APPLIED
MATHEMATICAL
PROGRAMMING

APPLIED MATHEMATICAL PROGRAMMING

STEPHEN P. BRADLEY
Graduate School of Business Administration, Harvard University

ARNOLDO C. HAX
Alfred P. Sloan School of Management, Massachusetts Institute of Technology

THOMAS L. MAGNANTI
Alfred P. Sloan School of Management, Massachusetts Institute of Technology

ADDISON-WESLEY PUBLISHING COMPANY
Reading, Massachusetts
Menlo Park, California · London · Amsterdam · Don Mills, Ontario · Sydney

Part of the research involved in the preparation of this book was supported by the Office of Naval Research under contract N00014–75–C–0556.

ISBN 0-201-00464-X
FGHIJKLMN-MA-89876543210

To our wives,
Edeltraud, Neva, and Beverly,
for their patience and encouragement,
and to George B. Dantzig,
for his continuing inspiration

Preface

Applied Mathematical Programming is designed as a text for a first course in mathematical programming aimed at integrating methods and applications. The objective of the book is to present material that leads to a greater understanding of applied problems and to an ability to structure and carry out the implementation of projects that utilize mathematical programming models.

Today, mathematical programming is one of the most widely used forms of modeling. Since its initial development, it has attracted the attention of managers, engineers, economists, and social scientists as one of the most powerful and flexible tools available to support decision making. The result has been applications of mathematical programming that are both numerous and varied. In this book our aim is to provide students, as well as practitioners in these fields, with a broad exposure to the primary topics of mathematical programming, by blending methodology and applications.

The book presents in a unifying manner the methodology of mathematical programming, an overview of mathematical programming in practice, and a series of selected projects reporting on the actual implementation of mathematical programming models in business and the public sector.

WHAT THE BOOK COVERS

The book surveys at an introductory level the most important methodology of mathematical programming. Linear programming is covered in some depth including recognition and formulation of problems, techniques for solving problems, sensitivity analysis and parametric procedures, and duality theory and its implications. Special linear models such as games, network models, and large-scale systems are also presented. The basic linear optimization procedures are extended to include the more general issues of integer and mixed-integer programming, as well as dynamic and nonlinear programming. Throughout the book emphasis is on pedagogical exposition as opposed to exhaustive survey of the theory. We have made a careful choice of material, and attempted to cover a number of important topics that would not ordinarily be dealt with in an introductory text. At all times, however, our concern is to make these techniques available to a broad class of readers who may not be primarily interested in the mathematics. Our approach is to tie the more difficult but important areas to simpler issues already covered in the text.

The book also provides an introduction to the issues of application and implementation. One chapter covers a general discussion of the decision-making process, the steps in model formulation, and the role of the computer in solving mathematical-programming models. In addition, four selected projects are included where mathematical-programming models are at the center of a complex decision-making system. The emphasis in these chapters is on actual systems or studies that have been implemented in large companies or government agencies. The particular projects were selected to cover a reasonable scope of applications as well as a variety of mathematical-programming techniques. The applications presented include the integration of long- and short-range planning in the aluminum industry, optimization of the composition and mission of the U.S. Merchant Marine Fleet, design of a job shop on a Navy tender, and a system for bond portfolio management. The techniques employed in the projects include linear programming and sensitivity analysis, mixed-integer programming, simulation, and important aspects of large-scale systems and computer implementation. In these chapters we attempt to give a feeling for the variety of methods that have been applied, as well as the flavor of the difficulties encountered in the implementation process.

FOR WHOM IS THE BOOK INTENDED?

The level of the book is that of the advanced undergraduate or first-year graduate in business, economics, or engineering, or of the undergraduate in applied mathematics. More precisely, the book should be part of the basic foundation of all students and practitioners alike who are working in the field of management science/operations research. There are essentially no prerequisites for using the book. It is definitely not necessary that a student be familiar with linear algebra or calculus. A level of mathematical sophistication at roughly that of two undergraduate years of mathematics should suffice. The book is intended to serve as a text for one- or two-semester courses entitled "Introduction to Mathematical Programming" or "Linear Programming," as well as the first semester of two-semester courses entitled "Management Science" or "Operations Research."

We should point out that no use is made of vectors and matrices in the methodology chapters of the book. It is our contention that complicated topics can be effectively developed without resorting to a notation that will be difficult and unfamiliar to some readers. Whenever possible, numerical and geometrical examples are used to illustrate important points. Additional material utilizing vectors and matrices, as well as material deemed too difficult to explain at the introductory level, is included in three technical appendices to the book. With the addition of these appendices, the book can be used for more advanced one- or two-semester courses in "Mathematical Programming." Throughout, the emphasis of the book is on that material which it is most important for the management science/operations research practitioner to understand and be able to apply.

WHAT DISTINGUISHES THIS BOOK FROM OTHERS?

We believe the book has three salient features. First, our pedagogical approach reflects teaching methods that we have tested for many years in our classrooms. The essence of the approach is inductive—to move from simple examples that highlight concepts and methods in concrete settings, to general principles and procedures. In this we have used as elementary a mathematical treatment as possible.

Second, we have devoted a significant portion of the book to in-depth treatments of actual applications. These chapters are intended to communicate to the reader the problems associated with the development of mathematical-programming models to support business and public-sector decisions.

Third, at the end of each chapter we have presented a comprehensive collection of exercises to allow the students to test their understanding of the basic concepts and methods presented. These exercises cover a wide spectrum of topics at various levels of difficulty—ranging from simple illustrations with which to practice the mechanics of solution procedures, to realistic formulations of complex applications in a variety of settings. Moreover, several computer-based exercises are included, dealing with model implementation as well as interpretation of results generated by standard computer systems.

HOW TO USE THE BOOK

As we have indicated, the book can be used at various levels depending on the depth desired, the time available, and the background of the students.

The foundations of linear programming are presented in Chapters 1 through 4. This material can be treated in great detail or it can be reviewed at a relatively fast pace, depending on whether or not the students have been previously exposed to linear programming. Moreover, the material of these chapters can be combined with Appendix B if matrix notation is desirable for a more advanced mathematical treatment of the subject.

Chapter 5 contains some useful taxonomies of the decision-making process, that help in presenting mathematical-programming models in proper perspective as decision-support tools. We have found it useful at times to discuss these taxonomies in the first introductory class, particularly when dealing with business students. The additional material of Chapter 5, dealing with the steps in model formulation and the role of computers in implementing mathematical-programming models, provides a useful foundation for introducing the students to a discussion of practical applications.

Chapters 6 and 7 discuss actual projects. Chapter 6 is particularly effective in illustrating the development of a hierarchical planning system to provide guidance for strategic and tactical planning decisions. Moreover, this chapter addresses basic issues in complex model formulation and transmits some of the flavor of "the art of modeling." Chapter 7 presents a simple mathematical program where the major lesson relies on a fairly comprehensive sensitivity analysis. Either one of these chapters, and preferably both, should be used if students are to be prepared to use mathematical-programming models realistically.

Chapters 8, 9, 11, 12, and 13 provide extensions of linear programming to different methodologies including network models, integer programming, dynamic programming, large-scale systems, and nonlinear programming. The introductory sections of these chapters can be used to present a broad view of these general mathematical-programming techniques. These chapters can be covered completely only in a two-semester course or in an advanced one-semester course devoted to extensions of linear programming.

Chapters 10 and 14 deal with applications of some of the more advanced topics. Therefore, their discussion is dependent on having selected the relevant methodological topics in advance. Chapter 10 describes a project that uses mixed-integer programming in a simulation environment. Chapter 14 addresses a bond portfolio-selection problem by means of Dantzig–Wolfe decomposition and dynamic-programming models.

ACKNOWLEDGMENTS

First and foremost we would like to extend our deep appreciation and recognition to Professor George B. Dantzig. Following George Dantzig's contributions to mathematical programming is like following the historical development of this field. In the last three decades, his name has been associated with most of the major achievements related to mathematical programming. This is even more of an accomplishment when one considers the impressive intellectual calibre of the researchers who have been attracted to this field.

The authors of this book had the unique experience of having been students of Professor Dantzig during his tenure at the University of California at Berkeley and later at Stanford University. Undoubtedly, a careful reader will detect the strong influence he has had on us, and will recognize much of George Dantzig's approach to mathematical programming in these pages. We hope that what we have finally accomplished gives to our students a little of what George Dantzig has given to us.

However strong the initial inspiration, there are more practical aspects to the writing of a book. Since the book has evolved through extensive classroom use, not only by us but by many of our colleagues, we owe a substantial debt to those who have helped by teaching their courses from early drafts of the manuscript. In addition to our own teaching, Gordon Kaufman, John D. C. Little, Roy Marsten, Steven Alter, and Gabriel Handler have tested the material at M.I.T.; Ronald Frank and John Mulvey at Harvard; Roger Glassey at the University of California, Berkeley; and

Bruce Golden at Boston University. We would also like to thank Richard Grinold for his extensive comments on our first completed draft.

Further, through the writing of the book, a number of our students provided great assistance in developing exercises and reviewing various chapters. In this regard, we thank Arjang Assad, Raymond Coulombe, Kalyan Chatterjee, John Dallen, Raphael Lazimy, James Litchfield, Lawrence Menkl, Silvia Pariente, John Schmitz, and Ronald Zlatoper. In addition, we would like to single out one student, Dan Candea, for the important role he played in providing both critical comments and innovative exercises in an untiring manner.

In the production of the book we owe our thanks to Deborah Cohen and Martha Laisne, among others, for typing the manuscript, and our gratitude again to Martha for an outstanding job of copy editing and proofreading.

Finally, we would like to thank the Division of Research of the Harvard Business School, the Sloan School of Management of M.I.T., and the Office of Naval Research, through the auspices of Dr. Marvin Denicoff, for their financial support throughout this effort.

Cambridge   *S.P.B.*
December 1976   *A.C.H.*
                                                                                                 T.L.M.

Contents

Mathematical Programming: An Overview

1

Management science is characterized by a scientific approach to managerial decision making. It attempts to apply mathematical methods and the capabilities of modern computers to the difficult and unstructured problems confronting modern managers. It is a young and novel discipline. Although its roots can be traced back to problems posed by early civilizations, it was not until World War II that it became identified as a respectable and well defined body of knowledge. Since then, it has grown at an impressive pace, unprecedented for most scientific accomplishments; it is changing our attitudes toward decision-making, and infiltrating every conceivable area of application, covering a wide variety of business, industrial, military, and public-sector problems.

Management science has been known by a variety of other names. In the United States, *operations research* has served as a synonym and it is used widely today, while in Britain *operational research* seems to be the more accepted name. Some people tend to identify the scientific approach to managerial problem-solving under such other names as systems analysis, cost–benefit analysis, and cost-effectiveness analysis. We will adhere to *management science* throughout this book.

Mathematical programming, and especially linear programming, is one of the best developed and most used branches of management science. It concerns the optimum allocation of limited resources among competing activities, under a set of constraints imposed by the nature of the problem being studied. These constraints could reflect financial, technological, marketing, organizational, or many other considerations. In broad terms, mathematical programming can be defined as a mathematical representation aimed at programming or planning the best possible allocation of scarce resources. When the mathematical representation uses linear functions exclusively, we have a linear-programming model.

In 1947, George B. Dantzig, then part of a research group of the U.S. Air Force known as Project SCOOP (Scientific Computation Of Optimum Programs), developed the *simplex method* for solving the general linear-programming problem.

The extraordinary computational efficiency and robustness of the simplex method, together with the availability of high-speed digital computers, have made linear programming the most powerful optimization method ever designed and the most widely applied in the business environment.

Since then, many additional techniques have been developed, which relax the assumptions of the linear-programming model and broaden the applications of the mathematical-programming approach. It is this spectrum of techniques and their effective implementation in practice that are considered in this book.

1.1 AN INTRODUCTION TO MANAGEMENT SCIENCE

Since mathematical programming is only a tool of the broad discipline known as management science, let us first attempt to understand the management-science approach and identify the role of mathematical programming within that approach.

It is hard to give a noncontroversial definition of management science. As we have indicated before, this is a rather new field that is renewing itself and changing constantly. It has benefited from contributions originating in the social and natural sciences, econometrics, and mathematics, much of which escape the rigidity of a definition. Nonetheless it is possible to provide a general statement about the basic elements of the management-science approach.

> Management science is characterized by the use of *mathematical models* in providing guide-lines to managers for making effective decisions within the state of the current information, or in seeking further information if current knowledge is insufficient to reach a proper decision.

There are several elements of this statement that are deserving of emphasis. First, the essence of management science is the model-building approach—that is, an attempt to capture the most significant features of the decision under consideration by means of a mathematical abstraction. Models are simplified representations of the real world. In order for models to be useful in supporting management decisions, they have to be simple to understand and easy to use. At the same time, they have to provide a complete and realistic representation of the decision environment by incorporating all the elements required to characterize the essence of the problem under study. This is not an easy task but, if done properly, it will supply managers with a formidable tool to be used in complex decision situations.

Second, through this model-design effort, management science tries to provide guidelines to managers or, in other words, to increase managers' understanding of the consequences of their actions. There is never an attempt to replace or substitute for managers, but rather the aim is to *support* management actions. It is critical, then, to recognize the strong interaction required between managers and models. Models can expediently and effectively account for the many interrelationships that might be present among the alternatives being considered, and can explicitly evaluate the economic consequences of the actions available to managers within the constraints imposed by the existing resources and the demands placed upon the use of those resources. Managers, on the other hand, should formulate the basic questions to be addressed by the model, and then interpret the model's results in light of their own experience and intuition, recognizing the model's limitations. The complementarity between the superior computational capabilities provided by the model and the higher judgmental capabilities of the human decision-maker is the key to a successful management-science approach.

Finally, it is the complexity of the decision under study, and not the tool being used to investigate the decision-making process, that should determine the amount of information needed to handle that decision effectively. Models have been criticized for creating unreasonable requirements for information. In fact, this is not necessary. Quite to the contrary, models can be constructed within the current state of available information and they can be used to evaluate whether or not it is economically desirable to gather additional information.

The subject of proper model design and implementation will be covered in detail in Chapter 5.

1.2 MODEL CLASSIFICATION

The management-science literature includes several approaches to classifying models. We will begin with a categorization that identifies broad types of models according to the degree of realism that they achieve in representing a given problem. The model categories can be illustrated as shown in Fig. 1.1.

Operational Exercise

The first model type is an *operational exercise*. This modeling approach operates directly with the real environment in which the decision under study is going to take place. The modeling effort merely involves designing a set of experiments to be conducted in that environment, and measuring and interpreting the results of those experiments. Suppose, for instance, that we would like to determine what mix of several crude oils should be blended in a given oil refinery to satisfy, in the most

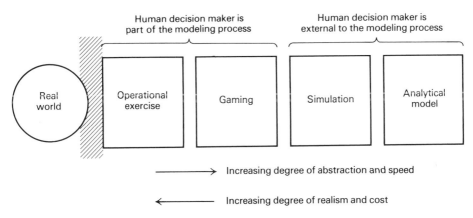

Fig. 1.1   Types of model representation.

effective way, the market requirements for final products to be delivered from that refinery. If we were to conduct an operational exercise to support that decision, we would try different quantities of several combinations of crude oil types directly in the actual refinery process, and observe the resulting revenues and costs associated with each alternative mix. After performing quite a few trials, we would begin to develop an understanding of the relationship between the crude oil input and the net revenue obtained from the refinery process, which would guide us in identifying an appropriate mix.

In order for this approach to operate successfully, it is mandatory to design experiments to be conducted carefully, to evaluate the experimental results in light of errors that can be introduced by measurement inaccuracies, and to draw inferences about the decisions reached, based upon the limited number of observations performed. Many statistical and optimization methods can be used to accomplish these tasks properly. The essence of the operational exercise is an inductive learning process, characteristic of empirical research in the natural sciences, in which generalizations are drawn from particular observations of a given phenomenon.

Operational exercises contain the highest degree of realism of any form of modeling approach, since hardly any external abstractions or oversimplifications are introduced other than those connected with the interpretation of the observed results and the generalizations to be drawn from them. However, the method is exceedingly, usually prohibitively, expensive to implement. Moreover, in most cases it is impossible to exhaustively analyze the alternatives available to the decision-maker. This can lead to severe *suboptimization* in the final conclusions. For these reasons, operational exercises seldom are used as a pure form of modeling practice. It is important to recognize, however, that direct observation of the actual environment underlies most model conceptualizations and also constitutes one of the most important sources of data. Consequently, even though they may not be used exclusively, operational exercises produce significant contributions to the improvement of managerial decision-making.

Gaming

The second type of model in this classification is *gaming*. In this case, a model is constructed that is an abstract and simplified representation of the real environment. This model provides a responsive mechanism to evaluate the effectiveness of proposed alternatives, which the decision-maker must supply in an organized and sequential fashion. The model is simply a device that allows the decision-maker to test the performance of the various alternatives that seem worthwhile to pursue. In addition, in a gaming situation, all the human interactions that affect the decision environment are allowed to participate actively by providing the inputs they usually are responsible for in the actual realization of their activities. If a gaming approach is used in our previous example, the refinery process would be represented by a computer or mathematical model, which could assume any kind of structure.

The model should reflect, with an acceptable degree of accuracy, the relationships between the inputs and outputs of the refinery process. Subsequently, all the personnel who participate in structuring the decision process in the management of the refinery would be allowed to interact with the model. The production manager would establish production plans, the marketing manager would secure contracts and develop marketing strategies, the purchasing manager would identify prices and sources of crude oil and develop acquisition programs, and so forth. As before, several combinations of quantities and types of crude oil would be tried, and the resulting revenues and cost figures derived from the model would be obtained, to guide us in formulating an optimal policy. Certainly, we have lost some degree of realism in our modeling approach with respect to the operational exercise, since we are operating with an abstract environment, but we have retained some of the human interactions of the real process. However, the *cost of processing each alternative* has been reduced, and the *speed of measuring the performance* of each alternative has been increased.

Gaming is used mostly as a learning device for developing some appreciation for those complexities inherent in a decision-making process. Several management games have been designed to illustrate how marketing, production, and financial decisions interact in a competitive economy.

Simulation

Simulation models are similar to gaming models except that all human decision-makers are removed from the modeling process. The model provides the means to evaluate the performance of a number of alternatives, supplied externally to the model by the decision-maker, without allowing for human interactions at intermediate stages of the model computation.

Like operational exercises and gaming, simulation models neither generate alternatives nor produce an optimum answer to the decision under study. These types of models are inductive and empirical in nature; they are useful only to assess the performance of alternatives identified previously by the decision-maker.

If we were to conduct a simulation model in our refinery example, we would program in advance a large number of combinations of quantities and types of crude oil to be used, and we would obtain the net revenues associated with each alternative without any external inputs of the decision-makers. Once the model results were produced, new runs could be conducted until we felt that we had reached a proper understanding of the problem on hand.

Many simulation models take the form of computer programs, where logical arithmetic operations are performed in a prearranged sequence. It is not necessary, therefore, to define the problem exclusively in analytic terms. This provides an added flexibility in model formulation and permits a high degree of realism to be achieved, which is particularly useful when uncertainties are an important aspect of the decision.

Analytical Model

Finally, the fourth model category proposed in this framework is the *analytical model*. In this type of model, the problem is represented completely in mathematical terms, normally by means of a criterion or objective, which we seek to maximize or minimize, subject to a set of mathematical constraints that portray the conditions under which the decisions have to be made. The model computes an optimum solution, that is, one that satisfies all the constraints and gives the best possible value of the objective function.

In the refinery example, the use of an analytical model implies setting up as an objective the *maximization of the net revenues* obtained from the refinery operation as a function of the types and quantities of the crude oil used. In addition, the technology of the refinery process, the final product requirements, and the crude oil availabilities must be represented in mathematical terms to define the constraints of our problem. The solution to the model will be the exact amount of each available crude-oil type to be processed that will maximize the net revenues within the proposed constraint set. Linear programming has been, in the last two decades, the indisputable analytical model to use for this kind of problem.

Analytical models are normally the least expensive and easiest models to develop. However, they introduce the highest degree of simplification in the model representation. As a rule of thumb, it is better to be as much to the right as possible in the model spectrum (no political implication intended!), provided that the resulting degree of realism is appropriate to characterize the decision under study.

Most of the work undertaken by management scientists has been oriented toward the development and implementation of analytical models. As a result of this effort, many different techniques and methodologies have been proposed to address specific kinds of problems. Table 1.1 presents a classification of the most important types of analytical and simulation models that have been developed.

Table 1.1 Classification of Analytical and Simulation Models

	Strategy evaluation	*Strategy generation*

Certainty	Deterministic simulation
	Econometric models
	Systems of simultaneous
	    equations
	Input-output models
		Linear programming
		Network models
		Integer and mixed-integer
		    programming
		Nonlinear programming
		Control theory

Uncertainty	Monte Carlo simulation Econometric models Stochastic processes Queueing theory Reliability theory	Decision theory Dynamic programming Inventory theory Stochastic programming Stochastic control theory

Statistics and subjective assessment are used in all models to determine values for parameters of the models and limits on the alternatives.

The classification presented in Table 1.1 is not rigid, since strategy evaluation models are used for improving decisions by trying different alternatives until one is determined that appears "best." The important distinction of the proposed classification is that, for strategy evaluation models, the user must first choose and construct the alternative and then evaluate it with the aid of the model. For strategy generation models, the alternative is not completely determined by the user; rather, the class of alternatives is determined by establishing constraints on the decisions, and then an algorithmic procedure is used to automatically generate the "best" alternative within that class. The horizontal classification should be clear, and is introduced because the inclusion of uncertainty (or not) generally makes a substantial difference in the type and complexity of the techniques that are employed. Problems involving uncertainty are inherently more difficult to formulate well and to solve efficiently.

This book is devoted to mathematical programming—a part of management science that has a common base of theory and a large range of applications. Generally, mathematical programming includes all of the topics under the heading of strategy generation except for decision theory and control theory. These two topics are entire disciplines in themselves, depending essentially on different underlying theories and techniques. Recently, though, the similarities between mathematical programming and control theory are becoming better understood, and these disciplines are beginning to merge. In mathematical programming, the main body of material that has been developed, and more especially applied, is under the assumption of certainty. Therefore, we concentrate the bulk of our presentation on the topics in the upper righthand corner of Table 1.1. The critical emphasis in the book is on developing those principles and techniques that lead to *good formulations* of actual decision problems and *solution procedures* that are efficient for solving these formulations.

1.3 FORMULATION OF SOME EXAMPLES

In order to provide a preliminary understanding of the types of problems to which mathematical programming can be applied, and to illustrate the kind of rationale that should be used in formulating linear-programming problems, we will present in this section three highly simplified examples and their corresponding linear-programming formulations.

Charging a Blast Furnace   An iron foundry has a firm order to produce 1000 pounds of castings containing at least 0.45 percent manganese and between 3.25 percent and 5.50 percent silicon. As these particular castings are a special order, there are no suitable castings on hand. The castings sell for $0.45 per pound. The foundry has three types of pig iron available in essentially unlimited amounts, with the following properties:

	Type of pig iron		
	A	B	C
Silicon	4 %	1 %	0.6%
Manganese	0.45%	0.5%	0.4%

Further, the production process is such that pure manganese can also be added directly to the melt. The costs of the various possible inputs are:

Pig A	$21/thousand pounds
Pig B	$25/thousand pounds
Pig C	$15/thousand pounds
Manganese	$ 8/pound.

It costs 0.5 cents to melt down a pound of pig iron. Out of what inputs should the foundry produce the castings in order to maximize profits?

The first step in formulating a linear program is to define the *decision variables* of the problem. These are the elements under the control of the decision-maker, and their values determine the solution of the model. In the present example, these variables are simple to identify, and correspond to the number of pounds of pig A, pig B, pig C, and pure manganese to be used in the production of castings. Specifically, let us denote the decision variables as follows:

$$x_1 = \text{Thousands of pounds of pig iron A,}$$
$$x_2 = \text{Thousands of pounds of pig iron B,}$$
$$x_3 = \text{Thousands of pounds of pig iron C,}$$
$$x_4 = \text{Pounds of pure manganese.}$$

The next step to be carried out in the formulation of the problem is to determine the criterion the decision-maker will use to evaluate alternative solutions to the problem. In mathematical-programming terminology, this is known as the *objective function.* In our case, we want to maximize the total profit resulting from the production of 1000 pounds of castings. Since we are producing exactly 1000 pounds of castings, the total income will be the selling price per pound times 1000 pounds. That is:

$$\text{Total income} = 0.45 \times 1000 = 450.$$

To determine the total cost incurred in the production of the alloy, we should add the melting cost of $0.005/pound to the corresponding cost of each type of pig iron used. Thus, the relevant unit cost of the pig iron, in dollars per thousand pounds, is:

Pig iron A	$21 + 5 = 26,$
Pig iron B	$25 + 5 = 30,$
Pig iron C	$15 + 5 = 20.$

Therefore, the total cost becomes:

$$\text{Total cost} = 26x_1 + 30x_2 + 20x_3 + 8x_4, \tag{1}$$

and the total profit we want to maximize is determined by the expression:

$$\text{Total profit} = \text{Total income} - \text{Total cost.}$$

Thus,

$$\text{Total profit} = 450 - 26x_1 - 30x_2 - 20x_3 - 8x_4. \tag{2}$$

It is worthwhile noticing in this example that, since the amount of castings to be produced was fixed in advance, equal to 1000 pounds, the maximization of the total profit, given by Eq. (2), becomes completely equivalent to the minimization of the total cost, given by Eq. (1).

We should now define the *constraints* of the problem, which are the restrictions imposed upon the values of the decision variables by the characteristics of the problem under study. First, since the producer does not want to keep any supply of the castings on hand, we should make the total amount to be produced exactly equal to 1000 pounds; that is,

$$1000x_1 + 1000x_2 + 1000x_3 + x_4 = 1000. \tag{3}$$

Next, the castings should contain at least 0.45 percent manganese, or 4.5 pounds in the 1000 pounds of castings to be produced. This restriction can be expressed as follows:

$$4.5x_1 + 5.0x_2 + 4.0x_3 + x_4 \geq 4.5. \tag{4}$$

The term $4.5x_1$ is the pounds of manganese contributed by pig iron A since each 1000 pounds of this pig iron contains 4.5 pounds of manganese. The x_2, x_3, and x_4 terms account for the manganese contributed by pig iron B, by pig iron C, and by the addition of pure manganese.

Similarly, the restriction regarding the silicon content can be represented by the following inequalities:

$$40x_1 + 10x_2 + 6x_3 \geq 32.5, \tag{5}$$

$$40x_1 + 10x_2 + 6x_3 \leq 55.0. \tag{6}$$

Constraint (5) establishes that the minimum silicon content in the castings is 3.25 percent, while constraint (6) indicates that the maximum silicon allowed is 5.5 percent.

Finally, we have the obvious nonnegativity constraints:

$$x_1 \geq 0, \qquad x_2 \geq 0, \qquad x_3 \geq 0, \qquad x_4 \geq 0.$$

If we choose to minimize the total cost given by Eq. (1), the resulting linear programming problem can be expressed as follows:

Minimize $z = \quad 26x_1 + \quad 30x_2 + \quad 20x_3 + \quad 8x_4,$

subject to:

$$
\begin{aligned}
1000x_1 + 1000x_2 + 1000x_3 + \quad x_4 &= 1000, \\
4.5x_1 + \quad 5.0x_2 + \quad 4.0x_3 + \quad x_4 &\geq 4.5, \\
40x_1 + \quad 10x_2 + \quad 6x_3 \quad\quad &\geq 32.5, \\
40x_1 + \quad 10x_2 + \quad 6x_3 \quad\quad &\leq 55.0, \\
x_1 \geq 0, \quad x_2 \geq 0, \quad x_3 \geq 0, \quad x_4 &\geq 0.
\end{aligned}
$$

Often, this algebraic formulation is represented in *tableau form* as follows:

	Decision variables				Relation	Requirements
	x_1	x_2	x_3	x_4		
Total lbs.	1000	1000	1000	1	$=$	1000
Manganese	4.5	5	4	1	$\geq$	4.5
Silicon lower	40	10	6	0	$\geq$	32.5
Silicon upper	40	10	6	0	$\leq$	55.0
Objective	26	30	20	8	$=$	z(min)
	0.779	0	0.220	0.111		25.54

Optimal solution

The bottom line of the tableau specifies the *optimal solution* to this problem. The solution includes the values of the decision variables as well as the minimum cost attained, $25.54 per thousand pounds; this solution was generated using a commercially available linear-programming computer system. The underlying solution procedure, known as the *simplex method*, will be explained in Chapter 2.

It might be interesting to see how the solution and the optimal value of the objective function is affected by changes in the cost of manganese. In Fig. 1.2 we give the optimal value of the objective function as this cost is varied. Note that if the cost of manganese rises above $9.86/lb., then *no* pure manganese is used. In the range from $0.019/lb. to $9.86/lb., the values of the decision variables remain unchanged. When manganese becomes extremely *inexpensive*, less than $0.019/lb., a great deal of manganese is used, in conjunction with only one type of pig iron.

Similar analyses can be performed to investigate the behavior of the solution as other parameters of the problem (for example, minimum allowed silicon content)

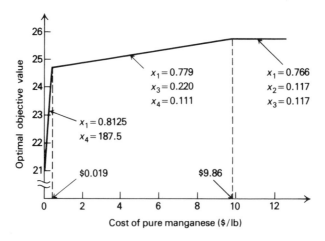

Fig. 1.2 Optimal cost of castings as a function of the cost of pure manganese.

are varied. These results, known as parametric analysis, are reported routinely by commercial linear-programming computer systems. In Chapter 3 we will show how to conduct such analyses in a comprehensive way.

Portfolio Selection A portfolio manager in charge of a bank portfolio has $10 million to invest. The securities available for purchase, as well as their respective quality ratings, maturities, and yields, are shown in Table 1.2.

Table 1.2

Bond name	Bond type	Quality scales		Years to maturity	Yield to maturity	After-tax yield
		Moody's	Bank's			
A	Municipal	Aa	2	9	4.3%	4.3%
B	Agency	Aa	2	15	5.4	2.7
C	Government	Aaa	1	4	5.0	2.5
D	Government	Aaa	1	3	4.4	2.2
E	Municipal	Ba	5	2	4.5	4.5

The bank places the following policy limitations on the portfolio manager's actions:

1. Government and agency bonds must total at least $4 million.

2. The average quality of the portfolio cannot exceed 1.4 on the bank's quality scale. (Note that a low number on this scale means a high-quality bond.)

3. The average years to maturity of the portfolio must not exceed 5 years.

Assuming that the objective of the portfolio manager is to maximize after-tax earnings and that the tax rate is 50 percent, what bonds should he purchase? If it became possible to borrow up to $1 million at 5.5 percent before taxes, how should his selection be changed?

Leaving the question of borrowed funds aside for the moment, the decision variables for this problem are simply the dollar amount of each security to be purchased:

x_A = Amount to be invested in bond A; in millions of dollars.

x_B = Amount to be invested in bond B; in millions of dollars.

x^C = Amount to be invested in bond C; in millions of dollars.

x_D = Amount to be invested in bond D; in millions of dollars.

x_E = Amount to be invested in bond E; in millions of dollars.

We must now determine the form of the objective function. Assuming that all securities are purchased at par (face value) and held to maturity and that the income on municipal bonds is tax-exempt, the after-tax earnings are given by:

$$z = 0.043x_A + 0.027x_B + 0.025x_C + 0.022x_D + 0.045x_E.$$

Now let us consider each of the restrictions of the problem. The portfolio manager has only a total of ten million dollars to invest, and therefore:

$$x_A + x_B + x_C + x_D + x_E \leq 10.$$

Further, of this amount at least \$4 million must be invested in government and agency bonds. Hence,

$$x_B + x_C + x_D \geq 4.$$

The average quality of the portfolio, which is given by the ratio of the total quality to the total value of the portfolio, must not exceed 1.4:

$$\frac{2x_A + 2x_B + x_C + x_D + 5x_E}{x_A + x_B + x_C + x_D + x_E} \leq 1.4.$$

Note that the inequality is less-than-or-equal-to, since a low number on the bank's quality scale means a high-quality bond. By clearing the denominator and rearranging terms, we find that this inequality is clearly equivalent to the linear constraint:

$$0.6x_A + 0.6x_B - 0.4x_C - 0.4x_D + 3.6x_E \leq 0.$$

The constraint on the average maturity of the portfolio is a similar ratio. The average maturity must not exceed five years:

$$\frac{9x_A + 15x_B + 4x_C + 3x_D + 2x_E}{x_A + x_B + x_C + x_D + x_E} \leq 5,$$

which is equivalent to the linear constraint:

$$4x_A + 10x_B - x_C - 2x_D - 3x_E \leq 0.$$

Note that the two ratio constraints are, in fact, *nonlinear* constraints, which would require sophisticated computational procedures if included in this form. However, simply multiplying both sides of each ratio constraint by its denominator (which must be nonnegative since it is the sum of nonnegative variables) transforms this nonlinear constraint into a simple linear constraint. We can summarize our formulation in tableau form, as follows:

	x_A	x_B	x_C	x_D	x_E	Relation	Limits
Cash	1	1	1	1	1	$\leq$	10
Governments		1	1	1		$\geq$	4
Quality	0.6	0.6	-0.4	-0.4	3.6	$\leq$	0
Maturity	4	10	-1	-2	-3	$\leq$	0
Objective	0.043	0.027	0.025	0.022	0.045	$=$	z(max)
	3.36	0	0	6.48	0.16		0.294

Optimal solution

The values of the decision variables and the optimal value of the objective function are again given in the last row of the tableau.

Now consider the additional possibility of being able to borrow up to $1 million at 5.5 percent before taxes. Essentially, we can increase our cash supply above ten million by borrowing at an after-tax rate of 2.75 percent. We can define a new decision variable as follows:

$$y = \text{amount borrowed in millions of dollars.}$$

There is an upper bound on the amount of funds that can be borrowed, and hence

$$y \leq 1.$$

The cash constraint is then modified to reflect that the total amount purchased must be less than or equal to the cash that can be made available including borrowing:

$$x_A + x_B + x_C + x_D + x_E \leq 10 + y.$$

Now, since the borrowed money costs 2.75 percent after taxes, the new after-tax earnings are:

$$z = 0.043x_A + 0.027x_B + 0.025x_C + 0.022x_D + 0.045x_E - 0.0275y.$$

We summarize the formulation when borrowing is allowed and give the solution in tableau form as follows:

	x_A	x_B	x_C	x_D	x_E	y	Relation	Limits
Cash	1	1	1	1	1	-1	$\leq$	10
Borrowing						1	$\leq$	1
Governments		1	1	1			$\geq$	4
Quality	0.6	0.6	-0.4	-0.4	3.6		$\leq$	0
Maturity	4	10	-1	-2	-3		$\leq$	0
Objective	0.043	0.027	0.025	0.022	0.045	-0.0275	$=$	z(max)
	3.70	0	0	7.13	0.18	1		0.296

Optimal solution

Production and Assembly   A division of a plastics company manufactures three basic products: sporks, packets, and school packs. A spork is a plastic utensil which purports to be a combination spoon, fork, and knife. The packets consist of a spork, a napkin, and a straw wrapped in cellophane. The school packs are boxes of 100 packets with an additional 10 loose sporks included.

Production of 1000 sporks requires 0.8 standard hours of molding machine capacity, 0.2 standard hours of supervisory time, and $2.50 in direct costs. Production of 1000 packets, including 1 spork, 1 napkin, and 1 straw, requires 1.5 standard hours of the packaging-area capacity, 0.5 standard hours of supervisory time, and $4.00 in direct costs. There is an unlimited supply of napkins and straws. Production of 1000 school packs requires 2.5 standard hours of packaging-area capacity, 0.5 standard hours of supervisory time, 10 sporks, 100 packets, and $8.00 in direct costs.

Any of the three products may be sold in unlimited quantities at prices of $5.00, $15.00, and $300.00 per thousand, respectively. If there are 200 hours of production time in the coming month, what products, and how much of each, should be manufactured to yield the most profit?

The first decision one has to make in formulating a linear programming model is the selection of the proper variables to represent the problem under consideration. In this example there are at least two different sets of variables that can be chosen as decision variables. Let us represent them by x's and y's and define them as follows:

x_1 = Total number of sporks produced in thousands,

x_2 = Total number of packets produced in thousands,

x_3 = Total number of school packs produced in thousands,

and

y_1 = Total number of sporks sold as sporks in thousands,

y_2 = Total number of packets sold as packets in thousands,

y_3 = Total number of school packs sold as school packs in thousands.

We can determine the relationship between these two groups of variables. Since each packet needs one spork, and each school pack needs ten sporks, the total number of sporks sold as sporks is given by:

$$y_1 = x_1 - x_2 - 10x_3. \tag{7}$$

Similarly, for the total number of packets sold as packets we have:

$$y_2 = x_2 - 100x_3. \tag{8}$$

Finally, since all the school packs are sold as such, we have:

$$y_3 = x_3. \tag{9}$$

From Eqs. (7), (8), and (9) it is easy to express the x's in terms of the y's, obtaining:

$$x_1 = y_1 + y_2 + 110y_3, \tag{10}$$

$$x_2 = y_2 + 100y_3, \tag{11}$$

$$x_3 = y_3. \tag{12}$$

As a matter of exercise, let us formulate the linear program corresponding to the present example in two forms: first, using the x's as decision variables, and second, using the y's as decision variables.

The objective function is easily determined in terms of both y- and x-variables by using the information provided in the statement of the problem with respect to the selling prices and the direct costs of each of the units produced. The total profit is given by:

$$\text{Total profit} = 5y_1 + 15y_2 + 300y_3 - 2.5x_1 - 4x_2 - 8x_3. \tag{13}$$

Equations (7), (8), and (9) allow us to express this total profit in terms of the x-variables alone. After performing the transformation, we get:

$$\text{Total profit} = 2.5x_1 + 6x_2 - 1258x_3. \tag{14}$$

Now we have to set up the restrictions imposed by the maximum availability of 200 hours of production time. Since the sporks are the only items requiring time in the injection-molding area, and they consume 0.8 standard hours of production per 1000 sporks, we have:

$$0.8x_1 \leqq 200. \tag{15}$$

For packaging-area capacity we have:

$$1.5x_2 + 2.5x_3 \leqq 200, \tag{16}$$

while supervisory time requires that

$$0.2x_1 + 0.5x_2 + 0.5x_3 \leqq 200. \tag{17}$$

In addition to these constraints, we have to make sure that the number of sporks, packets, and school packs sold as such (i.e., the y-variables) are nonnegative. Therefore, we have to add the following constraints:

$$x_1 - x_2 - 10x_3 \geqq 0, \tag{18}$$
$$x_2 - 100x_3 \geqq 0, \tag{19}$$
$$x_3 \geqq 0. \tag{20}$$

Finally, we have the trivial nonnegativity conditions

$$x_1 \geqq 0, \qquad x_2 \geqq 0, \qquad x_3 \geqq 0. \tag{21}$$

Note that, besides the nonnegativity of all the variables, which is a condition always implicit in the methods of solution in linear programming, this form of stating the problem has generated six constraints, (15) to (20). If we expressed the problem in terms of the y-variables, however, conditions (18) to (20) correspond merely to the nonnegativity of the y-variables, and these constraints are automatically guaranteed because the y's are nonnegative and the x's expressed by (10), (11), and (12), in terms of the y's, are the sum of nonnegative variables, and therefore the x's are always nonnegative.

By performing the proper changes of variables, it is easy to see that the linear-programming formulation in terms of the y-variables is given in tableau form as follows:

	y_1	y_2	y_3	*Relation*	*Limits*
Molding	0.8	0.8	88	$\leqq$	200
Packaging		1.5	152.5	$\leqq$	200
Supervisory	0.2	0.7	72.5	$\leqq$	200
Objective	2.5	8.5	-383	$=$	z(max)
	116.7	133.3	0		1425

Optimal solution

Since the computation time required to solve a linear programming problem increases roughly with the cube of the number of rows of the problem, in this example the y's constitute better decision variables than the x's. The values of the x's are easily determined from Eqs. (10), (11), and (12).

1.4 A GEOMETRICAL PREVIEW

Although in this introductory chapter we are not going to discuss the details of computational procedures for solving mathematical programming problems, we can gain some useful insight into the characteristics of the procedures by looking at the geometry of a few simple examples. Since we want to be able to draw simple graphs depicting various possible situations, the problem initially considered has only two decision variables.

The Problem   Suppose that a custom molder has one injection-molding machine and two different dies to fit the machine. Due to differences in number of cavities and cycle times, with the first die he can produce 100 cases of six-ounce juice glasses in six hours, while with the second die he can produce 100 cases of ten-ounce fancy cocktail glasses in five hours. He prefers to operate only on a schedule of 60 hours of production per week. He stores the week's production in his own stockroom where he has an effective capacity of 15,000 cubic feet. A case of six-ounce juice glasses requires 10 cubic feet of storage space, while a case of ten-ounce cocktail glasses requires 20 cubic feet due to special packaging. The contribution of the six-ounce juice glasses is \$5.00 per case; however, the only customer available will not accept more than 800 cases per week. The contribution of the ten-ounce cocktail glasses is \$4.50 per case and there is no limit on the amount that can be sold. How many cases of each type of glass should be produced each week in order to maximize the total contribution?

Formulation of the Problem

We first define the decision variables and the units in which they are measured. For this problem we are interested in knowing the optimal number of cases of each type of glass to produce per week. Let

x_1 = Number of cases of six-ounce juice glasses produced per week (in hundreds of cases per week), and

x_2 = Number of cases of ten-ounce cocktail glasses produced per week (in hundreds of cases per week).

The objective function is easy to establish since we merely want to maximize the contribution to overhead, which is given by:

$$\text{Contribution} = 500x_1 + 450x_2,$$

since the decision variables are measured in hundreds of cases per week. We can write the constraints in a straightforward manner. Because the custom molder is limited to 60 hours per week, and production of six-ounce juice glasses requires 6 hours per hundred cases while production of ten-ounce cocktail glasses requires 5 hours per hundred cases, the constraint imposed on production capacity in units of production hours per week is:

$$6x_1 + 5x_2 \leqq 60.$$

Now since the custom molder has only 15,000 cubic feet of effective storage space and a case of six-ounce juice glasses requires 10 cubic feet, while a case of ten-ounce cocktail glasses requires 20 cubic feet, the constraint on storage capacity, in units of hundreds of cubic feet of space per week, is:

$$10x_1 + 20x_2 \leqq 150.$$

Finally, the demand is such that no more than 800 cases of six-ounce juice glasses can be sold each week. Hence,

$$x_1 \leqq 8.$$

Since the decision variables must be nonnegative,

$$x_1 \geqq 0, \qquad x_2 \geqq 0$$

and we have the following linear program:

$$\text{Maximize } z = 500x_1 + 450x_2,$$

subject to:

$$
\begin{array}{llll}
6x_1 + & 5x_2 & \leqq 60 & \text{(production hours)} \\
10x_1 + & 20x_2 & \leqq 150 & \text{(hundred sq. ft. storage)} \\
x_1 & & \leqq \ 8 & \text{(sales limit 6 oz. glasses)} \\
x_1 \geqq 0, & x_2 \geqq 0. &
\end{array}
$$

Graphical Representation of the Decision Space

We now look at the constraints of the above linear-programming problem. All of the values of the decision variables x_1 and x_2 that simultaneously satisfy these constraints can be represented geometrically by the shaded area given in Fig. 1.3. Note that each line in this figure is represented by a constraint expressed as an equality. The arrows associated with each line show the direction indicated by the inequality sign in each constraint. The set of values of the decision variables x_1 and x_2 that simultaneously satisfy all the constraints indicated by the shaded area are the feasible production possibilities or *feasible solutions* to the problem. Any production alternative not within the feasible region must violate at least one of the constraints of the problem. Among these feasible production alternatives, we want to find the values of the decision variables x_1 and x_2 that maximize the resulting contribution to overhead.

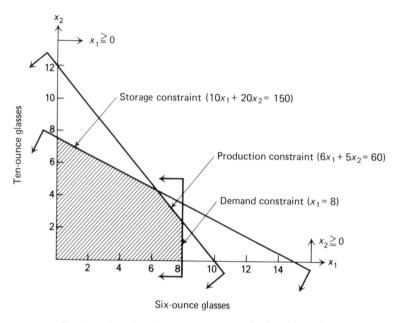

Fig. 1.3   Graphical representation of the feasible region.

Finding an Optimal Solution

To find an optimal solution we first note that any point in the interior of the feasible region cannot be an optimal solution since the contribution can be increased by increasing either x_1 or x_2 or both. To make this point more clearly, let us rewrite the objective function

$$z = 500x_1 + 450x_2,$$

in terms of x_2 as follows:

$$x_2 = \frac{1}{450}z - \left(\frac{500}{450}\right)x_1.$$

If z is held fixed at a given constant value, this expression represents a straight line, where $\frac{1}{450}z$ is the intercept with the x_2 axis (i.e., the value of x_2 when $x_1 = 0$), and $-\frac{500}{450}$ is the slope (i.e., the change in the value of x_2 corresponding to a unit increase in the value of x_1). Note that the slope of this straight line is constant, independent of the value of z. As the value of z increases, the resulting straight lines move parallel to themselves in a northeasterly direction away from the origin (since the intercept $\frac{1}{450}z$ increases when z increases, and the slope is constant at $-\frac{500}{450}$). Figure 1.4 shows some of these parallel lines for specific values of z. At the point labeled P1, the line intercepts the farthest point from the origin within the feasible region, and the contribution z cannot be increased any more. Therefore, point P1 represents the optimum solution.

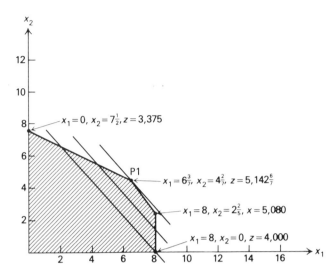

Fig. 1.4 Finding the optimal solution.

Since reading the graph may be difficult, we can compute the values of the decision variables by recognizing that point P1 is determined by the intersection of the production-capacity constraint and the storage-capacity constraint. Solving these constraints,

$$6x_1 + 5x_2 = 60,$$
$$10x_1 + 20x_2 = 150,$$

yields $x_1 = 6\frac{3}{7}$, $x_2 = 4\frac{2}{7}$; and substituting these values into the objective function yields $z = 5142\frac{6}{7}$ as the maximum contribution that can be attained.

Note that the optimal solution is at a corner point, or *vertex*, of the feasible region. This turns out to be a general property of linear programming: if a problem has an optimal solution, there is always a vertex that is optimal. The *simplex method* for finding an optimal solution to a general linear program exploits this property by starting at a vertex and moving from vertex to vertex, improving the value of the objective function with each move. In Fig. 1.4, the values of the decision variables and the associated value of the objective function are given for each vertex of the feasible region. Any procedure that starts at one of the vertices and looks for an improvement among adjacent vertices would also result in the solution labeled P1.

An optimal solution of a linear program in its simplest form gives the value of the criterion function, the levels of the decision variables, and the amount of slack or surplus in the constraints. In the custom-molder example, the criterion was *maximum contribution*, which turned out to be $z = \$5142\frac{6}{7}$; the levels of the decision variables are $x_1 = 6\frac{3}{7}$ hundred cases of six-ounce juice glasses and $x_2 = 4\frac{2}{7}$ hundred cases of ten-ounce cocktail glasses. Only the constraint on demand for six-ounce juice glasses has slack in it, since the custom molder could have chosen to make an additional $1\frac{4}{7}$ hundred cases if he had wanted to decrease the production of ten-ounce cocktail glasses appropriately.

Shadow Prices on the Constraints

Solving a linear program usually provides more information about an optimal solution than merely the values of the decision variables. Associated with an optimal solution are *shadow prices* (also referred to as *dual variables, marginal values,* or *pi values*) for the constraints. The shadow price on a particular constraint represents the change in the value of the objective function per unit increase in the righthand-side value of that constraint. For example, suppose that the number of hours of molding-machine capacity was increased from 60 hours to 61 hours. What is the change in the value of the objective function from such an increase? Since the constraints on production capacity and storage capacity remain binding with this increase, we need only solve

$$6x_1 + 5x_2 = 61,$$
$$10x_1 + 20x_2 = 150,$$

to find a new optimal solution. The new values of the decision variables are $x_1 = 6\frac{5}{7}$ and $x_2 = 4\frac{1}{7}$, and the new value of the objective function is:

$$z = 500x_1 + 450x_2 = 500(6\tfrac{5}{7}) + 450(4\tfrac{1}{7}) = 5,221\tfrac{3}{7}.$$

The shadow price associated with the constraint on production capacity then becomes:

$$5221\tfrac{3}{7} - 5142\tfrac{6}{7} = 78\tfrac{4}{7}.$$

The shadow price associated with production capacity is $\$78\frac{4}{7}$ per additional hour of production time. This is important information since it implies that it would be profitable to invest up to $\$78\frac{4}{7}$ each week to increase production time by one hour. Note that the units of the shadow price are determined by the ratio of the units of the objective function and the units of the particular constraint under consideration.

We can perform a similar calculation to find the shadow price $2\frac{6}{7}$ associated with the storage-capacity constraint, implying that an addition of one hundred cubic feet of storage capacity is worth $\$2\frac{6}{7}$. The shadow price associated with the demand for six-ounce juice glasses clearly must be zero. Since currently we are not producing up to the 800-case limit, increasing this limit will certainly not change our decision and therefore will not be of any value.

Finally, we must consider the shadow prices associated with the nonnegativity constraints. These shadow prices often are called the *reduced costs* and usually are reported separately from the shadow prices on the other constraints; however, they have the identical interpretation. For our problem, increasing either of the nonnegativity constraints separately will not affect the optimal solution, so the values of the shadow prices, or reduced costs, are zero for both nonnegativity constraints.

Objective and Righthand-Side Ranges

The data for a linear program may not be known with certainty or may be subject to change. When solving linear programs, then, it is natural to ask about the *sensitivity* of the optimal solution to *variations in the data*. For example, over what range can a particular objective-function coefficient vary without changing the optimal solution?

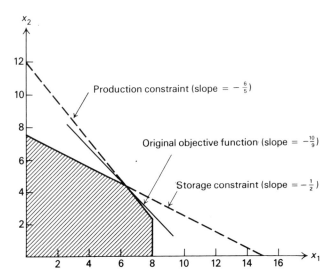

Fig. 1.5   Range on the slope of the objective function.

It is clear from Fig. 1.5 that some variation of the contribution coefficients is possible without a change in the optimal levels of the decision variables. Throughout our discussion of shadow prices, we assumed that the *constraints* defining the optimal solution did not change when the values of their righthand sides were varied. Further, when we made changes in the righthand-side values we made them one at a time, leaving the remaining coefficients and values in the problem unchanged. The question naturally arises, over what range can a particular righthand-side value change without changing the shadow prices associated with that constraint? These questions of simple one-at-a-time changes in either the objective-function coefficients or the right-hand-side values are determined easily and therefore usually are reported in any computer solution.

Changes in the Coefficients of the Objective Function

We will consider first the question of making one-at-a-time changes in the coefficients of the objective function. Suppose we consider the contribution per one hundred cases of six-ounce juice glasses, and determine the range for that coefficient such that the optimal solution remains unchanged. From Fig. 1.5, it should be clear that the optimal solution remains unchanged as long as the slope of the objective function lies between the slope of the constraint on production capacity and the slope of the constraint on storage capacity.

We can determine the range on the coefficient of contribution from six-ounce juice glasses, which we denote by c_1, by merely equating the respective slopes. Assuming the remaining coefficients and values in the problem remain unchanged, we must have:

$$\text{Production slope} \leqq \text{Objective slope} \leqq \text{Storage slope.}$$

Since $z = c_1 x_1 + 450 x_2$ can be written as $x_2 = (z/450) - (c_1/450)x_1$, we see, as before, that the objective slope is $-(c_1/450)$. Thus

$$-\frac{6}{5} \leqq -\frac{c_1}{450} \leqq -\frac{1}{2}$$

or, equivalently,

$$225 \leqq c_1 \leqq 540,$$

where the current value of $c_1 = 500$.

Similarly, by holding c_1 fixed at 500, we can determine the range of the coefficient of contribution from ten-ounce cocktail glasses, which we denote by c_2:

$$-\frac{6}{5} \leqq -\frac{500}{c_2} \leqq -\frac{1}{2}$$

or, equivalently,

$$416\tfrac{2}{3} \leqq c_2 \leqq 1000,$$

where the current value of $c_2 = 450$. The objective ranges are therefore the range over which a particular objective coefficient can be varied, all other coefficients and values in the problem remaining unchanged, and have the optimal solution (i.e., levels of the decision variables) remain unchanged.

From Fig. 1.5 it is clear that the same binding constraints will define the optimal solution. Although the levels of the decision variables remain unchanged, the *value* of the objective function, and therefore the shadow prices, will change as the objective-function coefficients are varied.

It should now be clear that an optimal solution to a linear program is not always unique. If the objective function is parallel to one of the binding constraints, then

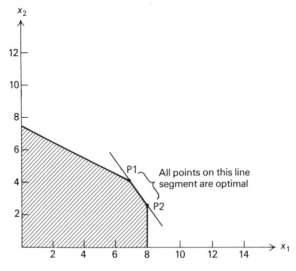

Fig. 1.6 Objective function coincides with a constraint.

there is an entire set of optimal solutions. Suppose that the objective function were

$$z = 540x_1 + 450x_2.$$

It would be parallel to the line determined by the production-capacity constraint; and all levels of the decision variables lying on the line segment joining the points labeled P1 and P2 in Fig. 1.6 would be optimal solutions.

Changes in the Righthand-Side Values of the Constraints

Now consider the question of making one-at-a-time changes in the righthand-side values of the constraints. Suppose that we want to find the range on the number of hours of production capacity that will leave all of the shadow prices unchanged. The essence of our procedure for computing the shadow prices was to assume that the constraints defining the optimal solution would remain the same even though a righthand-side value was being changed. First, let us consider increasing the number of hours of production capacity. How much can the production capacity be increased and still give us an increase of $78\frac{4}{7}$ per hour of increase? Looking at Fig. 1.7, we see that we cannot usefully increase production capacity beyond the point where storage capacity and the limit on demand for six-ounce juice glasses become binding. This point is labeled P3 in Fig. 1.7. Any further increase in production hours would be worth zero since they would go unused. We can determine the number of hours of production capacity corresponding to the point labeled P3, since this point is characterized by $x_1 = 8$ and $x_2 = 3\frac{1}{2}$. Hence, the upper bound on the range of the righthand-side value for production capacity is $6(8) + 5(3\frac{1}{2}) = 65\frac{1}{2}$ hours.

Alternatively, let us see how much the capacity can be *decreased* before the shadow prices change. Again looking at Fig. 1.7, we see that we can decrease production capacity to the point where the constraint on storage capacity and the nonnegativity

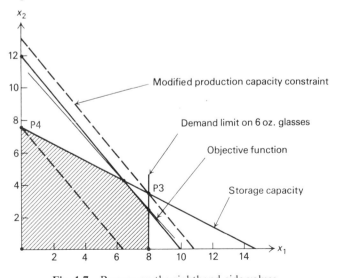

Fig. 1.7   Ranges on the righthand-side values.

constraint on ten-ounce cocktail glasses become binding. This point is labeled P4 in Fig. 1.7 and corresponds to only $37\frac{1}{2}$ hours of production time per week, since $x_1 = 0$ and $x_2 = 7\frac{1}{2}$. Any further decreases in production capacity beyond this point would result in lost contribution of $90 per hour of further reduction. This is true since at this point it is optimal just to produce as many cases of the ten-ounce cocktail glasses as possible while producing no six-ounce juice glasses at all. Each hour of reduced production time now causes a reduction of $\frac{1}{5}$ of one hundred cases of ten-ounce cocktail glasses valued at $450, i.e., $\frac{1}{5}(450) = 90$. Hence, the range over which the shadow prices remain unchanged is the range over which the optimal solution is defined by the same binding constraints. If we take the righthand-side value of production capacity to be b_1, the range on this value, such that the shadow prices will remain unchanged, is:

$$37\tfrac{1}{2} \leqq b_1 \leqq 65\tfrac{1}{2},$$

where the current value of production capacity $b_1 = 60$ hours. It should be emphasized again that this righthand-side range assumes that all other righthand-side values and all variable coefficients in the problem remain unchanged.

In a similar manner we can determine the ranges on the righthand-side values of the remaining constraints:

$$128 \leqq b_2 \leqq 240,$$
$$6\tfrac{3}{7} \leqq b_3.$$

Observe that that there is no upper bound on six-ounce juice-glass demand b_3, since this constraint is nonbinding at the current optional solution P1.

We have seen that both cost and righthand-side ranges are valid if the coefficient or value is varied by itself, all other variable coefficients and righthand-side values being held constant. The objective ranges are the ranges on the coefficients of the objective function, varied one at a time, such that the levels of the decision variables remain unchanged in the optimal solution. The righthand-side ranges are the ranges on the righthand-side values, varied one at a time, such that the shadow prices associated with the optimal solution remain unchanged. In both instances, the ranges are defined so that the binding constraints at the optimal solution remain unchanged.

Computational Considerations

Until now we have described a number of the properties of an optimal solution to a linear program, assuming first that there was such a solution and second that we were able to find it. It could happen that a linear program has *no* feasible solution. An infeasible linear program might result from a poorly formulated problem, or from a situation where requirements exceed the capacity of the existing available resources. Suppose, for example, that an additional constraint was added to the model imposing a minimum on the number of production hours worked. However, in recording the data, an error was made that resulted in the following constraint being included in the formulation:

$$6x_1 + 5x_2 \geqq 80.$$

The graphical representation of this error is given in Fig. 1.8.

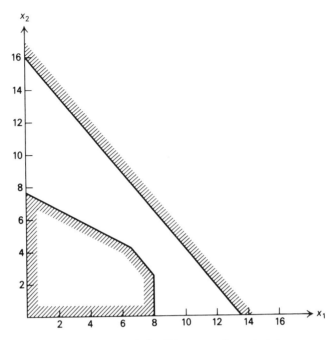

Fig. 1.8   An infeasible group of constraints.

The shading indicates the direction of the inequalities associated with each constraint. Clearly, there are no points that satisfy all the constraints simultaneously, and the problem is therefore infeasible. Computer routines for solving linear programs must be able to tell the user when a situation of this sort occurs. In general, on large problems it is relatively easy to have infeasibilities in the initial formulation and not know it. Once these infeasibilities are detected, the formulation is corrected.

Another type of error that can occur is somewhat less obvious but potentially more costly. Suppose that we consider our original custom-molder problem but that a control message was typed into the computer incorrectly so that we are, in fact, attempting to solve our linear program with "*greater* than or equal to" constraints instead of "*less* than or equal to" constraints. We would then have the following linear program:

Maximize $z = 500x_1 + 450x_2,$

subject to:

$$6x_1 + 5x_2 \geqq 60,$$
$$10x_1 + 20x_2 \geqq 150,$$
$$x_1 \qquad \geqq 8,$$
$$x_1 \geqq 0, \qquad x_2 \geqq 0.$$

The graphical representation of this error is given in Fig. 1.9.

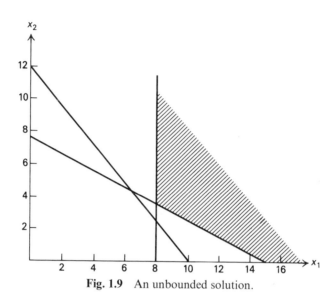

Fig. 1.9   An unbounded solution.

Clearly, the maximum of the objective function is now unbounded. That is, we apparently can make our contribution arbitrarily large by increasing either x_1 or x_2. Linear-programming solution procedures also detect when this kind of error has been made, and automatically terminate the calculations indicating the direction that produces the unbounded objective value.

Integer Solutions

In linear programming, the decision space is *continuous*, in the sense that fractional answers always are allowed. This is contrasted with discrete or integer programming, where integer values are required for some or all variables. In our custom-molding example, if the customer will accept each product only in even hundred-case lots, in order to ease his reordering situation, and if, further, we choose not to store either product from one week to the next, we must seek an *integer solution* to our problem. Our initial reaction is to round off our continuous solution, yielding

$$x_1 = 6, \qquad x_2 = 4,$$

which in this case is feasible, since we are rounding down. The resulting value of contribution is $z = \$4800$. Is this the optimal integer solution? We clearly could increase the total contribution if we could round either x_1 or x_2 *up* instead of down. However, neither the point $x_1 = 7$, $x_2 = 4$, nor the point $x_1 = 6$, $x_2 = 5$ is within the feasible region.

Another alternative would be to start with our trial solution $x_1 = 6$, $x_2 = 4$, and examine "nearby" solutions such as $x_1 = 5$, $x_2 = 5$, which turns out to be feasible and which has a contribution of $z = \$4750$, not as high as our trial solution. Another "nearby" solution is $x_1 = 7$, $x_2 = 3$, which is also feasible and has a contribution $z = \$4850$. Since this integer solution has a higher contribution than any previous integer solution, we can use it as our new trial solution. It turns out that the *optimal integer solution* for our problem is:

$$x_1 = 8, \qquad x_2 = 2,$$

with a contribution of $z = \$4900$. It is interesting to note that this solution is not particularly "nearby" the optimal *continuous* solution $x_1 = 6\frac{3}{7}, x_2 = 4\frac{2}{7}$.

Basically, the integer-programming problem is inherently difficult to solve and falls in the domain of *combinatorial analysis* rather than simple linear programming. Special algorithms have been developed to find optimal integer solutions; however, the size of problem that can be solved successfully by these algorithms is an order of magnitude smaller than the size of linear programs that can easily be solved. Whenever it is possible to avoid integer variables, it is usually a good idea to do so. Often what at first glance seem to be integer variables can be interpreted as production or operating *rates*, and then the integer difficulty disappears. In our example, if it is not necessary to ship in even hundred-case lots, or if the odd lots are shipped the following week, then it still is possible to produce at rates $x_1 = 6\frac{3}{7}$ and $x_2 = 4\frac{2}{7}$ hundred cases per week. Finally, in any problem where the numbers involved are large, rounding to a feasible integer solution usually results in a good approximation.

Nonlinear Functions

In linear programming the variables are continuous and the functions involved are linear. Let us now consider the decision problem of the custom molder in a slightly altered form. Suppose that the interpretation of production in terms of rates is acceptable, so that we need not find an integer solution; however, we now have a *nonlinear* objective function. We assume that the injection-molding machine is fairly old and that operating near capacity results in a reduction in contribution per unit for each additional unit produced, due to higher maintenance costs and downtime. Assume that the reduction is $0.05 per case of six-ounce juice glasses and $0.04 per case of ten-ounce cocktail glasses. In fact, let us assume that we have fit the following functions to per-unit contribution for each type of glass:

$$z_1(x_1) = 60 - 5x_1 \qquad \text{for six-ounce juice glasses,}$$
$$z_2(x_2) = 80 - 4x_2 \qquad \text{for ten-ounce cocktail glasses.}$$

The resulting total contribution is then given by:

$$(60 - 5x_1)x_1 + (80 - 4x_2)x_2.$$

Hence, we have the following nonlinear programming problem to solve:

$$\text{Maximize } z = 60x_1 - 5x_1^2 + 80x_2 - 4x_2^2,$$

subject to:

$$6x_1 + 5x_2 \leq 60,$$
$$10x_1 + 20x_2 \leq 150,$$
$$x_1 \qquad\quad \leq 8,$$
$$x_1 \geq 0, \qquad x_2 \geq 0.$$

This situation is depicted in Fig. 1.10. The curved lines represent lines of constant contribution.

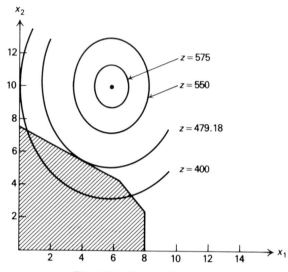

Fig. 1.10   The nonlinear program.

Note that the optimal solution is no longer at a corner point of the feasible region. This property alone makes finding an optimal solution much more difficult than in linear programming. In this situation, we cannot merely move from vertex to vertex, looking for an improvement at each iteration. However, this particular problem has the property that, if you have a trial solution and cannot find an improving direction to move in, then *the trial solution is an optimal solution.* It is this property that is generally exploited in computational procedures for nonlinear programs.

1.5 A CLASSIFICATION OF MATHEMATICAL PROGRAMMING MODELS

We are now in a position to provide a general statement of the mathematical programming problem and formally summarize definitions and notation introduced previously. Let us begin by giving a formal representation of the general linear programming model.

In mathematical terms, the linear programming model can be expressed as the maximization (or minimization) of an *objective* (or criterion) *function*, subject to a given set of linear *constraints*. Specifically, the linear programming problem can be described as finding the values of n decision variables, $x_1, x_2, \ldots, x_n$, such that they maximize the objective function z where

$$z = c_1 x_1 + c_2 x_2 + \cdots + c_n x_n, \tag{22}$$

subject to the following constraints:

$$
\begin{aligned}
a_{11} x_1 + a_{12} x_2 + \cdots + a_{1n} x_n &\quad b_1, \\
a_{21} x_1 + a_{22} x_2 + \cdots + a_{2n} x_n &\le b_2, \\
\vdots \qquad\qquad\qquad \vdots & \\
a_{m1} x_1 + a_{m2} x_2 + \cdots + a_{mn} x_n &\le b_m,
\end{aligned}
\tag{23}
$$

and, usually,

$$x_1 \geqq 0, \qquad x_2 \geqq 0, \qquad \ldots, \qquad x_n \geqq 0, \qquad (24)$$

where c_j, a_{ij}, and b_i are given constants.

It is easy to provide an immediate interpretation to the general linear-programming problem just stated in terms of a production problem. For instance, we could assume that, in a given production facility, there are n possible products we may manufacture; for each of these we want to determine the level of production which we shall designate by $x_1, x_2, \ldots, x_n$. In addition, these products compete for m limited resources, which could be manpower availability, machine capacities, product demand, working capital, and so forth, and are designated by $b_1, b_2, \ldots, b_m$. Let a_{ij} be the amount of resource i required by product j and let c_j be the unit profit of product j. Then the linear-programming model seeks to determine the production quantity of each product in such a way as to maximize the total resulting profit z (Eq. 22), given that the available resources should not be exceeded (constraints 23), and that we can produce only *positive* or *zero* amounts of each product (constraints 24).

Linear programming is not restricted to the structure of the problem presented above. First, it is perfectly possible to minimize, rather than maximize, the objective function. In addition, "greater than or equal to" or "equal to" constraints can be handled simultaneously with the "less than or equal to" constraints presented in constraints (23). Finally, some of the variables may assume both positive and negative values.

There is some technical terminology associated with mathematical programming, informally introduced in the previous section, which we will now define in more precise terms. Values of the decision variables $x_1, x_2, \ldots, x_n$ that satisfy all the constraints of (23) and (24) simultaneously are said to form a *feasible solution* to the linear programming problem. The set of *all* values of the decision variables characterized by constraints (23) and (24) form the *feasible region* of the problem under consideration. A feasible solution that in addition optimizes the objective function (22) is called an *optimum feasible solution*.

As we have seen in the geometric representation of the problem, solving a linear program can result in three possible situations.

i) The linear program could be infeasible, meaning that there are no values of the decision variables $x_1, x_2, \ldots, x_n$ that simultaneously satisfy all the constraints of (23) and (24).

ii) It could have an unbounded solution, meaning that, if we are maximizing, the value of the objective function can be increased indefinitely without violating any of the constraints. (If we are minimizing, the value of the objective function may be decreased indefinitely.)

iii) In most cases, it will have at least one finite optimal solution and often it will have multiple optimal solutions.

The *simplex method* for solving linear programs, which will be discussed in Chapter 2, provides an efficient procedure for constructing an optimal solution, if one exists, or for determining whether the problem is infeasible or unbounded.

Note that, in the linear programming formulation, the decision variables are allowed to take any continuous value. For instance, values such that $x_1 = 1.5$, $x_2 = 2.33$, are perfectly acceptable as long as they satisfy constraints (23) and (24). An important extension of this linear programming model is to require that all or some of the decision variables be restricted to be integers.

Another fundamental extension of the above model is to allow the objective function, or the constraints, or both, to be nonlinear functions. The general nonlinear programming model can be stated as finding the values of the decision variables x_1, $x_2, \ldots, x_n$ that maximize the objective function z where

$$z = f_0(x_1, x_2, \ldots, x_n), \tag{25}$$

subject to the following constraints:

$$
\begin{aligned}
f_1(x_1, x_2, \ldots, x_n) &\leq b_1, \\
f_2(x_1, x_2, \ldots, x_n) &\leq b_2, \\
&\vdots \\
f_m(x_1, x_2, \ldots, x_n) &\leq b_m,
\end{aligned}
\tag{26}
$$

and sometimes

$$x_1 \geq 0, \qquad x_2 \geq 0, \qquad \ldots, \qquad x_n \geq 0. \tag{27}$$

Often in nonlinear programming the righthand-side values are included in the definition of the function $f_i(x_1, x_2, \ldots, x_n)$, leaving the righthand side zero. In order to solve a nonlinear programming problem, some assumptions must be made about the shape and behavior of the functions involved. We will leave the specifics of these assumptions until later. Suffice it to say that the nonlinear functions must be rather well-behaved in order to have computationally efficient means of finding a solution.

Optimization models can be subject to various classifications depending on the point of view we adopt. According to the number of time periods considered in the model, optimization models can be classified as *static* (single time period) or *multistage* (multiple time periods). Even when all relationships are linear, if several time periods are incorporated in the model the resulting linear program could become prohibitively large for solution by standard computational methods. Fortunately, in most of these cases, the problem exhibits some form of special structure that can be adequately exploited by the application of special types of mathematical programming methods. Dynamic programming, which is discussed in Chapter 11, is one approach for solving multistage problems. Further, there is a considerable research effort underway today, in the field of large-scale linear programming, to develop special algorithms to deal with multistage problems. Chapter 12 addresses these issues.

Another important way of classifying optimization models refers to the behavior of the parameters of the model. If the parameters are known constants, the optimization model is said to be *deterministic*. If the parameters are specified as uncertain quantities, whose values are characterized by probability distributions, the optimization model is said to be *stochastic*. Finally, if some of the parameters are allowed to vary systematically, and the changes in the optimum solution corresponding to changes in those parameters are determined, the optimization model is said to be *parametric*. In general, stochastic and parametric mathematical programming give rise to much more difficult problems than deterministic mathematical programming. Although important theoretical and practical contributions have been made in the areas of stochastic and parametric programming, there are still no effective general procedures that cope with these problems. Deterministic linear programming, however, can be efficiently applied to very large problems of up to 5000 rows and an almost unlimited number of variables. Moreover, in linear programming, sensitivity analysis and parametric programming can be conducted effectively after obtaining the deterministic optimum solution, as will be seen in Chapter 3.

A third way of classifying optimization models deals with the behavior of the variables in the optimal solution. If the variables are allowed to take any value that satisfies the constraints, the optimization model is said to be *continuous*. If the variables are allowed to take on only discrete values, the optimization model is called *integer* or *discrete*. Finally, when there are some integer variables and some continuous variables in the problem, the optimization model is said to be *mixed*. In general, problems with integer variables are significantly more difficult to solve than those with continuous variables. Network models, which are discussed in Chapter 8, are a class of linear programming models that are an exception to this rule, as their special structure results in integer optimal solutions. Although significant progress has been made in the general area of mixed and integer linear programming, there is still no algorithm that can efficiently solve all general medium-size integer linear programs in a reasonable amount of time though, for special problems, adequate computational techniques have been developed. Chapter 9 comments on the various methods available to solve integer programming problems.

EXERCISES

1. Indicate graphically whether each of the following linear programs has a feasible solution. Graphically determine the optimal solution, if one exists, or show that none exists.

a)   Maximize $z = x_1 + 2x_2$,

subject to:
$$x_1 - 2x_2 \leq 3,$$
$$x_1 + x_2 \leq 3,$$
$$x_1 \geq 0, \qquad x_2 \geq 0.$$

b)   Minimize $z = x_1 + x_2$,

subject to:
$$x_1 - x_2 \leq 2,$$
$$x_1 - x_2 \geq -2,$$
$$x_1 \geq 0, \qquad x_2 \geq 0.$$

c) Redo (b) with the objective function

$$\text{Maximize } z = x_1 + x_2.$$

d) $\text{Maximize } z = 3x_1 + 4x_2,$

subject to:

$$x_1 - 2x_2 \geq 4,$$
$$x_1 + x_2 \leq 3,$$
$$x_1 \geq 0, \qquad x_2 \geq 0.$$

2. Consider the following linear program:

$$\text{Maximize } z = 2x_1 + x_2,$$

subject to:

$$12x_1 + 3x_2 \leq 6,$$
$$-3x_1 + x_2 \leq 7,$$
$$x_2 \leq 10,$$
$$x_1 \geq 0, \qquad x_2 \geq 0.$$

a) Draw a graph of the constraints and shade in the feasible region. Label the vertices of this region with their coordinates.
b) Using the graph obtained in (a), find the optimal solution and the maximum value of the objective function.
c) What is the slack in each of the constraints?
d) Find the shadow prices on each of the constraints.
e) Find the ranges associated with the two coefficients of the objective function.
f) Find the righthand-side ranges for the three constraints.

3. Consider the bond-portfolio problem formulated in Section 1.3. Reformulate the problem restricting the bonds available only to bonds A and D. Further add a constraint that the holdings of municipal bonds must be less than or equal to $3 million.

a) What is the optimal solution?
b) What is the shadow price on the municipal limit?
c) How much can the municipal limit be relaxed before it becomes a nonbinding constraint?
d) Below what interest rate is it favorable to borrow funds to increase the overall size of the portfolio?
e) Why is this rate less than the earnings rate on the portfolio as a whole?

4. A liquor company produces and sells two kinds of liquor: blended whiskey and bourbon. The company purchases intermediate products in bulk, purifies them by repeated distillation, mixes them, and bottles the final product under their own brand names. In the past, the firm has always been able to sell all that it produced.

The firm has been limited by its machine capacity and available cash. The bourbon requires 3 machine hours per bottle while, due to additional blending requirements, the blended whiskey requires 4 hours of machine time per bottle. There are 20,000 machine hours available in the current production period. The direct operating costs, which are principally for labor and materials, are $3.00 per bottle of bourbon and $2.00 per bottle of blended whiskey. The working capital available to finance labor and material is $4000;

however, 45% of the bourbon sales revenues and 30% of the blended-whiskey sales revenues from production in the current period will be collected during the current period and be available to finance operations. The selling price to the distributor is $6 per bottle of bourbon and $5.40 per bottle of blended whiskey.

a) Formulate a linear program that maximizes contribution subject to limitations on machine capacity and working capital.

b) What is the optimal production mix to schedule?

c) Can the selling prices change without changing the optimal production mix?

d) Suppose that the company could spend $400 to repair some machinery and increase its available machine hours by 2000 hours. Should the investment be made?

e) What interest rate could the company afford to pay to borrow funds to finance its operations during the current period?

5. The truck-assembly division of a large company produces two different models: the Aztec and the Bronco. Their basic operation consists of separate assembly departments: drive-train, coachwork, Aztec final, and Bronco final. The drive-train assembly capacity is limited to a total of 4000 units per month, of either Aztecs or Broncos, since it takes the same amount of time to assemble each. The coachwork capacity each month is either 3000 Aztecs or 6000 Broncos. The Aztecs, which are vans, take twice as much time for coachwork as the Broncos, which are pickups. The final assembly of each model is done in separate departments because of the parts availability system. The division can do the final assembly of up to 2500 Aztecs and 3000 Broncos each month.

The profit per unit on each model is computed by the firm as follows:

	Aztec		Bronco	
Selling Price		$4200		$4000
Material cost	$2300		$2000	
Labor cost	400	2700	450	2450
Gross Margin		$1500		$1550
Selling & Administrative*	$ 210		$ 200	
Depreciation†	60		180	
Fixed overhead†	50		150	
Variable overhead	590	910	750	1280
Profit before taxes		$ 590		$ 270

* 5% of selling price.
† Allocated according to planned production of 1000 Aztecs and 3000 Broncos per month for the coming year.

a) Formulate a linear program to aid management in deciding how many trucks of each type to produce per month.

b) What is the proper objective function for the division?

c) How many Aztecs and Broncos should the division produce each month?

6. Suppose that the division described in Exercise 5 now has the opportunity to increase its drive-train capacity by subcontracting some of this assembly work to a nearby firm. The

drive-train assembly department cost breakdown per unit is as follows:

	Aztec	*Bronco*
Direct labor	$ 80	$ 60
Fixed overhead	20	60
Variable overhead	200	150
	$300	$270

The subcontractor will pick up the parts and deliver the assembled drive-trains back to the division. What is the maximum amount that the division would be willing to pay the sub-contractor for assembly of each type of drive-train?

7. Suppose that the division described in Exercises 5 and 6 has decided against subcontracting the assembly of drive-trains but now is considering assembling drive-trains on overtime. If there is a 50% overtime labor premium on each drive-train assembled on overtime and increased fixed overhead of $150,000 per month, should the division go on overtime drive-train production? Formulate a linear program to answer this question, but do not solve explicitly.

8. A manufacturer of wire cloth products can produce four basic product lines: (1) industrial wire cloth; (2) insect screen; (3) roofing mesh; and (4) snow fence. He can sell all that he can produce of each product line for the next year.

The production process for each product line is essentially the same. Aluminum wire is purchased in large diameters of approximately 0.375 inches and drawn down to finer diameters of 0.009 to 0.018 inches. Then the fine wire is woven on looms, in much the same manner as textiles. Various types of different wire meshes are produced, depending on product line. For example, industrial wire cloth consists of meshes as fine as 30 wires per inch, while snow fence has approximately 6 wires per inch. The production process is limited by both wire-drawing capacity and weaving capacity, as well as the availability of the basic raw material, large-diameter aluminum wire.

For the next planning period, there are 600 hours of wire-drawing machine capacity, 1000 hours of loom capacity, and 15 cwt (hundred weight) of large-diameter aluminum wire. The four product lines require the following inputs to make a thousand square feet of output:

	Aluminum wire (cwt)	*Wire drawing* (100's of hrs)	*Weaving* (100's of hrs)
Industrial cloth	1	1	2
Insect screen	3	1	1
Roofing mesh	3	2	1.5
Snow fence	2.5	1.5	2

The contributions from industrial cloth, insect screen, roofing mesh, and snow fence are 2.0, 3.0, 4.2, and 4.0, respectively, in 100's of dollars per thousand square feet.

a) Formulate a linear program to maximize the firm's contribution from the four product lines.

b) Since roofing mesh and snow fence are really new product lines, analyze the problem graphically, considering *only* industrial wire cloth and insect screen.

   i) What is the optimal solution using only these two product lines?
   ii) Over what range of contribution of screen cloth will this solution remain optimal?
   iii) Determine the shadow prices on the resources for this solution. Give an economic interpretation of these shadow prices.
   iv) Suppose that the manufacturer is considering adding wire-drawing capacity. Over what range of wire-drawing capacity will the shadow price on this constraint remain unchanged? In this range, what happens to the other shadow prices?

c) Now, considering the *full* line of four products, answer the following, making use of the information developed in (b).

   i) Will it pay to transfer resources into the production of roofing mesh or snow fence?
   ii) Without performing the calculations, what will be the form of the optimal solution of the full product line? What is the largest number of product lines that will be produced?
   iii) Suppose that in the future some new product line is developed. Is it possible that one of the product lines *not* presently included in the optimal solution will then be included? Explain.

9. The Candid Camera Company manufactures three lines of cameras: the Cub, the Quickiematic and the VIP, whose contributions are $3, $9, and $25, respectively. The distribution center requires that at least 250 Cubs, 375 Quickiematics, and 150 VIPs be produced each week.

   Each camera requires a certain amount of time in order to: (1) manufacture the body parts; (2) assemble the parts (lenses are purchased from outside sources and can be ignored in the production scheduling decision); and (3) inspect, test, and package the final product. The Cub takes 0.1 hours to manufacture, 0.2 hours to assemble, and 0.1 hours to inspect, test, and package. The Quickiematic needs 0.2 hours to manufacture, 0.35 hours to assemble, and 0.2 hours for the final set of operations. The VIP requires 0.7, 0.1, and 0.3 hours, respectively. In addition, there are 250 hours per week of manufacturing time available, 350 hours of assembly, and 150 hours total to inspect, test, and package.

   Formulate this scheduling problem as a linear program that maximizes contribution.

10. A leather-goods factory manufactures five styles of handbags, whose variable contributions are $30, $40, $45, $25, and $60 per dozen, respectively. The products must pass through four work centers and the man-hours available in each are: clicking (700), paring (600), stitching (400), and finishing (900). Hourly requirements for each dozen handbags are:

Style	Click	Pare	Stitch	Finish
341	3	8	2	6
262	4	3	1	0
43	2	2	0	2
784	2	1	3	4
5-A	5	4	4	3

To prevent adding to inventory levels already on hand, the production manager, after reviewing the weekly sales forecasts, has specified that no more than 100, 50, 90, 70, and 30 dozen of each style respectively may be produced. Each handbag is made from five materials as specified in the following table:

Style	Leather	Fabric	Backing	Lining	Accessories
341	0	1	4	2	3
262	4	0	7	4	4
43	5	7	6	2	0
784	6	4	1	1	2
5-A	2	3	3	0	4
Total available	300	400	1000	900	1600

Formulate a linear program for next week's optimum manufacturing schedule if overall contribution is to be maximized.

11. A corporation that produces gasoline and oil specialty additives purchases three grades of petroleum distillates, A, B, and C. The company then combines the three according to specifications of the maximum or minimum percentages of grades A or C in each blend.

Mixture	Max % allowed for Additive A	Min % allowed for Additive C	Selling price $/gallon
Deluxe	60%	20%	7.9
Standard	15%	60%	6.9
Economy	—	50%	5.0

Supplies of the three basic additives and their costs are:

Distillate	Maximum quantity available per day (gals)	Cost $/gallon
A	4000	0.60
B	5000	0.52
C	2500	0.48

Show how to formulate a linear program to determine the production policy that will maximize profits.

12. Universal Aviation currently is investigating the possibility of branching out from its passenger service into the small-plane air-freight business. With $4,000,000 available to invest in the purchase of new twin-engined cargo aircraft, Universal is considering three types of planes. Aircraft A, costing $80,000, has a ten-ton payload and is expected to cruise at 350 knots, while airplane B can haul 20 tons of goods at an average speed of 300 knots. Aircraft B will cost $130,000. The third aircraft is a modified form of B with provisions for a copilot, a 300-knot cruising speed, a reduced capacity of 18 tons, and a cost of $150,000.

Plane A requires one pilot and, if flown for three shifts, could average 18 hours a day in the air, as could aircraft B. While transports B and C both require a crew of two, C could average 21 hours of flying per day on a three-shift pilot basis due to superior loading equipment. Universal's operations department currently estimates that 150 pilot-shifts will be available for each day's cargo operations. Limitations on maintenance facilities indicate that no more than thirty planes can be purchased. The contributions of planes A, B, and C per ton-mile are, respectively, $1, $8, and $120. 3,500,000 ton-miles of shipping must be completed each day, but deliveries not completed by the in-house fleet can still be subcontracted to outside air carriers at a contribution of $0.20 per ton mile. What "mix" of aircraft should be purchased if the company desires to maximize its contribution per day? (*Note.* Consider a knot = 1 mile/hour.)

13. A mobile-home manufacturer in Indiana channels its mobile-home units through distribution centers located in Elkhart, Ind., Albany, N.Y., Camden, N.J., and Petersburg, Va. An examination of their shipping department records indicates that, in the upcoming quarter, the distribution centers will have in inventory 30, 75, 60, and 35 mobile homes, respectively. Quarterly orders submitted by dealerships serviced by the distribution centers require the following numbers of mobile home units for the upcoming quarter:

	Number of units		Number of units
Dealer A	25	Dealer D	25
Dealer B	40	Dealer E	50
Dealer C	15	Dealer F	45

Transportation costs (in dollars per unit) between each distribution center and the dealerships are as shown in the table below.

Distribution centers	Dealers					
	A	B	C	D	E	F
Elkhart	75	65	175	90	110	150
Albany	90	30	45	50	105	130
Camden	40	55	35	80	70	75
Petersburg	95	150	100	115	55	55

a) Formulate this problem as a linear-programming problem with the objective of minimizing the transportation costs from the distribution centers to the dealerships.

b) Suppose Dealer E had placed an order for 65 units, assuming all other data remain unchanged. How does this affect the problem? (Note that total supply is less than total demand.)

14. A strategic planner for an airline that flies to four different cities from its Boston base owns 10 large jets (B707's), 15 propeller-driven planes (Electra's), and two small jets (DC9's).

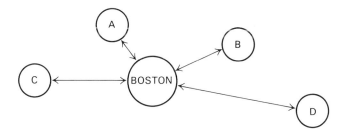

Assuming constant flying conditions and passenger usage, the following data is available.

	City	Trip cost*	Trip revenue	Average flying time (hours)
B707	A	$ 6,000	$ 5,000	1
	B	7,000	7,000	2
	C	8,000	10,000	5
	D	10,000	18,000	10
Electra	A	$ 1,000	$ 3,000	2
	B	2,000	4,000	4
	C	4,000	6,000	8
	D	——	——	20
DC9	A	2,000	$ 4,000	1
	B	3,500	5,500	2
	C	6,000	8,000	6
	D	10,000	14,000	12

* Data is for a round trip.

Formulate constraints to take into account the following:

i) city D must be served twice daily; cities A, B, and C must be served four times daily;
ii) limitation on number of planes available, assuming that each plane can fly at most 18 hours/day.

Formulate objective functions for:

i) cost minimization;
ii) profit maximization;
iii) fleet flying-time minimization.

Indicate when a continuous linear-programming formulation is acceptable, and when an integer-programming formulation is required.

15. The Temporary Help Company must provide secretaries to its clients over the next year on the following estimated schedule: spring, 6000 secretary-days; summer, 7500 secretary-days; fall, 5500 secretary-days; and winter, 9000 secretary-days. A secretary must be trained for 5 days before becoming eligible for assignment to clients.

There are 65 working days in each quarter, and at the beginning of the spring season there are 120 qualified secretaries on the payroll. The secretaries are paid by the company and not the client; they earn a salary of $800 a month. During each quarter, the company loses 15% of its personnel (including secretaries trained in the previous quarter).

Formulate the problem as a linear-programming problem. (*Hint*: Use x_t as the number of secretaries hired at the beginning of season t, and S_t as the total number of secretaries at the beginning of season t.)

16. An imaginary economy has six distinct geographic regions; each has its own specific economic functions, as follows:

Region	Function
A	Food producing
B	Manufacturing—Machinery
C	Manufacturing—Machinery and consumer durables
D	Administrative
E	Food-producing
F	Manufacturing—Consumer durables and nondurables

The regions also have the following annual requirements (all quantities measured in tons):

Region	Food	Machinery	Consumer durables	Consumer nondurables
A	5	30	20	10
B	15	100	40	30
C	20	80	50	40
D	30	10	70	60
E	10	60	30	20
F	25	60	60	50

Using the national railroad, shipping costs are $1/ton per 100 miles for all hauls over 100 miles. Within 100 miles, all goods are carried by truck at a cost of $1.25/ton per 100 miles ($1/ton minimum charge). The distances (in miles) between regions are as follows:

	A	B	C	D	E	F
A	—					
B	500	—				
C	200	400	—			
D	75	500	150	—		
E	600	125	350	550	—	
F	300	200	100	400	300	—

Assume producing regions can meet all requirements, but that, due to government regulation, food production in sector A is restricted to half that of sector E.

Formulate a linear program that will meet the requirements and minimize the total transportation costs in the economy.

17. The Environmental Protection Agency (EPA) wants to restrict the amount of pollutants added by a company to the river water. The concentrations of phenol and nitrogen in the water are to be restricted to, respectively, P and N lbs. MG (million gallons) on a daily basis. The river has a flow of M MG/day. The company diverts a portion of the river water, adds the pollutants, namely, phenol and nitrogen, to it, and sends the water back to the river. The company has four possible ways to treat the water it uses before returning it to the river. The characteristics of each treatment are given in the following table.

(lbs. of pollutant)/MG after treatment

Treatment	1	2	3	4
Phenol	P_1	P_2	P_3	P_4
Nitrogen	N_1	N_2	N_3	N_4
Cost/MG	c_1	c_2	c_3	c_4

Assume: (i) that the river is initially free of pollutants; (ii) that addition of pollutants does not affect the amount of water flow; and (iii) that the company has to process at least K (MG/day) of river water.

a) Set up a linear program to solve for the amount of water to be processed by each treatment, so that total cost of treatment is minimized.

b) How does the formulation change if the EPA regulations apply not to total river concentration downstream from the plant, but rather to the concentration of effluent from the plant?

18. Suppose the company of Exercise 17 has three plants, A, B, and C, in a river network as shown below:

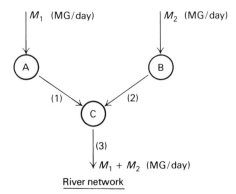

River network

River water can be clean enough for (1) swimming, or (2) to support bio-life (but not clean enough for swimming). Maximum permissible concentrations of pollutants for both cases are given in the table below.

Maximum permissible concentrations (lbs./MG)

	Swimming	Biological life
Phenol	P_S	P_B
Nitrogen	N_S	N_B

Assume again that each plant has four different alternatives for water treatment, as in the table of Exercise 17, and that A, B, and C must process at least K_A, K_B, K_C MG/day, respectively. It is necessary that water in sections (1) and (3) be clean enough for swimming and that water in section (2) be clean enough to support bio-life.

Further, assume: (i) that the river is free of pollutants upstream from plants A and B; and (ii) that adding pollutants does not affect the amount of water flow.

a) Formulate a linear program, including plants A, B, and C, that minimizes total treatment costs.

b) How would the formulation change if all the plants did not necessarily have the same capabilities and cost structure for each treatment? Indicate modifications only—do not reformulate completely.

19. "Chemico" company produces 3 products from a certain mineral. The production process consists of 2 stages: mining the crude mineral, and processing the mineral to produce 3 different products.

The crude mineral can be mined by one of 5 methods. The quality of the mineral depends on the method used and is measured by the percentage of the quantity of the final products that are produced from 1 ton of crude mineral. Furthermore, the distribution of products obtained from 1 ton of crude mineral depends on the mining method used.

A limited amount of the crude mineral can be mined by each method. In addition, the capacity of the production process is limited to 850 tons of crude mineral. The table on page 41 gives the relevant data.

Mine	Mining method	Mining cost (per ton of final products) ($)	No. of hours per 1 ton of final products	Quality	Production capacity (tons of crude material)	Production distribution		
						Product I	Product II	Product III
A	1	250	50	65%	300	20%	30%	50%
A	2	320	62	80%	280	25%	35%	40%
B	3	260	63	76%	400	15%	45%	40%
B	4	210	55	76%	250	7%	24%	69%
B	5	350	60	78%	500	35%	40%	25%

Note that the mining costs are per 1 ton of final products, and not per 1 ton of crude mineral. The production capacity of each mining method is in terms of crude material.

There are 2 mines: Mine A uses methods 1 and 2, whereas Mine B uses methods 3, 4, and 5. The total number of work hours available in Mines A and B are 13,000 and 15,000, respectively (for the planning period). It is possible to transfer workers from A to B and vice versa, but it costs $2 to transfer 1 work hour, and 10% of the transferred work hours are wasted.

The company sells its products in 5 markets. The selling price depends on the market, and in each market the amount that can be sold is limited. The relevant data are given in a second table:

Market	Product	Price per 1 ton ($)	Max. amount (tons)
1	I	1550	150
2	I	1600	100
3	II	1400	200
4	III	1000	50
5	III	850	400

Management wants to plan the production for the next period such that the profit will be maximized. They want to know how many work hours to assign to each mine, what amount to sell in each market, and what is the expected profit. Formulate the problem as a linear program.

20. From past data, the production manager of a factory knows that, by varying his production rate, he incurs additional costs. He estimates that his cost per unit increases by $0.50 when production is increased from one month to the next. Similarly, reducing production increases costs by $0.25 per unit. A smooth production rate is obviously desirable.

Sales forecasts for the next twelve months are (in thousands):

July	4	October	12	January	20	April	6
August	6	November	16	February	12	May	4
September	8	December	20	March	8	June	4

June's production schedule already has been set at 4000 units, and the July 1 inventory level is projected to be 2000 units. Storage is available for only 10,000 units at any one time. Ignoring inventory costs, formulate a production schedule for the coming year that will minimize the cost of changing production rates while meeting all sales demands. (*Hint*: Express the change in production from month t to month $t + 1$ in terms of nonnegative variables x_t^+ and x_t^- as $x_t^+ - x_t^-$. Variable x_t^+ is the increase in production and x_t^- the decrease. It is possible for both x_t^+ and x_t^- to be positive in the optimal solution?)

21. Videocomp, Inc., a new manufacturer of cathode ray tube (CRT) devices for computer applications, is planning to enlarge its capacity over the next two years. The company's primary objective is to grow as rapidly as possible over these two years to make good on its marketing claims.

The CRT's are produced in sets of 200 units on modular assembly lines. It takes three months to produce a set of 200 units from initial chemical coating to final assembly and testing. To ensure quality control, none of the units in a set is shipped until the entire set has been completed. Videocomp has three modular assembly lines and thus currently can produce up to 600 units in a quarter. Each set of 200 units requires $15,000 at the beginning of the quarter when production is initiated for purchasing component parts and paying direct labor expenses. Each set produces revenue of $35,000, of which 20% is received at the time of shipment and the remaining 80% a full three months later.

Videocomp has negotiated the terms for adding modular assembly lines with a number of contractors and has selected two possible contractors. The first contractor requires an investment of $60,000 paid in advance and guarantees that the assembly line will be completed in three months. For the same assembly line, the second contractor requires an investment of $20,000 in advance and an additional investment of $20,000 upon completion; however, his completion time is six months.

The present assets of Videocomp for investment in new modular assembly lines and financing current operations are $150,000. No further funds will be made available except those received from sales of CRT's. However, as the market is expanding rapidly, all CRT's produced can be sold immediately.

Formulate a linear program to maximize Videocomp's productive capacity at the end of two years using eight planning periods of three months' duration each.

22. A rent-a-car company operates a rental-agent training program, which students complete in one month. The teachers in this program are trained rental agents, at a ratio of one for every fifteen students. Experience has shown that twenty-five students must be hired for every twenty who successfully complete the program. The demand for rental cars is seasonal and, for the next six months, requires rental agents as follows:

January	135	April	170
February	125	May	160
March	150	June	180

As of 1 January, 145 trained rental agents are available for counter work or teaching, and the personnel records indicate that 8% of the trained rental agents leave the company at the end of each month. Payroll costs per month are:

Student	$350
Trained agent renting or teaching	$600
Trained agent idle	$500

Company policy forbids firing because of an excess of agents.

Formulate a linear program that will produce the minimum-cost hiring and training schedule that meets the demand requirements, assuming that average labor turnover rates prevail and ignoring the indivisibility of agents. Assume also that in June the training school closes for vacation.

23. The Radex Electronics Corporation is a medium-size electronics firm, one division of which specializes in made-to-order radar and ship-to-shore radio equipment. Radex has been awarded an Army contract to assemble and deliver a number of special radar units that are small enough and light enough to be carried on a man's back. Radex receives $500 for each

radar unit and the delivery schedule is as follows:

Shipping date	Requirements	Cumulative shipments
1 April	300	300
1 May	400	700
1 June	300	1000

The actual cumulative shipments cannot exceed this schedule, since the Army, in order to have time for effective testing before accepting all units, will refuse to accept any faster delivery. Further, it is permissible in the terms of the contract to ship the radar units late; however, a penalty cost of $50 is assessed for each radar unit that is shipped late. If this division of Radex produces faster than the given schedule, then it must purchase storage space in the company warehouse at $10 per assembled radar unit on hand on the first of the month (after the shipment is sent out).

Production requirements are as follows: One radar unit requires 43 standard hours of assembly labor; one trained man (with more than one month's experience), working at 100% efficiency, produces 172 standard hours of output per month, regular time; one new man (with less than one month's experience), rated at 75% efficiency, produces 129 standard hours of output per month.

The labor requirements are as follows: Employees are paid $5.00 per hour of regular time and $7.50 per hour of overtime; a maximum of 35 hours of overtime is allowed per month, but new men may not work overtime; each man on the payroll is guaranteed 172 regular-time hours (that is, $860 per month); at the end of the month, five percent of the labor force quits; hiring costs are $200 per man, and all hiring is done on the first of the month.

On 1 March, Radex begins work on the government contract with 90 trained men on the payroll, and new men can be hired immediately. On completion of the contract 1 June, Radex will begin work on a new contract requiring 200 trained men.

Construct the coefficient matrix for a linear program to optimize the operations of this division of the Radex Electronics Corporation over the indicated time period. *Briefly* define the variables and explain the significance of the equations. (*Suggestion*: Set up the block of the coefficient matrix corresponding to the March decisions first, then indicate how additional blocks should be constructed.)

24. Construct the coefficient matrix, define each of the variables, and explain briefly the significance of each equation or inequality in the linear-programming model for optimal operation, during March and April, of the hydroelectric power system described as follows:

The system consists of two dams and their associated reservoirs and power plants on a river. The important flows of power and water are shown in the accompanying diagram.

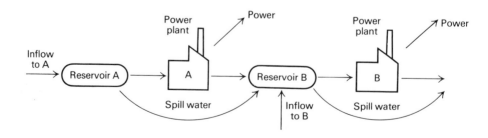

In the following table, all quantities measuring water are in units of 10^3 acre-feet (KAF). Power is measured in megawatt hours (MWH).

	A	B	Units
Storage capacity	2,000	1,500	KAF
Minimum allowable level	1,200	800	KAF
Predicted inflow:			
March	200	40	KAF
April	130	15	KAF
March 1 level	1,900	850	KAF
Water-power conversion	400	200	MWH/KAF
Power-plant capacity	60,000	35,000	MWH/per month

Power can be sold at $5.00 per MWH for up to 50,000 MWH each month, and excess power above that figure can be sold for $3.50 per MWH.

Assume flow rates in and out through the power plants are constant within the month. If the capacity of the reservoir is exceeded, the excess water runs down the spillway and by-passes the power plant. A consequence of these assumptions is that the maximum and minimum water-level constraints need to be satisfied only at the end of the month. (*Suggestion*: First set up the model for the March operating decisions, then modify it as necessary to extend the planning horizon to the end of April.)

25. A problem common to many different industries entails "trim" losses in cutting rolls of paper, textiles, foil, or other material in the process of filling the orders of its customers. The problem arises from the fact that, due to production economies, a factory normally produces rolls of material in standard widths (for example, 100 inches) of fixed length (say, 500 feet). Customers who order from the factory, however, usually require rolls of smaller width for the purposes of their own industrial uses. The rolls are cut on a large cutting machine, the knives of which can be set for virtually any combination of widths so long as the combined total does not exceed the width of the roll. Thus the problem becomes one of assigning the orders in such a manner that the number of standard rolls used to fill the orders is minimized. All wasted material, or "trim loss," represents a loss to the firm. This loss can sometimes be alleviated, however, through recycling or selling as "scrap" or "seconds."

For purposes of illustration, assume that a factory produces newsprint in standard rolls, each having a width of 100 inches, and a fixed length of 500 feet. The factory must fill the following orders: 75 rolls of 24-inch width; 50 rolls of 40-inch width; and 110 rolls of 32-inch width. For simplicity, assume that the factory delivers all orders (no matter what the width) in the standard length of 500 feet. Further, assume that there are on hand as many standard rolls as necessary, and that only the widths on order are cut. Set up the problem as a linear program with integer variables that minimizes the trim losses.

(*Hint*: Completely itemize the number of possible ways n in which a 100-inch roll can be cut into combinations of 24-, 40-, and 32-inch widths; i.e., one 24-inch roll, one 40-inch roll, one 32-inch roll, with 4 inches of trim waste. Then let the decision variable x_i represent the number of rolls cut as combination i, $i = 1, 2, \ldots, n$. For simplicity in itemizing the possible combinations, assume that each standard roll is cut into as many smaller rolls as possible. Thus, if any smaller rolls are produced in excess of the number ordered, they are counted as waste.)

26. For the trim problem described in Exercise 25, assume that the factory has two cutting machines available; machine #1 is capable of cutting a standard roll of 100 inches, and

machine #2 is capable of cutting a standard roll of 90 inches. Each machine can be used to cut no more than 400 rolls. The following orders must be filled: 225 rolls of 40-inch width; 180 rolls of 32-inch width; and 300 rolls of 24-inch width. Using applicable assumptions and hints given in the statement of Exercise 25, formulate this problem as a linear program to allocate the orders to the two cutting machines so as to minimize trim losses.

27. The U.S. Supreme Court decision of 1954 on *de jure* segregation of schools, and recent decisions denying *de facto* segregation and barring "freedom-of-choice" pupil assignments, have forced school districts to devise plans for integrating public schools. Finding a feasible method of achieving racially balanced schools is, at best, difficult. A great number of factors must be taken into consideration.

 For a given school district the following is known. There are G grades and J schools. Each school has a capacity C_{jg} for grade g. In each of I neighborhoods in the district, there is student population S_{ig} for neighborhood i and grade g. Let the distance from neighborhood i to school j be represented by d_{ij}.

a) Formulate a model to assign all students to schools, minimizing total distance.

b) Now let S_{ikg} = the number of students in neighborhood i of race k and grade g; a_k = the maximum percent of racial group k assigned to a school; and b_k = the minimum percent of racial group k. Reformulate the model while also satisfying the racial-balance constraints.

c) Minimizing total distance might lead to some students having to travel great distances by bus, while others would be within walking distance. Reformulate the problems in (a) and (b) to minimize the maximum distance traveled by any student.

28. The selling prices of a number of houses in a particular section of the city overlooking the bay are given in the following table, along with the size of the lot and its elevation:

Selling price P_i	Lot size (sq. ft.) L_i	Elevation (feet) E_i
$155,000	$12,000	350
120,000	10,000	300
100,000	9,000	100
70,000	8,000	200
60,000	6,000	100
100,000	9,000	200

A real-estate agent wishes to construct a model to forecast the selling prices of other houses in this section of the city from their lot sizes and elevations. The agent feels that a linear model of the form

$$P = b_0 + b_1 L + b_2 E$$

would be reasonably accurate and easy to use. Here b_1 and b_2 would indicate how the price varies with lot size and elevation, respectively, while b_0 would reflect a base price for this section of the city.

 The agent would like to select the "best" linear model in some sense, but he is unsure how to proceed. If he knew the three parameters b_0, b_1, and b_2, the six observations in the table would each provide a forecast of the selling price as follows:

$$\hat{P}_i = b_0 + b_1 L_i + b_2 E_i \qquad i = 1, 2, \ldots, 6.$$

However, since b_0, b_1, and b_2 cannot, in general, be chosen so that the actual prices P_i are exactly equal to the forecast prices $\hat{P}_i$ for all observations, the agent would like to minimize the absolute value of the residuals $R_i = P_i - \hat{P}_i$. Formulate mathematical programs to find the "best" values of b_0, b_1, and b_2 by minimizing each of the following criteria:

a) $\sum_{i=1}^{6} (P_i - \hat{P}_i)^2$,   Least squares

b) $\sum_{i-1}^{6} |P_i - \hat{P}_i|$,   Linear absolute residual

c) $\underset{1 \leq i \leq 6}{\text{Max}} |P_i - \hat{P}_i|$,   Maximum absolute residual

(*Hint*: (b) and (c) can be formulated as linear programs. How should (a) be solved?)

29. A firm wishes to maximize its total funds accumulated after N time periods. It receives d_i dollars in external income in period i, where $i = 1, 2, \ldots, N$, and can generate additional income by making any of 10 investments in each time period. An investment of \$1 in the jth investment in period i produces nonnegative income $(a^j_{i,\,i+1}), (a^j_{i,\,i+2}), \ldots, (a^j_{i,\,N})$ in periods $i + 1, i + 2, \ldots, N$ respectively. For example, $a^3_{1,\,N}$ is the return in period N for investing \$1 in the third investment opportunity in period 1. The total cash available for investment in each period equals d_i, plus yield in period i from previous investments, plus savings of money from previous periods. Thus, cash can be saved, invested, or partially saved and the rest invested.

a) Formulate a linear program to determine the investment schedule that maximizes the accumulated funds in time period N.

b) How does the formulation change if the investment opportunities vary (in number and kind) from period to period?

c) What happens if savings earn 5% per time period?

ACKNOWLEDGMENTS

Exercise 4 is based on the Rectified Liquor case, written by V. L. Andrews. Exercises 5, 6, and 7 are inspired by the Sherman Motor Company case, written by Charles J. Christenson, which in turn is adapted from an example used by Robert Dorfman in "Mathematical or 'Linear' Programming: a Nonmathematical Approach," *American Economic Review*, December 1953. Exercises 17, 18, and 24 are variations of problems used by Professor C. Roger Glassey of the University of California, Berkeley.

Solving Linear Programs

2

In this chapter, we present a systematic procedure for solving linear programs. This procedure, called the *simplex method*, proceeds by moving from one feasible solution to another, at each step improving the value of the objective function. Moreover, the method terminates after a finite number of such transitions.

Two characteristics of the simplex method have led to its widespread acceptance as a computational tool. First, the method is robust. It solves *any* linear program; it detects redundant constraints in the problem formulation; it identifies instances when the objective value is unbounded over the feasible region; and it solves problems with one or more optimal solutions. The method is also self-initiating. It uses itself either to generate an appropriate feasible solution, as required, to start the method, or to show that the problem has no feasible solution. Each of these features will be discussed in this chapter.

Second, the simplex method provides much more than just optimal solutions. As byproducts, it indicates how the optimal solution varies as a function of the problem data (cost coefficients, constraint coefficients, and righthand-side data). This information is intimately related to a linear program called the *dual* to the given problem, and the simplex method automatically solves this dual problem along with the given problem. These characteristics of the method are of primary importance for applications, since data rarely is known with certainty and usually is approximated when formulating a problem. These features will be discussed in detail in the chapters to follow.

Before presenting a formal description of the algorithm, we consider some examples. Though elementary, these examples illustrate the essential algebraic and geometric features of the method and motivate the general procedure.

2.1 SIMPLEX METHOD—A PREVIEW

Optimal Solutions

Consider the following linear program:

$$\text{Maximize } z = 0x_1 + 0x_2 - 3x_3 - x_4 + 20, \qquad\qquad \text{(Objective 1)}$$

subject to:

$$x_1 \qquad\quad - 3x_3 + 3x_4 = 6, \qquad\qquad\qquad (1)$$
$$x_2 - 8x_3 + 4x_4 = 4, \qquad\qquad\qquad (2)$$
$$x_j \geq 0 \qquad (j = 1, 2, 3, 4).$$

Note that as stated the problem has a very special form. It satisfies the following:

1. All decision variables are constrained to be nonnegative.

2. All constraints, except for the nonnegativity of decision variables, are stated as equalities.

3. The righthand-side coefficients are all nonnegative.

4. One decision variable is isolated in each constraint with a $+1$ coefficient (x_1 in constraint (1) and x_2 in constraint (2)). The variable isolated in a given constraint does not appear in any other constraint, and appears with a zero coefficient in the objective function.

A problem with this structure is said to be in *canonical form*. This formulation might appear to be quite limited and restrictive; as we will see later, however, *any* linear programming problem can be transformed so that it is in canonical form. Thus, the following discussion is valid for linear programs in general.

Observe that, given any values for x_3 and x_4, the values of x_1 and x_2 are determined uniquely by the equalities. In fact, setting $x_3 = x_4 = 0$ immediately gives a feasible solution with $x_1 = 6$ and $x_2 = 4$. Solutions such as these will play a central role in the simplex method and are referred to as *basic feasible solutions*. In general, given a canonical form for any linear program, a basic feasible solution is given by setting the variable isolated in constraint j, called the jth *basic-variable*, equal to the righthand side of the jth constraint and by setting the remaining variables, called *nonbasic*, all to zero. Collectively the basic variables are termed a *basis*.*

In the example above, the basic feasible solution $x_1 = 6, x_2 = 4, x_3 = 0, x_4 = 0$, is optimal. For any other feasible solution, x_3 and x_4 must remain nonnegative. Since their coefficients in the objective function are negative, if either x_3 or x_4 is positive, z will be less than 20. Thus the maximum value for z is obtained when $x_3 = x_4 = 0$.

---

* We have introduced the new terms *canonical, basis,* and *basic variable* at this early point in our discussion because these terms have been firmly established as part of linear-programming vernacular. *Canonical* is a word used in many contexts in mathematics, as it is here, to mean "a special or standard representation of a problem or concept," usually chosen to facilitate study of the problem or concept. *Basis* and *basic* are concepts in linear algebra; our use of these terms agrees with linear-algebra interpretations of the simplex method that are discussed formally in Appendix A.

To summarize this observation, we state the:

Optimality Criterion.   Suppose that, in a maximization problem, every nonbasic variable has a nonpositive coefficient in the objective function of a canonical form. Then the basic feasible solution given by that canonical form maximizes the objective function over the feasible region.

Unbounded Objective Value

Next consider the example just discussed but with a new objective function:

$$\text{Maximize } z = 0x_1 + 0x_2 + 3x_3 - x_4 + 20, \qquad\qquad \text{(Objective 2)}$$

subject to:

$$x_1 \qquad\quad - 3x_3 + 3x_4 = 6, \qquad\qquad (1)$$
$$x_2 - 8x_3 + 4x_4 = 4, \qquad\qquad (2)$$
$$x_j \geq 0 \quad (j = 1, 2, 3, 4).$$

Since x_3 now has a positive coefficient in the objective function, it appears promising to increase the value of x_3 as much as possible. Let us maintain $x_4 = 0$, increase x_3 to a value t to be determined, and update x_1 and x_2 to preserve feasibility. From constraints (1) and (2),

$$x_1 \qquad\qquad = 6 + 3t,$$
$$x_2 \qquad\qquad = 4 + 8t,$$
$$z = 20 + 3t.$$

No matter how large t becomes, x_1 and x_2 remain nonnegative. In fact, as t approaches $+\infty$, z approaches $+\infty$. In this case, the objective function is unbounded over the feasible region.

The same argument applies to any linear program and provides the:

Unboundedness Criterion.   Suppose that, in a maximization problem, some non-basic variable has a positive coefficient in the objective function of a canonical form. If that variable has negative or zero coefficients in all constraints, then the objective function is unbounded from above over the feasible region.

Improving a Nonoptimal Solution

Finally, let us consider one further version of the previous problem:

$$\text{Maximize } z = 0x_1 + 0x_2 - 3x_3 + x_4 + 20, \qquad\qquad \text{(Objective 3)}$$

subject to:

$$x_1 \qquad\quad - 3x_3 + 3x_4 = 6, \qquad\qquad (1)$$
$$x_2 - 8x_3 + 4x_4 = 4, \qquad\qquad (2)$$
$$x_j \geq 0 \quad (j = 1, 2, 3, 4).$$

Now as x_4 increases, z increases. Maintaining $x_3 = 0$, let us increase x_4 to a value t, and update x_1 and x_2 to preserve feasibility. Then, as before, from constraints (1) and (2),

$$x_1 \qquad\qquad\qquad\qquad = \ 6 - 3t,$$
$$x_2 \qquad\qquad\qquad = \ 4 - 4t,$$
$$z = 20 + \ t.$$

If x_1 and x_2 are to remain nonnegative, we require:

$$6 - 3t \geq 0, \qquad \text{that is, } t \leq \tfrac{6}{3} = 2$$

and

$$4 - 4t \geq 0, \qquad \text{that is, } t \leq \tfrac{4}{4} = 1.$$

Therefore, the largest value for t that maintains a feasible solution is $t = 1$. When $t = 1$, the new solution becomes $x_1 = 3$, $x_2 = 0$, $x_3 = 0$, $x_4 = 1$, which has an associated value of $z = 21$ in the objective function.

Note that, in the new solution, x_4 has a positive value and x_2 has become zero. Since nonbasic variables have been given zero values before, it appears that x_4 has replaced x_2 as a basic variable. In fact, it is fairly simple to manipulate Eqs. (1) and (2) algebraically to produce a new canonical form, where x_1 and x_4 become the basic variables. If x_4 is to become a basic variable, it should appear with coefficient $+1$ in Eq. (2), and with zero coefficients in Eq. (1) and in the objective function. To obtain a $+1$ coefficient in Eq. (2), we divide that equation by 4, changing the constraints to read:

$$x_1 \qquad\quad - 3x_3 + 3x_4 = \quad 6, \qquad\qquad (1)$$
$$\tfrac{1}{4}x_2 - 2x_3 + \ x_4 = \quad 1. \qquad\qquad (2')$$

Now, to eliminate x_4 from the first constraint, we may multiply Eq. (2') by 3 and subtract it from constraint (1), giving:

$$x_1 - \tfrac{3}{4}x_2 + 3x_3 \qquad\quad = \quad 3, \qquad\qquad (1')$$
$$\tfrac{1}{4}x_2 - 2x_3 + \ x_4 = \quad 1. \qquad\qquad (2')$$

Finally, we may rearrange the objective function and write it as:

$$(-z) \qquad\quad - 3x_3 + \ x_4 = \ -20 \qquad\qquad (3)$$

and use the same technique to eliminate x_4; that is, multiply (2') by -1 and add to Eq. (3) giving:

$$(-z) - \tfrac{1}{4}x_2 - \ x_3 \qquad\quad = \ -21.$$

Collecting these equations, the system becomes:

Maximize $z = 0x_1 - \tfrac{1}{4}x_2 - \ x_3 + 0x_4 + 21$,

subject to:

$$x_1 - \tfrac{3}{4}x_2 + 3x_3 \qquad\quad = \quad 3, \qquad\qquad (1')$$
$$\tfrac{1}{4}x_2 - 2x_3 + \ x_4 = \quad 1, \qquad\qquad (2')$$
$$x_j \geq 0 \qquad (j = 1, 2, 3, 4).$$

Now the problem is in canonical form with x_1 and x_4 as basic variables, and z has increased from 20 to 21. Consequently, we are in a position to reapply the arguments of this section, beginning with this improved solution. In this case, the new canonical form satisfies the optimality criterion since all nonbasic variables have nonpositive coefficients in the objective function, and thus the basic feasible solution $x_1 = 3, x_2 = 0, x_3 = 0, x_4 = 1$, is optimal.

The procedure that we have just described for generating a new basic variable is called *pivoting*. It is the essential computation of the simplex method. In this case, we say that we have just pivoted on x_4 in the second constraint. To appreciate the simplicity of the pivoting procedure and gain some additional insight, let us see that it corresponds to nothing more than elementary algebraic manipulations to re-express the problem conveniently.

First, let us use constraint (2) to solve for x_4 in terms of x_2 and x_3, giving:

$$x_4 = \tfrac{1}{4}(4 - x_2 + 8x_3) \qquad \text{or} \qquad x_4 = 1 - \tfrac{1}{4}x_2 + 2x_3. \tag{2'}$$

Now we will use this relationship to substitute for x_4 in the objective equation:

$$z = 0x_1 + 0x_2 - 3x_3 + (1 - \tfrac{1}{4}x_2 + 2x_3) + 20,$$
$$z = 0x_1 - \tfrac{1}{4}x_2 - x_3 + 0x_4 + 21,$$

and also in constraint (1)

$$x_1 \quad - 3x_3 + 3(1 - \tfrac{1}{4}x_2 + 2x_3) = 6,$$

or, equivalently,

$$x_1 - \tfrac{3}{4}x_2 + 3x_3 = 3. \tag{1'}$$

Note that the equations determined by this procedure for eliminating variables are the same as those given by pivoting. We may interpret pivoting the same way, even in more general situations, as merely rearranging the system by solving for one variable and then substituting for it. We pivot because, for the new basic variable, we want a $+1$ coefficient in the constraint where it replaces a basic variable, and 0 coefficients in all other constraints and in the objective function.

Consequently, after pivoting, the form of the problem has been altered, but the modified equations still represent the original problem and have the same feasible solutions and same objective value when evaluated at any given feasible solution.

Indeed, the substitution is merely the familiar variable-elimination technique from high-school algebra, known more formally as Gauss–Jordan elimination.

In summary, the basic step for generating a canonical form with an improved value for the objective function is described as:

Improvement Criterion. Suppose that, in a maximization problem, some nonbasic variable has a positive coefficient in the objective function of a canonical form. If that variable has a positive coefficient in some constraint, then a new basic feasible solution may be obtained by pivoting.

Recall that we chose the constraint to pivot in (and consequently the variable to drop from the basis) by determining *which basic variable* first goes to zero as we increase the nonbasic variable x_4. The constraint is selected by taking the ratio of the righthand-side coefficients to the coefficients of x_4 in the constraints, i.e., by performing the *ratio test*:

$$\min \left\{ \tfrac{6}{3}, \tfrac{4}{4} \right\}.$$

Note, however, that if the coefficient of x_4 in the second constraint were -4 instead of $+4$, the values for x_1 and x_2 would be given by:

$$x_1 \quad = 6 - 3t,$$
$$x_2 = 4 + 4t,$$

so that as $x_4 = t$ increases from 0, x_2 never becomes zero. In this case, we would increase x_4 to $t = \tfrac{6}{3} = 2$. This observation applies in general for any number of constraints, so that we need never compute ratios for nonpositive coefficients of the variable that is coming into the basis, and we establish the following criterion:

> **Ratio and Pivoting Criterion.** When improving a given canonical form by introducing variable x_s into the basis, pivot in a constraint that gives the minimum ratio of righthand-side coefficient to corresponding x_s coefficient. Compute these ratios only for constraints that have a positive coefficient for x_s.

Observe that the value t of the variable being introduced into the basis is the minimum ratio. This ratio is zero if the righthand side is zero in the pivot row. In this instance, a new basis will be obtained by pivoting, but the values of the decision variables remain unchanged since $t = 0$.

As a final note, we point out that a linear program may have multiple optimal solutions. Suppose that the optimality criterion is satisfied and a nonbasic variable has a zero objective-function coefficient in the final canonical form. Since the value of the objective function remains unchanged for increases in that variable, we obtain an alternative optimal solution whenever we can increase the variable by pivoting.

Geometrical Interpretation

The three simplex criteria just introduced algebraically may be interpreted geometrically. In order to represent the problem conveniently, we have plotted the feasible region in Figs. 2.1(a) and 2.1(b) in terms of only the nonbasic variables x_3 and x_4. The values of x_3 and x_4 contained in the feasible regions of these figures satisfy the equality constraints and ensure nonnegativity of the basic and nonbasic variables:

$$x_1 \quad = 6 + 3x_3 - 3x_4 \geqq 0, \tag{1}$$
$$x_2 = 4 + 8x_3 - 4x_4 \geqq 0, \tag{2}$$
$$x_3 \geqq 0, \qquad x_4 \geqq 0.$$

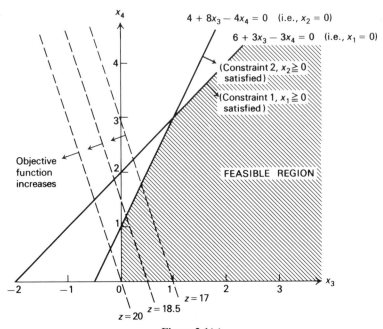

Figure 2.1(a)

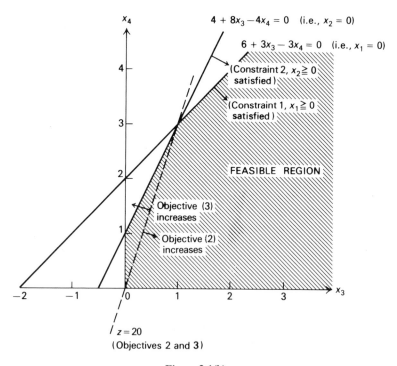

Figure 2.1(b)

Consider the objective function that we used to illustrate the optimality criterion,

$$z = -3x_3 - x_4 + 20. \qquad \text{(Objective 1)}$$

For any value of z, say $z = 17$, the objective function is represented by a straight line in Fig. 2.1(a). As z increases to 20, the line corresponding to the objective function moves parallel to itself across the feasible region. At $z = 20$, it meets the feasible region only at the point $x_3 = x_4 = 0$; and, for $z > 20$, it no longer touches the feasible region. Consequently, $z = 20$ is optimal.

The unboundedness criterion was illustrated with the objective function:

$$z = 3x_3 - x_4 + 20, \qquad \text{(Objective 2)}$$

which is depicted in Fig. 2.1(b). Increasing x_3 while holding $x_4 = 0$ corresponds to moving outward from the origin (i.e., the point $x_3 = x_4 = 0$) along the x_3-axis. As we move along the axis, we never meet either constraint (1) or (2). Also, as we move along the x_3-axis, the value of the objective function is increasing to $+\infty$.

The improvement criterion was illustrated with the objective function

$$z = -3x_3 + x_4 + 20, \qquad \text{(Objective 3)}$$

which also is shown in Fig. 2.1(b). Starting from $x_3 = 0$, $x_4 = 0$, and increasing x_4 corresponds to moving from the origin along the x_4-axis. In this case, however, we encounter constraint (2) at $x_4 = t = 1$ and constraint (3) at $x_4 = t = 2$. Consequently, to maintain feasibility in accordance with the ratio test, we move to the intersection of the x_4-axis and constraint (2), which is the optimal solution.

2.2 REDUCTION TO CANONICAL FORM

To this point we have been solving linear programs posed in canonical form with (1) nonnegative variables, (2) equality constraints, (3) nonnegative righthand-side coefficients, and (4) one basic variable isolated in each constraint. Here we complete this preliminary discussion by showing how to transform any linear program to this canonical form.

1. *Inequality constraints.*

In Chapter 1, the blast-furnace example contained the two constraints:

$$40x_1 + 10x_2 + 6x_3 \leqq 55.0,$$

$$40x_1 + 10x_2 + 6x_3 \geqq 32.5.$$

The lefthand side in these constraints is the silicon content of the 1000-pound casting being produced. The constraints specify the quality requirement that the silicon content must be between 32.5 and 55.0 pounds. To convert these constraints to equality form, introduce two new nonnegative variables (the blast-furnace example already includes a variable denoted x_4) defined as:

$$x_5 = 55.0 - 40x_1 - 10x_2 - 6x_3,$$
$$x_6 = 40x_1 + 10x_2 + 6x_3 - 32.5.$$

Variable x_5 measures the amount that the actual silicon content falls *short* of the maximum content that can be added to the casting, and is called a *slack variable*; x_6 is the amount of silicon in *excess* of the minimum requirement and is called a *surplus variable*. The constraints become:

$$40x_1 + 10x_2 + 6x_3 + x_5 = 55.0,$$
$$40x_1 + 10x_2 + 6x_3 - x_6 = 32.5.$$

Slack or surplus variables can be used in this way to convert any inequality to equality form.

2. *Free variables*

To see how to treat free variables, or variables unconstrained in sign, consider the basic balance equation of inventory models:

$$x_t \quad + \quad I_{t-1} \quad = \quad d_t \quad + \quad I_t.$$

$$\begin{pmatrix} \text{Production} \\ \text{in period } t \end{pmatrix} \quad \begin{pmatrix} \text{Inventory} \\ \text{from period } (t-1) \end{pmatrix} \quad \begin{pmatrix} \text{Demand in} \\ \text{period } t \end{pmatrix} \quad \begin{pmatrix} \text{Inventory at} \\ \text{end of period } t \end{pmatrix}$$

In many applications, we may assume that demand is known and that production x_t must be nonnegative. Inventory I_t may be positive or negative, however, indicating either that there is a surplus of goods to be stored or that there is a shortage of goods and some must be produced later. For instance, if $d_t - x_t - I_{t-1} = 3$, then $I_t = -3$ units must be produced later to satisfy current demand. To formulate models with free variables, we introduce two nonnegative variables I_t^+ and I_t^-, and write

$$I_t = I_t^+ - I_t^-$$

as a substitute for I_t everywhere in the model. The variable I_t^+ represents positive inventory on hand and I_t^- represents backorders (i.e., unfilled demand). Whenever $I_t \geq 0$, we set $I_t^+ = I_t$ and $I_t^- = 0$, and when $I_t < 0$, we set $I_t^+ = 0$ and $I_t^- = -I_t$. The same technique converts any free variable into the difference between two nonnegative variables. The above equation, for example, is expressed with non-negative variables as:

$$x_t + I_{t-1}^+ - I_{t-1}^- - I_t^+ + I_t^- = d_t.$$

Using these transformations, any linear program can be transformed into a linear program with nonnegative variables and equality constraints. Further, the model can be stated with only nonnegative righthand-side values by multiplying by -1 any constraint with a negative righthand side. Then, to obtain a canonical form, we must make sure that, in each constraint, one basic variable can be isolated with a $+1$ coefficient. Some constraints already will have this form. For example, the slack variable x_5 introduced previously into the silicon equation,

$$40x_1 + 10x_2 + 6x_3 + x_5 = 55.0,$$

appears in no other equation in the model. It can function as an intial basic variable for this constraint. Note, however, that the surplus variable x_6 in the constraint

$$40x_1 + 10x_2 + 6x_3 - x_6 = 32.5$$

does not serve this purpose, since its coefficient is -1.

3. *Artificial variables.*

There are several ways to isolate basic variables in the constraints where one is not readily apparent. One particularly simple method is just to add a new variable to any equation that requires one. For instance, the last constraint can be written as:

$$40x_1 + 10x_2 + 6x_3 - x_6 + x_7 = 32.5,$$

with nonnegative basic variable x_7. This new variable is completely fictitious and is called an *artificial variable*. Any solution with $x_7 = 0$ is feasible for the original problem, but those with $x_7 > 0$ are not feasible. Consequently, we should attempt to drive the artificial variable to zero. In a minimization problem, this can be accomplished by attaching a high unit cost M (>0) to x_7 in the objective function (for maximization, add the penalty $-Mx_7$ to the objective function). For M sufficiently large, x_7 will be zero in the final linear programming solution, so that the solution satisfies the original problem constraint without the artificial variable. If $x_7 > 0$ in the final tableau, then there is no solution to the original problem where the artificial variables have been removed; that is, we have shown that the problem is infeasible.

   Let us emphasize the distinction between artificial and slack variables. Whereas slack variables have meaning in the problem formulation, artificial variables have no significance; they are merely a mathematical convenience useful for initiating the simplex algorithm.

   This procedure for penalizing artificial variables, called the *big M method*, is straightforward conceptually and has been incorporated in some linear programming systems. There are, however, two serious drawbacks to its use. First, we don't know *a priori* how large M must be for a given problem to ensure that all artificial variables are driven to zero. Second, using large numbers for M may lead to numerical difficulties on a computer. Hence, other methods are used more commonly in practice.

   An alternative to the big M method that is often used for initiating linear programs is called the *phase I–phase II procedure* and works in two stages. Phase I determines a canonical form for the problem by solving a linear program related to the original problem formulation. The second phase starts with this canonical form to solve the original problem.

To illustrate the technique, consider the linear program:

Maximize $z = -3x_1 + 3x_2 + 2x_3 - 2x_4 - x_5 + 4x_6$,

subject to:

$$
\begin{aligned}
x_1 - x_2 + x_3 - x_4 - 4x_5 + 2x_6 - x_7 \quad\quad + x_9 \quad\quad\quad\quad\quad &= 4, \\
-3x_1 + 3x_2 + x_3 - x_4 - 2x_5 \quad\quad\quad\quad\quad + x_8 \quad\quad\quad\quad\quad &= 6, \\
- x_3 + x_4 \quad\quad + x_6 \quad\quad\quad\quad\quad + x_{10} \quad\quad\quad &= 1, \\
x_1 - x_2 + x_3 - x_4 - x_5 \quad\quad\quad\quad\quad\quad\quad\quad\quad\quad + x_{11} &= 0,
\end{aligned}
$$

$x_j \geqq 0 \quad (j = 1, 2, \ldots, 11)$.

Artificial variables
added

Assume that x_8 is a slack variable, and that the problem has been augmented by the introduction of artificial variables x_9, x_{10}, and x_{11} in the first, third and fourth constraints, so that x_8, x_9, x_{10}, and x_{11} form a basis. The following elementary, yet important, fact will be useful:

Any feasible solution to the augmented system with all artificial variables equal to zero provides a feasible solution to the original problem. Conversely, every feasible solution to the original problem provides a feasible solution to the augmented system by setting all artificial variables to zero.

Next, observe that since the artificial variables x_9, x_{10}, and x_{11} are all nonnegative, they are all zero only when their sum $x_9 + x_{10} + x_{11}$ is zero. For the basic feasible solution just derived, this sum is 5. Consequently, the artificial variables can be eliminated by ignoring the original objective function for the time being and minimizing $x_9 + x_{10} + x_{11}$ (i.e., minimizing the sum of all artificial variables). Since the artificial variables are all nonnegative, minimizing their sum means driving their sum towards zero. If the minimum sum is 0, then the artificial variables are all zero and a feasible, but not necessarily optimal, solution to the original problem has been obtained. If the minimum is greater than zero, then every solution to the augmented system has $x_9 + x_{10} + x_{11} > 0$, so that *some* artificial variable is still positive. In this case, the original problem has no feasible solution.

The essential point to note is that minimizing the infeasibility in the augmented system is a linear program. Moreover, adding the artificial variables has isolated one basic variable in each constraint. To complete the canonical form of the phase I linear program, we need to eliminate the basic variables from the phase I objective function. Since we have presented the simplex method in terms of maximizing an objective function, for the phase I linear program we will maximize w defined to be *minus* the sum of the artificial variables, rather than minimizing their sum directly. The canonical form for the phase I linear program is then determined simply by adding the artificial variables to the w equation. That is, we add the first, third, and fourth constraints in the previous problem formulation to:

$$(-w) - x_9 - x_{10} - x_{11} = 0,$$

and express this equation as:

$$w = 2x_1 - 2x_2 + x_3 - x_4 - 5x_5 + 3x_6 - x_7 + 0x_9 + 0x_{10} + 0x_{11} - 5.$$

The artificial variables now have zero coefficients in the phase I objective.

Note that the initial coefficients for the nonartificial variable x_j in the w equation is the sum of the coefficients of x_j from the equations with an artificial variable (see Fig. 2.2).

If $w = 0$ is the solution to the phase I problem, then all artificial variables are zero. If, in addition, every artificial variable is nonbasic in this optimal solution, then basic variables have been determined from the original variables, so that a canonical form has been constructed to initiate the original optimization problem. (Some artificial variables may be basic at value zero. This case will be treated in Section 2.5.) Observe that the unboundedness condition is unnecessary. Since the artificial variables are nonnegative, w is bounded from above by zero (for example, $w = -x_9 - x_{10} - x_{11} \leqq 0$) so that the unboundedness condition will never apply.

To recap, artificial variables are added to place the linear program in canonical form. Maximizing w either

i) gives max $w < 0$. The original problem is infeasible and the optimization terminates; or

ii) gives max $w = 0$. Then a canonical form has been determined to initiate the original problem. Apply the optimality, unboundedness, and improvement criteria to the original objective function z, starting with this canonical form.

In order to reduce a general linear-programming problem to canonical form, it is convenient to perform the necessary transformations according to the following sequence:

1. Replace each decision variable unconstrained in sign by a difference between two nonnegative variables. This replacement applies to all equations including the objective function.

2. Change inequalities to equalities by the introduction of slack and surplus variables. For $\geqq$ inequalities, let the nonnegative *surplus variable* represent the amount by which the lefthand side exceeds the righthand side; for $\leqq$ inequalities, let the nonnegative *slack variable* represent the amount by which the righthand side exceeds the lefthand side.

3. Multiply equations with a negative righthand side coefficient by -1.

4. Add a (nonnegative) artificial variable to any equation that does not have an isolated variable readily apparent, and construct the phase I objective function.

To illustrate the orderly application of these rules we provide, in Fig. 2.2, a full example of reduction to canonical form. The succeeding sets of equations in this table represent the stages of problem transformation as we apply each of the steps indicated above. We should emphasize that at each stage the form of the given problem is exactly equivalent to the original problem.

Problem

Maximize $z = -3y_1 + 2y_2 - y_3 + 4y_4$,

subject to:

$$
\begin{aligned}
y_1 + y_2 - 4y_3 + 2y_4 &\geqq 4, \\
-3y_1 + y_2 - 2y_3 &\leqq 6, \\
y_2 - y_4 &= -1, \\
y_1 + y_2 - y_3 &= 0, \\
y_3 \geqq 0, \quad y_4 &\geqq 0.
\end{aligned}
$$

STEP 1 REDUCTION

COMMENTS

Maximize $z = -3x_1 + 3x_2 + 2x_3 - 2x_4 - x_5 + 4x_6$,

Substitute
$$x_1 - x_2 = y_1, \qquad x_5 = y_3$$
$$x_3 - x_4 = y_2, \qquad x_6 = y_4.$$

subject to:

$$
\begin{aligned}
x_1 - x_2 + x_3 - x_4 - 4x_5 + 2x_6 &\leqq 4, \\
-3x_1 + 3x_2 + x_3 - x_4 - 2x_5 &\geqq 6, \\
x_3 - x_4 - x_6 &= -1, \\
x_1 - x_2 + x_3 - x_4 - x_5 &= 0, \\
x_j \geqq 0 \quad (j = 1, 2, \ldots, 6).
\end{aligned}
$$

STEP 2 REDUCTION

Maximize $z = -3x_1 + 3x_2 + 2x_3 - 2x_4 - x_5 + 4x_6$,

subject to:

$$
\begin{aligned}
x_1 - x_2 + x_3 - x_4 - 4x_5 + 2x_6 - x_7 \qquad &= 4, && \text{Introduce surplus variable, } x_7. \\
-3x_1 + 3x_2 + x_3 - x_4 - 2x_5 \qquad\quad + x_8 &= 6, && \text{Introduce slack variable, } x_8. \\
x_3 - x_4 - x_6 &= -1, \\
x_1 - x_2 + x_3 - x_4 - x_5 &= 0, \\
x_j \geqq 0 \quad (j = 1, 2, \ldots, 8).
\end{aligned}
$$

STEP 3 REDUCTION

Maximize $z = -3x_1 + 3x_2 + 2x_3 - 2x_4 - x_5 + 4x_6,$

subject to:

$$
\begin{aligned}
x_1 - x_2 + x_3 - x_4 - 4x_5 + 2x_6 - x_7 \qquad\quad &= 4, \\
-3x_1 + 3x_2 + x_3 - x_4 - 2x_5 \qquad\quad + x_8 &= 6, \\
- x_3 + x_4 \qquad + x_6 \qquad\qquad\quad &= 1, \\
x_1 - x_2 + x_3 - x_4 - x_5 \qquad\qquad\qquad\quad &= 0,
\end{aligned}
$$

$x_j \geqq 0 \qquad (j = 1, 2, \ldots, 8).$

The third constraint was multiplied by -1.

STEP 4 REDUCTION

Maximize $z = -3x_1 + 3x_2 + 2x_3 - 2x_4 - x_5 + 4x_6,$

subject to:

$$
\begin{aligned}
x_1 - x_2 + x_3 - x_4 - 4x_5 + 2x_6 - x_7 \quad + x_9 \qquad\qquad\qquad &= 4, \\
-3x_1 + 3x_2 + x_3 - x_4 - 2x_5 \qquad + x_8 \qquad\qquad\qquad\qquad &= 6, \\
- x_3 + x_4 \quad + x_6 \qquad\qquad + x_{10} \qquad\quad &= 1, \\
x_1 - x_2 + x_3 - x_4 - x_5 \qquad\qquad\qquad\qquad + x_{11} &= 0,
\end{aligned}
$$

$x_j \geqq 0 \qquad (j = 1, 2, \ldots, 11).$

Artificial variables, x_9, x_{10}, x_{11}, added.

PHASE I OBJECTIVE

Maximize $w =$
$= +2x_1 - 2x_2 + x_3 - x_4 - 5x_5 + 3x_6 - x_7 \quad - x_9 - x_{10} - x_{11} - 5.$

Constraints 1, 3, and 4 added to w objective.

Fig. 2.2   Reduction to canonical form.

2.3 SIMPLEX METHOD—A FULL EXAMPLE

The simplex method for solving linear programs is but one of a number of methods, or algorithms, for solving optimization problems. By an algorithm, we mean a systematic procedure, usually iterative, for solving a class of problems. The simplex method, for example, is an algorithm for solving the class of linear-programming problems. Any finite optimization algorithm should terminate in one, and only one, of the following possible situations:

1. by demonstrating that there is no feasible solution;
2. by determining an optimal solution; or
3. by demonstrating that the objective function is unbounded over the feasible region.

We will say that an algorithm solves a problem if it always satisfies one of these three conditions. As we shall see, a major feature of the simplex method is that it solves any linear-programming problem.

Most of the algorithms that we are going to consider are iterative, in the sense that they move from one decision point $x_1, x_2, \ldots, x_n$ to another. For these algorithms, we need:

i) a starting point to initiate the procedure;

ii) a termination criterion to indicate when a solution has been obtained; and

iii) an improvement mechanism for moving from a point that is not a solution to a better point.

Every algorithm that we develop should be analyzed with respect to these three requirements.

In the previous section, we discussed most of these criteria for a sample linear-programming problem. Now we must extend that discussion to give a formal and general version of the simplex algorithm. Before doing so, let us first use the improvement criterion of the previous section iteratively to solve a complete problem. To avoid needless complications at this point, we select a problem that does not require artificial variables.

Simple Example. The owner of a shop producing automobile trailers wishes to determine the best mix for his three products: flat-bed trailers, economy trailers, and luxury trailers. His shop is limited to working 24 days/month on metalworking and 60 days/month on woodworking for these products. The following table indicates production data for the trailers.

	Usage per unit of trailer			Resource availabilities
	Flat-bed	Economy	Luxury	
Metalworking days	$\frac{1}{2}$	2	1	24
Woodworking days	1	2	4	60
Contribution ($ × 100)	6	14	13	

Let the decision variables of the problem be:

$$x_1 = \text{Number of flat-bed trailers produced per month,}$$
$$x_2 = \text{Number of economy trailers produced per month,}$$
$$x_3 = \text{Number of luxury trailers produced per month.}$$

Assuming that the costs for metalworking and woodworking capacity are fixed, the problem becomes:

Maximize $z = 6x_1 + 14x_2 + 13x_3$,

subject to:

$$\tfrac{1}{2}x_1 + 2x_2 + x_3 \leq 24,$$
$$x_1 + 2x_2 + 4x_3 \leq 60,$$
$$x_1 \geq 0, \qquad x_2 \geq 0, \qquad x_3 \geq 0.$$

Letting x_4 and x_5 be slack variables corresponding to unused hours of metalworking and woodworking capacity, the problem above is equivalent to the linear program:

Maximize $z = 6x_1 + 14x_2 + 13x_3$,

subject to:

$$\tfrac{1}{2}x_1 + 2x_2 + x_3 + x_4 = 24,$$
$$x_1 + 2x_2 + 4x_3 + x_5 = 60,$$
$$x_j \geq 0 \qquad (j = 1, 2, \ldots, 5).$$

This linear program is in canonical form with basic variables x_4 and x_5. To simplify our exposition and to more nearly parallel the way in which a computer might be used to solve problems, let us adopt a tabular representation of the equations instead of writing them out in detail. Tableau 1 corresponds to the given canonical form. The first two rows in the tableau are self-explanatory; they simply represent the constraints, but with the variables detached. The third row represents the z-equation, which may be rewritten as:

$$(-z) + 6x_1 + 14x_2 + 13x_3 = 0.$$

By convention, we say that $(-z)$ is the basic variable associated with this equation. Note that no formal column has been added to the tableau for the $(-z)$-variable.

Tableau 1

Basic variables	Current values	x_1	x_2	x_3	x_4	x_5	Equation identification and transformations	Ratio test
x_4	24	$\tfrac{1}{2}$	②	1	1		1	24/2
x_5	60	1	2	4		1	2	60/2
$(-z)$	0	$+6$	$+14$	$+13$			3	

↑

The data to the right of the tableau is not required for the solution. It simply identifies the rows and summarizes calculations. The arrow below the tableau indicates the variable being introduced into the basis; the circled element of the tableau indicates the pivot element; and the arrow to the left of the tableau indicates the variable being removed from the basis.

By the improvement criterion, introducing either x_1, x_2, or x_3 into the basis will improve the solution. The simplex method selects the variable with best payoff per unit (largest objective coefficient), in this case x_2. By the ratio test, as x_2 is increased, x_4 goes to zero before x_5 does; we should pivot in the first constraint. After pivoting, x_2 replaces x_4 in the basis and the new canonical form is as given in Tableau 2.

Tableau 2

Basic variables	Current values	x_1	x_2	x_3	x_4	x_5	Equation identification and transformations	Ratio test
x_2	12	$\frac{1}{4}$	1	$\frac{1}{2}$	$\frac{1}{2}$		$\boxed{4} = \frac{1}{2}\boxed{1}$	$12/(1/2)$
x_5	36	$\frac{1}{2}$		③	-1	1	$\boxed{5} = \boxed{2} - 2\boxed{4}$	$36/3$
$(-z)$	-168	$+\frac{5}{2}$		$+6$	-7		$\boxed{6} = \boxed{3} - 14\boxed{4}$	

(← points to x_5 row; ↑ under x_3)

Next, x_3 is introduced in place of x_5 (Tableau 3).

Tableau 3

Basic variables	Current values	x_1	x_2	x_3	x_4	x_5	Equation identification and transformations	Ratio test
x_2	6	⑥$\frac{1}{6}$	1		$\frac{2}{3}$	$-\frac{1}{6}$	$\boxed{7} = \boxed{4} - \frac{1}{2}\boxed{8}$	$6/(1/6)$
x_3	12	$\frac{1}{6}$		1	$-\frac{1}{3}$	$\frac{1}{3}$	$\boxed{8} = \frac{1}{3}\boxed{5}$	$12/(1/6)$
$(-z)$	-240	$+\frac{3}{2}$			-5	-2	$\boxed{9} = \boxed{6} - 6\boxed{8}$	

(← points to x_2 row; ↑ under x_1)

Finally, x_1 is introduced in place of x_2 (Tableau 4).

Tableau 4

Basic variables	Current values	x_1	x_2	x_3	x_4	x_5	Equation identification and transformations
x_1	36	1	6		4	-1	$\boxed{10} = 6\boxed{7}$
x_3	6		-1	1	-1	$\frac{1}{2}$	$\boxed{11} = \boxed{8} - \frac{1}{6}\boxed{10}$
$(-z)$	-294		-9		-11	$-\frac{1}{2}$	$\boxed{12} = \boxed{9} - \frac{3}{2}\boxed{10}$

Tableau 4 satisfies the optimality criterion, giving an optimal contribution of $29,400 with a monthly production of 36 flat-bed trailers and 6 luxury trailers.

Note that in this example, x_2 entered the basis at the first iteration, but does not appear in the optimal basis. In general, a variable might be introduced into (and dropped from) the basis several times. In fact, it is possible for a variable to enter the basis at one iteration and drop from the basis at the very next iteration.

Variations

The simplex method changes in minor ways if the canonical form is written differently. Since these modifications appear frequently in the management-science literature, let us briefly discuss these variations. We have chosen to consider the maximizing form of the linear program (max z) and have written the objective function in the canonical form as:

$$(-z) + c_1x_1 + c_2x_2 + \cdots + c_nx_n = -z_0,$$

so that the current solution has $z = z_0$. We argued that, if all $c_j \leq 0$, then $z = z_0 + c_1x_1 + c_2x_2 + \cdots + c_nx_n \geq z_0$ for any feasible solution, so that the current solution is optimal. If instead, the objective equation is written as:

$$(z) + c_1'x_1 + c_2'x_2 + \cdots + c_n'x_n = z_0,$$

where $c_j' = -c_j$, then z is maximized if each coefficient $c_j' \geq 0$. In this case, the variable with the most negative coefficient $c_j' < 0$ is chosen to enter the basis. All other steps in the method are unaltered.

The same type of association occurs for the minimizing objective:

$$\text{Minimize } z = c_1x_1 + c_2x_2 + \cdots + c_nx_n.$$

If we write the objective function as:

$$(-z) + c_1x_1 + c_2x_2 + \cdots + c_nx_n = -z_0,$$

then, since $z = z_0 + c_1x_1 + \cdots + c_nx_n$, the current solution is optimal if every $c_j \geq 0$. The variable x_s to be introduced is selected from $c_s = \min c_j < 0$, and every other step of the method is the same as for the maximizing problem. Similarly, if the objective function is written as:

$$(z) + c_1'x_1 + c_2'x_2 + \cdots + c_n'x_n = z_0,$$

where $c_j' = -c_j$, then, for a minimization problem, we introduce the variable with the most positive c_j' into the basis.

Note that these modifications affect only the way in which the variable entering the basis is determined. The pivoting computations are not altered.

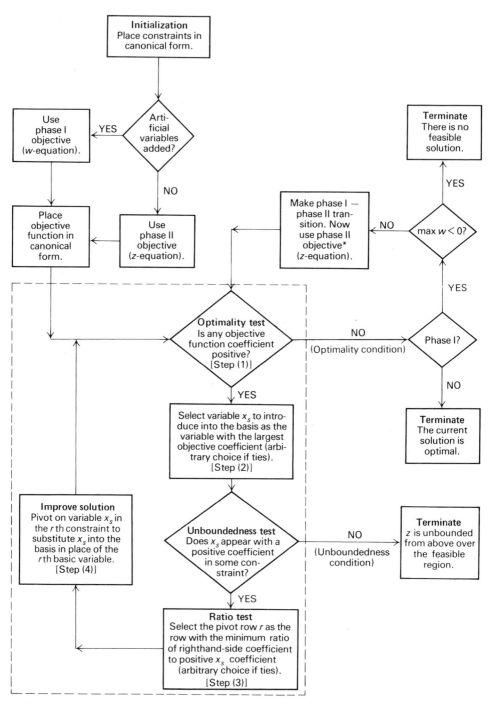

Fig. 2.3 Simplex phase I–phase II maximization procedure.

---

* See Section 2.5.

Given these variations in the selection rule for incoming variables, we should be wary of memorizing formulas for the simplex algorithm. Instead, we should be able to argue as in the previous example and as in the simplex preview. In this way, we maintain flexibility for almost any application and will not succumb to the rigidity of a "formula trap."

2.4 FORMAL PROCEDURE

Figure 2.3 summarizes the simplex method in flow-chart form. It illustrates both the computational steps of the algorithm and the interface between phase I and phase II. The flow chart indicates how the algorithm is used to show that the problem is infeasible, to find an optimal solution, or to show that the objective function is unbounded over the feasible region. Figure 2.4 illustrates this algorithm for a phase I–phase II example by solving the problem introduced in Section 2.2 for reducing a problem to canonical form. The remainder of this section specifies the computational steps of the flow chart in algebraic terms.

At any intermediate step during phase II of the simplex algorithm, the problem is posed in the following canonical form:

$$
\begin{array}{ll}
x_1 & + \bar{a}_{1,m+1}x_{m+1} + \cdots + \bar{a}_{1s}x_s + \cdots + \bar{a}_{1n}x_n = \bar{b}_1, \\
\quad x_2 & + \bar{a}_{2,m+1}x_{m+1} + \cdots \qquad\qquad\qquad + \bar{a}_{2n}x_n = \bar{b}_2, \\
\qquad \ddots & \qquad\qquad \vdots \qquad\qquad\qquad\qquad\qquad \vdots \qquad \vdots \\
\qquad\quad x_r & + \bar{a}_{r,m+1}x_{m+1} + \cdots + \boxed{\bar{a}_{rs}}x_s + \cdots + \bar{a}_{rn}x_n = \bar{b}_r, \\
\qquad\qquad \ddots & \qquad\qquad \vdots \qquad\qquad\qquad\qquad\qquad \vdots \qquad \vdots \\
\qquad\qquad\quad x_m + \bar{a}_{m,m+1}x_{m+1} + \cdots + \bar{a}_{ms}x_s + \cdots + \bar{a}_{mn}x_n = \bar{b}_m, \\
(-z) & \quad + \bar{c}_{m+1}x_{m+1} \quad + \cdots + \bar{c}_s x_s + \cdots + \bar{c}_n x_n = -\bar{z}_0,
\end{array}
$$

$$x_j \geq 0 \qquad (j = 1, 2, \ldots, n).$$

Originally, this canonical form is developed by using the procedures of Section 2.2. The data $\bar{a}_{ij}$, $\bar{b}_i$, $\bar{z}_0$, $\bar{w}_0$, and $\bar{c}_j$ are known. They are either the original data (without bars) or that data as updated by previous steps of the algorithm. We have assumed (by reindexing variables if necessary) that $x_1, x_2, \ldots, x_m$ are the basic variables. Also, since this is a canonical form, $\bar{b}_i \geq 0$ for $i = 1, 2, \ldots, m$.

Simplex Algorithm (Maximization Form)

STEP (0)   The problem is initially in canonical form and all $\bar{b}_i \geq 0$.

STEP (1)   If $\bar{c}_j \leq 0$ for $j = 1, 2, \ldots, n$, then *stop*; we are optimal. If we continue then there exists some $\bar{c}_j > 0$.

Initial tableau

| | | | | | | | | | | Artificial variables | | |
Basic variables	Current values	x_1	x_2	x_3	x_4	x_5	x_6	x_7	x_8	x_9	x_{10}	x_{11}
x_9	4	1	−1	1	−1	−4	2	−1		1		
x_8	6	−3	3	1	−1	−2	0	0	1			
x_{10}	1	0	0	−1	1	0	①	0			1	
x_{11}	0	1	−1	1	−1	−1	0	0				1
$(-z)$	0	−3	3	2	−2	−1	4	0				
$(-w)$	5	2	−2	1	−1	−5	3	−1				

Equation identification and transformations: 1, 2, 3, 4, 5, 6

Ratio test: 4/2, 1/1

Tableau 2

| | | | | | | | | | | Artificial variables | | |
Basic variables	Current values	x_1	x_2	x_3	x_4	x_5	x_6	x_7	x_8	x_9	x_{10}	x_{11}
x_9	2	1	−1	3	−3	−4		−1		1	−2	
x_8	6	−3	3	1	−1	−2		0	1		0	
x_6	1	0	0	−1	−1	0	1	0			1	
x_{11}	0	1	−1	①	−1	−1		0			0	1
$(-z)$	−4	−3	3	6	−6	−1		0			−4	
$(-w)$	2	2	−2	4	−4	−5		−1			−3	

Equation identification and transformations:

$7 = 1 - 2|3$
$8 = 2$
$9 = 3$
$10 = 4$
$11 = 5 - 4|3$
$12 = 6 - 3|3$

Ratio test: 2/3, 6/1, 0/1

Tableau 3

| | | | | | | | | | | Artificial variables | | |
Basic variables	Current values	x_1	x_2	x_3	x_4	x_5	x_6	x_7	x_8	x_9	x_{10}	x_{11}
x_9	2	−2	②		0	−1		−1		1	−2	−3
x_8	6	−4	4		0	−1		0	1		0	−1
x_6	1	1	−1	1	0	−1	1	0			1	1
x_3	0	1	−1		−1	−1		0			0	1
$(-z)$	−4	−9	9		0	5		0			−4	−6
$(-w)$	2	−2	2		0	−1		−1			−3	−4

Equation identification and transformations:

$13 = 7 - 3|10$
$14 = 8 - |$
$15 = 9 + |$
$16 = 10$
$17 = 11 - 6|10$
$18 = 12 - 4|10$

Ratio test: 2/2, 6/4

Tableau 4

2/1

Basic variables	Current values	x_1	x_2	x_3	x_4	x_5	x_6	x_7	x_8	x_9	x_{10}	x_{11}
x_2	1	-1	1		0	$-\frac{1}{2}$		$-\frac{1}{2}$		$-\frac{1}{2}$	-1	$-\frac{3}{2}$
x_8	2	0			0	$\boxed{1}$		2	1	-2	4	5
x_6	2	0			0	$-\frac{3}{2}$	1	$-\frac{1}{2}$		$-\frac{1}{2}$	0	$-\frac{1}{2}$
x_3	1	0		1	-1	$-\frac{3}{2}$		$-\frac{1}{2}$		$-\frac{1}{2}$	-1	$-\frac{1}{2}$
$(-z)$	-13	0			0	$\frac{19}{2}$		$\frac{9}{2}$		$-\frac{9}{2}$	5	$\frac{15}{2}$
$(-w)$	0	0			0	0		0		-1	-1	-1

$\leftarrow$

End of phase I. All artificial variables are nonbasic, so proceed with phase II, dropping the w-equation and maintaining $x_9 = x_{10} = x_{11} = 0$ (i.e., never introduce an artificial variable into the basis).

Final tableau

Basic variables	Current values	x_1	x_2	x_3	x_4	x_5	x_6	x_7	x_8	x_9	x_{10}	x_{11}
x_2	2	-1	1		0			$\frac{1}{2}$	$\frac{1}{2}$	$-\frac{1}{2}$	1	1
x_5	2	0			0	1		2	1	-2	4	5
x_6	5	0			0		1	$\frac{5}{2}$	$\frac{3}{2}$	$-\frac{5}{2}$	6	7
x_3	4	0		1	-1			$\frac{5}{2}$	$\frac{3}{2}$	$-\frac{5}{2}$	5	7
$(-z)$	-32	0			0			$-\frac{29}{2}$	$-\frac{19}{2}$	$\frac{29}{2}$	-33	-40

End of phase II. Substituting for the original variables, the optimal solution is $y_1 = -2$, $y_2 = 4$, $y_3 = 2$, $y_4 = 5$, max $z = 32$.

Fig. 2.4 Phase I–phase II illustration.

$\boxed{19} = \frac{1}{2}\boxed{13}$

$\boxed{20} = \boxed{14} - 4\boxed{19}$

$\boxed{21} = \boxed{15} + \boxed{19}$

$\boxed{22} = \boxed{16} + \boxed{19}$

$\boxed{23} = \boxed{17} - 9\boxed{19}$

$\boxed{24} = \boxed{18} - 2\boxed{19}$

$\boxed{25} = \boxed{19} + \frac{1}{2}\boxed{20}$

$\boxed{26} = \boxed{20}\boxed{20}$

$\boxed{27} = \boxed{21} + \frac{3}{2}\boxed{20}$

$\boxed{28} = \boxed{22} + \frac{3}{2}\boxed{20}$

$\boxed{29} = \boxed{23} - \frac{19}{2}\boxed{20}$

STEP (2)   Choose the column to pivot in (i.e., the variable to introduce into the basis) by:

$$\bar{c}_s = \max_j \{\bar{c}_j \mid \bar{c}_j > 0\}.*$$

If $\bar{a}_{is} \leq 0$ for $i = 1, 2, \ldots, m$, then *stop*; the primal problem is unbounded. If we continue, then $\bar{a}_{is} > 0$ for some $i = 1, 2, \ldots, m$.

STEP (3)   Choose row r to pivot in (i.e., the variable to drop from the basis) by the ratio test:

$$\frac{\bar{b}_r}{\bar{a}_{rs}} = \min_i \left\{ \frac{\bar{b}_i}{\bar{a}_{is}} \mid \bar{a}_{is} > 0 \right\}.$$

STEP (4)   Replace the basic variable in row r with variable s and re-establish the canonical form (i.e., pivot on the coefficient $\bar{a}_{rs}$).

STEP (5)   Go to step (1).

These steps are the essential computations of the simplex method. They apply to either the phase I or phase II problem. For the phase I problem, the coefficients $\bar{c}_j$ are those of the phase I objective function.

The only computation remaining to be specified formally is the effect that pivoting in step (4) has on the problem data. Recall that we pivot on coefficient $\bar{a}_{rs}$ merely to isolate variable x_s with a $+1$ coefficient in constraint r. The pivot can be viewed as being composed of two steps:

i) normalizing the rth constraint so that x_s has a $+1$ coefficient, and

ii) subtracting multiples of the normalized constraint from the other equations in order to eliminate variable x_s.

These steps are summarized pictorially in Fig. 2.5.

The last tableau in Fig. 2.5 specifies the new values for the data. The new righthand-side coefficients, for instance, are given by:

$$\bar{b}_r^{\text{new}} = \frac{\bar{b}_r}{\bar{a}_{rs}} \quad \text{and} \quad \bar{b}_i^{\text{new}} = \bar{b}_i - \bar{a}_{is} \left(\frac{\bar{b}_r}{\bar{a}_{rs}} \right) \geq 0 \quad \text{for } i \neq r.$$

Observe that the new coefficients for the variable x_r being removed from the basis summarize the computations. For example, the coefficient of x_r in the first row of the final tableau is obtained from the first tableau by subtracting $\bar{a}_{1s}/\bar{a}_{rs}$ times the rth row from the first row. The coefficients of the other variables in the first row of the third tableau can be obtained from the first tableau by performing this same calculation. This observation can be used to partially streamline the computations of the simplex method. (See Appendix B for details.)

---

* The vertical bar within braces is an abbreviation for the phrase "such that."

x_1 $\cdots$ x_r $\cdots$ x_m	x_{m+1} $\cdots$ x_s $\cdots$ x_n	
1	$\bar{a}_{1,m+1}$ $\cdots$ $\bar{a}_{1s}$ $\cdots$ $\bar{a}_{1n}$	$\bar{b}_1$
$\ddots$	$\vdots$ $\qquad$ $\vdots$	$\vdots$
1	$\bar{a}_{r,m+1}$ $\cdots$ $\boxed{\bar{a}_{rs}}$ $\cdots$ $\bar{a}_{rn}$	$\bar{b}_r$
$\ddots$	$\vdots$ $\qquad$ $\vdots$	$\vdots$
1	$\bar{a}_{m,m+1}$ $\cdots$ $\bar{a}_{ms}$ $\cdots$ $\bar{a}_{mn}$	$\bar{b}_m$
	$\bar{c}_{m+1}$ $\cdots$ $\bar{c}_s$ $\cdots$ $\bar{c}_n$	$-\bar{z}_0$

$\downarrow$ Normalization

x_1 $\cdots$ x_r $\cdots$ x_m	x_{m+1} $\cdots$ x_s $\cdots$ x_n	
1	$\bar{a}_{1,m+1}$ $\cdots$ $\bar{a}_{1s}$ $\cdots$ $\bar{a}_{1n}$	$\bar{b}_1$
$\ddots$	$\vdots$ $\qquad$ $\vdots$	$\vdots$
$\left(\dfrac{1}{\bar{a}_{rs}}\right)$	$\left(\dfrac{\bar{a}_{r,m+1}}{\bar{a}_{rs}}\right)$ $\cdots$ 1 $\cdots$ $\left(\dfrac{\bar{a}_{rn}}{\bar{a}_{rs}}\right)$	$\left(\dfrac{\bar{b}_r}{\bar{a}_{rs}}\right)$
$\ddots$	$\vdots$ $\qquad$ $\vdots$	$\vdots$
1	$\bar{a}_{m,m+1}$ $\cdots$ $\bar{a}_{ms}$ $\cdots$ $\bar{a}_{mn}$	$\bar{b}_m$
	$\bar{c}_{m+1}$ $\cdots$ $\bar{c}_s$ $\cdots$ $\bar{c}_n$	$-\bar{z}_0$

$\downarrow$ Elimination of x_s

1 $\quad$ $-\left(\dfrac{\bar{a}_{1s}}{\bar{a}_{rs}}\right)$	$\bar{a}_{1,m+1} - \bar{a}_{1s}\left(\dfrac{\bar{a}_{r,m+1}}{\bar{a}_{rs}}\right)$ $\cdots$ 0 $\cdots$ $\bar{a}_{1n} - \bar{a}_{1s}\left(\dfrac{\bar{a}_{rn}}{\bar{a}_{rs}}\right)$	$\bar{b}_1 - \bar{a}_{1s}\left(\dfrac{\bar{b}_r}{\bar{a}_{rs}}\right)$
$\ddots$	$\vdots$ $\qquad$ $\vdots$	$\vdots$
$\left(\dfrac{1}{\bar{a}_{rs}}\right)$	$\left(\dfrac{\bar{a}_{r,m+1}}{\bar{a}_{rs}}\right)$ $\cdots$ 1 $\cdots$ $\left(\dfrac{\bar{a}_{rn}}{\bar{a}_{rs}}\right)$	$\dfrac{\bar{b}_r}{\bar{a}_{rs}}$
$\ddots$	$\vdots$ $\qquad$ $\vdots$	$\vdots$
$-\left(\dfrac{\bar{a}_{ms}}{\bar{a}_{rs}}\right)$ $\quad$ 1	$\bar{a}_{m,m+1} - \bar{a}_{ms}\left(\dfrac{\bar{a}_{r,m+1}}{\bar{a}_{rs}}\right)$ $\cdots$ 0 $\cdots$ $\bar{a}_{mn} - \bar{a}_{ms}\left(\dfrac{\bar{a}_{rn}}{\bar{a}_{rs}}\right)$	$\bar{b}_m - \bar{a}_{ms}\left(\dfrac{\bar{b}_r}{\bar{a}_{rs}}\right)$
$-\left(\dfrac{\bar{c}_s}{\bar{a}_{rs}}\right)$	$\bar{c}_{m+1} - \bar{c}_s\left(\dfrac{\bar{a}_{r,m+1}}{\bar{a}_{rs}}\right)$ $\cdots$ 0 $\cdots$ $\bar{c}_n - \bar{c}_s\left(\dfrac{\bar{a}_{rn}}{\bar{a}_{rs}}\right)$	$-\bar{z}_0 - \bar{c}_s\left(\dfrac{\bar{b}_r}{\bar{a}_{rs}}\right)$

Fig. 2.5   Algebra for a pivot operation.

Note also that the new value for z will be given by:

$$\bar{z}_0 + \left(\frac{\bar{b}_r}{\bar{a}_{rs}}\right)\bar{c}_s.$$

By our choice of the variable x_s to introduce into the basis, $\bar{c}_s > 0$. Since $\bar{b}_r \geqq 0$ and $\bar{a}_{rs} > 0$, this implies that $z^{\text{new}} \geqq z^{\text{old}}$. In addition, if $\bar{b}_r > 0$, then z^{new} is strictly greater than z^{old}.

Convergence

Though the simplex algorithm has solved each of our previous examples, we have yet to show that it solves *any* linear program. A formal proof requires results from linear algebra, as well as further technical material that is presented in Appendix B. Let us outline a proof assuming these results. We assume that the linear program has n variables and m equality constraints.

First, note that there are only a finite number of bases for a given problem, since a basis contains m variables (one isolated in each constraint) and there are a finite number of variables to select from. A standard result in linear algebra states that, once the basic variables have been selected, all the entries in the tableau, including the objective value, are determined uniquely. Consequently, there are only a finite number of canonical forms as well. If the objective value *strictly* increases after every pivot, the algorithm never repeats a canonical form and must determine an optimal solution after a *finite* number of pivots (any nonoptimal canonical form is transformed to a new canonical form by the simplex method).

This argument shows that the simplex method solves linear programs as long as the objective value strictly increases after each pivoting operation. As we have just seen, each pivot affects the objective function by adding a multiple of the pivot equation to the objective function. The current value of the z-equation increases by a multiple of the righthand-side coefficient; if this coefficient is positive (not zero), the objective value increases. With this in mind, we introduce the following definition:

> A canonical form is called *nondegenerate* if each righthand-side coefficient is strictly positive. The linear-programming problem is called nondegenerate if, starting with an initial canonical form, every canonical form determined by the algorithm is nondegenerate.

In these terms, we have shown that the simplex method solves every non-degenerate linear program using a finite number of pivoting steps. When a problem is degenerate, it is possible to perturb the data slightly so that every righthand-side coefficient remains positive and again show that the method works. Details are given in Appendix B. A final note is that, empirically, the finite number of iterations mentioned here to solve a problem frequently lies between 1.5 and 2 times the number of constraints (i.e., between $1.5m$ and $2m$).

Applying this perturbation, if required, to both phase I and phase II, we obtain the essential property of the simplex method.

> ***Fundamental Property of the Simplex Method.*** The simplex method (with perturbation if necessary) solves any given linear program in a finite number of iterations. That is, in a finite number of iterations, it shows that there is no feasible solution; finds an optimal solution; or shows that the objective function is unbounded over the feasible region.

Although degeneracy occurs in almost every problem met in practice, it rarely causes any complications. In fact, even without the perturbation analysis, the simplex method never has failed to solve a practical problem, though problems that are highly degenerate with many basic variables at value zero frequently take more computational time than other problems.

Applying this fundamental property to the phase I problem, we see that, if a problem is feasible, the simplex method finds a basic feasible solution. Since these solutions correspond to corner or extreme points of the feasible region, we have the

> **Fundamental Property of Linear Equations.** If a set of linear equations in non-negative variables is feasible, then there is an extreme-point solution to the equations.

2.5 TRANSITION FROM PHASE I TO PHASE II

We have seen that, if an artificial variable is positive at the end of phase I, then the original problem has no feasible solution. On the other hand, if all artificial variables are nonbasic at value zero at the end of phase I, then a basic feasible solution has been found to initiate the original optimization problem. Section 2.4 furnishes an example of this case. Suppose, though, that when phase I terminates, all artificial variables are zero, but that some artificial variable remains in the basis. The following example illustrates this possibility.

Problem. Find a canonical form for x_1, x_2, and x_3 by solving the phase I problem (x_4, x_5, and x_6 are artificial variables):

Maximize $w = \qquad\qquad -x_4 - x_5 - x_6,$

subject to:

$$
\begin{aligned}
x_1 - 2x_2 + x_4 &= 2, \\
x_1 - 3x_2 - x_3 + x_5 &= 1, \\
x_1 - x_2 + ax_3 + x_6 &= 3, \\
x_j \geq 0 \quad (j = 1, 2, \ldots, 6).
\end{aligned}
$$

To illustrate the various terminal conditions, the coefficient of x_3 is unspecified in the third constraint. Later it will be set to either 0 or 1. In either case, the pivoting sequence will be the same and we shall merely carry the coefficient symbolically.

Putting the problem in canonical form by eliminating x_4, x_5, and x_6 from the objective function, the simplex solution to the phase I problem is given in Tableaus 1 through 3.

Tableau 1

Basic variables	Current values	x_1	x_2	x_3	x_4	x_5	x_6
x_4	2	1	-2	0	1		
x_5	1	①	-3	-1		1	
x_6	3	1	-1	a			1
$(-w)$	$+6$	$+3$	-6	$(a-1)$			

↑

Tableau 2

Basic variables	Current values	x_1	x_2	x_3	x_4	x_5	x_6
x_4	1		①	1	1	-1	
x_1	1	1	-3	-1		1	
x_6	2		2	$a+1$		-1	1
$(-w)$	$+3$		$+3$	$a+2$		-3	

↑

Tableau 3

Basic variables	Current values	x_1	x_2	x_3	x_4	x_5	x_6
x_2	1		1	1	1	-1	
x_1	4	1		2	3	-2	
x_6	0			$(a-1)$	-2	1	1
$(-w)$	0			$a-1$	-3	0	

For $a = 0$ or 1, phase I is complete since $\bar{c}_3 = a - 1 \leqq 0$, but with x_6 still part of the basis. Note that in Tableau 2, either x_4 or x_6 could be dropped from the basis. We have arbitrarily selected x_4. (A similar argument would apply if x_6 were chosen.)

First, assume $a = 0$. Then we can introduce x_3 into the basis in place of the artificial variable x_6, pivoting on the coefficient $a - 1$ or x_3 in the third constraint, giving Tableau 4.

Tableau 4

Basic variables	Current values	x_1	x_2	x_3	x_4	x_5	x_6
x_2	1		1		-1	0	1
x_1	4	1			-1	0	2
x_3	0			1	2	-1	-1
$(-w)$	0				-1	-1	-1

Note that we have pivoted on a negative coefficient here. Since the righthand-side element of the third equation is zero, dividing by a negative pivot element will not make the resulting righthand-side coefficient negative. Dropping x_4, x_5, and x_6, we obtain the desired canonical form. Note that x_6 is now set to zero and is nonbasic.

Next, suppose that $a = 1$. The coefficient $(a - 1)$ in Tableau 3 is zero, so we cannot pivot x_3 into the basis as above. In this case, however, dropping artificial variables x_4 and x_5 from the system, the third constraint of Tableau 3 reads $x_6 = 0$.

Consequently, even though x_6 is a basic variable, in the canonical form for the original problem it will always remain at value zero during phase II. Thus, throughout phase II, a feasible solution to the original problem will be maintained as required. When more than one artificial variable is in the optimal basis for phase I, these techniques can be applied to each variable.

For the general problem, the transition rule from phase I to phase II can be stated as:

> **Phase I–Phase II Transition Rule.** Suppose that artificial variable x_i is the ith basic variable at the end of Phase I (at value zero). Let $\bar{a}_{ij}$ be the coefficient of the nonartificial variable x_j in the ith constraint of the final tableau. If some $\bar{a}_{ij} \neq 0$, then pivot on any such $\bar{a}_{ij}$, introducing x_j into the basis in place of x_i. If all $\bar{a}_{ij} = 0$, then maintain x_i in the basis throughout phase II by including the ith constraint, which reads $x_i = 0$.

As a final note, observe that if all $\bar{a}_{ij} = 0$ above, then constraint i is a redundant constraint in the original system, for, by adding multiples of the other equation to constraint i via pivoting, we have produced the equation (ignoring artificial variables):

$$0x_1 + 0x_2 + \cdots + 0x_n = 0.$$

For example, when $a = 1$ for the problem above, (constraint 3) = 2 times (constraint 1) − (constraint 2), and is redundant.

Phase I–Phase II Example

$$\text{Maximize } z = -3x_1 \qquad + x_3,$$

subject to:

$$
\begin{aligned}
x_1 + x_2 + x_3 + x_4 &= 4, \\
-2x_1 + x_2 - x_3 \qquad &= 1, \\
3x_2 + x_3 + x_4 &= 9, \\
x_j \geq 0 \qquad (j &= 1, 2, 3, 4).
\end{aligned}
$$

Adding artificial variables x_5, x_6, and x_7, we first minimize $x_5 + x_6 + x_7$ or, equivalently, maximize $w = -x_5 - x_6 - x_7$. The iterations are shown in Fig. 2.6. The first tableau is the phase I problem statement. Basic variables x_5, x_6 and x_7 appear in the objective function and, to achieve the initial canonical form, we must add the constraints to the objective function to eliminate these variables.

Tableaus 2 and 3 contain the phase I solution. Tableau 4 gives a feasible solution to the original problem. Artificial variable x_7 remains in the basis and is eliminated by pivoting on the -1 coefficient for x_4. This pivot replaces $x_7 = 0$ in the basis by $x_4 = 0$, and gives a basis from the original variables to initiate phase II.

Tableaus 5 and 6 give the phase II solution.

Tableau 1 (Phase I—problem statement)

Basic variables	Current values	x_1	x_2	x_3	x_4	x_5	x_6	x_7		Equation identification and transformations	Ratio test
						Artificial variables					
x_5	4	1	1	1	1	1				[1]	
x_6	1	−2	1	−1	0		1			[2]	
x_7	9	0	3	1	1			1		[3]	
$(-z)$	0	−3	0	1	0					[4]	
$(-w)$	0	0	0	0	0	−1	−1	−1		[5]	

Tableau 2 (Phase I—initial canonical form)

Basic variables	Current values	x_1	x_2	x_3	x_4	x_5	x_6	x_7		Equation identification and transformations	Ratio test
x_5	4	1	1	1	1	1				[1]	4/1
x_6	1	−2	(1)	−1	0		1			[2]	1/1
x_7	9	0	3	1	1			1		[3]	9/3
$(-z)$	0	−3	0	1	0					[4]	
$(-w)$	14	−1	5	1	2					[6] = [5] + [1] + [2] + [3]	

↓   ↑

Tableau 3

Basic variables	Current values	x_1	x_2	x_3	x_4	x_5	x_6	x_7		Equation identification and transformations	Ratio test
x_5	3	(3)		2	1	1	−1			[7] = [1] − [2]	3/3
x_2	1	−2	1	−1	0		1			[8] = [2]	
x_7	6	6		4	1		−3	1		[9] = [3] − 3[2]	6/6
$(-z)$	0	−3		1	0		0			[10] = [4]	
$(-w)$	9	9		6	2		−5			[11] = [6] − 5[2]	

↑

Tableau 4

Basic variables	Current values	x_1	x_2	x_3	x_4	x_5	x_6	x_7
x_1	1	1		$\frac{2}{3}$	$\frac{1}{3}$	$\frac{1}{3}$	$-\frac{1}{3}$	
x_2	3		1	$\frac{1}{3}$	$\frac{2}{3}$	$\frac{2}{3}$	$-\frac{1}{3}$	
x_7	0			0	(-1)	-2	-1	1
$(-z)$	3			3	1	1	-1	
$(-w)$	0			0	-1	-3	-2	

$\boxed{12} = \frac{1}{3}\boxed{7}$
$\boxed{13} = \boxed{8} + 2\,\boxed{12} \quad = \boxed{9} \quad -2\,\boxed{7}$
$\boxed{14} = \boxed{9} - 6\,\boxed{12} \quad = \boxed{9}$
$\boxed{15} = \boxed{10} + 3\,\boxed{12} = \boxed{10} \quad +\boxed{7}$
$\boxed{16} = \boxed{11} - 9\,\boxed{12} = \boxed{11} \quad -3\,\boxed{7}$
(Pivot to remove artificial variable from the basis.)

Tableau 5 (start phase II)

Basic variables	Current values	x_1	x_2	x_3	x_4	x_5	x_6	x_7
x_1	1	1		$\left(\frac{2}{3}\right)$		$-\frac{1}{3}$	$-\frac{2}{3}$	$\frac{1}{3}$
x_2	3		1	$-\frac{1}{3}$		$\frac{2}{3}$	$-\frac{1}{3}$	$\frac{2}{3}$
x_4	0			0	1	2	1	-1
$(-z)$	3			3		-1	-2	1

$\boxed{17} = \boxed{12} - \frac{1}{3}\boxed{19}$
$\boxed{18} = \boxed{13} - \frac{2}{3}\boxed{19}$
$\boxed{19} = -1\,\boxed{14}$
$\boxed{20} = \boxed{15} - \boxed{19}$

Tableau 6

Basic variables	Current values	x_1	x_2	x_3	x_4	x_5	x_6	x_7
x_3	$\frac{3}{2}$	$\frac{3}{2}$		1		$-\frac{1}{2}$	-1	$\frac{1}{2}$
x_2	$\frac{5}{2}$	$-\frac{1}{2}$	1			$-\frac{1}{2}$	0	$\frac{1}{2}$
x_4	0	0			1	2	1	-1
$(-z)$	$-\frac{3}{2}$	$-\frac{9}{2}$				$\frac{1}{2}$	1	$-\frac{1}{2}$

$\boxed{21} = \frac{3}{2}\boxed{17}$
$\boxed{22} = \boxed{18} - \frac{1}{3}\boxed{21}$
$\boxed{23} = \boxed{19}$
$\boxed{24} = \boxed{20} - 3\,\boxed{21}$

$1(2/3)$
$3(1/3)$

Optimal solution $x_1 = 0,\ x_2 = \frac{5}{2},\ x_3 = \frac{3}{2},\ x_4 = 0,\ z = \frac{3}{2}$

Fig. 2.6 Phase I–phase II example.

2.6 LINEAR PROGRAMS WITH BOUNDED VARIABLES

In most linear-programming applications, many of the constraints merely specify upper and lower bounds on the decision variables. In a distribution problem, for example, variables x_j representing inventory levels might be constrained by storage capacities u_j and by predetermined safety stock levels ℓ_j so that $\ell_j \leq x_j \leq u_j$. We are led then to consider linear programs with *bounded variables*:

$$\text{Maximize } z = \sum_{j=1}^{n} c_j x_j,$$

subject to:

$$\sum_{j=1}^{n} a_{ij} x_j = b_i, \qquad (i = 1, 2, \ldots, m) \tag{3}$$

$$\ell_j \leq x_j \leq u_j, \qquad (j = 1, 2, \ldots, n). \tag{4}$$

The lower bounds ℓ_j may be $-\infty$ and/or the upper bounds u_j may be $+\infty$, indicating respectively that the decision variable x_j is unbounded from below or from above. Note that when each $\ell_j = 0$ and each $u_j = +\infty$, this problem reduces to the linear-programming form that has been discussed up until now.

The bounded-variable problem can be solved by the simplex method as discussed thus far, by adding slack variables to the upper-bound constraints and surplus variables to the lower-bound constraints, thereby converting them to equalities. This approach handles the bounding constraints explicitly. In contrast, the approach proposed in this section modifies the simplex method to consider the bounded-variable constraints implicitly. In this new approach, pivoting calculations are computed only for the equality constraints (3) rather than for the entire system (3) and (4). In many instances, this reduction in pivoting calculations will provide substantial computational savings. As an extreme illustration, suppose that there is one equality constraint and 1000 nonnegative variables with upper bounds. The simplex method will maintain 1001 constraints in the tableau, whereas the new procedure maintains only the single equality constraint.

We achieve these savings by using a canonical form with one basic variable isolated in each of the equality constraints, as in the usual simplex method. However, basic feasible solutions now are determined by setting nonbasic variables to either their lower or upper bound. This method for defining basic feasible solutions extends our previous practice of setting nonbasic variables to their lower bounds of zero, and permits us to assess optimality and generate improvement procedures much as before.

Suppose, for example, that x_2 and x_4 are nonbasic variables constrained by:

$$4 \leq x_2 \leq 15,$$

$$2 \leq x_4 \leq 5;$$

and that

$$z = 4 - \tfrac{1}{4}x_2 + \tfrac{1}{2}x_4,$$
$$x_2 = 4,$$
$$x_4 = 5,$$

in the current canonical form. In any feasible solution, $x_2 \geq 4$, so $-\tfrac{1}{4}x_2 \leq -1$; also, $x_4 \leq 5$, so that $\tfrac{1}{2}x_4 \leq \tfrac{1}{2}(5) = 2\tfrac{1}{2}$. Consequently,

$$z = 4 - \tfrac{1}{4}x_2 + \tfrac{1}{2}x_4 \leq 4 - 1 + 2\tfrac{1}{2} = 5\tfrac{1}{2}$$

for any feasible solution. Since the current solution with $x_2 = 4$ and $x_4 = 5$ attains this upper bound, it must be optimal. In general, the current canonical form represents the optimal solution whenever nonbasic variables at their lower bounds have nonpositive objective coefficients, and nonbasic variables at their upper bound have nonnegative objective coefficients.

> **Bounded Variable Optimality Condition.** In a maximization problem in canonical form, if every nonbasic variable at its lower bound has a nonpositive objective coefficient, and every nonbasic variable at its upper bound has a nonnegative objective coefficient, then the basic feasible solution given by that canonical form maximizes the objective function over the feasible region.

Improving a nonoptimal solution becomes slightly more complicated than before. If the objective coefficient $\bar{c}_j$ of nonbasic variable x_j is positive and $x_j = \ell_j$, then we increase x_j; if $\bar{c}_j < 0$ and $x_j = u_j$, we decrease x_j. In either case, the objective value is improving.

When changing the value of a nonbasic variable, we wish to maintain feasibility. As we have seen, for problems with only nonnegative variables, we have to test, via the ratio rule, to see when a basic variable first becomes zero. Here we must consider the following contingencies:

i) the nonbasic variable being changed reaches its upper or lower bound; or

ii) some basic variable reaches either its upper or lower bound.

In the first case, no pivoting is required. The nonbasic variable simply changes from its lower to upper bound, or upper to lower bound, and remains nonbasic. In the second case, pivoting is used to remove the basic variable reaching either its lower or upper bound from the basis.

These ideas can be implemented on a computer in a number of ways. For example, we can keep track of the lower bounds throughout the algorithm; or every lower bound

$$x_j \geq \ell_j$$

can be converted to zero by defining a new variable

$$x_j'' = x_j - \ell_j \geq 0,$$

and substituting $x_j'' + \ell_j$ for x_j everywhere throughout the model. Also, we can always redefine variables so that every nonbasic variable is at its lower bound. Let

Tableau 1

Basic variables	Current values	x_1	x_2	x_3	x_4	x_5	x_6	Ratio test
x_1	15	1				4	1	$15/4$
x_2	8		1			6	2	$8/6$
x_3	4			1		-7	-2	$(15-4)/7$
x_4	2				1	-1	-1	$(5-2)/1$
$(-z)$	0					2	1	
Upper bounds	15	15	15	15	5	1	8	

$\uparrow$ (under x_5)

$\begin{cases} t_1 = 8/6 \\ t_2 = 11/7 \\ u_5 = 1 \end{cases}$ (Upper bound substitution)

Upper bound ratios:
$15/4$, $8/6$, $(15-4)/7$, $(5-2)/1$

Tableau 2

Basic variables	Current values	x_1	x_2	x_3	x_4	x_5'	x_6	Ratio test
x_1	11	1				-4	1	$11/1$
x_2	2		1			-6	②	$2/2$
x_3	11			1		7	-2	$(15-11)/2$
x_4	3				1	1	-1	$(5-3)/1$
$(-z)$	-2					-2	1	
Upper bounds	15	15	15	15	5	1	8	

$\downarrow$ (at left) $\uparrow$ (under x_6)

$\begin{cases} t_1 = 1 \\ t_2 = 2 \\ u_6 = 8 \end{cases}$ (Usual pivot)

Tableau 3

Basic variables	Current values	x_1	x_2	x_3	x_4	x_5'	x_6	Ratio test
x_1	10	1	$-\frac{1}{2}$			-1		$(15-10)/1$
x_6	1		$\frac{1}{2}$			-3	1	$(8-1)/3$
x_3	13		1	1		1		$13/1$
x_4	4		$\frac{1}{2}$		1	-2		$(5-4)/2$
$(-z)$	-3		$-\frac{1}{2}$			1		
Upper bounds	15	15	15	15	5	1	8	

$\uparrow$ (under x_5')

$\begin{cases} t_1 = 13 \\ t_2 = \frac{1}{2} \\ u_5 = 1 \end{cases}$ (Upper bound substitution for x_4)

80

Tableau 4

Basic variables	Current values	x_1	x_2	x_3	x_4'	x_5'	x_6	
x_1	10	1	$-\frac{1}{2}$			-1		$(15 - 10)/1$
x_6	1		$\frac{1}{2}$			-3	1	$(8 - 1)/3$
x_3	13		1	1		1		$13/1$
x_4'	1		$-\frac{1}{2}$		1	②		$1/2$
$(-z)$	-3		$-\frac{1}{2}$			1		
Upper bounds		15	15	15	5	1	8	

$$\left\{\begin{aligned} t_1 &= \tfrac{1}{2} \\ t_2 &= \tfrac{8}{3} \\ u_5 &= 1 \end{aligned}\right.$$
(Usual pivot)

Tableau 5

Basic variables	Current values	x_1	x_2	x_3	x_4'	x_5'	x_6
x_1	$10\frac{1}{2}$	1	$-\frac{3}{4}$		$\frac{1}{2}$		
x_6	$2\frac{1}{2}$		$-\frac{1}{4}$		$\frac{3}{2}$		1
x_3	$12\frac{1}{2}$		$\frac{5}{4}$	1	$-\frac{1}{2}$		
x_5'	$\frac{1}{2}$		$-\frac{1}{4}$		$-\frac{1}{2}$	1	
$(-z)$	$-3\frac{1}{2}$		$-\frac{1}{4}$		$-\frac{1}{2}$		
Upper bounds		15	15	15	5	1	8

Fig. 2.7 Simplex method with bounded variables.

Optimal solution: $x_1 = 10\frac{1}{2}$, $x_2 = 0$, $x_3 = 12\frac{1}{2}$, $x_4 = 5 - x_4' = 5$, $x_5 = 1 - x_5' = \frac{1}{2}$, $x_6 = 2\frac{1}{2}$.

x'_j denote the slack variable for the upper-bound constraint $x_j \leq u_j$; that is,

$$x_j + x'_j = u_j.$$

Whenever x_j is nonbasic at its upper bound u_j, the slack variable $x'_j = 0$. Consequently, substituting $u_j - x'_j$ for x_j in the model makes x'_j nonbasic at value zero in place of x_j. If, subsequently in the algorithm, x'_j becomes nonbasic at its upper bound, which is also u_j, we can make the same substitution for x'_j, replacing it with $u_j - x_j$, and x_j will appear nonbasic at value zero. These transformations are usually referred to as the *upper-bounding substitution*.

The computational steps of the upper-bounding algorithm are very simple to describe if both of these transformations are used. Since all nonbasic variables (either x_j or x'_j) are at value zero, we increase a variable for maximization as in the usual simplex method if its objective coefficient is positive. We use the usual ratio rule to determine at what value t_1 for the incoming variable, a basic variable first reaches zero. We also must ensure that variables do not exceed their upper bounds. For example, when increasing nonbasic variable x_s to value t, in a constraint with x_1 basic, such as:

$$x_1 - 2x_s = 3,$$

we require that:

$$x_1 = 3 + 2t \leq u_1 \qquad \left(\text{that is, } t \leq \frac{u_1 - 3}{2}\right).$$

We must perform such a check in every constraint in which the incoming variable has a negative coefficient; thus $x_s \leq t_2$ where:

$$t_2 = \min_i \left\{ \frac{u_k - \bar{b}_i}{-\bar{a}_{is}} \,\middle|\, \bar{a}_{is} < 0 \right\},$$

and u_k is the upper bound for the basic variable x_k in the ith constraint, $\bar{b}_i$ is the current value for this variable, and $\bar{a}_{is}$ are the constraint coefficients for variable x_s. This test might be called the *upper-bounding ratio test*. Note that, in contrast to the usual ratio test, the upper-bounding ratio uses negative coefficients $\bar{a}_{is} < 0$ for the nonbasic variable x_s being increased.

In general, the incoming variable x_s (or x'_s) is set to:

$$x_s = \min\{u_s, t_1, t_2\}.$$

If the minimum is

i) u_s, then the upper bounding substitution is made for x_s (or x'_s);

ii) t_1, then a usual simplex pivot is made to introduce x_s into the basis;

iii) t_2, then the upper bounding substitution is made for the basic variable x_k (or x'_k) reaching its upper bound and x_s is introduced into the basis in place of x'_k (or x_k) by a usual simplex pivot.

The procedure is illustrated in Fig. 2.7. Tableau 1 contains the problem formulation, which is in canonical form with x_1, x_2, x_3, and x_4 as basic variables and x_5

and x_6 as nonbasic variables at value zero. In the first iteration, variable x_5 increases, reaches its upper bound, and the upper bounding substitution $x_5' = 1 - x_5$ is made. Note that, after making this substitution, the variable x_5' has coefficients opposite in sign from the coefficients of x_5. Also, in going from Tableau 1 to Tableau 2, we have updated the current value of the basic variables by multiplying the upper bound of x_5, in this case $u_5 = 1$, by the coefficients of x_5 and moving these constants to the righthand side of the equations.

In Tableau 2, variable x_6 increases, basic variable x_2 reaches zero, and a usual simplex pivot is performed. After the pivot, it is attractive to increase x_5' in Tableau 3. As x_5' increases basic variable x_4 reaches its upper bound at $x_4 = 5$ and the upper-bounding substitution $x_4' = 5 - x_4$ is made. Variable x_4' is isolated as the basic variable in the fourth constraint in Tableau 4 (by multiplying the constraint by -1 after the upper-bounding substitution); variable x_5' then enters the basis in place of x_4'. Finally, the solution in Tableau 5 is optimal, since the objective coefficients are nonpositive for the nonbasic variables, each at value zero.

EXERCISES

1. Given:

$$
\begin{aligned}
x_1 \quad\quad\quad + 2x_4 &= 8, \\
x_2 \quad\quad + 3x_4 &= 6, \\
x_3 + 8x_4 &= 24, \\
+ 10x_4 &= -32, \\
\end{aligned}
$$

$-z$

$$x_1 \geqq 0, \quad x_2 \geqq 0, \quad x_3 \geqq 0, \quad x_4 \geqq 0.$$

a) What is the optimal solution of this problem?

b) Change the coefficient of x_4 in the z-equation to -3. What is the optimal solution now?

c) Change the signs on all x_4 coefficients to be negative. What is the optimal solution now?

2. Consider the linear program:

$$\text{Maximize } z = \quad 9x_2 + \quad x_3 \quad\quad - 2x_5 - x_6,$$

subject to:

$$
\begin{aligned}
5x_2 + 50x_3 + x_4 + \quad x_5 \quad\quad &= 10, \\
x_1 - 15x_2 + 2x_3 \quad\quad\quad\quad\quad &= 2, \\
x_2 + \quad x_3 \quad\quad + \quad x_5 + x_6 &= 6, \\
x_j \geqq 0 \quad (j = 1, 2, \ldots, 6).
\end{aligned}
$$

a) Find an initial basic feasible solution, specify values of the decision variables, and tell which are basic.

b) Transform the system of equations to the canonical form for carrying out the simplex routine.

c) Is your initial basic feasible solution optimal? Why?

d) How would you select a column in which to pivot in carrying out the simplex algorithm?
e) Having chosen a pivot column, now select a row in which to pivot and describe the selection rule. How does this rule guarantee that the new basic solution is feasible? Is it possible that no row meets the criterion of your rule? If this happens, what does this indicate about the original problem?
f) Without carrying out the pivot operation, compute the new basic feasible solution.
g) Perform the pivot operation indicated by (d) and (e) and check your answer to (f). Substitute your basic feasible solution in the original equations as an additional check.
h) Is your solution optimal now? Why?

3. a) Reduce the following system to canonical form. Identify slack, surplus, and artificial variables.

$$-2x_1 + x_2 \leq 4 \qquad (1)$$
$$3x_1 + 4x_2 \geq 2 \qquad (2)$$
$$5x_1 + 9x_2 = 8 \qquad (3)$$
$$x_1 + x_2 \geq 0 \qquad (4)$$
$$2x_1 + x_2 \geq -3 \qquad (5)$$
$$-3x_1 - x_2 \leq -2 \qquad (6)$$
$$3x_1 + 2x_2 \leq 10 \qquad (7)$$
$$x_1 \geq 0, \qquad x_2 \geq 0.$$

b) Formulate phase I objective functions for the following systems with $x_1 \geq 0$ and $x_2 \geq 0$:

i) expressions 2 and 3 above.

ii) expressions 1 and 7 above.

iii) expressions 4 and 5 above.

4. Consider the linear program

$$\text{Maximize } z = x_1,$$

subject to:

$$-x_1 + x_2 \leq 2,$$
$$x_1 + x_2 \leq 8,$$
$$-x_1 + x_2 \geq -4,$$
$$x_1 \geq 0, \qquad x_2 \geq 0.$$

a) State the above in canonical form.
b) Solve by the simplex method.
c) Solve geometrically and also trace the simplex procedure steps graphically.
d) Suppose that the objective function is changed to $z = x_1 + cx_2$. Graphically determine the values of c for which the solution found in parts (b) and (c) remains optimal.
e) Graphically determine the shadow price corresponding to the third constraint.

5. The bartender of your local pub has asked you to assist him in finding the combination of mixed drinks that will maximize his revenue. He has the following bottles available:

1 quart (32 oz.) Old Cambridge (a fine whiskey—cost $8/quart)
1 quart Joy Juice (another fine whiskey—cost $10/quart)
1 quart Ma's Wicked Vermouth ($10/quart)
2 quarts Gil-boy's Gin ($6/quart)

Since he is new to the business, his knowledge is limited to the following drinks:

Whiskey Sour	2 oz. whiskey	Price $1
Manhattan	2 oz. whiskey	$2
	1 oz. vermouth	
Martini	2 oz. gin	$2
	1 oz. vermouth	
Pub Special	2 oz. gin	$3
	2 oz. whiskey	

    Use the simplex method to maximize the bar's profit. (Is the cost of the liquor relevant in this formulation?)

6. A company makes three lines of tires. Its four-ply biased tires produce $6 in profit per tire, its fiberglass belted line $4 a tire, and its radials $8 a tire. Each type of tire passes through three manufacturing stages as a part of the entire production process.

    Each of the three process centers has the following hours of available production time per day:

Process	Hours
1 Molding	12
2 Curing	9
3 Assembly	16

The time required in each process to produce one hundred tires of each line is as follows:

Tire	Hours per 100 units		
	Molding	Curing	Assembly
Four-ply	2	3	2
Fiberglass	2	2	1
Radial	2	1	3

Determine the optimum product mix for each day's production, assuming all tires are sold.

7. An electronics firm manufactures printed circuit boards and specialized electronics devices. Final assembly operations are completed by a small group of trained workers who labor simultaneously on the products. Because of limited space available in the plant, no more then ten assemblers can be employed. The standard operating budget in this functional department allows a maximum of $9000 per month as salaries for the workers.

    The existing wage structure in the community requires that workers with two or more years of experience receive $1000 per month, while recent trade-school graduates will work for only $800. Previous studies have shown that experienced assemblers produce $2000 in "value added" per month while new-hires add only $1800. In order to maximize the value added by the group, how many persons from each group should be employed? Solve graphically and by the simplex method.

8. The processing division of the Sunrise Breakfast Company must produce one ton (2000 pounds) of breakfast flakes per day to meet the demand for its Sugar Sweets cereal. Cost per pound of the three ingredients is:

$$
\begin{array}{ll}
\text{Ingredient A} & \$4 \text{ per pound} \\
\text{Ingredient B} & \$3 \text{ per pound} \\
\text{Ingredient C} & \$2 \text{ per pound}
\end{array}
$$

Government regulations require that the mix contain at least 10% ingredient A and 20% ingredient B. Use of more than 800 pounds per ton of ingredient C produces an unacceptable taste.

Determine the minimum-cost mixture that satisfies the daily demand for Sugar Sweets. Can the bounded-variable simplex method be used to solve this problem?

9. Solve the following problem using the two phases of the simplex method:

$$\text{Maximize } z = 2x_1 + x_2 + x_3,$$

subject to:

$$
\begin{array}{rcl}
2x_1 + 3x_2 - x_3 & \leq & 9, \\
2x_2 + x_3 & \geq & 4, \\
x_1 \qquad\;\; + x_3 & = & 6,
\end{array}
$$

$$x_1 \geq 0, \qquad x_2 \geq 0, \qquad x_3 \geq 0.$$

Is the optimal solution unique?

10. Consider the linear program:

$$\text{Maximize } z = -3x_1 + 6x_2,$$

subject to:

$$
\begin{array}{rcl}
5x_1 + 7x_2 & \leq & 35, \\
-x_1 + 2x_2 & \leq & 2,
\end{array}
$$

$$x_1 \geq 0, \qquad x_2 \geq 0.$$

a) Solve this problem by the simplex method. Are there alternative optimal solutions? How can this be determined at the final simplex iteration?

b) Solve the problem graphically to verify your answer to part (a).

11. Solve the following problem using the simplex method:

$$\text{Minimize } z = x_1 - 2x_2 - 4x_3 + 2x_4,$$

subject to:

$$
\begin{array}{rcl}
x_1 \qquad\;\; - 2x_3 & \leq & 4, \\
x_2 \qquad\quad - x_4 & \leq & 8, \\
-2x_1 + x_2 + 8x_3 + x_4 & \leq & 12,
\end{array}
$$

$$x_1 \geq 0, \qquad x_2 \geq 0, \qquad x_3 \geq 0, \qquad x_4 \geq 0.$$

12. a) Set up a linear program that will determine a feasible solution to the following system
     of equations and inequalities if one exists. *Do not solve* the linear program.

$$x_1 - 6x_2 + x_3 - x_4 = 5,$$
$$- 2x_2 + 2x_3 - 3x_4 \geqq 3,$$
$$3x_1 \qquad - 2x_3 + 4x_4 = -1,$$
$$x_1 \geqq 0, \qquad x_3 \geqq 0, \qquad x_4 \qquad 0.$$

     b) Formulate a phase I linear program to find a feasible solution to the system:

$$3x_1 + 2x_2 - x_3 \qquad \leqq -3,$$
$$- x_1 - x_2 + 2x_3 \qquad \leqq -1,$$
$$x_1 \geqq 0, \qquad x_2 \geqq 0, \qquad x_3 \geqq 0.$$

     Show, from the phase I objective function, that the system contains no feasible solution
     (no pivoting calculations are required).

13. The tableau given below corresponds to a maximization problem in decision variables
    $x_j \geqq 0 \ (j = 1, 2, \ldots, 5)$:

Basic variables	Current values	x_1	x_2	x_3	x_4	x_5
x_3	4	-1	a_1	1		
x_4	1	a_2	-4		1	
x_5	b	a_3	3			1
$(-z)$	-10	c	-2			

State conditions on all five unknowns a_1, a_2, a_3, b, and c, such that the following statements
are true.

   a) The current solution is optimal. There are multiple optimal solutions.
   b) The problem is unbounded.
   c) The problem is infeasible.
   d) The current solution is not optimal (assume that $b \geqq 0$). Indicate the variable that
      enters the basis, the variable that leaves the basis, and what the total change in profit
      would be for one iteration of the simplex method for all values of the unknowns that
      are not optimal.

14. Consider the linear program:

   Maximize $z = \alpha x_1 + 2x_2 + x_3 - 4x_4,$

   subject to:

$$x_1 + x_2 \qquad - x_4 = 4 + 2\Delta \qquad (1)$$
$$2x_1 - x_2 + 3x_3 - 2x_4 = 5 + 7\Delta \qquad (2)$$
$$x_1 \geqq 0, \qquad x_2 \geqq 0, \qquad x_3 \geqq 0, \qquad x_4 \geqq 0,$$

   where α and Δ are viewed as parameters.

a) Form two new constraints as $(1') = (1) + (2)$ and $(2') = -2(1) + (2)$. Solve for x_1 and x_2 from $(1')$ and $(2')$, and substitute their values in the objective function. Use these transformations to express the problem in canonical form with x_1 and x_2 as basic variables. $3 = f(x_2, x_4, \alpha, \Delta)$

b) Assume $\Delta = 0$ (constant). For what values of α are x_1 and x_2 *optimal basic variables* in the problem? $3 \le \alpha \le 4$

c) Assume $\alpha = 3$. For what values of Δ do x_1 and x_2 form an optimal basic feasible solution? $-1 \le \Delta \le 1$

15. Let

$$(-w) + d_1 x_1 + d_2 x_2 + \cdots + d_m x_m = 0 \qquad (*)$$

be the phase I objective function when phase I terminates for maximizing w. Discuss the following two procedures for making the phase I to II transition when an artificial variable remains in the basis at value zero. Show, using either procedure, that every basic solution determined during phase II will be feasible for the *original* problem formulation.

a) Multiply each coefficient in $(*)$ by -1. Initiate phase II with the original objective function, but maintain $(*)$ in the tableau as a new constraint with (w) as the basic variable.

b) Eliminate $(*)$ from the tableau and at the same time eliminate from the problem any variable x_j with $d_j < 0$. Any artificial variable in the optimal phase I basis is now treated as though it were a variable from the original problem.

16. In our discussion of reduction to canonical form, we have replaced variables unconstrained in sign by the difference between two nonnegative variables. This exercise considers an alternative transformation that does not introduce as many new variables, and also a simplex-like procedure for treating free variables directly without any substitutions. For concreteness, suppose that y_1, y_2, and y_3 are the only unconstrained variables in a linear program.

a) Substitute for y_1, y_2, and y_3 in the model by:

$$y_1 = x_1 - x_0,$$
$$y_2 = x_2 - x_0,$$
$$y_3 = x_3 - x_0,$$

with $x_0 \geq 0$, $x_1 \geq 0$, $x_2 \geq 0$, and $x_3 \geq 0$. Show that the models are equivalent before and after these substitutions.

b) Apply the simplex method directly with y_1, y_2, and y_3. When are these variables introduced into the basis at positive levels? At negative levels? If y_1 is basic, will it ever be removed from the basis? Is the equation containing y_1 as a basic variable used in the ratio test? Would the simplex method be altered in any other way?

17. Apply the phase I simplex method to find a feasible solution to the problem:

$$x_1 - 2x_2 + x_3 = 2,$$
$$-x_1 + 3x_2 + x_3 = 1,$$
$$2x_1 - 3x_2 + 4x_3 = 7,$$

$$x_1 \geq 0, \qquad x_2 \geq 0, \qquad x_3 \geq 0.$$

Does the termination of phase I show that the system contains a redundant equation? How?

18. Frequently, linear programs are formulated with *interval constraints* of the form:

$$5 \leq 6x_1 - x_2 + 3x_3 \leq 8.$$

a) Show that this constraint is equivalent to the constraints

$$6x_1 - x_2 + 3x_3 + x_4 = 8,$$
$$0 \leq x_4 \leq 3.$$

b) Indicate how the general *interval linear program*

$$\text{Maximize } z = \sum_{j=1}^{n} c_j x_j,$$

subject to:

$$b_i' \leq \sum_{j=1}^{n} a_{ij} x_j \leq b_i \qquad (i = 1, 2, \ldots, m),$$
$$x_j \geq 0 \qquad (j = 1, 2, \ldots, n),$$

can be formulated as a bounded-variable linear program with m equality constraints.

19. a) What is the solution to the linear-programming problem:

$$\text{Maximize } z = c_1 x_1 + c_2 x_2 + \cdots + c_n x_n,$$

subject to:

$$a_1 x_1 + a_2 x_2 + \cdots + a_n x_n \leq b,$$
$$0 \leq x_j \leq u_j \qquad (j = 1, 2, \ldots, n),$$

with bounded variables and one additional constraint? Assume that the constants c_j, a_j, and u_j for $j = 1, 2, \ldots, n$, and b are all positive and that the problem has been formulated so that:

$$\frac{c_1}{a_1} \geq \frac{c_2}{a_2} \geq \cdots \geq \frac{c_n}{a_n}.$$

b) What will be the steps of the simplex method for bounded variables when applied to this problem (in what order do the variables enter and leave the basis)?

20. a) Graphically determine the steps of the simplex method for the problem:

$$\text{Maximize } 8x_1 + 6x_2,$$

subject to:

$$3x_1 + 2x_2 \leq 28,$$
$$5x_1 + 2x_2 \leq 42,$$
$$x_1 \qquad\quad \leq 8,$$
$$x_2 \leq 8,$$
$$x_1 \geq 0, \qquad x_2 \geq 0.$$

Indicate on the sketch the basic variables at each iteration of the simplex algorithm in terms of the given variables and the slack variables for the four less-than-or-equal-to constraints.

b) Suppose that the bounded-variable simplex method is applied to this problem. Specify how the iterations in the solution to part (a) correspond to the bounded-variable simplex method. Which variables from x_1, x_2, and the slack variable for the first two constraints, are basic at each iteration? Which nonbasic variables are at their upper bounds?

c) Solve the problem algebraically, using the simplex algorithm for bounded variables.

Sensitivity Analysis

3

We have already been introduced to sensitivity analysis in Chapter 1 via the geometry of a simple example. We saw that the values of the decision variables and those of the slack and surplus variables remain unchanged even though some coefficients in the objective function are varied. We also saw that varying the righthand-side value for a particular constraint alters the optimal value of the objective function in a way that allows us to impute a per-unit value, or *shadow price*, to that constraint. These shadow prices and the shadow prices on the implicit nonnegativity constraints, called *reduced costs*, remain unchanged even though some of the righthand-side values are varied. Since there is always some uncertainty in the data, it is useful to know over what range and under what conditions the components of a particular solution remain unchanged. Further, the sensitivity of a solution to changes in the data gives us insight into possible technological improvements in the process being modeled. For instance, it might be that the available resources are not balanced properly and the primary issue is not to resolve the most effective allocation of these resources, but to investigate what additional resources should be acquired to eliminate possible bottlenecks. Sensitivity analysis provides an invaluable tool for addressing such issues.

There are a number of questions that could be asked concerning the sensitivity of an optimal solution to changes in the data. In this chapter we will address those that can be answered most easily. Every commercial linear-programming system provides this elementary sensitivity analysis, since the calculations are easy to perform using the tableau associated with an optimal solution. There are two variations in the data that invariably are reported: objective function and righthand-side ranges. The objective-function ranges refer to the range over which an individual coefficient of the objective function can vary, without changing the basis associated with an optimal solution. In essence, these are the ranges on the objective-function coefficients over which we can be sure the values of the decision variables in an optimal solution will remain unchanged. The righthand-side ranges refer to the range over which an individual righthand-side value can vary, again without changing the basis associated with an optimal solution. These are the ranges on the righthand-side values over which we can be sure the values of the shadow prices and reduced costs will remain unchanged. Further, associated with each range is information concerning how the basis would change if the range were *exceeded*. These concepts will become clear if we deal with a specific example.

3.1 AN EXAMPLE FOR ANALYSIS

We will consider for concreteness the custom-molder example from Chapter 1; in order to increase the complexity somewhat, let us add a third alternative to the production possibilities. Suppose that, besides the six-ounce juice glasses x_1 and the ten-ounce cocktail glasses x_2, our molder is approached by a new customer to produce a champagne glass. The champagne glass is not difficult to produce except that it must be molded in two separate pieces—the bowl with stem and then the base. As a result, the production time for the champagne glass is 8 hours per hundred cases, which is greater than either of the other products. The storage space required for the champagne glasses is 1000 cubic feet per hundred cases; and the contribution is $6.00 per case, which is higher than either of the other products. There is no limit on the demand for champagne glasses. Now what is the optimal product mix among the three alternatives?

The formulation of the custom-molding example, including the new activity of producing champagne glasses, is straightforward. We have exactly the same capacity limitations—hours of production capacity, cubic feet of warehouse capacity, and limit on six-ounce juice-glass demand—and one additional decision variable for the production of champagne glasses. Letting

x_1 = Number of cases of six-ounce juice glasses, in hundreds;

x_2 = Number of cases of ten-ounce cocktail glasses, in hundreds;

x_3 = Number of cases of champagne glasses, in hundreds;

and measuring the contribution in hundreds of dollars, we have the following formulation of our custom-molder example:

Maximize $z = \ 5x_1 + 4.5x_2 + \ 6x_3,$ (hundreds of dollars)

subject to:

$$6x_1 + \ 5x_2 + \ 8x_3 \leqq \ 60,$$ (production capacity; hours)

$$10x_1 + 20x_2 + 10x_3 \leqq 150,$$ (warehouse capacity;   (1) hundreds of sq. ft.)

$$x_1 \qquad\qquad\qquad \leqq \ 8,$$ (demand for 6 oz. glasses; hundreds of cases)

$$x_1 \geqq 0, \qquad x_2 \geqq 0, \qquad x_3 \geqq 0.$$

If we add one slack variable in each of the less-than-or-equal-to constraints, the problem will be in the following canonical form for performing the simplex method:

$$6x_1 + 5x_2 + 8x_3 + x_4 \qquad\qquad\qquad = 60, \qquad (2)$$
$$10x_1 + 20x_2 + 10x_3 \qquad + x_5 \qquad\qquad = 150, \qquad (3)$$
$$x_1 \qquad\qquad\qquad\qquad + x_6 \qquad = 8, \qquad (4)$$
$$5x_1 + 4.5x_2 + 6x_3 \qquad\qquad\qquad - z = 0. \qquad (5)$$

The corresponding initial tableau is shown in Tableau 1.

Tableau 1

Basic variables	Current values	x_1	x_2	x_3	x_4	x_5	x_6
x_4	60	6	5	8	1		
x_5	150	10	20	10		1	
x_6	8	1	0	0			1
$(-z)$	0	5	4.5	6			

After applying the simplex method as described in Chapter 2, we obtain the the final tableau shown in Tableau 2.

Tableau 2

Basic variables	Current values	x_1	x_2	x_3	x_4	x_5	x_6
x_2	$4\frac{2}{7}$		1	$-\frac{2}{7}$	$-\frac{1}{7}$	$\frac{3}{35}$	
x_6	$1\frac{4}{7}$			$-\frac{11}{7}$	$-\frac{2}{7}$	$\frac{1}{14}$	1
x_1	$6\frac{3}{7}$	1		$\frac{11}{7}$	$\frac{2}{7}$	$-\frac{1}{14}$	
$(-z)$	$-51\frac{3}{7}$			$-\frac{4}{7}$	$-\frac{11}{14}$	$-\frac{1}{35}$	

Since the final tableau is in canonical form and all objective-function coefficients of the nonbasic variables are currently nonpositive, we know from Chapter 2 that we have the optimal solution, consisting of $x_1 = 6\frac{3}{7}$, $x_2 = 4\frac{2}{7}$, $x_6 = 1\frac{4}{7}$, and $z = 51\frac{3}{7}$.

In this chapter we present results that depend only on the initial and final tableaus of the problem. Specifically, we wish to analyze the effect on the optimal solution of changing various elements of the problem data without re-solving the linear program or having to remember any of the intermediate tableaus generated in solving the problem by the simplex method. The type of results that can be derived in this way are conservative, in the sense that they provide sensitivity analysis for changes in the problem data small enough so that the same decision variables remain basic, but not for larger changes in the data. The example presented in this section will be used to motivate the discussions of sensitivity analysis throughout this chapter.

3.2 SHADOW PRICES, REDUCED COSTS, AND NEW ACTIVITIES

In our new variation of the custom-molder example, we note that the new activity of producing champagne glasses is not undertaken at all. An immediate question arises, could we have known this without performing the simplex method on the entire problem? It turns out that a proper interpretation of the shadow prices in the Chapter 1 version of the problem would have told us that producing champagne glasses would not be economically attractive. However, let us proceed more slowly. Recall the definition of the shadow price associated with a particular constraint.

> **Definition.** The *shadow price* associated with a particular constraint is the change in the optimal value of the objective function per unit increase in the righthand-side value for that constraint, all other problem data remaining unchanged.

In Chapter 1 we implied that the shadow prices were readily available when a linear program is solved. Is it then possible to determine the shadow prices from the final tableau easily? The answer is yes, in general, but let us consider our example for concreteness.

Suppose that the production capacity in the first constraint of our model

$$6x_1 + 5x_2 + 8x_3 + x_4 = 60 \tag{6}$$

is increased from 60 to 61 hours. We then essentially are procuring one additional unit of production capacity at no cost. We can obtain the same result algebraically by allowing the slack variable x_4 to take on negative values. If x_4 is replaced by $x_4 - 1$ (i.e., from its optimum value $x_4 = 0$ to $x_4 = -1$), Eq. (6) becomes:

$$6x_1 + 5x_2 + 8x_3 + x_4 = 61,$$

which is exactly what we intended.

Since x_4 is a slack variable, it does not appear in any other constraint of the original model formulation, nor does it appear in the objective function. Therefore, this replacement does not alter any other righthand-side value in the original problem formulation. What is the contribution to the optimal profit of this additional unit of capacity? We can resolve this question by looking at the objective function of the final tableau, which is given by:

$$z = 0x_1 + 0x_2 - \tfrac{4}{7}x_3 - \tfrac{11}{14}x_4 - \tfrac{1}{35}x_5 + 0x_6 + 51\tfrac{3}{7}. \tag{7}$$

The optimality conditions of the simplex method imply that the optimal solution is determined by setting the nonbasic variables $x_3 = x_4 = x_5 = 0$, which results in a profit of $51\tfrac{3}{7}$. Now, if we are allowed to make $x_4 = -1$, the profit increases by $\tfrac{11}{14}$ hundred dollars for each additional unit of capacity available. This, then, is the marginal value, or shadow price, for production hours.

The righthand side for every constraint can be analyzed in this way, so that the shadow price for a particular constraint is merely the negative of the coefficient of the appropriate slack (or artificial) variable in the objective function of the final tableau. For our example, the shadow prices are $\frac{11}{14}$ hundred dollars per hour of production capacity, $\frac{1}{35}$ hundred dollars per hundred cubic feet of storage capacity, and zero for the limit on six-ounce juice-glass demand. It should be understood that the shadow prices are associated with the constraints of the problem and not the variables. They are in fact the marginal worth of an additional unit of a particular righthand-side value.

So far, we have discussed shadow prices for the explicit structural constraints of the linear-programming model. The nonnegativity constraints also have a shadow price, which, in linear-programming terminology, is given the special name of reduced cost.

Definition. The *reduced cost* associated with the nonnegativity constraint for each variable is the shadow price of that constraint (i.e., the corresponding change in the objective function per unit increase in the lower bound of the variable).

The reduced costs can also be obtained directly from the objective equation in the final tableau. In our example, the final objective form is

$$z = 0x_1 + 0x_2 - \tfrac{4}{7}x_3 - \tfrac{11}{14}x_4 - \tfrac{1}{35}x_5 + 0x_6 + 51\tfrac{3}{7}. \tag{8}$$

Increasing the righthand side of $x_3 \geq 0$ by one unit to $x_3 \geq 1$ forces champagne glasses to be used in the final solution. From (8), the optimal profit decreases by $-\frac{4}{7}$. Since the basic variables have values $x_1 = 6\frac{3}{7}$ and $x_2 = 4\frac{2}{7}$, increasing the righthand sides of $x_1 \geq 0$ and $x_2 \geq 0$ by a small amount does not affect the optimal solution, so their reduced costs are zero. Consequently, in every case, the shadow price for the nonnegativity constraint on a variable is the objective coefficient for this variable in the final canonical form. For basic variables, these reduced costs are zero.

Alternatively, the reduced costs for all decision variables can be computed directly from the shadow prices on the structural constraints and the objective-function coefficients. In this view, the shadow prices are thought of as the opportunity costs associated with diverting resources away from the optimal production mix. For example, consider x_3. Since the new activity of producing champagne glasses requires 8 hours of production capacity per hundred cases, whose opportunity cost is $\frac{11}{14}$ hundred dollars per hour, and 10 hundred cubic feet of storage capacity per hundred cases, whose opportunity cost is $\frac{1}{35}$ hundred dollars per hundred cubic feet, the resulting total opportunity cost of producing one hundred cases of champagne glasses is:

$$(\tfrac{11}{14})8 + (\tfrac{1}{35})10 = \tfrac{46}{7} = 6\tfrac{4}{7}.$$

Now the contribution per hundred cases is only 6 hundred dollars so that producing any champagne glasses is not as attractive as producing the current levels of six-ounce juice glasses and ten-ounce cocktail glasses. In fact, if resources were diverted from the current optimal production mix to produce champagne glasses, the optimal value of the objective function would be reduced by $\frac{4}{7}$ hundred dollars per hundred cases of champagne glasses produced. This is exactly the reduced cost associated with variable x_3. This operation of determining the reduced cost of an activity from the shadow price and the objective function is generally referred to as *pricing out an activity.*

Given the reduced costs, it becomes natural to ask how much the contribution of the new activity would have to increase to make producing champagne glasses attractive? Using the opportunity-cost interpretation, the contribution clearly would have to be \$$6\frac{4}{7}$ in order for the custom-molder to be indifferent to transferring resources to the production of champagne glasses. Since the reduced cost associated with the new activity $6 - 6\frac{4}{7} = -\frac{4}{7}$ is negative, the new activity will not be introduced into the basis. If the reduced cost had been positive, the new activity would have been an attractive candidate to introduce into the basis.

The shadow prices determined for the Chapter 1 version of the custom-molder example are the same as those determined here, since the optimal solution is un-changed by the introduction of the new activity of producing champagne glasses. Had the new activity been priced out at the outset, using the shadow prices determined in Chapter 1, we would have immediately discovered that the opportunity cost of diverting resources from the current solution to the new activity exceeded its potential contribution. There would have been no need to consider the new activity further. This is an important observation, since it implies that the shadow prices provide a mechanism for screening new activities that were not included in the initial model formulation. In a maximization problem, if any new activity prices out negatively using the shadow prices associated with an optimal solution, it may be immediately dropped from consideration. If, however, a new activity prices out positively with these shadow prices, it must be included in the problem formulation and the new optimal solution determined by pivoting.

General Discussion

The concepts presented in the context of the custom-molder example can be applied to any linear program. Consider a problem in initial canonical form:

$$
\begin{array}{rcl|c}
 & & & \text{Shadow price} \\
\hline
a_{11}x_1 + a_{12}x_2 + \cdots + a_{1n}x_n + x_{n+1} & = & b_1 & y_1 \\
a_{21}x_1 + a_{22}x_2 + \cdots + a_{2n}x_n \quad\quad + x_{n+2} & = & b_2 & y_2 \\
\vdots & & \vdots & \vdots \\
a_{m1}x_1 + a_{m2}x_2 + \cdots + a_{mn}x_n + \quad\quad \cdots + x_{n+m} & = & b_m & y_m \\
\hline
(-z) + c_1x_1 + c_2x_2 + \cdots + c_nx_n + 0x_{n+1} + 0x_{n+2} + \cdots + 0x_{n+m} & = & 0 &
\end{array}
$$

The variables $x_{n+1}, x_{n+2}, \ldots, x_{n+m}$ are either slack variables or artificial variables that have been introduced in order to transform the problem into canonical form.

Assume that the optimum solution to this problem has been found and the corresponding final form of the objective function is:

$$(-z) + \bar{c}_1 x_1 + \bar{c}_2 x_2 + \cdots + \bar{c}_n x_n + \bar{c}_{n+1} x_{n+1}$$
$$+ \bar{c}_{n+2} x_{n+2} + \cdots + \bar{c}_{n+m} x_{n+m} = -\bar{z}_0. \quad (9)$$

As we have indicated before, $\bar{c}_j$ is the reduced cost associated with variable x_j. Since (9) is in canonical form, $\bar{c}_j = 0$ if x_j is a basic variable. Let y_i denote the shadow price for the ith constraint. The arguments from the example problem show that the negative of the final objective coefficient of the variable x_{n+i} corresponds to the shadow price associated with the ith constraint. Therefore:

$$\bar{c}_{n+1} = -y_1, \qquad \bar{c}_{n+2} = -y_2, \qquad \ldots, \qquad \bar{c}_{n+m} = -y_m. \quad (10)$$

Note that this result applies whether the variable x_{n+i} is a slack variable (i.e., the ith constraint is a less-than-or-equal-to constraint), or whether x_{n+i} is an artificial variable (i.e., the ith constraint is either an equality or a greater-than-or-equal-to constraint).

We now shall establish a fundamental relationship between shadow prices, reduced costs, and the problem data. Recall that, at each iteration of the simplex method, the objective function is transformed by subtracting from it a multiple of the row in which the pivot was performed. Consequently, the final form of the objective function could be obtained by subtracting multiples of the original constraints from the original objective function. Consider first the final objective coefficients associated with the original basic variables $x_{n+1}, x_{n+2}, \ldots, x_{n+m}$. Let $\pi_1, \pi_2, \ldots, \pi_n$ be the multiples of each row that are subtracted from the original objective function to obtain its final form (9). Since x_{n+i} appears only in the ith constraint and has a $+1$ coefficient, we should have:

$$\bar{c}_{n+i} = 0 - 1\pi_i.$$

Combining this expression with (10), we obtain:

$$\bar{c}_{n+i} = -\pi_i = -y_i.$$

Thus the shadow prices y_i are the multiples π_i.

Since these multiples can be used to obtain every objective coefficient in the final form (9), the reduced cost $\bar{c}_j$ of variable x_j is given by:

$$\bar{c}_j = c_j - \sum_{i=1}^{m} a_{ij} y_i \qquad (j = 1, 2, \ldots, n), \quad (11)$$

and the current value of the objective function is:

$$-\bar{z}_0 = -\sum_{i=1}^{m} b_i y_i$$

or, equivalently,

$$\bar{z}_0 = \sum_{i=1}^{m} b_i y_i \quad (12)$$

Expression (11) links the shadow prices to the reduced cost of each variable, while (12) establishes the relationship between the shadow prices and the optimum value of the objective function.

Expression (11) also can be viewed as a mathematical definition of the shadow prices. Since $\bar{c}_j = 0$ for the m basic variables of the optimum solution, we have:

$$0 = c_j - \sum_{i=1}^{m} a_{ij}y_i \qquad \text{for } j \text{ basic.}$$

This is a system of m equations in m unknowns that uniquely determines the values of the shadow prices y_i.

3.3 VARIATIONS IN THE OBJECTIVE COEFFICIENTS

Now let us consider the question of how much the objective-function coefficients can vary without changing the values of the decision variables in the optimal solution. We will make the changes one at a time, holding all other coefficients and righthand-side values constant. The reason for this is twofold: first, the calculation of the range on one coefficient is fairly simple and therefore not expensive; and second, describing ranges when more than two coefficients are simultaneously varied would require a system of equations instead of a simple interval.

We return to consideration of the objective coefficient of x_3, a nonbasic variable in our example. Clearly, if the contribution is reduced from $6 per case to something less it would certainly not become attractive to produce champagne glasses. If it is now not attractive to produce champagne glasses, then *reducing* the contribution from their production only makes it less attractive. However, if the contribution from production of champagne glasses is increased, presumably there is some level of contribution such that it becomes attractive to produce them. In fact, it was argued above that the opportunity cost associated with diverting resources from the optimal production schedule was merely the shadow price associated with a resource multiplied by the amount of the resource consumed by the activity. For this activity, the opportunity cost is $6\frac{4}{7}$ per case compared to a contribution of $6 per case. If the contribution were increased above the break-even opportunity cost, then it would become attractive to produce champagne glasses.

Let us relate this to the procedures of the simplex method. Suppose that we increase the objective function coefficient of x_3 in the original problem formulation (5) by Δc_3, giving us:

$$5x_1 + 4.5x_2 + \underbrace{(6 + \Delta c_3)x_3}_{= c_3^{\text{new}}} - z = 0.$$

In applying the simplex method, multiples of the rows were subtracted from the objective function to yield the final system of equations. Therefore, the objective function in the final tableau will remain unchanged except for the addition of $\Delta c_3 x_3$. The modified final tableau is given in Tableau 3.

Tableau 3

Basic variables	Current values	x_1	x_2	x_3	x_4	x_5	x_6
x_2	$4\frac{2}{7}$		1	$-\frac{2}{7}$	$-\frac{1}{7}$	$\frac{3}{35}$	
x_6	$1\frac{4}{7}$			$-\frac{11}{7}$	$-\frac{2}{7}$	$\frac{1}{14}$	1
x_1	$6\frac{3}{7}$	1		$\frac{11}{7}$	$\frac{2}{7}$	$-\frac{1}{14}$	
$(-z)$	$-51\frac{3}{7}$			$-\frac{4}{7}+\Delta c_3$	$-\frac{11}{14}$	$-\frac{1}{35}$	

Now x_3 will become a candidate to enter the optimal solution at a positive level, i.e., to enter the basis, only when its objective-function coefficient is positive. The optimal solution remains unchanged so long as:

$$-\tfrac{4}{7} + \Delta c_3 \leqq 0 \qquad \text{or} \qquad \Delta c_3 \leqq \tfrac{4}{7}.$$

Equivalently, we know that the range on the original objective-function coefficient of x_3, say c_3^{new}, must satisfy

$$-\infty < c_3^{\text{new}} \leqq 6\tfrac{4}{7}$$

if the optimal solution is to remain unchanged.

Next, let us consider what happens when the objective-function coefficient of a basic variable is varied. Consider the range of the objective-function coefficient of variable x_1. It should be obvious that if the contribution associated with the production of six-ounce juice glasses is reduced sufficiently, we will stop producing them. Also, a little thought tells us that if the contribution were *increased* sufficiently, we might end up producing *only* six-ounce juice glasses. To understand this mathematically, let us start out as we did before by adding Δc_1 to the objective-function coefficient of x_1 in the original problem formulation (5) to yield the following modified objective function:

$$(5 + \Delta c_1)x_1 + 4.5x_2 + 6x_3 - z = 0.$$

If we apply the same logic as in the case of the nonbasic variable, the result is Tableau 4.

Tableau 4

Basic variables	Current values	x_1	x_2	x_3	x_4	x_5	x_6
x_2	$4\frac{2}{7}$		1	$-\frac{2}{7}$	$-\frac{1}{7}$	$\frac{3}{35}$	
x_6	$1\frac{4}{7}$			$-\frac{11}{7}$	$-\frac{2}{7}$	$\frac{1}{14}$	1
x_1	$6\frac{3}{7}$	1		$\frac{11}{7}$	$\frac{2}{7}$	$-\frac{1}{14}$	
$(-z)$	$-51\frac{3}{7}$	Δc_1		$-\frac{4}{7}$	$-\frac{11}{14}$	$-\frac{1}{35}$	

However, the simplex method requires that the final system of equations be in canonical form with respect to the basic variables. Since the basis is to be unchanged, in order to make the coefficient of x_1 zero in the final tableau we must subtract Δc_1 times row 3 from row 4 in Tableau 4. The result is Tableau 5.

Tableau 5

Basic variables	Current values	x_1	x_2	x_3	x_4	x_5	x_6
x_2	$4\frac{2}{7}$		1	$\frac{2}{7}$	$-\frac{1}{7}$	$\frac{3}{35}$	
x_6	$1\frac{4}{7}$			$-\frac{11}{7}$	$-\frac{2}{7}$	$\frac{1}{14}$	1
x_1	$6\frac{3}{7}$	1		$\frac{11}{7}$	$\frac{2}{7}$	$-\frac{1}{14}$	
$(-z)$	$-51\frac{3}{7} - 6\frac{3}{7}\Delta c_1$			$-\frac{4}{7} - \frac{11}{7}\Delta c_1$	$-\frac{11}{14} - \frac{2}{7}\Delta c_1$	$-\frac{1}{35} + \frac{1}{14}\Delta c_1$	

By the simplex optimality criterion, all the objective-function coefficients in the final tableau must be nonpositive in order to have the current solution remain unchanged. Hence, we must have:

$$-\tfrac{4}{7} - \tfrac{11}{7}\Delta c_1 \leq 0 \qquad \text{(that is, } \Delta c_1 \geq -\tfrac{4}{11}\text{),}$$
$$-\tfrac{11}{14} - \tfrac{2}{7}\Delta c_1 \leq 0 \qquad \text{(that is, } \Delta c_1 \geq -\tfrac{11}{4}\text{),}$$
$$-\tfrac{1}{35} + \tfrac{1}{14}\Delta c_1 \leq 0 \qquad \text{(that is, } \Delta c_1 \leq +\tfrac{2}{5}\text{);}$$

and, taking the most limiting inequalities, the bounds on Δc_1 are:

$$-\tfrac{4}{11} \leq \Delta c_1 \leq \tfrac{2}{5}.$$

If we let $c_1^{\text{new}} = c_1 + \Delta c_1$ be the objective-function coefficient of x_1 in the initial tableau, then:

$$4\tfrac{7}{11} \leq c_1^{\text{new}} \leq 5\tfrac{2}{5},$$

where the current value of $c_1 = 5$.

It is easy to determine which variables will enter and leave the basis when the new cost coefficient reaches either of the extreme values of the range. When $c_1^{\text{new}} = 5\frac{2}{5}$, the objective coefficient of x_5 in the final tableau becomes 0; thus x_5 enters the basis for any further increase of c_1^{new}. By the usual ratio test of the simplex method,

$$\text{Min}_i \left\{ \frac{\bar{b}_i}{\bar{a}_{is}} \,\middle|\, \bar{a}_{is} > 0 \right\} = \text{Min} \left\{ \frac{4\frac{2}{7}}{\frac{3}{35}}, \frac{1\frac{4}{7}}{\frac{1}{14}} \right\} = \text{Min } \{50, 22\},$$

and the variable x_6, which is basic in row 2, leaves the basis when x_5 is introduced. Similarly, when $c_1^{\text{new}} = 4\frac{7}{11}$, the objective coefficient of x_3 in the final tableau becomes 0, and x_3 is the entering variable. In this case, the ratio test shows that x_1 leaves the basis.

General Discussion

To determine the ranges of the cost coefficients in the optimum solution of any linear program, it is useful to distinguish between nonbasic variables and basic variables.

If x_j is a nonbasic variable and we let its objective-function coefficient c_j be changed by an amount Δc_j with all other data held fixed, then the current solution remains unchanged so long as the new reduced cost $\bar{c}_j^{\text{new}}$ remains nonnegative, that is,

$$\bar{c}_j^{\text{new}} = c_j + \Delta c_j - \sum_{i=1}^{m} a_{ij} y_i = \bar{c}_j + \Delta c_j \leq 0.$$

The range on the variation of the objective-function coefficient of a nonbasic variable is then given by:

$$-\infty < \Delta c_j \leqq -\overline{c}_j, \tag{13}$$

so that the range on the objective-function coefficient $c_j^{\text{new}} = c_j + \Delta c_j$ is:

$$-\infty < c_j^{\text{new}} \leqq c_j - \overline{c}_j.$$

If x_r is a basic variable in row k and we let its original objective-function coefficient c_r be changed by an amount Δc_r with all other data held fixed, then the coefficient of the variable x_r in the final tableau changes to:

$$\overline{c}_r^{\text{new}} = c_r + \Delta c_r - \sum_{i=1}^{m} a_{ir} y_i = \overline{c}_r + \Delta c_r.$$

Since x_r is a basic variable, $\overline{c}_r = 0$; so, to recover a canonical form with $\overline{c}_r^{\text{new}} = 0$, we subtract Δc_r times the kth constraint in the final tableau from the final form of the objective function, to give new reduced costs for all nonbasic variables,

$$\overline{c}_j^{\text{new}} = \overline{c}_j - \Delta c_r \overline{a}_{kj}. \tag{14}$$

Here $\overline{a}_{kj}$ is the coefficient of variable x_j in the kth constraint of the final tableau. Note that $\overline{c}_j^{\text{new}}$ will be zero for all basic variables.

The current basis remains optimal if $\overline{c}_j^{\text{new}} \leqq 0$. Using this condition and (14), we obtain the range on the variation of the objective-function coefficient:

$$\underset{j}{\text{Max}} \left\{ \frac{\overline{c}_j}{\overline{a}_{kj}} \,\middle|\, \overline{a}_{kj} > 0 \right\} \leqq \Delta c_r \leqq \underset{j}{\text{Min}} \left\{ \frac{\overline{c}_j}{\overline{a}_{kj}} \,\middle|\, \overline{a}_{kj} < 0 \right\}. \tag{15}$$

The range on the objective-function coefficient $c_r^{\text{new}} = c_r + \Delta c_r$ of the basic variable x_r is determined by adding c_r to each bound in (15).

The variable transitions that occur at the limits of the cost ranges are easy to determine. For nonbasic variables, the entering variable is the one whose cost is being varied. For basic variables, the entering variable is the one giving the limiting value in (15). The variable that leaves the basis is then determined by the minimum-ratio rule of the simplex method. If x_s is the entering variable, then the basic variable in row r, determined by:

$$\frac{\overline{b}_r}{\overline{a}_{rs}} = \underset{i}{\text{Min}} \left\{ \frac{\overline{b}_i}{\overline{a}_{is}} \,\middle|\, \overline{a}_{is} > 0 \right\},$$

is dropped from the basis.

Since the calculation of these ranges and the determination of the variables that will enter and leave the basis if a range is exceeded are computationally inexpensive to perform, this information is invariably reported on any commercially available computer package. These computations are easy since no iterations of the simplex method need be performed. It is necessary only (1) to check the entering variable conditions of the simplex method to determine the ranges, as well as the variable that enters the basis, and (2) to check the leaving variable condition (i.e., the minimum-ratio rule) of the simplex method to compute the variable that leaves the basis.

3.4 VARIATIONS IN THE RIGHTHAND-SIDE VALUES

Now let us turn to the questions related to the righthand-side ranges. We already have noted that a righthand-side range is the interval over which an individual righthand-side value can be varied, all the other problem data being held constant, such that variables that constitute the basis remain the same. Over these ranges, the *values* of the decision variables are clearly modified. Of what use are these righthand-side ranges? Any change in the righthand-side values that keep the current basis, and therefore the canonical form, unchanged has no effect upon the objective-function coefficients. Consequently, the righthand-side ranges are such that the *shadow prices* (which are the negative of the coefficients of the slack or artificial variables in the final tableau) and the *reduced costs* remain unchanged for variations of a single value within the stated range.

We first consider the righthand-side value for the demand limit on six-ounce juice glasses in our example. Since this constraint is not binding, the shadow price associated with it is zero and it is simple to determine the appropriate range. If we add an amount Δb_3 to the righthand side of this constraint (4), the constraint changes to:

$$x_1 + x_6 = 8 + \Delta b_3.$$

In the original problem formulation, it should be clear that, since x_6, the slack in this constraint, is a basic variable in the final system of equations, x_6 is merely increased or decreased by Δb_3. In order to keep the current solution feasible, x_6 must remain greater than or equal to zero. From the final tableau, we see that the current value of $x_6 = 1\frac{4}{7}$; therefore x_6 remains in the basis if the following condition is satisfied:

$$x_6 = 1\tfrac{4}{7} + \Delta b_3 \geqq 0.$$

This implies that:

$$\Delta b_3 \geqq -1\tfrac{4}{7}$$

or, equivalently, that:

$$b_3^{\text{new}} = 8 + \Delta b_3 \geqq 6\tfrac{3}{7}.$$

Now let us consider changing the righthand-side value associated with the storage-capacity constraint of our example. If we add Δb_2 to the righthand-side value of constraint (3), this constraint changes to:

$$10x_1 + 20x_2 + 10x_3 + x_5 = 150 + \Delta b_2.$$

In the original problem formulation, as was previously remarked, changing the righthand-side value is essentially equivalent to decreasing the value of the slack variable x_5 of the corresponding constraint by Δb_2; that is, substituting $x_5 - \Delta b_2$ for x_5 in the original problem formulation. In this case, x_5, which is zero in the final solution, is changed to $x_5 = -\Delta b_2$. We can analyze the implications of this increase in the righthand-side value by using the relationships among the variables represented by the final tableau. Since we are allowing only one value to change at a time, we will maintain the remaining nonbasic variables, x_3 and x_4, at zero level, and we let $x_5 = -\Delta b_2$. Making these substitutions in the final tableau provides the following relationships:

$$x_2 - \tfrac{3}{35}\Delta b_2 = 4\tfrac{2}{7},$$
$$x_6 - \tfrac{1}{14}\Delta b_2 = 1\tfrac{4}{7},$$
$$x_1 + \tfrac{1}{14}\Delta b_2 = 6\tfrac{3}{7}.$$

In order for the current basis to remain optimal, it need only remain feasible, since the reduced costs will be unchanged by any such variation in the righthand-side value. Thus,

$$x_2 = 4\tfrac{2}{7} + \tfrac{3}{35}\Delta b_2 \geq 0 \qquad \text{(that is, } \Delta b_2 \geq -50\text{)},$$
$$x_6 = 1\tfrac{4}{7} + \tfrac{1}{14}\Delta b_2 \geq 0 \qquad \text{(that is, } \Delta b_2 \geq -22\text{)},$$
$$x_1 = 6\tfrac{3}{7} - \tfrac{1}{14}\Delta b_2 \geq 0 \qquad \text{(that is, } \Delta b_2 \leq 90\text{)},$$

which implies:

$$-22 \leq \Delta b_2 \leq 90,$$

or, equivalently,

$$128 \leq b_2^{\text{new}} \leq 240,$$

where the current value of $b_2 = 150$ and $b_2^{\text{new}} = 150 + \Delta b_2$.

Observe that these computations can be carried out directly in terms of the final tableau. When changing the ith righthand side by Δb_i, we simply substitute $-\Delta b_i$ for the slack variable in the corresponding constraint and update the current values of the basic variables accordingly. For instance, the change of storage capacity just considered is accomplished by setting $x_5 = -\Delta b_2$ in the final tableau to produce Tableau 6 with modified current values.

Tableau 6

Basic variables	Current values	x_1	x_2	x_3	x_4	x_5	x_6
x_2	$4\tfrac{2}{7} + \tfrac{3}{35}\Delta b_2$		1	$-\tfrac{2}{7}$	$-\tfrac{1}{7}$	$\tfrac{3}{35}$	
x_6	$1\tfrac{4}{7} + \tfrac{1}{14}\Delta b_2$			$-\tfrac{11}{7}$	$-\tfrac{2}{7}$	$\tfrac{1}{14}$	1
x_1	$6\tfrac{3}{7} - \tfrac{1}{14}\Delta b_2$	1		$\tfrac{11}{7}$	$\tfrac{2}{7}$	$-\tfrac{1}{14}$	
$(-z)$	$-51\tfrac{3}{7} - \tfrac{1}{35}\Delta b_2$			$-\tfrac{4}{7}$	$-\tfrac{11}{14}$	$-\tfrac{1}{35}$	

Note that the change in the optimal value of the objective function is merely the shadow price on the storage-capacity constraint multiplied by the increased number of units of storage capacity.

Variable Transitions

We have just seen how to compute righthand-side ranges so that the current basis remains optimal. For example, when changing demand on six-ounce juice glasses, we found that the basis remains optimal if

$$b_3^{\text{new}} \geq 6\tfrac{3}{7},$$

and that for $b_3^{new} < 6\frac{3}{7}$, the basic variable x_6 becomes negative. If b_3^{new} were reduced below $6\frac{3}{7}$, what change would take place in the basis? First, x_6 would have to leave the basis, since otherwise it would become negative. What variable would then enter the basis to take its place? In order to have the new basis be an optimal solution, the entering variable must be chosen so that the reduced costs are not allowed to become positive.

Regardless of which variable enters the basis, the entering variable will be isolated in row 2 of the final tableau to replace x_6, which leaves the basis. To isolate the entering variable, we must perform a pivot operation, and a multiple, say t, of row 2 in the final tableau will be subtracted from the objective-function row. Assuming that b_3^{new} were set equal to $6\frac{3}{7}$ in the initial tableau, the results are given in Tableau 7.

Tableau 7

Basic variables	Current values	x_1	x_2	x_3	x_4	x_5	x_6
x_2	$4\frac{2}{7}$		1	$-\frac{2}{7}$	$-\frac{1}{7}$	$\frac{3}{35}$	
x_6	0			$-\frac{11}{7}$	$-\frac{2}{7}$	$\frac{1}{14}$	1
x_1	$6\frac{3}{7}$	1		$\frac{11}{7}$	$\frac{2}{7}$	$-\frac{1}{14}$	
$(-z)$	$-51\frac{3}{7}$			$-\frac{4}{7}+\frac{11}{7}t$	$-\frac{11}{14}+\frac{2}{7}t$	$-\frac{1}{35}-\frac{1}{14}t$	$-t$

In order that the new solution be an optimal solution, the coefficients of the variables in the objective function of the final tableau must be nonpositive; hence,

$$-\tfrac{4}{7}+\tfrac{11}{7}t \leq 0 \qquad \text{(that is, } t \leq \tfrac{4}{11}\text{)},$$
$$-\tfrac{11}{14}+\tfrac{2}{7}t \leq 0 \qquad \text{(that is, } t \leq \tfrac{11}{4}\text{)},$$
$$-\tfrac{1}{35}-\tfrac{1}{14}t \leq 0 \qquad \text{(that is, } t \geq -\tfrac{2}{5}\text{)},$$
$$-t \leq 0 \qquad \text{(that is, } t \geq 0\text{)},$$

which implies:

$$0 \leq t \leq \tfrac{4}{11}.$$

Since the coefficient of x_3 is most constraining on t, x_3 will enter the basis. Note that the range on the righthand-side value and the variable transitions that would occur if that range were exceeded by a small amount are easily computed. However, the pivot operation actually introducing x_3 into the basis and eliminating x_6 need not be performed.

As another example, when the change Δb_2 in storage capacity reaches -22 in Tableau 6, then x_6, the slack on six-ounce juice glasses, reaches zero and will drop from the basis, and again x_3 enters the basis. When $\Delta b_2 = 90$, though, then x_1, the production of six-ounce glasses, reaches zero and will drop from the basis. Since x_1 is a basic variable in the third constraint of Tableau 6, we must pivot in the third row, subtracting t times this row from the objective function. The result is given in Tableau 8.

Tableau 8

Basic variables	Current values	x_1	x_2	x_3	x_4	x_5	x_6
x_2	12		1	$-\frac{2}{7}$	$-\frac{1}{7}$	$\frac{3}{35}$	
x_6	8			$-\frac{11}{7}$	$-\frac{2}{7}$	$\frac{1}{14}$	1
x_1	0	1		$\frac{11}{7}$	$\frac{2}{7}$	$-\frac{1}{14}$	
$(-z)$	-54	$-t$		$-\frac{4}{7} - \frac{11}{7}t$	$-\frac{11}{7} - \frac{2}{7}t$	$-\frac{1}{35} + \frac{1}{14}t$	

The new objective-function coefficients must be nonpositive in order for the new basis to be optimal; hence,

$$-t \leq 0 \qquad \text{(that is, } t \geq 0\text{)},$$
$$-\tfrac{4}{7} - \tfrac{11}{7}t \leq 0 \qquad \text{(that is, } t \geq -\tfrac{4}{11}\text{)},$$
$$-\tfrac{11}{7} - \tfrac{2}{7}t \leq 0 \qquad \text{(that is, } t \geq -\tfrac{11}{2}\text{)},$$
$$-\tfrac{1}{35} + \tfrac{1}{14}t \leq 0 \qquad \text{(that is, } t \leq \tfrac{2}{5}\text{)},$$

which implies $0 \leq t \leq \tfrac{2}{5}$. Consequently, x_5 enters the basis. The implication is that, if the storage capacity were increased from 150 to 240 cubic feet, then we would produce only the ten-ounce cocktail glasses, which have the highest contribution per hour of production time.

General Discussion

In the process of solving linear programs by the simplex method, the initial canonical form:

$$a_{11}x_1 + a_{12}x_2 + \cdots + a_{1n}x_n + x_{n+1} \qquad\qquad\qquad = b_1,$$
$$a_{21}x_1 + a_{22}x_2 + \cdots + a_{2n}x_n \qquad\quad + x_{n+2} \qquad\qquad = b_2,$$
$$\vdots \qquad\quad \vdots \qquad\qquad \vdots \qquad\qquad\qquad \ddots \qquad\qquad \vdots$$
$$a_{m1}x_1 + a_{m2}x_2 + \cdots + a_{mn}x_n \qquad\qquad\qquad\quad + x_{n+m} = b_m$$

$$\underbrace{\qquad\qquad\qquad\qquad\qquad}_{\text{Original basic variables}}$$

is transformed into a canonical form:

$$\bar{a}_{11}x_1 + \bar{a}_{12}x_2 + \cdots + \bar{a}_{1n}x_n + \beta_{11}x_{n+1} + \beta_{12}x_{n+2} + \cdots + \beta_{1m}x_{n+m} = \bar{b}_1,$$
$$\bar{a}_{21}x_1 + \bar{a}_{22}x_2 + \cdots + \bar{a}_{2n}x_n + \beta_{21}x_{n+1} + \beta_{22}x_{n+2} + \cdots + \beta_{2m}x_{n+m} = \bar{b}_2,$$
$$\vdots \qquad\qquad\qquad\qquad\qquad\qquad\qquad\qquad\qquad\qquad\qquad \vdots$$
$$\bar{a}_{m1}x_1 + \bar{a}_{m2}x_2 + \cdots + \bar{a}_{mn}x_n + \beta_{m1}x_{n+1} + \beta_{m2}x_{n+2} + \cdots + \beta_{mn}x_{n+m} = \bar{b}_m.$$

Of course, because this is a canonical form, the updated data $\bar{a}_{ij}$ and β_{ij} will be structured so that one basic variable is isolated in each constraint. Since the updated coefficients of the initial basic (slack or artificial) variables $x_{n+1}, x_{n+2}, \ldots, x_{n+m}$ in the final tableau play such an important role in sensitivity analysis, we specifically denote these values as β_{ij}.

We can change the coefficient b_k of the kth righthand side in the initial tableau by Δb_k with all the other data held fixed, simply by substituting $x_{n+k} - \Delta b_k$ for x_{n+k} in the original tableau. To see how this change affects the updated righthand-side coefficients, we make the same substitution in the final tableau. Only the terms $\beta_{ik} x_{n+k}$ for $i = 1, 2, \ldots, m$ change in the final tableau. They become $\beta_{ik}(x_{n+k} - \Delta b_k) = \beta_{ik} x_{n+k} - \beta_{ik}\Delta b_k$. Since $\beta_{ik}\Delta b_k$ is a constant, we move it to the righthand side to give modified righthand-side values:

$$\bar{b}_i + \beta_{ik}\Delta b_k \qquad (i = 1, 2, \ldots, m).$$

As long as all of these values are nonnegative, the basis specified by the final tableau remains optimal, since the reduced costs have not been changed. Consequently, the current basis is optimal whenever $\bar{b}_i + \beta_{ik}\Delta b_k \geq 0$ for $i = 1, 2, \ldots, m$ or, equivalently,

$$\operatorname*{Max}_i \left\{ \frac{-\bar{b}_i}{\beta_{ik}} \,\middle|\, \beta_{ik} > 0 \right\} \leq \Delta b_k \leq \operatorname*{Min}_i \left\{ \frac{-b_i}{\beta_{ik}} \,\middle|\, \beta_{ik} < 0 \right\}. \tag{16}$$

The lower bound disappears if all $\beta_{ik} \leq 0$, and the upper bound disappears if all $\beta_{ik} \geq 0$.

When Δb_k reaches either its upper or lower bound in Eq. (16), any further increase (or decrease) in its value makes one of the updated righthand sides, say $\bar{b}_r + \beta_{rk}\Delta b_k$, negative. At this point, the basic variable in row r leaves the basis, and we must pivot in row r in the final tableau to find the variable to be introduced in its place. Since pivoting subtracts a multiple t of the pivot row from the objective equation, the new objective equation has coefficients:

$$\bar{c}_j - t\bar{a}_{rj} \qquad (j = 1, 2, \ldots, n). \tag{17}$$

For the new basis to be optimal, each of these coefficients must be nonpositive. Since $\bar{c}_j = 0$ for the basic variable being dropped and its coefficient in constraint r is $\bar{a}_{rj} = 1$, we must have $t \geq 0$. For any nonnegative t, the updated coefficient $\bar{c}_j - t\bar{a}_{rj}$ for any other variable remains nonpositive if $\bar{a}_{rj} \geq 0$. Consequently we need only consider $\bar{a}_{rj} < 0$, and t is given by

$$t = \operatorname*{Min}_j \left\{ \frac{\bar{c}_j}{\bar{a}_{rj}} \,\middle|\, \bar{a}_{rj} < 0 \right\}. \tag{18}$$

The index s giving this minimum has $\bar{c}_s - t\bar{a}_{rs} = 0$, and the corresponding variable x_s can become the new basic variable in row r by pivoting on $\bar{a}_{rs}$. Note that this pivot is made on a negative coefficient.

Since the calculation of the righthand-side ranges and the determination of the variables that will enter and leave the basis when a range is exceeded are computationally easy to perform, this information is reported by commercially available computer packages. For righthand-side ranges, it is necessary to check only the feasibility conditions to determine the ranges as well as the variable that leaves, and it is necessary to check only the entering variable condition (18) to complete the variable transitions. This condition will be used again in Chapter 4, since it is the minimum-ratio rule of the so-called dual simplex method.

3.5 ALTERNATIVE OPTIMAL SOLUTIONS AND SHADOW PRICES

In many applied problems it is important to identify alternative optimal solutions when they exist. When there is more than one optimal solution, there are often good external reasons for preferring one to the other; therefore it is useful to be able to easily determine alternative optimal solutions.

As in the case of the objective function and righthand-side ranges, the final tableau of the linear program tells us something conservative about the possibility of alternative optimal solutions. First, if all reduced costs of the nonbasic variables are strictly negative (positive) in a maximization (minimization) problem, then there is *no* alternative optimal solution, because introducing any variable into the basis at a positive level would reduce (increase) the value of the objective function. On the other hand, if one or more of the reduced costs are zero, there *may* exist alternative optimal solutions. Suppose that, in our custom-molder example, the contribution from champagne glasses, x_3, had been $6\frac{4}{7}$. From Section 3.1 we know that the reduced cost associated with this activity would be zero. The final tableau would look like Tableau 9.

Tableau 9

Basic variables	Current values	x_1	x_2	x_3	x_4	x_5	x_6	Ratio test
x_2	$4\frac{2}{7}$		1	$-\frac{2}{7}$	$-\frac{1}{7}$	$\frac{3}{35}$		
x_6	$1\frac{4}{7}$			$-\frac{11}{7}$	$-\frac{2}{7}$	$\frac{1}{14}$	1	
x_1	$6\frac{3}{7}$	1		$\frac{11}{7}$	$\frac{2}{7}$	$-\frac{1}{14}$		$6\frac{3}{7}/\frac{11}{7}$
$(-z)$	$-51\frac{3}{7}$			0	$-\frac{11}{14}$	$-\frac{1}{35}$		

$\uparrow$

This would imply that x_3, production of champagne glasses, could be introduced into the basis without changing the value of the objective function. The variable that would be dropped from the basis would be determined by the usual minimum-ratio rule:

$$\operatorname*{Min}_{i}\left\{\frac{\bar{b}_i}{\bar{a}_{i3}}\,\middle|\,\bar{a}_{i3}>0\right\} = \operatorname{Min}\left\{\frac{4\frac{2}{7}}{\frac{2}{7}},\frac{6\frac{3}{7}}{\frac{11}{7}}\right\} = \operatorname{Min}\left\{14,\tfrac{45}{11}\right\}.$$

In this case, the minimum ratio occurs for row 3 so that x_1 leaves the basis. The alternative optimal solution is then found by completing the pivot that introduces x_3 and eliminates x_1. The values of the new basic variables can be determined from the final tableau as follows:

$$x_3 = \tfrac{45}{11},$$
$$x_2 = 4\tfrac{2}{7} - \tfrac{2}{7}x_3 = 4\tfrac{2}{7} - \tfrac{2}{7}\left(\tfrac{45}{11}\right) = 3\tfrac{9}{77},$$
$$x_6 = 1\tfrac{4}{7} + \tfrac{11}{7}x_3 = 1\tfrac{4}{7} + \tfrac{11}{7}\left(\tfrac{45}{11}\right) = 8,$$
$$x_1 = 6\tfrac{3}{7} - \tfrac{11}{7}x_3 = 6\tfrac{3}{7} - \tfrac{11}{7}\left(\tfrac{45}{11}\right) = 0,$$
$$z = 51\tfrac{3}{7}.$$

Under the assumption that the contribution from champagne-glass production is $6\frac{4}{7}$, we have found an alternative optimal solution. It should be pointed out that any weighted combination of these two solutions is then also an alternative optimal solution.

In general, we can say that there *may* exist an alternative optimal solution to a linear program if one or more of the reduced costs of the nonbasic variables are zero. There *does* exist an alternative optimal solution if one of the nonbasic variables with zero reduced cost can be introduced into the basis at a positive level. In this case, any weighted combination of these solutions is also an alternative optimal solution. However, if it is not possible to introduce a new variable at a positive level, then no such alternative optimal solution exists even though some of the nonbasic variables have zero reduced costs. Further, the problem of finding *all* alternative optimal solutions cannot be solved by simply considering the reduced costs of the final tableau, since there can in general exist alternative optimal solutions that cannot be reached by a *single* pivot operation.

Independent of the question of whether or not alternative optimal solutions exist in the sense that different values of the decision variables yield the same optimal value of the objective function, there may exist alternative optimal shadow prices. The problem is completely analogous to that of identifying alternative optimal solutions. First, if all righthand-side values in the final tableau are positive, then there do not exist alternative optimal shadow prices. Alternatively, if one or more of these values are zero, then there *may* exist alternative optimal shadow prices. Suppose, in our custom-molder example, that the value of storage capacity had been 128 hundred cubic feet rather than 150. Then from Tableau 6 in Section 3.4, setting $\Delta b_2 = -22$, we illustrate this situation in Tableau 10.

Tableau 10

Basic variables	Current values	x_1	x_2	x_3	x_4	x_5	x_6
x_2	$\frac{12}{5}$		1	$-\frac{2}{7}$	$-\frac{1}{7}$	$\frac{3}{35}$	
x_6	0			$-\frac{11}{7}$	$-\frac{2}{7}$	$\frac{1}{14}$	1
x_1	8	1		$\frac{11}{7}$	$\frac{2}{7}$	$-\frac{1}{14}$	
$(-z)$	$-50\frac{4}{5}$			$-\frac{4}{7}$	$-\frac{11}{14}$	$-\frac{2}{70}$	

Since the righthand-side value in row 2 of Tableau 10 is zero, it is possible to drop x_6, the basic variable in row 2 of the final tableau, from the basis as long as there is a variable to introduce into the basis. The variable to be introduced into the basis can be determined from the entering variable condition, Eq. (18):

$$\underset{j}{\text{Min}}\left\{\frac{\bar{c}_j}{\bar{a}_{2j}}\,\middle|\,\bar{a}_{2j} < 0\right\} = \text{Min}\left\{\frac{-\frac{4}{7}}{-\frac{11}{7}}, \frac{-\frac{11}{14}}{-\frac{2}{7}}\right\} = \text{Min}\left\{\frac{4}{11}, \frac{11}{4}\right\}.$$

In this case the minimum ratio implies that the production of champagne glasses, x_3, enters the basis. The values of the reduced costs for the new basis can be determined by subtracting $\frac{4}{11}$ times row 2 in the final tableau from the objective function in the final tableau. Thus we have:

$$\bar{c}_3 = -\tfrac{4}{7} + \tfrac{11}{7}(\tfrac{4}{11}) = 0,$$
$$\bar{c}_4 = -\tfrac{11}{14} + \tfrac{2}{7}(\tfrac{4}{11}) = -\tfrac{15}{22},$$
$$\bar{c}_5 = -\tfrac{1}{35} - \tfrac{1}{14}(\tfrac{4}{11}) = -\tfrac{3}{55},$$
$$\bar{c}_6 = -\tfrac{4}{11}.$$

Since the shadow prices are the negative of the objective-function coefficients of the slack variables in the final tableau, the alternative optimal shadow prices in this case are $\tfrac{15}{22}$, $\tfrac{3}{55}$, and $\tfrac{4}{11}$ for the three constraints, respectively.

In general we can say that there *may* exist alternative optimal shadow prices for a linear program if one or more of the righthand-side values in the final tableau are zero. There *does* exist an alternative set of optimal shadow prices if the variable to enter the basis determined by the minimum-ratio rule has a strictly negative reduced cost. As in the case of alternative optimal solutions, any weighted combination of these sets of shadow prices is also an alternative set of optimal shadow prices. Finding all such sets of alternative optimal shadow prices is of the same degree of complexity as finding all alternative optimal solutions, since in general there can exist alternative sets of shadow prices that cannot be reached in a *single* pivot. Finally, it should be pointed out that all four cases can take place: unique solution with unique shadow prices; unique solution with alternative optimal shadow prices; alternative optimal solutions with unique shadow prices; and alternative optimal solutions and alternative shadow prices simultaneously.

3.6 THE COMPUTER OUTPUT—AN EXAMPLE

We have remarked that any commercially available linear-programming package routinely reports a great deal of information besides the values of the decision variables at the optimal solution. In this section we illustrate the typical computer output for the variation of the custom-molder example presented in the first few sections of this chapter. The results shown in Fig. 3.1 were obtained on a commercial time-sharing system. Note that the output includes a tabular display of the problem as stated in Eq. (1).

The first four numbered sections of this output should be compared to Tableau 2 for this example, given in Section 3.1. The optimal value of the objective function is $z = 51\tfrac{3}{7}$ hundred dollars, while the associated values of the decision variables are as follows: production of six-ounce juice glasses $x_1 = 6\tfrac{3}{7}$ hundred cases, and production of ten-ounce cocktail glasses $x_2 = 4\tfrac{2}{7}$ hundred cases. Note that there is slack in the constraint on demand for six-ounce juice glasses of $1\tfrac{4}{7}$ hundred cases, which corresponds to variable x_6. Finally, champagne glasses are not produced, and hence $x_3 = 0$. Note that the reduced cost associated with production of champagne glasses is $-\tfrac{4}{7}$, which is the amount the objective function would decrease per hundred cases if champagne glasses were in fact produced.

In our discussion of shadow prices in Section 3.2, it is pointed out that the shadow prices are the negative of the reduced costs associated with the slack variables.

```
TITLE: CUSTOM MOLDER
PROCEED, DISPLAY, OR REJECT? DISPLAY

FORMAT? 0

OBJECTIVES:

                 SIX-OZ    TEN-OZ    CHAMP

CONTRIB          5.000     4.500     6.000

CONSTRAINTS:

                 SIX-OZ    TEN-OZ    CHAMP    RELATION   RHS

PROD-HR          6.000     5.000     8.000      LE      60.00
STORAGE          10.00     20.00     10.00      LE      150.0
DEMAND           1.000     .0000     .0000      LE      8.000

PROCEED OR REJECT? PROCEED

PARAMETRICS? NO

MAXIMIZE OR MINIMIZE? MAX

OPTIMAL SOLUTION FOUND.
     CONTRIB        51.4286

OUTPUT OPTION? ALL

ALL ITEMS NOT LISTED IN SECTIONS 1 - 4 HAVE THE VALUE ZERO.

*1* DECISION VARIABLES
  1. SIX-OZ    6.42857
  2. TEN-OZ    4.28571

*2* SLACK(+) AND SURPLUS(-) IN CONSTRAINTS
  3. +DEMAND   1.57143

*3* SHADOW PRICES FOR CONSTRAINTS
  1. PROD-HR   .785714
  2. STORAGE   .285714E-01

*4* REDUCED COSTS FOR DECISION VARIABLES
  3. CHAMP     -.571429

*5* RANGES ON COEFFICIENTS OF OBJECTIVE CONTRIB
     VARIABLE   LOWER BOUND  CURRENT VALUE  UPPER BOUND
  1. SIX-OZ     4.6364       5.0000         5.4000
  2. TEN-OZ     4.1667       4.5000         6.5000
  3. CHAMP      UNBOUNDED    6.0000         6.5714

*6* RANGES ON VALUES OF RIGHT-HAND-SIDE RHS
     CONSTRNT   LOWER BOUND  CURRENT VALUE  UPPER BOUND
  1. PROD-HR    37.500       60.000         65.500
  2. STORAGE    128.00       150.00         240.00
  3. DEMAND     6.4286       8.0000         UNBOUNDED

*7* VARIABLE TRANSITIONS RESULTING FROM RANGING OBJECTIVE CONTRIB
     VARIABLE        LOWER BOUND           UPPER BOUND
                  VAR. IN   VAR. OUT    VAR. IN    VAR. OUT
  1. SIX-OZ       CHAMP     SIX-OZ      +STORAGE   +DEMAND
  2. TEN-OZ       +STORAGE  +DEMAND     CHAMP      SIX-OZ
  3. CHAMP                              CHAMP      SIX-OZ

*8* VARIABLE TRANSITIONS RESULTING FROM RANGING RHS RHS
     CONSTRNT        LOWER BOUND           UPPER BOUND
                  VAR. IN   VAR. OUT    VAR. IN    VAR. OUT
  1. PROD-HR      +STORAGE  SIX-OZ      CHAMP      +DEMAND
  2. STORAGE      CHAMP     +DEMAND     +STORAGE   SIX-OZ
  3. DEMAND       CHAMP     +DEMAND

OUTPUT OPTION? NO
```

Fig. 3.1 Solution of the custom-molder example.

Thus the shadow price on production capacity is $\frac{11}{14}$ hundred dollars per hour of production time, and the shadow price on storage capacity is $\frac{1}{35}$ hundred dollars per hundred square feet of storage space. The shadow price on six-ounce juice glasses demand is zero, since there remains unfulfilled demand in the optimal solution. It is intuitive that, in general, either the shadow price associated with a constraint is nonzero or the slack (surplus) in that constraint is nonzero, but both will not simultaneously be nonzero. This property is referred to as *complementary slackness*, and is discussed in detail in Chapter 4.

Sections ∗5∗ and ∗7∗ of the computer output give the ranges on the coefficients of the objective function and the variable transitions that take place. (This material is discussed in Section 3.3.) Note that the range on the nonbasic variable x_3, production of champagne glasses, is one-sided. Champagne glasses are not currently produced when their contribution is $6 per case, so that, if their contribution were reduced, we would certainly not expect this to change. However, if the contribution from the production of champagne glasses is increased to $6\frac{4}{7}$ per case, their production becomes attractive, so that variable x_3 would enter the basis and variable x_1, production of six-ounce juice glasses, would leave the basis. Consider now the range on the coefficient of x_1, production of six-ounce juice glasses, where the current contribution is $5 per case. If this contribution were raised to $5\frac{2}{5}$ per case, the slack in storage capacity would enter the basis, and the slack in juice-glass demand would leave. This means we would meet all of the juice-glass demand and storage capacity would no longer be binding. On the other hand, if this contribution were reduced to $4\frac{4}{7}$ per case, variable x_3, production of champagne glasses, would enter the basis and variable x_1, production of six-ounce juice glasses, would leave. In this instance, juice glasses would no longer be produced.

Sections ∗6∗ and ∗8∗ of the computer output give the ranges on the righthand-side values and the variable transitions that result. (This material is discussed in Section 3.4.) Consider, for example, the range on the righthand-side value of the constraint on storage capacity, where the current storage capacity is 150 hundred square feet. If this value were increased to 240 hundred square feet, the slack in storage capacity would enter the basis and variable x_1, production of six-ounce glasses, would leave. This means we would no longer produce six-ounce juice glasses, but would devote all our production capacity to ten-ounce cocktail glasses. If the storage capacity were reduced to 128 hundred square feet, we would begin producing champagne glasses and the slack in the demand for six-ounce juice glasses would leave the basis. The other ranges have similar interpretations.

Thus far we have covered information that is available routinely when a linear program is solved. The format of the information varies from one computer system to another, but the information available is always the same. The role of computers in solving linear-programming problems is covered more extensively in Chapter 5.

3.7 SIMULTANEOUS VARIATIONS WITHIN THE RANGES

Until now we have described the sensitivity of an optimal solution in the form of ranges on the objective-function coefficients and righthand-side values. These ranges were shown to be valid for changes in *one* objective-function coefficient or righthand-side value, while the remaining problem data are held fixed. It is then natural to ask what can be said about simultaneously changing *more* than one coefficient or value within the ranges. In the event that simultaneous changes are made *only* in objective-function coefficients of *nonbasic* variables and righthand-side values of *nonbinding* constraints within their appropriate ranges, the basis will remain unchanged. Unfortunately, it is not true, in general, that when the simultaneous variations within the ranges involve basic variables or binding constraints the basis will remain unchanged. However, for both the ranges on the objective-function coefficients when basic variables are involved, and the righthand-side values when binding constraints are involved, there is a conservative bound on these simultaneous changes that we refer to as the "100 percent rule."

Let us consider first, simultaneous changes in the righthand-side values involving binding constraints for which the basis, and therefore the shadow prices and reduced costs, remain unchanged. The righthand-side ranges, as discussed in Section 3.4, give the range over which *one* particular righthand-side value may be varied, with all other problem data being held fixed, such that the basis remains unchanged. As we have indicated, it is not true that simultaneous changes of more than one righthand-side value within these ranges will leave the optimal basis unchanged. However, it turns out that if these simultaneous changes are made in such a way that the sum of the fractions of allowable range utilized by these changes is less than or equal to one, the optimal basis will be unchanged.

Let us consider our custom-molder example, and make simultaneous changes in the binding production and storage capacity constraints. The righthand-side range on production capacity is 37.5 to 65.5 hundred hours, with the current value being 60. The righthand-side range on storage capacity is 128 to 240 hundred cubic feet, with the current value being 150. Although it is not true that the optimal basis remains unchanged for simultaneous changes in the current righthand-side values anywhere within these ranges, it is true that the optimal basis remains unchanged for any simultaneous change that is a weighted combination of values within these ranges. Figure 3.2 illustrates this situation. The horizontal and vertical lines in the figure are the ranges for production and storage capacity, respectively. The four-sided figure includes all weighted combinations of these ranges and is the space over which simultaneous variations in the values of production and storage capacity can be made while still ensuring that the basis remains unchanged. If we consider moving from the current righthand-side values of 60 and 150 to b_1^{new} and b_2^{new} respectively, where $b_1^{new} \leq 60$ and $b_2^{new} \geq 150$, we can ensure that the basis remains unchanged if

$$\frac{60 - b_1^{new}}{60 - 37.5} + \frac{b_2^{new} - 150}{240 - 150} \leqq 1.$$

As long as the sum of the fractions formed by the ratio of the change to the maximum possible change in that direction is less than or equal to one, the basis remains unchanged. Hence, we have the 100 percent rule. Since the basis remains unchanged, the shadow prices and reduced costs also are unchanged.

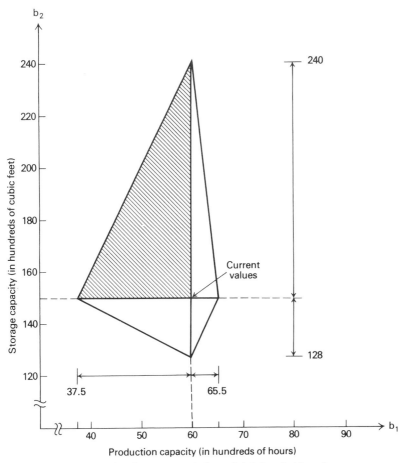

Fig. 3.2 Simultaneous variation of righthand-side values.

A similar situation exists in the case of simultaneous variations in the objective-function coefficients when basic variables are involved. It is not true that the basis remains unchanged for simultaneous changes in these coefficients anywhere within their individual ranges. However, it is true that the optimal basis remains unchanged for any simultaneous change in the objective-function coefficients that is a weighted combination of the values within these ranges. If, for example, we were to increase all cost coefficients simultaneously, the new optimal basis would remain unchanged so long as the new values c_1^{new}, c_2^{new}, and c_3^{new} satisfy:

$$\frac{c_1^{new} - 5}{5.4 - 5} + \frac{c_2^{new} - 4.5}{6.5 - 4.5} + \frac{c_3^{new} - 6}{6.5714 - 6} \leqq 1.$$

Again, the sum of the fractions formed by the ratio of the change in the coefficient to the maximum possible change in that direction must be less than or equal to one.

General Discussion

Let us analyze the 100 percent rule more formally. If we first consider simultaneous variations in the righthand-side values, the 100 percent rule states that the basis remains unchanged so long as the sum of the fractions, corresponding to the percent of maximum change in each direction, is less than or equal to one. To see that this must indeed be true, we look at weighted combinations of the solution for the current values and the solutions corresponding to particular boundaries of the ranges. In Fig. 3.2, the shaded area contains all weighted combinations of the current values, the lower bound on b_1, and the upper bound on b_2.

Let $x_j^0, j = 1, 2, \ldots, n$, be an optimal solution to the given linear program, and let a_{ij} be the coefficients of the initial tableau corresponding to the basic variables. Then,

$$\sum_{j_B} a_{ij}x_j^0 = b_i \qquad (i = 1, 2, \ldots, m), \tag{19}$$

where j_B indicates that the sum is taken only over basic variables. Further, let x_j^k, $j = 1, 2, \ldots, n$, be the optimal solution at either the upper or lower limit of the range when changing b_k alone, depending on the direction of the variation being considered. Since the basis remains unchanged for these variations, these solutions satisfy:

$$\sum_{j_B} a_{ij}x_j^k = b_i' = \begin{cases} b_k + \Delta b_k^{\max} & \text{for } i = k, \\ b_i & \text{for } i \neq k. \end{cases} \tag{20}$$

It is now easy to show that any nonnegative, weighted combination of these solutions must be an optimal feasible solution to the problem with simultaneous variations. Let λ_0 be the weight associated with the optimal solution corresponding to the current values, Eq. (19), and let λ_k be the weight associated with the optimal solution corresponding to the variation in b_k in Eq. (20). Consider all solutions that are nonnegative weighted combinations of these solutions, such that:

$$\sum_{k=0}^{m} \lambda_k = 1. \tag{21}$$

The corresponding weighted solution must be nonnegative; that is,

$$x_j = \sum_{k=0}^{m} \lambda_k x_j^k \geq 0 \qquad (j = 1, 2, \ldots, n), \tag{22}$$

since both λ_k and x_j^k are nonnegative. By multiplying the ith constraint of Eq. (19) by λ_0 and the ith constraint of Eq. (20) by λ_k and adding, we have

$$\sum_{k=0}^{m} \lambda_k \left(\sum_{j_B} a_{ij}x_j^k \right) = \sum_{k=0}^{m} \lambda_k b_i'.$$

Since the righthand-side reduces to:

$$\sum_{k=0}^{m} \lambda_k b'_i = \sum_{k \neq i} \lambda_k b_k + \lambda_i(b_i + \Delta b_i^{\max}) = b_i + \lambda_i \Delta b_i^{\max},$$

we can rewrite this expression by changing the order of summation as:

$$\sum_{j_B} a_{ij} \left(\sum_{k=0}^{m} \lambda_k x_j^k \right) = b_i + \lambda_i \Delta b_i^{\max} \qquad (i = 1, 2, \ldots, m). \qquad (23)$$

Expressions (22) and (23) together show that the weighted solution $x_j, j = 1, 2, \ldots, n$, is a feasible solution to the righthand-side variations indicated in Eq. (20) and has the same basis as the optimal solution corresponding to the current values in Eq. (19). This solution must also be optimal, since the operations carried out do not change the objective-function coefficients.

Hence, the basis remains optimal so long as the sum of the weights is one, as in Eq. (21). However, this is equivalent to requiring that the weights $\lambda_k, k = 1, 2, \ldots, m$, corresponding to the solutions associated with the ranges, while *excluding* the solution associated with the current values, satisfy:

$$\sum_{k=1}^{m} \lambda_k \leqq 1. \qquad (24)$$

The weight λ_0 on the solution corresponding to the current values is then determined from Eq. (21). Expression (24) can be seen to be the 100 percent rule by defining:

$$\Delta b_k \equiv \lambda_k \Delta b_k^{\max}$$

or, equivalently,

$$\lambda_k = \frac{\Delta b_k}{\Delta b_k^{\max}}$$

which, when substituted in Eq. (24) yields:

$$\sum_{k=1}^{m} \frac{\Delta b_k}{\Delta b_k^{\max}} \leqq 1. \qquad (25)$$

Since it was required that $\lambda_k \geqq 0$, Δb_k and $\Delta b_k^{\max}$ must have the same sign. Hence, the fractions in the 100 percent rule are the ratio of the actual change in a particular direction to the maximum possible change in that direction.

A similar argument can be made to establish the 100 percent rule for variations in the objective-function coefficients. The argument will not be given, but the form of the rule is:

$$\sum_{k=1}^{n} \frac{\Delta c_k}{\Delta c_k^{\max}} \leqq 1. \qquad (26)$$

3.8 PARAMETRIC PROGRAMMING

In the sensitivity analysis discussed thus far, we have restricted our presentation to changes in the problem data that can be made without changing the optimal basis. Consequently, what we have been able to say is fairly conservative. We did go so far as to indicate the variable that would enter the basis and the variable that would leave the basis when a boundary of a range was encountered. Further, in the case of alternative optimal solutions and alternative optimal shadow prices, the indicated pivot was completed at least far enough to exhibit the particular alternative. One important point in these discussions was the ease with which we could determine the pivot to be performed at a boundary of a range. This seems to indicate that it is relatively easy to make systematic calculations beyond the indicated objective-function or righthand-side ranges. This, in fact, is the case; and the procedure by which these systematic calculations are made is called *parametric programming*.

Preview

Consider once again our custom-molder example, and suppose that we are about to negotiate a contract for storage space and are interested in knowing the optimal value of the objective function for all values of storage capacity. We know from Section 3.2 that the shadow price on storage capacity is $\frac{1}{35}$ hundred dollars per hundred cubic feet, and that this shadow price holds over the range of 128 to 240 hundred cubic feet. As long as we are negotiating within this range, we know the worth of an additional hundred cubic feet of storage capacity. We further know that if we go above a storage capacity of 240, x_5, the slack variable in the storage-capacity constraint, enters the basis and x_1, the production of six-ounce juice glasses, leaves the basis. Since slack in storage capacity exists beyond 240, the shadow price beyond 240 must be zero. If we go below 128, we know that x_3, production of champagne glasses, will enter the basis and the slack in six-ounce juice-glass demand leaves the basis. However, we do not know the shadow price on storage capacity below 128.

To determine the shadow price on storage capacity below 128, we need to perform the indicated pivot and exhibit the new canonical form. Once we have the new canonical form, we immediately know the new shadow prices and can easily compute the new righthand-side ranges such that these shadow prices remain unchanged. The new shadow price on storage capacity turns out to be $\frac{3}{55}$ and this holds over the range 95 to 128. We can continue in this manner until the optimal value of the objective function for all possible values of storage capacity is determined.

Since the shadow price on a particular constraint is the change in the optimal value of the objective function per unit increase in the righthand-side value of that constraint, the optimal value of the objective function within some range must be a linear function of the righthand-side value with a slope equal to the corresponding shadow price. In Fig. 3.3, the optimal value of the objective function is plotted versus the available storage capacity. Note that this curve consists of a number of straight-line segments. The slope of each straight-line segment is equal to the shadow price on storage capacity, and the corresponding interval is the righthand-side range for

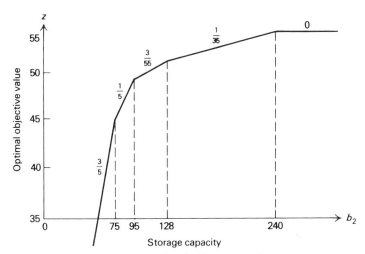

Fig. 3.3 Righthand-side parametrics.

storage capacity. For example, the slope of this curve is $\frac{1}{35}$ over the interval 128 to 240 hundred cubic feet.

In order to determine the curve given in Fig. 3.3, we essentially need to compute the optimal solution for all possible values of storage capacity. It is intuitive that this may be done efficiently since the breakpoints in the curve correspond to changes in the basis, and the optimal solution need only be calculated at these breakpoints. From a given solution the next breakpoint may be determined, since this is the same as computing a righthand-side range.

Now let us fix the storage capacity again at 150 hundred cubic feet and consider solving our custom-molder example as a function of the contribution from six-ounce juice glasses. Since the basis will remain unchanged for variations within the objective-function ranges, we might expect results similar to those obtained by varying a right-hand-side value. In Fig. 3.4, the optimal value of the objective function is plotted versus the contribution from six-ounce juice glasses. Note that this curve also consists of a number of straight-line segments. These segments correspond to ranges on the objective-function coefficient of six-ounce juice glasses such that the optimal basis does not change. Since the curve is a function of the contribution from production of six-ounce juice glasses, and the basis remains unchanged over each interval, the slope of the curve is given by the value of x_1 in the appropriate range.

In general then, it is straightforward to find the optimal solution to a linear program as a function of any *one* parameter: hence, the name *parametric programming*. In the above two examples, we first found the optimal solution as a function of the righthand-side value b_2, and then found the optimal solution as a function of the objective-function coefficient c_1. The procedure used in these examples is easily generalized to include simultaneous variation of more than one coefficient, as long as the variation is made a function of *one* parameter.

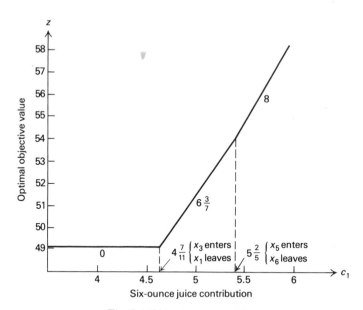

Fig. 3.4 Objective parametrics.

Righthand-Side Parametrics

To illustrate the general procedure in detail, consider the trailer-production problem introduced in Chapter 2. That example included two constraints, a limit of 24 days/month on metalworking capacity and a limit of 60 days/month on woodworking capacity. Suppose that, by reallocating floor space and manpower in the workshop, we can exchange any number of days of woodworking capacity for the same number of days of metalworking capacity. After such an exchange, the capacities will become $(24 + \theta)$ days/month for metalworking capacity and $(60 - \theta)$ days/month for woodworking. The initial tableau for this problem is then given in Tableau 11. What is the optimal contribution to overhead that the firm can obtain for each value of θ? In particular, what value of θ provides the greatest contribution to overhead?

Tableau 11

Basic variables	Current values	x_1	x_2	x_3	x_4	x_5
x_4	$24 + \theta$	$\frac{1}{2}$	2	1	1	
x_5	$60 - \theta$	1	2	4		1
$(-z)$	0	$+6$	$+14$	$+13$		

We can answer these questions by performing parametric righthand-side analysis, starting with the final linear-programming tableau that was determined in Chapter 2 and is repeated here as Tableau 12.

Tableau 12

	Basic variables	Current values	x_1	x_2	x_3	x_4	x_5
at $\theta = -7\frac{1}{5}$	x_1	$36 + 5\theta$	1	6		4	$\ominus 1$
at $\theta = 4$	x_3	$6 - \frac{3}{2}\theta$		$\ominus 1$	1	-1	$\frac{1}{2}$
	$(-z)$	$-294 - 10\frac{1}{2}\theta$		-9		-11	$-\frac{1}{2}$

$$\uparrow \qquad\qquad\qquad\qquad \uparrow$$
$$\text{at } \theta = 4 \qquad\qquad \text{at } \theta = -7\frac{1}{5}$$

In Tableau 12 we have introduced the changes in the current values of the basic variables implied by the parametric variation. These are obtained in the usual way. Increasing metalworking capacity by θ units is equivalent to setting its slack variable $x_4 = -\theta$, and decreasing woodworking capacity by θ is equivalent to setting its slack variable $x_5 = +\theta$. Making these substitutions simultaneously in the tableau and moving the parameter θ to the righthand side gives the current values specified in Tableau 12.

Since the reduced costs are not affected by the value of θ, we see that the basis remains optimal in the final tableau so long as the basic variables x_1 and x_2 remain nonnegative. Hence,

$$x_1 = 36 + 5\theta \geq 0 \qquad \text{(that is, } \theta \geq -7\frac{1}{5}\text{),}$$
$$x_2 = 6 - \frac{3}{2}\theta \geq 0 \qquad \text{(that is, } \theta \leq 4\text{),}$$

which implies that the optimal contribution is given by

$$z = 294 + 10\frac{1}{2}\theta \qquad \text{for } -7\frac{1}{5} \leq \theta \leq 4.$$

At $\theta = 4$, variable x_3 becomes zero, while at $\theta = -7\frac{1}{5}$, variable x_1 becomes zero, and in each case a pivot operation is indicated. As we saw in Section 3.5 on variable transitions associated with righthand-side ranges, the variable to be introduced at these values of θ is determined by pivoting on the negative coefficient in the appropriate pivot row r that gives the minimum ratio as follows:

$$\frac{\bar{c}_s}{\bar{a}_{rs}} = \text{Min}_j \left\{ \frac{\bar{c}_j}{\bar{a}_{rj}} \,\middle|\, \bar{a}_{rj} < 0 \right\}.$$

At $\theta = -7\frac{1}{5}$, only variable x_5 has a negative coefficient in the pivot row, so we pivot to introduce x_5 into the basis in place of x_1. The result is given in Tableau 13.

Tableau 13

	Basic variables	Current values	x_1	x_2	x_3	x_4	x_5
	x_5	$-36 - 5\theta$	-1	-6		-4	1
at $\theta = -24$	x_3	$24 + \theta$	$\frac{1}{2}$	2	1	1	
	$(-z)$	$-312 - 13\theta$	$-\frac{1}{2}$	-12		-13	

Tableau 13 is optimal for:

$$x_5 = -36 - 50 \geq 0 \qquad \text{(that is, } \theta \leq -7\tfrac{1}{5}\text{)},$$

and

$$x_3 = \quad 24 + \quad \theta \geq 0 \qquad \text{(that is, } \theta \geq -24\text{)},$$

and the optimal contribution is then:

$$z = 312 + 13\theta \qquad \text{for } -24 \leq \theta \leq -7\tfrac{1}{5}.$$

At $\theta = -24$, x_3 is zero and becomes a candidate to drop from the basis. We would like to replace x_3 in the basis with another decision variable for $\theta < -24$. We cannot perform any such pivot, however, because there is no negative constraint coefficient in the pivot row. The row reads:

$$\tfrac{1}{2}x_1 + 2x_2 + x_3 + x_4 = 24 + \theta.$$

For $\theta < -24$, the righthand side is negative and the constraint cannot be satisfied by the nonnegative variables x_1, x_2, x_3, and x_4. Consequently, the problem is infeasible for $\theta < -24$. This observation reflects the obvious fact that the capacity $(24 + \theta)$ in the first constraint of the original problem formulation becomes negative for $\theta < -24$ and so the constraint is infeasible.

Having now investigated the problem behavior for $\theta \leq 4$, let us return to Tableau 12 with $\theta = 4$. At this value, x_2 replaces x_3 in the basis. Performing the pivot gives Tableau 14.

Tableau 14

	Basic variables	Current values	x_1	x_2	x_3	x_4	x_5
at $\theta = 18$ ←	x_1	$72 - 4\theta$	1		6	-2	2
	x_2	$-6 + \tfrac{3}{2}\theta$		1	-1	1	$-\tfrac{1}{2}$
	$(-z)$	$-348 + 3\theta$			-9	-2	-5

The basic variables are nonnegative in Tableau 14 for $4 \leq \theta \leq 18$ with optimal objective value $z = 348 - 3\theta$. At $\theta = 18$, we must perform another pivot to introduce x_4 into the basis in place of x_1, giving Tableau 15.

Tableau 15

	Basic variable	Current values	x_1	x_2	x_3	x_4	x_5
	x_4	$20 - 36$	$-\tfrac{1}{2}$		-3	1	-1
at $\theta = 60$ ←	x_2	$30 - \tfrac{1}{2}\theta$	$\tfrac{1}{2}$	1	2		$\tfrac{1}{2}$
	$(-z)$	$-420 + 70$	-1		-15		-7

The basic variables are nonnegative in Tableau 15 for $18 \leq \theta \leq 60$ with optimal objective value $z = 420 - 7\theta$. At $\theta = 60$, no variable can be introduced into the basis for the basic variable x_2, which reaches zero. This is caused by an infeasibility in the original problem formulation, since the woodworking capacity $60 - \theta$ becomes negative for any value of θ in excess of 60.

By collecting the various pieces of the optimal contribution as a function of the parameter θ, we determine the graph given in Fig. 3.5. The highest contribution that can be obtained is \$33,600 per month, which occurs at $\theta = 4$, with

$$24 + \theta = 28 \text{ days/month of metalworking capacity,}$$

and

$$60 - \theta = 56 \text{ days/month of woodworking capacity.}$$

By exchanging 4 days of capacity, we can increase contribution by more than 15 percent from its original value of \$29,400 per month. This increase must be weighed against the costs incurred in the changeover.

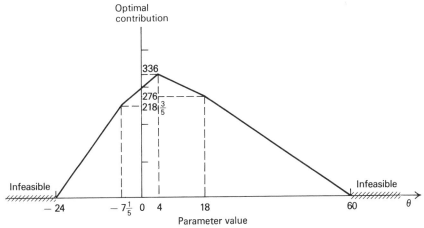

Fig. 3.5 Parametric righthand-side analysis.

Objective-Function Parametrics

To illustrate parametric programming of the objective-function coefficients, we will consider the trailer-production example once again. However, this time we will keep the capacities unchanged and vary the contribution coefficients in the objective function. Suppose that our trailer manufacturer is entering a period of extreme price competition and is considering changing his prices in such a way that his contribution would become:

$$c_1 = 6 + \theta,$$
$$c_2 = 14 + 2\theta,$$
$$c_3 = 13 + 2\theta.$$

How does the optimal production strategy change when the contribution is reduced (i.e., as θ becomes more and more negative)?

The initial tableau for this example is then Tableau 16, and the final tableau, assuming $\theta = 0$, is Tableau 17.

Tableau 16

Basic variables	Current values	x_1	x_2	x_3	x_4	x_5
x_4	24	$\frac{1}{2}$	2	1	1	
x_5	60	1	2	4		1
$(-z)$	0	$6 + \theta$	$14 + 2\theta$	$13 + 2\theta$		

Tableau 17

Basic variables	Current values	x_1	x_2	x_3	x_4	x_5
x_1	36	1	⑥		4	-1
x_3	6		-1	1	-1	$\frac{1}{2}$
$(-z)$	$-294 - 48\theta$		$(-9 - 2\theta)$		$(-11 - 2\theta)$	$-\frac{1}{2}$

$$\uparrow$$
$$\text{at } \theta = -\tfrac{9}{2}$$

As θ is allowed to decrease, the current solution remains optimal so long as $\theta \geq -\frac{9}{2}$. At this value of θ, the reduced cost of x_2 becomes zero, indicating that there may be alternative optimal solutions. In fact, since a pivot is possible in the first row, we can determine an alternative optimal solution where x_2 enters the basis and x_1 leaves. The result is given in Tableau 18.

Tableau 18

Basic variables	Current values	x_1	x_2	x_3	x_4	x_5
x_2	6	$\frac{1}{6}$	1		$\frac{2}{3}$	$-\frac{1}{6}$
x_3	12	$\frac{1}{6}$		1	$-\frac{1}{3}$	⑤ $\frac{1}{3}$
$(-z)$	$-240 - 36\theta$	$(\frac{3}{2} + \frac{1}{3}\theta)$			$(-5 - \frac{2}{3}\theta)$	$(-2 - \frac{1}{3}\theta)$

$$\uparrow$$
$$\text{at } \theta = -6$$

The new basis, consisting of variables x_2 and x_3, remains optimal so long as $-6 \leq \theta \leq -4\frac{1}{2}$. Once $\theta = -6$, the reduced cost of x_5 is zero, so that x_5 becomes a candidate to enter the basis. A pivot can be performed so that x_5 replaces x_3 in the basis. The result is given in Tableau 19.

Tableau 19

Basic variables	Current values	x_1	x_2	x_3	x_4	x_5
x_2	12	$\frac{1}{4}$	1	$\frac{1}{2}$	$\textcircled{\tfrac{1}{2}}$	
x_5	36	$\frac{1}{2}$		3	-1	1
$(-z)$	$-168 - 240$	$(\frac{5}{2} + \frac{1}{2}\theta)$		$(6 + \theta)$	$(-7 - \theta)$	

$$\uparrow$$
$$\text{at } \theta = -7$$

The new basis, consisting of the variables x_2 and x_5, remains optimal so long as $-7 \le \theta \le -6$. At $\theta = -7$, x_4 enters the basis, replacing x_2, and the resulting tableau is identical to Tableau 16 that has the slack basis x_4 and x_5 and so will not be repeated here.

In Fig. 3.6, we plot the optimal value of the objective function as a function of θ. As θ is decreased to $-4\frac{1}{2}$, the contributions of the three products become $1\frac{1}{2}$, 5, and 4, respectively, and x_1 is replaced in the basis by x_2. As we continue to decrease θ to -6, the contributions become 0, 2, and 1, and x_2 replaces x_3 in the basis. It is interesting to note that x_2 is not in the optimal basis with $\theta = 0$, but eventually is the only product produced. At $\theta = 0$, even though x_2 has the highest contribution, it is not produced because a combination of x_1 and x_3 is a better match with the productive resources available. As the relative contributions are reduced, x_2 eventually becomes more attractive than x_1 and finally is the only product manufactured.

The algorithm for objective-function parametrics is clearly straightforward to apply and consists of computing a sequence of optimal bases by the primal simplex method. At each stage, a range is computed on the parameter θ such that the current

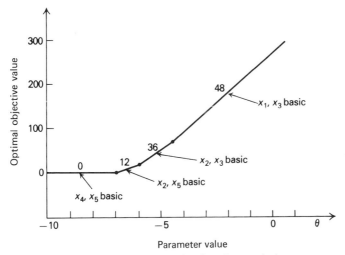

Fig. 3.6 Parametric objective-function analysis.

basis remains optimal. The variable transition at the boundary of this range is easily determined, and one pivot operation determines a new basis that is optimal for some adjacent range.

EXERCISES

1. Outdoors, Inc. has, as one of its product lines, lawn furniture. They currently have three items in that line: a lawn chair, a standard bench, and a table. These products are produced in a two-step manufacturing process involving the tube-bending department and the welding department. The time required by each item in each department is as follows:

	Product			Present capacity
	Lawn chair	Bench	Table	
Tube bending	1.2	1.7	1.2	1000
Welding	0.8	0	2.3	1200

The contribution that Outdoors, Inc. receives from the manufacture and sale of one unit of each product is $3 for a chair, $3 for a bench, and $5 for a table.

The company is trying to plan its production mix for the current selling season. It feels that it can sell any number it produces, but unfortunately production is further limited by available material, because of a prolonged strike. The company currently has on hand 2000 lbs. of tubing. The three products require the following amounts of this tubing: 2 lbs. per chair, 3 lbs. per bench, and 4.5 lbs. per table.

In order to determine the optimal product mix, the production manager has formulated the linear program shown in Fig. E3.1 and obtained the results shown in Fig. E3.2.

a) What is the optimal production mix? What contribution can the firm anticipate by producing this mix?

b) What is the value of one unit more of tube-bending time? of welding time? of metal tubing?

c) A local distributor has offered to sell Outdoors, Inc. some additional metal tubing for $0.60/lb. Should Outdoors buy it? If yes, how much would the firm's contribution increase if they bought 500 lbs. and used it in an optimal fashion?

d) If Outdoors, Inc. feels that it must produce at least 100 benches to round out its product line, what effect will that have on its contribution?

	Chair	Bench	Table	Relation	Limit
D-Tube	1.2	1.7	1.2	$\leq$	1000
D-Weld	0.8	0	2.3	$\leq$	1200
TSuply	2.0	3.0	4.5	$\leq$	2000
Contrib.	3.00	3.00	5.00	$=$	z (max)

Fig. E3.1 Formulation of Outdoors, Inc.

```
TITLE: OUTDOORS
PROCEED, DISPLAY, OR REJECT? PROCEED

MAXIMIZE OR MINIMIZE? MAX

OPTIMAL SOLUTION FOUND.
      CONTRIB.    2766.67

OUTPUT OPTION? EXTEND

ALL ITEMS NOT LISTED IN SECTIONS 1 - 4 HAVE THE VALUE ZERO.

*1* DECISION VARIABLES
   1. CHAIR    700.000
   3. TABLE    133.333

*2* SLACK(+) AND SURPLUS(-) IN CONSTRAINTS
   2. +D-WELD    333.333

*3* SHADOW PRICES FOR CONSTRAINTS
   1. D-TUBE    1.16667
   3. TSUPLY     .800000

*4* REDUCED COSTS FOR DECISION VARIABLES
   2. BENCH    -1.38333

*5* RANGES ON COEFFICIENTS OF OBJECTIVE CONTRIB.
      VARIABLE    LOWER BOUND   CURRENT VALUE   UPPER BOUND
   1. CHAIR        2.2222         3.0000         5.0000
   2. BENCH       UNBOUNDED       3.0000         4.3833
   3. TABLE        3.0000         5.0000         6.7500

*6* RANGES ON VALUES OF RIGHT-HAND-SIDE LIMIT
      CONSTRNT    LOWER BOUND   CURRENT VALUE   UPPER BOUND
   1. D-TUBE       533.33         1000.0         1200.0
   2. D-WELD       866.67         1200.0        UNBOUNDED
   3. TSUPLY       1666.7         2000.0         2555.6

OUTPUT OPTION? NO
```

Fig. E3.2 Solution of Outdoors, Inc.

e) The R&D department has been redesigning the bench to make it more profitable. The new design will require 1.1 hours of tube-bending time, 2.0 hours of welding time, and 2.0 lbs. of metal tubing. If it can sell one unit of this bench with a unit contribution of $3, what effect will it have on overall contribution?

f) Marketing has suggested a new patio awning that would require 1.8 hours of tube-bending time, 0.5 hours of welding time, and 1.3 lbs. of metal tubing. What contribution must this new product have to make it attractive to produce this season?

g) Outdoors, Inc. has a chance to sell some of its capacity in tube bending at cost + $1.50/ hour. If it sells 200 hours at that price, how will this affect contribution?

h) If the contribution on chairs were to decrease to $2.50, what would be the optimal production mix and what contribution would this production plan give?

2. A commercial printing firm is trying to determine the best mix of printing jobs it should seek, given its current capacity constraints in its four capital-intensive departments: typesetting, camera, pressroom, and bindery. It has classified its commercial work into three classes: A, B, and C, each requiring different amounts of time in the four major departments.

The production requirements in hours per unit of product are as follows:

| | Class of work | | |
Department	A	B	C
Typesetting	0	2	3
Camera	3	1	3
Pressroom	3	6	2
Bindery	5	4	0

Assuming these units of work are produced using regular time, the contribution to overhead and profit is $200 for each unit of Class A work, $300 for each unit of Class B work, and $100 for each unit of Class C work.

The firm currently has the following regular-time capacity available in each department for the next time period: typesetting, 40 hours; camera, 60 hours; pressroom, 200 hours; bindery, 160 hours. In addition to this regular time, the firm could utilize an overtime shift in typesetting, which would make available an additional 35 hours in that department. The premium for this overtime (i.e., incremental costs in addition to regular time) would be $4/hour.

Since the firm wants to find the optimal job mix for its equipment, management assumes it can sell all it produces. However, to satisfy long-established customers, management decides to produce at least 10 units of each class of work in each time period.

Assuming that the firm wants to maximize its contribution to profit and overhead, we can formulate the above situation as a linear program, as follows:

Decision variables:

X_{AR} = Number of units of Class A work produced on regular time;

X_{BR} = Number of units of Class B work produced on regular time;

X_{CR} = Number of units of Class C work produced on regular time;

X_{BO} = Number of units of Class B work produced on overtime typesetting;

X_{CO} = Number of units of Class C work produced on overtime typesetting.

Objective function:

Maximize $\qquad z = 200X_{AR} + 300X_{BR} + 100X_{CR} + 292X_{BO} + 88X_{CO},$

Constraints:

Regular Typesetting		$2X_{BR} +$	$3X_{CR}$		$\leq 40,$
Overtime Typesetting				$2X_{BO} + 3X_{CO} \leq 35,$	
Camera	$3X_{AR} +$	$X_{BR} +$	$3X_{CR} +$	$X_{BO} + 3X_{CO} \leq 60,$	
Pressroom	$3X_{AR} +$	$6X_{BR} +$	$2X_{CR} +$	$6X_{BO} + 2X_{CO} \leq 200,$	
Bindery	$5X_{AR} +$	$4X_{BR}$		$+ \quad 4X_{BO}$	$\leq 160,$
Class A—minimum	X_{AR}				$\geq 10,$
Class B—minimum		X_{BR}		$+ \quad X_{BO}$	$\geq 10,$
Class C—minimum			X_{CR}	$+ \quad X_{CO} \geq 10,$	

$$X_{AR} \geq 0, \quad X_{BR} \geq 0, \quad X_{CR} \geq 0, \quad X_{BO} \geq 0, \quad X_{CO} \geq 0.$$

```
TITLE: COMMERCIAL PRINTING
PROCEED, DISPLAY, OR REJECT? PROCEED

MAXIMIZE OR MINIMIZE? MAX

OPTIMAL SOLUTION FOUND.
      CONTRIB      10110.0

OUTPUT OPTION? EXTENDED

ALL ITEMS NOT LISTED IN SECTIONS 1 - 4 HAVE THE VALUE ZERO.

*1* DECISION VARIABLES
   1. CLASS-AR   12.5000
   2. CLASS-BR   20.0000
   4. CLASS-BO   2.50000
   5. CLASS-CO   10.0000

*2* SLACK(+) AND SURPLUS(-) IN CONSTRAINTS
   4. +PRESS     7.50000
   5. +BINDERY   7.50000
   6. -CLA-MIN   2.50000
   7. -CLB-MIN   12.5000

*3* SHADOW PRICES FOR CONSTRAINTS
   1. TYPE-REG   116.667
   2. TYPE-OVT   112.667
   3. CAMERA     66.6667
   8. CLC-MIN    -450.000

*4* REDUCED COSTS FOR DECISION VARIABLES
   3. CLASS-CR   .000000

*5* RANGES ON COEFFICIENTS OF OBJECTIVE CONTRIB
      VARIABLE   LOWER BOUND   CURRENT VALUE   UPPER BOUND
   1. CLASS-AR   .19073E-05    200.00          876.00
   2. CLASS-BR   300.00        300.00          UNBOUNDED
   3. CLASS-CR   UNBOUNDED     100.00          100.00
   4. CLASS-BO   66.667        292.00          292.00
   5. CLASS-CO   88.000        88.000          538.00

*6* RANGES ON VALUES OF RIGHT-HAND-SIDE CAPACITY
      CONSTRNT   LOWER BOUND   CURRENT VALUE   UPPER BOUND
   1. TYPE-REG   15.000        40.000          43.000
   2. TYPE-OVT   30.000        35.000          38.000
   3. CAMERA     82.500        90.000          94.500
   4. PRESS      192.50        200.00          UNBOUNDED
   5. BINDERY    152.50        160.00          UNBOUNDED
   6. CLA-MIN    UNBOUNDED     10.000          12.500
   7. CLB-MIN    UNBOUNDED     10.000          22.500
   8. CLC-MIN    9.1176        10.000          11.667

OUTPUT OPTION? NO
```

Fig. E3.3 Solution for commercial printing firm.

The solution of the firm's linear programming model is given in Fig. E3.3.

a) What is the optimal production mix?

b) Is there any unused production capacity?

c) Is this a unique optimum? Why?

d) Why is the shadow price of regular typesetting different from the shadow price of overtime typesetting?

e) If the printing firm has a chance to sell a new type of work that requires 0 hours of type-setting, 2 hours of camera, 2 hours of pressroom, and 1 hour of bindery, what contribution is required to make it attractive?

f) Suppose that both the regular and overtime typesetting capacity are reduced by 4 hours. How does the solution change? (*Hint*: Does the basis change in this situation?)

3. Jean-Pierre Leveque has recently been named the Minister of International Trade for the new nation of New France. In connection with this position, he has decided that the welfare of the country (and his performance) could best be served by maximizing the net dollar value of the country's exports for the coming year. (The net dollar value of exports is defined as exports *less* the cost of all materials imported by the country.)

The area that now constitutes New France has traditionally made three products for export: steel, heavy machinery, and trucks. For the coming year, Jean-Pierre feels that they could sell all that they could produce of these three items at existing world market prices of $900/unit for steel, $2500/unit for machinery, and $3000/unit for trucks.

In order to produce one unit of steel with the country's existing technology, it takes 0.05 units of machinery, 0.08 units of trucks, two units of ore purchased on the world market for $100/unit, and other imported materials costing $100. In addition, it takes .5 man-years of labor to produce each unit of steel. The steel mills of New France have a maximum usable capacity of 300,000 units/year.

To produce one unit of machinery requires .75 units of steel, 0.12 units of trucks, and 5 man-years of labor. In addition, $150 of materials must be imported for each unit of machinery produced. The practical capacity of the country's machinery plants is 50,000 units/year.

In order to produce one unit of trucks, it takes one unit of steel, 0.10 units of machinery, three man-years of labor, and $500 worth of imported materials. Existing truck capacity is 550,000 units/year.

The total manpower available for production of steel, machinery, and trucks is 1,200,000 men/year.

To help Jean-Pierre in his planning, he had one of his staff formulate the model shown in Fig. E3.4 and solved in Fig. E3.5. Referring to these two exhibits, he has asked you to help him with the following questions:

a) What is the optimal production and export mix for New France, based on Fig. E3.5? What would be the net dollar value of exports under this plan?

b) What do the first three constraint equations (STEEL, MACHIN, and TRUCK) represent? Why are they equality constraints?

c) The optimal solution suggests that New France produce 50,000 units of machinery. How are those units to be utilized during the year?

d) What would happen to the value of net exports if the world market price of steel increased to $1225/unit and the country chose to export one unit of steel?

e) If New France wants to identify other products it can profitably produce and export, what characteristics should those products have?

f) There is a chance that Jean-Pierre may have $500,000 to spend on expanding capacity. If this investment will buy 500 units of truck capacity, 1000 units of machine capacity, or 300 units of steel capacity, what would be the best investment?

g) If the world market price of the imported materials needed to produce one unit of trucks were to increase by $400, what would be the optimal export mix for New France, and what would be the dollar value of their net exports?

Variables:

X_1 = Steel production for export (EXSTEE),

X_2 = Machinery production for export (EXMACH),

X_3 = Truck production for export (EXTRUC),

X_4 = Total steel production (TOSTEE),

X_5 = Total machinery production (TOMACH),

X_6 = Total truck production (TOTRUC).

Constraints:

Steel output	(STEEL)
Machinery output	(MACHIN)
Truck output	(TRUCK)
Steel capacity	(CAPSTE)
Machinery capacity	(CAPMAC)
Truck capacity	(CAPTRU)
Manpower available	(AVAMAN)

```
TITLE: NEW FRANCE INTERNATIONAL TRADE
PROCEED, DISPLAY, OR REJECT? DISPLAY


OBJECTIVES:

              EXSTEE     EXMACH     EXTRUC     TOSTEE    TOMACH     TOTRUC

EXPORTS       900.0      2500.      3000.      -300.0    -150.0     -500.0


CONSTRAINTS:

              EXSTEE     EXMACH     EXTRUC     TOSTEE      TOMACH     TOTRUC
              RELATION   CAPACITY

STEEL         -1.000     .0000      .0000      1.000       -.7500     -1.000
              EQ         .0000
MACHIN        .0000      -1.000     .0000      -.5000E-01  1.000      -.1000
              EQ         .0000
TRUCK         .0000      .0000      -1.000     -.8000E-01  -.1200     1.000
              EQ         .0000
CAPSTE        .0000      .0000      .0000      1.000       .0000      .0000
              LE         .3000E+06
CAPMAC        .0000      .0000      .0000      .0000       1.000      .0000
              LE         .5000E+05
CAPTRU        .0000      .0000      .0000      .0000       .0000      1.000
              LE         .5500E+06
AVAMAN        .0000      .0000      .0000      .5000       5.000      3.000
              LE         .1200E+07
```

Fig. E3.4 Formulation of optimal production and export for New France.

```
TITLE: NEW FRANCE INTERNATIONAL TRADE
PROCEED, DISPLAY, OR REJECT? PROCEED

MAXIMIZE OR MINIMIZE? MAX

OPTIMAL SOLUTION FOUND.
      EXPORTS      0.490625E+09

OUTPUT OPTION? EXTENDED

ALL ITEMS NOT LISTED IN SECTIONS 1 - 4 HAVE THE VALUE ZERO.

*1* DECISION VARIABLES
    2. EXMACH     8750.00
    3. EXTRUC     232500.
    4. TOSTEE     300000.
    5. TOMACH     50000.0
    6. TOTRUC     262500.

*2* SLACK(+) AND SURPLUS(-) IN CONSTRAINTS
    6. +CAPTRU    287500.
    7. +AVAMAN    12500.0

*3* SHADOW PRICES FOR CONSTRAINTS
    1. STEEL      -2250.00
    2. MACHIN     -2500.00
    3. TRUCK      -3000.00
    4. CAPSTE     1585.00
    5. CAPMAC     302.500

*4* REDUCED COSTS FOR DECISION VARIABLES
    1. EXSTEE     -1350.00

*5* RANGES ON COEFFICIENTS OF OBJECTIVE EXPORTS
       VARIABLE  LOWER BOUND  CURRENT VALUE  UPPER BOUND
    1. EXSTEE    UNBOUNDED    900.00         2250.0
    2. EXMACH    2218.6       2500.0         13067.
    3. EXTRUC    1650.0       3000.0         3347.7
    4. TOSTEE    -1885.0      -300.00        UNBOUNDED
    5. TOMACH    -452.50      -150.00        UNBOUNDED
    6. TOTRUC    -1850.0      -500.00        -96.667

*6* RANGES ON VALUES OF RIGHT-HAND-SIDE CAPACITY
       CONSTRNT  LOWER BOUND  CURRENT VALUE  UPPER BOUND
    1. STEEL     -4166.7      .00000         .23250E+06
    2. MACHIN    UNBOUNDED    .00000         8750.0
    3. TRUCK     UNBOUNDED    .00000         .23250E+06
    4. CAPSTE    47283.       .30000E+06     .30357E+06
    5. CAPMAC    41860.       50000.         54545.
    6. CAPTRU    .26250E+06   .55000E+06     UNBOUNDED
    7. AVAMAN    .11875E+07   .12000E+07     UNBOUNDED

OUTPUT OPTION? NO
```

Fig. E3.5 Solution of New France model.

h) The Minister of Defense has recently come to Jean-Pierre and said that he would like to stockpile (inventory) an additional 10,000 units of steel during the coming year. How will this change the constraint equation STEEL, and what impact will it have on net dollar exports?

i) A government R&D group has recently come to Jean-Pierre with a new product, Product X, that can be produced for export with 1.5 man-years of labor and 0.3 units of machinery for each unit produced. What must Product X sell for on the world market to make it attractive for production?

j) How does this particular formulation deal with existing inventories at the start of the year and any desired inventories at the end of the year?

4. Another member of Jean-Pierre's staff has presented an alternative formulation of New France's planning problem as described in Exercise 3, which involves only three variables. This formulation is as follows:

$$Y_1 = \text{Total steel production,}$$
$$Y_2 = \text{Total machinery production,}$$
$$Y_3 = \text{Total truck production.}$$

$$\text{Maximize } z = 900(Y_1 - 0.75Y_2 - Y_3) - 300Y_1 + 2500(Y_2 - 0.05Y_1 - 0.10Y_3)$$
$$- 150Y_2 + 3000(Y_3 - 0.80Y_1 - 0.12Y_2) - 500Y_3,$$

subject to:

$$Y_1 \leq 300{,}000,$$
$$Y_2 \leq 50{,}000,$$
$$Y_3 \leq 550{,}000,$$
$$0.5Y_1 + 5Y_2 + 3Y_3 \leq 1{,}200{,}000, \qquad Y_1, Y_2, Y_3 \geq 0.$$

a) Is this formulation equivalent to the one presented in Fig. E3.4 of Exercise 3? How would the optimal solution here compare with that found in Fig. E3.5?

b) If we had the optimal solution to this formulation in terms of total production, how would we find the optimal exports of each product?

c) What assumptions does this formulation make about the quantities and prices of products that can be exported and imported?

d) New France is considering the production of automobiles. It will take 0.5 units of steel, 0.05 units of machinery, and 2.0 man-years to produce one unit of automobiles. Imported materials for this unit will cost $250 and the finished product will sell on world markets at a price of $2000. Each automobile produced will use up .75 units of the country's limited truck capacity. How would you alter this formulation to take the production of automobiles into account?

5. Returning to the original problem formulation shown in Fig. E3.4 of Exercise 3, Jean-Pierre feels that, with existing uncertainties in international currencies, there is some chance that the New France dollar will be revalued upward relative to world markets. If this happened, the cost of imported materials would go down by the same percent as the devaluation, and the market price of the country's exports would go up by the same percent as the revaluation. Assuming that the country can always sell all it wishes to export, how much of a revaluation could occur before the optimal solution in Fig. E3.5 would change?

6. After paying its monthly bills, a family has $100 left over to spend on leisure. The family enjoys two types of activities: (i) eating out; and (ii) other entertainment, such as seeing movies and sporting events. The family has determined the relative value (utility) of these two activities—each dollar spent on eating out provides 1.2 units of value for each dollar spent on other entertainment. Suppose that the family wishes to determine how to spend this money to maximize the value of its expenditures, but that no more than $70 can be spent on eating out and no more than $50 on other entertainment (if desired, part of the money

can be saved). The family also would like to know:

a) How the total value of its expenditures for leisure would change if there were only $99 to spend.

b) How the total value would change if $75 could be spent on eating out.

c) Whether it would *save* any money if each dollar of savings would provide 1.1 units of value for each dollar spent on other entertainment.

Formulate the family's spending decision as a linear program. Sketch the feasible region, and answer each of the above questions graphically. Then solve the linear program by the simplex method, identify shadow prices on the constraints, and answer each of these questions, using the shadow prices.

7. Consider the linear program:

$$\text{Maximize } z = x_1 + 3x_2,$$

subject to:

$$
\begin{array}{lll}
x_1 + x_2 \le 8 & \text{(resource 1)}, \\
-x_1 + x_2 \le 4 & \text{(resource 2)}, \\
x_1 \le 6 & \text{(resource 3)}, \\
\end{array}
$$

$$x_1 \ge 0, \qquad x_2 \ge 0.$$

a) Determine graphically:

   i) the optimal solution;

   ii) the shadow prices on the three constraints;

   iii) the range on the objective coefficient of each variable, holding the coefficient of the other variable at its current value, for which the solution to part (i) remains optimal;

   iv) the range on the availability of each resource, holding the availability of the other resources at their current values, for which the shadow prices in part (ii) remain optimal.

b) Answer the same question, using the following optimal tableau determined by the simplex method:

Basic variables	Current values	x_1	x_2	x_3	x_4	x_5
x_1	2	1		$\frac{1}{2}$	$-\frac{1}{2}$	
x_2	6		1	$\frac{1}{2}$	$\frac{1}{2}$	
x_5	4			$-\frac{1}{2}$	$\frac{1}{2}$	1
$(-z)$	-20			-2	-1	

8. Determine the variable transitions for the previous problem, when each objective coefficient or righthand-side value is varied by itself to the point where the optimal basis no longer remains optimal. Carry out this analysis using the optimal tableau, and interpret the transitions graphically on a sketch of the feasible region.

9. A wood-products company produces four types of household furniture. Three basic operations are involved: cutting, sanding, and finishing. The plant's capacity is such that there is a limit of 900 machine hours for cutting, 800 hours for sanding, and 480 hours for finishing.

The firm's objective is to maximize profits. The initial tableau is as follows:

Basic variables	Current values	x_1	x_2	x_3	x_4	x_5	x_6	x_7
x_5	480	2	8	4	2	1		
x_6	800	5	4	8	5		1	
x_7	900	7	8	3	5			1
$(-z)$	0	$+90$	$+160$	$+40$	$+100$			

Using the simplex algorithm, the final tableau is found to be:

Basic variables	Current values	x_1	x_2	x_3	x_4	x_5	x_6	x_7
x_2	25	0	1	$\frac{1}{8}$		$\frac{5}{32}$	$-\frac{1}{16}$	
x_4	140	1		$\frac{3}{2}$	1	$-\frac{1}{8}$	$\frac{1}{4}$	
x_7	0	2		$-\frac{11}{2}$		$-\frac{5}{8}$	$-\frac{3}{4}$	1
$(-z)$	$-18{,}000$	-10		-130		$-12\frac{1}{2}$	-15	

a) What are the shadow prices on each of the constraints?
b) What profit for x_3 would justify its production?
c) What are the limits on sanding capacity that will allow the present basic variables to stay in the optimal solution?
d) Suppose management had to decide whether or not to introduce a new product requiring 20 hours of cutting, 3 hours of sanding, and 2 hours of finishing, with an expected profit of $120. Should the product be produced?
e) If another saw to perform cutting can be rented for $10/hour, should it be procured? What about a finisher at the same price? If either is rented, what will be the gain from the first hour's use?

10. The Reclamation Machining Company makes nuts and bolts from scrap material supplied from the waste products of three steel-using firms. For each 100 pounds of scrap material provided by firm A, 10 cases of nuts and 4 cases of bolts can be made, with a contribution of $46. 100 pounds from firm B results in 6 cases of nuts, 10 cases of bolts, and a contribution of $57. Use of 100 pounds of firm C's material will produce 12 cases of nuts, 8 of bolts, and a contribution of $60. Assuming Reclamation can sell only a maximum of 62 cases of nuts and 60 of bolts, the final tableau for a linear-programming solution of this problem is as follows:

Basic variables	Current values	x_1	x_2	x_3	x_4	x_5
x_1	3.421	1		0.947	0.132	-0.079
x_2	4.632		1	0.421	-0.052	0.132
$(-z)$	-421.368			-7.759	-3.053	-3.868

a) What is the optimal solution?
b) Is the solution unique? Why?
c) For each of the three sources, determine the interval of contribution for which the solution in part (a) remains optimal.

d) What are the shadow prices associated with the optimal solution in (a), and what do they mean?

e) Give an interval for each sales limitation such that the shadow prices in (d) remain unchanged.

11. Consider the linear program:

$$\text{Maximize } z = 2x_1 + x_2 + 10x_3,$$

subject to:

$$x_1 - x_2 + 3x_3 \qquad\quad = 10,$$
$$x_2 + x_3 + x_4 = 6,$$
$$x_j \geq 0 \qquad (j = 1, 2, 3, 4).$$

The optimal tableau is:

Basic variables	Current values	x_1	x_2	x_3	x_4
x_3	4	$\frac{1}{4}$		1	$\frac{1}{4}$
x_2	2	$-\frac{1}{4}$	1		$\frac{3}{4}$
$(-z)$	-24	$-\frac{1}{4}$			$-3\frac{1}{4}$

a) What are the optimal shadow prices for the two constraints? Can the optimal shadow price for the first constraint be determined easily from the final objective coefficient for x_1? (*Hint*: The initial problem formulation is in canonical form if the objective coefficient of x_1 is changed to $0x_1$.)

b) Suppose that the initial righthand-side coefficient of the first constraint is changed to $10 + \delta$. For what values of δ do x_2 and x_3 form an optimal basis for the problem?

12. Consider the following linear program:

$$\text{Minimize } z = 2x_i + x_2 + 2x_3 - 3x_4,$$

subject to:

$$8x_1 - 4x_2 - x_3 + 3x_4 \leq 10,$$
$$2x_1 + 3x_2 + x_3 - x_4 \leq 7,$$
$$- 2x_2 - x_3 + 4x_4 \leq 12,$$
$$x_1 \geq 0, \quad x_2 \geq 0, \quad x_3 \geq 0, \quad x_4 \geq 0.$$

After several iterations of the simplex algorithm, the following tableau has been determined:

Basic variables	Current values	x_1	x_2	x_3	x_4	x_5	x_6	x_7
x_5	11	10		$\frac{1}{2}$		1	1	$-\frac{1}{2}$
x_2	4	$\frac{4}{5}$	1	$\frac{3}{10}$			$\frac{2}{5}$	$\frac{1}{10}$
x_4	5	$\frac{2}{5}$		$-\frac{1}{10}$	1		$\frac{1}{5}$	$\frac{3}{10}$
$(-z)$	-11	$\frac{12}{5}$		$\frac{7}{5}$			$\frac{1}{5}$	$\frac{4}{5}$

a) What are the values of the decision variables? How do you know this is the optimal solution?

b) For each nonbasic variable of the solution, find the interval for its objective-function coefficient such that the solution remains optimal.

c) Do the same for each basic variable.

d) Determine the interval for each righthand-side value such that the optimal basis determined remains unchanged. What does this imply about the shadow prices? reduced costs?

13. A cast-iron foundry is required to produce 1000 lbs. of castings containing at least 0.35 percent manganese and not more than 3.2 percent silicon. Three types of pig iron are available in unlimited amounts, with the following properties:

	Type of pig iron		
	A	B	C
Silicon	4%	1%	5%
Manganese	0.35%	0.4%	0.3%
Cost/1000 lbs.	$28	$30	$20

Assuming that pig iron is melted with other materials to produce cast iron, a linear-programming formulation that minimizes cost is as follows:

Minimize $z = 28x_1 + 30x_2 + 20x_3$,

subject to:

$$4x_1 + x_2 + 5x_3 \leq 3.2 \quad \text{(lb. Si} \times 10),$$
$$3.5x_1 + 4x_2 + 3x_3 \geq 3.5 \quad \text{(lb. Mn)},$$
$$x_1 + x_2 + x_3 = 1 \quad \text{(lb.} \times 10^3),$$

$$x_1 \geq 0, \quad x_2 \geq 0, \quad x_3 \geq 0.$$

Initial tableau

					Surplus	Slack	Artificial	
Basic variables	Current values	x_1	x_2	x_3	x_4	x_5	v_1	v_2
x_5	3.2	4	1	5	0	1		
v_1	3.5	3.5	4	3	-1		1	
v_2	1	1	1	1	0			1
$(-z)$	0	28	30	20	0			

Final tableau

Basic variables	Current values	x_1	x_2	x_3	x_4	x_5	v_1	v_2
x_5	0.2	1			-4	1	4	-17
x_3	0.5	0.5	1		-1		1	-3
x_2	0.5	0.5		1	1		-1	4
$(-z)$	-25	3			10		-10	10

a) At what cost does pig type A become a candidate for entry into the optimal basis? What activity would it replace?

b) How much can we afford to pay for pure manganese to add to the melt?

c) How much can the manganese requirement be reduced without changing the basis? What are the values of the other basic variables when this happens?

d) How much can the cost of pig type B change without changing the optimal basis? What are the new basic variables when such a change occurs?

e) How can the final tableau be optimal if the reduced cost of v_1 is -10?

14. The Classic Stone Cutter Company produces four types of stone sculptures: figures, figurines, free forms, and statues. Each product requires the following hours of work for cutting and chiseling stone and polishing the final product:

	Type of product			
Operation	Figures	Figurines	Free Forms	Statues
Cutting	30	5	45	60
Chiseling	20	8	60	30
Polishing	0	20	0	120
Contribution/unit	$280	$40	$500	$510

The last row in the above table specifies the contribution to overhead for each product.

   Classic's current work force has production capacity to allocate 300 hours to cutting, 180 hours to chiseling, and 300 hours to polishing in any week. Based upon these limitations, it finds its weekly production schedule from the linear-programming solution given by the following tableau.

Basic variables	Current values	Figures x_1	Figurines x_2	Forms x_3	Statues x_4	Cutting slack, hours x_5	Chiseling slack, hours x_6	Polishing slack, hours x_7
Statues	2		$-\frac{7}{15}$	-6	1	$\frac{1}{15}$	$-\frac{1}{10}$	
Figures	6	1	$\frac{11}{10}$	$\frac{15}{2}$		$-\frac{1}{10}$	$\frac{1}{5}$	
Slack	60		76	360		-8	12	1
$-$Contrib.	-2700		-30	-70		-6	-5	

a) Determine a range on the cutting capacity such that the current solution remains optimal.

b) Busts have the following characteristics:

Cutting	15 hrs.
Chiseling	10 hrs.
Polishing	20 hrs.
Contribution/unit	$240

Should Classic maintain its present line or expand into busts?

c) Classic can buy 5 hours of cutting capacity and 5 hours of chiseling capacity from an outside contractor at a total cost of $75. Should Classic make the purchase or not?

d) By how much does the contribution from free forms have to be increased to make free forms profitable to produce?

e) Give a range on the contribution from figures such that the current solution remains optimal. What activities enter the basis at the bounds of this range?

15. The Concrete Products Corporation has the capability of producing four types of concrete blocks. Each block must be subjected to four processes: batch mixing, mold vibrating, inspection, and yard drying. The plant manager desires to maximize profits during the next month. During the upcoming thirty days, he has 800 machine hours available on the batch mixer, 1000 hours on the mold vibrator, and 340 man-hours of inspection time. Yard-drying time is unconstrained. The production director has formulated his problem as a linear program with the following initial tableau:

Basic variables	Current values	x_1	x_2	x_3	x_4	x_5	x_6	x_7
x_5	800	1	2	10	16	1		
x_6	1000	1.5	2	4	5		1	
x_7	340	0.5	0.6	1	2			1
$(-z)$	0	8	14	30	50			

where x_1, x_2, x_3, x_4 represent the number of pallets of the four types of blocks. After solving by the simplex method, the final tableau is:

Basic variables	Current values	x_1	x_2	x_3	x_4	x_5	x_6	x_7
x_2	200		1	$+11$	19	15	-1	
x_1	400	1		-12	-22	-2	2	
x_7	20			0.4	1.6	0.1	-0.4	1
$(-z)$	6000			-28	-40	-5	-2	

a) By how much must the profit on a pallet of number 3 blocks be increased before it would be profitable to manufacture them?

b) What minimum profit on x_2 must be realized so that it remains in the production schedule?

c) If the 800 machine-hours capacity on the batch mixer is uncertain, for what range of machine hours will it remain feasible to produce blocks 1 and 2?

d) A competitor located next door has offered the manager additional batch-mixing time at a rate of $4.00 per hour. Should he accept this offer?

e) The owner has approached the manager with a thought about producing a new type of concrete block that would require 4 hours of batch mixing, 4 hours of molding, and 1 hour of inspection per pallet. What should be the profit per pallet if block number 5 is to be included in the optimal schedule?

16. The linear-programming program:

$$\text{Maximize } z = \qquad 4x_4 + 2x_5 - 3x_6,$$

subject to:

$$
\begin{aligned}
x_1 \qquad\qquad + x_4 - x_5 + 4x_6 &= 2,\\
x_2 \quad + x_4 + x_5 - 2x_6 &= 6,\\
x_3 - 2x_4 + x_5 - 3x_6 &= 6,\\
x_j \geq 0 \qquad (j = 1, 2, \ldots, 6)
\end{aligned}
$$

has an optimal canonical form given by:

Basic variables	Current values	x_1	x_2	x_3	x_4	x_5	x_6
x_4	4	$\frac{1}{2}$	$\frac{1}{2}$		1		1
x_5	2	$-\frac{1}{2}$	$\frac{1}{2}$			1	-3
x_3	12	$\frac{3}{2}$	$\frac{1}{2}$	1			2
$(-z)$	-20	-1	-3				-1

Answer the following questions. All parts are independent and refer to changes in the *original* problem formulation as stated above. That is, do not consider parts (a), (b), and (c) together, but refer to the original problem formulation in each part.

a) Suppose that the objective function changes to

$$z = 4x_4 + 2x_5 + 0x_6.$$

Find the new optimal solution. Is the solution to the new problem unique? Explain your answer.

b) Consider the parametric-programming problem with the objective function:

$$\text{Maximize } z = (4 + 2\theta)x_4 + (2 - 4\theta)x_5 + (-3 + \theta)x_6.$$

For what values $\underline{\theta} \leq \theta \leq \overline{\theta}$ of θ do the variables x_4, x_5, and x_3 remain in the optimal basis (e.g., if $\underline{\theta} = -1$ and $\overline{\theta} = 2$, then the interval is $-1 \leq \theta \leq 2$). What are the variable transitions at $\theta = \underline{\theta}$ and $\theta = \overline{\theta}$?

c) Consider the parametric programming problem with the righthand sides

$$
\begin{aligned}
2 + 4\theta,\\
6 - 2\theta,\\
6 - \theta.
\end{aligned}
$$

For what values $\underline{\theta} \leq \theta \leq \overline{\theta}$ of θ do the variables x_4, x_5, and x_3 remain in the optimal basis? What are the variable transitions at $\theta = \underline{\theta}$ and $\theta = \overline{\theta}$?

17. The Massachusetts Electric Company is planning additions to its mix of nuclear and conventional electric power plants to cover a ten-year horizon. It measures the output of plants in terms of equivalent conventional plants. Production of electricity must satisfy the state's

additional requirements over the planning horizon for guaranteed power (kW), peak power (kW), and annual energy (MWh). The state's additional requirements, and the contribution of each type of plant to these requirements, are as follows:

	Requirements		
	Guaranteed power (kW)	Peak power (kW)	Annual energy (MWh)
Conventional	1	3	1
Nuclear	1	1	4
Additional requirements	20	30	40

The costs for each type of plant include fixed investment costs and variable operating costs per year. The conventional plants have very high operating costs, since they burn fossil fuel, while the nuclear plants have very high investment costs. These costs are as follows:

Type of plant	Investment costs (millions)	Annual operating costs (millions)
Conventional	$ 30	$20
Nuclear	100	5

For simplicity, we will assume that the annual operating costs are an infinite stream discounted at a rate of $(1/(1 + r))$ per year. The present value of this infinite stream of operating costs is:

$$PV = \sum_{n=1}^{\infty} \left(\frac{1}{1 + r}\right)^n c = \left[\frac{1}{1 - [1/(1 + r)]} - 1\right] c = \frac{1}{r} c$$

and the term $1/r$ is sometimes referred to as the *coefficient of capitalization*.

a) Formulate a linear program to decide what mix of conventional and nuclear power plants to build, assuming that you must meet or exceed the state's three requirements and you want to minimize the *total* investment plus discounted operating costs.

b) There has been a debate at Massachusetts Electric as to what discount rate r to use in making this calculation. Some have argued for a low "social" rate of about 2 or 3 percent, while others have argued for the private "cost of capital" rate of from 10 to 15 percent. Graphically determine the optimal solution for all values of the discount rate r, to aid in settling this question. Comment on the implications of the choice of discount rate.

c) Another difficulty is that there may not be sufficient investment funds to make the decisions independently of an investment limitation. Assuming the "social" discount rate of 2 percent, find the optimal solution for all conceivable budget levels for total investment. What is the impact of limiting the budget for investment?

18. Consider the parametric-programming problem:

$$\text{Maximize } z = (-3 + 3\theta)x_1 + (1 - 2\theta)x_2,$$

subject to:

$$-2x_1 + x_2 \leq 2,$$
$$x_1 - 2x_2 \leq 2,$$
$$x_1 - x_2 \leq 4,$$
$$x_1 \geq 0, \qquad x_2 \geq 0.$$

Letting x_3, x_4, and x_5 be slack variables for the constraints, we write the optimal canonical form at $\theta = 0$ as:

Basic variables	Current values	x_1	x_2	x_3	x_4	x_5
x_2	2	-2	1	1		
x_4	6	-3		2	1	
x_5	6	-1		1		1
$(-z)$	-2	-1		-1		

a) Place the objective function in this tableau in canonical form for values of θ other than 0. For what values of θ is this canonical form optimal?

b) What are the variable transitions when this canonical form is no longer optimal? Does the problem become unbounded at either of the transition points?

c) Use the parametric programming algorithm to find the optimal solution for all values of θ. Plot the optimal objective value as a function of θ.

d) Graph the feasible region and interpret the parametric algorithm on the graph.

19. Consider the following parametric linear program:

$$z^*(\theta) = \text{Max } x_1 + 2x_2,$$

subject to:

$$2x_1 + x_2 \leq 14,$$
$$x_1 + x_2 \leq 10,$$
$$x_1 \leq 1 + 2\theta,$$
$$x_2 \leq 8 - \theta,$$
$$x_1 \geq 0, \qquad x_2 \geq 0.$$

a) Graph the feasible region and indicate the optimal solution for $\theta = 0, \theta = 1$, and $\theta = 2$.

b) For what values of θ is this problem feasible?

c) From the graphs in part (a), determine the optimal solution for all values of θ. Graph the optimal objective value $z^*(\theta)$.

d) Starting with the optimal canonical form given below for $\theta = 0$, use the parametric simplex algorithm to solve for all values of θ.

Basic variables	Current values	x_1	x_2	x_3	x_4	x_5	x_6
x_3	$4 - 3\theta$			1		-2	-1
x_4	$1 - \theta$				1	-1	-1
x_1	$1 + 2\theta$	1				1	
x_2	$8 - \theta$		1				1
$(-z)$	-17					-1	-2

20. Consider the linear program:

$$\text{Minimize } z = -10x_1 + 16x_2 - x,$$

subject to:

$$x_1 - 2x_2 + x_3 \leq 2 + 2\theta,$$
$$x_1 - x_2 \leq 4 + \theta,$$
$$x_1 \geq 0, \qquad x_2 \geq 0, \qquad x_3 \geq 0,$$

where θ is a parameter.

a) For $\theta = 0$

   i) Solve the linear program.
   ii) What are the optimal shadow prices?
   iii) Suppose that the constant on the righthand side of the second inequality is changed from 4 to 6. What is the new optimal value of z?

b) For what values of θ does the optimal basis to part a(i) remain optimal?

c) Solve the linear program for all values of θ.

21. When discussing parametric analysis in Section 3.8, we considered reallocating floor space for a trailer-production problem to trade off woodworking capacity for metalworking capacity. The trade-off was governed by the parameter θ in the following tableau:

Basic variables	Current values	x_1	x_2	x_3	x_4	x_5	
x_4	$24 + \theta$	$\frac{1}{2}$	2	1	1		Woodworking capacity
x_5	$60 - \theta$	1	2	4		1	Metalworking capacity
$(-z)$	0	6	14	13			

We found that the following tableau was optimal for values of θ in the range $0 \leq \theta \leq 4$:

Basic variables	Current values	x_1	x_2	x_3	x_4	x_5
x_1	$35 + 5\theta$	1	6		4	-1
x_3	$6 - \frac{3}{2}\theta$		-1	1	-1	$\frac{1}{2}$
$(-z)$	$-294 - 10\frac{1}{2}\theta$		-9		-11	$-\frac{1}{2}$

The optimal solution at $\theta = 4$ is $x_1 = 56, x_2 = x_3 = x_4 = x_5 = 0$, and $z = 336$. At $\theta = 5$, we found that the optimal basic variables are $x_1 = 52, x_2 = 1.5$, and that the optimal objective value is $z = 348 - 3(5) = 333$. Therefore as θ increases from 4 to 5,

$$\Delta z = 333 - 336 = -3.$$

The previous optimal tableau tells us that for $0 \leq \theta \leq 4$, woodworking capacity is worth \$11/day and metalworking capacity is worth \$0.50/day. Increasing θ from 4 to 5 increases woodworking capacity by one day. Using the prices \$11/day and \$0.50/day, the change in the optimal objective value should be:

$$\Delta z = 11(1) + 0.50(-1) = 10.50.$$

We have obtained two different values for the change in the optimal objective value, $\Delta z = -3$ and $\Delta z = 10.50$. Reconcile the difference between these values.

22. An investor has \$5000 and two potential investments. Let x_j for $j = 1$ and $j = 2$ denote her allocation to investment j in thousands of dollars. From historical data, investments 1 and 2 are known to have an expected annual return of 20 and 16 percent, respectively. Also, the total risk involved with investments 1 and 2, as measured by the variance of total return, is known to be given by $2x_1 + x_2 + (x_1 + x_2)$, so that risk increases with total investment $(x_1 + x_2)$ and with the amount of each individual investment. The investor would like to maximize her return and at the same time minimize her risk. Figure E3.6 illustrates the conflict between these two objectives. Point A in this figure corresponds to investing nothing, point B to investing all \$5000 in the second alternative, and point C to investing completely in the first alternative. Every other point in the shaded region of the figure corresponds to some other investment strategy; that is, to a feasible solution to the constraints:

$$
\begin{aligned}
x_1 + x_2 &\leq 5, \\
x_1 \geq 0, \quad x_2 &\geq 0.
\end{aligned}
\tag{27}
$$

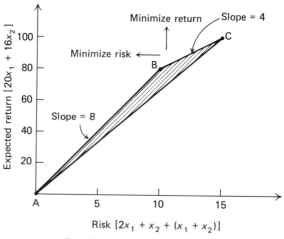

Fig. E3.6 Risk-return trade-off.

To deal with the conflicting objectives, the investor decides to combine return and risk into a single objective function.

$$\text{Objective} = \text{Return} - \theta \,(\text{Risk})$$
$$= 20x_1 + 16x_2 - \theta[2x_1 + x_2 + (x_1 + x_2)]. \tag{28}$$

She uses the parameter θ in this objective function to weigh the trade-off between the two objectives.

Because the "most appropriate" trade-off parameter is difficult to assess, the investor would like to maximize the objective function (28) subject to the constraints (27), for all values of θ, and then use secondary considerations, not captured in this simple model, to choose from the alternative optimal solutions.

a) Use parametric linear programming to solve the problem for all nonnegative values of θ.
b) Plot the optimal objective value, the expected return, and the risk as a function of θ.
c) Interpret the solutions on Fig. E3.6.
d) Suppose that the investor can save at 6% with no risk. For what values of θ will she save any money using the objective function (28)?

23. An alternative model for the investment problem discussed in the previous exercise is:

$$\text{Maximize } z = 20x_1 + 16x_2,$$

subject to:

$$x_1 + x_2 \leq 5,$$
$$2x_1 + x_2 + (x_1 + x_2) \leq \gamma,$$
$$x_1 \geq 0, \qquad x_2 \geq 0.$$

In this model, the investor maximizes her expected return, constraining the risk that she is willing to incur by the parameter γ.

a) Use parametric linear programming to solve the problem for all nonnegative values of γ.
b) Plot the expected return and risk as a function of γ.
c) Interpret the solutions on Fig. E3.6 of Exercise 22.

24. The shadow-price concept has been central to our development of sensitivity analysis in this chapter. In this exercise, we consider how changing the initial problem formulation alters the development. Suppose that the initial formulation of the model is given by:

$$\text{Maximize } z = \sum_{j=1}^{n} c_j x_j,$$

subject to:

$$\sum_{j=1}^{n} a_{ij} x_j \leq b_i \qquad (i = 1, 2, \ldots, m),$$
$$x_j \geq 0 \qquad (j = 1, 2, \ldots, n).$$

Solving this problem by the simplex method determines the shadow prices for the constraints as $y_1, y_2, \ldots, y_m$.

a) Suppose that the first constraint were multiplied by 2 and stated instead as:

$$\sum_{j=1}^{n} (2a_{1j})x_j \leq (2b_1).$$

Let $\hat{y}_1$ denote the value of the shadow price for this constraint. How is $\hat{y}_1$ related to y_1?

b) What happens to the value of the shadow prices if every coefficient c_j is multiplied by 2 in the original problem formulation?

c) Suppose that the first variable x_1 in the model is rescaled by a factor of 3. That is, let $x'_1 = 3x_1$ and replace x_1 everywhere in the model by $(x'_1/3)$. Do the shadow prices change? Would it be possible for x'_1 to appear in an optimal basis, if x_1 could not appear in an optimal basis of the original formulation for the problem?

d) Do the answers to parts (a), (b), and (c) change if the original problem is stated with all equality constraints:

$$\sum_{j=1}^{n} a_{ij}x_j = b_i \qquad (i = 1, 2, \ldots, m)?$$

25. Shortly after the beginning of the oil embargo, the Eastern District Director of the Government Office of Fuel Allocation was concerned that he would soon have to start issuing monthly allocations specifying the amounts of heating oil that each refinery in the district would send to each city to ensure that every city would receive its quota.

Since different refineries used a variety of alternative sources of crude oil, both foreign and domestic, the cost of products at each refinery varied considerably. Consequently, under current government regulations, the prices that could be charged for heating oil would vary from one refinery to another. To avoid political criticism, it was felt that, in making allocations of supplies from the refineries to the cities, it was essential to maintain a reasonably uniform average refinery price for each city. In fact, the Director felt that the average refinery price paid by any city should not be more than 3% above the overall average.

Another complication had arisen in recent months, since some of the emergency allocations of supplies to certain cities could not be delivered due to a shortage in tanker capacity. If deliveries of the allocated supplies were to be feasible, the limited transportation facilities on some of the shipping routes would have to be recognized. Finally, it would be most desirable to maintain distribution costs as low as possible under any allocation plan, if charges of government inefficiency were to be avoided.

Data for a simplified version of the problem, with only three refineries and four cities, are given in Fig. E3.7. The allocation problem can be formulated as the linear program shown in Fig. E3.8. The decision variables are defined as follows:

$A - n =$ Barrels of heating oil (in thousands) to be supplied from Refinery A to City n where $n = 1, 2, 3,$ or 4.

$B - n =$ Barrels of heating oil (in thousands) to be supplied from Refinery B to City n where $n = 1, 2, 3,$ or 4.

$C - n =$ Barrels of heating oil (in thousands) to be supplied by Refinery C to City n where $n = 1, 2, 3,$ or 4.

PMAX $=$ A value higher than the average refinery price paid by any of the four cities.

Refinery	December availability (barrels)	Price at refinery ($ per barrel)
A. Norton	95,000	10.53
B. Chatham	63,000	9.39
C. Eastport	116,000	12.43
Average Refinery Price $11.07		

City	December quotas (barrels)
1. Westville	55,000
2. Bridgeton	73,000
3. Brookfield	105,000
4. Dorval	38,000

	Shipping costs ($ per barrel) City			
Refinery	1	2	3	4
A	0.10	0.16	0.32	0.28
B	0.20	0.34	0.30	0.12
C	0.34	0.38	0.22	0.18

	capacities (thousands of barrels per month) City			
Refinery	1	2	3	4
A	*	25	*	*
B	*	20	20	*
C	25	*	*	*

* No effective limit.

Fig. E3.7 Data for heating oil allocation problem.

There are four types of constraints that must be satisfied:

1. Supply constraints at each refinery.
2. Quota restrictions for each city.
3. Average price restrictions which impose an upper limit to the average refinery price which will be paid by any city.
4. Shipping capacity limits on some of the routes.

All the constraints are straightforward except possibly the average price reductions. To restrict the average refinery price paid by City 1 to be less than the variable PMAX, the following expression is used:

$$\frac{(10.53)(A - 1) + (9.39)(B - 1) + (12.43)(C - 1)}{55} \leqq PMAX$$

or

$$0.1915(A - 1) + 0.1701(B - 1) + 0.2260(C - 1) - PMAX \leqq 0.$$

Such a restriction is included for each city, to provide a uniform upper bound on average prices. The value of PMAX is limited by the prescribed maximum of $11.40 = 11.07 \times 1.03$ by the constraint

$$PMAX \leqq 11.40.$$

	A-1	A-2	A-3	A-4	B-1	B-2	B-3	B-4	C-1	C-2	C-3	C-4	PMAX	Relation	Righthand-side	
REF-A	1.0000	1.0000	1.0000	1.0000										≤	95.0000	Supply Constraints
REF-B					1.0000	1.0000	1.0000	1.0000						≤	63.0000	
REF-C									1.0000	1.0000	1.0000	1.0000		≤	116.0000	
CIT-1	1.0000				1.0000				1.0000					=	55.0000	Quota Restrictions
CIT-2		1.0000				1.0000				1.0000				=	73.0000	
CIT-3			1.0000				1.0000				1.0000			=	105.0000	
CIT-4				1.0000				1.0000				1.0000		=	38.0000	
AP-CIT-1	0.1915			1.701	0.1286				0.2260				-1.0000	≤	0.0000	Average Price Restrictions
AP-CIT-2		0.1442	0.1003			0.0894				0.1703			-1.0000	≤	0.0000	
AP-CIT-3			0.2771	0.2771			0.2471				0.1184		-1.0000	≤	0.0000	
AP-CIT-4												0.3271	-1.0000	≤	0.0000	
PRICELIM													1.0000	≤	11.4000	
ROUTE-A2		1.0000												≤	25.0000	Shipping Capacity Limits
ROUTE-B2						1.0000								≤	20.0000	
ROUTE-B3							1.0000							≤	20.0000	
ROUTE-C1									1.0000					≤	25.0000	
Shipcost	0.1000	0.1600	0.3200	0.2800	0.2000	0.3400	0.3000	0.1200	0.3400	0.3800	0.2200	0.1800	0.0000	=	z(min) Objective	

Fig. E3.8 Formulation of the heating-oil allocation model.

The computer analysis of the linear programming model is given in Fig. E3.9

a) Determine the detailed allocations that might be used for December.
b) Evaluate the average refinery prices paid by each city under these allocations.
c) Determine the best way to utilize an additional tanker-truck capacity of 10,000 barrels per month.
d) Discuss the inefficiency in distribution costs resulting from the average refinery price restrictions.
e) Evaluate the applicability of the model in the context of a full allocation system.
f) Discuss any general insights into the fuel allocation problem provided by this model.

```
TITLE: FUEL ALLOCATION
PROCEED, DISPLAY, OR REJECT? PROCEED

MAXIMIZE OR MINIMIZE? MIN

OPTIMAL SOLUTION FOUND.
     SHIPCOST      61.6563

OUTPUT OPTION? USUAL

ALL ITEMS NOT LISTED IN SECTIONS 1 - 4 HAVE THE VALUE ZERO.

*1* DECISION VARIABLES
   1. A-1       54.4253
   2. A-2       15.6023
   3. A-3       24.9724
   5. B-1        .574694
   6. B-2       14.9803
   7. B-3       20.0000
   8. B-4       27.4450
  10. C-2       42.4174
  11. C-3       60.0276
  12. C-4       10.5550
  13. PMAX      11.4000

*2* SLACK(+) AND SURPLUS(-) IN CONSTRAINTS
   3. +REF-C     3.00000
  11. +AP-CIT-4 1.16580
  13. +ROUTE-A2 9.39768
  14. +ROUTE-B2 5.01967
  16. +ROUTE-C1 25.0000

*3* SHADOW PRICES FOR CONSTRAINTS
   1. REF-A      -.232518
   2. REF-B      -.600000E-01
   4. CIT-1       .341718
   5. CIT-2       .461679
   6. CIT-3      2.39515
   7. CIT-4       .180000
   8. AP-CIT-1  -.480411E-01
   9. AP-CIT-2  -.479616
  10. AP-CIT-3 -18.3712
  12. PRICELIM -18.8988
  15. ROUTE-B3  -.392764

*4* REDUCED COSTS FOR DECISION VARIABLES
   4. A-4        .332518
   9. C-1        .913941E-02

OUTPUT OPTION?
```

Fig. E3.9 Solution of the allocation model. (*Continued on next page.*)

```
*5* RANGES ON COEFFICIENTS OF OBJECTIVE SHIPCOST
        VARIABLE    LOWER BOUND   CURRENT VALUE    UPPER BOUND
     1. A-1          .27482E-01      .10000          .10935
     2. A-2          .15065          .16000          .23252
     3. A-3          .74861E-01      .32000          .13630E+06
     4. A-4         -.52518E-01      .28000          UNBOUNDED
     5. B-1          UNBOUNDED       .20000          .27252
     6. B-2          .32000          .34000          .35494
     7. B-3          UNBOUNDED       .30000          .69276
     8. B-4          .10559          .12000          .14000
     9. C-1          .33086          .34000          UNBOUNDED
    10. C-2          .18615          .38000          .40000
    11. C-3         -.13630E+06      .22000          .55252
    12. C-4          .16000          .18000          .19441
    13. PMAX         UNBOUNDED       .00000          18.899


*6* RANGES ON VALUES OF RIGHT-HAND-SIDE RHS
        CONSTRNT    LOWER BOUND   CURRENT VALUE    UPPER BOUND
     1. REF-A         92.000        95.000          104.40
     2. REF-B         60.000        63.000          73.555
     3. REF-C        113.00        116.00           UNBOUNDED
     4. CIT-1         46.660        55.000          58.000
     5. CIT-2         70.415        73.000          74.229
     6. CIT-3        103.56        105.00           106.23
     7. CIT-4         27.445        38.000          41.000
     8. AP-CIT-1      -.86750        .00000         14.186
     9. AP-CIT-2      -.20932        .00000          .44014
    10. AP-CIT-3      -.14516        .00000          .17010
    11. AP-CIT-4     -1.1658         .00000         UNBOUNDED
    12. PRICELIM      11.315        11.400          11.568
    13. ROUTE-A2      15.602        25.000          UNBOUNDED
    14. ROUTE-B2      14.980        20.000          UNBOUNDED
    15. ROUTE-B3      14.994        20.000          25.865
    16. ROUTE-C1     -.23842E-06    25.000          UNBOUNDED

OUTPUT OPTION? NO
```

Figure E3.9 (*Cont.*)

26. The Krebs Wire Company is an intermediate processor that purchases uncoated wire in standard gauges and then applies various coatings according to customer specification. Krebs Wire has essentially two basic products—standard inexpensive plastic and the higher quality Teflon. The two coatings come in a variety of colors but these are changed easily by introducing different dyes into the basic coating liquid.

The production facilities at Krebs Wire consist of two independent wire trains, referred to as the Kolbert and Loomis trains. Both the standard plastic-coated and the quality Teflon-coated wire can be produced on either process train; however, production of Teflon-coated wire is a slower process due to drying requirements. The different production rates in tons per day are given below:

Process train	Plastic	Teflon
Kolbert	40 tons/day	35 tons/day
Loomis	50 tons/day	42 tons/day

It has been traditional at Krebs Wire to view production rates in terms of daily tonnage, as opposed to reels per day or other production measures. The respective contributions in dollars per day are:

Process train	Plastic	Teflon
Kolbert	525 $/day	546 $/day
Loomis	580 $/day	590 $/day

Planning at Krebs Wire is usually done on a monthly basis. However, since most employee vacations are scheduled over the two summer months, management feels that production planning for the two summer months should be combined to facilitate vacation scheduling. Each month the process trains must be shut down for scheduled maintenance, so that the total days available for production per month are as follows:

Process train	July	August
Kolbert	26 days	26 days
Loomis	28 days	27 days

The scheduling process is further complicated by the fact that, over the two summer months, the total amount of time available for production is limited to 102 machine days due to vacation schedules.

The amounts of wire that the management feels it can sell in the coming two months are:

Product	July	August
Plastic	1200 tons	1400 tons
Teflon	800 tons	900 tons

Both types of wire may be stored for future delivery. Space is available in Krebs' own warehouse, which has a capacity of 20 tons. The inventory and carrying costs in dollars per ton for wire produced in July and delivered in August are:

Product	Inventory and carrying costs
Plastic	1.00 $/ton
Teflon	1.20 $/ton

Due to a commitment of warehouse capacity to other products in September, it is not possible to stock any wire in inventory at the end of August.

To help in planning production for the two summer months, Krebs Wire management has formulated and solved a linear program (see Figs. E3.10 and E3.11) in order to determine the production schedule that will maximize contribution from these products.

	JULY				INV-P	INV-T	AUGUST				Rela-tion	Limit	Constraint name
	K-P	L-P	K-T	L-T			K-P	L-P	K-T	L-T			
	525	580	546	590	−1.0	−1.2	525	580	546	590			
K-Days-July	1	0	1	0	0	0	0	0	0	0	≤	26	K-Days-July
L-Days-July	0	1	0	1	0	0	0	0	0	0	≤	28	L-Days-July
P-Dem-July	40	50	0	0	−1	0	0	0	0	0	≤	1200	P-Dem-July
T-Dem-July	0	0	35	42	0	−1	0	0	0	0	≤	800	T-Dem-July
Warehouse	0	0	0	0	1	1	0	0	0	0	≤	20	Warehouse
K-Days-August	0	0	0	0	0	0	1	0	1	0	≤	26	K-Days-August
L-Days-August	0	0	0	0	0	0	0	1	0	1	≤	27	L-Days-August
P-Dem-August	0	0	0	0	1	0	40	50	0	0	≤	1400	P-Dem-August
T-Dem-August	0	0	0	0	0	1	0	0	35	42	≤	900	T-Dem-August
Max-days-total	1	1	1	1	0	0	1	1	1	1	≤	102	Max-days-total
Contribution											=	z (Max)	Contribution

INDEX

K : Kolbert
L : Loomis
P : Plastic
T : Teflon
INV : Inventory
DEM : Demand maximum
Max-days-total : The maximum number of days available for production due to vacation constraints.

Fig. 3.10 Formulation of the Krebs Wire linear program.

```
OPTIMAL SOLUTION FOUND.
     CONTRIB      56839.0

OUTPUT OPTION? EXTENDED

ALL ITEMS NOT LISTED IN SECTIONS 1 - 4 HAVE THE VALUE ZERO.

*1* DECISION VARIABLES
    1. K-P-J     26.0000
    2. L-P-J      3.20000
    4. L-T-J     19.5238
    6. INV-T     20.0000
    7. K-P-A      .857143
    8. L-P-A     27.0000
    9. K-T-A     25.1429

*2* SLACK(+) AND SURPLUS(-) IN CONSTRAINTS
    2. +L-DAY-J   5.27619
    8. +P-DEM-A  15.7143
   10. +M-DAY-T   .276191

*3* SHADOW PRICES FOR CONSTRAINTS
    1. K-DAY-J   61.0000
    3. P-DEM-J   11.6000
    4. T-DEM-J   14.0476
    5. WAR-CAP   12.2476
    6. K-DAY-A  525.000
    7. L-DAY-A  580.000
    9. T-DEM-A    .600000

*4* REDUCED COSTS FOR DECISION VARIABLES
    3. K-T-J     -6.66667
    5. INV-P     -1.64762
   10. L-T-A    -15.2000

*5* RANGES ON COEFFICIENTS OF OBJECTIVE CONTRIB
        VARIABLE   LOWER BOUND   CURRENT VALUE   UPPER BOUND
    1. K-P-J        518.33         525.00        UNBOUNDED
    2. L-P-J         .00000        580.00          588.33
    3. K-T-J       UNBOUNDED       546.00          552.67
    4. L-T-J        582.00         590.00        UNBOUNDED
    5. INV-P       UNBOUNDED      -1.0000           .64762
    6. INV-T       -2.8476        -1.2000        UNBOUNDED
    7. K-P-A        467.33         525.00          537.67
    8. L-P-A        564.80         580.00        UNBOUNDED
    9. K-T-A        533.33         546.00          603.67
   10. L-T-A       UNBOUNDED       590.00          605.20

*6* RANGES ON VALUES OF RIGHT-HAND-SIDE CONSTRAI
        CONSTRNT   LOWER BOUND   CURRENT VALUE   UPPER BOUND
    1. K-DAY-J      19.405         26.000          27.381
    2. L-DAY-J      22.724         28.000        UNBOUNDED
    3. P-DEM-J      1040.0         1200.0          1213.8
    4. T-DEM-J     -20.000         800.00          811.60
    5. WAR-CAP       .00000        20.000          31.600
    6. K-DAY-A      25.143         26.000          26.276
    7. L-DAY-A       .00000        27.000          27.276
    8. P-DEM-A      1384.3         1400.0        UNBOUNDED
    9. T-DEM-A      886.25         900.00          930.00
   10. M-DAY-T      101.72         102.00        UNBOUNDED
```

Fig. E3.11 Solution of the Krebs Wire model.

a) What does the output in Fig. E3.12 tell the production manager about the details of scheduling his machines?   ⌐|⌐

b) There is the possibility that some employees might be persuaded to take their vacations in June or September. Should this be encouraged?

c) The solution in Fig. E3.12 suggests that Teflon should be made only on the Loomis train in July and only on the Kolbert train in August. Why?

d) Should Krebs Wire lease additional warehouse capacity at a cost of $2.00 above the inventory and carrying costs? If so, how would the optimal solution change?

e) The sales manager feels that future sales might be affected if the firm could not meet demand for plastic-coated wire in August. What, if anything, should be done?

f) One of Krebs' customers has requested a special run of twenty tons of corrosion-resistant wire to be delivered at the end of August. Krebs has made this product before and found that it can be produced only on the Loomis machine, due to technical restrictions. The Loomis machine can produce this special wire at a rate of 40 tons per day, and the customer will pay $12 per ton. Krebs cannot start production before the 1st of August due to a shortage of raw materials. Should the firm accept the order?

27. Consider the computer output for the Krebs Wire case in Exercise 26. What does the "100 percent rule" tell us in each of the following situations?

a) The objective-function coefficients changed as follows:

K-T-J from 546 to 550,
L-T-A from 590 to 600.

b) The objective-function coefficients changed as follows:

L-P-J from 580 to 585,
L-T-J from 590 to 585.

c) The objective-function coefficients changed as follows:

L-P-J from 580 to 585,
L-T-J from 590 to 588.

d) The righthand-side values changed as follows:

L-Day-J from 28 to 25,
War-Cap from 20 to 24.

e) The righthand-side values changed as follows:

L-Day-J from 28 to 23,
P-Dem-A from 1400 to 1385.

28. Mr. Watson has 100 acres that can be used for growing corn or soybeans. His yield is 95 bushels per acre per year of corn or 60 bushels of soybeans. Any fraction of the 100 acres can be devoted to growing either crop. Labor requirements are 4 hours per acre per year, plus 0.70 hour per bushel of corn and 0.15 hour per bushel of soybeans. Cost of seed, fertilizer, and so on is 24 cents per bushel of corn and 40 cents per bushel of soybeans. Corn can be sold for $1.90 per bushel, and soybeans for $3.50 per bushel. Corn can be purchased for $3.00 per bushel, and soybeans for $5.00 per bushel.

In the past, Mr. Watson has occasionally raised pigs and calves. He sells the pigs or calves when they reach the age of one year. A pig sells for $80 and a calf for $160. One pig requires 20 bushels of corn or 25 bushels of soybeans, plus 25 hours of labor and 25 square feet of floor space. One calf requires 50 bushels of corn or 20 bushels of soybeans, 80 hours of labor, and 30 square feet of floor space.

Mr. Watson has 10,000 square feet of floor space. He has available per year 2000 hours of his own time and another 4000 hours from his family. He can hire labor at $3.00 per hour. However, for each hour of hired labor, 0.15 hour of his time is required for supervision.

Mr. Watson's son is a graduate student in business, and he has formulated a linear program to show his father how much land should be devoted to corn and soybeans and in addition, how many pigs and/or calves should be raised to maximize profit.

In Fig. 3.12, Tableau 1 shows an initial simplex tableau for Watson's farm using the definitions of variables and constraints given below, and Tableau 2 shows the results of several iterations of the simplex algorithm.

Variables

1. Grow 1 acre of corn
2. Grow 1 acre of soybeans
3. Buy 1 bu. of corn
4. Buy 1 bu. of soybeans
5. Sell 1 bu. of corn
6. Sell 1 bu. of soybeans
7. Raise 1 pig on corn
8. Raise 1 pig on soybeans
9. Raise 1 calf on corn
10. Raise 1 calf on soybeans
11. Hire 1 hour of labor
12–15. Slack variables

Constraints

1. Acres of land
2. Bushels of corn
3. Bushels of soybeans
4. Hundreds of sq. ft. floor space
5. Hundreds of labor hours
6. Hundreds of farmer hours

Objective

Dollars of cost to be minimized.

a) What is the optimal solution to the farmer's problem?
b) What are the binding constraints and what are the shadow prices on these constraints? [Hint: The initial tableau is not quite in canonical form.]
c) At what selling price for corn does the raising of corn become attractive? At this price + $0.05, what is an optimal basic solution?
d) The farmer's city nephew wants a summer job, but because of his inexperience he requires 0.2 hours of supervision for each hour he works. How much can the farmer pay him and break even? How many hours can the nephew work without changing the optimal basis? What activity leaves the basis if he works more than that amount?
e) One of the farmer's sons wants to leave for the city. How much can the old man afford to pay him to stay? Since that is not enough, he goes, reducing the family labor pool by 2000 hours/year. What is the optimal program now?
f) How much can the selling price of soybeans increase without changing the basis? Decrease? For both of these basis changes, what activity leaves the basis? Are these basis changes intuitively obvious?
g) Does there exist an alternative optimal solution to the linear program? Alternative optimal shadow prices? If so, how can they be found?

Tableau 1

	x_1	x_2	x_3	x_4	x_5	x_6	x_7	x_8	x_9	x_{10}	x_{11}	x_{12}	x_{13}	x_{14}	x_{15}	Relation	RHS
1. Land	1	1										1				=	100
2. Corn	−95		−1												1	=	0
3. Soybeans		−60		−1												=	0
4. Space					1	1	20	25	50	20						=	100
5. Labor	0.705	0.13					0.25	0.25	0.30	0.30	−0.85		1			=	60
6. Farmer							0.25	0.25	0.80	0.80	0.15			1		=	20
$	22.8	24	3.00	5.00	−1.90	−3.50	−80	−80	−160	−160	300					=	0

Tableau 2

	x_1 x_2	x_3	x_4	x_5	x_6	x_7	x_8	x_9	x_{10}	x_{11}	x_{12}	x_{13}	x_{14} x_{15}	Relation	RHS
1. Land	1	−0.0105		0.0105		0.2105	0.3125	0.5264	1	−1.065	1	1.25		=	100
2. Corn		−0.00756		0.00756		0.4638		1.378			−0.1625	−25.0		=	58.75
3. Soybeans		−0.4803	−1	0.4803	1	3.355	18.75	4.014		21.25	63.25			=	4825
4. Space		0.0023		−0.0023		0.1109	0.1562	−0.1134		0.319	0.0488	−0.375	1	=	82.375
5. Labor	1	0.0105		−0.0105		−0.2105		−0.5264						=	0
6. Farmer										0.15			1	=	20
$		0.1212	1.5	0.9788		5.701	35.625	73.94		204.38	171.375	112.50		=	23387.5

Fig. E3.12 Initial and final tableaus for Watson's model.

29. The initial data tableau, in canonical form, for a linear program to be minimized is given below:

Basic variables	Current values	x_1	x_2	x_3	x_4	x_5	x_6	x_7
x_5	3	1	1	1	1	1		
x_6	2	3	4	1	1		1	
x_7	1	1	3	2	1			1
$(-z)$	0	-4	-6	-1	-3			

A standard commercial linear-programming code employing the *revised simplex method* would have the following information at its disposal: (1) the above initial data tableau; (2) the current values of the basic variables and the row in which each is basic; and (3) the current coefficients of the initial unit columns. Items (2) and (3) are given below:

Basic variables	Current values	x_5	x_6	x_7
x_5	$\frac{12}{5}$	1	$-\frac{2}{5}$	$\frac{1}{5}$
x_1	$\frac{2}{5}$		$\frac{3}{5}$	$-\frac{4}{5}$
x_2	$\frac{1}{5}$		$-\frac{1}{5}$	$\frac{3}{5}$
$(-z)$	$\frac{14}{5}$		$\frac{6}{5}$	$\frac{2}{5}$

a) What is the current basic feasible solution?

b) We can define *simplex multipliers* to be the shadow prices associated with the current basic solution even if the solution is not optimal. What are the values of the simplex multipliers associated with the current solution?

c) The current solution is optimal if the reduced costs of the nonbasic variables are nonnegative. Which variables are nonbasic? Determine the reduced cost of the nonbasic variables and show that the current solution is *not* optimal.

d) Suppose that variable x_4 should now be introduced into the basis. To determine the variable to drop from the basis, we use the minimum-ratio rule, which requires that we know not only the current righthand-side values but also the coefficients of x_4 in the current tableau. These coefficients of x_4 need to be computed.

   In performing the simplex method, multiples of the initial tableau have been added to and subtracted from one another to produce the final tableau. The coefficients in the current tableau of the initial unit columns summarize these operations.

   i) What multiple of rows 1, 2, and 3, when added together, *must* produce the current row 1 (even though we do not know all the current coefficients in row 1)? The current row 2? The current row 3?

   ii) Using the rationale of (i) determine the coefficients of x_4 in the current tableau. Note that it is unnecessary to determine any of the other unknown columns.

   iii) How should the pivot operation be performed to update the tableau consisting of only x_5, x_6, and x_7?

   You have now completed an iteration of the simplex algorithm using *only* (1) the initial data, (2) the current values of the basic variables and the row in which each is basic, and (3) the current coefficients of the initial unit columns. This is the essence of the revised simplex method. (See Appendix B for further details.)

ACKNOWLEDGMENTS

Exercises 1 and 2 are due to Steven C. Wheelwright of the Harvard Business School.

Exercises 3, 4, and 5 are based on the Land of Milk and Honey case, written by Steven Wheelwright, which in turn is based on a formulation exercise from *Quantitative Analysis of Business Decisions*, by H. Bierman, C. P. Bonnini, and W. H. Hausman, Third Edition, Richard D. Irwin, Inc., 1969.

Exercises 13 and 28 are variations of problems used by C. Roger Glassey of the University of California, Berkeley, and Exercise 28 is in turn based on a formulation exercise from *Linear Programming*, by G. Hadley, Addison-Wesley Publishing Company, Inc., 1963.

Exercise 17 is inspired by the French Electric Power Industry case written by John E. Bishop, based on "Application of Linear Programming to Investments in the Electric Power Industry," by P. Massé and R. Gibrat, which appeared in *Management Science*, 1957.

Exercise 25 is based on the Holden Consulting Company case written by Basil A. Kalyman.

Exercise 26 is based on the Krebs Wire Company case written by Ronald S. Frank, based on a case of one of the authors.

Duality in Linear Programming

4

In the preceding chapter on sensitivity analysis, we saw that the shadow-price interpretation of the optimal simplex multipliers is a very useful concept. First, these shadow prices give us directly the marginal worth of an additional unit of any of the resources. Second, when an activity is "priced out" using these shadow prices, the opportunity cost of allocating resources to that activity relative to other activities is determined. Duality in linear programming is essentially a unifying theory that develops the relationships between a given linear program and another related linear program stated in terms of variables with this shadow-price interpretation. The importance of duality is twofold. First, fully understanding the shadow-price interpretation of the optimal simplex multipliers can prove very useful in understanding the implications of a particular linear-programming model. Second, it is often possible to solve the related linear program with the shadow prices as the variables in place of, or in conjunction with, the original linear program, thereby taking advantage of some computational efficiencies. The importance of duality for computational procedures will become more apparent in later chapters on network-flow problems and large-scale systems.

4.1 A PREVIEW OF DUALITY

We can motivate our discussion of duality in linear programming by considering again the simple example given in Chapter 2 involving the firm producing three types of automobile trailers. Recall that the decision variables are:

$$x_1 = \text{number of flat-bed trailers produced per month,}$$

$$x_2 = \text{number of economy trailers produced per month,}$$

$$x_3 = \text{number of luxury trailers produced per month.}$$

The constraining resources of the production operation are the metalworking and woodworking capacities measured in days per month. The linear program to maximize contribution to the firm's overhead (in hundreds of dollars) is:

$$\text{Maximize } z = 6x_1 + 14x_2 + 13x_3,$$

subject to:

$$\tfrac{1}{2}x_1 + 2x_2 + x_3 \leq 24,$$
$$x_1 + 2x_2 + 4x_3 \leq 60, \tag{1}$$
$$x_1 \geq 0, \qquad x_2 \geq 0, \qquad x_3 \geq 0.$$

After adding slack variables, the initial tableau is stated in canonical form in Tableau 1.

Tableau 1

Basic variables	Current values	x_1	x_2	x_3	x_4	x_5
x_4	24	$\tfrac{1}{2}$	2	1	1	
x_5	60	1.	2	4		1
$(-z)$	0	6	14	13		

In Chapter 2, the example was solved in detail by the simplex method, resulting in the final tableau, repeated here as Tableau 2.

Tableau 2

Basic variables	Current values	x_1	x_2	x_3	x_4	x_5
x_1	36	1	6		4	-1
x_3	6		-1	1	-1	$\tfrac{1}{2}$
$(-z)$	-294		-9		-11	$-\tfrac{1}{2}$

As we saw in Chapter 3, the shadow prices, y_1 for metalworking capacity and y_2 for woodworking capacity, can be determined from the final tableau as the negative of the reduced costs associated with the slack variables x_4 and x_5. Thus these shadow prices are $y_1 = 11$ and $y_2 = \tfrac{1}{2}$, respectively.

We can interpret the shadow prices in the usual way. One additional day of metalworking capacity is worth \$1100, while one additional day of woodworking capacity is worth only \$50. These values can be viewed as the breakeven rents that the firm could pay per day for additional capacity of each type. If additional capacity could be rented for *less* than its corresponding shadow price, it would be profitable to expand capacity in this way. Hence, in allocating the scarce resources to the production activities, we have determined shadow prices for the resources, which are the values imputed to these resources at the margin.

Let us examine some of the economic properties of the shadow prices associated with the resources. Recall, from Chapter 3, Eq. (11), that the reduced costs are given in terms of the shadow prices as follows:

$$\bar{c}_j = c_j - \sum_{i=1}^{m} a_{ij} y_i \qquad (j = 1, 2, \ldots, n).$$

Since a_{ij} is the amount of resource i used per unit of activity j, and y_i is the imputed value of that resource, the term

$$\sum_{i=1}^{m} a_{ij} y_i$$

is the total value of the resources used per unit of activity j. It is thus the marginal resource cost for using that activity. If we think of the objective coefficients c_j as being marginal revenues, the reduced costs $\bar{c}_j$ are simply net marginal revenues (i.e., marginal revenue minus marginal cost).

For the basic variables x_1 and x_3, the reduced costs are zero,

$$\bar{c}_1 = 6 - 11(\tfrac{1}{2}) - \tfrac{1}{2}(1) = 0,$$
$$\bar{c}_3 = 13 - 11(1) - \tfrac{1}{2}(4) = 0.$$

The values imputed to the resources are such that the net marginal revenue is zero on those activities operated at a positive level. That is, for any production activity at positive level, *marginal revenue must equal marginal cost.*

The situation is much the same for the nonbasic variables x_2, x_4, and x_5, with corresponding reduced costs:

$$\bar{c}_2 = 14 - 11(2) - \tfrac{1}{2}(2) = -9,$$
$$\bar{c}_4 = 0 - 11(1) - \tfrac{1}{2}(0) = -11,$$
$$\bar{c}_5 = 0 - 11(0) - \tfrac{1}{2}(1) = -\tfrac{1}{2}.$$

The reduced costs for all nonbasic variables are negative. The interpretation is that, for the values imputed to the scarce resources, *marginal revenue is less than marginal cost* for these activities, so they should not be pursued. In the case of x_2, this simply means that we should not produce *any* economy trailers. The cases of x_4 and x_5 are somewhat different, since slack variables represent unused capacity. Since the marginal revenue of a slack activity is zero, its reduced cost equals *minus its marginal cost,* which is just the shadow price of the corresponding capacity constraint, as we have seen before.

The above conditions interpreted for the reduced costs of the decision variables are the familiar optimality conditions of the simplex method. Economically we can see why they must hold. If marginal revenue exceeds marginal cost for any activity, then the firm would improve its contribution to overhead by *increasing* that activity. If, however, marginal cost exceeds marginal revenue for an activity operated at a positive level, then the firm would increase its contribution by *decreasing* that activity. In either case, a new solution could be found that is an improvement on the current solution. Finally, as we have seen in Chapter 3, those nonbasic variables with zero reduced costs represent possible alternative optimal solutions.

Until now we have used the shadow prices mainly to impute the marginal resource cost associated with each activity. We then selected the best activities for the firm to pursue by comparing the marginal revenue of an activity with its marginal resource cost. In this case, the shadow prices are interpreted as the opportunity costs associated with consuming the firm's resources. If we now value the firm's total resources at these prices, we find their value,

$$v = 11(24) + \tfrac{1}{2}(60) = 294,$$

is exactly equal to the optimal value of the objective function of the firm's decision problem. The implication of this valuation scheme is that the firm's metalworking and woodworking capacities have an imputed worth of \$264 and \$30, respectively. Essentially then, the shadow prices constitute an *internal* pricing system for the firm's resources that:

1. permits the firm to select which activity to pursue by considering only the marginal profitability of its activities; and
2. allocates the contribution of the firm to its resources at the margin.

Suppose that we consider trying to determine directly the shadow prices that satisfy these conditions, without solving the firm's production-decision problem. The shadow prices must satisfy the requirement that marginal revenue be less than or equal to marginal cost for all activities. Further, they must be nonnegative since they are associated with less-than-or-equal-to constraints in a maximization decision problem. Therefore, the unknown shadow prices y_1 on metalworking capacity and y_2 on woodworking capacity must satisfy:

$$\tfrac{1}{2}y_1 + \ y_2 \geqq \ 6,$$
$$2y_1 + 2y_2 \geqq 14,$$
$$y_1 + 4y_2 \geqq 13,$$
$$y_1 \geqq 0, \qquad y_2 \geqq 0.$$

These constraints require that the shadow prices be chosen so that the net marginal revenue for each activity is nonpositive. If this were not the case, an improvement in the firm's total contribution could be made by changing the choice of production activities.

Recall that the shadow prices were interpreted as breakeven rents for capacity at the margin. Imagine for the moment that the firm does not own its productive capacity but has to rent it. Now consider any values for the shadow prices, or rental rates, that satisfy the above constraints, say $y_1 = 4$ and $y_2 = 4$. The total worth of the rented capacities evaluated at these shadow prices is $v = 24(4) + 60(4) = 336$, which is greater than the maximum contribution of the firm. Since the imputed value of the firm's resources is derived solely from allocating the firm's contribution to its resources, $v = 336$ is too high a total worth to impute to the firm's resources. The firm clearly could not break even if it had to rent its production capacity at such rates.

If we think of the firm as renting all of its resources, then surely it should try to rent them at least cost. This suggests that we might determine the appropriate values of the shadow prices by minimizing the total rental cost of the resources, subject to the above constraints. That is, solve the following linear program:

Minimize $v = 24y_1 + 60y_2$,

subject to:

$$\frac{1}{2}y_1 + \quad y_2 \geq 6,$$
$$2y_1 + 2y_2 \geq 14,$$
$$y_1 + 4y_2 \geq 13, \tag{2}$$
$$y_1 \geq 0, \quad y_2 \geq 0.$$

If we solve this linear program by the simplex method, the resulting optimal solution is $y_1 = 11$, $y_2 = \frac{1}{2}$, and $v = 294$. These are exactly the desired values of the shadow prices, and the value of v reflects that the firm's contribution is fully allocated to its resources. Essentially, the linear program (2), in terms of the shadow prices, determines rents for the resources that would allow the firm to break even, in the sense that its total contribution would exactly equal the total rental value of its resources. However, the firm in fact owns its resources, and so the shadow prices are interpreted as the breakeven rates for renting *additional* capacity.

Thus, we have observed that, by solving (2), we can determine the shadow prices of (1) directly. Problem (2) is called the *dual* of Problem (1). Since Problem (2) has a name, it is helpful to have a generic name for the original linear program. Problem (1) has come to be called the *primal*.

In solving any linear program by the simplex method, we also determine the shadow prices associated with the constraints. In solving (2), the shadow prices associated with its constraints are $u_1 = 36$, $u_2 = 0$, and $u_3 = 6$. However, these shadow prices for the constraints of (2) are exactly the optimal values of the decision variables of the firm's allocation problem. Hence, in solving the dual (2) by the simplex method, we apparently have solved the primal (1) as well. As we will see later, this will always be the case since "the dual of the dual is the primal." This is an important result since it implies that the dual may be solved instead of the primal whenever there are computational advantages.

Let us further emphasize the implications of solving these problems by the simplex method. The optimality conditions of the simplex method require that the reduced costs of basic variables be zero. Hence,

$$\text{if } \hat{x}_1 > 0, \quad \text{then } \bar{c}_1 = 6 - \tfrac{1}{2}\hat{y}_1 - \hat{y}_2 = 0;$$
$$\text{if } \hat{x}_3 > 0, \quad \text{then } \bar{c}_3 = 13 - \hat{y}_1 - 4\hat{y}_2 = 0.$$

These equations state that, if a decision variable of the primal is positive, then the corresponding constraint in the dual must hold with equality. Further, the optimality conditions require that the nonbasic variables be zero (at least for those variables with negative reduced costs); that is,

$$\text{if } \bar{c}_2 = 14 - 2\hat{y}_1 - 2\hat{y}_2 < 0, \quad \text{then } \hat{x}_2 = 0.$$

These observations are often referred to as *complementary slackness* conditions since, if a variable is positive, its corresponding (complementary) dual constraint holds with equality while, if a dual constraint holds with strict inequality, then the corresponding (complementary) primal variable must be zero.

These results are analogous to what we have seen in Chapter 3. If some shadow price is positive, then the corresponding constraint must hold with equality; that is,

$$\text{if } \hat{y}_1 > 0, \quad \text{then } \tfrac{1}{2}\hat{x}_1 + 2\hat{x}_2 + \hat{x}_3 = 24;$$
$$\text{if } \hat{y}_2 > 0, \quad \text{then } \hat{x}_1 + 2\hat{x}_2 + 4\hat{x}_3 = 60.$$

Further, if a constraint of the primal is not binding, then its corresponding shadow price must be zero. In our simple example there do not happen to be any nonbinding constraints, other than the implicit nonnegativity constraints. However, the reduced costs have the interpretation of shadow prices on the nonnegativity constraints, and we see that the reduced costs of x_1 and x_3 are appropriately zero.

In this chapter we develop these ideas further by presenting the general theory of duality in linear programming.

4.2 DEFINITION OF THE DUAL PROBLEM

The duality principles we have illustrated in the previous sections can be stated formally in general terms. Let the primal problem be:

Primal

$$\text{Maximize } z = \sum_{j=1}^{n} c_j x_j,$$

subject to:

$$\sum_{j=1}^{n} a_{ij} x_j \leq b_i \qquad (i = 1, 2, \ldots, m), \tag{3}$$

$$x_j \geq 0 \qquad (j = 1, 2, \ldots, n),$$

Associated with this primal problem there is a corresponding dual problem given by:

Dual

$$\text{Minimize } v = \sum_{i=1}^{m} b_i y_i,$$

subject to:

$$\sum_{i=1}^{m} a_{ij} y_i \geq c_j \qquad (j = 1, 2, \ldots, n), \tag{4}$$

$$y_i \geq 0 \qquad (i = 1, 2, \ldots, m).$$

These primal and dual relationships can be conveniently summarized as in Fig. 4.1.

Dual \ Primal variables	$x_1 \geq 0$	$x_2 \geq 0$	$x_3 \geq 0$	$\cdots$	$x_n \geq 0$	Primal relation	Min v
$y_1 \geq 0$	a_{11}	a_{12}	a_{13}	$\cdots$	a_{1n}	$\leq$	b_1
$y_2 \geq 0$	a_{21}	a_{22}	a_{23}	$\cdots$	a_{2n}	$\leq$	b_2
$\vdots$	$\vdots$	$\vdots$	$\vdots$		$\vdots$	$\vdots$	$\vdots$
$y_m \geq 0$	a_{m1}	a_{m2}	a_{m3}	$\cdots$	a_{mn}	$\leq$	b_m
Dual Relation	II∨	II∨	II∨		II∨		
Max z	c_1	c_2	c_3	$\cdots$	c_n		

Fig. 4.1 Primal and dual relationships.

Without the variables $y_1, y_2, \ldots, y_m$, this tableau is essentially the tableau form utilized in Chapters 2 and 3 for a linear program. The first m rows of the tableau correspond to the constraints of the primal problem, while the last row corresponds to the objective function of the primal problem. If the variables $x_1, x_2, \ldots, x_n$, are ignored, the columns of the tableau have a similar interpretation for the dual problem. The first n columns of the tableau correspond to the constraints of the dual problem, while the last column corresponds to the objective function of the dual problem. Note that there is one dual variable for each explicit constraint in the primal, and one primal variable for each explicit constraint in the dual. Moreover, the dual constraints are the familiar optimality condition of "pricing out" a column. They state that, at optimality, no activity should appear to be profitable from the standpoint of its reduced cost; that is,

$$\bar{c}_j = c_j - \sum_{i=1}^{m} a_{ij} y_i \leq 0.$$

To illustrate some of these relationships, let us consider an example formulated in Chapter 1. Recall the portfolio-selection problem, where the decision variables are the amounts to invest in each security type:

Maximize $z = 0.043x_A + 0.027x_B + 0.025x_C + 0.022x_D + 0.045x_E$,

subject to:

Cash	$x_A +$	$x_B +$	$x_C +$	$x_D +$	$x_E \leq 10,$
Governments		$x_B +$	$x_C +$	x_D	$\geq 4,$
Quality	$0.6x_A +$	$0.6x_B -$	$0.4x_C -$	$0.4x_D +$	$3.6x_E \leq 0,$
Maturity	$4x_A +$	$10x_B -$	$x_C -$	$2x_D -$	$3x_E \leq 0,$

$$x_A \geq 0, \quad x_B \geq 0, \quad x_C \geq 0, \quad x_D \geq 0, \quad x_E \geq 0.$$

The dual of this problem can be found easily by converting it to the standard primal formulation given in (3). This is accomplished by multiplying the second constraint by -1, thus changing the "greater than or equal to" constraint to a "less than or equal to" constraint. The resulting primal problem becomes:

Maximize $z = 0.043x_A + 0.027x_B + 0.025x_C + 0.022x_D + 0.045x_E,$

subject to:

$$
\begin{array}{rcrcrcrcrcl}
x_A & + & x_B & + & x_C & + & x_D & + & x_E & \leq & 10, \\
 & - & x_B & - & x_C & - & x_D & & & \leq & -4, \\
0.6x_A & + & 0.6x_B & - & 0.4x_C & - & 0.4x_D & + & 3.6x_E & \leq & 0, \\
4x_A & + & 10x_B & - & x_C & - & 2x_D & - & 3x_E & \leq & 0, \\
\end{array}
$$

$$x_A \geq 0, \qquad x_B \geq 0, \qquad x_C \geq 0, \qquad x_D \geq 0, \qquad x_E \geq 0.$$

According to expression (4), the corresponding dual problem is:

Minimize $v = 10y_1 - 4y_2,$

subject to:

$$
\begin{array}{rcrcrcrcl}
y_1 & & & + & 0.6y_3 & + & 4y_4 & \geq & 0.043, \\
y_1 & - & y_2 & + & 0.6y_3 & + & 10y_4 & \geq & 0.027, \\
y_1 & - & y_2 & - & 0.4y_3 & - & y_4 & \geq & 0.025, \\
y_1 & - & y_2 & - & 0.4y_3 & - & 2y_4 & \geq & 0.022, \\
y_1 & & & + & 3.6y_3 & - & 3y_4 & \geq & 0.045, \\
\end{array}
$$

$$y_1 \geq 0, \qquad y_2 \geq 0, \qquad y_3 \geq 0, \qquad y_4 \geq 0.$$

By applying the simplex method, the optimal solution to both primal and dual problems can be found to be:

Primal: $x_A = 3.36,$ $x_B = 0,$ $x_C = 0,$ $x_D = 6.48,$
$x_E = 0.16,$ and $z = 0.294;$

Dual: $y_1 = 0.0294,$ $y_2 = 0,$ $y_3 = 0.00636,$
$y_4 = 0.00244,$ and $v = 0.294.$

As we have seen before, the optimal values of the objective functions of the primal and dual solutions are equal. Furthermore, an optimal dual variable is nonzero only if its associated constraint in the primal is binding. This should be intuitively clear, since the optimal dual variables are the shadow prices associated with the constraints. These shadow prices can be interpreted as values imputed to the scarce resources (binding constraints), so that the value of these resources equals the value of the primal objective function.

To further develop that the optimal dual variables are the shadow prices discussed in Chapter 3, we note that they satisfy the optimality conditions of the simplex method. In the final tableau of the simplex method, the reduced costs of the basic variables must be zero. As an example, consider basic variable x_A. The reduced cost of x_A in the final tableau can be determined as follows:

$$\bar{c}_A = c_A - \sum_{i=1}^{5} a_{iA} y_i$$
$$= 0.043 - 1(0.0294) - 0(0) - 0.6(0.00636) - 4(0.00244) = 0.$$

For nonbasic variables, the reduced cost in the final tableau must be nonpositive in order to ensure that no improvements can be made. Consider nonbasic variable x_B, whose reduced cost is determined as follows:

$$\bar{c}_B = c_B - \sum_{i=1}^{5} a_{iB} y_i$$
$$= 0.027 - 1(0.0294) - 1(0) - 0.6(0.00636) - 10(0.00244) = -0.0306.$$

The remaining basic and nonbasic variables also satisfy these optimality conditions for the simplex method. Therefore, the optimal dual variables must be the shadow prices associated with an optimal solution.

Since any linear program can be put in the form of (3) by making simple transformations similar to those used in this example, then any linear program must have a dual linear program. In fact, since the dual problem (4) is a linear program, it must also have a dual. For completeness, one would hope that the dual of the dual is the primal (3), which is indeed the case. To show this we need only change the dual (4) into the form of (3) and apply the definition of the dual problem. The dual may be reformulated as a maximization problem with less-than-or-equal-to constraints, as follows:

$$\text{Maximize } v' = \sum_{i=1}^{m} - b_i y_i,$$

subject to:

$$\sum_{i=1}^{m} - a_{ij} y_i \leqq - c_j \qquad (j = 1, 2, \ldots, n), \tag{5}$$

$$y_i \geqq 0 \qquad (i = 1, 2, \ldots, m).$$

Applying the definition of the dual problem and letting the dual variables be x_j, $j = 1, 2, \ldots, n$, we have

$$\text{Minimize } z' = \sum_{j=1}^{n} - c_j x_j,$$

subject to:

$$\sum_{j=1}^{n} - a_{ij} x_j \geqq - b_i \qquad (i = 1, 2, \ldots, m), \tag{6}$$

$$x_j \geqq 0 \qquad (j = 1, 2, \ldots, n),$$

which, by multiplying the constraints by minus one and converting the objective function to maximization, is clearly equivalent to the primal problem. Thus, the *dual of the dual is the primal.*

4.3 FINDING THE DUAL IN GENERAL

Very often linear programs are encountered in equality form with nonnegative variables. For example, the canonical form, which is used for computing a solution by the simplex method, is in equality form. It is of interest, then, to find the dual of the equality form:

$$\text{Maximize } z = \sum_{j=1}^{n} c_j x_j,$$

subject to:

$$\sum_{j=1}^{n} a_{ij} x_j = b_i \qquad (i = 1, 2, \ldots, m), \tag{7}$$

$$x_j \geq 0 \qquad (j = 1, 2, \ldots, n).$$

A problem in equality form can be transformed into inequality form by replacing each equation by two inequalities. Formulation (7) can be rewritten as

$$\text{Maximize } z = \sum_{j=1}^{n} c_j x_j,$$

subject to:

$$\left.\begin{aligned}\sum_{j=1}^{n} a_{ij} x_j &\leq b_i \\ \sum_{j=1}^{n} -a_{ij} x_j &\leq -b_i\end{aligned}\right\} \qquad (i = 1, 2, \ldots, m),$$

$$x_j \geq 0 \qquad (j = 1, 2, \ldots, n).$$

The dual of the equality form can then be found by applying the definition of the dual to this problem. Letting y_i^+ and y_i^- $(i = 1, 2, \ldots, m)$ be the dual variables associated with the first m and second m constraints, respectively, from expression (4), we find the dual to be:

$$\text{Minimize } v = \sum_{i=1}^{m} b_i y_i^+ + \sum_{i=1}^{m} - b_i y_i^-,$$

subject to:

$$\sum_{i=1}^{m} a_{ij} y_i^+ + \sum_{i=1}^{m} - a_{ij} y_i^- \geq c_j \qquad (j = 1, 2, \ldots, n),$$

$$y_i^+ \geq 0, \qquad y_i^- \geq 0 \qquad (i = 1, 2, \ldots, m).$$

Collecting terms, we have:

$$\text{Minimize } v = \sum_{i=1}^{m} b_i(y_i^+ - y_i^-),$$

subject to:

$$\sum_{i=1}^{m} a_{ij}(y_i^+ - y_i^-) \geqq c_j \qquad (j = 1, 2, \ldots, n),$$

$$y_i^+ \geqq 0, \qquad y_i^- \geqq 0 \qquad (i = 1, 2, \ldots, m).$$

Letting $y_i = y_i^+ - y_i^-$, and noting that y_i is unrestricted in sign, gives us the dual of the equality form (7):

$$\text{Minimize } v = \sum_{i=1}^{m} b_i y_i,$$

subject to:

$$\sum_{i=1}^{m} a_{ij} y_i \geqq c_j \qquad (j = 1, 2, \ldots, n), \tag{8}$$

$$y_i \text{ unrestricted} \qquad (i = 1, 2, \ldots, m).$$

Note that the dual variables associated with equality constraints are unrestricted.

There are a number of relationships between primal and dual, depending upon whether the primal problem is a maximization or a minimization problem and upon the types of constraints and restrictions on the variables. To illustrate some of these relationships, let us consider a general maximization problem as follows:

$$\text{Maximize } z = \sum_{j=1}^{n} c_j x_j,$$

subject to:

$$\sum_{j=1}^{n} a_{ij} x_j \leqq b_i \qquad (i = 1, 2, \ldots, m'),$$

$$\sum_{j=1}^{n} a_{ij} x_j \geqq b_i \qquad (i = m' + 1, m' + 2, \ldots, m''), \tag{9}$$

$$\sum_{j=1}^{n} a_{ij} x_j = b_i \qquad (i = m'' + 1, m'' + 2, \ldots, m),$$

$$x_j \geqq 0 \qquad (j = 1, 2, \ldots, n).$$

We can change the general primal problem to equality form by adding the appropriate slack and surplus variables, as follows:

$$\text{Maximize } z = \sum_{j=1}^{n} c_j x_j,$$

subject to:

$$\sum_{j=1}^{n} a_{ij} x_j + x_{n+i} = b_i \qquad (i = 1, 2, \ldots, m'),$$

$$\sum_{j=1}^{n} a_{ij} x_j - x_{n+i} = b_i \qquad (i = m' + 1, m' + 2, \ldots, m''),$$

$$\sum_{j=1}^{n} a_{ij} x_j = b_i \qquad (i = m'' + 1, m'' + 2, \ldots, m),$$

$$x_j \geq 0 \qquad (j = 1, 2, \ldots, n + m'').$$

Letting y_i, y_i', and y_i'' be dual variables associated respectively with the three sets of equations, the dual of (9) is then

$$\text{Minimize } v = \sum_{i=1}^{m'} b_i y_i + \sum_{i=m'+1}^{m''} b_i y_i' + \sum_{i=m''+1}^{m} b_i y_i'',$$

subject to:

$$\sum_{i=1}^{m'} a_{ij} y_i + \sum_{i=m'+1}^{m''} a_{ij} y_i' + \sum_{i=m''+1}^{m} a_{ij} y_i'' \geq c_j \qquad (10)$$

$$y_i \geq 0,$$

$$-y_i' \geq 0,$$

where y_i'' is unrestricted in sign and the last inequality could be written $y_i' \leq 0$. Thus, if the primal problem is a maximization problem, the dual variables associated with the less-than-or-equal-to constraints are nonnegative, the dual variables associated with the greater-than-or-equal-to constraints are nonpositive, and the dual variables associated with the equality constraints are unrestricted in sign.

These conventions reflect the interpretation of the dual variables as shadow prices of the primal problem. A less-than-or-equal-to constraint, normally representing a scarce resource, has a positive shadow price, since the expansion of that resource generates additional profits. On the other hand, a greater-than-or-equal-to constraint usually represents an external requirement (e.g., demand for a given commodity). If that requirement increases, the problem becomes more constrained; this produces a decrease in the objective function and thus the corresponding constraint has a negative shadow price. Finally, changes in the righthand side of an equality constraint might produce either negative or positive changes in the value of the objective function. This explains the unrestricted nature of the corresponding dual variable.

Let us now investigate the duality relationships when the primal problem is cast in *minimization*, rather than maximization form:

$$\text{Minimize } z = \sum_{j=1}^{n} c_j x_j,$$

subject to:

$$\sum_{j=1}^{n} a_{ij} x_j \leqq b_i \qquad (i = 1, 2, \ldots, m'),$$

$$\sum_{j=1}^{n} a_{ij} x_j \geqq b_i \qquad (i = m' + 1, m' + 2, \ldots, m''), \tag{11}$$

$$\sum_{j=1}^{n} a_{ij} x_j = b_i \qquad (i = m'' + 1, m'' + 2, \ldots, m),$$

$$x_j \geqq 0 \qquad (j = 1, 2, \ldots, n).$$

Since the dual of the dual is the primal, we know that the dual of a minimization problem will be a maximization problem. The dual of (11) can be found by performing transformations similar to those conducted previously. The resulting dual of (11) is the following maximization problem:

$$\text{Maximize } v = \sum_{i=1}^{m'} b_i y_i + \sum_{i=m'+1}^{m''} b_i y_i' + \sum_{i=m''+1}^{m} b_i y_i'',$$

subject to:

$$\sum_{i=1}^{m'} a_{ij} y_i + \sum_{i=m'+1}^{m''} a_{ij} y_i' + \sum_{i=m''+1}^{m} a_{ij} y_i'' \geqq c_j \tag{12}$$

$$y_i \leqq 0 \qquad (i = 1, 2, \ldots, m'),$$

$$y_i' \geqq 0 \qquad (i = m' + 1, m' + 2, \ldots, m'').$$

Observe that now the sign of the dual variables associated with the inequality constraints has changed, as might be expected from the shadow-price interpretation. In a cost-minimization problem, increasing the available resources will tend to decrease the total cost, since the constraint has been relaxed. As a result, the dual variable associated with a less-than-or-equal-to constraint in a minimization problem is nonpositive. On the other hand, increasing requirements could only generate a cost increase. Thus, a greater-than-or-equal-to constraint in a minimization problem has an associated nonnegative dual variable.

The primal and dual problems that we have just developed illustrate one further duality correspondence. If (12) is considered as the primal problem and (11) as its dual, then unrestricted variables in the primal are associated with equality constraints in the dual.

We now can summarize the general duality relationships. Basically we note that equality constraints in the primal correspond to unrestricted variables in the dual, while inequality constraints in the primal correspond to restricted variables in the dual, where the sign of the restriction in the dual depends upon the combination of objective-function type and constraint relation in the primal. These various correspondences are summarized in Table 4.1. The table is based on the assumption that the primal is a maximization problem. Since the dual of the dual is the primal, we can interchange the words primal and dual in Table 4.1 and the correspondences will still hold.

Table 4.1

Primal (Maximize)	Dual (Minimize)
ith constraint $\leq$	ith variable ≥ 0
ith constraint $\geq$	ith variable ≤ 0
ith constraint $=$	ith variable unrestricted
jth variable ≥ 0	jth constraint $\geq$
jth variable ≤ 0	jth constraint $\leq$
jth variable unrestricted	jth constraint $=$

4.4 THE FUNDAMENTAL DUALITY PROPERTIES

In the previous sections of this chapter we have illustrated many duality properties for linear programming. In this section we formalize some of the assertions we have already made. The reader not interested in the theory should skip over this section entirely, or merely read the statements of the properties and ignore the proofs. We consider the primal problem in inequality form so that the primal and dual problems are symmetric. Thus, any statement that is made about the primal problem immediately has an analog for the dual problem, and conversely. For convenience, we restate the primal and dual problems.

Primal

$$\text{Maximize } z = \sum_{j=1}^{n} c_j x_j,$$

subject to:

$$\sum_{j=1}^{n} a_{ij} x_j \leq b_i \qquad (i = 1, 2, \ldots, m), \tag{13}$$

$$x_j \geq 0 \qquad (j = 1, 2, \ldots, n).$$

Dual

$$\text{Minimize } v = \sum_{i=1}^{m} b_i y_i,$$

subject to:

$$\sum_{i=1}^{m} a_{ij} y_i \geq c_j \qquad (j = 1, 2, \ldots, n), \tag{14}$$

$$y_i \geq 0 \qquad (i = 1, 2, \ldots, m).$$

The first property is referred to as "weak duality" and provides a bound on the optimal value of the objective function of either the primal or the dual. Simply stated, the value of the objective function for any feasible solution to the primal maximization problem is bounded from above by the value of the objective function for any feasible solution to its dual. Similarly, the value of the objective function for its dual is bounded from below by the value of the objective function of the primal. Pictorially, we might represent the situation as follows:

Dual feasible	$\downarrow$ v decreasing
Primal feasible	$\uparrow$ z increasing

The sequence of properties to be developed will lead us to the "strong duality" property, which states that the optimal values of the primal and dual problems are in fact equal. Further, in developing this result, we show how the solution of one of these problems is readily available from the solution of the other.

Weak Duality Property. If $\bar{x}_j$, $j = 1, 2, \ldots, n$, is a feasible solution to the primal problem and $\bar{y}_i$, $i = 1, 2, \ldots, m$, is a feasible solution to the dual problem, then

$$\sum_{j=1}^{n} c_j \bar{x}_j \leq \sum_{i=1}^{m} b_i \bar{y}_i.$$

The weak duality property follows immediately from the respective feasibility of the two solutions. Primal feasibility implies:

$$\sum_{j=1}^{n} a_{ij} \bar{x}_j \leq b_i \quad (i = 1, 2, \ldots, m) \quad \text{and} \quad \bar{x}_j \geq 0 \quad (j = 1, 2, \ldots, n),$$

while dual feasibility implies

$$\sum_{j=1}^{n} a_{ij} \bar{y}_i \geq c_j \quad (j = 1, 2, \ldots, n) \quad \text{and} \quad \bar{y}_i \geq 0 \quad (i = 1, 2, \ldots, m).$$

Hence, multiplying the ith primal constraint by $\bar{y}_i$ and adding yields:

$$\sum_{i=1}^{m} \sum_{j=1}^{n} a_{ij} \bar{x}_j \bar{y}_i \leq \sum_{i=1}^{m} b_i \bar{y}_i,$$

while multiplying the jth dual constraint by $\bar{x}_j$ and adding yields:

$$\sum_{j=1}^{n} \sum_{i=1}^{m} a_{ij} \bar{y}_i \bar{x}_j \geq \sum_{j=1}^{n} c_j \bar{x}_j.$$

Since the lefthand sides of these two inequalities are equal, together they imply the desired result that

$$\sum_{j=1}^{n} c_j \bar{x}_j \leq \sum_{i=1}^{m} b_i \bar{y}_i.$$

There are a number of direct consequences of the weak duality property. If we have feasible solutions to the primal and dual problems such that their respective objective functions are equal, then these solutions are optimal to their respective problems. This result follows immediately from the weak duality property, since a dual feasible solution is an upper bound on the optimal primal solution and this bound is attained by the given feasible primal solution. The argument for the dual problem is analogous. Hence, we have an optimality property of dual linear programs.

Optimality Property. If $\hat{x}_j, j = 1, 2, \ldots, n$, is a feasible solution to the primal problem and $\hat{y}_i, i = 1, 2, \ldots, m$, is a feasible solution to the dual problem, and, further,

$$\sum_{j=1}^{n} c_j \hat{x}_j = \sum_{i=1}^{m} b_i \hat{y}_i,$$

then $\hat{x}_j, j = 1, 2, \ldots, n$, is an optimal solution to the primal problem and $\hat{y}_i$, $i = 1, 2, \ldots, m$, is an optimal solution to the dual problem.

Furthermore, if one problem has an unbounded solution, then the dual of that problem is infeasible. This must be true for the primal since any feasible solution to the dual would provide an upper bound on the primal objective function by the weak duality theorem; this contradicts the fact that the primal problem is unbounded. Again, the argument for the dual problem is analogous. Hence, we have an unboundedness property of dual linear programs.

Unboundedness Property. If the primal (dual) problem has an unbounded solution, then the dual (primal) problem is infeasible.

We are now in a position to give the main result of this section, the "strong duality" property. The importance of this property is that it indicates that we may in fact solve the dual problem in place of or in conjunction with the primal problem. The proof of this result depends merely on observing that the shadow prices determined by solving the primal problem by the simplex method give a dual feasible solution, satisfying the optimality property given above.

Strong Duality Property. If the primal (dual) problem has a finite optimal solution, then so does the dual (primal) problem, and these two values are equal. That is, $\hat{z} = \hat{v}$ where

$$\hat{z} = \text{Max} \sum_{j=1}^{n} c_j x_j, \qquad\qquad \hat{v} = \text{Min} \sum_{i=1}^{m} b_i y_i,$$

subject to: $\qquad\qquad\qquad\qquad\qquad$ subject to:

$$\sum_{j=1}^{n} a_{ij} x_j \leq b_i, \qquad\qquad\qquad \sum_{i=1}^{m} a_{ij} y_i \geq c_j,$$

$$x_j \geq 0; \qquad\qquad\qquad\qquad\qquad y_i \geq 0.$$

Let us see how to establish this property. We can convert the primal problem to the equivalent equality form by adding slack variables as follows:

$$\text{Maximize } z = \sum_{j=1}^{n} c_j x_j,$$

subject to:

$$\sum_{j=1}^{n} a_{ij} x_j + x_{n+i} = b_i \qquad (i = 1, 2, \ldots, m),$$

$$x_j \geq 0 \qquad (j = 1, 2, \ldots, n + m).$$

Suppose that we have applied the simplex method to the linear program and $\hat{x}_j$, $j = 1, 2, \ldots, n$, is the resulting optimal solution. Let $\hat{y}_i$, $i = 1, 2, \ldots, m$, be the shadow prices associated with the optimal solution. Recall that the shadow prices associated with the original constraints are the multiples of those constraints which, when subtracted from the original form of the objective function, yield the form of the objective function in the final tableau [Section 3.2, expression (11)]. Thus the following condition holds:

$$-z + \sum_{j=1}^{n} \overline{c}_j x_j = -\sum_{i=1}^{m} b_i \hat{y}_i, \tag{15}$$

where, due to the optimality criterion of the simplex method, the reduced costs satisfy:

$$\overline{c}_j = c_j - \sum_{i=1}^{m} a_{ij} \hat{y}_i \leq 0 \qquad (j = 1, 2, \ldots, n), \tag{16}$$

and

$$\overline{c}_j = 0 - \hat{y}_i \leq 0 \qquad (j = n + 1, n + 2, \ldots, n + m). \tag{17}$$

Conditions (16) and (17) imply that $\hat{y}_i$, for $i = 1, 2, \ldots, m$, constitutes a feasible solution to the dual problem. When x_j is replaced by the optimum value $\hat{x}_j$ in expression (15), the term

$$\sum_{j=1}^{n} \overline{c}_j \hat{x}_j$$

is equal to zero, since $\overline{c}_j = 0$ when $\hat{x}_j$ is basic, and $\hat{x}_j = 0$ when $\hat{x}_j$ is nonbasic. Therefore, the maximum value of z, say $\hat{z}$, is given by:

$$-\hat{z} = -\sum_{i=1}^{m} b_i \hat{y}_i.$$

Moreover, since $\hat{x}_j$, for $j = 1, 2, \ldots, n$, is an optimum solution to the primal problem,

$$\sum_{j=1}^{n} c_j \hat{x}_j = \hat{z} = \sum_{i=1}^{m} b_i \hat{y}_i.$$

This is the *optimality property* for the primal feasible solution $\hat{x}_j$, $j = 1, 2, \ldots, n$, and the dual feasible solution $\hat{y}_i$, $i = 1, 2, \ldots, m$, so they are optimal for their respective problems. (The argument in terms of the dual problem is analogous.)

It should be pointed out that it is *not* true that if the primal problem is infeasible, then the dual problem is unbounded. In this case the dual problem may be either unbounded or infeasible. All four combinations of feasibility and infeasibility for primal and dual problems may take place. An example of each is indicated in Table 4.2.

Table 4.2

1	*Primal feasible*	*Dual feasible*
	Maximize $z = 2x_1 + x_2$,	Minimize $v = 4y_1 + 2y_2$,
	subject to:	subject to:
	$\quad x_1 + x_2 \leqq 4,$	$\quad y_1 + y_2 \geqq 2,$
	$\quad x_1 - x_2 \leqq 2,$	$\quad y_1 - y_2 \geqq 1,$
	$\quad x_1 \geqq 0, \quad x_2 \geqq 0.$	$\quad y_1 \geqq 0, \quad y_2 \geqq 0.$
2	*Primal feasible and unbounded*	*Dual infeasible*
	Maximize $z = 2x_1 + x_2$,	Minimize $v = 4y_1 + 2y_2$,
	subject to:	subject to:
	$\quad x_1 - x_2 \leqq 4,$	$\quad y_1 + y_2 \geqq 2,$
	$\quad x_1 - x_2 \leqq 2,$	$\quad -y_1 - y_2 \geqq 1,$
	$\quad x_1 \geqq 0, \quad x_2 \geqq 0.$	$\quad y_1 \geqq 0, \quad y_2 \geqq 0.$
3	*Primal infeasible*	*Dual feasible and unbounded*
	Maximize $z = 2x_1 + x_2$,	Minimize $v = -4y_1 + 2y_2$,
	subject to:	subject to:
	$\quad -x_1 - x_2 \leqq -4,$	$\quad -y_1 + y_2 \geqq 2,$
	$\quad x_1 + x_2 \leqq 2,$	$\quad -y_1 + y_2 \geqq 1,$
	$\quad x_1 \geqq 0, \quad x_2 \geqq 0.$	$\quad y_1 \geqq 0, \quad y_2 \geqq 0.$
4	*Primal infeasible*	*Dual infeasible*
	Maximize $z = 2x_1 + x_2$,	Minimize $v = -4y_1 + 2y_2$,
	subject to:	subject to:
	$\quad -x_1 + x_2 \leqq -4,$	$\quad -y_1 + y_2 \geqq 2,$
	$\quad x_1 - x_2 \leqq 2,$	$\quad y_1 - y_2 \geqq 1,$
	$\quad x_1 \geqq 0, \quad x_2 \geqq 0.$	$\quad y_1 \geqq 0, \quad y_2 \geqq 0.$

In example (2) of Table 4.2 it should be clear that x_2 may be increased indefinitely without violating feasibility while at the same time making the objective function arbitrarily large. The constraints of the dual problem for this example are clearly infeasible since $y_1 + y_2 \geq 2$ and $y_1 + y_2 \leq -1$ cannot simultaneously hold. A similar observation is made for example (3), except that the primal problem is now infeasible while the dual variable y_1 may be increased indefinitely. In example (4), the fact that neither primal nor dual problem is feasible can be checked easily by multiplying the first constraint of each by minus one.

4.5 COMPLEMENTARY SLACKNESS

We have remarked that the duality theory developed in the previous section is a unifying theory relating the optimal solution of a linear program to the optimal solution of the dual linear program, which involves the shadow prices of the primal as decision variables. In this section we make this relationship more precise by defining the concept of complementary slackness relating the two problems.

    Complementary Slackness Property. If, in an optimal solution of a linear program, the value of the dual variable (shadow price) associated with a constraint is nonzero, then that constraint must be satisfied with equality. Further, if a constraint is satisfied with strict inequality, then its corresponding dual variable must be zero.

    For the primal linear program posed as a maximization problem with less-than-or-equal-to constraints, this means:

i) $\text{if } \hat{y}_i > 0, \quad \text{then } \sum_{j=1}^{n} a_{ij}\hat{x}_j = b_i;$

ii) $\text{if } \sum_{j=1}^{n} a_{ij}\hat{x}_j < b_i, \quad \text{then } \hat{y}_i = 0.$

We can show that the complementary-slackness conditions follow directly from the strong duality property just presented. Recall that, in demonstrating the weak duality property, we used the fact that:

$$\sum_{j=1}^{n} c_j\hat{x}_j \leq \sum_{i=1}^{m}\sum_{j=1}^{n} a_{ij}\hat{x}_j\hat{y}_i \leq \sum_{i=1}^{m} b_i\hat{y}_i \tag{18}$$

for any $\hat{x}_j$, $j = 1, 2, \ldots, n$, and $\hat{y}_i$, $i = 1, 2, \ldots, m$, feasible to the primal and dual problems, respectively. Now, since these solutions are not only feasible but optimal to these problems, equality must hold throughout. Hence, considering the righthand relationship in (18), we have:

$$\sum_{i=1}^{m}\sum_{j=1}^{n} a_{ij}\hat{x}_j\hat{y}_i = \sum_{i=1}^{m} b_i\hat{y}_i,$$

which implies:

$$\sum_{i=1}^{m}\left[\sum_{j=1}^{n}a_{ij}\hat{x}_j - b_i\right]\hat{y}_i = 0.$$

Since the dual variables $\hat{y}_i$ are nonnegative and their coefficients

$$\sum_{j=1}^{n}a_{ij}\hat{x}_j - b_i$$

are nonpositive by primal feasibility, this condition can hold only if each of its terms is equal to zero; that is,

$$\left[\sum_{j=1}^{n}a_{ij}\hat{x}_j - b_i\right]\hat{y}_i = 0 \qquad (i = 1, 2, \ldots, m).$$

These latter conditions are clearly equivalent to (i) and (ii) above.

For the dual linear program posed as a minimization problem with greater-than-or-equal-to constraints, the complementary-slackness conditions are the following:

iii) $$\qquad\qquad \text{if } \hat{x}_j > 0, \quad \text{then } \sum_{i=1}^{m}a_{ij}y_i = c_j,$$

iv) $$\qquad\qquad \text{if } \sum_{i=1}^{m}a_{ij}\hat{y}_i > c_j, \quad \text{then } \hat{x}_j = 0.$$

These conditions also follow directly from the strong duality property by an argument similar to that given above. By considering the lefthand relationship in (18), we can easily show that

$$\left[\sum_{i=1}^{m}a_{ij}y_i - c_j\right]x_j = 0 \qquad (j = 1, 2, \ldots, n),$$

which is equivalent to (iii) and (iv).

The complementary-slackness conditions of the primal problem have a fundamental economic interpretation. If the shadow price of the ith resource (constraint) is strictly positive in the optimal solution $\hat{y}_i > 0$, then we should require that all of this resource be consumed by the optimal program; that is,

$$\sum_{j=1}^{n}a_{ij}\hat{x}_j = b_i.$$

If, on the other hand, the ith resource is not fully used; that is,

$$\sum_{j=1}^{n}a_{ij}\hat{x}_j < b_i,$$

then its shadow price should be zero, $\hat{y}_i = 0$.

The complementary-slackness conditions of the dual problem are merely the optimality conditions for the simplex method, where the reduced cost $\bar{c}_j$ associated with any variable must be nonpositive and is given by

$$\bar{c}_j = c_j - \sum_{i=1}^{m} a_{ij}\hat{y}_i \leq 0 \qquad (j = 1, 2, \ldots, n).$$

If $\hat{x}_j > 0$, then $\hat{x}_j$ must be a basic variable and its reduced cost is defined to be zero. Thus,

$$c_j = \sum_{i=1}^{m} a_{ij}\hat{y}_i.$$

If, on the other hand,

$$c_j - \sum_{i=1}^{m} a_{ij}\hat{y}_i < 0,$$

then $\hat{x}_j$ must be nonbasic and set equal to zero in the optimal solution; $\hat{x}_j = 0$.

We have shown that the strong duality property implies that the complementary-slackness conditions must hold for both the primal and dual problems. The converse of this also is true. If the complementary-slackness conditions hold for both problems, then the strong duality property holds. To see this, let $\hat{x}_j$, $j = 1, 2, \ldots, n$, and $\hat{y}_i$, $i = 1, 2, \ldots, m$, be feasible solutions to the primal and dual problems, respectively. The complementary-slackness conditions (i) and (ii) for the primal problem imply:

$$\sum_{i=1}^{m} \left[\sum_{j=1}^{n} a_{ij}\hat{x}_j - b_i \right] \hat{y}_i = 0,$$

while the complementary-slackness conditions (iii) and (iv) for the dual problem imply:

$$\sum_{j=1}^{n} \left[\sum_{i=1}^{m} a_{ij}\hat{y}_i - c_j \right] \hat{x}_j = 0.$$

These two equations together imply:

$$\sum_{j=1}^{n} c_j\hat{x}_j = \sum_{i=1}^{m}\sum_{j=1}^{n} a_{ij}\hat{x}_j\hat{y}_i = \sum_{i=1}^{m} b_i\hat{y}_i,$$

and hence the values of the primal and dual objective functions are equal. Since these solutions are feasible to the primal and dual problems respectively, the *optimality property* implies that these solutions are optimal to the primal and dual problems. We have, in essence, shown that the complementary-slackness conditions holding for both the primal and dual problems is equivalent to the *strong duality property*. For this reason, the complementary-slackness conditions are often referred to as the *optimality conditions*.

> ***Optimality Conditions.*** If $\hat{x}_j$, $j = 1, 2, \ldots, n$, and $\hat{y}_i$, $i = 1, 2, \ldots, m$, are feasible solutions to the primal and dual problems, respectively, then they are optimal solutions to these problems if, and only if, the complementary-slackness conditions hold for both the primal and the dual problems.

4.6 THE DUAL SIMPLEX METHOD

One of the most important impacts of the general duality theory presented in the previous section has been on computational procedures for linear programming. First, we have established that the dual can be solved in place of the primal whenever there are advantages to doing so. For example, if the number of constraints of a problem is much greater than the number of variables, it is usually wise to solve the dual instead of the primal since the solution time increases much more rapidly with the number of constraints in the problem than with the number of variables. Second, new algorithms have been developed that take advantage of the duality theory in more subtle ways. In this section we present the dual simplex method. The steps of the algorithm result from applying the primal simplex method directly to the dual problem; however, the algorithm is carried out in terms of the ordinary primal tableau. We have already seen the essence of the dual simplex method in Section 3.5, on righthand-side ranging, where the variable transitions at the boundaries of each range are essentially computed by the dual simplex method. Further, in Section 3.8, on parametric programming of the righthand side, the dual simplex method is the cornerstone of the computational approach. In this section we formalize the general algorithm.

Recall the canonical form employed in the simplex method:

$$
\begin{aligned}
x_1 \quad\quad\quad & + \bar{a}_{1,m+1}x_{m+1} + \cdots + \bar{a}_{1,n}x_n = \bar{b}_1, \\
x_2 \quad\quad & + \bar{a}_{2,m+1}x_{m+1} + \cdots + \bar{a}_{2,n}x_n = \bar{b}_2, \\
& \quad\vdots \quad\quad\quad\quad\quad\quad \vdots \quad\quad\quad \vdots \\
x_m + \bar{a}_{m,m+1}x_{m+1} & + \cdots + \bar{a}_{m,n}x_n = \bar{b}_m, \\
(-z) \quad\quad & + \bar{c}_{m+1}x_{m+1} + \cdots + \bar{c}_n x_n = -\bar{z}_0.
\end{aligned}
$$

The conditions for $x_1, x_2, \ldots, x_m$ to constitute an optimal basis for a maximization problem are:

i) $\qquad\qquad\qquad\qquad \bar{c}_j \leq 0 \qquad (j = 1, 2, \ldots, n),$

ii) $\qquad\qquad\qquad\qquad \bar{b}_i \geq 0 \qquad (i = 1, 2, \ldots, m).$

We could refer to condition (i) as primal optimality (or equivalently, dual feasibility) and condition (ii) as primal feasibility. In the primal simplex method, we move from basic feasible solution to adjacent basic feasible solution, increasing (not decreasing) the objective function at each iteration. Termination occurs when the primal optimality conditions are satisfied. Alternatively, we could maintain primal optimality (dual feasibility) by imposing (i) and terminating when the primal feasibility conditions (ii) are satisfied. This latter procedure is referred to as the *dual simplex method* and in fact results from applying the simplex method to the dual problem. In Chapter 3, on sensitivity analysis, we gave a preview of the dual simplex method when we determined the variable to leave the basis at the boundary of a righthand-side range. In fact, the dual simplex method is most useful in applications when a problem has been solved and a subsequent change on the righthand side makes the optimal solution no longer primal feasible, as in the case of parametric programming of the righthand-side values.

The rules of the dual simplex method are identical to those of the primal simplex algorithm, except for the selection of the variable to leave and enter the basis. At each iteration of the dual simplex method, we require that:

$$\bar{c}_j = c_j - \sum_{i=1}^{m} y_i a_{ij} \leqq 0;$$

and since $y_i \geqq 0$ for $i = 1, 2, \ldots, m$, these variables are a dual feasible solution. Further, at each iteration of the dual simplex method, the most negative $\bar{b}_i$ is chosen to determine the pivot row, corresponding to choosing the most positive $\bar{c}_j$ to determine the pivot column in the primal simplex method.

Prior to giving the formal rules of the dual simplex method, we present a simple example to illustrate the essence of the procedure. Consider the following maximization problem with nonnegative variables given in Tableau 3. This problem is in "dual canonical form" since the optimality conditions are satisfied, but the basic variables are not yet nonnegative.

Tableau 3

Basic variables	Current values	x_1	x_2	x_3	x_4
x_3	-1	-1	-1	1	
x_4	-2	-2	-3		1
$(-z)$	0	-3	-1		

In the dual simplex algorithm, we are attempting to make all variables nonnegative. The procedure is the opposite of the primal method in that it first selects the variable to drop from the basis and then the new variable to introduce into the basis in its place. The variable to drop is the basic variable associated with the constraint with the most negative righthand-side value; in this case x_4. Next we have to determine the entering variable. We select from only those nonbasic variables that have a negative coefficient in the pivot row, since then, after pivoting, the righthand side becomes positive, thus eliminating the primal infeasibility in the constraint. If all coefficients are positive, then the primal problem is clearly infeasible, because the sum of nonnegative terms can never equal the negative righthand side.

In this instance, both nonbasic variables x_1 and x_2 are candidates. Pivoting will subtract some multiple, say t, of the pivot row containing x_4 from the objective function, to give:

$$(-3 + 2t)x_1 + (-1 + 3t)x_2 - tx_4 - z = 2t.$$

Since we wish to maintain the optimality conditions, we want the coefficient of each variable to remain nonpositive, so that:

$$-3 + 2t \leqq 0 \quad \text{(that is, } t \leqq \tfrac{3}{2}),$$
$$-1 + 3t \leqq 0 \quad \text{(that is, } t \leqq \tfrac{1}{3}),$$
$$- \ t \leqq 0 \quad \text{(that is, } t \geqq 0).$$

Setting $t = \frac{1}{3}$ preserves the optimality conditions and identifies the new basic variable with a zero coefficient in the objective function, as required to maintain the dual canonical form.

These steps can be summarized as follows: the variable to leave the basis is chosen by:

$$\bar{b}_r = \underset{i}{\text{Min}} \{\bar{b}_i\} = \text{Min} \{-1, -2\} = \bar{b}_2.$$

The variable to enter the basis is determined by the dual ratio test:

$$\frac{\bar{c}_s}{\bar{a}_{rs}} = \underset{j}{\text{Min}} \left\{ \frac{\bar{c}_j}{|\bar{a}_{rj}|} \Big| \bar{a}_{rj} < 0 \right\} = \text{Min} \left\{ \frac{-3}{-2}, \frac{-1}{-3} \right\} = \frac{\bar{c}_2}{\bar{a}_{22}}.$$

Pivoting in x_2 in the second constraint gives the new canonical form in Tableau 4.

Tableau 4

Basic variables	Current values	x_1	x_2	x_3	x_4
x_3	$-\frac{1}{3}$	$-\frac{1}{3}$		1	$-\frac{1}{3}$
x_2	$\frac{2}{3}$	$\frac{2}{3}$	1		$-\frac{1}{3}$
$(-z)$	$\frac{2}{3}$	$-\frac{7}{3}$			$-\frac{1}{3}$

Clearly, x_3 is the leaving variable since $\bar{b}_1 = -\frac{1}{3}$ is the only negative righthand-side coefficient; and x_4 is the entering variable since $\bar{c}_+/\bar{a}_{14}$ is the minimum dual ratio in the first row. After pivoting, the new canonical form is given in Tableau 5

Tableau 5

Basic variables	Current values	x_1	x_2	x_3	x_4
x_4	1	1		-3	1
x_2	1	1	1	-1	
$(-z)$	1	-2		-1	

Since $\bar{b}_i \geq 0$ for $i = 1, 2$, and $\bar{c}_j \leq 0$ for $j = 1, 2, \dots, 4$, we have the optimal solution.

Following is a formal statement of the procedure. The proof of it is analogous to that of the primal simplex method and is omitted.

Dual Simplex Algorithm

STEP (0) The problem is initially in canonical form and all $\bar{c}_j \leq 0$.

STEP (1) If $\bar{b}_i \geq 0$, $i = 1, 2, \dots, m$, then *stop*, we are optimal. If we continue, then there exists some $\bar{b}_i < 0$.

STEP (2) Choose the row to pivot in (i.e., variable to drop from the basis) by:

$$\bar{b}_r = \underset{i}{\text{Min}} \{\bar{b}_i | \bar{b}_i < 0\}.$$

If $\bar{a}_{rj} \geqq 0$, $j = 1, 2, \ldots, n$, then *stop*; the primal problem is infeasible (dual unbounded). If we continue, then there exists $\bar{a}_{rj} < 0$ for some $j = 1, 2, \ldots, n$.

STEP (3) Choose column s to enter the basis by:

$$\frac{\bar{c}_s}{\bar{a}_{rs}} = \operatorname*{Min}_{j} \left\{ \frac{\bar{c}_j}{\bar{a}_{rj}} \middle| \bar{a}_{rj} < 0 \right\}.$$

STEP (4) Replace the basic variable in row r with variable s and reestablish the canonical form (i.e., pivot on the coefficient $\bar{a}_{rs}$).

STEP (5) Go to Step (1).

Step (3) is designed to maintain $\bar{c}_j \leqq 0$ at each iteration, and Step (2) finds the most promising candidate for an improvement in feasibility.

4.7 PRIMAL-DUAL ALGORITHMS

There are many other variants of the simplex method. Thus far we have discussed only the primal and dual methods. There are obvious generalizations that combine these two methods. Algorithms that perform both primal and dual steps are referred to as primal-dual algorithms and there are a number of such algorithms. We present here one simple algorithm of this form called the *parametric primal-dual*. It is most easily discussed in an example.

$$x_1 \geqq 0, \qquad x_2 \geqq 0,$$
$$x_1 + x_2 \leqq 6,$$
$$-x_1 + 2x_2 \leqq -\tfrac{1}{2},$$
$$x_1 - 3x_2 \leqq -1,$$
$$-2x_1 + 3x_2 = z(\text{max}).$$

The above example can easily be put in canonical form by addition of slack variables. However, neither primal feasibility nor primal optimality conditions will be satisfied. We will arbitrarily consider the above example as a function of the parameter θ in Tableau 6.

Tableau 6

Basic variables	Current values	x_1	x_2	x_3	x_4	x_5
x_3	6	1	1	1		
x_4	$-\tfrac{1}{2} + \theta$	-1	2		1	
x_5	$-1 + \theta$	1	-3			1
$(-z)$	0	-2	$(3 - \theta)$			

Clearly, if we choose θ large enough, this system of equations satisfies the primal feasibility and primal optimality conditions. The idea of the parametric primal-dual algorithm is to choose θ large initially so that these conditions are satisfied, and attempt to reduce θ to zero through a sequence of pivot operations. If we start with $\theta = 4$ and let θ approach zero, primal optimality is violated when $\theta < 3$. If we were to reduce θ below 3, the objective-function coefficient of x_2 would become positive. Therefore we perform a primal simplex pivot by introducing x_2 into the basis. We determine the variable to leave the basis by the minimum-ratio rule of the primal simplex method:

$$\frac{\bar{b}_r}{\bar{a}_{rs}} = \underset{i}{\text{Min}} \left\{ \frac{\bar{b}_i}{\bar{a}_{is}} \,\middle|\, \bar{a}_{is} > 0 \right\} = \left\{ \frac{6}{1}, \frac{2\frac{1}{2}}{2} \right\} = \frac{\bar{b}_2}{\bar{a}_{22}}.$$

x_4 leaves the basis and the new canonical form is then shown in Tableau 7.

Tableau 7

Basic variables	Current values	x_1	x_2	x_3	x_4	x_5
x_3	$6\frac{1}{4} - \frac{1}{2}\theta$	$\frac{3}{2}$		1	$-\frac{1}{2}$	
x_2	$-\frac{1}{4} + \frac{1}{2}\theta$	$-\frac{1}{2}$	1		$\frac{1}{2}$	
x_5	$-\frac{7}{4} + \frac{5}{2}\theta$	$-\frac{1}{2}$			$\frac{3}{2}$	1
$(-z)$	$-(3 - \theta)(-\frac{1}{4} + \frac{1}{2}\theta)$	$(-\frac{1}{2} - \frac{1}{2}\theta)$			$(-\frac{3}{2} + \frac{1}{2}\theta)$	

The optimality conditions are satisfied for $\frac{7}{10} \leq \theta \leq 3$. If we were to reduce θ below $\frac{7}{10}$, the righthand-side value of the third constraint would become negative. Therefore, we perform a dual simplex pivot by dropping x_5 from the basis. We determine the variable to enter the basis by the rules of the dual simplex method:

$$\underset{j}{\text{Min}} \left\{ \frac{\bar{c}_j}{\bar{a}_{3j}} \,\middle|\, \bar{a}_{3j} < 0 \right\} = \left\{ \frac{-\frac{1}{2} + -\frac{1}{2}\theta}{-\frac{1}{2}} \right\} = \frac{\bar{c}_1}{\bar{a}_{31}}.$$

After the pivot is performed the new canonical form is given in Tableau 8.

Tableau 8

Basic variables	Current values	x_1	x_2	x_3	x_4	x_5
x_3	$1 + 7\theta$			1	4	3
x_2	$\frac{3}{2} - 2\theta$		1		-1	-1
x_1	$\frac{7}{2} - 5\theta$	1			-3	-2
$(-z)$	$-(3 - \theta)(-\frac{1}{4} + \frac{1}{2}\theta)$ $+ (\frac{1}{2} + \frac{1}{2}\theta)(\frac{7}{2} - 5\theta)$				$(-3 - \theta)$	$(-1 - \theta)$

As we continue to decrease θ to zero, the optimality conditions remain satisfied. Thus the optimal final tableau for this example is given by setting θ equal to zero.

Primal-dual algorithms are useful when simultaneous changes of both righthand-side and cost coefficients are imposed on a previous optimal solution, and a new optimal solution is to be recovered. When parametric programming of both the objective-function coefficients and the righthand-side values is performed simultaneously, a variation of this algorithm is used. This type of parametric programming is usually referred to as the *rim problem*. (See Appendix B for further details.)

4.8 MATHEMATICAL ECONOMICS

As we have seen in the two previous sections, duality theory is important for developing computational procedures that take advantage of the relationships between the primal and dual problems. However, there is another less obvious area that also has been heavily influenced by duality, and that is mathematical economics.

In the beginning of this chapter, we gave a preview of duality that interpreted the dual variables as shadow prices imputed to the firm's resources by allocating the profit (contribution) of the firm to its resources at the margin. In a perfectly competitive economy, it is assumed that, if a firm could make profits in excess of the value of its resources, then some other firm would enter the market with a lower price, thus tending to eliminate these excess profits. The duality theory of linear programming has had a significant impact on mathematical economics through the interpretation of the dual as the price-setting mechanism in a perfectly competitive economy. In this section we will briefly sketch this idea.

Suppose that a firm may engage in any n production activities that consume and/or produce m resources in the process. Let $x_j \geq 0$ be the level at which the jth activity is operated, and let c_j be the revenue per unit (minus means cost) generated from engaging in the jth activity. Further, let a_{ij} be the amount of the ith resource consumed (minus means produced) per unit level of operation of the jth activity. Assume that the firm starts with a position of b_i units of the ith resource and may buy or sell this resource at a price $y_i \geq 0$ determined by an external market. Since the firm generates revenues and incurs costs by engaging in production activities and by buying and selling resources, its profit is given by:

$$\sum_{j=1}^{n} c_j x_j + \sum_{i=1}^{m} y_i \left(b_i - \sum_{j=1}^{n} a_{ij} x_j \right), \tag{19}$$

where the second term includes revenues from selling excess resources and costs of buying additional resources.

Note that if $b_i > \sum_{j=1}^{n} a_{ij} x_j$, the firm sells $b_i - \sum_{j=1}^{n} a_{ij} x_j$ units of resource i to the marketplace at a price y_i. If, however, $b_i < \sum_{j=1}^{n} a_{ij} x_j$, then the firm buys $\sum_{j=1}^{n} a_{ij} x_j - b_i$ units of resource i from the marketplace at a price y_i.

Now assume that the market mechanism for setting prices is such that it tends to minimize the profits of the firm, since these profits are construed to be at the expense of someone else in the market. That is, given x_j for $j = 1, 2, \ldots, n$, the market reacts to minimize (19). Two consequences immediately follow. First, consuming any resource that needs to be purchased from the marketplace, that is,

$$b_i - \sum_{j=1}^{n} a_{ij}x_j < 0,$$

is clearly uneconomical for the firm, since the market will tend to set the price of the resource arbitrarily high so as to make the firm's profits arbitrarily small (i.e., the firm's losses arbitrarily large). Consequently, our imaginary firm will always choose its production activities such that:

$$\sum_{j=1}^{n} a_{ij}x_j \leqq b_i \qquad (i = 1, 2, \dots, m).$$

Second, if any resource were not completely consumed by the firm in its own production and therefore became available for sale to the market, that is,

$$b_i - \sum_{j=1}^{n} a_{ij}x_j > 0,$$

this "malevolent market" would set a price of zero for that resource in order to minimize the firm's profit. Therefore, the second term of (19) will be zero and the firm's decision problem of choosing production activities so as to maximize profits simply reduces to:

$$\text{Maximize } \sum_{j=1}^{n} c_j x_j,$$

subject to:

$$\sum_{j=1}^{n} a_{ij}x_j \leqq b_i \qquad (i = 1, 2, \dots, m), \tag{20}$$

$$x_j \geqq 0 \qquad (j = 1, 2, \dots, n),$$

the usual primal linear program.

Now let us look at the problem from the standpoint of the market. Rearranging (19) as follows:

$$\sum_{j=1}^{n} \left(c_j - \sum_{i=1}^{m} a_{ij}y_i \right) x_j + \sum_{i=1}^{m} b_i y_i, \tag{21}$$

we can more readily see the impact of the firm's decision on the market. Note that the term $\sum_{i=1}^{m} a_{ij}y_i$ is the market opportunity cost for the firm using the resources $a_{1j}, a_{2j}, \dots, a_{mj}$, in order to engage in the jth activity at unit level. Again, two consequences immediately follow. First, if the market sets the prices so that the revenue from engaging in an activity exceeds the market cost, that is,

$$c_j - \sum_{i=1}^{m} a_{ij}y_i > 0,$$

then the firm would be able to make arbitrarily large profits by engaging in the activity at an arbitrarily high level, a clearly unacceptable situation from the standpoint of the market. The market instead will always choose to set its prices such that:

$$\sum_{i=1}^{m} a_{ij} y_i \geq c_j \qquad (j = 1, 2, \ldots, n).$$

Second, if the market sets the price of a resource so that the revenue from engaging in that activity does not exceed the potential revenue from the sale of the resources directly to the market, that is,

$$c_j - \sum_{i=1}^{m} a_{ij} y_i < 0,$$

then the firm will not engage in that activity at all. In this latter case, the opportunity cost associated with engaging in the activity is in excess of the revenue produced by engaging in the activity. Hence, the first term of (21) will always be zero, and the market's "decision" problem of choosing the prices for the resources so as to minimize the firm's profit reduces to:

$$\text{Minimize} \sum_{i=1}^{m} b_i y_i,$$

subject to:

$$\sum_{i=1}^{m} a_{ij} y_i \geq c_j \qquad (j = 1, 2, \ldots, n), \tag{22}$$

$$y_i \geq 0 \qquad (i = 1, 2, \ldots, m).$$

This linear program (22) is the *dual* of (20).

The questions that then naturally arise are "When do these problems have solutions?" and "What is the relationship between these solutions?" In arriving at the firm's decision problem and the market's "decision" problem, we assumed that the firm and the market would interact in such a manner that an equilibrium would be arrived at, satisfying:

$$\left(b_i - \sum_{j=1}^{n} a_{ij} \hat{x}_j \right) \hat{y}_i = 0 \qquad (i = 1, 2, \ldots, m),$$

$$\left(c_j - \sum_{i=1}^{m} a_{ij} \hat{y}_i \right) \hat{x}_j = 0 \qquad (j = 1, 2, \ldots, n). \tag{23}$$

These equations are just the *complementary-slackness conditions* of linear programming. The first condition implies that either the *amount* of resource i that is unused (slack in the ith constraint of the primal) is zero, or the *price* of resource i is zero. This is intuitively appealing since, if a firm has excess of a particular resource, then the market should not be willing to pay anything for the surplus of that resource since the market wishes to minimize the firm's profit. There may be a nonzero market price on a resource only if the firm is consuming all of that resource that is available. The second condition implies that either the amount of excess profit on the jth activity (slack in the jth constraint of the dual) is zero or the level of activity j is zero. This is also appealing from the standpoint of the perfectly competitive market, which acts to eliminate any excess profits.

If we had an equilibrium satisfying (23), then, by equating (19) and (21) for this equilibrium, we can quickly conclude that the extreme values of the primal and dual problems are equal; that is,

$$\sum_{j=1}^{n} c_j \hat{x}_j = \sum_{i=1}^{m} b_i \hat{y}_i.$$

Observe that this condition has the usual interpretation for a firm operating in a perfectly competitive market. It states that the maximum profit that the firm can make equals the market evaluation of its initial endowment of resources. That is, the firm makes no excess profits. The important step is to answer the question of when such equilibrium solutions exist. As we have seen, if the primal (dual) has a finite optimal solution, then so does the dual (primal), and the optimal values of these objective functions are equal. This result is just the strong duality property of linear programming.

4.9 GAME THEORY

The example of the perfectly competitive economy given in the previous section appears to be a game of some sort between the firm and the malevolent market. The firm chooses its strategy to maximize its profits while the market behaves ("chooses" its strategy) in such a way as to minimize the firm's profits. Duality theory is, in fact, closely related to game theory, and in this section we indicate the basic relationships.

In many contexts, a decision-maker does not operate in isolation, but rather must contend with other decision-makers with conflicting objectives. Consider, for example, advertising in a competitive market, portfolio selection in the presence of other investors, or almost any public-sector problem with its multifaceted implications. Each instance contrasts sharply with the optimization models previously discussed, since they concern a single decision-maker, be it an individual, a company, a government, or, in general, any group acting with a common objective and common constraints.

Game theory is one approach for dealing with these "multiperson" decision problems. It views the decision-making problem as a *game* in which each decision-maker, or *player*, chooses a *strategy* or an action to be taken. When all players have selected a strategy, each individual player receives a *payoff*. As an example, consider the advertising strategies of a two-firm market. There are two players, firm R (row player) and firm C (column player). The alternatives open to each firm are its advertising possibilities; payoffs are market shares resulting from the combined advertising selections of both firms. The payoff table in Tableau 9 summarizes the situation.

Tableau 9   Market Share of Firm R

Firm R alternatives \ Firm C alternatives	Advertising campaign 1	Advertising campaign 2	Advertising campaign 3
Advertising campaign 1	30%	40%	60%
Advertising campaign 2	20%	10%	30%

Since we have assumed a two-firm market, firm R and firm C share the market, and firm C receives whatever share of the market R does not. Consequently, firm R would like to maximize the payoff entry from the table and firm B would like to minimize this payoff. Games with this structure are called *two-person, zero-sum games.* They are *zero-sum*, since the gain of one player is the loss of the other player.

To analyze the game we must make some behavioral assumptions as to how the players will act. Let us suppose, in this example, that both players are conservative, in the sense that they wish to assure themselves of their possible payoff level regardless of the strategy adopted by their opponent. In selecting its alternative, firm R chooses a row in the payoff table. The worst that can happen from its viewpoint is for firm C to select the minimum column entry in that row. If firm R selects its first alternative, then it can be assured of securing 30% of the market, but no more, whereas if it selects its second alternative it is assured of securing 10%, but no more. Of course, firm R, wishing to secure as much market share as possible, will select alternative 1, to obtain the maximum of these security levels. Consequently, it selects the alternative giving the maximum of the column minimum, known as a *maximin* strategy. Similarly, firm C's security levels are given by the maximum row entries; if it selects alternative 1, it can be assured of losing only 30% of the market, and no more, and so forth. In this way, firm C is led to a *minimax* strategy of selecting the alternative that minimizes its security levels of maximum row entries (see Tableau 10).

Tableau 10   Market Share of Firm R

Firm R \ Firm C	Alternative 1	Alternative 2	Alternative 3	Security level for firm R
Alternative 1	30	40	60	30 ⎫ maximin = 30
Alternative 2	20	10	30	10 ⎭
Security level for firm C	30	40	60	

minimax = 30

For the problem at hand, the maximin and minimax are both 30 and we say that the problem has a *saddlepoint.* We might very well expect both players to select the alternatives that lead to this common value—both selecting alternative 1. Observe that, by the way we arrived at this value, the saddlepoint is an *equilibrium* solution in the sense that neither player will move unilaterally from this point. For instance, if firm R adheres to alternative 1, then firm C cannot improve its position by moving to either alternative 2 or 3 since then firm R's market share increases to either 40%

or 60%. Similarly, if firm C adheres to alternative 1, then firm R as well will not be induced to move from its saddlepoint alternative, since its market share drops to 20% if it selects alternative 2.

The situation changes dramatically if we alter a single entry in the payoff table (see Tableau 11).

Tableau 11   Market Share of Firm R

Firm R \ Firm C	Alternative 1	Alternative 2	Alternative 3	Security level for firm R
Alternative 1	30 ←	40 ↑	60	30 ⎱ maximin = 30
Alternative 2	↓ 60 →	10	30	10 ⎰
Security level for firm C	60	40	60	

minimax = 40

Now the security levels for the two players do not agree, and moreover, given any choice of decisions by the two firms, one of them can always improve its position by changing its strategy. If, for instance, both firms choose alternative 1, then firm R increases its market share from 30% to 60% by switching to alternative 2. After this switch though, it is attractive for firm C then to switch to its second alternative, so that firm R's market share decreases to 10%. Similar movements will continue to take place as indicated by the arrows in the table, and no single choice of alternatives by the players will be "stable."

Is there any way for the firms to do better in using *randomized strategies*? That is, instead of choosing a strategy outright, a player selects one according to some preassigned probabilities. For example, suppose that firm C selects among its alternatives with probabilities x_1, x_2, and x_3, respectively. Then the expected market share of firm R is

$$30x_1 + 40x_2 + 60x_3 \qquad \text{if firm R selects alternative 1,}$$

or

$$60x_1 + 10x_2 + 30x_3 \qquad \text{if firm R selects alternative 2.}$$

Since any gain in market share by firm R is a loss to firm C, firm C wants to make the expected market share of firm R as small as possible, i.e., maximize its own expected market share. Firm C can minimize the maximum expected market share of firm R by solving the following linear program.

Minimize v,

subject to:

$$
\begin{aligned}
30x_1 + 40x_2 + 60x_3 - v &\leq 0, \\
60x_1 + 10x_2 + 30x_3 - v &\leq 0, \\
x_1 + x_2 + x_3 &= 1, \\
x_1 \geq 0, \quad x_2 \geq 0, \quad x_3 &\geq 0.
\end{aligned}
\qquad (24)
$$

The first two constraints limit firm R's expected market share to be less than or equal to v for each of firm R's pure strategies. By minimizing v, firm C limits the expected market share of firm R as much as possible. The third constraint simply states that the chosen probabilities must sum to one. The solution to this linear program is $x_1 = \frac{1}{2}$, $x_2 = \frac{1}{2}$, $x_3 = 0$, and $v = 35$. By using a randomized strategy (i.e., selecting among the first two alternatives with equal probability), firm C has improved its security level. Its expected market share has increased, since the expected market share of firm R has decreased from 40 percent to 35 percent.

Let us now see how firm R might set probabilities y_1 and y_2 on its alternative selections to achieve its best security level. When firm R weights its alternatives by y_1 and y_2, it has an expected market share of:

$$30y_1 + 60y_2 \quad \text{if firm C selects alternative 1,}$$
$$40y_1 + 10y_2 \quad \text{if firm C selects alternative 2,}$$
$$60y_1 + 30y_2 \quad \text{if firm C selects alternative 3.}$$

Firm R wants its market share as large as possible, but takes a conservative approach in maximizing its minimum expected market share from these three expressions. In this case, firm R solves the following linear program:

Maximize w,

subject to:

$$30y_1 + 60y_2 - w \geq 0,$$
$$40y_1 + 10y_2 - w \geq 0,$$
$$60y_1 + 30y_2 - w \geq 0, \qquad \qquad (25)$$
$$y_1 + \quad y_2 \qquad \quad = 1,$$
$$y_1 \geq 0, \qquad y_2 \geq 0.$$

The first three constraints require that, regardless of the alternative selected by firm C, firm R's market share will be at least w, which is then maximized. The fourth constraint again states that the probabilities must sum to one. In this case, firm R acts optimally by selecting its first alternative with probability $y_1 = \frac{5}{6}$ and its second alternative with probability $y_2 = \frac{1}{6}$, giving an expected market share of 35 percent.

Note that the security levels resulting from each linear program are identical. On closer examination we see that (25) is in fact the dual of (24)! To see the duality correspondence, first convert the inequality constraints in (24) to the standard ($\geq$) for a minimization problem by multiplying by -1. Then the dual constraint derived from "pricing-out" x_1, for example, will read $-30y_1 - 60y_2 + w \leq 0$, which is the first constraint in (25). In this simple example, we have shown that two-person, zero-sum games reduce to primal and dual linear programs. This is true in general, so that the results presented for duality theory in linear programming may be used to draw conclusions about two-person, zero-sum games. Historically, this took place in the reverse, since game theory was first developed by John von Neumann in 1928 and then helped motivate duality theory in linear programming some twenty years later.

General Discussion

The general situation for a two-person, zero-sum game has the same characteristics as our simple example. The payoff table in Tableau 12 and the conservative assumption on the player's behavior lead to the primal and dual linear programs discussed below.

Tableau 12   Payoff to Row Player ($-$Payoff to Column Player)

Row player alternatives \ Column player alternatives	1	2	3	$\cdots$	n
1	a_{11}	a_{12}	a_{13}	$\cdots$	a_{1n}
2	a_{21}	a_{22}	a_{23}		a_{2n}
3	a_{31}	a_{32}	a_{33}		a_{3n}
$\vdots$	$\vdots$	$\vdots$	$\vdots$		$\vdots$
m	a_{m1}	a_{m2}	a_{m3}	$\cdots$	a_{mn}

The column player must solve the linear program:

Minimize v,

subject to:

$$a_{i1}x_1 + a_{i2}x_2 + \cdots + a_{in}x_n - v \leq 0 \qquad (i = 1, 2, \ldots, m),$$
$$x_1 + x_2 + \cdots + x_n = 1,$$

$$x_j \geq 0 \qquad (j = 1, 2, \ldots, n);$$

and the row player the dual linear program:

Maximize w,

subject to:

$$a_{1j}y_1 + a_{2j}y_2 + \cdots + a_{mn}y_m - w \geq 0 \qquad (j = 1, 2, \ldots, n),$$
$$y_1 + y_2 + \cdots + y_m = 1,$$

$$y_i \geq 0 \qquad (i = 1, 2, \ldots, m).$$

The optimal values for $x_1, x_2, \ldots, x_n$, and $y_1, y_2, \ldots, y_m$, from these problems are the probabilities that the players use to select their alternatives. The optimal solutions give (min v) = (max w) and, as before, these solutions provide a stable equilibrium, since neither player has any incentive for improved value by unilaterally altering its optimal probabilities.

More formally, if the row player uses probabilities y_i and the column player uses probabilities x_j, then the expected payoff of the game is given by:

$$\sum_{i=1}^{m} \sum_{j=1}^{n} y_i a_{ij} x_j.$$

By complementary slackness, the optimal probabilities $\hat{y}_i$ and $\hat{x}_j$ of the dual linear programs satisfy:

$$\hat{y}_i = 0 \qquad \text{if } \sum_{j=1}^{n} a_{ij} \hat{x}_j < \hat{v}$$

and

$$\hat{y}_i > 0 \qquad \text{only if } \sum_{j=1}^{n} a_{ij} \hat{x}_j = \hat{v}.$$

Consequently, multiplying the ith inequality in the primal by $\hat{y}_i$, and adding gives

$$\sum_{i=1}^{m} \sum_{j=1}^{n} \hat{y}_i a_{ij} \hat{x}_j = \sum_{i=1}^{m} \hat{y}_i \hat{v} = \hat{v}, \tag{26}$$

showing that the primal and dual solutions $\hat{x}_j$ and $\hat{y}_i$ lead to the payoff $\hat{v}$. The last equality uses the fact that the probabilities $\hat{y}_i$ sum to 1.

Next, consider any other probabilities y_i for the row player. Multiplying the ith primal equation

$$\sum_{j=1}^{n} a_{ij} \hat{x}_j \leq \hat{v}$$

by y_i and adding, we find that:

$$\sum_{i=1}^{m} \sum_{j=1}^{n} y_i a_{ij} \hat{x}_j \leq \sum_{i=1}^{m} y_i \hat{v} = \hat{v}. \tag{27}$$

Similarly, multiplying the jth dual constraint

$$\sum_{i=1}^{m} \hat{y}_i a_{ij} \geq \hat{w}$$

by any probabilities x_j and adding gives:

$$\sum_{i=1}^{m} \sum_{j=1}^{n} \hat{y}_i a_{ij} x_j \geq \sum_{j=1}^{n} \hat{w} x_j = \hat{w}. \tag{28}$$

Since $\hat{v} = \hat{w}$ by linear programming duality theory, Eqs. (26), (27), and (28) imply that:

$$\sum_{i=1}^{m} \sum_{j=1}^{n} y_i a_{ij} \hat{x}_j \leq \sum_{i=1}^{m} \sum_{j=1}^{n} \hat{y}_i a_{ij} \hat{x}_j \leq \sum_{i=1}^{m} \sum_{j=1}^{n} \hat{y}_i a_{ij} x_j.$$

This expression summarizes the equilibrium condition. By unilaterally altering its selection of probabilities $\hat{y}_i$ to y_i, the row player cannot increase its payoff beyond $\hat{v}$.

Similarly, the column player cannot *reduce* the row player's payoff *below* $\hat{v}$ by unilaterally changing *its* selection of probabilities $\hat{x}_j$ to x_j. Therefore, the probabilities $\hat{y}_i$ and $\hat{x}_j$ acts as an equilibrium, since neither player has an incentive to move from these solutions.

EXERCISES

1. Find the dual associated with each of the following problems:

a)   Minimize $z = 3x_1 + 2x_2 - 3x_3 + 4x_4$,

subject to:

$$x_1 - 2x_2 + 3x_3 + 4x_4 \leq \quad 3,$$
$$x_2 + 3x_3 + 4x_4 \geq -5,$$
$$2x_1 - 3x_2 - 7x_3 - 4x_4 = \quad 2,$$
$$x_1 \geq 0, \qquad x_4 \leq 0.$$

b)   Maximize $z = 3x_1 + 2x_2$,

subject to:

$$x_1 + 3x_2 \leq 3,$$
$$6x_1 - x_2 = 4,$$
$$x_1 + 2x_2 \leq 2,$$
$$x_1 \geq 0, \qquad x_2 \geq 0.$$

2. Consider the linear-programming problem:

Maximize $z = 2x_1 + x_2 + 3x_3 + x_4$,

subject to:

$$x_1 + x_2 + x_3 + x_4 \leq \quad 5,$$
$$2x_1 - x_2 + 3x_3 \qquad = -4,$$
$$x_1 \qquad - x_3 + x_4 \geq \quad 1,$$
$$x_1 \geq 0, \qquad x_3 \geq 0, \qquad x_2 \text{ and } x_4 \text{ unrestricted.}$$

a) State this problem with equality constraints and nonnegative variables.
b) Write the dual to the given problem and the dual to the transformed problem found in part (a). Show that these two dual problems are equivalent.

3. The initial and final tableaus of a linear-programming problem are as follows:

Initial Tableau

Basic variables	Current values	x_1	x_2	x_3	x_4	x_5	x_6
x_5	6	4	9	7	10	1	
x_6	4	1	1	3	40		1
$(-z)$	0	12	20	18	40		

Final Tableau

Basic variables	Current values	x_1	x_2	x_3	x_4	x_5	x_6
x_1	$\frac{4}{3}$	1	$\frac{7}{3}$	$\frac{5}{3}$		$\frac{4}{15}$	$-\frac{1}{15}$
x_4	$\frac{1}{15}$		$-\frac{1}{30}$	$\frac{1}{30}$	1	$-\frac{1}{150}$	$\frac{2}{75}$
$(-z)$	$-\frac{56}{3}$		$-\frac{20}{3}$	$-\frac{10}{3}$		$-\frac{44}{15}$	$-\frac{4}{15}$

a) Find the optimum solution for the dual problem.

b) Verify that the values of the shadow prices are the dual feasible, i.e., that they satisfy the following relationship:

$$\bar{c}_j = c_j - \sum_{i=1}^{m} a_{ij}\bar{y}_i \leqq 0,$$

where the terms with bars refer to data in the final tableau and the terms without bars refer to data in the initial tableau.

c) Verify the complementary-slackness conditions.

4. In the second exercise of Chapter 1, we graphically determined the shadow prices to the following linear program:

Maximize $z = 2x_1 + x_2$,

subject to:

$$12x_1 + x_2 \leqq 6,$$
$$-3x_1 + x_2 \leqq 7,$$
$$x_2 \leqq 10,$$
$$x_1 \geqq 0, \qquad x_2 \geqq 0.$$

a) Formulate the dual to this linear program.

b) Show that the shadow prices solve the dual problem.

5. Solve the linear program below as follows: First, solve the dual problem graphically. Then use the solution to the dual problem to determine which variables in the primal problem are zero in the optimal primal solution. [*Hint*: Invoke complementary slackness.] Finally, solve for the optimal basic variables in the primal, using the primal equations.

Primal

Maximize $-4x_2 + 3x_3 + 2x_4 - 8x_5$,

subject to:

$$3x_1 + x_2 + 2x_3 + x_4 = 3,$$
$$x_1 - x_2 + x_4 - x_5 \geqq 2,$$
$$x_j \geqq 0 \qquad (j = 1, 2, 3, 4, 5).$$

6. A dietician wishes to design a minimum-cost diet to meet minimum daily requirements for calories, protein, carbohydrate, fat, vitamin A and vitamin B dietary needs. Several different foods can be used in the diet, with data as specified in the following table.

	Content and costs per pound consumed						
	Food 1	Food 2	....	Food j	....	Food n	Daily requirements
Calories	a_{11}	a_{12}		a_{1j}		a_{1n}	b_1
Protein (grams)	a_{21}	a_{22}		a_{2j}		a_{2n}	b_2
Carbohydrate (grams)	a_{31}	a_{32}		a_{3j}		a_{3n}	b_3
Fat (grams)	a_{41}	a_{42}		a_{4j}		a_{4n}	b_4
Vitamin A (milligrams)	a_{51}	a_{52}		a_{5j}		a_{5n}	b_5
Vitamin B (milligrams)	a_{61}	a_{62}		a_{6j}		a_{6n}	b_6
Costs (dollars)	c_1	c_2		c_j		c_n	

a) Formulate a linear program to determine which foods to include in the minimum cost diet. (More than the minimum daily requirements of any dietary need can be consumed.)

b) State the dual to the diet problem, specifying the units of measurement for each of the dual variables. Interpret the dual problem in terms of a druggist who sets prices on the dietary needs in a manner to sell a dietary pill with b_1, b_2, b_3, b_4, b_5, and b_6 units of the given dietary needs at maximum profit.

7. In order to smooth its production scheduling, a footwear company has decided to use a simple version of a linear cost model for aggregate planning. The model is:

$$\text{Minimize } z = \sum_{i=1}^{N} \sum_{t=1}^{T} (v_i X_{it} + c_i I_{it}) + \sum_{t=1}^{T} (rW_t + oO_t),$$

subject to:

$$X_{it} + I_{i,t-1} - I_{it} = d_{it} \qquad \begin{cases} t = 1, 2, \ldots, T \\ i = 1, 2, \ldots, N \end{cases}$$

$$\sum_{i=1}^{N} k_i X_{it} - W_t - O_t = 0 \qquad t = 1, 2, \ldots, T$$

$$0 \le W_t \le (rm) \qquad t = 1, 2, \ldots, T$$

$$-pW_t + O_t \le 0 \qquad t = 1, 2, \ldots, T$$

$$X_{it}, I_{it} \ge 0 \qquad \begin{cases} i = 1, 2, \ldots, N \\ t = 1, 2, \ldots, T \end{cases}$$

with parameters:

v_i = Unit production cost for product i in each period,

c_i = Inventory-carrying cost per unit of product i in each period,

r = Cost per man-hour of regular labor,

o = Cost per man-hour of overtime labor,

d_{it} = Demand for product i in period t,

k_i = Man-hours required to produce one unit of product i,

(rm) = Total man-hours of regular labor available in each period,

p = Fraction of labor man-hours available as overtime,

T = Time horizon in periods,

N = Total number of products.

The decision variables are:

X_{it} = Units of product i to be produced in period t,

I_{it} = Units of product i to be left over as inventory at the end of period t,

W_t = Man-hours of regular labor used during period (fixed work force),

O_t = Man-hours of overtime labor used during period t.

The company has two major products, boots and shoes, whose production it wants to schedule for the next three periods. It costs \$10 to make a pair of boots and \$5 to make a pair of shoes. The company estimates that it costs \$2 to maintain a pair of boots as inventory through the end of a period and half this amount for shoes. Average wage rates, including benefits, are three dollars an hour with overtime paying double. The company prefers a constant labor force and estimates that regular time will make 2000 man-hours available per period. Workers are willing to increase their work time up to 25% for overtime compensation. The demand for boots and shoes for the three periods is estimated as:

Period	Boots	Shoes
1	300 pr.	3000 pr.
2	600 pr.	5000 pr.
3	900 pr.	4000 pr.

a) Set up the model using 1 man-hour and $\frac{1}{2}$ man-hour as the effort required to produce a pair of boots and shoes, respectively.

b) Write the dual problem.

c) Define the physical meaning of the dual objective function and the dual constraints.

8. In capital-budgeting problems within the firm, there is a debate as to whether or not the appropriate objective function should be discounted. The formulation of the capital-budgeting problem without discounting is as follows:

Maximize v_N,

subject to:

$$\sum_{j=1}^{J} (-c_{ij}x_j) \leqq f_i \qquad (i = 0, 1, 2, \ldots, N-1) \qquad y_i$$

$$\sum_{j=1}^{J} (-c_{Nj}x_j) + v_N \leqq f_N, \qquad\qquad y_N$$

$$0 \leqq x_j \leqq u_j \qquad (j = 1, 2, \ldots, J),$$

Shadow prices

where c_{ij} is the cash outflow ($c_{ij} < 0$) or inflow ($c_{ij} > 0$) in period i for project j; the right-hand-side constant f_i is the net exogenous funds made available to ($f_i > 0$) or withdrawn from ($f_i < 0$) a division of the firm in period i; the decision variable x_j is the level of investment in project j; u_j is an upper bound on the level of investment in project j; and v_N is a variable measuring the value of the holdings of the division at the end of the planning horizon.

If the undiscounted problem is solved by the bounded variable simplex method, the optimal solution is x_j^* for $j = 1, 2, \ldots, J$ with associated shadow prices (dual variables) y_i^* for $i = 0, 1, 2, \ldots, N$.

a) Show that $y_N^* = 1$.

b) The discount factor for time i is defined to be the present value (time $i = 0$) of one dollar received at time i. Show that the discount factor for time i, ρ_i^* is given by:

$$\rho_i^* = \frac{y_i^*}{y_0^*} \qquad (i = 0, 1, 2, \ldots, N).$$

(Assume that the initial budget constraint is binding so that $y_0^* > 0$.)

c) Why should $\rho_0^* = 1$?

9. The *discounted* formulation of the capital-budgeting problem described in the previous exercise can be stated as:

$$\text{Maximize} \sum_{j=1}^{J} \left(\sum_{i=0}^{N} \rho_i c_{ij} \right) x_j,$$

subject to: Shadow
                                                                                                 prices

$$\sum_{j=1}^{J} (-c_{ij} x_j) \leqq f_i \qquad (i = 0, 1, 2, \ldots, N), \qquad \lambda_i$$

$$0 \leqq x_j \leqq u_j \qquad (j = 1, 2, \ldots, J),$$

where the objective-function coefficient

$$\sum_{i=0}^{N} \rho_i c_{ij}$$

represents the discounted present value of the cash flows from investing at unit level in project j, and λ_i for $i = 1, 2, \ldots, N$ are the shadow prices associated with the funds-flow constraints.

Suppose that we wish to solve the above discounted formulation of the capital-budgeting problem, using the discount factors determined by the optimal solution of the previous exercise, that is, setting:

$$\rho_i = \rho_i^* = \frac{y_i^*}{y_0^*} \qquad (i = 0, 1, 2, \ldots, N).$$

Show that the optimal solution x_j^* for $j = 1, 2, \ldots, J$, determined from the undiscounted case in the previous exercise, is also optimal to the above discounted formulation, assuming:

$$\rho_i = \rho_i^* \qquad (i = 0, 1, 2, \ldots, N).$$

[*Hint*: Write the optimality conditions for the discounted problem, using shadow-price values $\lambda_i^* = 0$ for $i = 1, 2, \ldots, N$. Does x_j^* $(j = 1, 2, \ldots, J)$ satisfy these conditions? Do the shadow prices on the upper bounding constraints for the discounted model differ from those of the undiscounted model?]

10. An alternative formulation of the undiscounted problem developed in Exercise 8 is to maximize the total *earnings* on the projects rather than the horizon value. Earnings of a project are defined to be the net cash flow from a project over its lifetime. The alternative formulation is then:

$$\text{Maximize} \sum_{j=1}^{J} \left(\sum_{i=0}^{N} c_{ij} \right) x_j,$$

subject to:

	Shadow prices

$$\sum_{j=1}^{J} (-c_{ij}x_j) \leq f_i \qquad (i = 0, 1, 2, \ldots, N), \qquad y_i'$$

$$0 \leq x_j \leq u_j \qquad (j = 1, 2, \ldots, J).$$

Let x_j^* for $j = 1, 2, \ldots, J$ solve this earnings formulation, and suppose that the funds constraints are binding at all times $i = 0, 1, \ldots, N$ for this solution.

a) Show that x_j^* for $j = 1, 2, \ldots, J$ also solves the horizon formulation given in Exercise 8.

b) Denote the optimal shadow prices of the funds constraints for the earnings formulation as y_i' for $i = 0, 1, 2, \ldots, N$. Show that $y_i = 1 + y_i'$ for $i = 0, 1, 2, \ldots, N$ are optimal shadow prices for the funds constraints of the horizon formulation.

11. Suppose that we now consider a variation of the horizon model given in Exercise 8, that explicitly includes one-period borrowing b_i at rate r_b and one-period lending ℓ_i at rate r_ℓ. The formulation is as follows:

$$\text{Maximize} \ v_N,$$

subject to:

	Shadow prices

$$\sum_{j=1}^{J} (-c_{0j}x_j) + \ell_0 - b_0 \leq f_0, \qquad y_0$$

$$\sum_{j=1}^{J} (-c_{ij}x_j) - (1 + r_\ell)\ell_{i-1} + \ell_i + (1 + r_b)b_{i-1} - b_i \leq f_i \qquad (i = 1, 2, \ldots, N-1), \qquad y_i$$

$$\sum_{j=1}^{J} (-c_{Nj}x_j) - (1 + r_\ell)\ell_{N-1} + (1 - r_b)b_{N-1} + v_N \leq f_N, \qquad y_N$$

$$b_i \leq B_i \qquad (i = 0, 1, \ldots, N-1), \qquad w_i$$

$$0 \leq x_j \leq u_j \qquad (j = 1, 2, \ldots, J),$$

$$b_i \geq 0, \quad \ell_i \geq 0 \qquad (i = 0, 1, \ldots, N-1).$$

a) Suppose that the optimal solution includes lending in every time period; that is, $\ell_i^* > 0$ for $i = 0, 1, \ldots, N-1$. Show that the present value of a dollar in period i is

$$\rho_i = \left(\frac{1}{1 + r_\ell} \right)^i.$$

b) Suppose that there is no upper bound on borrowing; that is, $B_i = +\infty$ for $i = 0, 1, \ldots, N - 1$, and that the borrowing rate equals the lending rate, $r = r_b = r_\ell$. Show that

$$\frac{y_{i-1}}{y_i} = 1 + r,$$

and that the present value of a dollar in period i is

$$\rho_i = \left(\frac{1}{1 + r}\right)^i.$$

c) Let w_i be the shadow prices on the upper-bound constraints $b_i \leq B_i$, and let $r = r_b = r_\ell$. Show that the shadow prices satisfy:

$$1 + r \leq \frac{y_{i-1}}{y_i} \leq 1 + r + w_{i-1}.$$

d) Assume that the borrowing rate is greater than the lending rate, $r_b > r_\ell$; show that the firm will not borrow and lend in the same period if it uses this linear-programming model for its capital budgeting decisions.

12. As an example of the present-value analysis given in Exercise 8, consider four projects with cash flows and upper bounds as follows:

Project	End of year 0	End of year 1	End of year 2	Upper bound
A	−1.00	0.60	0.60	∞
B	−1.00	1.10	0	500
C	0	−1.00	1.25	∞
D	−1.00	0	1.30	∞

A negative entry in this table corresponds to a cash outflow and a positive entry to a cash inflow.

The horizon-value model is formulated below, and the optimal solution and associated shadow prices are given:

x_A	x_B	x_C	x_D	v	Relation	RHS	
1	1		1		$\leq$	1000	1.35
−0.60	−1.1	1			$\leq$	0	1.25	Shadow
−0.60		−1.25	−1.30	1	$\leq$	0	1.00	prices
	1				$\leq$	500	0.025
				1	$=$	z(max)	
500	500	850	0	−1362.5			

Solution

a) Explain exactly how *one* additional dollar at the end of year 2 would be invested to return the shadow price 1.25. Similarly, explain how to use the projects in the portfolio to achieve a return of 1.35 for an additional dollar at the end of year 0.

b) Determine the discount factors for each year, from the shadow prices on the constraints.

c) Find the discounted present value $\sum_{i=0}^{N} p_i c_{ij}$ of each of the four projects. How does the sign of the discounted present value of a project compare with the sign of the reduced cost for that project? What is the relationship between the two? Can discounting interpreted in this way be used for decision-making?

d) Consider the two additional projects with cash flows and upper bounds as follows:

Project	End of year 0	End of year 1	End of year 2	Upper bound
E	-1.00	0.75	0.55	∞
F	-1.00	0.30	1.10	∞

What is the discounted present value of each project? Both appear promising. Suppose that funds are transferred from the current portfolio into project E, will project F still appear promising? Why? Do the discount factors change?

13. In the exercises following Chapter 1, we formulated the absolute-value regression problem:

$$\text{Minimize} \sum_{i=1}^{m} |y_i - x_{i1}\beta_1 - x_{i2}\beta_2 - \cdots - x_{in}\beta_n|$$

as the linear program:

$$\text{Minimize} \sum_{i=1}^{m} (P_i + N_i),$$

subject to:

$$x_{i1}\beta_1 + x_{i2}\beta_2 + \cdots + x_{in}\beta_n + P_i - N_i = y_i \qquad \text{for } i = 1, 2, \ldots, m,$$
$$P_i \geq 0, \qquad N_i \geq 0 \qquad \text{for } i = 1, 2, \ldots, m.$$

In this formulation, the y_i are measurements of the dependent variable (e.g., income), which is assumed to be explained by independent variables (e.g., level of education, parents' income, and so forth), which are measured as $x_{i1}, x_{i2}, \ldots, x_{in}$. A linear model assumes that y depends linearly upon the β's, as:

$$\hat{y}_i = x_{i1}\beta_1 + x_{i2}\beta_2 + \cdots + x_{in}\beta_n. \tag{29}$$

Given any choice of the parameters $\beta_1, \beta_2, \ldots, \beta_n, \hat{y}_i$ is an estimate of y_i. The above formulation aims to minimize the deviations of the estimates of $\hat{y}_i$ from y_i as measured by the sum of absolute values of the deviations. The variables in the linear-programming model are the parameters $\beta_1, \beta_2, \ldots, \beta_n$ as well as the P_i and N_i. The quantities $y_i, x_{i1}, x_{i2}, \ldots, x_{in}$ are known data found by measuring several values for the dependent and independent variables for the linear model (29).

In practice, the number of observations m frequently is much larger than the number of parameters n. Show how we can take advantage of this property by formulating the dual to the above linear program in the dual variables $u_1, u_2, \ldots, u_m$. How can the special structure of the dual problem be exploited computationally?

14. The following tableau is in canonical form for maximizing z, except that one righthand-side value is negative.

Basic variables	Current values	x_1	x_2	x_3	x_4	x_5	x_6
x_1	14	1			2	3	-2
x_2	6		1		1	-2	-2
x_3	-10			1	-1	-2	0
$(-z)$	-40				-1	-4	-2

However, the reduced costs of the nonbasic variables all satisfy the primal optimality conditions. Find the optimal solution to this problem, using the dual simplex algorithm to find a feasible canonical form while maintaining the primal optimality conditions.

15. In Chapter 2 we solved a two-constraint linear-programming version of a trailer-production problem:

$$\text{Maximize } 6x_1 + 14x_2 + 13x_3,$$

subject to:

$$\tfrac{1}{2}x_1 + 2x_2 + 4x_3 \leq 24 \quad \text{(Metalworking capacity)},$$
$$x_1 + 2x_2 + 4x_3 \leq 60 \quad \text{(Woodworking capacity)},$$
$$x_1 \geq 0, \quad x_2 \geq 0, \quad x_3 \geq 0,$$

obtaining an optimal tableau:

Basic variables	Current values	x_1	x_2	x_3	x_4	x_5
x_1	36	1	6		4	-1
x_3	6		-1	1	-1	$\tfrac{1}{2}$
$(-z)$	-294		-9		-11	$-\tfrac{1}{2}$

Suppose that, in formulating this problem, we ignored a constraint limiting the time available in the shop for inspecting the trailers.

a) If the solution $x_1 = 36, x_2 = 0$, and $x_3 = 6$ to the original problem satisfies the inspection constraint, is it necessarily optimal for the problem when we impose the inspection constraint?

b) Suppose that the inspection constraint is

$$x_1 + x_2 + x_3 + x_6 = 30,$$

where x_6 is a nonnegative slack variable. Add this constraint to the optimal tableau with x_6 as its basic variable and pivot to eliminate the basic variables x_1 and x_3 from this constraint. Is the tableau now in dual canonical form?

c) Use the dual simplex method to find the optimal solution to the trailer-production problem with the inspection constraint given in part (b).

d) Can the ideas used in this example be applied to solve a linear program whenever a new constraint is added after the problem has been solved?

16. Apply the dual simplex method to the following tableau for maximizing z with nonnegative decision variables $x_1, x_2, \ldots, x_5$.

Basic variables	Current values	x_1	x_2	x_3	x_4	x_5
x_1	-3	1		-1	2	-2
x_2	7		1	3	-4	8
$(-z)$	-5			-1	-5	-6

Is the problem feasible? How can you tell?

17. Consider the linear program:

$$\text{Minimize } z = 2x_1 + x_2,$$

subject to:

$$-4x_1 + 3x_2 - x_3 \geq 16,$$

$$x_1 + 6x_2 + 3x_3 \geq 12,$$

$$x_i \geq 0 \qquad \text{for } i = 1, 2, 3.$$

a) Write the associated dual problem.
b) Solve the primal problem, using the dual simplex algorithm.
c) Utilizing the final tableau from part (b), find optimal values for the dual variables y_1 and y_2. What is the corresponding value of the dual objective function?

18. For what values of the parameter θ is the following tableau in canonical form for maximizing the objective value z?

Basic variables	Current values	x_1	x_2	x_3	x_4	x_5
x_1	$-2 + \theta$	1		-1	2	-3
x_2	1		1	1	0	1
$(-z)$	-20			$3 - \theta$	$4 - \theta$	-6

Starting with this tableau, use the parametric primal-dual algorithm to solve the linear program at $\theta = 0$.

19. After solving the linear program:

$$\text{Maximize } z = 5x_1 + 7x_2 + 2x_3,$$

subject to:

$$2x_1 + 3x_2 + x_3 + x_4 = 5,$$

$$\tfrac{1}{2}x_1 + x_2 + x_5 = 1,$$

$$x_j \geq 0 \qquad (j = 1, 2, \ldots, 5),$$

and obtaining the optimal canonical form

$b_1 = {}^-2 + \theta$

Basic variables	Current values	x_1	x_2	x_3	x_4	x_5
x_3	1		−1	1	1	−4
x_1	2	1	2		0	2
$(-z)$	−12		−1		−2	−2

(handwritten: $-2 \ge 1$ in Current values; $-1+4$ under x_2; $c_2 = 3-\theta$ in left margin)

we discover that the problem was formulated improperly. The objective coefficient for x_2 should have been 11 (not 7) and the righthand side of the first constraint should have been 2 (not 5).

a) How do these modifications in the problem formulation alter the data in the tableau above? *(handwritten: $b_1 = 1-3 = -2$ $\bar{c}_2 = -1+4 \le 3$)*

x_2 rentie $\theta < 3$
prochain dual
b) How can we use the parametric primal-dual algorithm to solve the linear program after these data changes, starting with x_1 and x_3 as basic variables? *(handwritten: $\bar{c}_2 < 3-\theta$ $\bar{b}_1 \sim -2+\theta$)*

c) Find an optimal solution to the problem after the data changes are made, using the parametric primal-dual algorithm.

20. Rock, Paper, and Scissors is a game in which two players simultaneously reveal no fingers (rock), one finger (paper), or two fingers (scissors). The payoff to player 1 for the game is governed by the following table:

Player 1 \ Player 2	Rock	Paper	Scissors
Rock	0	−1	1
Paper	1	0	−1
Scissors	−1	1	0

Payoff to player 1
(or minus payoff to player 2)

Note that the game is void if both players select the same alternative: otherwise, rock breaks scissors and wins, scissors cut paper and wins, and paper covers rock and wins.

Use the linear-programming formulation of this game and linear-programming duality theory, to show that both players' optimal strategy is to choose each alternative with probability $\frac{1}{3}$.

21. Solve for an optimal strategy for both player 1 and player 2 in a zero-sum two-person game with the following payoff table:

Player 1 \ Player 2	Alternative 1	Alternative 2	Alternative 3
Alternative 1	1	4	3
Alternative 2	5	2	3

Payoff to player 1
(or minus payoff to player 2)

Is the optimal strategy of each player unique?

22. In a game of tic-tac-toe, the first player has three different choices at the first move: the center square, a corner square, or a side square. The second player then has different alternatives depending upon the move of the first player. For instance, if the first player chooses a corner square, the second player can select the center square, the opposite corner, an adjacent corner, an opposite side square, or an adjacent side square. We can picture the possible outcomes for the game in a decision tree as follows:

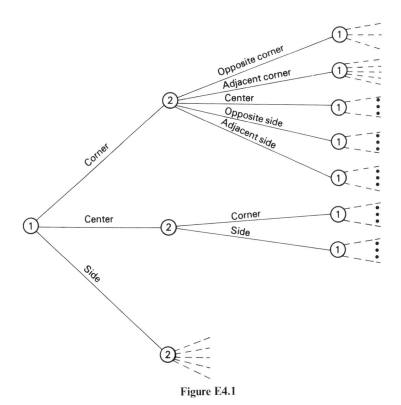

Figure E4.1

Nodes ① correspond to decision points for the first player; nodes ② are decision points for the second player. The tree is constructed in this manner by considering the options available to each player at a given turn, until either one player has won the game (i.e., has selected three adjacent horizontal or vertical squares or the squares along one of the diagonals), or all nine squares have been selected. In the first case, the winning player receives 1 point; in the second case the game is a draw and neither player receives any points.

The decision-tree formulation of a game like that illustrated here is called an *extensive form* formulation. Show how the game can be recast in the linear-programming form (called a *normal form*) discussed in the text. Do not attempt to enumerate every strategy for both players. Just indicate how to construct the payoff tables.

[*Hint*: Can you extract a strategy for player 1 by considering his options at each decision point? Does his strategy depend upon how player 2 reacts to player 1's selections?]

The remaining exercises extend the theory developed in this chapter, providing new results and relating some of the duality concepts to other ideas that arise in mathematical programming.

23. Consider a linear program with bounded variables:

$$\text{Maximize } z = \sum_{j=1}^{n} c_j x_j,$$

subject to:

Dual variables

$$\sum_{j=1}^{n} a_{ij} x_j = b_i \quad (i = 1, 2, \ldots, m), \qquad y_i$$

$$x_j \leq u_j \quad (j = 1, 2, \ldots, n), \qquad w_j$$

$$x_j \geq 0 \quad (j = 1, 2, \ldots, n),$$

where the upper bounds u_j are positive constants. Let y_i for $i = 1, 2, \ldots, m$ and w_j for $j = 1, 2, \ldots, m$ be variables in the dual problem.

a) Formulate the dual to this bounded-variable problem.

b) State the optimality conditions (that is, primal feasibility, dual feasibility, and complementary slackness) for the problem.

c) Let $\bar{c}_j = c_j - \sum_{i=1}^{m} y_i a_{ij}$ denote the reduced costs for variable x_j determined by pricing out the a_{ij} constraints and not the upper-bounding constraints. Show that the optimality conditions are equivalent to the bounded-variable optimality conditions

$$\bar{c}_j \leq 0 \quad \text{if } x_j = 0,$$
$$\bar{c}_j \geq 0 \quad \text{if } x_j = u_j,$$
$$\bar{c}_j = 0 \quad \text{if } 0 < x_j < u_j,$$

given in Section 2.6 of the text, for any feasible solution x_j $(j = 1, 2, \ldots, n)$ of the primal problem.

24. Let x_j^* for $j = 1, 2, \ldots, n$ be an optimal solution to the linear program with

$$\text{Maximize } \sum_{j=1}^{n} c_j x_j,$$

subject to:

Shadow prices

$$\sum_{j=1}^{n} a_{ij} x_j \leq b_i \quad (i = 1, 2, \ldots, m), \qquad y_i$$

$$\sum_{j=1}^{n} e_{ij} x_j \leq d_i \quad (i = 1, 2, \ldots, q), \qquad w_i$$

$$x_j \geq 0 \quad (j = 1, 2, \ldots, n),$$

two groups of constraints, the a_{ij} constraints and the e_{ij} constraints. Let y_i for $i = 1, 2, \ldots, m$ and w_i for $i = 1, 2, \ldots, q$ denote the shadow prices (optimal dual variables) for the constraints.

a) Show that x_j^* for $j = 1, 2, \ldots, n$ also solves the linear program:

$$\text{Maximize } \sum_{j=1}^{n} \left[c_j - \sum_{i=1}^{m} y_i a_{ij} \right] x_j,$$

subject to:

$$\sum_{j=1}^{n} e_{ij}x_j \leq d_i \qquad (i = 1, 2, \ldots, q),$$

$$x_j \geq 0 \qquad (j = 1, 2, \ldots, n),$$

in which the a_{ij} constraints have been incorporated into the objective function as in the method of Lagrange multipliers. [*Hint:* Apply the optimality conditions developed in Section 4.5 to the original problem and this "Lagrangian problem."]

b) Illustrate the property from part (a) with the optimal solution $x_1^* = 2$, $x_2^* = 1$, and shadow prices $y_1 = \frac{1}{2}$, $w_1 = \frac{1}{2}$, $w_2 = 0$, to the linear program:

Maximize x_1,

subject to:

$$x_1 + x_2 \leq 3 \qquad [a_{ij} \text{ constraint}],$$

$$\left.\begin{array}{r} x_1 - x_2 \leq 1 \\ x_2 \leq 2 \end{array}\right\} \quad [e_{ij} \text{ constraints}],$$

$$x_1 \geq 0, \qquad x_2 \geq 0.$$

c) Show that an optimal solution $\bar{x}_j$ for $j = 1, 2, \ldots, n$ for the "Lagrangian problem" from part (a) need not be optimal for the original problem [see the example in part (b)]. Under what conditions is an optimal solution to the Lagrangian problem optimal in the original problem?

d) We know that if the "Lagrangian problem" has an optimal solution, then it has an extreme-point solution. Does this imply that there is an extreme point to the Lagrangian problem that is optimal in the original problem?

25. The payoff matrix (a_{ij}) for a two-person zero-sum game is said to be *skew symmetric* if the matrix has as many rows as columns and $a_{ij} = -a_{ji}$ for each choice of i and j. The payoff matrix for the game Rock, Paper and Scissors discussed in Exercise 20 has this property.

a) Show that if players 1 and 2 use the same strategy,

$$x_1 = y_1, \quad x_2 = y_2, \quad \ldots, \quad x_n = y_n,$$

then the payoff

$$\sum_{i=1}^{n} \sum_{j=1}^{n} y_i a_{ij} x_j$$

from the game is zero.

b) Given any strategy $x_1, x_2, \ldots, x_n$ for player 2, i.e., satisfying:

$$a_{11}x_1 + a_{12}x_2 + \cdots + a_{1n}x_n \geq w,$$

$$a_{21}x_1 + a_{22}x_2 + \cdots + a_{2n}x_n \geq w,$$

$$\vdots \qquad\qquad \vdots$$

$$a_{n1}x_1 + a_{n2}x_2 + \cdots + a_{nn}x_n \geq w,$$

$$x_1 + x_2 + \cdots + x_n = 1,$$

multiply the first n constraints by $y_1 = x_1$, $y_2 = x_2, \ldots$, and $y_n = x_n$, respectively, and add. Use this manipulation, together with part (a), to show that $w \leq 0$. Similarly, show that the value to the player 1 problem satisfies $v \geq 0$. Use linear-programming duality to conclude that $v = w = 0$ is the value of the game.

26. In Section 4.9 we showed how to use linear-programming duality theory to model the column and row players' decision-making problems in a zero-sum two-person game. We would now like to exhibit a stronger connection between linear-programming duality and game theory. Consider the linear-programming dual problems

$$\text{Maximize } z = \sum_{j=1}^{n} c_j x_j, \qquad\qquad \text{Minimize } w = \sum_{i=1}^{m} y_i b_i,$$

subject to:

subject to:

$$\sum_{j=1}^{n} a_{ij} x_j \leq b_i, \qquad\qquad \sum_{i=1}^{m} y_i a_{ij} \geq c_j,$$

$$x_j \geq 0, \qquad\qquad y_i \geq 0,$$

where the inequalities apply for $i = 1, 2, \ldots, m$ and $j = 1, 2, \ldots, n$; in addition, consider a zero-sum two-person game with the following payoff table:

$\bar{y}_1$	$\bar{y}_2$	$\cdots$	$\bar{y}_m$	$\bar{x}_1$	$\bar{x}_2$	$\cdots$	$\bar{x}_m$	t
0	0	$\cdots$	0	a_{11}	a_{12}	$\cdots$	a_{1n}	$-b_1$
0	0	$\cdots$	0	a_{21}	a_{22}	$\cdots$	a_{2n}	$-b_2$
$\vdots$				$\vdots$				$\vdots$
0	0	$\cdots$	0	a_{m1}	a_{m2}	$\cdots$	a_{mn}	$-b_m$
$-a_{11}$	$-a_{21}$	$\cdots$	$-a_{m1}$	0	0	$\cdots$	0	c_1
$-a_{12}$	$-a_{22}$	$\cdots$	$-a_{m2}$	0	0	$\cdots$	0	c_2
$\vdots$				$\vdots$				$\vdots$
$-a_{1n}$	$-a_{2n}$	$\cdots$	$-a_{mn}$	0	0	$\cdots$	0	c_n
b_1	b_2	$\cdots$	b_m	$-c_1$	$-c_2$	$\cdots$	$-c_n$	0

The quantities $\bar{x}_j$, $\bar{y}_i$, and t shown above the table are the column players' selection probabilities.

a) Is this game skew symmetric? What is the value of the game? (Refer to the previous exercise.)

b) Suppose that x_j for $j = 1, 2, \ldots, n$ and y_i for $i = 1, 2, \ldots, m$ solve the linear-programming primal and dual problems. Let

$$t = \frac{1}{\sum_{j=1}^{n} x_j + \sum_{i=1}^{m} y_i + 1},$$

let $\bar{x}_j = t x_j$ for $j = 1, 2, \ldots, n$, and let $\bar{y}_i = t y_i$ for $i = 1, 2, \ldots, m$. Show that $\bar{x}_j$, $\bar{y}_i$, and t solve the given game.

c) Let $\bar{x}_j$ for $j = 1, 2, \ldots, n$, $\bar{y}_i$ for $i = 1, 2, \ldots, m$, and t solve the game, and suppose that $t > 0$. Show that

$$x_j = \frac{\bar{x}_j}{t} \quad \text{and} \quad y_i = \frac{\bar{y}_i}{t}$$

solve the primal and dual linear programs. (See *Hint.*)

[*Hint*: Use the value of the game in parts (b) and (c), together with the conditions for strong duality in linear programming.]

27. Suppose we approach the solution of the usual linear program:

$$\text{Maximize} \sum_{j=1}^{n} c_j x_j,$$

subject to:

$$\sum_{j=1}^{n} c_{ij} x_j \leq b_i \qquad (i = 1, 2, \dots, m), \tag{30}$$

$$x_j \geq 0 \qquad (j = 1, 2, \dots, n),$$

by the method of Lagrange multipliers. Define the Lagrangian maximization problem as follows:

$$L(\lambda_1, \lambda_2, \dots, \lambda_m) = \text{Maximize} \left\{ \sum_{j=1}^{n} c_j x_j - \sum_{i=1}^{m} \lambda_i \left(\sum_{j=1}^{n} c_{ij} x_j - b_i \right) \right\}$$

subject to:

$$x_j \geq 0 \qquad (j = 1, 2, \dots, n).$$

Show that the dual of (30) is given by:

$$\text{Minimize } L(\lambda_1, \lambda_2, \dots, \lambda_m),$$

subject to:

$$\lambda_i \geq 0 \qquad (i = 1, 2, \dots, m).$$

[*Note*: The Lagrangian function $L(\lambda_1, \lambda_2, \dots, \lambda_m)$ may equal $+\infty$ for certain choices of the Lagrange multipliers $\lambda_1, \lambda_2, \dots, \lambda_m$. Would these values for the Lagrange multipliers ever be selected when minimizing $L(\lambda_1, \lambda_2, \dots, \lambda_m)$?]

28. Consider the bounded-variable linear program:

$$\text{Maximize} \sum_{j=1}^{n} c_j x_j,$$

subject to:

$$\sum_{j=1}^{n} a_{ij} x_j \leq b_i \qquad (i = 1, 2, \dots, m),$$

$$\ell_j \leq x_j \leq u_j \qquad (j = 1, 2, \dots, n).$$

Define the Lagrangian maximization problem by:

$$L(\lambda_1, \lambda_2, \dots, \lambda_m) = \text{Maximize} \left\{ \sum_{j=1}^{n} c_j x_j - \sum_{i=1}^{m} \lambda_i \left(\sum_{j=1}^{n} a_{ij} x_j - b_i \right) \right\}$$

subject to:

$$\ell_j \leq x_j \leq u_j \qquad (j = 1, 2, \dots, n).$$

Show that the primal problem is bounded above by $L(\lambda_1, \lambda_2, \dots, \lambda_m)$ so long as $\lambda_i \geq 0$ for $i = 1, 2, \dots, m$.

29. Consider the bounded-variable linear program and the corresponding Lagrangian maximization problem defined in Exercise 28.

   a) Show that:

$$\text{Maximize} \sum_{j=1}^{n} c_j x_j \qquad = \qquad \text{Minimize } L(\lambda_1, \lambda_2, \ldots, \lambda_m),$$

   subject to: subject to:

$$\sum_{j=1}^{n} a_{ij} x_j \leq b_i, \qquad\qquad\qquad \lambda_i \geq 0,$$

$$\ell_j \leq x_j \leq u_j,$$

   which is a form of strong duality.

   b) Why is it that $L(\lambda_1, \lambda_2, \ldots, \lambda_m)$ in this case does not reduce to the usual linear-programming dual problem?

30. Suppose that we define the function:

$$f(x_1, x_2, \ldots, x_n; y_1, y_2, \ldots, y_m) = \sum_{j=1}^{n} c_j x_j - \sum_{i=1}^{m} \sum_{j=1}^{n} a_{ij} x_j y_j + \sum_{i=1}^{m} b_i y_i.$$

   Show that the minimax property

$$\begin{array}{cc} \text{Max} & \text{Min} \\ x_j \geq 0 & y_i \geq 0 \end{array} f(x_1, \ldots, x_n; y_1, \ldots, y_m) = \begin{array}{cc} \text{Min} & \text{Max} \\ y_i \geq 0 & x_j \geq 0 \end{array} f(x_1, \ldots, x_n; y_1, \ldots, y_m)$$

   implies the strong duality property.

ACKNOWLEDGMENTS

In Exercise 6, the pill interpretation of the diet problem appears in *The Theory of Linear Economic Models,* by David Gale, McGraw-Hill Book Company, Inc., 1960.

Exercises 8 and 11 are based on material contained in *Mathematical Programming and the Analysis of Capital Budgeting Problems,* by H. Martin Weingartner, Prentice-Hall, Inc. 1963.

Exercise 12 is based on the Gondon Enterprises case written by Paul W. Marshall and Colin C. Blaydon.

Mathematical Programming
in Practice

5

In management science, as in most sciences, there is a natural interplay between theory and practice. Theory provides tools for applied work and suggests viable approaches to problem solving, whereas practice adds focus to the theory by suggesting areas for theoretical development in its continual quest for problem-solving capabilities. It is impossible, then, to understand fully either theory or practice in isolation. Rather, they need to be considered as *continually interacting* with one another.

Having established linear programming as a foundation for mathematical programming theory, we now are in a position to appreciate certain aspects of implementing mathematical programming models. In the next three chapters, we address several issues of mathematical-programming applications, starting with a general discussion and followed by specific applications. By necessity, at this point, the emphasis is on linear programming; however, most of our comments apply equally to other areas of mathematical programming, and we hope they will provide a useful perspective from which to view forthcoming developments in the text.

This chapter begins by presenting two frameworks that characterize the decision-making process. These frameworks suggest how modeling approaches to problem solving have evolved; specify those types of problems for which mathematical programming has had most impact; and indicate how other techniques can be integrated with mathematical-programming models. The remainder of the chapter concentrates on mathematical programming itself in terms of problem formulation and implementation, including the role of the computer. Finally, an example is presented to illustrate much of the material.

5.1 THE DECISION-MAKING PROCESS

Since management science basically aims to improve the quality of decision-making by providing managers with a better understanding of the consequences of their decisions, it is important to spend some time reflecting upon the nature of the decision-making process and evaluating the role that quantitative methods, especially mathematical programming, can play in increasing managerial effectiveness.

There are several ways to categorize the decisions faced by managers. We would like to discuss two frameworks, in particular, since they have proved to be extremely helpful in generating better insights into the decision-making process, and in defining the characteristics that a sound decision-support system should possess. Moreover, each framework leads to new perceptions concerning model formulation and the role of mathematical programming.

Anthony's Framework: Strategic, Tactical, and Operational Decisions

The first of these frameworks was proposed by Robert N. Anthony.* He classified decisions in three categories: strategic planning, tactical planning, and operations control. Let us briefly comment on the characteristics of each of these categories and review their implications for a model-based approach to support management decisions.

The examples given to illustrate specific decisions belonging to each category are based primarily on the production and distribution activities of a manufacturing firm. This is done simply for consistency and convenience; the suggested framework is certainly appropriate for dealing with broader kinds of decisions.

a) *Strategic Planning*

Strategic planning is concerned mainly with establishing managerial policies and with developing the necessary resources the enterprise needs to satisfy its external requirements in a manner consistent with its specific goals. Examples of strategic decisions are major capital investments in new production capacity and expansions of existing capacity, merger and divestiture decisions, determination of location and size of new plants and distribution facilities, development and introduction of new products, and issuing of bonds and stocks to secure financial resources.

These decisions are extremely important because, to a great extent, they are responsible for maintaining the competitive capabilities of the firm, determining its rate of growth, and eventually defining its success or failure. An essential characteristic of these strategic decisions is that they have long-lasting effects, thus mandating long planning horizons in their analysis. This, in turn, requires the consideration of uncertainties and risk attitudes in the decision-making process. Moreover, strategic decisions are resolved at fairly high managerial levels, and are affected by information that is both external and internal to the firm. Thus, any form of rational analysis of these decisions necessarily has a very broad scope, requiring information to be processed in a very aggregate form so that all the dimensions of the problem can be included and so that top managers are not distracted by unnecessary operational details.

---

* Robert N. Anthony, *Planning and Control Systems: A Framework for Analysis*, Harvard University Graduate School of Business Administration, Boston, 1965.

b) *Tactical Planning*

Once the physical facilities have been decided upon, the basic problem to be resolved is the effective allocation of resources (e.g., production, storage, and distribution capacities; work-force availabilities; marketing, financial, and managerial resources) to satisfy demand and technological requirements, taking into account the costs and revenues associated with the operation of the resources available to the firm. When we are dealing with several plants, many distribution centers, and regional and local warehouses, having products that require complex multistage fabrication and assembly processes, and serving broad market areas affected by strong randomness and seasonalities in their demand patterns, these decisions are far from simple. They usually involve the consideration of a medium-range time horizon divided into several periods, and require significant aggregation of the information to be processed. Typical decisions to be made within this context are utilization of regular and overtime work force, allocation of aggregate capacity resources to product families, definition of distribution channels, selection of transportation and transshipment alternatives, and allocation of advertising and promotional budgets.

c) *Operations Control*

After making an aggregate allocation of the resources of the firm, it is necessary to deal with the day-to-day operational and scheduling decisions. This requires the complete *dis*aggregation of the information generated at higher levels into the details consistent with the managerial procedures followed in daily activities. Typical decisions at this level are the assignment of customer orders to individual machines, the sequencing of these orders in the work shop, inventory accounting and inventory control activities, dispatching, expediting and processing of orders, vehicular scheduling, and credit granting to individual customers.

Table 5.1 Distinct Characteristics of Strategic, Tactical, and Operational Decisions

Characteristics	*Strategic planning*	*Tactical planning*	*Operations control*
Objective	Resource acquisition	Resource utilization	Execution
Time horizon	Long	Middle	Short
Level of management involvement	Top	Medium	Low
Scope	Broad	Medium	Narrow
Source of information	(External	& Internal)	Internal
Level of detail of information	Highly aggregate	Moderately aggregate	Low
Degree of uncertainty	High	Moderate	Low
Degree of risk	High	Moderate	Low

These three types of decisions differ markedly in various dimensions. The nature of these differences, expressed in relative terms, can be characterized as in Table 5.1.

Implications of Anthony's Framework: A Hierarchical Integrative Approach

There are significant conclusions that can be drawn from Anthony's classification, regarding the nature of the model-based decision-support systems. First, strategic, tactical, and operational decisions cannot be made in isolation because they interact strongly with one another. Therefore, an integrated approach is required in order to avoid suboptimization. Second, this approach, although essential, cannot be made without decomposing the elements of the problem in some way, within the context of a hierarchical system that links higher-level decisions with lower-level ones in an effective manner. Decisions that are made at higher levels of the system provide constraints for lower-level decision making, and decision-makers must recognize their impact upon lower-level operations.

This hierarchical approach recognizes the distinct characteristics of the type of management participation, the scope of the decision, the level of aggregation of the required information, and the time frame in which the decision is to be made. In our opinion, it would be a serious mistake to attempt to deal with all these decisions at once via a monolithic system (or model). Even if computer and methodological capabilities would permit the solution of large, detailed integrated models—clearly not the case today—that approach is inappropriate because it is not responsive to management needs at each level of the organization, and would prevent interactions between models and managers at each organization echelon.

In designing a system to support management decision-making, it is imperative, therefore, to identify ways in which the decision process can be partitioned, to select adequate models to deal with the individual decisions at each hierarchical level, to design linking mechanisms for the transferring of the higher-level results to the lower hierarchical levels, which include means to disaggregate information, and to provide quantitative measures to evaluate the resulting deviations from optimal performance at each level.

Mathematical programming is suited particularly well for supporting tactical decisions. This category of decisions, dealing with allocation of resources through a middle-range time horizon, lends itself quite naturally to representation by means of mathematical-programming models. Typically, tactical decisions generate models with a large number of variables and constraints due to the complex interactions among the choices available to the decision-maker. Since these choices are hard to evaluate on merely intuitive grounds, a decision-maker could benefit greatly from a model-based support effort. Historically, mathematical programming has been the type of model used most widely in this capacity, and has contributed a great deal to improving the quality of decision-making at the tactical level.

As we indicated before, tactical decisions are affected by only moderate uncertainties. This characteristic is useful for the application of linear programming, since models of this kind do not handle uncertainties directly. The impact of moderate uncertainties, however, can be assessed indirectly by performing sensitivity analysis. Furthermore, sensitivity tests and shadow price information allow the decision-maker to evaluate how well the resources of the firm are balanced.

For example, a tactical linear-programming model designed to support production-planning decisions might reveal insufficient capacity in a given stage of the production process. The shadow price associated with that capacity constraint provides a local indication of the payoff to be obtained from a unit increase in that limited capacity.

The role of mathematical programming in supporting strategic and operational decisions has been more limited. The importance of uncertainties and risk in strategic decisions, and the great amount of detailed information necessary to resolve operational problems work against the direct use of mathematical programming at these levels. In the decision-making hierarchy, mathematical-programming models become the links between strategic and operational decisions. By carefully designing a sequence of model runs, the decision-maker can identify bottlenecks or idle facilities; and this information provides useful feedback for strategic decisions dealing with acquisition or divestment of resources. On the other hand, by contributing to the optimum allocation of the aggregate resources of the firm, mathematical-programming models generate the broad guidelines for detailed implementation. Operational decisions are reduced, therefore, to producing the appropriate disaggregation of the tactical plans suggested by the mathematical-programming model against the day-to-day requirements of the firm.

The design of a hierarchical system to support the overall managerial process is an art that demands a great deal of pragmatism and experience. In Chapter 6 we describe an integrated system to deal with strategic and tactical decisions in the aluminum industry. The heart of the system is formed by two linear-programming models that actively interact with one another. In Chapter 10 we analyze a hierarchical system to decide on the design of a job-shop facility. In that case a mixed-integer programming model and a simulation model represent tactical and operational decisions, respectively. Chapter 7 presents another practical application of linear programming, stressing the role of sensitivity analysis in coping with future uncertainties. Chapter 14 discusses the use of decomposition for bond-portfolio decisions.

Simon's Framework: Programmed and Nonprogrammed Decisions

A second decision framework that is useful in analyzing the role of models in managerial decision-making was proposed by Herbert A. Simon.* Simon distinguishes two types of decisions: *programmed* and *nonprogrammed*, which also are referred to as *structured* and *unstructured* decisions, respectively. Although decisions cover a continuous spectrum between these two extremes, it is useful first to analyze the characteristics of these two kinds.

---

* Herbert A. Simon, *The Shape of Automation for Men and Management*, Harper and Row, 1965.

a) *Programmed Decisions*

Programmed decisions are those that occur routinely and repetitively. They can be structured into specific procedural instructions, so they can be delegated without undue supervision to the lower echelons of the organization. As Simon put it, programmed decisions are normally the domain of clerks. This does not necessarily imply that high-level managers do not have their share of programmed decision-making; it simply indicates that the incidence of programmed decisions increases the lower we go in the hierarchy of the organization.

b) *Nonprogrammed Decisions*

Nonprogrammed decisions are complex, unique, and unstructured. They usually do not lend themselves to a well defined treatment, but rather require a large amount of good judgment and creativity in order for them to be handled effectively. Normally, top managers are responsible for the more significant nonprogrammed decisions of any organization.

Implications of Simon's Framework: The Degree of Automation of the Decision-Making Process

Simon's framework is also very helpful in identifying the role of models in the decision-making process. Its implications are summarized in Table 5.2 which shows the contribution of both conventional and modern methods to support programmed and nonprogrammed decisions.

Table 5.2 The Implications of Simon's Taxonomy

Type of decision	Conventional methods	Modern methods
Programmed (Structured) Routine Repetitive	Organizational Structure Procedures and regulations Habit-forming	Electronic data processing Mathematical models
Nonprogrammed (Unstructured) Unique Complex	Policies Judgment and intuition Rules of thumb Training and promotion	Hierarchical system design Decision theory Heuristic problem solving

A major issue regarding programmed decisions is how to develop a systematic approach to cope with routine situations that an organization faces on a repetitive basis. A traditional approach is the preparation of written procedures and regulations that establish what to do under normal operating conditions, and that signal higher-management intervention whenever deviations from current practices occur. If properly designed and implemented, these procedures tend to create desirable patterns of behavior and habits among the personnel dealing with routine decisions. Control mechanisms normally are designed to motivate people, to measure the effectiveness of the decisions, and to take corrective actions if necessary.

What allows the proper execution and management of programmed decisions is the organizational structure of the firm, which breaks the management functions into problems of smaller and smaller scope. At the lower echelons of the organization, most of the work assumes highly structured characteristics and, therefore, can be delegated easily to relatively unskilled personnel.

During the last twenty years, we have witnessed a tremendous change in the way programmed decisions are made. First, the introduction of computers has created new capabilities to store, retrieve, and process huge amounts of information. When these capabilities are used intelligently, significant improvements can be made in the quality of decision-making at all levels. Second, the data bank that usually is developed in the preliminary stages of computer utilization invites the introduction of models to support management decisions. In many situations, these models have been responsible for adding to the structure of a given decision. Inventory control, production planning, and capital budgeting are just a few examples of managerial functions that were thought of as highly nonprogrammed, but now have become significantly structured.

Conventional methods for dealing with nonprogrammed decisions rely quite heavily on the use of judgment. Policies and rules of thumb sometimes can be formulated to guide the application of sound judgment and lessen the chances that poor judgment is exercised. Since good judgment seems to be the most precious element for securing high-quality decision-making in nonprogrammed decisions, part of the conventional efforts are oriented toward the recognition of those individuals who possess this quality. Management-development programs attempt to stimulate the growth of qualified personnel, and promotions tend to raise the level of responsibility of those who excel in their managerial duties.

One of the greatest disappointments for the advocates of management science is that, in its present form, it has not had a strong impact in dealing with nonprogrammed decisions. Broad corporate models have been constructed to help the coordination of the high-level managerial functions. Also, hierarchical systems are beginning to offer meaningful ways to approach problems of the firm in which various echelons of managers are involved. In addition, disciplines like decision theory and heuristic problem-solving have made contributions to the effective handling of nonprogrammed decisions. However, it is fair to say that we are quite short of achieving all the potentials of management science in this area of decision-making. This situation is changing somewhat with all the interest in unstructured social problems such as energy systems, population control, environmental issues, and so forth.

We can conclude, therefore, that the more structured the decision, the more it is likely that a meaningful model can be developed to support that decision.

5.2 STAGES OF FORMULATION, SOLUTION, AND IMPLEMENTATION

Having seen where mathematical programming might be most useful and indicated its interplay with other managerial tools, we will now describe an orderly sequence of steps that can be followed for a systematic formulation, solution, and implementation of a mathematical-programming model. These steps could be applied to the development of any management-science model. However, due to the nature of this book, we will limit our discussions to the treatment of mathematical-programming models.

Although the practical applications of mathematical programming cover a broad range of problems, it is possible to distinguish five general stages that the solution of any mathematical-programming problem should follow.

A. Formulating the model.

B. Gathering the data.

C. Obtaining an optimal solution.

D. Applying sensitivity analysis.

E. Testing and implementing the solution.

Obviously, these stages are not defined very clearly, and they normally overlap and interact with each other. Nevertheless we can analyze, in general terms, the main characteristics of each stage.

A) Formulating the Model

The first step to be taken in a practical application is the development of the model. The following are elements that define the model structure:

a) *Selection of a Time Horizon*

One of the first decisions the model designer has to make, when applying mathematical programming to a planning situation, is the selection of the time horizon (also referred to as planning horizon, or cutoff date). The time horizon indicates how long we have to look into the future to account for all the significant factors of the decision under study. Its magnitude reflects the future impact of the decision under consideration. It might cover ten to twenty years in a major capital-investment decision, one year in an aggregate production-planning problem, or just a few weeks in a detailed scheduling issue.

Sometimes it is necessary to divide the time horizon into several time periods. This is done in order to identify the dynamic changes that take place throughout the time horizon. For example, in a production-planning problem it is customary to divide the one-year time horizon into time periods of one month's duration to reflect fluctuations in the demand pattern due to product seasonalities. However, if the model is going to be updated and resolved at the beginning of each month with a rolling one-year time horizon, it could be advantageous to consider unequal time periods; for example, to consider time periods of one month's duration during the first three months, and then extend the duration of the time periods to three months, up to the end of the time horizon. This generally will not adversely affect the quality of the decisions and will reduce the computational burden on the model significantly.

b) *Selection of Decision Variables and Parameters*

The next step in formulating the mathematical-programming model is to identify the decision variables, which are those factors under the control of the decision-maker, and the parameters, which are beyond the control of the decision-maker and are imposed by the external environment.

The decision variables are the answers the decision-maker is seeking. If we are dealing with production-planning models, some relevant decision variables might be the amount to be manufactured of each product at each time period, the amount of inventory to accumulate at each time period, and the allotment of man-hours of regular and overtime labor at each time period.

On some occasions, a great amount of ingenuity is required to select those decision variables that most adequately describe the problem being examined. In some instances it is possible to decrease the number of constraints drastically or to transform an apparent nonlinear problem into a linear one, by merely defining the decision variables to be used in the model formulation in a different way.

The parameters represent those factors which affect the decision but are not controllable directly (such as prices, costs, demand, and so forth). In deterministic mathematical-programming models, all the parameters are assumed to take fixed, known values, where estimates are provided via point forecasts. The impact of this assumption can be tested by means of sensitivity analysis. Examples of some of the parameters associated with a production-planning problem are: product demands, finished product prices and costs, productivity of the manufacturing process, and manpower availability.

The distinction made between parameters and decision variables is somewhat arbitrary, and one could argue that, for a certain price, most parameters can be controlled to some degree by the decision-maker. For instance, the demand for products can be altered by advertising and promotional campaigns; costs and prices can be increased or decreased within certain margins, and so on. We always can start, however, from a reference point that defines the appropriate values for the parameters, and insert as additional decision variables those actions the decision-maker can make (like promotions or advertising expenditures) to create changes in the initial values of the parameters. Also, shadow-price information can be helpful in assessing the consequences of changing the values of the initial parameters used in the model.

c) *Definition of the Constraints*

The constraint set reflects relationships among decision variables and parameters that are imposed by the characteristics of the problem under study (e.g., the nature of the production process, the resources available to the firm, and financial, marketing, economical, political, and institutional considerations). These relationships should be expressed in a precise, quantitative way. The nature of the constraints will, to a great extent, determine the computational difficulty of solving the model.

It is quite common, in the initial representation of the problem, to overlook some vital constraints or to introduce some errors into the model description, which will lead to unacceptable solutions. However, the mathematical programming solution of the ill-defined model provides enough information to assist in the detection of these errors and their prompt correction. The problem has to be reformulated and a new cycle has to be initiated.

d) *Selection of the Objective Function*

Once the decision variables are established, it is possible to determine the objective function to be minimized or maximized, provided that a measure of performance (or effectiveness) has been established and can be associated with the values that the decision variables can assume. This measure of performance provides a selection criterion for evaluating the various courses of action that are available in the situation being investigated. The most common index of performance selected in business applications is *dollar value*; thus, we define the objective function as the minimization of cost or the maximization of profit. However, other objective functions could become more relevant in some instances. Examples of alternative objectives are:

> Maximize total production, in units.
> Minimize production time.
> Maximize share of market for all or some products.
> Maximize total sales, in dollars or units.
> Minimize changes of production pattern.
> Minimize the use of a particular scarce (or expensive) commodity.

The definition of an acceptable objective function might constitute a serious problem in some situations, especially when social and political problems are involved. In addition, there could be conflicting objectives, each important in its own right, that the decision-maker wants to fulfill. In these situations it is usually helpful to define multiple objective functions and to solve the problem with respect to each one of them separately, observing the values that all the objective functions assume in each solution. If no one of these solutions appears to be acceptable, we could introduce as additional constraints the minimum desirable performance level of each of the objective functions we are willing to accept, and solve the problem again, having as an objective the most relevant of those objective functions being considered. Sequential tests and sensitivity analysis could be quite valuable in obtaining satisfactory answers in this context.

Another approach available to deal with the problem of multiple objectives is to unify all the conflicting criteria into a single objective function. This can be accomplished by attaching weights to the various measures of performance established by the decision-maker, or by directly assessing a multiattribute preference (or utility) function of the decision-maker. This approach, which is conceptually extremely attractive but requires a great deal of work in implementation, is the concern of a discipline called *decision theory* (or decision analysis) and is outside the scope of our present work.*

---

* For an introduction to decision theory, the reader is referred to Howard Raiffa, *Decision Analysis—Introductory Lectures on Choices under Uncertainty*, Addison-Wesley, 1970.

B) Gathering the Data

Having defined the model, we must collect the data required to define the parameters of the problem. The data involves the objective-function coefficients, the constraint coefficients (also called the matrix coefficients) and the righthand side of the mathematical-programming model. This stage usually represents one of the most time-consuming and costly efforts required by the mathematical-programming approach.

C) Obtaining an Optimal Solution

Because of the lengthy calculations required to obtain the optimal solution of a mathematical-programming model, a digital computer is invariably used in this stage of model implementation. Today, all the computer manufacturers offer highly efficient codes to solve linear-programming models. These codes presently can handle general linear-programming problems of up to 4000 rows, with hundreds of thousands of decision variables, and are equipped with sophisticated features that permit great flexibility in their operation and make them extraordinarily accurate and effective. Instructions for use of these codes are provided by the manufacturers; they vary slightly from one computer firm to another.

Recently, very efficient specialized codes also have become available for solving mixed-integer programming problems, separable programming problems, and large-scale problems with specific structures (using generalized upper-bounding techniques, network models, and partitioning techniques). Models of this nature and their solution techniques will be dealt with in subsequent chapters.

When dealing with large models, it is useful, in utilizing computer programs, to input the required data automatically. These programs, often called matrix generators, are designed to cover specific applications. Similarly, computer programs often are written to translate the linear-programming output, usually too technical in nature, into meaningful managerial reports ready to be used by middle and top managers. In the next section, dealing with the role of the computer in solving mathematical-programming models, we will discuss these issues more extensively.

D) Applying Sensitivity Analysis

One of the most useful characteristics of linear-programming codes is their capability to perform sensitivity analysis on the optimum solutions obtained for the problem originally formulated. These postoptimum analyses are important for several reasons:

a) *Data Uncertainty*

Much of the information that is used in formulating the linear program is uncertain. Future production capacities and product demand, product specifications and requirements, cost and profit data, among other information, usually are evaluated through projections and average patterns, which are far from being known with complete accuracy. Therefore, it is often significant to determine how sensitive the optimum solution is to changes in those quantities, and how the optimum solution varies when actual experience deviates from the values used in the original model.

b) *Dynamic Considerations*

Even if the data were known with complete certainty, we would still want to perform sensitivity analysis on the optimum solution to find out how the recommended courses of action should be modified after some time, when changes most probably will have taken place in the original specification of the problem. In other words, instead of getting merely a static solution to the problem, it is usually desirable to obtain at least some appreciation for a dynamic situation.

c) *Input Errors*

Finally, we want to inquire how errors we may have committed in the original formulation of the problem may affect the optimum solution.

In general, the type of changes that are important to investigate are changes in the objective-function coefficients, in the righthand-side elements, and in the matrix coefficients. Further, it is sometimes necessary to evaluate the impact on the objective function of introducing new variables or new constraints into the problem. Although it is often impossible to assess all of these changes simultaneously, good linear-programming codes provide several means of obtaining pertinent information about the impact of these changes with a minimum of extra computational work.

Further discussions on this topic, including a description of the types of output generated by computer codes, are presented in the next section of this chapter.

E) Testing and Implementing the Solution

The solution should be tested fully to ensure that the model clearly represents the real situation. We already have pointed out the importance of conducting sensitivity analysis as part of this testing effort. Should the solution be unacceptable, new refinements have to be incorporated into the model and new solutions obtained until the mathematical-programming model is adequate.

When testing is complete, the model can be implemented. Implementation usually means solving the model with real data to arrive at a decision or set of decisions. We can distinguish two kinds of implementation: a single or one-time use, such as a plant location problem; and continual or repeated use of the model, such as production planning or blending problems. In the latter case, routine systems should be established to provide input data to the linear-programming model, and to transform the output into operating instructions. Care must be taken to ensure that the model is flexible enough to allow for incorporating changes that take place in the real operating system.

5.3 THE ROLE OF THE COMPUTER

Solving even small mathematical-programming problems demands a prohibitive amount of computational effort, if undertaken manually. Thus, all practical applications of mathematical programming require the use of computers.

In this section we examine the role of digital computers in solving mathematical-programming models. The majority of our remarks will be directed to linear-programming models, since they have the most advanced computer support and are the models used most widely in practice.

The task of solving linear programs is greatly facilitated by the very efficient linear-programming codes that most computer manufacturers provide for their various computer models. In a matter of one or two days, a potential user can become familiar with the program instructions (control language or control commands) that have to be followed in order to solve a given linear-programming problem by means of a specific commercial code. The ample availability of good codes is one of the basic reasons for the increasing impact of mathematical programming on management decision-making.

Commercial mathematical-programming codes have experienced an interesting evolution during the last decade. At the beginning, they were simple and rigid computer programs capable of solving limited-size models. Today, they are sophisticated and highly flexible information-processing systems, with modules that handle incoming information, provide powerful computational algorithms, and report the model results in a way that is acceptable to a manager or an application-oriented user. Most of the advances experienced in the design of mathematical-programming codes have been made possible by developments in software and hardware technology, as well as break-throughs in mathematical-programming theory. The continual improvement in mathematical-programming computational capabilities is part of the trend to reduce running times and computational costs, to facilitate solutions of larger problems (by using network theory and large-scale system theory), and to extend the capabilities beyond the solution of linear-programming models into nonlinear and integer-programming problems.

The field of mathematical-programming computer systems is changing rapidly and the number of computer codes currently available is extremely large. For these reasons, we will not attempt to cover the specific instructions of any linear-programming code in detail. Rather, we will concentrate on describing the basic features common to most good commercial systems, emphasizing those important concepts the user has to understand in order to make sound use of the currently existing codes.

Input Specifications

The first task the user is confronted with, when solving a linear-programming model by means of a commercial computer code, is the specification of the model structure and the input of the corresponding values of the parameters. There are several options available to the user in the performance of this task. We will examine some of the basic features common to most good codes in dealing with input descriptions.

In general, a linear-programming problem has the following structure:

$$\text{Maximize (or Minimize) } z = \sum_{j=1}^{n} c_j x_j, \tag{1}$$

subject to:

$$\sum_{j=1}^{n} a_{ij}x_j \ (\geq, =, \leq) \ b_i, \qquad (i = 1, 2, \ldots, m), \quad (2)$$

$$x_j \geq 0, \qquad \text{for some or all } j, \qquad\qquad (3)$$

$$x_j \text{ free}, \qquad \text{for some or all } j, \qquad\qquad (4)$$

$$x_j \geq \ell_j, \quad (\ell_j \neq 0), \qquad \text{for some or all } j, \quad (5)$$

$$x_j \leq u_j, \quad (u_j \neq \infty), \qquad \text{for some or all } j. \quad (6)$$

This model formulation permits constraints to be specified either as "greater than or equal to" ($\geq$) inequalities, "less than or equal to" ($\leq$) inequalities, or simple equalities ($=$). It also allows for some variables to be nonnegative, some to be unconstrained in sign, and some to have lower and/or upper bounds.

a) Reading the Data

The most important elements the user has to specify in order to input the problem data are:

Names of the decision variables, constraints, objective function(s), and righthand side(s). Names have to be given to each of this model's elements so that they can be identified easily by the user. The names cannot exceed a given number of characters (normally 6 or 8), depending on the code being used. Moreover, it is important to adopt proper mnemonics in assigning names to these elements for easy identification of the actual meaning of the variables and constraints. For instance, instead of designating by X37 and X38 the production of regular and white-wall tires during the month of January, it would be better to designate them by RTJAN and WTJAN, respectively.

Numerical values of the parameters a_{ij}, b_i, and c_j. Only nonzero coefficients have to be designated. This leads to significant savings in the amount of information to be input, since a great percentage of coefficients are zero in most linear-programming applications. The specification of a nonzero coefficient requires three pieces of information: the row name, the column name, and the numerical value corresponding to the coefficient.

Nature of the constraint relationship. The user should indicate whether a given constraint has a $=$, $\geq$, or $\leq$ relationship. Also, many systems permit the user to specify a range on a constraint consisting of both the $\geq$ and $\leq$ relationships.

Free and bounded variables. Almost all codes assume the variables to be nonnegative unless specified otherwise. This reduces the amount of data to be input, since in most practical applications the variables are indeed nonnegative. Thus the user need only specify those variables that are "free" (i.e., bounded neither from above nor from below), and those variables that have lower and upper bounds. The lower and upper bounds may be either positive or negative numbers.

Nature of the optimization command. Instructions have to establish whether the objective function is to be minimized or maximized. Moreover, most linear-programming codes allow the user to specify, if so desired, several objective functions and righthand sides. The code then proceeds to select one objective function and one righthand side at a time, in a sequence prescribed by the user, and the optimum solution is found for each of the resulting linear programs. This is a practical and effective way to perform sensitivity analysis. Whenever several objective functions and righthand sides are input, the user should indicate which one of these elements to use for a given optimization run.

Title of problem and individual runs.   It is useful to assign a title to the overall problem and to each of the sensitivity runs that might be performed. These titles have to be supplied externally by the user.

     The information given to the computer is stored in the main memory core or, if the problem is extremely large, in auxiliary memory devices. Needless to say, the computational time required to solve a problem increases significantly whenever auxiliary memory is required, due to the added information-processing time. Within the state of the current computer technology, large computers (e.g., IBM 370/195) can handle problems with up to 4000 constraints without using special large-scale theoretical approaches.

b) *Retrieving a Summary of the Input Data*

Most codes permit the user to obtain summarized or complete information on the input data. This information is useful in identifying possible errors due to careless data specification, or to inappropriate model formulation. In addition, the input summary can be of assistance in detecting any special structure of the model that can be exploited for more efficient computational purposes.

     The following are examples of types of summary information on the data input that are available in many codes:

     Number of rows and columns in the coefficient matrix,

     Number of righthand sides and objective functions,

     Density of the coefficient matrix (i.e., the percentage of nonzero coefficients in the total number of coefficients),

     Number of nonzero elements in each row and column,

     Value of the largest and smallest element in the coefficient matrix, righthand side(s), and objective function(s),

     Printout of all the equations in expanded form,

     Picture of the full coefficient matrix or a condensed form of the coefficient matrix.

c) *Scaling the Input Data*

The simultaneous presence of both large and small numbers in the matrix of coefficients should be avoided whenever possible, because it tends to create problems of numerical instability. This can be avoided by changing the unit of measure of a given variable (say, from lbs to 100 lbs, or from $1000 to $)—this change will reduce (or increase) the numerical values on a given column; or by dividing (or multiplying) a row by a constant, thus reducing (or increasing) the numerical values of the coefficients belonging to that row.

d) *Matrix Generators*

When dealing with large linear-programming models, the task of producing the original matrix of coefficients usually requires an extraordinary amount of human effort and elapsed time. Moreover, the chance of performing these tasks without errors is almost nil. For example, a problem consisting of 500 rows and 2000 variables, with a matrix density of 10 percent, has $500 \times 2000 \times 0.10 = 100{,}000$ nonzero coefficients that need to be input. As we indicated before, most linear-programming codes require three pieces of information to be specified when reporting a nonzero coefficient; these are the corresponding row name, column name, and coefficient value. Thus, in our example the user will need to input 300,000 elements. If the input operations are to be performed manually, the chance of absolute freedom from error is extremely low.

Moreover, in most practical applications, there is a great need for restructuring the matrix of coefficients due to dynamic changes in the problem under study, to model-formulation alternatives to be analyzed, and to sensitivity runs to be performed. A matrix generator helps address these difficulties by providing an effective mechanism to produce and revise the coefficient matrix.

A matrix generator is a computer program that allows the user to input the linear-programming data in a convenient form by prescribing only a minimum amount of information. Normally, it takes advantage of the repetitive structure that characterizes most large linear-programming models, where many of the coefficients are $+1$, -1, or zero. The matrix generator fills in those coefficients automatically, leaving to the user the task of providing only the input of those coefficients the program cannot anticipate or compute routinely. Multistage or multiperiod models have the tendency to repeat a given submatrix of coefficients several times. The matrix generator automatically duplicates this submatrix as many times as necessary to produce the final model structure. Matrix generators also provide consistency checks on input data to help detect possible sources of error.

Matrix generators also can be used to compute the values of some of the model coefficients directly from raw data provided by the user (e.g., the program might calculate the variable unit cost of an item by first evaluating the raw material costs, labor costs, and variable overhead). Moreover, the matrix generator can provide validity checks to ensure that no blunt errors have been introduced in the data. Simple checks consist of counting the number of nonzero coefficients in some or all the rows and columns, in analyzing the sign of some critical variable coefficients, and so forth.

Sophisticated mathematical-programming codes have general-purpose matrix generators as part of their system. Instances are also quite common where the user can prepare his own special-purpose matrix generator to be used as part of the linear-programming computations.

Solution Techniques

There has been considerable concern in the development of mathematical-programming systems to reduce the computational time required to solve a linear-programming problem by improving the techniques used in the optimization stage of the process. This has been accomplished primarily by means of refinements in the matrix-inversion procedures (derived from matrix triangularization concepts), improvements in the path-selection criterion (by allowing multiple pricing of several columns to be considered as possible incoming variables), and use of more efficient presolution techniques (by providing *crashing* options to be used in order to obtain an initial feasible solution, or by permitting the user to specify starting columns in the initial basis).

Other important issues associated with the use of computers in solving linear-programming problems are the presence of numerical errors and the tolerances allowed in defining the degree of accuracy sought in performing the computations. Most codes permit the user to exercise some options with regard to these issues. We will comment briefly on some of the most important features of those options.

A. *Reinversion*

Since the simplex method is a numeric computational procedure, it is inevitable for roundoff errors to be produced and accumulated throughout the sequence of iterations required to find the optimum solution. These errors could create infeasibility or nonoptimal conditions; that is, when the final solution is obtained and the corresponding values of the variables are substituted in the original set of constraints, either the solution might not satisfy the original requirements, or the pricing-out computations might generate some nonzero reduced costs for the basic variables.

To address these problems, most codes use the original values of the coefficient matrix to reinvert the basis periodically, thus maintaining its accuracy. The frequency of reinversion can be specified externally by the user; otherwise, it will be defined internally by the mathematical-programming system. Some computer codes fix the automatic reinversion frequency at a constant number of iterations performed; e.g., the basis is reinverted every 50 to 100 iterations. Other codes define the reinversion frequency as a multiple of the number of rows in the problem to be optimized; e.g. the basis is reinverted every 1.25 to 1.40m iterations, where m is the number of constraints.

Most codes compute a basis reinversion when the final solution has been obtained to assure its feasibility and optimality.

B. *Tolerances*

In order to control the degree of accuracy obtained in the numerical computations, most linear-programming codes permit the user to specify the tolerances he is willing to accept. Feasibility tolerances are designed to check whether or not the values of the nonbasic variables and the reduced cost of the basic variables are zero. Unless otherwise specified by the user, these checks involve internally set-up tolerance limits. Typical limits are -0.000001 and $+0.000001$.

There are also pivot rejection tolerances that prevent a coefficient very close to zero from becoming a pivot element in a simplex iteration. Again, 10^{-6} is a normal tolerance to use for this purpose, unless the user specifies his own limits.

C. *Errors*

As we have indicated before, errors usually are detected when the optimum values of the variables are substituted in the original set of constraints. *Row errors* measure the difference between the computed values of the lefthand sides of the constraints, and the original righthand-side values, i.e., for equality constraints:

$$\sum_{j=1}^{n} a_{ij}x_j^0 - b_i = i\text{th row error},$$

where the x_j^0's are the corresponding optimum values of the variable. Similarly, there are *column errors*, which are calculated by performing similar computations with the dual problem, i.e., for a basic column:

$$\sum_{i=1}^{m} a_{ij}y_i^0 - c_i = j\text{th column error},$$

where the y_i^0's are the corresponding optimum values for the dual variables. Some codes have an internally determined or externally specified error frequency, which dictates how often errors are calculated and checked against the feasibility tolerances. If the computed errors exceed the prescribed tolerances, a *basis reinversion* is called for automatically. Restoring feasibility might demand a Phase I computation, or a dual simplex computation.

Many codes offer options to the user with regard to the output of the error calculations. The user might opt for a detailed printout of all column and row errors, or might be satisfied with just the maximum column and row error.

Whenever extra accuracy is required, some codes allow *double precision* to be used. This means that the space reserved in the computer for the number of significant figures to represent the value of each variable is doubled. Computing with the additional significant digits provides additional precision in the calculations required by the solution procedure.

Output Specifications

Once the optimum solution has been obtained, most computer codes provide the user with a great amount of information that describes in detail the specific values of the optimum solution and its sensitivity to changes in some of the parameters of the original linear-programming problem. We will review some of the most important features associated with output specifications.

a) *Standard Output Reports*

Typical information produced by standard commercial codes might include:

Optimum value of the objective function.   This is obtained by substituting the optimum values of the decision variables in the original objective function. When the model has been constructed properly, the value of the objective function is a finite number. If no feasible solution exists, the computer program will indicate that the problem is infeasible, and no specific value of the objective function will be given. The remaining alternative occurs when the objective function can be increased (if we are maximizing) or decreased (if we are minimizing) indefinitely. In this situation, the program will report an unbounded solution, which is a clear indication in any practical application that an error has been made in the model formulation.

Optimum values of the decision variables.   For each of the decision variables the program specifies its optimum value. Except for degenerate conditions, all the basic variables assume positive values. Invariably, all nonbasic variables have zero values. There are as many basic variables as constraints in the original problem.

Slacks and surpluses in the constraints.   "Less than or equal to" ($\leq$) constraints might have positive slacks associated with them. Correspondingly, "greater than or equal to" ($\geq$) constraints might have positive surpluses associated with them. The program reports the amount of these slacks and surpluses in the optimum solution. Normally, a slack corresponds to an unused amount of a given resource, and a surplus corresponds to an excess above a minimum requirement.

Shadow prices for constraints.   The shadow price associated with a given constraint corresponds to the change in the objective function when the original righthand side of that constraint is increased by one unit. Shadow prices usually can be interpreted as marginal costs (if we are minimizing) or marginal profits (if we are maximizing). Constraints that have positive slacks or surpluses have zero shadow prices.

Reduced costs for decision variables.   Reduced costs can be interpreted as the shadow prices corresponding to the nonnegativity constraints. All basic variables have zero reduced costs. The reduced cost associated with a nonbasic corresponds to the change in the objective function whenever the corresponding value of the nonbasic variable is increased from 0 to 1.

Ranges on coefficients of the objective function.   Ranges are given for each decision variable, indicating the lower and upper bound the cost coefficient of the variable can take without changing the current value of the variable in the optimum solution.

Ranges on coefficients of the righthand side.   Ranges are given for the righthand-side element of each constraint, indicating the lower and upper value the righthand-side coefficient of a given constraint can take without affecting the value of the shadow price associated with that constraint.

Variable transitions resulting from changes in the coefficients of the objective function. Whenever a coefficient of the objective function is changed beyond the range prescribed above, a change of the basis will take place. This element of the output report shows, for each variable, what variable will leave the basis and what new variable will enter the basis if the objective-function coefficient of the corresponding variable were to assume a value beyond its current range. If there is no variable to drop, the problem becomes unbounded.

Variable transition resulting from changes in the coefficient of the righthand side.
Similarly, whenever a coefficient of the righthand side of a constraint is changed
beyond the range prescribed above, a change in the current basis will occur. This
portion of the report shows, for each constraint, which variable will leave the basis
and which new variable will enter the basis if the righthand-side coefficient of the
corresponding constraint were to assume a value beyond its current range. If there
is no variable to enter, the problem becomes infeasible.

In the next section we present a simple illustration of these output reports.

b) *Sensitivity and Parametric Analyses*

In addition to providing the information just described, most codes allow the user
to perform a variety of sensitivity and parametric runs, which permit an effective
analysis of the changes resulting in the optimum solution when the original specifications of the problem are modified. Quite often the user is not interested in obtaining
a single solution to a problem, but wants to examine thoroughly a series of solutions
to a number of cases. Some of the system features to facilitate this process are:

Multiple objective functions and righthand sides. We noticed before that provisions
are made in many codes for the user to input several objective functions. The problem
is solved with one objective function at a time, the optimum solution of one problem
serving as an initial solution for the new problem. Sometimes only a few iterations
are required to determine the new optimum, so that this process is quite effective.
In this fashion, changes of the optimal solution with changes in the cost or revenue
structure of the model can be assessed very rapidly. Similar options are available
for processing the problem with one righthand side at a time, through a sequence of
righthand sides provided by the user. Some codes use the dual simplex method
for this purpose.

Parametric Variation. Another way to assess sensitivity analysis with regard to
objective functions and righthand sides is to allow continuous changes to occur from
a specified set of coefficients for the objective function or the righthand side to another
specified set. This continuous parametrization exhaustively explores the pattern of
the solution sensitivity in a very efficient manner. Some codes allow for a joint parametrization of cost coefficients and righthand-side elements.

Revisions of the original model formulation. Finally, many codes allow revisions to
be incorporated in the model structure without requiring a complete reprocessing
of the new problem. These revisions might effect changes in specific parameters of
the model, as well as introducing new variables and new constraints.

c) *Report generators*

Given the massive amount of highly technical information produced by a linear-
programming code, it is most desirable to translate the output into a report directed
to an application-oriented user. This is almost mandatory when using linear pro-

gramming as an operational tool to support routine managerial decision-making. The most sophisticated mathematical-programming systems contain capabilities for the generation of general-purpose reports. Special-purpose reports easily can be programmed externally in any one of the high-level programming languages (like FORTRAN, APL, or BASIC).

5.4 A SIMPLE EXAMPLE

The basic purpose of this section is to illustrate, via a very small and simple example, the elements contained in a typical computer output of a linear-programming model. The example presented is a multistage production-planning problem. In order to simplify our discussion and to facilitate the interpretation of the computer output, we have limited the size of the problem by reducing to a bare minimum the number of machines, products, and time periods being considered. This tends to limit the degree of realism of the problem, but greatly simplifies the model formulation. For some rich and realistic model-formulation examples, the reader is referred to the exercises at the end of this chapter, and to Chapters 6, 7, 10, and 14.

A Multistage Planning Problem

An automobile tire company has the ability to produce both nylon and fiberglass tires. During the next three months they have agreed to deliver tires as follows:

Date	Nylon	Fiberglass
June 30	4,000	1,000
July 31	8,000	5,000
August 31	3,000	5,000
Total	15,000	11,000

The company has two presses, a Wheeling machine and a Regal machine, and appropriate molds that can be used to produce these tires, with the following production hours available in the upcoming months:

	Wheeling machine	Regal machine
June	700	1500
July	300	400
August	1000	300

The production rates for each machine-and-tire combination, in terms of *hours per tire*, are as follows:

	Wheeling machine	Regal machine
Nylon	0.15	0.16
Fiberglass	0.12	0.14

The variable costs of producing tires are $5.00 per operating hour, regardless of which machine is being used or which tire is being produced. There is also an inventory-carrying charge of $0.10 per tire per month. Material costs for the nylon and fiberglass tires are $3.10 and $3.90 per tire, respectively. Finishing, packaging and shipping costs are $0.23 per tire. Prices have been set at $7.00 per nylon tire and $9.00 per fiberglass tire.

The following questions have been raised by the production manager of the company:

a) How should the production be scheduled in order to meet the delivery requirements at minimum cost?

b) What would be the total contribution to be derived from this optimum schedule?

c) A new Wheeling machine is due to arrive at the beginning of September. For a $200 fee, it would be possible to expedite the arrival of the machine to August 2, making available 172 additional hours of Wheeling machine time in August. Should the machine be expedited?

d) When would it be appropriate to allocate time for the yearly maintenance checkup of the two machines?

Model Formulation

We begin by applying the steps in model formulation recommended in Section 5.2.

Selection of a Time Horizon

In this particular situation the time horizon covers three months, divided into three time periods of a month's duration each. More realistic production-planning models normally have a full-year time horizon.

Selection of Decision Variables and Parameters

We can determine what decision variables are necessary by defining exactly what information the plant foreman must have in order to schedule the production. Essentially, he must know the number of each type of tire to be produced on each machine in each month and the number of each type of tire to place in inventory at the end of each month. Hence, we have the following decision variables:

$W_{n,t}$ = Number of nylon tires to be produced on the Wheeling machine during month t;

$R_{n,t}$ = Number of nylon tires to be produced on the Regal machine during month t;

$W_{g,t}$ = Number of fiberglass tires to be produced on the Wheeling machine in month t;

$R_{g,t}$ = Number of fiberglass tires to be produced on the Regal machine in month t;

$I_{n,t}$ = Number of nylon tires put into inventory at the end of month t;

$I_{g,t}$ = Number of fiberglass tires put into inventory at the end of month t.

In general there are six variables per time period and since there are three months under consideration, we have a total of eighteen variables. However, it should be clear that it would never be optimal to put tires into inventory at the end of August since all tires must be *delivered* by then. Hence, we can ignore the inventory variables for August.

The parameters of the problem are represented by the demand requirements, the machine availabilities, the machine productivity rates, and the cost and revenue information. All these parameters are assumed to be known deterministically.

Definition of the Constraints

There are two types of constraint in this problem representing production-capacity available and demand requirements at each month.

Let us develop the constraints for the month of June. The production-capacity constraints can be written in terms of production hours on each machine. For the Wheeling machine in June we have:

$$0.15W_{n,1} + 0.12W_{g,1} \leq 700,$$

while, for the Regal machine in June, we have:

$$0.16R_{n,1} + 0.14R_{g,1} \leq 1500.$$

The production constraints for future months differ only in the available number of hours of capacity for the righthand side.

Now consider the demand constraints for June. For each type of tire produced in June we must meet the demand requirement and then put any excess production into inventory. The demand constraint for nylon tires in June is then:

$$W_{n,1} + R_{n,1} - I_{n,1} = 4000,$$

while, for fiberglass tires for June, it is:

$$W_{g,1} + R_{g,1} - I_{g,1} = 1000.$$

In July, however, the tires put into inventory in June are available to meet demand. Hence, the demand constraint for nylon tires in July is:

$$I_{n,1} + W_{n,2} + R_{n,2} - I_{n,2} = 8000,$$

while for fiberglass tires in July it is:

$$I_{g,1} + W_{g,2} + R_{g,2} - I_{g,2} = 5000.$$

In August it is clear that tires will *not* be put into inventory at the end of the month, so the demand constraint for nylon tires in August is:

$$I_{n, 2} + W_{n, 3} + R_{n, 3} = 3000,$$

while for fiberglass tires in August it is:

$$I_{g, 2} + W_{g, 3} + R_{g, 3} = 5000.$$

Finally, we have the nonnegativity constraints on all of the decision variables:

$$
\begin{array}{lll}
W_{n, t} \geq 0, & W_{g, t} \geq 0, & (t = 1, 2, 3); \\
R_{n, t} \geq 0, & R_{g, t} \geq 0, & (t = 1, 2, 3); \\
I_{n, t} \geq 0, & I_{g, t} \geq 0, & (t = 1, 2).
\end{array}
$$

Selection of the Objective Function

The total revenues to be obtained in this problem are fixed, because we are meeting all the demand requirements, and maximization of profit becomes equivalent to minimization of cost. Also, the material-cost component is fixed, since we know the total amount of each product to be produced during the model time horizon. Thus, a proper objective function to select is the minimization of the variable relevant cost components: variable production costs plus inventory-carrying costs.

Now, since each kind of tire on each machine has a different production rate, the cost of producing a tire on a particular machine will vary, even though the variable cost per hour is constant for each tire-and-machine combination. The variable production cost per tire for the fiberglass tires made on the Regal machine can be determined by multiplying the production rate (in hours/tire) by the variable production cost (in $/hour) resulting in $(0.14)(5) = \$0.70$/tire. The remaining costs for producing each tire on each machine can be computed similarly, yielding

	Wheeling machine	Regal machine
Nylon	0.75	0.80
Fiberglass	0.60	0.70

Given the inventory-carrying cost of $0.10 per tire per month, we have the following objective function for minimizing costs:

$$\sum_{t=1}^{3} (0.75 W_{n, t} + 0.80 R_{n, t} + 0.60 W_{g, t} + 0.70 R_{g, t} + 0.10 I_{n, t} + 0.10 I_{g, t});$$

and we understand that $I_{n,3} = 0$ and $I_{g,3} = 0$.

The formulation of this problem is summarized in Table 5.3. This problem is what we call a multistage model, because it contains more than one time period. Note that the constraints of one time period are linked to the constraints of another only by the inventory variables. This type of problem structure is very common in mathematical programming. Note that there are very few elements different from 0, 1, and -1 in the tableau given in Table 5.3. This problem structure can be exploited easily in the design of a matrix generator, to provide the input for the linear-programming computation.

Table 5.3 Multistage Planning Model*

		June					July						August					RHS	
		Nylon		Glass		Inventory		Nylon		Glass		Inventory		Nylon		Glass			
		$W_{n,1}$	$R_{n,1}$	$W_{g,1}$	$R_{g,1}$	$I_{n,1}$	$I_{g,1}$	$W_{n,2}$	$R_{n,2}$	$W_{g,2}$	$R_{g,2}$	$I_{n,2}$	$I_{g,2}$	$W_{n,3}$	$R_{n,3}$	$W_{g,3}$	$R_{g,3}$		
Machine time constraints	June Wheeling	0.15		0.12														$\leq$	700
	June Regal		0.16		0.14													$\leq$	1500
	July Wheeling							0.15		0.12								$\leq$	300
	July Regal								0.16		0.14							$\leq$	400
	Aug. Wheeling													0.15		0.12		$\leq$	1000
	Aug. Regal														0.16		0.14	$\leq$	300
Demand constraints	June Nylon	1	1			−1												=	4000
	June Glass			1	1		−1											=	1000
	July Nylon					1		1	1			−1						=	8000
	July Glass						1			1	1		−1					=	5000
	Aug. Nylon											1		1	1			=	3000
	Aug. Glass												1			1	1	=	5000
Objective		0.75	0.80	0.60	0.70	0.10	0.10	0.75	0.80	0.60	0.70	0.10	0.10	0.75	0.80	0.60	0.70		Minimum

Hours available: 700, 1500, 300, 400, 1000, 300

Number of tires demanded: 4000, 1000, 8000, 5000, 3000, 5000

* All blanks are zeros.

Computer Results

We will now present the computer output obtained by solving the linear-programming model set forth in Table 5.3 by means of an interactive system operated from a computer terminal. The notation describing the decision variables has been changed slightly, in order to facilitate computer printouts. For example:

$$WN-T = \text{Number of nylon tires produced}$$
$$\text{on the Wheeling machine in month T.}$$

Similar interpretations can be given to variables WG–T (number of fiberglass tires on the Wheeling machine), RN–T, and RG–T (number of nylon and fiberglass tires, respectively, produced on the Regal machine at month T). IN–T and IG–T denote the number of nylon and fiberglass tires, respectively, left over at the end of period T.

The production constraints are represented by W–T and R–T, meaning the hours of Wheeling and Regal machine availability at period T. N–T and G–T stand for the demand at period T of nylon and fiberglass tires, respectively.

Figure 5.1 is the computer output of the problem. The reader should reflect about the meaning of each of the elements of the output. The output provides exactly the same information discussed under the title Standard Output Reports in Section 5.3. The reader is referred to that section for a detailed explanation of each output element.

Answering the Proposed Questions

With the aid of the optimum solution of the linear-programming model, we can answer the questions that were formulated at the beginning of this problem.

a) *Production Scheduling*

Examination of the optimum values of the decision variables in the linear-programming solution reveals that the appropriate production schedule should be:

		Wheeling machine	*Regal machine*
June	No. of nylon tires	1867	7633
	No. of fiberglass tires	3500	0
	Hrs. of unused capacity	0	279
July	No. of nylon tires	0	2500
	No. of fiberglass tires	2500	0
	Hrs. of unused capacity	0	0
August	No. of nylon tires	2667	333
	No. of fiberglass tires	5000	0
	Hrs. of unused capacity	0	247

The unused hours of each machine are the slack variables of the computer output.

```
TITLE: RUBICON RUBBER
PROCEED, DISPLAY, OR REJECT? PRO

MAXIMIZE OR MINIMIZE? MIN

OPTIMAL SOLUTION FOUND.
     COST          19173.3

OUTPUT OPTION? USUAL

ALL ITEMS NOT LISTED IN SECTIONS 1 - 4 HAVE THE VALUE ZERO.

*1* DECISION VARIABLES
    1. WN-1      1866.67
    2. RN-1      7633.33
    3. WG-1      3500.00
    5. IN-1      5500.00
    6. IG-1      2500.00
    8. RN-2      2500.00
    9. WG-2      2500.00
   13. WN-3      2666.67
   14. RN-3      333.333
   15. WG-3      5000.00

*2* SLACK(+) AND SURPLUS(-) IN CONSTRAINTS
    2. +R-1      278.667
    6. +R-3      246.667

*3* SHADOW PRICES FOR CONSTRAINTS
    1. W-1       -.333333
    3. W-2       -1.16667
    4. R-2       -.625000
    5. W-3       -.333333
    7. N-1        .800000
    8. G-1        .640000
    9. N-2        .900000
   10. G-2        .740000
   11. N-3        .800000
   12. G-3        .640000

*4* REDUCED COSTS FOR DECISION VARIABLES
    4. RG-1       .600000E-01
    7. WN-2       .250000E-01
   10. RG-2      8.71000
   11. IN-2       .200000
   12. IG-2       .200000
   16. RG-3       .600000E-01

OUTPUT OPTION? 5 7

*5* RANGES ON COEFFICIENTS OF OBJECTIVE COST
     VARIABLE  LOWER BOUND  CURRENT VALUE  UPPER BOUND
    1. WN-1      .67500        .75000        .77500
    2. RN-1      .75000        .80000        .87500
    3. WG-1      .58000        .60000        .66000
    4. RG-1      .64000        .70000       UNBOUNDED
    5. IN-1      .45714E-03    .10000        .12500
    6. IG-1      .80000E-01    .10000       8.8100
    7. WN-2      .72500        .75000       UNBOUNDED
    8. RN-2     UNBOUNDED      .80000        .89954
    9. WG-2     UNBOUNDED      .60000        .62000
   10. RG-2    -8.0100         .70000       UNBOUNDED
   11. IN-2     -.10000        .10000       UNBOUNDED
   12. IG-2     -.10000        .10000       UNBOUNDED
   13. WN-3      .67500        .75000        .80000
   14. RN-3      .75000        .80000        .87500
   15. WG-3     UNBOUNDED      .60000        .66000
   16. RG-3      .64000        .70000       UNBOUNDED
```

Fig. 5.1 Computer printout of solution of the problem. (*Cont. on page 236.*)

```
*7* VARIABLE TRANSITIONS RESULTING FROM RANGING OBJECTIVE COST
    VARIABLE           LOWER BOUND              UPPER BOUND
                     VAR. IN    VAR. OUT     VAR. IN    VAR. OUT
  1. WN-1            RG-1       WG-1         WN-2       WN-1
  2. RN-1            +W-1       +R-1         RG-1       WG-1
  3. WG-1            WN-2       WN-1         RG-1       WG-1
  4. RG-1            RG-1       WG-1
  5. IN-1            RG-2       +R-1         WN-2       WN-1
  6. IG-1            WN-2       WN-1         RG-2       +R-1
  7. WN-2            WN-2       WN-1
  8. RN-2                                    RG-2       +R-1
  9. WG-2                                    WN-2       WN-1
 10. RG-2            RG-2       +R-1
 11. IN-2            IN-2       RN-3
 12. IG-2            IG-2       RN-3
 13. WN-3            RG-3       RN-3         +W-3       +R-3
 14. RN-3            +W-3       +R-3         RG-3       RN-3
 15. WG-3                                    RG-3       RN-3
 16. RG-3            RG-3       RN-3
```

OUTPUT OPTION? 6 8

```
*6* RANGES ON VALUES OF RIGHT-HAND-SIDE RHS1
     CONSTRNT   LOWER BOUND   CURRENT VALUE   UPPER BOUND
  1.   W-1        438.75         700.00         1845.0
  2.   R-1       1221.3         1500.0        UNBOUNDED
  3.   W-2         38.750        300.00         600.00
  4.   R-2        121.33         400.00        1280.0
  5.   W-3        768.75        1000.0         1050.0
  6.   R-3         53.333        300.00        UNBOUNDED
  7.   N-1      -3633.3         4000.0         5741.7
  8.   G-1      -2500.0         1000.0         3177.1
  9.   N-2       2500.0         8000.0         9741.7
 10.   G-2       2500.0         5000.0         7177.1
 11.   N-3       2666.7         3000.0         4541.7
 12.   G-3       4583.3         5000.0         6927.1
```

```
*8* VARIABLE TRANSITIONS RESULTING FROM RANGING RHS RHS1
     CONSTRNT           LOWER BOUND              UPPER BOUND
                     VAR. IN    VAR. OUT     VAR. IN    VAR. OUT
  1.   W-1           +W-3       +R-1         +W-1       RN-1
  2.   R-1           +W-3       +R-1
  3.   W-2           +W-3       +R-1         WN-2       IG-1
  4.   R-2           +W-3       +R-1         RG-2       IN-1
  5.   W-3           IN-2       +R-3         +W-3       RN-3
  6.   R-3           IN-2       +R-3
  7.   N-1           +W-1       RN-1         +W-3       +R-1
  8.   G-1           WN-2       WG-1         +W-3       +R-1
  9.   N-2           RG-2       IN-1         +W-3       +R-1
 10.   G-2           WN-2       IG-1         +W-3       +R-1
 11.   N-3           +W-3       RN-3         IN-2       +R-3
 12.   G-3           +W-3       RN-3         IN-2       +R-3
```

OUTPUT OPTION? NO

PARAMETRICS? NO

OPTION? TER

Fig. 5.1 (*Concluded*)

The resulting inventory at the end of each month for the two types of products is as follows:

	June	July	August
Inventory of nylon tires	5500	0	0
Inventory of fiberglass tires	2500	0	0

b) *Summary of Costs and Revenues*

Total costs. The total costs are the variable production costs, the inventory costs, the cost of raw materials, and the finishing, packaging, and shipping costs.

The variable production and inventory costs are obtained directly from the optimum value of the objective function of the linear-programming model. These costs are equal to $19,173.30.

Raw-material costs are $3.10 per nylon tire and $3.90 per fiberglass tire. The total production of nylon and fiberglass tires is given in the delivery schedule. The total material costs are therefore:

Raw material cost for nylon tires	$3.10 \times 15,000 = $46,500
Raw material cost for fiberglass tires	$3.90 \times 11,000 = $42,900
Total raw material cost	$89,400

The finishing, packaging, and shipping costs are $0.23 per tire. Since we are producing 26,000 tires, this cost amounts to $0.23 \times 26,000 = $5,980.

Thus the total costs are:

Variable production and inventory	$19,173.30
Raw material	89,400.00
Finishing, packaging and shipping	5,980.00
Total cost	$114,553.30

Total revenues. Prices per tire are $7.00 for nylon and $9.00 for fiberglass. Therefore, the total revenue is:

Nylon revenues	$7.00 \times 15,000 = $105,000
Fiberglass revenues	$9.00 \times 11,000 = $99,000
Total revenues	$204,000

Contribution. The contribution to the company is:

Total revenues	$204,000.00
Total cost	114,553.30
Total contribution	$ 89,446.70

From this contribution we should subtract the corresponding overhead cost to be absorbed by this contract, in order to obtain the *net contribution before taxes* to the company.

c) *Expediting of the New Machine*

The new machine is a Wheeling machine, which is preferred to the Regal equipment. The question is *can* it be used, and *how much*? Even if the machine were fully utilized, the hourly marginal cost would be:

$$\frac{\$200}{172 \text{ hrs}} = \$1.16/\text{hr}.$$

Examination of the shadow price for Wheeling machines in August reveals that it would be worth only $0.33 to have an additional hour of time on Wheeling equipment. We should therefore recommend *against* expediting the additional machine.

d) *Maintenance Schedule*

We are not told in the problem statement the amount of time required to service a given machine, or whether maintenance can just as well be performed later, or how much it would cost to do it at night and on weekends. We therefore cannot tell the maintenance department exactly what to do. We can, however, tell them that the Wheeling machines are completely used, but 278 hours are available on the Regal machines in June and 246 in August. These hours would be "free."

The shadow prices show that, by adjusting the production schedule, Wheeling machine time could be made available in June and August at a cost of $0.33/hr. During June we can have, at this cost, a total of 261.25 hours (the difference between the current availability, 700 hours, and the lower bound of the range for the W–1 righthand-side coefficient, 438.75). During August we will have available at $0.33/hr. 231.25 hours (the difference between the current availability, 1000 hours, and the lower bound of the range for the W–3 righthand-side coefficient, 768.75).

EXERCISES

1. The Pearce Container Corporation manufactures glass containers for a variety of products including soft drinks, beer, milk, catsup and peanut butter. Pearce has two products that account for 34% of its sales: container Type 35 and container Type 42. Pearce is interested in using a linear program to allocate production of these two containers to the five plants at which they are manufactured. The following is the estimated demand for the two types of containers for each quarter of next year: (1 unit = 100,000 containers)

Type	Plant	1st Qtr.	2nd Qtr.	3rd Qtr.	4th Qtr.
35	1	1388	1423	1399	1404
35	2	232	257	256	257
35	3	661	666	671	675
35*	4	31	32	34	36
42	1	2842	2787	3058	3228
42	2	2614	2551	2720	2893
42	3	1341	1608	1753	1887
42	4	1168	1165	1260	1305
42	5	1106	1138	1204	1206

* Plant 5 does not produce Type 35.

Pearce has ten machines at the five plants; eight of the machines can produce both Types 35 and 42, but two of the machines can produce only Type 42. Because the ten machines were purchased singly over a long period of time, no two machines have the same production rate or variable costs. After considerable research, the following information was gathered regarding the production of 1 unit (100,000 containers) by each of the ten machines:

Plant	Machine	Type	Cost	Machine-days	Man-days
1	D5	35	760	0.097	0.0194
		42	454	0.037	0.0037
	D6	35	879	0.103	0.0206
		42	476	0.061	0.0088
	C4	35	733	0.080	0.0204
		42	529	0.068	0.0083
2	T	42	520	0.043	0.0109
	U2	35	790	0.109	0.0145
		42	668	0.056	0.0143
3	K4	35	758	0.119	0.0159
		42	509	0.061	0.0129
	J6	35	799	0.119	0.0159
		42	521	0.056	0.0118
	70	35	888	0.140	0.0202
		42	625	0.093	0.0196
4	1	35	933	0.113	0.0100
		42	538	0.061	0.0081
5	V1	42	503	0.061	0.0135

During production a residue from the glass containers is deposited on the glass machines; during the course of a year machines are required to be shut down in order to clean off the residue. However, four of the machines are relatively new and will not be required to be shut down at all during the year; these are machines: C4, D5, K4, and V1. The following table shows the production days available for each machine by quarters:

Machine	1st Qtr.	2nd Qtr.	3rd Qtr.	4th Qtr.
C4	88	89	89	88
D5	88	89	89	88
D6	72	63	58	65
U2	81	88	87	55
T	88	75	89	88
K4	88	89	89	88
J6	37	89	39	86
70	54	84	85	73
1	42	71	70	68
V1	88	89	89	88
Days in quarter	88	89	89	88

In order to meet demands most efficiently, Pearce ships products from plant to plant. This transportation process ensures that the most efficient machines will be used first. However, transportation costs must also be considered. The following table shows transportation costs for shipping one unit between the five plants; the cost is the same for Type 35 and Type 42; Type 35 is not shipped to or from Plant 5.

Inter-plant transport (100,000 containers)

	To Plant				
From Plant	*1*	*2*	*3*	*4*	*5*
1	—	226	274	933	357
2	226	—	371	1022	443
3	274	371	—	715	168
4	941	1032	715	—	715
5	357	443	168	715	—

It is possible to store Type 35 and Type 42 containers almost indefinitely without damage to the containers. However, there is limited storage space at the five plants. Also, due to storage demands for other products, the available space varies during each quarter. The space available for Types 35 and 42 is as follows:

Inventory capacity (100,000 bottles = 1 unit)

Plant	*1st Qtr.*	*2nd Qtr.*	*3rd Qtr.*	*4th Qtr.*
1	376	325	348	410
2	55	48	62	58
3	875	642	573	813
4	10	15	30	24
5	103	103	103	103

It was found that there was no direct cost for keeping the containers in inventory other than using up storage space. However, there were direct costs for handling items, i.e., putting in *and* taking out of inventory. The costs for handling containers was as follows by type and by plant for one unit:

Handling costs (in $/unit)

Plant	*Type 35*	*Type 42*
1	85	70
2	98	98
3	75	75
4	90	80
5	—	67

a) Apply the stages of model formulation discussed in Section 5.2 to the Pearce Container Corp. problem. Precisely interpret the decision variables, the constraints, and the objective function to be used in the linear-programming model.

b) Indicate how to formulate the linear program mathematically. It is not necessary to write out the entire initial tableau of the model.

c) Determine the number of decision variables and constraints involved in the model formulation.

2. The Maynard Wire Company was founded in 1931 primarily to capitalize on the telephone company's expanding need for high-quality color-coded wire. As telephone services were rapidly expanding at that time, the need for quality color-coded wire was also expanding. Since then, the Maynard Wire Company has produced a variety of wire coatings, other wire products, and unrelated molded-plastic components. Today a sizable portion of its business remains in specially coated wire. Maynard Wire has only one production facility for coated wire, located in eastern Massachusetts, and has sales over much of the northeastern United States.

Maynard Wire is an intermediate processor, in that it purchases uncoated wire in standard gauges and then applies the various coatings that its customers desire. Basically there are only two types of coatings requested—standard inexpensive plastic and the higher-quality Teflon. The two coatings then come in a variety of colors, achieved by putting special dyes in the basic coating liquid. Since changing the color of the coating during the production process is a simple task, Maynard Wire has essentially two basic products.

Planning at Maynard Wire is done on a quarterly basis, and for the next quarter the demands for each type of wire in tons per month are:

Product	July	August	September
Plastic coated	1200	1400	1300
Quality Teflon	800	900	1150

The production of each type of wire must then be scheduled to minimize the cost of meeting this demand.

The production process at Maynard Wire is very modern and highly automated. The uncoated wire arrives in large reels, which are put on spindles at one end of the plant. The uncoated wire is continuously drawn off each successive reel over some traverse guides and through a coating bath containing either the standard plastic or the more expensive Teflon. The wire then is pulled through an extruder, so that the coating liquid adheres evenly to the wire, which then continues through a sequence of four electric drying ovens to harden the coating. Finally, the wire is reeled up on reels similar to those it arrived on. Different dyes are added to the coating liquid during the process to produce the various colors of wire ordered.

Maynard Wire has two, basically independent, wire trains within the plant, one engineered by the Kolbert Engineering Corporation and the other purchased secondhand from the now defunct Loomis Wire Company. Both the standard plastic and the quality Teflon types of wire can be produced on either process train. The production rates in tons per day are:

Process train	Plastic	Quality Teflon
Kolbert	40	35
Loomis	50	42

Producing the quality Teflon wire is a slower process due to the longer drying time required. The associated variable operating costs for the month of July in dollars per day are:

Process train	Plastic	Quality Teflon
Kolbert	100	102
Loomis	105	108

However, because each month the process trains must be shut down for scheduled maintenance, there are fewer days available for production than days in the month. The process-train availabilities in days per month are:

Process train	July	August	September
Kolbert	26	26	29
Loomis	28	27	30

Both types of wire may be stored for future delivery. Space is available in Maynard Wire's own warehouse, but only up to 100 tons. Additional space is available essentially without limit on a leased basis. The warehousing costs in dollars per ton between July and August are:

Product	Warehouse	Leased
Plastic	8.00	12.00
Quality Teflon	9.00	13.00

A linear program has been formulated and solved that minimizes the total discounted manufacturing and warehousing costs. Future manufacturing and warehousing costs have been discounted at approximately ten percent per month. The MPS input format*, picture, and solution of the model are presented (see Figs. E5.1 and E5.2). Also, there is a parametric righthand-side study that increases the demand for standard plastic-coated wire in September from 1300 to 1600 tons. Finally, there is a parametric cost run that varies the warehousing cost for quality Teflon-coated wire from $8.00 to $12.00.

Typical rows and columns of the linear program are defined as follows:

Rows
COST   Objective function
DCOST   Change row for PARAOBJ
1P   Demand for plastic-coated wire in month 1
1K   Process-train availability in month 1
2WS   Warehouse limitation in month 2

Columns
1K-P   Production of plastic-coated wire on Kolbert train in month 1
WP12   Warehousing plastic-coated wire from the end of month 1 to the end of month 2
LP12   Leasing warehouse space for plastic-coated wire from the end of month 1 to the end of month 2
RHS1   Righthand side
RHS2   Change column for PARARHS

---

* MPS stands for Mathematical Programming System. It is a software package that IBM has developed to solve general linear-programming models.

ROWS		COMPUTER INPUT
N	COST	
N	DCOST	We first present the basic input of the linear programming model.
E	1P	
E	1Q	The ROWS listing provides the names given to every row in the linear program-
L	1K	ming model. The first two rows are, respectively, the original cost objective
L	1L	function and the elements to be added to the objective function later in order
E	2P	to perform sensitivity analysis. The first letter in each heading specifies the
E	2Q	nature of the constraint represented by the corresponding row.
L	2K	(N = unrestricted; E = equality; L = less-than-or-equal-to constraint; G = greater-
L	2L	than-or-equal-to constraint.)
E	3P	
E	3Q	
L	3K	
L	3L	
L	2WS	
L	3WS	

COLUMNS			
	1K-P	COST	100.0
	1K-P	1P	40.0
	1K-P	1K	1.0
	1K-Q	COST	102.0
	1K-Q	1Q	35.0
	1K-Q	1K	1.0
	1L-P	COST	105.0
	1L-P	1P	50.0
	1L-P	1L	1.0
	1L-Q	COST	108.0
	1L-Q	1Q	42.0
	1L-Q	1L	1.0
	WP12	COST	8.0
	WP12	1P	-1.0
	WP12	2P	1.0
	WP12	2WS	1.0
	WP13	COST	8.0
	WP13	1P	-1.0
	WP13	3P	1.0
	WP13	2WS	1.0
	WP13	3WS	1.0
	WQ12	COST	9.0
	WQ12	DCOST	1.0
	WQ12	1Q	-1.0
	WQ12	2Q	1.0
	WQ12	2WS	1.0
	WQ13	COST	9.0
	WQ13	DCOST	1.0
	WQ13	1Q	-1.0
	WQ13	3Q	1.0
	WQ13	2WS	1.0
	WQ13	3WS	1.0
	LP12	COST	12.0
	LP12	1P	-1.0
	LP12	2P	1.0
	LP13	COST	14.0
	LP13	1P	-1.0
	LP13	3P	1.0
	LQ12	COST	13.0
	LQ12	1Q	-1.0
	LQ12	2Q	1.0
	LQ13	COST	15.0
.	LQ13	1Q	-1.0
	LQ13	3Q	1.0
	2K-P	COST	90.0
	2K-P	2P	40.0
	2K-P	2K	1.0
	2K-Q	COST	92.0
	2K-Q	2Q	35.0
	2K-Q	2K	1.0

Under the COLUMNS heading every nonzero coefficient is identified by indicating the column name, the row name, and the corresponding numerical value of the coefficient.

Fig. E5.1 Model of program for manufacturing and warehousing costs. (*Continued on page 244.*)

2L-P	COST	95.0
2L-P	2P	50.0
2L-P	2L	1.0
2L-Q	COST	98.0
2L-Q	2Q	42.0
2L-Q	2L	1.0
WP23	COST	7.0
WP23	2P	-1.0
WP23	3P	1.0
WP23	3WS	1.0
WQ23	COST	8.0
WQ23	DCOST	1.0
WQ23	2Q	-1.0
WQ23	3Q	1.0
WQ23	3WS	1.0
LP23	COST	11.0
LP23	2P	-1.0
LP23	3P	1.0
LQ23	COST	12.0
LQ23	2Q	-1.0
LQ23	3Q	1.0
3K-P	COST	80.0
3K-P	3P	40.0
3K-P	3K	1.0
3K-Q	COST	82.0
3K-Q	3Q	35.0
3K-Q	3K	1.0
3L-P	COST	85.0
3L-P	3P	50.0
3L-P	3L	1.0
3L-Q	COST	88.0
3L-Q	3Q	42.0
3L-Q	3L	1.0
RHS		
RHS1	1P	1200.0
RHS1	1Q	800.0
RHS1	1K	26.0
RHS1	1L	28.0
RHS1	2P	1400.0
RHS1	2Q	900.0
RHS1	2K	26.0
RHS1	2L	27.0
RHS1	3P	1300.0
RHS1	3Q	1150.0
RHS1	3K	29.0
RHS1	3L	30.0
RHS1	2WS	100.0
RHS1	3WS	100.0
RHS2	3P	1.0
RHS2	3P	1.0
RHS3	1P	1200.0
RHS3	1Q	800.0
RHS3	1K	26.0
RHS3	1L	28.0
RHS3	2P	1400.0
RHS3	2Q	900.0
RHS3	2K	26.0
RHS3	2L	27.0
RHS3	3P	1600.0
RHS3	3Q	1150.0
RHS3	3K	29.0
RHS3	3L	30.0
RHS3	2WS	100.0
RHS3	3WS	100.0
ENDATA		

Under the RHS heading every nonzero right-hand-side value is given. In our case there are three different righthand-side values that will be presented one at a time to perform the required sensitivity analysis.

The ENDATA command instructs the computer program that all the required data has been specified.

Fig. E5.1 (*Continued*)

CONTROL PROGRAM COMPILER

```
0001                    PROGRAM
0002                    TITLE('MAYNARD WIRE COMPANY')
0003                    INITIALZ
0060                    MOVE(XDATA,'MAYNARD')
0061                    MOVE(XPBNAME,'PBFILE')
0062                    CONVERT('SUMMARY')
0063                    SETUP('MIN')
0064                    MOVE(XOBJ,'COST')
0065                    MOVE(XRHS,'RHS1')
0066                    PICTURE
0067                    PRIMAL
0068                    SOLUTION
0069                    RANGE
0070                    MOVE(XCHCOL,'RHS2')
0071                    XPARAM=0.0
0072                    XPARDELT=50.0
0073                    XPARMAX=300.0
0074                    PARARHS
0075                    SOLUTION
0076                    MOVE(XRHS,'RHS3')
0077                    MOVE(XCHROW,'DCOST')
0078                    XPARAM=0.0
0079                    XPARDELT=1.0
0080                    XPARMAX=4.0
0081                    PARAOBJ
0082                    SOLUTION
0083                    EXIT
0084                    PEND
```

The CONTROL PROGRAM COMPILER represents the commands which are given to solve the problem with the data set specified before. They pertain to the type of analysis and output information to be obtained from the computer.

```
             1 1 1 1 W W W W L L L L 2 2 2 2 W W L L 3 3 3 3 R R R
             K K L L P P Q Q P P Q Q K K L L P Q P Q K K L L H H H
             - - - - 1 1 1 1 1 1 1 1 - - - - 2 2 2 2 - - - - S S S
             P Q P Q 2 3 2 3 2 3 2 3 P Q P Q 3 3 3 3 P Q P Q 1 2 3

COST   N     B C C C A A A A B B B B B B B B A A B B B B B B
DCOST  N                 1 1                           1
1P     E     B    B  -1 -1      -1 -1                              D    D
1Q     E        B    B    -1 -1      -1 -1                         C    C
1K     L     1 1                                                  B    B
1L     L        1 1                                               B    B
2P     E           1       1          B    B  -1  -1              D    D
2Q     E             1       1          B    B  -1  -1            C    C
2K     L                              1 1                         B    B
2L     L                                1 1                       B    B
3P     E           1       1                    1 1  B    B   D 1 D
3Q     E             1       1                    1 1  B   B  D   D
3K     L                                              1 1         B    B
3L     L                                                1 1 B     B    B
2WS    L        1 1 1 1                                           B    B
3WS    L        1 1                          1 1                  B    B
```

This exhibit is a pictorial representation of the initial tableau showing the position and magnitude of the nonzero coefficients in the tableau. The following conventions have been used:

FROM	NOTATION	TO
l	A	l0
l0	B	l00
l00	C	l000
l000	D	l0000

Fig. E5.1 (*Concluded*)

MAYNARD WIRE COMPANY

SECTION 1 - ROWS   FOR: COST AND RHS 1

NUMBER	...ROW..	AT	...ACTIVITY...	SLACK ACTIVITY	..LOWER LIMIT.	..UPPER LIMIT.	.DUAL ACTIVITY
1	COST	BS	15013.12143	15013.12143-	NONE	NONE	1.00000
2	DCOST	BS	33.75000	33.75000-	NONE	NONE	.
3	1P	EQ	1200.00000	.	1200.00000	1200.00000	2.38800-
4	1Q	EQ	800.00000	.	800.00000	800.00000	2.91429-
5	1K	BS	19.02143	6.97857	NONE	26.00000	.
6	1L	UL	28.00000		NONE	28.00000	14.40000
7	2P	EQ	1400.00000	.	1400.00000	1400.00000	10.37500-
8	2Q	EQ	900.00000	.	900.00000	900.00000	11.91429-
9	2K	UL	26.00000	.	NONE	26.00000	325.00000
10	2L	UL	27.00000	.	NONE	27.00000	423.75000
11	3P	EQ	1300.00000	.	1300.00000	1300.00000	1.90800-
12	3Q	EQ	1150.00000	.	1150.00000	1150.00000	2.34286-
13	3K	BS	28.05714	.94286	NONE	29.00000	.
14	3L	UL	30.00000	.	NONE	30.00000	10.40000
15	2WS	BS	33.75000	66.25000	NONE	100.00000	.
16	3WS	BS	33.75000	100.00000	NONE	100.00000	.

SECTION 2 - COLUMNS

NUMBER	.COLUMN.	AT	...ACTIVITY...	..INPUT COST..	..LOWER LIMIT.	..UPPER LIMIT.	.REDUCED COST.
17	1K-P	LL	.	100.00000	.	NONE	4.48000
18	1K-Q	BS	19.02143	102.00000	.	NONE	.
19	1L-P	BS	24.00000	105.00000	.	NONE	.
20	1L-Q	BS	4.00000	108.00000	.	NONE	.
21	WP12	LL	.	8.00000	.	NONE	.01300
22	WP13	LL	.	8.00000	.	NONE	8.48000
23	WQ12	BS	33.75000	9.00000	.	NONE	.
24	WQ13	LL	.	9.00000	.	NONE	9.57143
25	LP12	LL	.	12.00000	.	NONE	4.01300
26	LP13	LL	.	14.00000	.	NONE	14.48000
27	LQ12	LL	.	13.00000	.	NONE	4.00000
28	LQ13	LL	.	15.00000	.	NONE	15.57143
29	2K-P	BS	1.25000	90.00000	.	NONE	.
30	2K-Q	BS	24.75000	92.00000	.	NONE	.
31	2L-P	BS	27.00000	95.00000	.	NONE	.
32	2L-Q	LL	.	98.00000	.	NONE	21.35000
33	WP23	LL	.	7.00000	.	NONE	15.46700
34	WQ23	LL	.	8.00000	.	NONE	17.57143
35	LP23	LL	.	11.00000	.	NONE	19.46700
36	LQ23	LL	.	12.00000	.	NONE	21.57143
37	3K-P	LL	.	80.00000	.	NONE	3.68000
38	3K-Q	BS	28.05714	82.00000	.	NONE	.
39	3L-P	BS	26.00000	85.00000	.	NONE	.
40	3L-Q	BS	4.00000	88.00000	.	NONE	.

(Cont.)

Fig. E5.2 Solution of program.

PARARHS   OBJ = COST   RHS = RHS1   CHCOL = RHS2   PARAM = .

TIME = 0.34 MINS.

ITER NUMBER M	NUMBER NONOPT	VECTOR OUT	VECTOR IN	REDUCED COST	FUNCTION VALUE	PARAM VALUE
13	0	13	24	9.57143	15088.1	39.2857

SECTION 1 - ROWS   FOR: COST AND RHS1   50.0+ RHS2 [1]

NUMBER	...ROW..	AT	...ACTIVITY...	SLACK ACTIVITY	..LOWER LIMIT.	..UPPER LIMIT.	.DUAL ACTIVITY
1	COST	BS	15194.66429	15194.66429-	NONE	NONE	1.00000
2	DCOST	BS	42.75000	42.75000-	NONE	NONE	.
3	1P	EQ	1200.00000	.	1200.00000	1200.00000	2.38800-
4	1Q	EQ	800.00000	.	800.00000	800.00000	2.91429-
5	1K	BS	19.27857	6.72143	NONE	26.00000	.
6	1L	UL	28.00000	.	NONE	28.00000	14.40000
7	2P	EQ	1400.00000	.	1400.00000	1400.00000	10.37500-
8	2Q	EQ	900.00000	.	900.00000	900.00000	11.91429-
9	2K	UL	26.00000	.	NONE	26.00000	325.00000
10	2L	UL	27.00000	.	NONE	27.00000	423.75000
11	3P	EQ	1350.00000	.	1350.00000	1350.00000	9.94800-
12	3Q	EQ	1150.00000	.	1150.00000	1150.00000	11.91429-
13	3K	UL	29.00000	.	NONE	29.00000	335.00000
14	3L	UL	30.00000	.	NONE	30.00000	412.40000
15	2WS	BS	42.75000	57.25000	NONE	100.00000	.
16	3WS	BS	9.00000	91.00000	NONE	100.00000	.

SECTION 2 - COLUMNS

NUMBER	.COLUMN.	AT	...ACTIVITY...	..INPUT COST..	..LOWER LIMIT.	..UPPER LIMIT.	.REDUCED COST.
17	1K-P	LL	.	100.00000	.	NONE	4.48000
18	1K-Q	BS	19.27857	102.00000	.	NONE	.
19	1L-P	BS	24.00000	105.00000	.	NONE	.
20	1L-Q	BS	4.00000	108.00000	.	NONE	.01300
21	WP12	LL	.	8.00000	.	NONE	.44000
22	WP13	LL	.	8.00000	.	NONE	.
23	WQ12	BS	33.75000	9.00000	.	NONE	.
24	WQ13	BS	9.00000	9.00000	.	NONE	.
25	LP12	LL	.	12.00000	.	NONE	4.01300
26	LP13	LL	.	14.00000	.	NONE	6.44000
27	LQ12	LL	.	13.00000	.	NONE	4.00000
28	LQ13	LL	.	15.00000	.	NONE	6.00000
29	2K-P	BS	1.25000	90.00000	.	NONE	.
30	2K-Q	BS	24.75000	92.00000	.	NONE	.
31	2L-P	BS	27.00000	95.00000	.	NONE	.
32	2L-Q	LL	.	98.00000	.	NONE	21.35000
33	WP23	LL	.	7.00000	.	NONE	7.42700
34	WQ23	BS	8.00000	8.00000	.	NONE	8.00000
35	LP23	LL	.	11.00000	.	NONE	11.42700
36	LQ23	LL	.	12.00000	.	NONE	12.00000
37	3K-P	LL	.	80.00000	.	NONE	17.08000
38	3K-Q	BS	29.00000	82.00000	.	NONE	.
39	3L-P	BS	27.00000	85.00000	.	NONE	.
40	3L-Q	BS	3.00000	88.00000	.	NONE	.

1 the reader should refer to the picture of the initial tableau to visualize the type of analysis being performed (What is the resulting righthand-side of the problem formed by RHS1 + 50 RHS2?)

Fig. E5.2 (*Cont.*)

MAYNARD WIRE COMPANY

SECTION 1 - ROWS   FOR: COST AND RHS1 + 100.0 RHS2

NUMBER	...ROW..	AT	...ACTIVITY...	SLACK ACTIVITY	..LOWER LIMIT.	..UPPER LIMIT.	.DUAL ACTIVITY
1	COST	BS	15692.06429	15692.06429-	NONE	NONE	1.00000
2	DCOST	BS	84.75000	84.75000-	NONE	NONE	.
3	1P	EQ	1200.00000	.	1200.00000	1200.00000	2.38800-
4	1Q	EQ	800.00000	.	800.00000	800.00000	2.91429-
5	1K	BS	20.47857	5.52143	NONE	26.00000	.
6	1L	UL	28.00000	.	NONE	28.00000	14.40000
7	2P	EQ	1400.00000	.	1400.00000	1400.00000	10.37500-
8	2Q	EQ	900.00000	.	900.00000	900.00000	11.91429-
9	2K	UL	26.00000	.	NONE	26.00000	325.00000
10	2L	UL	27.00000	.	NONE	27.00000	423.75000
11	3P	EQ	1400.00000	.	1400.00000	1400.00000	9.94800-
12	3Q	EQ	1150.00000	.	1150.00000	1150.00000	11.91429-
13	3K	UL	29.00000	.	NONE	29.00000	335.00000
14	3L	UL	30.00000	.	NONE	30.00000	412.40000
15	2WS	BS	84.75000	15.25000	NONE	100.00000	.
16	3WS	BS	51.00000	49.00000	NONE	100.00000	.

SECTION 2 - COLUMNS

NUMBER	.COLUMN.	AT	...ACTIVITY...	..INPUT COST..	..LOWER LIMIT.	..UPPER LIMIT.	.REDUCED COST.
17	1K-P	LL	.	100.00000	.	NONE	4.48000
18	1K-Q	BS	20.47857	102.00000	.	NONE	.
19	1L-P	BS	24.00000	105.00000	.	NONE	.
20	1L-Q	BS	4.00000	108.00000	.	NONE	.
21	WP12	LL	.	8.00000	.	NONE	.01300
22	WP13	LL	.	8.00000	.	NONE	.44000
23	WQ12	BS	33.75000	9.00000	.	NONE	.
24	WQ13	BS	51.00000	9.00000	.	NONE	.
25	LP12	LL	.	12.00000	.	NONE	4.01300
26	LP13	LL	.	14.00000	.	NONE	6.44000
27	LQ12	LL	.	13.00000	.	NONE	4.00000
28	LQ13	LL	.	15.00000	.	NONE	6.00000
29	2K-P	BS	1.25000	90.00000	.	NONE	.
30	2K-Q	BS	24.75000	92.00000	.	NONE	.
31	2L-P	BS	27.00000	95.00000	.	NONE	.
32	2L-Q	LL	.	98.00000	.	NONE	21.35000
33	WP23	LL	.	7.00000	.	NONE	7.42700
34	WQ23	LL	.	8.00000	.	NONE	8.00000
35	LP23	LL	.	11.00000	.	NONE	11.42700
36	LQ23	LL	.	12.00000	.	NONE	12.00000
37	3K-P	LL	.	80.00000	.	NONE	17.08000
38	3K-Q	BS	29.00000	82.00000	.	NONE	.
39	3L-P	BS	28.00000	85.00000	.	NONE	.
40	3L-Q	BS	2.00000	88.00000	.	NONE	.

Fig. E5.2 (Cont.)

PARARHS	OBJ = COST	RHS = RHS1	CHCOL = RHS2	PARAM = 100.00000

TIME = 0.41 MINS.

ITER NUMBER	NUMBER NONOPT	VECTOR IN	VECTOR OUT	REDUCED COST	FUNCTION VALUE	PARAM VALUE
14	0	27	15	4.00000	15872.7	118.155

M

SECTION 1 - ROWS   FOR: COST AND RHS1 + 100.0 RHS2

NUMBER	...ROW...	AT	...ACTIVITY...	SLACK ACTIVITY	..LOWER LIMIT.	..UPPER LIMIT.	.DUAL ACTIVITY
1	COST	BS	16296.46429	16296.46429-	NONE	NONE	1.00000
2	DCOST	BS	100.00000	100.00000-	NONE	NONE	NONE
3	1P	EQ	1200.00000		1200.00000	1200.00000	2.38800-
4	1Q	EQ	800.00000		800.00000	800.00000	2.91429-
5	1K	BS	21.67857	4.32143	NONE	26.00000	14.40000
6	1L	UL	28.00000		NONE	28.00000	13.87500-
7	2P	EQ	1400.00000		1400.00000	1400.00000	15.91429-
8	2Q	EQ	900.00000		900.00000	900.00000	465.00000
9	2K	UL	26.00000		NONE	26.00000	598.75000
10	2L	UL	27.00000		NONE	27.00000	13.30800-
11	3P	EQ	1450.00000		1450.00000	1450.00000	15.91429-
12	3Q	EQ	1150.00000		1150.00000	1150.00000	475.00000
13	3K	UL	29.00000		NONE	29.00000	580.40000
14	3L	UL	30.00000		NONE	30.00000	4.00000
15	2WS	UL	100.00000		NONE	100.00000	
16	3WS	BS	93.00000	7.00000	NONE	100.00000	

SECTION 2 - COLUMNS

NUMBER	.COLUMN.	AT	...ACTIVITY...	..INPUT COST..	..LOWER LIMIT.	..UPPER LIMIT.	.REDUCED COST.
17	1K-P	LL	.	100.00000	.	NONE	4.48000
18	1K-Q	BS	21.67857	102.00000	.	NONE	.
19	1L-P	RS	24.00000	105.00000	.	NONE	.
20	1L-Q	BS	4.00000	108.00000	.	NONE	.
21	WP12	LL	.	8.00000	.	NONE	.51300
22	WP13	LL	.	8.00000	.	NONE	1.08000
23	WQ12	BS	7.00000	9.00000	.	NONE	.
24	WQ13	BS	93.00000	9.00000	.	NONE	.51300
25	LP12	LL	.	12.00000	.	NONE	.
26	LP13	LL	.	14.00000	.	NONE	3.08000
27	LQ12	BS	26.75000	13.00000	.	NONE	.
28	LQ13	LL	.	15.00000	.	NONE	2.00000
29	2K-P	BS	1.25000	90.00000	.	NONE	.
30	2K-Q	BS	24.75000	92.00000	.	NONE	.
31	2L-P	BS	27.00000	95.00000	.	NONE	.
32	2L-Q	LL	.	98.00000	.	NONE	28.35000
33	WP23	LL	.	7.00000	.	NONE	7.56700
34	WQ23	LL	.	8.00000	.	NONE	8.00000
35	LP23	LL	.	11.00000	.	NONE	11.56700
36	LQ23	LL	.	12.00000	.	NONE	12.00000
37	3K-P	LL	.	80.00000	.	NONE	22.68000
38	3K-Q	BS	29.00000	82.00000	.	NONE	.
39	3L-P	BS	29.00000	85.00000	.	NONE	.
40	3L-Q	BS	1.00000	88.00000	.	NONE	.

Fig. E5.2 (*Cont.*)

PARARHS   OBJ = COST   RHS = RHS1   CHCOL = RHS2   PARAM = 150.00000

TIME = 0.45 MINS.

ITER NUMBER	NUMBER NONOPT	VECTOR OUT	VECTOR IN	REDUCED COST	FUNCTION VALUE	PARAM VALUE
M 15	0	16	28	2.00000	16407.4	158.333

SECTION 1 - ROWS   FOR: COST AND RHS 1 + 200.0 RHS 2

NUMBER	..ROW..	AT	...ACTIVITY...	SLACK ACTIVITY	..LOWER LIMIT.	..UPPER LIMIT.	.DUAL ACTIVITY
1	COST	BS	17031.86429	17031.86429-	NONE	NONE	1.00000
2	DCOST	BS	100.00000	100.00000-	NONE	NONE	.
3	1P	EQ	1200.00000	.	1200.00000	1200.00000	2.38800-
4	1Q	EQ	800.00000	.	800.00000	800.00000	2.91429-
5	1K	BS	22.87857	3.12143	NONE	26.00000	.
6	1L	UL	28.00000	.	NONE	28.00000	14.40000
7	2P	EQ	1400.00000	.	1400.00000	1400.00000	13.87500-
8	2Q	EQ	900.00000	.	900.00000	900.00000	15.91429-
9	2K	UL	26.00000	.	NONE	26.00000	465.00000
10	2L	UL	27.00000	.	NONE	27.00000	598.75000
11	3P	EQ	1500.00000	.	1500.00000	1500.00000	14.98800-
12	3Q	EQ	1150.00000	.	1150.00000	1150.00000	17.91429-
13	3K	UL	29.00000	.	NONE	29.00000	545.00000
14	3L	UL	30.00000	.	NONE	30.00000	664.40000
15	2WS	UL	100.00000	.	NONE	100.00000	4.00000
16	3WS	UL	100.00000	.	NONE	100.00000	2.00000

SECTION 2 - COLUMNS

NUMBER	.COLUMN.	AT	...ACTIVITY...	..INPUT COST..	..LOWER LIMIT.	..UPPER LIMIT.	.REDUCED COST.
17	1K-P	LL	.	100.00000	.	NONE	4.48000
18	1K-Q	BS	22.87857	102.00000	.	NONE	.
19	1L-P	BS	24.00000	105.00000	.	NONE	.
20	1L-Q	BS	4.00000	108.00000	.	NONE	.
21	WP12	LL	.	8.00000	.	NONE	.51300
22	WP13	LL	.	8.00000	.	NONE	1.40000
23	WQ12	BS	.	9.00000	.	NONE	.
24	WQ13	BS	100.00000	9.00000	.	NONE	.
25	LP12	LL	.	12.00000	.	NONE	.
26	LP13	LL	.	14.00000	.	NONE	.51300
27	LQ12	BS	33.75000	13.00000	.	NONE	1.40000
28	LQ13	BS	35.00000	15.00000	.	NONE	.
29	2K-P	BS	1.25000	90.00000	.	NONE	.
30	2K-Q	BS	24.75000	92.00000	.	NONE	.
31	2L-P	BS	27.00000	95.00000	.	NONE	.
32	2L-Q	LL	.	98.00000	.	NONE	28.35000
33	WP23	LL	.	7.00000	.	NONE	7.88700
34	WQ23	LL	.	8.00000	.	NONE	8.00000
35	LP23	LL	.	11.00000	.	NONE	9.88700
36	LQ23	LL	.	12.00000	.	NONE	10.00000
37	3K-P	LL	.	80.00000	.	NONE	25.48000
38	3K-Q	BS	29.00000	82.00000	.	NONE	.
39	3L-P	BS	30.00000	85.00000	.	NONE	.
40	3L-Q	BS	30.00000	88.00000	.	NONE	.

Fig. E5.2 (Cont.)

MAYNARD WIRE COMPANY

PARARHS   OBJ = COST   RHS = RHS1   CHCOL = RHS2   PARAM = 200.00000

TIME =   0.49 MINS.

ITER NUMBER	NUMBER NONOPT	VECTOR IN	VECTOR OUT	REDUCED COST	FUNCTION VALUE	PARAM VALUE
M 16	0	37	40	25.4800	17031.9	200.000

SECTION 1 - ROWS   FOR: COST AND RHS 1 + 250.0 RHS 2

NUMBER	...ROW..	AT	...ACTIVITY...	SLACK ACTIVITY	..LOWER LIMIT.	..UPPER LIMIT.	.DUAL ACTIVITY
1	COST	BS	17813.11429	17813.11429-	NONE	NONE	1.00000
2	DCOST	BS	100.00000	100.00000-	NONE	NONE	.
3	1P	EQ	1200.00000	.	1200.00000	1200.00000	2.38800-
4	1Q	EQ	800.00000	.	800.00000	800.00000	2.91429-
5	1K	BS	24.12857	1.87143	NONE	26.00000	.
6	1L	UL	28.00000	.	NONE	28.00000	14.40000
7	2P	EQ	1400.00000	.	1400.00000	1400.00000	13.87500-
8	2Q	EQ	900.00000	.	900.00000	900.00000	15.91429-
9	2K	UL	26.00000	.	NONE	26.00000	465.00000
10	2L	UL	27.00000	.	NONE	27.00000	598.75000
11	3P	EQ	1550.00000	.	1550.00000	1550.00000	15.62500-
12	3Q	EQ	1150.00000	.	1150.00000	1150.00000	17.91429-
13	3K	UL	29.00000	.	NONE	29.00000	545.00000
14	3L	UL	30.00000	.	NONE	30.00000	696.25000
15	2WS	UL	100.00000	.	NONE	100.00000	4.00000
16	3WS	UL	100.00000	.	NONE	100.00000	2.00000

SECTION 2 - COLUMNS

NUMBER	.COLUMN.	AT	...ACTIVITY...	..INPUT COST..	..LOWER LIMIT.	..UPPER LIMIT.	.REDUCED COST.
17	1K-P	LL	.	100.00000	.	NONE	4.48000
18	1K-Q	BS	24.12857	102.00000	.	NONE	.
19	1L-P	BS	24.00000	105.00000	.	NONE	.
20	1L-Q	BS	4.00000	108.00000	.	NONE	.
21	WP12	LL	.	8.00000	.	NONE	.51300
22	WP13	LL	.	9.00000	.	NONE	.76300
23	WQ12	BS	.	9.00000	.	NONE	.
24	WQ13	BS	100.00000	12.00000	.	NONE	.
25	LP12	LL	.	13.00000	.	NONE	.51300
26	LP13	LL	.	14.00000	.	NONE	.76300
27	LQ12	BS	.	15.00000	.	NONE	.
28	LQ13	BS	35.75000		.	NONE	.
29	2K-P	BS	78.75000	90.00000	.	NONE	.
30	2K-Q	BS	1.25000	92.00000	.	NONE	.
31	2L-P	BS	24.75000	95.00000	.	NONE	.
32	2L-Q	LL	.	98.00000	.	NONE	28.35000
33	WP23	LL	.	7.00000	.	NONE	7.25000
34	WQ23	LL	.	8.00000	.	NONE	8.00000
35	LP23	LL	.	11.00000	.	NONE	9.25000
36	LQ23	LL	.	12.00000	.	NONE	10.00000
37	3K-P	BS	1.25000	80.00000	.	NONE	.
38	3K-Q	BS	27.75000	82.00000	.	NONE	.
39	3L-P	BS	30.00000	85.00000	.	NONE	.
40	3L-Q	LL	.	88.00000	.	NONE	31.85000

Fig. E5.2 (Cont.)

SECTION 1 — ROWS   FOR: COST AND RHS 1 + 300.0 RHS2

NUMBER	...ROW..	AT	...ACTIVITY...	SLACK ACTIVITY	..LOWER LIMIT.	..UPPER LIMIT.	.DUAL ACTIVITY
1	COST	BS	18594.36429	18594.36429—	NONE	NONE	1.00000
2	DCOST	BS	100.00000	100.00000—	NONE	NONE	
3	1P	EQ	1200.00000	.	1200.00000	1200.00000	2.38800—
4	1Q	EQ	800.00000	.	800.00000	800.00000	2.91429—
5	1K	BS	25.37857	.	NONE	26.00000	
6	1L	UL	28.00000	.62143	NONE	28.00000	14.40000
7	2P	EQ	1400.00000	.	1400.00000	1400.00000	13.87500—
8	2Q	EQ	900.00000	.	900.00000	900.00000	15.91429—
9	2K	UL	26.00000	.	NONE	26.00000	465.00000
10	2L	UL	27.00000	.	NONE	27.00000	598.75000
11	3P	EQ	1600.00000	.	1600.00000	1600.00000	15.62500—
12	3Q	EQ	1150.00000	.	1150.00000	1150.00000	17.91429—
13	3K	UL	29.00000	.	NONE	29.00000	545.00000
14	3L	UL	30.00000	.	NONE	30.00000	696.25000
15	2WS	UL	100.00000	.	NONE	100.00000	4.00000
16	3WS	UL	100.00000	.	NONE	100.00000	2.00000

SECTION 2 — COLUMNS

NUMBER	.COLUMN.	AT	...ACTIVITY...	..INPUT COST..	..LOWER LIMIT.	..UPPER LIMIT.	.REDUCED COST.
17	1K-P	LL	.	100.00000	.	NONE	4.48000
18	1K-Q	BS	25.37857	102.00000	.	NONE	.
19	1L-P	BS	24.00000	105.00000	.	NONE	.
20	1L-Q	BS	4.00000	108.00000	.	NONE	.
21	WP12	LL	.	8.00000	.	NONE	.51300
22	WP13	LL	.	8.00000	.	NONE	.76300
23	WQ12	BS	8.00000	9.00000	.	NONE	.
24	WQ13	BS	100.00000	9.00000	.	NONE	.
25	LP12	LL	.	12.00000	.	NONE	.51300
26	LP13	LL	.	14.00000	.	NONE	.76300
27	LQ12	BS	33.75000	13.00000	.	NONE	.
28	LQ13	BS	122.50000	15.00000	.	NONE	.
29	2K-P	BS	1.25000	90.00000	.	NONE	.
30	2K-Q	BS	24.75000	92.00000	.	NONE	.
31	2L-P	BS	27.00000	95.00000	.	NONE	.
32	2L-Q	LL	.	98.00000	.	NONE	28.35000
33	WP23	LL	.	7.00000	.	NONE	7.25000
34	WQ23	LL	.	8.00000	.	NONE	8.00000
35	LP23	LL	.	11.00000	.	NONE	9.25000
36	LQ23	LL	.	12.00000	.	NONE	10.00000
37	3K-P	BS	2.50000	80.00000	.	NONE	.
38	3K-Q	BS	26.50000	82.00000	.	NONE	.
39	3L-P	BS	30.00000	85.00000	.	NONE	.
40	3L-Q	LL	.	88.00000	.	NONE	31.85000

Fig. E5.2 (*Cont.*)

MAYNARD WIRE COMPANY

PARAOBJ	OBJ = COST	RHS = RHS3	CHROM = DCOST	PARAM =

TIME = 0.57 MINS.   Note: RHS3 = RHS1 + 300.0 RHS2

ITER	NUMBER NONOPT	VECTOR IN	VECTOR OUT	REDUCED COST	FUNCTION VALUE	PARAM VALUE
M 17	0	23	21	.51300	18645.7	.51300
M 18	0	24	22	.25000	18670.7	.76300

NO MAXIMUM FOR PARAMETER (ic. no basis change for increases in the parameter above this level)

SECTION 1 - ROWS   FOR: COST + 4.0 DCOST AND RHS3

NUMBER	..ROW..	AT	...ACTIVITY...	SLACK ACTIVITY	..LOWER LIMIT.	..UPPER LIMIT.	.DUAL ACTIVITY
1	COST	BS	18670.66429	18670.66429-	NONE	NONE	1.00000
2	DCOST	BS	.	.	NONE	NONE	4.00000
3	1P	EQ	1200.00000	.	1200.00000	1200.00000	2.38800-
4	1Q	EQ	800.00000	.	800.00000	800.00000	2.91429-
5	1K	BS	25.27857	.72143	NONE	26.00000	
6	1L	UL	28.00000	.	NONE	28.00000	14.40000-
7	2P	EQ	1400.00000	.	1400.00000	1400.00000	13.87500-
8	2Q	EQ	900.00000	.	900.00000	900.00000	15.91429-
9	2K	UL	26.00000	.	NONE	26.00000	465.00000-
10	2L	UL	27.00000	.	NONE	27.00000	598.75000
11	3P	EQ	1600.00000	.	1600.00000	1600.00000	15.62500-
12	3Q	EQ	1150.00000	.	1150.00000	1150.00000	17.91429-
13	3K	UL	29.00000	.	NONE	29.00000	545.00000
14	3L	UL	30.00000	.	NONE	30.00000	696.25000
15	2WS	UL	100.00000	.	NONF	100.00000	3.48700
16	3WS	UL	100.00000	.	NONE	100.00000	1.75000

SECTION 2 - COLUMNS

NUMBER	.COLUMN.	AT	...ACTIVITY...	..INPUT COST..	..LOWER LIMIT.	..UPPER LIMIT.	.REDUCED COST.
17	1K-P	LL	.	100.00000	.	NONE	4.48000
18	1K-Q	BS	25.27857	102.00000	.	NONE	.
19	1L-P	BS	26.00000	105.00000	.	NONE	.
20	1L-Q	BS	2.00000	108.00000	.	NONE	.
21	WP12	BS	.	8.00000	.	NONE	.
22	WP13	BS	100.00000	8.00000	.	NONE	.
23	WQ12	LL	.	13.00000	.	NONE	3.48700
24	WQ13	LL	.	13.00000	.	NONE	3.23700
25	LP12	LL	.	12.00000	.	NONE	.51300
26	LP13	LL	.	14.00000	.	NONE	.76300
27	LQ12	BS	33.75000	13.00000	.	NONE	.
28	LQ13	BS	135.00000	15.00000	.	NONE	.
29	2K-P	BS	1.25000	90.00000	.	NONE	.
30	2K-Q	BS	24.75000	92.00000	.	NONE	.
31	2L-P	BS	27.00000	95.00000	.	NONE	.
32	2L-Q	LL	.	98.00000	.	NONE	28.35000
33	WP23	LL	.	7.00000	.	NONE	7.00000
34	WQ23	LL	.	12.00000	.	NONE	11.25000
35	LP23	LL	.	11.00000	.	NONE	9.25000
36	LQ23	LL	.	12.00000	.	NONE	10.00000
37	3K-P	BS	29.00000	80.00000	.	NONE	.
38	3K-Q	BS	.	82.00000	.	NONE	.
39	3L-P	BS	30.00000	85.00000	.	NONE	.
40	3L-Q	LL	.	88.00000	.	NONE	31.85000

Fig. E5.2 (*Concluded*)

a) Explain the optimal policy for Maynard Wire Company when the objective function is COST and the righthand side is RHS1.

b) What is the resulting production cost for the 300-ton incremental production of plastic-coated wire in month 3?

c) How does the marginal production cost of plastic-coated wire vary when its demand in month 3 is shifted from 1300 to 1600 tons?

d) How does the operating strategy vary when the warehousing cost for quality Teflon-coated wire shifts from $8.00 to $12.00 with the demand for plastic-coated wire in month 3 held at 1600 tons?

3. Toys, Inc., is a small manufacturing company that produces a variety of children's toys. In the past, water pistols have been an exceptionally profitable item, especially the miniature type which can be hidden in the palm of one hand. However, children recently have been buying water rifles, which permit the stream of water to be projected much farther. Recognizing that this industry trend was not a short-lived fad, Toys, Inc., started to produce a line of water rifles.

After several months of production, Toys' General Manager, Mr. Whett, ran into a storage problem. The older and smaller water pistols had not occupied much space, but the larger water rifles were quite bulky. Consequently, Mr. Whett was forced to rent storage space in a public warehouse at 28¢ per case per month, plus 44¢ per case for cost of handling. This made Mr. Whett wonder whether producing water rifles was profitable. In addition, Mr. Whett wondered whether it might not be better to increase his production capacity so that fewer cases would be stored in the slack season.

Data:

The following information was obtained from Toys' accounting department:

Desired return on investment: 10% after taxes

Tax rate: 55% (including state taxes)

Variable manufacturing cost: $21.00/case, or $9.50/case after taxes

(Variable costs include all overhead and so-called "fixed" costs, except for the cost of production equipment. This seems appropriate, since the company has expanded so rapidly that "fixed" costs have actually been variable.)

Warehousing: $0.28/case per month, or $0.126/case per month after taxes

Handling: $0.44/case, or $0.198/case after taxes

Opportunity cost of tying up capital in inventory: ($21.00 × 10%) ÷ 12 months = $0.18/case per month

Selling price: $28.10/case, or $12.61 after taxes

Existing production capacity: 2750 cases per month

Cost of additional production capacity: $6400/year after taxes, for each additional 1000 cases/month. This figure takes into account the investment tax credit and the discounted value of the tax shield arising out of future depreciation.

The anticipated demand, based on past sales data, for the next 12 months is given below. The first month is October, and the units are in cases.

Month	Demand (Cases)
October	1490
November	2106
December	2777
January	843
February	1105
March	2932
April	1901
May	4336
June	4578
July	1771
August	4745
September	3216
Total	31800

Formulating the model:

The program variables are defined as follows:

PRD-1 to PRD12 identify the *produc*tion constraints for the 12 periods.

DEM-1 to DEM12 identify the *dem*and constraints for the 12 periods.

CAP-1 to CAP12 identify the *cap*acity constraints for the 12 periods.

CNG-1 to CNG12 identify the constraints describing *chang*es in inventory levels for the 12 periods.

X1 to X12 are the cases produced in each period. Associated with these variables are production costs of $9.50.

S1 to S12 are the cases sold in each period. Associated with these variables are revenues of $12.61 per case.

Y1 to Y12 are the cases in inventory at the beginning of the designated period. Associated with these variables are storage and cost of capital charges, totaling $0.306/case per month.

U1 to U12 are the unfilled demand in each period. No attempt has been made to identify a penalty cost associated with these variables.

Q1 to Q11 are the changes in inventory levels from one period to the next. Since a handling charge of $0.198/case is associated only with increases in inventory, these variables have been further designated as Q1+, Q1−, etc. to indicate increases (+) and decreases (−) in inventory levels.

+CAP-1 to +CAP12 are the slack variables supplied by the computer to represent unused production capacity in the designated period.

A typical production constraint, PRD-2, is shown below:

$$Y2 + X2 - S2 - Y3 = 0.$$

This expression indicates that the beginning inventory, plus the production, minus the sales must equal the ending inventory.

In the beginning of period 1 and at the end of period 12, the inventory level is set equal to zero. Hence, these equations become:

$$X1 - S1 - Y2 = 0 \quad \text{and} \quad Y12 + X12 - S12 = 0.$$

A typical demand constraint, DEM-2, is shown below:

$$S2 + U2 = 2106.$$

This expression indicates that the cases sold, plus the unfilled demand, must equal the total demand for that period.

A typical capacity constraint, CAP-2, is shown below:

$$X2 \leq 2750.$$

This inequality indicates that the maximum number of cases that can be produced in any given month is 2750.

And lastly, a typical inventory level constraint, CNG-2, is shown below:

$$Y3 - Y2 = Q2 = (Q2+) - (Q2-).$$

This expression indicates that Q2 must equal the change in inventory level that occurs during period 2.

Since there is no beginning inventory, the change in inventory level that occurs during period 1 must equal the beginning inventory for period 2. Hence,

$$Y2 = Q1+ \qquad Q1- \text{ must be zero, since negative inventories are impossible.}$$

The *objective function* is to maximize the contribution, which equals:

$$\$12.61 \sum_{i=1}^{12} Si \quad -9.50 \sum_{i=1}^{12} Xi \quad -0.306 \sum_{i=1}^{12} Yi \quad -0.198 \sum_{i=1}^{12} Qi+$$

The following 9 pages provide the output to be used in discussing the Toys, Inc. problem.

Page 257 gives the optimum solution for the initial problem statement. Capacity is fixed at 2750 cases per month at every time period.

Pages 258 and 259 contain, respectively, the cost ranges and righthand-side ranges associated with the optimum solution of the initial problem.

Pages 260 through 264 give details pertaining to a parametric analysis of the capacity availability. In each time period, the capacity is increased from its original value of 2750 to 2750 + THETA × 1375. The computer reports only solutions corresponding to a change of basis. Such changes have taken place at values of THETA equal to 0.235, 0.499, 1.153, 1.329, and 1.450, which are reported on pages 260 to 264, respectively.

Page 265 presents a parametric study for the cost associated with unfilled demand. The cost of unfilled demand is increased from its initial value of 0 to PHI × 1, for every demand period. Page 257 gives the optimum solution for PHI = 0; page 265 provides the optimum solution for PHI = 0.454. Beyond this value of PHI the solution does not change.

245800, BRADLEY TOYS, INC. PRODUCTION SCHEDULING

TOTAL	NO.	ETA	ROW	CURRENT	CHOSEN	VECTR	RHS	C/V	CURRENT D/J	
ITERS	ETAS	REC	IDENT.	VALUE	VECTOR	REMVD	NO.	NO.	THETA/PHI	OPTIMAL PRI!
54	54	0	OBJT1	90159.351			1		* * * *	

J(H)	BETA(H)	ROW(I)	PI(I)	B(I)
0 00000	90159.35199931	OBJT1	1.00000000	.
0 00000	466.00000000-	OBJT2	.	.
S1	1491.00000000	PRD-1	9.50000000-	.
S2	2106.00000000	PRD-2	9.50000000-	.
S3	2777.00000000	PRD-3	10.00400000-	.
S4	843.00000000	PRD-4	10.11200000-	.
S5	1105.00000000	PRD-5	10.41800000-	.
S6	2932.00000000	PRD-6	10.92200000-	.
S7	1901.00000000	PRD-7	11.03000000-	.
S8	4336.00000000	PRD-8	11.53400000-	.
S9	4578.00000000	PRD-9	11.84000000-	.
S10	1771.00000000	PRD10	11.94800000-	.
S11	4744.00000000	PRD11	12.45200000-	.
S12	2750.00000000	PRD12	12.61000000-	.
X1	1491.00000000	DEM-1	3.11000000	1491.00000000
X2	2343.00000000	DEM-2	3.11000000	2106.00000000
Q2+	237.00000000	DEM-3	2.60600000	2777.00000000
X4	2750.00000000	DEM-4	2.49800000	843.00000000
X5	2750.00000000	DEM-5	2.19200000	1105.00000000
Q5+	1645.00000000	DEM-6	1.68800000	2932.00000000
X7	2750.00000000	DEM-7	1.58000000	1901.00000000
Y7	3580.00000000	DEM-8	1.07600000	4336.00000000
Q4+	1907.00000000	DEM-9	.77000000	4578.00000000
X10	2750.00000000	DEM10	.66200000	1771.00000000
Y4	210.00000000	DEM11	.15800000	4744.00000000
U12	466.00000000	DEM12	.	3216.00000000
+ CAP-1	1259.00000000	+ CAP-1	.	2750.00000000
+ CAP-2	407.00000000	+ CAP-2	.	2750.00000000
X3	2750.00000000	+ CAP-3	.50400000	2750.00000000
Y10	1015.00000000	+ CAP-4	.61200000	2750.00000000
Q9-	1828.00000000	+ CAP-5	.91800000	2750.00000000
X6	2750.00000000	+ CAP-6	1.42200000	2750.00000000
Q8-	1586.00000000	+ CAP-7	1.53000000	2750.00000000
X8	2750.00000000	+ CAP-8	2.03400000	2750.00000000
X9	2750.00000000	+ CAP-9	2.34000000	2750.00000000
Q11-	1994.00000000	+ CAP10	2.44800000	2750.00000000
X11	2750.00000000	+ CAP11	2.95200000	2750.00000000
X12	2750.00000000	+ CAP12	3.11000000	2750.00000000
Y2	.	CNG-1	.10800000-	.
Y3	237.00000000	CNG-2	.19800000	.
Q3-	27.00000000	CNG-3	.	.
Y5	2117.00000000	CNG-4	.19800000	.
Y6	3762.00000000	CNG-5	.19800000	.
Q6-	182.00000000	CNG-6	.	.
Y8	4429.00000000	CNG-7	.19800000	.
Q7+	849.00000000	CNG-8	.	.
Y9	2843.00000000	CNG-9	.	.
Y11	1994.00000000	CNG10	.19800000	.
Q10+	979.00000000	CNG11	.	.

Fig. E5.3 Optimum solution for Toys, Inc., cost and righthand-side ranges; parametric RHS analysis, and parametric cost ranging. *(Cont.)*

245800, BRADLEY TOYS, INC. PRODUCTION SCHEDULING

COST RANGES

BASIS VECTOR	BETA VALUE	COST IN PROBLEM	LIM 1	LIMIT 2	INCOMING VECTOR AT LIM 1	AT LIM 2
S1	1491.0000	-12.610000	* * * *	-9.5000000	UNBOUNDED	U1
S2	2106.0000	-12.610000	* * * *	-9.5000000	UNBOUNDED	U2
S3	2777.0000	-12.610000	* * * *	-10.004000	UNBOUNDED	U3
S4	842.99999	-12.610000	* * * *	-10.112000	UNBOUNDED	U4
S5	1105.0000	-12.610000	* * * *	-10.418000	UNBOUNDED	U5
S6	2932.0000	-12.610000	* * * *	-10.922000	UNBOUNDED	U6
S7	1901.0000	-12.610000	* * * *	-11.030000	UNBOUNDED	U7
S8	4336.0000	-12.610000	* * * *	-11.534000	UNBOUNDED	U8
S9	4578.0000	-12.610000	* * * *	-11.840000	UNBOUNDED	U9
S10	1771.0000	-12.610000	* * * *	-11.948000	UNBOUNDED	U10
S11	4744.0000	-12.610000	* * * *	-12.452000	UNBOUNDED	U11
S12	2750.0000	-12.610000	-12.758000	-9.5000000	Y12	+ CAP12
X1	1491.0000	9.5000000	9.1940000	12.610000	Q1+	U1
X2	2343.0000	9.5000000	9.3520000	9.6579999	Y12	U11
Q2+	237.00000	.19800000	.05000000	.35600000	Y12	U11
X4	2750.0000	9.5000000	* * * *	10.112000	UNBOUNDED	+ CAP-4
X5	2750.0000	9.5000000	* * * *	10.418000	UNBOUNDED	+ CAP-5
Q5+	1645.0000	.19800000		1.1160000	Q5-	+ CAP-5
X7	2750.0000	9.5000000	* * * *	11.030000	UNBOUNDED	+ CAP-7
Y7	3580.0000	.30600000	.15800000	.46400000	Y12	U11
Q4+	1907.0000	.19800000		.80999999	Q4-	+ CAP-4
X10	2750.0000	9.5000000	* * * *	11.948000	UNBOUNDED	+ CAP10
Y4	210.00000	.30600000	.15800000	.46400000	Y12	U11
U12	466.00000		-3.1100000	-.14800000	+ CAP12	Y12
+ CAP-1	1259.0000		-3.1100000	.30600000	U1	Q1+
+ CAP-2	407.00000		-.15800000	.14800000	U11	Y12
X3	2750.0000	9.5000000	* * * *	10.004000	UNBOUNDED	+ CAP-3
Y10	1015.0000	.30600000	.15800000	.46400000	Y12	U11
Q9-	1828.0000		-.19800000	.77000000	Q9+	U9
X6	2750.0000	9.5000000	* * * *	10.922000	UNBOUNDED	+ CAP-6
Q8-	1586.0000		-.19800000	1.0760000	Q8+	U8
X8	2750.0000	9.5000000	* * * *	11.534000	UNBOUNDED	+ CAP-8
X9	2750.0000	9.5000000	* * * *	11.840000	UNBOUNDED	+ CAP-9
Q11-	1994.0000		-.19800000	.15800000	Q11+	U11
X11	2750.0000	9.5000000	* * * *	12.452000	UNBOUNDED	+ CAP11
X12	2750.0000	9.5000000	* * * *	12.610000	UNBOUNDED	+ CAP12
Y2		.30600000		* * * *	Q1+	UNBOUNDED
Y3	237.00000	.30600000	.15800000	.46400000	Y12	U11
Q3-	27.000000		-.19800000	2.6060000	Q3+	U3
Y5	2117.0000	.30600000	.15800000	.46400000	Y12	U11
Y6	3762.0000	.30600000	.15800000	.46400000	Y12	U11
Q6-	182.00000		-.19800000	1.6880000	Q6+	U6
Y8	4429.0000	.30600000	.15800000	.46400000	Y12	U11
Q7+	849.00000	.19800000		1.7280000	Q7-	+ CAP-7
Y9	2843.0000	.30600000	.15800000	.46400000	Y12	U11
Y11	1994.0000	.30600000	.15800000	.46400000	Y12	U11
Q10+	978.99999	.19800000		2.6460000	Q10-	+ CAP10

Fig. E5.3 (*Cont.*)

245800, BRADLEY TOYS, INC. PRODUCTION SCHEDULING

RIGHT HAND SIDE RANGES

ROW NAME	CURRENT RHS VAL	PI VALUE	MINIMUM VALUE	MAXIMUM VALUE	OUTGOING VECTOR AT MIN	AT MAX
PRD-1		-9.5000000	-1491.0000	1259.0000	X1	+ CAP-1
PRD-2		-9.5000000	-2343.0000	407.00000	X2	+ CAP-2
PRD-3		-10.004000	-27.000000	407.00000	Q3-	+ CAP-2
PRD-4		-10.112000	-210.00000	407.00000	Y4	+ CAP-2
PRD-5		-10.418000	-210.00000	407.00000	Y4	+ CAP-2
PRD-6		-10.922000	-182.00000	407.00000	Q6-	+ CAP-2
PRD-7		-11.030000	-210.00000	407.00000	Y4	+ CAP-2
PRD-8		-11.534000	-210.00000	407.00000	Y4	+ CAP-2
PRD-9		-11.840000	-210.00000	407.00000	Y4	+ CAP-2
PRD10		-11.948000	-210.00000	407.00000	Y4	+ CAP-2
PRD11		-12.452000	-210.00000	407.00000	Y4	+ CAP-2
PRD12		-12.610000	-466.00000	2750.0000	U12	S12
DEM-1	1491.000	3.1100000		2750.0000	S1	+ CAP-1
DEM-2	2106.000	3.1100000		2513.0000	S2	+ CAP-2
DEM-3	2777.000	2.6060000	2750.0000	3184.0000	Q3-	+ CAP-2
DEM-4	842.9999	2.4980000	632.99999	1250.0000	Y4	+ CAP-2
DEM-5	1105.000	2.1920000	895.00000	1512.0000	Y4	+ CAP-2
DEM-6	2932.000	1.6880000	2750.0000	3339.0000	Q6-	+ CAP-2
DEM-7	1901.000	1.5800000	1691.0000	2308.0000	Y4	+ CAP-2
DEM-8	4336.000	1.0760000	4126.0000	4743.0000	Y4	+ CAP-2
DEM-9	4578.000	.77000000	4368.0000	4985.0000	Y4	+ CAP-2
DEM10	1771.000	.66200000	1561.0000	2178.0000	Y4	+ CAP-2
DEM11	4744.000	.15800000	4534.0000	5150.9999	Y4	+ CAP-2
DEM12	3216.000		2750.0000	UNBOUNDED	U12	
+ CAP-1	2750.000		1491.0000	UNBOUNDED	+ CAP-1	
+ CAP-2	2750.000		2343.0000	UNBOUNDED	+ CAP-2	
+ CAP-3	2750.000	.50400000	2343.0000	2777.0000	+ CAP-2	Q3-
+ CAP-4	2750.000	.61200000	2343.0000	2960.0000	+ CAP-2	Y4
+ CAP-5	2750.000	.91800000	2343.0000	2960.0000	+ CAP-2	Y4
+ CAP-6	2750.000	1.4220000	2343.0000	2932.0000	+ CAP-2	Q6-
+ CAP-7	2750.000	1.5300000	2343.0000	2960.0000	+ CAP-2	Y4
+ CAP-8	2750.000	2.0340000	2343.0000	2960.0000	+ CAP-2	Y4
+ CAP-9	2750.000	2.3400000	2343.0000	2960.0000	+ CAP-2	Y4
CAP10	2750.000	2.4480000	2343.0000	2960.0000	+ CAP-2	Y4
CAP11	2750.000	2.9520000	2343.0000	2960.0000	+ CAP-2	Y4
+ CAP12	2750.000	3.1100000		3216.0000	S12	U12
CNG-1		-.10800000		237.00000	Y2	Q2+
CNG-2		.19800000	UNBOUNDED	237.00000		Q2+
CNG-3			-27.000000	UNBOUNDED	Q3-	
CNG-4		.19800000	UNBOUNDED	1907.0000		Q4+
CNG-5		.19800000	UNBOUNDED	1645.0000		Q5+
CNG-6			-182.00000	UNBOUNDED	Q6-	
CNG-7		.19800000	UNBOUNDED	849.00000		Q7+
CNG-8			-1586.0000	UNBOUNDED	Q8-	
CNG-9			-1828.0000	UNBOUNDED	Q9-	
CNG10		.19800000	UNBOUNDED	978.99999		Q10+
CNG11			-1994.0000	UNBOUNDED	Q11-	

Fig. E5.3 *(Cont.)*

PARAMETRIC RIGHT-HAND-SIDE RANGING

245800, BRADLEY TOYS, INC. PRODUCTION SCHEDULING

TOTAL NO. ITERS	ETA ETAS	ROW REC	IDENT.	CURRENT VALUE	CHOSEN VECTOR	VECTR REMVD	RHS NO.	C/V NO.	CURRENT THETA/PHI	D/J CURRENT RT.I
62	62	0	OBJT1	94366.246	Q4-	Q4+	1	2	.23481818	

J(H)	BETA(H)	ROW(I)	PI(I)	B(I)+T*C(I)
0 00000	94366.24799940	OBJT1	1.00000000	.
0 00000	.	OBJT2	.	.
S1	1491.00000000	PRD-1	9.50000000-	.
S2	2106.00000000	PRD-2	9.50000000-	.
S3	2777.00000000	PRD-3	9.50000000-	.
S4	843.00000000	PRD-4	9.50000000-	.
S5	1105.00000000	PRD-5	9.80600000-	.
S6	2932.00000000	PRD-6	10.11200000-	.
S7	1901.00000000	PRD-7	10.41800000-	.
S8	4336.00000000	PRD-8	10.92200000-	.
S9	4578.00000000	PRD-9	11.22800000-	.
S10	1771.00000000	PRD10	11.33600000-	.
S11	4744.00000000	PRD11	11.84000000-	.
S12	3216.00000000	PRD12	12.14600000-	.
X1	1491.00000000	DEM-1	3.11000000	1491.00000000
X2	2106.00000000	DEM-2	3.11000000	2106.00000000
+ CAP-4	2229.87500000	DEM-3	3.11000000	2777.00000000
X4	843.00000000	DEM-4	3.11000000	843.00000000
X5	3072.87500000	DEM-5	2.80400000	1105.00000000
Q5+	1967.87500000	DEM-6	2.49800000	2932.00000000
X7	3072.87500000	DEM-7	2.19200000	1901.00000000
Y7	2108.75000000	DEM-8	1.68800000	4336.00000000
Q4-	.	DEM-9	1.38200000	4578.00000000
X10	3072.87500000	DEM10	1.27400000	1771.00000000
Y12	143.12500000	DEM11	.77000000	4744.00000000
Q2-	.	DEM12	.46400000	3216.00000000
+ CAP-1	1581.87500000	+ CAP-1	.	3072.87500000
+ CAP-2	966.87500000	+ CAP-2	.	3072.87500000
X3	2777.00000000	+ CAP-3	.	3072.87500000
Y10	512.37500000	+ CAP-4	.	3072.87500000
Q9-	1505.12500000	+ CAP-5	.30600000	3072.87500000
X6	3072.87500000	+ CAP-6	.61200000	3072.87500000
Q8-	1263.12500000	+ CAP-7	.91800000	3072.87500000
X8	3072.87500000	+ CAP-8	1.42200000	3072.87500000
X9	3072.87500000	+ CAP-9	1.72800000	3072.87500000
Q11-	1671.12500000	+ CAP10	1.83600000	3072.87500000
X11	3072.87500000	+ CAP11	2.34000000	3072.87500000
X12	3072.87500000	+ CAP12	2.64600000	3072.87500000
Y2	.	CNG-1	.30600000-	.
+ CAP-3	295.87500000	CNG-2	.	.
Q3+	.	CNG-3	.19800000	.
Y5	.	CNG-4	.19800000	.
Y6	1967.87500000	CNG-5	.19800000	.
Q6+	140.87500000	CNG-6	.19800000	.
Y8	3280.62500000	CNG-7	.19800000	.
Q7+	1171.87500000	CNG-8	.	.
Y9	2017.50000000	CNG-9	.	.
Y11	1814.25000000	CNG10	.19800000	.
Q10+	1301.87500000	CNG11	.	.

Fig. E5.3 (*Cont.*)

245800, BRADLEY TOYS, INC. PRODUCTION SCHEDULING

TOTAL NO.	ETA	ROW	CURRENT	CHOSEN	VECTR	RHS	C/V	CURRENT D/J	
ITERS ETAS	REC	IDENT.	VALUE	VECTOR	REMVD	NO.	NO.	THETA/PHI	CURRENT RT.
66 66	0	OBJT1	96707.057	Q5-	Q5+	1	2	.49945454	

J(H)	BETA(H)	ROW(I)	PI(I)	B(I)+T*C(I)
0 00000	96707.05799949	OBJT1	1.00000000	.
0 00000	.	OBJT2	.	.
S1	1491.00000000	PRD-1	9.50000000-	.
S2	2106.00000000	PRD-2	9.50000000-	.
S3	2777.00000000	PRD-3	9.50000000-	.
S4	843.00000000	PRD-4	9.50000000-	.
S5	1105.00000000	PRD-5	9.50000000-	.
S6	2932.00000000	PRD-6	9.80600000-	.
S7	1901.00000000	PRD-7	10.11200000-	.
S8	4336.00000000	PRD-8	10.61600000-	.
S9	4578.00000000	PRD-9	10.92200000-	.
S10	1771.00000000	PRD10	9.50000000-	.
S11	4744.00000000	PRD11	10.00400000-	.
S12	3216.00000000	PRD12	9.50000000-	.
X1	1491.00000000	DEM-1	3.11000000	1491.00000000
X2	2106.00000000	DEM-2	3.11000000	2106.00000000
CAP-4	2593.75000000	DEM-3	3.11000000	2777.00000000
X4	843.00000000	DEM-4	3.11000000	843.00000000
X5	1105.00000000	DEM-5	3.11000000	1105.00000000
Q5-	.	DEM-6	2.80400000	2932.00000000
X7	3436.75000000	DEM-7	2.49800000	1901.00000000
Y7	504.75000000	DEM-8	1.99400000	4336.00000000
Q4-	.	DEM-9	1.68800000	4578.00000000
X10	3078.25000000	DEM10	3.11000000	1771.00000000
+ CAP12	220.75000000	DEM11	2.60600000	4744.00000000
Q2-	.	DEM12	3.11000000	3216.00000000
+ CAP-1	1945.75000000	+ CAP-1	.	3436.75000000
+ CAP-2	1330.75000000	+ CAP-2	.	3436.75000000
X3	2777.00000000	+ CAP-3	.	3436.75000000
+ CAP10	358.50000000	+ CAP-4	.	3436.75000000
Q9-	1141.25000000	+ CAP-5	.	3436.75000000
X6	3436.75000000	+ CAP-6	.30600000	3436.75000000
Q8-	899.25000000	+ CAP-7	.61200000	3436.75000000
X8	3436.75000000	+ CAP-8	1.11600000	3436.75000000
X9	3436.75000000	+ CAP-9	1.42200000	3436.75000000
Q11-	1307.25000000	+ CAP10	.	3436.75000000
X11	3436.75000000	+ CAP11	.50400000	3436.75000000
X12	3216.00000000	+ CAP12	.	3436.75000000
Y2	.	CNG-1	.30600000-	.
+ CAP-3	659.75000000	CNG-2	.	.
Q3+	.	CNG-3	.19800000	.
+ CAP-5	2331.75000000	CNG-4	.	.
Y6	.	CNG-5	.19800000	.
Q6+	504.75000000	CNG-6	.19800000	.
Y8	2040.50000000	CNG-7	.19800000	.
Q7+	1535.75000000	CNG-8	.	.
Y9	1141.25000000	CNG-9	.	.
Y11	1307.25000000	CNG10	.19800000	.
Q10+	1307.25000000	CNG11	.	.

Fig. E5.3 (*Cont.*)

245800, BRADLEY TOYS, INC. PRODUCTION SCHEDULING

TOTAL	NO.	ETA	ROW	CURRENT	CHOSEN	VECTR	PHS	C/V	CURRENT D/J	
ITERS	ETAS	REC	IDENT.	VALUE	VECTOR	REMVD	NO.	NO.	THETA/PHI	CURRENT RT.
70	70	0	OBJT1	98496.347	Q8+	Q8-	1	2	1.1534545	

J(H)		BETA(H)	ROW(I)	PI(I)	B(I)+T*C(I)
0	00000	98496.34799954	OBJT1	1.00000000	.
0	00000	.	OBJT2	.	.
	S1	1491.00000000	PRD-1	9.50000000-	.
	S2	2106.00000000	PRD-2	9.50000000-	.
	S3	2777.00000000	PRD-3	9.50000000-	.
	S4	843.00000000	PRD-4	9.50000000-	.
	S5	1105.00000000	PRD-5	9.50000000-	.
	S6	2932.00000000	PRD-6	9.50000000-	.
	S7	1901.00000000	PRD-7	9.50000000-	.
	S8	4336.00000000	PRD-8	10.00400000-	.
	S9	4578.00000000	PRD-9	10.31000000-	.
	S10	1771.00000000	PRD10	9.50000000-	.
	S11	4744.00000000	PRD11	10.00400000-	.
	S12	3216.00000000	PRD12	9.50000000-	.
	X1	1491.00000000	DEM-1	3.11000000	1491.00000000
	X2	2106.00000000	DEM-2	3.11000000	2106.00000000
+	CAP-4	3493.00000000	DEM-3	3.11000000	2777.00000000
	X4	843.00000000	DEM-4	3.11000000	843.00000000
	X5	1105.00000000	DEM-5	3.11000000	1105.00000000
	Q5-	.	DEM-6	3.11000000	2932.00000000
	X7	2143.00000000	DEM-7	3.11000000	1901.00000000
+	CAP-7	2193.00000000	DEM-8	2.60600000	4336.00000000
	Q4-	.	DEM-9	2.30000000	4578.00000000
	X10	2179.00000000	DEM10	3.11000000	1771.00000000
+	CAP12	1120.00000000	DEM11	2.60600000	4744.00000000
	Q2-	.	DEM12	3.11000000	3216.00000000
+	CAP-1	2845.00000000	+ CAP-1	.	4336.00000000
+	CAP-2	2230.00000000	+ CAP-2	.	4336.00000000
	X3	2777.00000000	+ CAP-3	.	4336.00000000
+	CAP10	2157.00000000	+ CAP-4	.	4336.00000000
	Q9-	242.00000000	+ CAP-5	.	4336.00000000
	X6	2932.00000000	+ CAP-6	.	4336.00000000
	Q8+	.	+ CAP-7	.	4336.00000000
	X8	4336.00000000	+ CAP-8	.50400000	4336.00000000
	X9	4336.00000000	+ CAP-9	.81000000	4336.00000000
	Q11-	408.00000000	+ CAP10	.	4336.00000000
	X11	4336.00000000	+ CAP11	.50400000	4336.00000000
	X12	3216.00000000	+ CAP12	.	4336.00000000
	Y2	.	CNG-1	.30600000-	.
+	CAP-3	1559.00000000	CNG-2	.	.
	Q3+	.	CNG-3	.19800000	.
+	CAP-5	3231.00000000	CNG-4	.	.
+	CAP-6	1404.00000000	CNG-5	.	.
	Q6-	.	CNG-6	.	.
	Y8	242.00000000	CNG-7	.19800000	.
	Q7+	242.00000000	CNG-8	.	.
	Y9	242.00000000	CNG-9	.	.
	Y11	408.00000000	CNG10	.19800000	..
	Q10+	408.00000000	CNG11	.	.

Fig. E5.3 (*Cont.*)

245800, BRADLEY TOYS, INC. PRODUCTION SCHEDULING

TOTAL	NO.	ETA	ROW	CURRENT	CHOSEN	VECTR	RHS	C/V	CURRENT D/J	
ITERS	ETAS	REC	IDENT.	VALUE	VECTOR	REMVD	NO.	NO.	THETA/PHI	CURRENT RT.
73	73	0	OBJT1	98814.335	Q8-	Q8+	1	2	1.3294545	

J(H)		BETA(H)	ROW(I)	PI(I)	B(I)+T*C(I)
0	00000	98814.33599955	OBJT1	1.00000000	.
0	00000	.	OBJT2		.
	S1	1491.00000000	PRD-1	9.50000000-	.
	S2	2106.00000000	PRD-2	9.50000000-	.
	S3	2777.00000000	PRD-3	9.50000000-	.
	S4	843.00000000	PRD-4	9.50000000-	.
	S5	1105.00000000	PRD-5	9.50000000-	.
	S6	2932.00000000	PRD-6	9.50000000-	.
	S7	1901.00000000	PRD-7	9.50000000-	.
	S8	4336.00000000	PRD-8	9.50000000-	.
	S9	4578.00000000	PRD-9	10.00400000-	.
	S10	1771.00000000	PRD10	9.50000000-	.
	S11	4744.00000000	PRD11	10.00400000-	.
	S12	3216.00000000	PRD12	9.50000000-	.
	X1	1491.00000000	DEM-1	3.11000000	1491.00000000
	X2	2106.00000000	DEM-2	3.11000000	2106.00000000
+	CAP-4	3735.00000000	DEM-3	3.11000000	2777.00000000
	X4	843.00000000	DEM-4	3.11000000	843.00000000
	X5	1105.00000000	DEM-5	3.11000000	1105.00000000
	Q5-	.	DEM-6	3.11000000	2932.00000000
	X7	1901.00000000	DEM-7	3.11000000	1901.00000000
+	CAP-7	2677.00000000	DEM-8	3.11000000	4336.00000000
	Q4-	.	DEM-9	2.60600000	4578.00000000
	X10	1937.00000000	DEM10	3.11000000	1771.00000000
+	CAP12	1362.00000000	DEM11	2.60600000	4744.00000000
	Q2-	.	DEM12	3.11000000	3216.00000000
+	CAP-1	3087.00000000	+ CAP-1	.	4578.00000000
+	CAP-2	2472.00000000	+ CAP-2	.	4578.00000000
	X3	2777.00000000	+ CAP-3	.	4578.00000000
+	CAP10	2641.00000000	+ CAP-4	.	4578.00000000
	Q9-	.	+ CAP-5	.	4578.00000000
	X6	2932.00000000	+ CAP-6	.	4578.00000000
	Q8-	.	+ CAP-7	.	4578.00000000
	X8	4336.00000000	+ CAP-8	.	4578.00000000
	X9	4578.00000000	+ CAP-9	.50400000	4578.00000000
	Q11-	166.00000000	+ CAP10	.	4578.00000000
	X11	4578.00000000	+ CAP11	.50400000	4578.00000000
	X12	3216.00000000	+ CAP12	.	4578.00000000
	Y2	.	CNG-1	.30600000-	.
+	CAP-3	1801.00000000	CNG-2	.	.
	Q3+	.	CNG-3	.19800000	.
+	CAP-5	3473.00000000	CNG-4	.	.
+	CAP-6	1646.00000000	CNG-5	.	.
	Q6-	.	CNG-6	.	.
+	CAP-8	242.00000000	CNG-7	.	.
	Q7-	.	CNG-8	.19800000	.
	Y9	.	CNG-9	.	.
	Y11	166.00000000	CNG10	.19800000	.
	Q10+	166.00000000	CNG11	.	.

Fig. E5.3 (*Cont.*)

245800, BRADLEY TOYS, INC. PRODUCTION SCHEDULING

TOTAL	NO.	ETA	ROW	CURRENT	CHOSEN	VECTR	RHS	C/V	CURRENT	D/J	
ITERS	ETAS	REC	IDENT.	VALUE	VECTOR	REMVD	NO.	NO.	THETA/PHI	THETA AT MA	
78	78	0	OBJT1	98897.999			1	2	1.4501818		

J(H)		BETA(H)	ROW(I)	PI(I)	B(I)+T*C(I)
0	00000	98897.99999956	OBJT1	1.00000000	.
0	00000	.	OBJT2		.
	S1	1491.00000000	PRD-1	9.50000000-	.
	S2	2106.00000000	PRD-2	9.50000000-	.
	S3	2777.00000000	PRD-3	9.50000000-	.
	S4	843.00000000	PRD-4	9.50000000-	.
	S5	1105.00000000	PRD-5	9.50000000-	.
	S6	2932.00000000	PRD-6	9.50000000-	.
	S7	1901.00000000	PRD-7	9.50000000-	.
	S8	4336.00000000	PRD-8	9.50000000-	.
	S9	4578.00000000	PRD-9	9.50000000-	.
	S10	1771.00000000	PRD10	9.50000000-	.
	S11	4744.00000000	PRD11	9.50000000-	.
	S12	3216.00000000	PRD12	9.50000000-	.
	X1	1491.00000000	DEM-1	3.11000000	1491.00000000
	X2	2106.00000000	DEM-2	3.11000000	2106.00000000
+	CAP-4	3901.00000000	DEM-3	3.11000000	2777.00000000
	X4	843.00000000	DEM-4	3.11000000	843.00000000
	X5	1105.00000000	DEM-5	3.11000000	1105.00000000
	Q5-	.	DEM-6	3.11000000	2932.00000000
	X7	1901.00000000	DEM-7	3.11000000	1901.00000000
+	CAP-7	2843.00000000	DEM-8	3.11000000	4336.00000000
	Q4-	.	DEM-9	3.11000000	4578.00000000
	X10	1771.00000000	DEM10	3.11000000	1771.00000000
+	CAP12	1528.00000000	DEM11	3.11000000	4744.00000000
	Q2-	.	DEM12	3.11000000	3216.00000000
+	CAP-1	3253.00000000	+ CAP-1	.	4744.00000000
+	CAP-2	2638.00000000	+ CAP-2	.	4744.00000000
	X3	2777.00000000	+ CAP-3	.	4744.00000000
+	CAP10	2973.00000000	+ CAP-4	.	4744.00000000
	Q9+	.	+ CAP-5	.	4744.00000000
	X6	2932.00000000	+ CAP-6	.	4744.00000000
	Q8-	.	+ CAP-7	.	4744.00000000
	X8	4336.00000000	+ CAP-8	.	4744.00000000
	X9	4578.00000000	+ CAP-9	.	4744.00000000
	Q11+	.	+ CAP10	.	4744.00000000
	X11	4744.00000000	+ CAP11	.	4744.00000000
	X12	3216.00000000	+ CAP12	.	4744.00000000
	Y2	.	CNG-1	.30600000-	.
+	CAP-3	1967.00000000	CNG-2	.	.
	Q3+	.	CNG-3	.19800000	.
+	CAP-5	3639.00000000	CNG-4	.	.
+	CAP-6	1812.00000000	CNG-5	.	.
	Q6-	.	CNG-6	.	.
+	CAP-8	408.00000000	CNG-7	.	.
	Q7-	.	CNG-8	.	.
+	CAP-9	166.00000000	CNG-9	.19800000	.
+	CAP11	.	CNG10	.	.
	Q10-	.	CNG11	.19800000	.

THETA UNBOUNDED ABOVE THIS VALUE

Fig. E5.3 (*Cont.*)

PARAMETRIC COST RANGING

245800, BRADLEY TOYS, INC. PRODUCTION SCHEDULING

TOTAL	NO.	ETA	ROW	CURRENT	CHOSEN	VECTR	RHS	C/V	CURRENT D/J	
ITERS	ETAS	REC	IDENT.	VALUE	VECTOR	REMVD	NO.	NO.	THETA/PHI	SOLUTION PR
107	107	0	OBJT1	90072.329	Q1+	U12	1	1	.45400000	

J(H)	BETA(H)	ROW(I)	PI(I)	B(I)
0 00000	90072.32999925	OBJT1	1.00000000	.
0 00000	.	OBJT2	.45400000	.
S1	1491.00000000	PRD-1	9.50000000-	.
S2	2106.00000000	PRD-2	9.80600000-	.
S3	2777.00000000	PRD-3	10.31000000-	.
S4	843.00000000	PRD-4	10.41800000-	.
S5	1105.00000000	PRD-5	10.72400000-	.
S6	2932.00000000	PRD-6	11.22800000-	.
S7	1901.00000000	PRD-7	11.33600000-	.
S8	4336.00000000	PRD-8	11.84000000-	.
S9	4578.00000000	PRD-9	12.14600000-	.
S10	1771.00000000	PRD10	12.25400000-	.
S11	4744.00000000	PRD11	12.75800000-	.
S12	3216.00000000	PRD12	13.06400000-	.
X1	1550.00000000	DEM-1	3.11000000	1491.00000000
X2	2750.00000000	DEM-2	2.80400000	2106.00000000
Y10	1481.00000000	DEM-3	2.30000000	2777.00000000
X4	2750.00000000	DEM-4	2.19200000	843.00000000
X5	2750.00000000	DEM-5	1.88600000	1105.00000000
Y6	4228.00000000	DEM-6	1.38200000	2932.00000000
X7	2750.00000000	DEM-7	1.27400000	1901.00000000
Q7+	849.00000000	DEM-8	.77000000	4336.00000000
Y5	2583.00000000	DEM-9	.46400000	4578.00000000
X10	2750.00000000	DEM10	.35600000	1771.00000000
Q10+	979.00000000	DEM11	.14800000-	4744.00000000
Y3	703.00000000	DEM12	.45400000-	3216.00000000
+ CAP-1	1200.00000000	+ CAP-1	.	2750.00000000
Y12	466.00000000	+ CAP-2	.30600000	2750.00000000
X3	2750.00000000	+ CAP-3	.81000000	2750.00000000
Q11-	1994.00000000	+ CAP-4	.91800000	2750.00000000
Y9	3309.00000000	+ CAP-5	1.22400000	2750.00000000
X6	2750.00000000	+ CAP-6	1.72800000	2750.00000000
Q8-	1586.00000000	+ CAP-7	1.83600000	2750.00000000
X8	2750.00000000	+ CAP-8	2.34000000	2750.00000000
X9	2750.00000000	+ CAP-9	2.64600000	2750.00000000
Y11	2460.00000000	+ CAP10	2.75400000	2750.00000000
X11	2750.00000000	+ CAP11	3.25800000	2750.00000000
X12	2750.00000000	+ CAP12	3.56400000	2750.00000000
Y2	59.00000000	CNG-1	.19800000	.
Q3-	27.00000000	CNG-2	.19800000	.
Q2+	644.00000000	CNG-3	.	.
Q9-	1828.00000000	CNG-4	.19800000	.
Q5+	1645.00000000	CNG-5	.19800000	.
Q6-	182.00000000	CNG-6	.	.
Y7	4046.00000000	CNG-7	.19800000	.
Y8	4895.00000000	CNG-8	.	.
Q4+	1907.00000000	CNG-9	.	.
Y4	676.00000000	CNG10	.19800000	.
Q1+	59.00000000	CNG11	.	.

PHI UNBOUNDED ABOVE THIS VALUE

Fig. E5.3 *(Concluded)*

Using the computer output supplied, answer the following questions.

a) Draw a graph depicting the following as a function of time, assuming a capacity of 2750 cases per month.

   i) Cases demanded
   ii) Cases produced
   iii) Cases in inventory
   iv) Cases of unfilled demand

Explain thoroughly what this graph implies about the optimal operations of Toys, Inc.

b) Give a complete economic interpretation of the dual variables.
c) Give a concise explanation of the righthand-side and cost ranging output.
d) Use the parametric programming of the righthand side as a basis for discussing the optimal production capacity.
e) Use the parametric programming of the cost function as a basis for discussing the "value" of goodwill loss associated with unfilled demand. (When demand is not met, we lose some goodwill of our customer. What is this loss worth?)

4. *Solving an LP by computer.* Your doctor has found in you a very rare disease, ordinarily incurable, but, in your case, he believes that perhaps something can be done by a series of very elaborate treatments coupled with a strict diet. The treatments are so expensive that it becomes necessary to minimize the cost of the diet.

    The diet must provide mimimum amounts of the following items: calories, calcium, vitamin A, riboflavin, and ascorbic acid. Your daily requirements for these items (in the above order) may be determined by reading off the numerical values corresponding to the first five letters of your name on Table 5.4. The following are the units used: 10^2 calories, 10^{-2} grams, 10^2 international units, 10^{-1} milligrams, and milligrams.

Table 5.4 Diet Requirements

	Diet	Product X		Diet	Product X
A	7	63	N	6	91
B	60	52	O	10	45
C	83	59	P	32	82
D	10	85	Q	51	98
E	39	82	R	47	67
F	59	58	S	20	97
G	38	50	T	66	28
H	30	69	U	78	54
I	65	44	V	81	33
J	27	26	W	81	59
K	91	30	X	61	61
L	68	43	Y	0	39
M	49	90	Z	86	83

Your choice of foods is somewhat limited because you find it financially advantageous to trade at a discount store that has little imagination. You can buy: (1) wheat flour (enriched), (2) evaporated milk, (3) cheddar cheese, (4) beef liver, (5) cabbage, (6) spinach, (7) sweet potatoes, (8) lima beans (dried).

The nutritional values per dollar spent have been tabulated by Dantzig (who regularly patronizes the store) in Table 5.5. In addition, the store features a grayish powder, Product X,

Table 5.5 Nutritive Values of Common Foods Per Dollar of Expenditure*

Commodity	Calories (1000)	Protein (grams)	Calcium (grams)	Iron (mg.)	Vitamin A (1000 I.U.)	Thiamine (mg.)	Riboflavin (mg.)	Niacin (mg.)	Ascorbic Acid (mg.)
1. Wheat flour (enriched)	44.7	1411	2.0	365	—	55.4	33.3	441	—
5. Corn meal	36.0	897	1.7	99	30.9	17.4	7.9	106	—
15. Evaporated milk (can)	8.4	422	15.1	9	26.0	3.0	23.5	11	60
17. Oleomargarine	20.6	17	.6	6	55.8	.2	—	—	—
19. Cheese (cheddar)	7.4	448	16.4	19	28.1	.8	10.3	4	—
21. Peanut butter	15.7	661	1.0	48	—	9.6	8.1	471	—
24. Lard	41.7	—	—	—	.2	—	.5	5	—
30. Liver (beef)	2.2	333	.2	139	169.2	6.4	50.8	316	525
34. Pork loin roast	4.4	249	.3	37	—	18.2	3.6	79	—
40. Salmon, pink (can)	5.8	705	6.8	45	3.5	1.0	4.9	209	—
45. Green beans	2.4	138	3.7	80	69.0	4.3	5.8	37	862
46. Cabbage	2.6	125	4.0	36	7.2	9.0	4.5	26	5369
50. Onions	5.8	166	3.8	59	16.6	4.7	5.9	21	1184
51. Potatoes	14.3	336	1.8	118	6.7	29.4	7.1	198	2522
52. Spinach	1.1	106	—	138	918.4	5.7	13.8	33	2755
53. Sweet potatoes	9.6	138	2.7	54	290.7	8.4	5.4	83	1912
64. Peaches, dried	8.5	87	1.7	173	86.8	1.2	4.3	55	57
65. Prunes, dried	12.8	99	2.5	154	85.7	3.9	4.3	65	257
68. Lima beans, dried	17.4	1055	3.7	459	5.1	26.9	38.2	93	—
69. Navy beans, dried	26.9	1691	11.4	792	—	38.4	24.6	217	—

* Source: G. B. Dantzig, Linear Programming and Extensions, Princeton University Press, Princeton, N.J., 1963.

sold in bulk, whose nutritional values per unit cost are also given in Table 5.4. The units (same order of items as before) are 10^3 calories/dollar, 10^{-1} grams/dollar, 10^3 international units/dollar, 10^{-1} milligrams/dollar, milligrams/dollar.

Your doctor has coded your diet requirements and the nutritional properties of Product X under the first five letters of your name in Table 5.4.

a) Find your minimum-cost diet and its cost.
b) How much would you be willing to pay for pure vitamin A? pure riboflavin?
c) A new food has come out (called GLUNK) having nutritional values per dollar of 83, 17, 25, 93, 07 (values are in the same order and same units as for Product X). Would you want to include the new food in your diet?
d) By how much would the cost of lima beans have to change before it would enter (or leave, as the case may be) your diet?
e) Over what range of values of cost of beef liver would your diet contain this wonderful food?
f) Suppose the cost of foods 1, 3, 5, 7, and 9 went up 10% but you continued on the diet found in (a). How much would you be willing to pay for pure vitamin A? pure riboflavin?
g) If the wheat flour were enriched by 10 units of vitamin A without additional cost, would this change your diet?

ACKNOWLEDGMENTS

Exercise 1 is based on Chapter 14 of *Applied Linear Programming*, by Norman J. Driebeck, Addison-Wesley Publishing Company, Inc., 1969.
Exercises 2 and 3 are based on cases with the same names written by one of the authors.
Exercise 4 is a variation of the diet problem used by John D. C. Little of the Massachusetts Institute of Technology.

Integration of Strategic and Tactical Planning in the Aluminum Industry

6

As we have indicated several times before, problem formulation usually is not straightforward; on the contrary, it requires a great deal of creativity on the part of the model builder. It also is often the case that a single model cannot provide all the support managers need in dealing with a complex set of decisions. This chapter describes a practical application to illustrate the complexities inherent in model design when both strategic and tactical decisions are involved. The emphasis is on the formulation and linking of two separate models representing distinct levels of decision making.

6.1 THE PLANNING APPROACH

Industrial logistics is concerned with the effective management of the total flow of materials, from the acquisition of raw materials to the delivery of finished goods to the final consumer. It is an important and complex field of management, which encompasses a large number of decisions and affects several organizational echelons. Included in the logistics process are decisions determining the resources of the firm (number and location of plants, number and location of warehouses, transportation facilities, communication equipment, data-processing facilities, and so forth) and the proper utilization of these resources (capacity planning, inventory control, production scheduling, dispatching, and so forth). Commonly, the resource-acquisition decisions are associated with the strategic-planning efforts of the firm, and the resource-utilization decisions are considered tactical-planning activities.

These two types of decisions, resource acquisition and resource utilization, differ in scope, level of management involvement, type of supporting information, and length of their planning horizons. If formal systems are to be designed to support these two widely different types of logistics decisions, it is logical to develop two distinct systems: one addressing the strategic-planning issues and one addressing the tactical-planning issues of the firm. However, the two systems should interact strongly with one another, since strategic decisions provide constraints that tactical planning decisions have to meet, and the execution of tactical decisions determines the resource requirements to be supplied by the higher-level strategic decisions. This suggests the development of a hierarchical planning system that is responsive to the organizational structure of the firm and defines a framework for the partitioning and linking of the planning activities.

It is the purpose of this chapter to describe an actual study dealing with the development of an integrated system to support the strategic and tactical planning of an aluminum company. The production of aluminum is a continuous process; therefore, it does not introduce the problems associated with discontinuities in lot-size production required in batch-processing operations. Linear programming becomes a very appropriate model to use in connection with the production-planning activities.

The heart of the planning system to be described is formed by two linear-programming models that interact with each other. One of these models addresses the long-range strategic issues associated with resource planning, while the other is an operational model oriented toward the tactical problems of short-range resource utilization.

We will first provide a brief background of the aluminum industry. Subsequently, we will describe the structure of both models and the way in which they are integrated to form a comprehensive planning system. Special attention should be paid to the different characteristics of the models in terms of time horizon and number of time periods, level of aggregation of the information processed, and the scope and level of management interaction.

6.2 THE ALUMINUM INDUSTRY AND SMELTER OPERATIONS

The production of aluminum from raw materials to finished goods is a fairly involved process but the smelting part of the process is straightforward (see Fig. 6.1). Calcined alumina is reduced to aluminum metal in electrolytic cells or "pots." The passing of electric current through the molten electrolyte causes an electrolytic action that reduces the alumina into two materials, molten aluminum and carbon dioxide. Periodically the aluminum is drawn off through a siphon into large crucibles, which are then used to transfer the molten aluminum to a holding furnace, where the blending or alloying of the metal takes place. Next the aluminum is cast into ingots of various sizes and shapes ranging from 5 lbs to 20,000 lbs depending on the type of equipment that will be used in processing the metal. The smelting operation is the major point of the process that is of concern to us in the present study.

The aluminum company where this project took place is one of the largest fully integrated aluminum companies in the world; the company supplies its products to approximately a hundred different countries. It has six smelters, all operating near

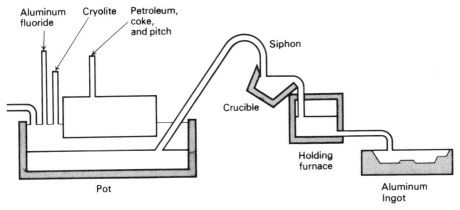

Fig. 6.1 The aluminum production process.

capacity, and the demand for ingot aluminum is expected to continue to grow. Of the six smelters, four are located in domestic locations and two are in foreign locations.

In addition to its own facilities, the company also is able to procure aluminum from external sources through "swapping" contracts. Swapping is the term used for the common practice in the aluminum industry of having a competitor produce one of *your* orders for a particular type of aluminum alloy and deliver it directly to *your* end-use customer under *your* label. The reason a company might want to do this may vary but the most common one is to save on transportation costs of the finished product by selecting a competitor's smelter that is in production closer to your customer area. These agreements generally are reciprocal in nature and are negotiated prior to the start of a year. They are actually a trade of material, since usually no money is exchanged and the major part of the negotiation concerns the amount of material that one company will produce for the other. All warrantees, complaints, and the like, will fall back on the company doing the contracting and not the company producing and delivering the aluminum. Such swapping agreements also are common in industries such as petroleum and fertilizer.

In view of the then current level of operations and the forecast growth in demand, the company was considering various alternatives for smelter-capacity expansion. One of the alternatives was the construction of a new smelter. A major concern of top management was the decision affecting the location, capacity, and the date for starting construction of the new smelter. In addition, a capacity-expansion program was required in the *existing* facilities. It was imperative to organize this effort in a well balanced form, allocating the resources of the company in those areas that offered the highest potential payoffs.

The first model to be presented is used to assess various options for capacity expansion at existing or new smelters, and for swapping contracts. It determines the sources of aluminum for meeting customer demands. A second model is used to assign incoming ingot orders to the various sources of supply, according to the capacity available at each source.

6.3 OVERVIEW OF THE STRATEGIC PLANNING MODEL

Objectives of the Strategic Logistics Model

An effective strategic model should be able to support the development of corporate logistics policies and to provide top managers with a better understanding of decisions on design of new production facilities, capacity expansion of existing facilities, acceptance of long-term contracts, and development of marketing and distribution strategies.

Specifically, our strategic-planning model was designed to assist managers in:

1. evaluating different options for increasing capacity at the existing aluminum smelters of the company;

2. measuring the economic consequences of installing a new smelter, whose size and location had yet to be determined;

3. defining the desired quantity and the price to be paid for various purchasing and swapping contracts;

4. setting general guidelines for the levels of operation at the company's smelters; and

5. assessing the attractiveness of each of the present market areas and defining a strategy for potential growth in these areas.

The Strategic Model—General Characteristics

In order to fulfill the objectives stated above, the planning model should have an overall corporate approach and should deal with aggregate information without going into details pertinent only for operating decisions. Due to the large number of interactions involved in the planning process and the continuous nature of the production activities, it was soon recognized that the most appropriate model to use in this kind of problem was a linear-programming model.

One of the first decisions that has to be made regarding the design of a model is the *total time horizon* covered and the *number of time periods* into which that time horizon is divided. In order to maintain as simple a model structure as possible, only one time period was included in the model. (This does not represent a shortcoming in an aggregated model such as this one, since multistage decisions affecting several years can be studied by changing the input to the one-time-period model properly in a *sequential* fashion.)

The model considers eleven different metal sources, six existing smelter locations, one new smelter, three swapping sources, and one other *possible* swapping source. The use of eleven sources makes it possible to (1) evaluate precisely what value to attach to current swapping agreements, (2) evaluate the marginal economic worth of the various alternative locations for a new smelter, and (3) allow for an unanticipated source of metal. At the same time, the model provides management with sufficient detailed information to answer questions regarding the operation of existing smelters (capacity planning, the level of operation desired, and so forth).

Customer areas and swapping destinations are broken into forty different market groups. These forty locations provide sufficient segmentation of the total market by allowing us to partition the market into *demand centers*, which have distinct transportation costs, customs duties, and in-transit inventory charges.

An analysis of the product line of the company revealed that, while the variety was extensive, an aggregation into only *eight major categories* provided sufficient detail for the strategic-planning model.

The primary constraints that are imposed on the planning process are the existing capacities of hot-metal and casting equipment, as well as the limits imposed on the purchasing or swapping contracts. In addition, the demand generated at each customer area should be satisfied with the present capacity; otherwise expansions of the current installations have to be made.

An important cost element in the production of aluminum ingot is the metal reduction cost. Large amounts of electricity are consumed in that process (approximately 8 KWH/lb), making the cost of electricity the single most important consideration in determining a smelter's location. The choice of location for a new smelter thus is limited to a small number of geographic areas where electrical costs are significantly low.

A second major cost factor is the cost of blending and casting the aluminum. A third relevant cost that affects decisions concerning either a new smelter location or order allocations among existing smelters is that for transportation–the total cost involved for both shipping the raw materials to a smelter and shipping the finished goods to a given customer. Because of the nature of the material (weight and volume) and the wide dispersion of the company's customers, transportation represents a large portion of the controllable variable cost of aluminum ingots. The problem is simplified in our case because the company owns its own shipping line for delivering raw materials to the smelters. The cost of transporting materials to the smelter then can be treated as a fixed cost in a first approximation. This assumption can be relaxed in a subsequent stage of the model development.

The final two major cost items to be considered when making logistic decisions are in-transit inventory and customs duties. While both of these costs are relatively small when compared with the other variable costs, they nonetheless are still large enough to warrant consideration.

Minimization of cost, instead of maximization of profit, was selected as the objective function because cost information is more readily available and prices depend on quantities purchased, type of contracts, and clients. Moreover, if demand *has* to be met, the resulting revenues are fixed and minimization of cost becomes equivalent to maximization of profit.

6.4 MATHEMATICAL FORMULATION OF THE STRATEGIC PLANNING MODEL

The Strategic Model—Notation

We now will describe the symbolic notation that is used in delineating the mathematical formulation of the strategic model.

Smelter and Purchase Source Locations

The letter "s" represents the location of smelters or purchase sources according to the following convention:

$$
\begin{aligned}
&s = 1, 2, \ldots, 6 &&\text{existing smelter locations}\\
&s = 7 &&\text{new smelter}\\
&s = 8, 9, 10 &&\text{swapping}\\
&s = 11 &&\text{other}
\end{aligned}
$$

Customer Areas

The letter "a" represents the forty locations of the various customer areas and swapping recipients.

Product Types

The letter "p" represents the eight different product types.

Data Requirements

r_{sp} = Reduction cost of product "p" at source "s", in $/ton;

c_{sp} = Casting cost of product "p" at source "s", in $/ton;

t_{sa} = Transportation cost from source "s" to customer area "a", in $/ton (the cost/ton is the same for all product types);

o_{sa} = Customs duty charged for the shipments from source "s" to customer area "a", in $/ton (the cost is the same for all product types);

ℓ_{sa} = Lead time and in-transit time required to ship from source "s" to customer area "a" (the time is the same for all product types);

h_p = Inventory-holding cost for product "p", in $/ton/day;

d_{ap} = Forecast demand for product "p" at customer location "a", in tons;

$\underline{m}_s, \overline{m}_s$ = Lower and upper bounds respectively, for the hot-metal capacity at source "s", in tons;

$\underline{e}_{sp}, \overline{e}_{sp}$ = Lower and upper bounds, respectively, for the casting-equipment capacity at source "s" for product "p", in tons.

Decision Variables

Q_{sap} = Quantity of product "p" to be shipped from source "s" to customer area "a", in tons;

M_s = Total hot-metal output at source "s", in tons;

D = Total overseas customs duties for all products shipped from all sources to all foreign customer areas, in dollars;

I_p = Total in-transit inventory costs for product "p", in dollars;

E_{sp} = Total amount of product "p" to be cast at source "s", in tons.

The last four decision variables (M_s, D, I_p, and E_{sp}) are introduced only for convenience in interpreting the results, as will be seen in the next section.

The Strategic Model—Formulation

Using the notation described above, we can now formulate the strategic model in mathematical terms.

The Objective Function—Logistics Cost

The objective of the model is the minimization of the total logistics cost incurred, which is represented by the following expression:

$$\text{Minimize cost} = \sum_s \sum_a \sum_p (r_{sp} + c_{sp} + t_{sa})Q_{sap} + \sum_p I_p + D.$$

The first term is the sum of reduction, casting, and transportation costs; the second term is the in-transit inventory cost; and the third term, D, is the total customs duties. (In-transit inventory costs and customs duties are defined in the constraint set as a function of the variable Q_{sap}.)

Metal-Supply Constraint at Sources

$$\sum_a \sum_p Q_{sap} - M_s = 0, \qquad s = 1, 2, \ldots, 11,$$

$$\underline{m}_s \leq M_s \leq \overline{m}_s, \qquad s = 1, 2, \ldots, 11.$$

The first equation merely states that the total amount shipped from location s to every customer, considering all products, should be equal to M_s, which is the total metal supply at smelter (or purchasing location) s. This equation serves to define the variable M_s.

The second constraint set represents the upper and lower bounds on the total metal supply at each smelter or puchase location. Recall that constraints of this type are handled implicitly rather than explicitly whenever a bounded-variable linear-programming code is used (see Chapter 2, Section 2.6). The upper and lower bounds define the maximum hot-metal capacity and the minimum economical operational level of the smelter, respectively. When dealing with swapping or puchasing locations, they provide the range in which purchasing or swapping agreements take place.

Shadow prices associated with the metal-supply constraints indicate whether expansion (or contraction) of a smelter hot-metal capacity or purchase contract are in order.

Equipment Casting Capacity at Smelters

$$\sum_a Q_{sap} - E_{sp} = 0 \qquad \begin{cases} s = 1, 2, \ldots, 11, \\ p = 1, 2, \ldots, 8. \end{cases}$$

$$\underline{e}_{sp} \leq E_{sp} \leq \bar{e}_{sp} \qquad \begin{cases} s = 1, 2, \ldots, 11, \\ p = 1, 2, \ldots, 8. \end{cases}$$

The first equation is used as a definition of variable E_{sp}. It indicates that E_{sp}, the total amount of product p cast at smelter s, must be equal to the total amount of product p shipped from location s to all customers.

The second set of constraints imposes lower and upper bounds on the amount of product p cast in smelter s. These bounds reflect maximum casting capacity and minimum economical levels of performance, respectively, and again do not add signifi- cant computational time to the solution of the model when a linear-programming code with bounded-variable provisions is used.

Shadow prices associated with the casting-equipment capacity constraint allow the efficiencies of the various casting equipment to be ranked and suggest expansion or replacement of current equipment.

Demand Constraints

$$\sum_s Q_{sap} = d_{ap} \qquad \begin{cases} a = 1, 2, \ldots, 40, \\ p = 1, 2, \ldots, 8. \end{cases}$$

This set of equations specifies that the amount of each product p received at customer region a, from all sources s, has to be equal to the demand of product p at customer region a.

The shadow prices of these constraints allow the relative attractiveness of each product group at each customer area to be defined and therefore serve as basic in- formation for marketing-penetration strategies. By ranking each market in accor- dance with the marginal returns to be derived by expanding its current requirements, priorities can be assigned that provide guidelines for marketing penetration. In addition, shadow prices indicate when swapping is of interest, since swapping means simply a trade-off between two customer areas. Ideally, we would like to swap an area with a very small marginal return with one that provides a very high marginal return.

Total Overseas Customs Duties

$$\sum_s \sum_a \sum_p o_{sa} Q_{sap} - D = 0.$$

This equation is used to define the total amount spent in customs duties, D, and permits the company to keep track of this expenditure without performing additional computations.

Total In-Transit Inventory Cost

$$\sum_s \sum_a h_p \ell_{sa} Q_{sap} - I_p = 0, \qquad p = 1, 2, \ldots, 8.$$

This equation records the total in-transit inventory cost for each product group p.

Nonnegativity of the Variables

All the variables should be nonnegative.

6.5 THE TACTICAL PLANNING MODEL

Objectives of the Tactical Model

The basic objective of the tactical model is to assist middle management in assigning ingot orders to the various possible sources of supply, in a way that is consistent with upper-level decisions, which are made with the help of the strategic planning model. The tactical model deals only with the order assignments to the four domestic smelters, although extensions of the model to incorporate the foreign smelters are straight-forward. The assignments are performed on a week-by-week basis for a four-week time horizon, followed by two months of planned operations based on orders actually received and forecasts for orders to be expected in those periods.

Each order refers to a demand for a single product type. If an original order contains requirements for more than one product, the order is broken into various individual single-product orders.

The model is intended to support management decisions in the following areas:

1. Assignment of ingot orders to specific casting machines at each smelter (initially only domestic smelters are considered);

2. Effective utilization of existing production equipment;

3. Assignment of labor crews to each production center;

4. Determination of aggregated inventory levels for each product type;

5. Specification of transportation requirements, in deciding ship reservations; and

6. Identification of operational bottlenecks, which could suggest capacity expansion opportunities.

Linking the Strategic and Tactical Models

Figure 6.2 is a diagram of the total logistics system, illustrating the relationships between the strategic and tactical models. It is important to notice the hierarchical nature of this approach, in which decisions made at the strategic level define some of the constraints that have to be observed at the tactical level. Specifically, the strategic model defines the capacity expansion that should take place in the hot-metal and casting facilities to cope with the increasing aluminum demand, including the location, size, and timing of construction of a new smelter. In addition, it fixes operational

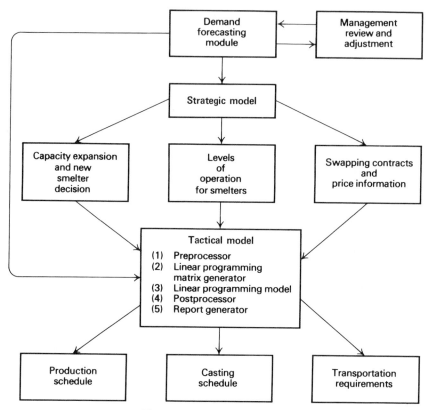

Fig. 6.2 The logistics system.

levels for hot-metal capacity, it defines swapping sources and the quantities to be swapped with these sources, and it prescribes allocation of market areas to metal sources. The strategic model essentially determines the capacity of all sources including new plants, old plants, and swapping points. The tactical model makes the most effective use of the existing production facilities by providing production, casting, and transportation schedules. Feedback from the actual implementation of the tactical-model decisions will in turn provide an important input to the strategic model to obtain a better balance between resource acquisition and resource utilization.

Segmentation of the Tactical Model

The design of the tactical model poses some serious difficulties. First, there are a number of institutional constraints resulting from the company's traditional practices in dealing with specific customers and from priority requirements of some orders, which force the schedule to allocate an order to a specific machine. These constraints must be considered in any realistic order-assignment procedure. Second, due to the large number of constraints and decision variables involved in the order-assignment process, it is mandatory to make every effort to produce a good starting solution to the model and to reduce, as much as possible, the number of constraints to be considered; otherwise the model would become computationally or economically infeasible to run and update every week. There were about one thousand outstanding customer orders to be assigned during the three-month time horizon considered in the tactical model. Thus, the dimension of the problem is such that linear programming becomes the only viable approach to consider. However, since splitting an order between two smelters was not acceptable from an operational point of view, and since the use of *integer* linear programming was out of the question due to the large model dimensions, extreme care had to be exercised to avoid order-splitting problems.

These considerations led to the design of an operating system composed of three segments: the *preprocessor*, which establishes a preliminary operation plan; the *linear-programming model*, which computes an optimal order assignment; and the *postprocessor*, which consolidates the orders that might be split and produces relevant management reports. Figure 6.3 illustrates how the system has been decomposed and how the three segments interact.

The first of these segments is a preprocessor subsystem. This is a computer program that performs the following functions:

1. determines the date by which the order should be completed at each smelter (known as the ex-mill date) to satisfy the promised delivery date to the client. The ex-mill date is determined by subtracting the transportation time, from the smelter to the customer location, from the promised delivery date;

2. accepts constraints on orders that must be processed at a given smelter because of purity specifications that can be met only by using a specific machine at that smelter;

3. accepts reservations for certain blocks of casting-capacity time that are required for some special purpose; and

4. accepts the demand-forecast estimates for the two look-ahead months (orders for the current month always will be known).

The preprocessor then constructs an initial order-assignment plan, based on the minimization of freight and in-transit inventory costs. This order-assignment plan does not consider any capacity constraints and thus represents an ideal plan for distribution. The ideal plan rarely would be feasible, but it constitutes an effective solution to initiate the linear-programming model that will be formulated in terms of order reassignments. Only freight and in-transit inventory costs are selected in this initial plan assignment, in order to simplify the cost computations. Freight and in-transit inventory costs are the predominant cost elements in the operating model. Other cost elements are incorporated into the subsequent linear-programming model. The total demand determined for each product type at each smelter during each time period is accumulated, for comparison with actual casting and hot-metal capacity limitations. The final output of the preprocessor system is a list for each time period,

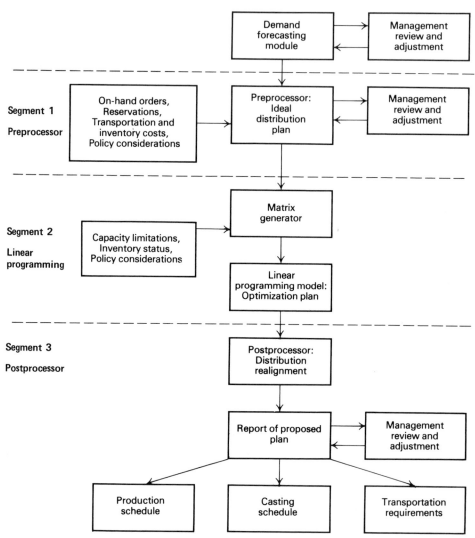

Fig. 6.3 General flow chart of the tactical system.

in order-number sequence, showing the freight-plus-in-transit-inventory cost for each order if it were shipped from each of the four smelters. Figure 6.4 is an example of this listing.

The second part of the system is the linear-programming model. This program accepts the output from the preprocessor program together with the operational constraints. The program outputs an optimum order assignment which minimizes the total logistics costs, taking into account all the constraints imposed by the production operations.

		Source cost (Freight and in-transit inventory cost)			
		Smelters			
Order number	Order quantity	I	II	III	IV
49001	350	10	12	27	28
49002	1000	25	20	41	44
49003	200	7	8	4	3
...	...	...	...	...	...
...	...	...	...	...	...
...	...	...	...	...	...
...	...	...	...	...	...
...	...	...	...	...	...
49999	290	18	16	10	8
Total assignment	6540	1554	1462	1632	1892

Note. The preprocessor produces one of these reports for each product group and for each time period. It assigns each order to minimize these costs, e.g., 49001 to smelter I, 49002 to smelter II.

Figure 6.4 Preprocessor order-assignment listing, for time period 1 and product group 1

The third segment of the system is a postprocessor program. This program is intended:

1. to consolidate those orders that may have been split between two or more smelters by the linear-programming model, since shipment from more than one source is, in general, not economical, because of the complexities created in controlling the order-processing procedures;

2. to provide the inventory policy the company should follow in order to minimize the cost of operating inventories (it should be noted that this is only for operating inventories; inventories carried for safety-stock purposes must be considered separately);

3. to provide the order-assignment output from the linear-programming model in an easily readable format;

4. to describe the marginal costs and shadow prices for the various distribution alternatives; and

5. to compute the size of the labor crews to be assigned in the coming weeks to each production center.

The main purposes of this segmented approach are to reduce the complexity and scope of the problem to a manageable size; to provide a means of seeing easily which orders could and should be changed from one time period to another; to greatly reduce the number of constraint equations in the linear-programming part of the model; to provide easy means of quickly viewing the variability of ideal demand for each of the smelters on a week-to-week basis; and to produce a report from the linear-programming output that managers can read and use for day-to-day planning operation.

There are two basic ways to reduce the number of constraint equations in the linear-programming model. One is simply to preassign some orders that must be produced on a particular machine for quality-control reasons. This preassigning is easily done in the preprocessor by reducing the capacities of the appropriate machines by the amount preassigned, and those orders will never reach the linear-programming model. The second way to reduce the size of the linear-programming model is by reducing the alternatives available for the reassignment of each order. If the choice for order assignment can be limited to only two smelters, then the corresponding constraint equation is reduced to a simple upper-bound type of constraint. The computational advantage of this approach will become clear when the linear-programming model is explained in detail.

Both the preprocessor and the postprocessor segments are relatively straightforward computer programs, which do not require much explanation. We now will review in detail the structure of the linear-programming model.

Tactical Linear-Programming Model—General Characteristics

As we indicated before, the model considers only the four smelters located in the U.S. It covers a three-month time horizon divided into six time periods of uneven duration. The first four time periods are one week each; the remaining two time periods cover the second and third months, respectively.

For scheduling purposes, each year is split into 13 periods, each of four weeks' duration. (This is an easy way to handle the uneven length of the calendar months.) A second convention adopted is the assignment of orders to time periods in terms of *when* the order must be completed at the smelter, rather than the *required delivery date* of the customer. This use of an ex-mill date (in lieu of a customer delivery date) is a common way to deal with the problem of deliveries to customer areas that are widely scattered.

An analysis of the product types revealed that while it is possible to use only eight different types for the strategic model, it is necessary to consider at least *thirteen* for the tactical model. The reason is the fact that, while some of the more subtle distinctions between product types can be ignored for long-range planning, they have to be considered for day-to-day assignments.

A number of capacity constraints must be considered in the program formulation. The first is that all the hot metal produced must be cast into some form, due to the impossibility of stocking hot metal as such. (This must be done even if there is insufficient current demand for the hot metal.) A second general constraint is that each product type requires a certain variety of production equipment. The total time required on a given machine must not exceed the capacity for that machine at the assigned smelter. This consideration is critical since, while some of the machinery is required for only one product type, other machines are jointly used for a number of different product types.

As indicated before, the preprocessor routine assigns orders based on minimizing freight and in-transit inventory costs without regard to the capacity constraints of the system. The preprocessor also lists the reassignment penalty associated with producing the order in an alternative smelter, measured by the extra freight and in-transit inventory cost that will result if a reassignment takes place. Another way to reassign an order is to produce it earlier than the time originally listed in the preprocessor program. This change results in an additional charge for inventory costs. The basic function of the linear-programming model is to decide on the best *reassignment* of orders, considering the capacity constraints at the smelters and *all* the logistics cost components (rather than only freight and in-transit inventory) as a basis for order allocation.

To account for the inherent uncertainties in each product demand, which cause unavoidable errors in the forecasts, and to prevent the model from exhausting the inventories at the end of each time period, constraints are imposed requiring the closing inventories to be at least equal to the safety stock associated with each product group.

The model is designed to minimize the total logistics cost. Major items of that cost are the order-reassignment penalties, the inventory-carrying charges, and the casting cost.

6.6 MATHEMATICAL FORMULATION OF THE TACTICAL PLANNING MODEL

The Tactical Linear-Programming Model—Notation

To facilitate the mathematical formulation of the model, a symbolic notation will be introduced to characterize its elements.

Time Periods

Each time period will be represented by the letter "t."

Smelter Locations

The four domestic smelters are represented by the subscript letter "s," where $s = 1, 2, 3$ or 4.

When considering transfers between smelters, instead of saying that the transfer will go from smelter "s" to smelter "s," the letter "ℓ" will be used to designate the receiving smelter. Thus the transfer will read as going from smelter "s" to smelter "ℓ."

Casting Machines

Each individual casting machine at each smelter is identified with a different number. This number is represented by the subscript "m" ($m = 1, 2, \ldots, 40$). Thus it is possible to assign each order to a specific machine in a given smelter, since this is the way in which orders are scheduled. The model handles forty different machines, ten in each of the four smelters.

Product Groups

The 13 different product groups are represented by the letter "p" ($p = 1, 2, \ldots, 13$). Combinations of product groups that require the use of some additional processes are represented by the letter "c" as a subscript. These product combinations introduce additional constraints imposed by secondary operations that require processing steps that use other than casting equipment (such as bundling, sawing, acid-dipping, and so forth). There are three such combinations.

Order Number

Each order is identified with a different number represented by the letter "k". This unique number is needed if we are to be able to identify where each order will be produced and whether an order has been split. It should be noted that no order contains more than one product group, and all the material in that order is to be delivered in only one time period.

Data Requirements

d_{tsp} Number of tons of product-group p assigned by the preprocessor to smelter s for production during time period t;

h_{ts} Number of tons of hot-metal capacity available at smelter s in time period t;

e_{tm} Number of hours available for the use of casting machine m during time period t;

r_{mp} Production rate of machine m for product-group p, in tons/hour;

g_{spt} Number of tons of safety stock of product group p to be carried at smelter s during time period t;

q_k Number of tons of aluminum contained in order k;

w_{tps} Upper bound of product-group p that can be produced at smelter s during time period t, in tons;

u_{cst} Upper bound of combination c that can be produced at smelter s in time period t, in tons;

$f_{s\ell k}$ Reassignment cost of transferring order k from smelter s to smelter ℓ, in dollars per ton;

a_{mp} Casting cost for the production of product-group p on machine m, in dollars per ton;

i_{1p} Weekly inventory-carrying cost per week for product-group p, in dollars per ton;

i_{2p} Monthly inventory-carrying cost for product-group p, in dollars per ton per month.

Decision Variables

P_{tmp} Number of tons of product group p to be produced on machine m during time period t;

I_{tps} Number of tons of inventory of product-group p at smelter location s at the end of time period t;

$R_{s\ell k}$ Number of tons of order k to be reassigned from smelter s to smelter ℓ.

The Tactical Linear-Programming Model—Formulation

With the notation defined above, we now can describe in exact mathematical terms the structure of the tactical linear-programming model.

The Objective Function—Logistics Cost

The objective function is to minimize the total logistics cost while fulfilling the operational constraints. The elements of the logistics cost are as follows:

$$\text{Minimize cost} = \sum_{s}\sum_{\ell \neq s}\sum_{k} f_{s\ell k} R_{s\ell k} \qquad \text{Order reassignment}$$

$$+ \sum_{t \leq 4}\sum_{p}\sum_{s} i_{1p} I_{tps} \qquad \text{Weekly inventory-carrying charges}$$

$$+ \sum_{t=5,6}\sum_{p}\sum_{s} i_{2p} I_{tps} \qquad \text{Monthly inventory-carrying charges}$$

$$+ \sum_{t}\sum_{m}\sum_{p} a_{mp} P_{tmp} \qquad \text{Casting}$$

Demand Constraint

$$\sum_{m\,\text{in}\,s} P_{tmp} + I_{(t-1)sp} - I_{tsp} + \sum_{k}\sum_{\ell \neq s} R_{s\ell k} - \sum_{k}\sum_{\ell \neq s} R_{\ell sk} = d_{tsp} \qquad \begin{array}{l}(t = 1, 2, \ldots, 6;\\ p = 1, 2, \ldots, 13;\\ s = 1, 2, 3, 4.)\end{array}$$

This equation indicates that, for each time period, for each smelter, and for each product group, the total production in that smelter, plus the initial available inventory, minus the ending inventory, plus the reassignments out of the smelter, minus the reassignments to that smelter, should be equal to the demand assigned to the smelter by the preprocessor program. It is the material-balance equation for each product group. The quantity d_{tsp} has been determined by the preprocessor.

Hot-Metal Constraint

$$\sum_{p\,\text{on}\,m}\sum_{m\,\text{in}\,s} P_{tmp} = h_{ts} \qquad \begin{cases} t = 1, 2, \ldots, 6;\\ s = 1, 2, 3, 4. \end{cases}$$

This equation states that the total casting production at a smelter during time period t must equal the hot metal available for that smelter. The quantity h_{ts} has been determined by the strategic model.

Casting-Machine Capacity

$$\sum_{p \text{ on } m} \frac{P_{tmp}}{r_{mp}} \leqq e_{tm} \qquad \begin{cases} t = 1, 2, \ldots, 6; \\ m = 1, 2, \ldots, 20. \end{cases}$$

This constraint merely states that the total machine time used cannot exceed the total machine time available for time period t. The conversion into hours for each machine (from tons) is based on the productivity factor for each machine. The summation is made over all product-groups p that require machine m. The quantity e_{tm} is suggested by the strategic model.

Production Bounds

$$\sum_{m \text{ in } s} P_{tmp} \leqq w_{tps} \qquad \begin{cases} t = 1, 2, \ldots, 6; \\ p = 1, 2, \ldots, 13; \\ s = 1, 2, 3, 4; \\ m = 1, 2, \ldots, 20. \end{cases}$$

The production of product-group p on machine m for time period t cannot exceed the capacity of that smelter for that product during that time period.

Combination Constraints

$$\sum_{p \text{ in } c} \sum_{m \text{ in } s} P_{tmp} \leqq u_{tsc} \qquad \begin{cases} t = 1, 2, \ldots, 6; \\ s = 1, 2, 3, 4; \\ c = 1, 2, 3. \end{cases}$$

The number of tons of the product-group p in the combination c that are produced on machine m in smelter s during time period t cannot exceed the capacity for combination c during time period t at smelter s. This is a secondary constraint involving the capacity of a process step for certain of the product groups.

Ending-Inventory Constraints

$$I_{tsp} \geqq g_{spt} \qquad \begin{cases} s = 1, 2, 3, 4; \\ t = 1, 2, \ldots, 6; \\ p = 1, 2, \ldots, 13. \end{cases}$$

The amount of inventory of product-group p that is available at smelter s at the end of time period t should be greater than or equal to the safety stock required for product group p at smelter s.

Reassignment Balance

$$\sum_{\ell \neq s} R_{s\ell k} \leqq q_k, \qquad \text{for all k.}$$

The amount of order k that is reassigned from smelter s to all other smelters ℓ cannot exceed the total order quantity.

Note that neither the time period nor the product type is specified, since order k contains only one product type and it has an ex-mill date within just one time period.

This set of equations explains the role that the preprocessor plays in reducing the number of constraints of the linear-programming model. Since there are about one thousand orders to be scheduled during the three-month time horizon, in theory we should have that many reassignment constraints. However, for most orders we need only consider two alternative smelter locations; the first one corresponds to the location to which the order has been preassigned, and the second one presents a reassignment alternative. This means that only one variable is required in most reassignment-balance constraints, making that constraint a simple upper bound for that variable. This type of constraint is handled with little additional computational time by an upper-bounded linear-programming code. In the subject company, if the order assignment is to be made properly, about one hundred orders might be assigned to more than two alternative smelter locations. Therefore, using the preprocessor reduced the total set of constraints from one thousand to one hundred. Moreover, even the general reassignment-balance constraint can be handled easily by a special linear-programming code known as *generalized upper bounding*, which is available on many commercial programming systems.

Given the tendency of the linear-programming solutions to drive toward extreme points, in practice very few orders are split. The problem of split orders is thus very easy to resolve at the postprocessor level.

Nonnegativity of the Variables

All of the variables should be nonnegative.

6.7 CONCLUSION

We have described, with a fair amount of detail, a formal, integrated system to deal with some important logistics decisions in an aluminum company. We purposely have emphasized the formulational aspects of this project in order to illustrate the modeling effort required in many mathematical-programming applications.

This project also can be viewed as an example of the *hierarchical planning approach*, which was discussed in Chapter 5.

EXERCISES

I. Strategic Planning Model

1. Time Horizon of the Strategic Planning Model

Make a list of the objectives of the strategic planning model. What is the time horizon required to address each one of these specific objectives? Do these time horizons have the same length? If not, how do you think a single model could deal with the different time horizons? What is the proposed time horizon of the strategic-planning model? How many time periods does the model consider? How will the model handle the dynamic changes throughout the time horizon required to deal with each of the specified objectives?

2. Objective Function of the Strategic Planning Model

Review the elements of the objective function of the strategic-planning model. Why does the model minimize cost rather than maximize profit? Are reduction, casting, and transportation costs really linear? How might any nonlinearities be modeled?

How do the costs associated with the new smelter enter into the objective function? What could you do if there are trade-offs between fixed cost and variable-reduction cost in the new smelter (i.e., if there is one option involving a small fixed cost but high variable-reduction cost, which should be compared against an option consisting of a high fixed cost that generates a smaller variable-reduction cost)?

Why does the model ignore the procurement and transportation cost of raw materials to the smelters? Why are customs duties included in the objective function? (Aren't customs duties unavoidable, anyhow?)

Why aren't inventory costs other than in-transit inventory costs included in the model? Why doesn't the model impose costs for safety stocks, cycle stocks, stock-piling costs, seasonal stocks, work-in-process inventory, and so on? Why aren't the in-transit inventory costs and customs duties expressed directly in terms of the quantities Q_{sap}?

3. Metal-Supply and Casting-Equipment Capacity Constraints

Consider the metal-supply constraint

$$\sum_a \sum_p Q_{sap} - M_s = 0, \qquad s = 1, 2, \ldots, 11,$$

$$M_s \leq \bar{m}_s, \qquad s = 1, 2, \ldots, 11, \tag{1}$$

$$M_s \geq \underline{m}_s, \qquad s = 1, 2, \ldots, 11. \tag{2}$$

In total, there are 33 constraints used to express the metal-supply availability. An equivalent form for expressing these conditions will be:

$$\sum_a \sum_p Q_{sap} \leq \bar{m}_s \qquad s = 1, 2, \ldots, 11,$$

$$\sum_a \sum_p Q_{sap} \geq \underline{m}_s \qquad s = 1, 2, \ldots, 11.$$

which gives a total of 22 constraints. Why do you think the model uses the first formulation, which appears to unnecessarily increase the numbers of constraints required?

Consider constraints (1). What are the possible signs (positive, zero, negative) of the shadow prices associated with these constraints? How would you interpret the shadow prices? Answer the same questions with regard to constraints (2). From the values of these shadow prices, when would you consider expanding the metal capacity of a given smelter? When would you consider closing down a given smelter?

Similar questions can be posed with regard to casting-equipment capacity.

4. Demand Constraints

How many demand constraints are there in the strategic-planning model? What implications does the number of these constraints impose with regard to aggregation of information into product types and market regions?

What are the possible signs (positive, zero, negative) of the shadow prices associated with these constraints? How would you use this shadow-price information to decide on market-penetration strategies, and on swapping agreements? How could you deal with uncertainties in the demand requirements?

5. Customs Duties and In-Transit Inventory Constraints

Can these constraints be eliminated from the model? How would you accomplish this?

6. Use of the Strategic-Planning Model to Support Managerial Objectives

Refer to the list of objectives stated for the strategic-planning model, and discuss in detail how you think the model should be used to provide managerial support on each one of the decisions implied by the stated objectives. In particular, analyze how the model should be used in connection with the decision affecting the size and location of the new smelter.

II. Tactical Planning Model

1. Time Horizon of the Tactical Planning Model

Discuss each objective of the tactical model. How long is the time horizon required to address each of the specific objectives? How many time periods does the model consider? Why are multiple time periods essential in the tactical model?

2. Segmentation of the Tactical Model

Discuss the proposed segmentation of the tactical model. What are the functions, inputs, and outputs of the demand-forecasting module, preprocessor, matrix generator, linear-programming optimization routine, postprocessor, and report generator? How does the preprocessor help in reducing the computational requirements of the linear-programming model?

3. Objective Function of the Tactical Linear-Programming Model

Why is cost minimization preferred over profit maximization? What cost elements are implicit in the order-reassignment cost? Why are inventory-carrying charges divided into weekly and monthly charges? What additional cost elements might you include in a tactical model such as this one?

4. Demand Constraints

Discuss each term in the demand constraint. How is the righthand-side element of these constraints determined? How many constraints are there? What are the implications of these constraints with regard to aggregation of information? How would you interpret the shadow prices associated with these constraints? What feedback would these shadow prices provide to the strategic-planning model?

5. Hot-Metal, Casting-Machine Capacity, Production Bounds, and Combination Constraints

Analyze the nature of these constraints. Discuss their shadow-price interpretations, and the feedback implications to the strategic-planning model.

6. Ending-Inventory Constraints

Discuss the nature of these constraints. How would you specify the value of the righthand sides of these constraints? Are there alternative ways of expressing inventory targets, particularly when only safety stocks are involved and the model is going to be updated every review period? How taxing are these constraints in terms of computational requirements?

7. Reassignment-Balance Constraints

These constraints are critical for understanding the computational economies introduced by the preprocessor. Discuss what happens when only one reassignment alternative is considered for every order; what computational implications does this have? Interpret the shadow price associated with these constraints. Indicate an alternative formulation of the tactical-planning model without using the preprocessor or reassignment-decision variables.

III. Interaction between Strategic and Tactical Models

Analyze the hierarchical nature of the proposed planning system. Which outputs of the strategic model are transferred to the tactical model? What feedback from the tactical model can be useful for defining new alternatives to be tested by the strategic model? How might the problem be approached by a single model incorporating both tactical and strategic decisions? What would be the advantages and disadvantages of such an approach?

IV. Data Requirements

How much data do the strategic and planning models require? What would you do to collect this data? Which elements of the data would be most costly to collect? Which must be estimated most carefully?

ACKNOWLEDGMENTS

This chapter is based on the technical paper "Integration of Strategic and Tactical Planning in the Aluminum Industry" by A. C. Hax, Operations Research Center, M.I.T. Working Paper 026–73, September 1973.

Planning the Mission and Composition of the U.S. Merchant Marine Fleet

7

As we have indicated in Chapter 3, it is quite easy to perform *sensitivity analyses* for linear-programming models. This is one of the most important features of linear programming, and has contributed significantly to its usefulness as a tool to support managerial decisions. The project described in this chapter provides a practical application where sensitivity analysis plays a major role in understanding the implications of a linear-programming model. The project aims to identify the mission and the optimum composition of the U.S. Merchant Marine Fleet. The fleet should be designed to carry fifteen percent of the U.S. foreign trade in the major dry and liquid bulk commodities—oil, coal, grains, phosphate rock, and ores of iron, aluminum, manganese, and chromium. The study was conducted to design a fleet intended to be operational by 1982. A linear-programming model was used to obtain preliminary guidance with regard to the best ship designs, the sizes, and the mission of the resulting fleet. Subsequently, extensive sensitivity analysis was employed to evaluate the changes resulting in the fleet composition when trade forecasts and port conditions were varied.

For a long time, ships flying the American flag have faced construction and operating costs higher than those of most of their foreign competitors. The gap has widened and, in recent years, shipbuilding in the U.S.A. has cost about twice, and wages for American crews four times, the figures for representative foreign equivalents. This situation has been recognized in legislation. The Merchant Marine Act of 1936 instituted cost-equalizing subsidies in order to maintain U.S.-flag general-cargo liner service. As a result, a substantial share of our foreign general-cargo liner trade is carried by ships built, registered, operated, and crewed by Americans. No similar assistance was available to bulkers and tankers, and so U.S.-flag ships have almost disappeared from such service, apart from certain protected trades. In addition, American shipyards have not had the market or the incentives to share in the modernization of facilities and the mass production of large ships that have been offered in Europe and Japan.

A program intended to cure this situation was announced in 1969. Some of the characteristics of the program were:

1. Subsidies to be extended to construction and operation of tankers and bulkers;
2. Design studies were commissioned for ships that could be mass-produced and that would be economical to build and operate; and
3. Funds were to be appropriated on a large enough scale to enable shipyards to pay off the large capital investment required for modernization.

This effort entailed building 300 ships over the ten years between 1972 and 1982, and in the latter years carrying at least 15 percent of the foreign trade of the U.S.A.

7.1 STRUCTURE OF THE PROBLEM

There are four elements that characterize the structure of this problem: the commodities to be hauled; the types of ships to be included in the overall fleet design; the missions assigned to the ships; and the physical constraints imposed by the loading and unloading capabilities of the ports. We begin by reviewing the impact that these elements have on the problem formulation.

Commodity Movements

Commodity movements are the basic element of the problem's structure. They are tonnages of the commodities in question to be moved in 1982 from a loading port to a discharge port, and they define the mission of the fleet to be optimized. It obviously was impossible to include in our analysis all the details of origins and destinations. Instead, representative (although specific) loading and discharging ports were used.

A trade forecast provided estimates of the origins and destinations of the commodities and their expected volumes of flow. For the purposes of this study, the target for the 1982 U.S. bulk fleet was set by the Maritime Administration at 15 percent of the total foreign trade of the U.S.A. This figure was not applied rigidly for all commodities, but was used as a guide to informed judgment. For example, the goals for individual commodities also were affected by projections of national and industrial behavior, which make U.S. penetration of certain trades quite unlikely. A typical mission for 1982, consisting of tonnages of seven major bulk commodities to be moved along 23 trade routes, is shown in Table 7.1.

Voyages

Voyages essentially consist of round trips from a port of origin with *at most two* destination ports of call, for example, Canada to Baltimore then back to Canada, or Guinea to New Orleans, New Orleans to Japan, and Japan back to Guinea. The primary leg is always loaded, while the return trip may be an empty ("ballast") leg. Backhauls, which are shipments to be obtained from a destination port and brought back to the port of origin, are of major economic importance in bulk shipping. As a result, our structure included voyages with up to two loaded legs and either a backhaul or ballast leg. After at most two loaded legs, a voyage was arbitrarily closed by a return to the port of origin, although this may or may not be the case in practice.

Table 7.1 A Typical Mission (Mission No. 4)

Commodity	Origin	Destination	Million tons per annum
Iron ore	Canada	Baltimore	0.85
	Canada	Philadelphia	0.85
	Peru	Baltimore & Philadelphia	1.1
	Venezuela	Baltimore & Philadelphia	1.6
	Brazil	Baltimore	0.5
	Liberia	Baltimore & Philadelphia	0.8
Bauxite	Surinam	New Orleans	1.9
	Guinea	New Orleans	1.0
Manganese/chrome ore	South Africa	Baltimore	0.8
Grain	New Orleans	East Coast India	0.6
	New Orleans	West Coast India	0.6
	New Orleans	Japan	1.3
	New Orleans	Brazil	1.2
	New Orleans	Rotterdam	2.8
	Portland	Japan	1.3
	Portland	Southeast Asia	1.3
Coal	Hampton Roads	Rotterdam	3.1
	Hampton Roads	Japan	1.1
Phosphate rock	Tampa	Rotterdam	1.6
Oil	Persian Gulf	Philadelphia	5.6
	Libya	Philadelphia	5.6
	Venezuela	Philadelphia	11.3
	Venezuela	New Orleans	3.4

Each commodity movement along a trade route generates at least one voyage; there may be two if alternative routings exist, e.g., New Orleans to Japan via the Panama Canal or via the Cape of Good Hope. Also, there may be no appropriate backhaul for some geographical region, or there may be one or several voyages combining a given commodity movement, with another as a backhaul. Hence the list ("menu") of voyages is much longer than the list of a mission's commodity movements.

The menu of voyages is put together from the mission of the fleet with a table of distances between ports. It is reasonably inclusive, rejecting only such unattractive backhauls as those with less than 50 percent of loaded miles per voyage.

Menu of Ships

The third structural element is the *menu of ships* from which fleets can be chosen to execute the mission. We consider three ship types: *tankers*, with pumps and small hatches, which can carry liquids and also (with some cost penalty) grains that can be handled pneumatically; *bulkers*, with large hatches, suitable for dry bulks (grains, ores, coal, phosphate rock); and *OBO*'s* with both pumps and large hatches, which can

---

* OBO stands for Oil–Bulk–Ore, a ship that can handle these three types of cargos.

lift any of the major bulk commodities, liquid or dry. Within these types, ships are defined by physical characteristics—cargo capacity, speed, dimensions (especially draft)—and have cost characteristics such as: *construction cost* as a function of number built, *fuel per day* at sea and in port, and *other operating costs.*

Port Constraints

Voyages also suggest the fourth structural element, *port constraints.* There are two kinds: loading and discharging rates, which control the port times, and their associated costs; and dimensions, especially depth of water, which limits the load that a ship can carry. Canals also form a type of port constraint; the Panama Canal, for instance, imposes a toll and a time delay on a ship passing through, and also limits all three dimensions (length, beam, and draft) of a large vessel. These constraints were enumerated from standard sources such as *Ports of the World*, from the Corps of Engineers, and from industry information. Projection of this information to 1982 was very uncertain and was the subject of sensitivity testing.

As an objective function in the optimization, it would have been most appropriate to use the *return on investment* of the fleet of the given mission. In the present study, however, there were several difficulties that prevented its use, most notably the need for a forecast of revenue rates in 1982. This would have been as large a task as our whole project. Therefore, it was decided to take as the criterion the *discounted present value* of the life-cycle costs of the fleet.

Each of the alternatives considered, then, had the following structure. We selected a fleet mission, a set of constraints, and menus of ships and of voyages that could fulfill the mission. The menus, of course, included more types of ships and voyages than ultimately were chosen. The optimum choice (types and numbers) of ships and voyages then was determined. Because of the target date (1982), there were many uncertainties in the trade forecasts, definition of missions, constraints, and details of ship-allocation practices. Therefore it was necessary to make a number of studies—sensitivity tests, in effect—with various missions and menus and constraints. Thus, we could explore the significance of the uncertainties. Nonlinearities relating ship-construction cost to number ordered also called for multiple studies.

There is a significant difference between ship optimization and fleet optimization. Both have been practiced for some time, manually or by computer. *Ship optimization* is usually done by the naval architect, who varies the characteristics of a fairly well-defined ship design so as to minimize cost or maximize profit in one or a few well-defined trade routes. *Fleet optimization* is more likely to be done by a ship operator, who optimizes the allocation of a number of vessels of various types of designs (in existence or to be built) to various trade routes or missions or contracts of affreightment. The result may be a plan of allocation, an evaluation of a mission or contract, or a choice among design for ships to be built. Our project was of the second type. It differs from ship optimization in that a range of ship designs and trades is studied as a whole, and it differs from the fleet optimization usually performed (say, by an oil company) in that the target date is well into the future and none of the fleet currently exists.

7.2 THE LINEAR-PROGRAMMING MODEL

A linear-programming model was developed to obtain the optimum composition and mission of the fleet. The basic characteristics of the model were as follows.

1. The mission was satisfied by allowing ships to make the required number of voyages per year.

2. These voyages may include one load-carrying leg and a return leg in ballast, or two load-carrying legs plus one or two in ballast.

3. The number of ships required depended on the number of voyages and their duration; the linear-programming solution normally involved fractional numbers of ships. This was not considered a source of difficulty. The missions were arbitrary, to a certain extent, and an adjustment to the number of ships, bringing them to integral values, could have been made by adjusting the size of the demands. The aim was to produce an informative set of values rather than a precise set of numbers that would perform a required mission exactly.

4. For each possible voyage, various ships may be employed; the role of the linear program was to select that combination of voyages and ships that executes the assigned mission at minimum life-cycle cost. Several computations were required in order to define data for the linear-programming model.

5. For each leg of a voyage, a calculation was needed to determine the maximum cargo that an eligible ship may carry. This was a function of the *ship's* characteristics, including physical constraints (especially draft restrictions) and *cargo* characteristics.

6. For each voyage, a calculation was required to ascertain the time taken by any eligible ship. The time *at sea* depended on the ship's speed; the time *in port* depended on the quantity of cargo and the rate of loading and discharging. The latter rates were considered to depend primarily on the port and the cargo in question, and differential loading or discharging rates between different ships were not considered.

7. As implied in the foregoing, the cost of the voyage by a candidate ship was required. This consisted primarily of the fuel cost while at sea, and the port costs, which were based on daily rates and hence depended on loading rates.

8. The capital costs of the ships and their annual fixed costs were included. The linear-programming model that was used assumes constant costs per ship, independent of the number of ships. In order to allow a varying cost structure, in which costs per ship decrease with both the number built and the modularity of design, an iterative procedure was followed. The iterative approach consisted of solving the problem under one set of cost assumptions and then re-doing the problem with adjusted costs if the resulting solution was not consistent with the initial cost estimates. As it was the aim of the study to investigate the effectiveness of different combinations of ships in a variety of situations rather than to produce an exact optimum, this approach proved entirely satisfactory. More formal procedures, employing integer programming, could have been used to address this issue.

7.3 MATHEMATICAL DESCRIPTION OF THE MODEL

Decision Variables

The decision variables adopted were those that defined the composition and employment of the optimum fleet and were denoted, respectively, by n_s and x_{sr}, where

n_s = Number of ships of type s;

x_{sr} = Number of voyages per annum assigned to ship s along route r.

The selection of these variables implied not only the specification of a variety of ship types considered appropriate for inclusion in the fleet, but also the determination of adequate voyages to comprehensively describe the mission to be accomplished by the fleet.

Time Horizon and Number of Time Periods

As stated before, the basic problem was to determine the optimum fleet composition for foreign trade in 1982. Thus, it was necessary to deal only with fairly aggregated information, without regard to detailed information relevant only to decisions affecting the operating schedule of the fleet. Consequently, in the model we considered only one time period, corresponding to the year 1982. No attempt was made to subdivide this year into shorter time intervals, since no great advantage would have been derived from such an approach. Neither did we attempt to examine explicitly the intervening years prior to 1982. That, again, could have been accomplished by using a model similar to that adopted, but incorporating each of the intervening years, in order to explore the *evolution* of the optimum fleet from the present until 1982. However, the development of a construction schedule for the fleet was not the objective of the study.

Constraints

There were two basic constraints imposed upon the fleet. The first one referred to the *mission of the fleet* and can be represented by the following set of equations:

$$\sum_s \sum_r V_{srk} x_{sr} = d_k,$$

for all commodity movements k, where:

V_{srk} = Maximum amount of commodity k that can be carried by ship type s along route r;

d_k = Total annual tonnage (forecast for 1982) of commodity k specified in the mission for the pertinent pair of ports (one for loading, one for discharging).

The summation on r is carried over all routes that connect a particular pair of ports. This constraint, therefore, merely states that the amount carried by all ships, following every possible route, should satisfy the demand in the mission for each commodity and each pair of loading and discharging ports, such as those shown in Table 7.1. The maximum tonnage of a given commodity that can be carried by a ship type was computed outside the model, and was a function of the cargo weight and volume capacities of the ship, the density of the cargo, the maximum ship draft allowed on each leg of each voyage, and further canal or port constraints.

The second set of constraints affecting the fleet referred to the *total time consumed* by the fleet in performing its mission. It simply indicated that the total time used by every ship type could not exceed the number of ship-days available during the year. Assuming that all ships were available 345 days per year, the constraints can be expressed as follows:

$$\sum_r t_{sr} x_{sr} - 345 n_s \leq 0,$$

for all ship types s, where

$$t_{sr} = \text{Voyage time for ship type s along route r.}$$

These voyage times also were evaluated outside the model, as the sum of times at sea and times in port. Times at sea were calculated from the ship speed and distance. A safety factor of 10 percent was included to allow for weather, mechanical, and other contingencies. The time in port was considered to be dependent on the type and quantity of cargo and the ports concerned. (Additional constraints could have been incorporated to introduce lower or upper bounds on the *number* of ships of any type.)

Objective Function

As indicated before, the objective function selected for our model was the minimization of the life-cycle cost of the fleet. One element of the life-cycle cost was the operating cost for a given ship type, which can be expressed by the following relationship:

$$C_s = \sum_r C_{sr} x_{sr},$$

for all ship types s, where

$$C_{sr} = \text{Variable operating cost incurred by ship type s along route r;}$$
$$C_s = \text{Total variable cost of ship type s.}$$

Now, if we let

$$a_s = \text{Annual fixed operating cost of ship type s,}$$

then the total annual operating cost for all ships is given by:

$$\sum_s (C_s + n_s a_s).$$

Allowing for an inflation rate of 4 percent and considering a discount rate α, a 25-year life for each ship, and a capital cost I_s for ship type s, the total life-cycle cost of the ship can be expressed as follows:

$$\text{Life-cycle cost} = \sum_s n_s I_s + \sum_s (C_s + n_s a_s) \sum_{t=1}^{25} \left(\frac{1.04}{1+\alpha}\right)^t$$

$$= \sum_s [\beta C_s + n_s(I_s + \beta a_s)],$$

where

$$\beta = \sum_{t=1}^{25} \frac{1.04^t}{(1+\alpha)^t}.$$

In the standard cases the discount rate α was set at 10 percent, giving a value of $\beta = 13.0682$.

The model was solved under a variety of assumptions on the data. The aim of these investigations was to produce a combination of ships and voyages that was optimum, not merely in the sense that it minimized the life-cycle cost of the fleet for given values for the problem data, but also that it produced a minimum or near-minimum cost over as wide a range of probable values for the data as possible. Be-cause of the long time horizon involved in the study, uncertainties in the data cannot be ignored, and must be addressed in a meaningful way. In particular, we would like to consider uncertainties inherent in the trade forecasts, the operating costs, and the physical constraints. Other problem parameters were subject to policy decisions, e.g., major port developments. By performing many sets of optimization computa-tions, it was possible to investigate how stable the solution was with respect to uncertainties in the numerical data.

7.4 BASIC FINDINGS

Ship Designs

The primary objective of this project was to provide guidance as to the optimum ship sizes and types. The relevant experiments and their output are discussed below.

Size and Standardization

A modular set consists of ships having the same bow, stern, machinery, and super-structure, but with options (tanker, bulker, OBO) for the cargo-containing midbody. The modular sets offer large economies for multiple production.

Table 7.2 summarizes studies of both ship size and standardization. The three sections of the table compare the optimum fleet and its total cost for a representative mission for three different ship menus:

1. a menu containing a wide range of ships, considered to be designed and built in small lots;

2. a menu consisting solely of a PanMax (maximum ship able to transit the Panama Canal) modular set "P" and a modular set "O" at about 110,000 dwt (dwt stands for the deadweight of the ship); and

Table 7.2 Standardization Economies

Menu 1—Without modular construction

Ship	dwt (thousands of tons)	Draft (feet)	Number used
FP (B)*	45	35	1.6
FI (B)	66	39	4.7
FO (B)	105	45	4.5
TP (T)*	73	44	14.4
YO (T)	111	45	18.9
MO (O)*	108	45	6.2

Cost: $2293 million

Menu 2—With modular construction (1)

Ship	dwt	Draft	Number used	
BP (B)	67	43	1.4	(a)†
TP (T)	73	44	14.4	
FO (B)	105	45	8.2	
YO (T)	111	45	18.9	(b)†
MO (O)	108	45	6.2	

Cost: $2267 million (saving: $26 million)

Menu 3—With modular construction (2)

Ship	dwt	Draft	Number used	
BP (B)	67	43	1.4	(a)
TP (T)	73	44	15.3	
BQ (B)	128	51	8.5	
TQ (T)	120	49	20.1	(c)†
OQ (O)	130	52	5.0	

Cost: $2356 million (extra cost: $63 million)

* (B) = Bulker; (T) = Tanker; (O) = OBO.
† (a) Modular set "P"; (b) Modular set "O"; (c) Modular set "Q".

  3. a menu of the PanMax modular set "P" and a larger modular set "Q" at 120,000–130,000 dwt.

When a number of ships are going to be constructed, it is possible to take advantage of a modular design, primarily with regard to large ship sizes.

Each ship type is represented by a code of two alphabetic characters. Its deadweight (in thousands of long tons) and design draft (feet) are given. The final column of each section of the table shows the number of the various types chosen in the optimum fleet, and the bottom of each section indicates the cost of the mission. The fractions of ships were not rounded off, since the purpose was guidance and analysis, rather than an exact fleet-construction program.

Comparing the first and second parts of the table shows that the voyages optimally assigned to nonmodular ships FP and FI in the first menu are shifted to larger and more costly modular ships in the second. The added cost of this change is more than offset by the savings in multiple modular construction, and the total cost drops by $26 million. If we insist on the larger modular set "Q", however, the third part of the table shows that the third menu produces costs $89 million higher than the second menu, so the smaller set "O" is unequivocally preferred. The difference between "O" and "Q" can also be looked at another way. Draft restrictions force the use of more ships (33.6 versus 32.8) of the larger "Q" sets than of the smaller "O" set. A higher total cost is therefore inescapable.

OBO's

The analysis that we have performed underestimates the attractiveness of the OBO design. We allow a payoff in return for the extra cost and flexibility of the OBO only within the deterministic confines of single backhaul voyages. That is, only if oil and a dry commodity are attractive for backhauling in a single voyage will the OBO even be considered, since an OBO costs more than a tanker or dry-bulker of similar size and speed. The flexibility of the OBO, however, permits the ship operator to move from any trade to any other in response to demand and good rates. The need for this flexibility might arise on successive voyages, or over a period of a year or more.

A fair evaluation of the OBO design could not be made straightforwardly, so instead we investigated the reverse question: What would be the extra cost if a certain number of OBO's (say, 15) were forced into the fleet, and then employed as near optimally as possible? Table 7.3 gives the result for two possible missions. This major shift in fleet composition cuts back only slightly on the tankers; dry-bulkers almost disappear; but the cost of the fleet rises by only about $20 million, or one percent. Such a cost could well be offset by only a slightly higher utilization during the life of the ships.

Table 7.3 Effect of Requiring at Least 15 OBO's (modular P, O ships)

Ship	dwt (thousands of tons)	Draft (feet)	Mission 3.3		Mission 4.3	
			No restriction on OBO's	15 OBO's	No restriction on OBO's	15 OBO's
Bulkers:						
BP	67	43	2.3	2.3	1.4	
FO	105	45	5.3	2.3	8.2	1.5
Tankers:						
TP	73	44			14.4	14.4
YO	111	45	29.3	29.3	18.9	18.2
OBO's:						
OP	67	43				1.4
MO	108	45	12.0	15.0	6.2	13.6
Life-cycle cost ($ millions)			2340	2362	2267	2284

7.5 SENSITIVITY ANALYSIS

Variations of Mission

A major change of mission, of course, will produce some change in fleet composition because of different requirements. It also will lead to a change of total cost because of new mission size. As an example, we generated a pair of basic missions, of which one, identified as "Mission No. 4," included a number of cargo types that were omitted from "Mission No. 3." The more inclusive mission called for a different number of ton-miles and had many combinations of a commodity with a particular pair of loading and discharge ports. Further, it introduced some special requirements, and also gave many new backhaul opportunities.

Despite this major change, only one new design was accepted by the linear program, although the *number* of ships was, of course, considerably changed. The two fleets are compared in Table 7.4. It can be seen that, of the ship menu of 28 designs, the less inclusive mission called for six designs, and the more inclusive one used the same six plus one other. The total cost was affected only slightly (less than 4 percent), and the change had the same sign as the change of ton-miles.

Table 7.4 Comparison of Two Basic Missions

	Mission identification number	
	No. 3	*No. 4*
Description of mission:		
Total tonnage, dry bulk (millions)	24.3	24.3
liquid bulk (millions)	25.9	25.9
Total ton-miles (billions)	290	260
Total combinations of commodities, loading ports, and discharging ports	15	23
Description of optimum fleet:		
Bulkers: FP (45,000 dwt)	2.7	1.6
FI (66,000 dwt)	3.2	4.7
FO (105,000 dwt)	1.5	3.7
BQ (128,000 dwt)	0.8	0.6
Tankers: TP (73,000 dwt)		14.4
YO (111,000 dwt)	29.3	18.9
OBO's: MO (108,000 dwt)	12.2	6.2
Total number of ships	49.7	50.1
Unused designs in ship menu	22	21
Approximate total cost* ($ millions)	2373	2293

* Adjusted for serial production and quantity discounts; not adjusted for modularity.

The foregoing major variation of missions tests the fleet's sensitivity to uncertainty as to the trades in which it will be engaged. We also made experiments with a larger number of smaller variations, which would correspond to uncertainties in the trade forecast of U.S.-flag share of trade. Each of the two basic missions was transformed into a set of five variant missions, by increasing or decreasing each tonnage in the basic mission by a random amount bounded by "high" and "low" trade forecasts. The optimum fleets for the variant missions then were determined. In the case of the variants of Mission No. 4, the *same seven designs* appear, plus (in two variants) a small fraction of a ship of an eighth design. Furthermore, the *number* of ships does not change greatly. On this basis, we conclude that the fleet is not very sensitive to uncertainties in the mission. Extended experiments on Mission No. 3 show even less sensitivity, since the same six designs were used throughout the basic mission and its variants.

The fleet's lack of sensitivity to variations of mission is a fact of profound importance. If the opposite had been found, the linear-programming model as formulated could not have been used as a basis for planning 1982's ships. The uncertainties in the employment of the fleet would have had to be included explicitly in any model in order to perform such advance-design work. Since the optimum ship types are not sensitive to the uncertainties of trade forecasting and market capture, we can have confidence that the design work suggested by the model is appropriate, although, as time proceeds, we may have to adjust the numbers to be built of the various types.

Oil Demand

At one point in marketing research, it was felt that the oil companies generally might have little interest in a U.S.-flag fleet. We therefore studied the effect of this possibility on the optimum fleet by making computations for a basic mission and for missions in which the oil demand was cut in half and finally reduced to zero. The fleet composition changed in a predictable manner. Tankers decreased, and OBO's disappeared, of course, since their pumps and piping no longer provided any advantages.

Backhauls

Several runs indicated the major importance of backhauls to the fleet. One pair is summarized in Table 7.5. If backhauls are used to the optimum extent, fleet life-cycle cost is cut by 15 percent, from $2.7 to $2.3 billion. The comparison is against a run in which no backhauls are permitted, but the mission and ship menu and the other data were otherwise identical. The number of ships also is cut by 20 percent. Naturally, the OBO design disappears in the no-backhaul case, as there is no payoff for its flexibility.

The conclusion is that good utilization is very important to the economic success of a bulk fleet. Backhauls must be found, and for the situations where they are elusive because of markets or geography, *crosshauls* of foreign-to-foreign cargoes will be needed.

Table 7.5 Effect of Utilizing Backhaul Opportunities (Mission No. 4)

| | | | Number of ships | |
| | dwt | Draft | Without | With |
Ships	(thousands of tons)	(feet)	backhaul	backhaul
Bulkers:				
FP*	45	35	6.1	1.6
BP*	67	43		
FG	50	35		
FI	66	39	1.9	4.7
HI	38	39		
GN	103	48		
FO	105	45	8.7	3.7
BQ	136	53	4.7	0.6
FQ	129	51		
GQ	132	52		
FT	152	55		
Tankers:				
UD*	35	37		
UG*	50	37		
YP*	69	42		
TP*	73	44		14.4
YG	50	35		
JI	65	38		
YI	65	38		
YO	103	44	38.1	18.9
TQ	120	49		
YT	142	49		
YU	150	52		
TY	300	80		
OBO's:				
NI*	62	41		
MP*	63	41		
OP*	67	43		
MG	51	36		
MI	69	40		
MO	110	46		6.2
OQ	139	54		
MQ	133	52		
NQ	136	53		
MT	157	57		
QT	170	63		
Total cost ($ millions)			2721	2293
Utilization			50%	67%

* Able to transit Panama Canal.

Grain Stevedoring

The costs of loading and discharging grain were subject to considerable uncertainty. We therefore made some sensitivity tests on these costs.

Carriage of grain in tankers imposes substantial extra stevedoring costs, compared with dry-bulk or OBO designs with large hatches. We received estimates of this cost ranging from 40¢/per long ton to $1.20 or $1.50, depending on the ports concerned and the particular experience of the estimator. Most of the runs reported here were made with the minimum penalty cost of 40¢/long ton, which had little effect against the low construction cost of tankers and the backhaul utilization by oil. We also tried calculations with 80¢ and $1.20 penalties, as well as a flat prohibition (which corresponds to an infinite penalty).

The estimated range of this penalty or extra cost was just about sufficient to drive most grain shipments out of tankers and into dry-bulkers or OBO's. In view of the scale of the penalty, it is clear that variations of the relative charter-market rates for tankers and dry-bulkers or OBO's will strongly affect the fraction of grain shipped in tankers. This aspect of tanker utilization, therefore, was extremely difficult to predict for 1982.

Port Constraints

It was very difficult to predict harbor depths for 1982; therefore, several experiments were made to explore the sensitivity of the fleet to these constraints. It turns out that this is an area of great sensitivity.

In discussion of our first round of results, the question was put to us: What would happen if all the ports in the study were deepened by 10 percent? At that time, the ship menu contained very few designs larger than those already used in the optimum solution. The linear program seemed unlikely to be able to take advantage of the greater draft limits, since the port constraints reflecting draft limits were not binding. We then looked at the effect of *decreasing* all depths by 10 percent. (The Panama Canal was held constant in the process.) Table 7.6 gives the result. The largest dry-bulker (BQ) and OBO (OQ) were forced out of the optimum fleet entirely, and the number of each remaining design was increased, because shallower drafts cut the payloads. The total number of ships in the fleet and the total cost both increased by over 15 percent, a substantial change.

Another study focused on oil imports. For the major portion of this tonnage, Philadelphia had been taken as a representative port with a depth of 50 ft in 1982 (38-ft draft limit). Much oil goes into the North Jersey area also, with a similar limitation. The large tonnage suggests the question: What if the Delaware were deepened? Two cases are reported in Table 7.7. The first column is the basic optimum fleet; the second shows the change if the Delaware were dredged to equal the Chesapeake (48.5-ft draft); and the third removes the limitation entirely by assumption of an *offshore discharge terminal*. The trend toward larger tankers and OBO's is obvious. More dry-bulkers are used in the offshore-terminal case because the very large

Table 7.6 Effect of Decreasing All Port Draft Limitations by 10 Percent

			Number of ships			
			Variant of Mission No. 4		Variant of Mission No. 3	
Ships	dwt (thousands of tons)	Draft (feet)	Standard drafts	Decreased drafts	Standard drafts	Decreased drafts
Bulkers:						
FP*	45	35	1.6	1.8	2.7	2.9
FI	66	39	2.0	2.3	0.1	
FO	105	45	2.5	8.2	1.0	6.3
BQ1	136	53	4.5		4.5	
Tankers:						
YP*	69	42	15.2	18.0		
YO1	103	44	21.5	24.9	32.6	38.0
OBO's:						
MO	110	46	1.4	2.1	7.7	8.5
OQ1	139	54	0.6			
Total ships in fleet			49.3	57.6	48.6	55.7
Total cost ($ millions)			2203	2550	2287	2643

* Able to transit Panama Canal.

tankers are unsuitable for any backhaul service. As the draft constraint is relaxed, the corresponding decreases in fleet cost are quite large, and they are even larger in a similar experiment using the other basic mission. These reductions should be compared with the costs of dredging, or of building the offshore terminal and attendant storage and pipelines. In that comparison, the savings should be multiplied by perhaps 5 or 6, since only about 15 percent of the U.S. trade is included in the mission, whereas the improvement would cut costs for all of the cargo shipments.

Note of Design and Operating Drafts

Draft limitations play a central role in this study of large dry- and liquid-bulk carriers, since the shipment sizes that are available are almost unconstrained. We therefore requested that ship costs and characteristics be supplied for conceptual designs optimized for a given draft, rather than for a given deadweight as is the usual custom. This objective function caused the ships to be just as long and as wide for their draft as is reasonable, and so they differ from the ordinary vessels designed without this purpose.

Another peculiarity also was observed in our fleet optimization: ships often are not fully loaded. This may arise from two causes. First, even on a voyage without backhaul, it is not novel to find that a fully loaded smaller ship is less economical

Table 7.7 Effect of East Coast Oil Draft Changes

| | dwt (thousands of tons) | Draft (feet) | Number of ships | | | | | |
| | | | Mission No. 4 | | | Mission No. 3 | | |
Ships			Standard draft	48.5-ft draft	No limit for oil	Standard draft	48.5-ft draft	No limit for oil
Bulkers:								
FP*	45	35	1.6	1.6	6.0	2.7	2.7	2.5
BP*	67	42						0.4
FG	50	35						
FI	66	39	4.7	3.4	1.9	3.2	3.9	1.1
HI	69	39						
GN	102	47						
FO	105	45	3.7	5.0	6.5	1.5	2.0	6.9
BQ	128	51	0.6	0.8	2.5	0.8	1.2	0.3
GQ	132	52						
FT	150	55						
Tankers:								
UD*	35	35						
UG*	50	37						
TP*	73	44	14.4	7.0				
YG	50	35						
JI	65	38						
YI	65	38						
YO	111	45	18.9	5.4	5.4	29.3	1.5	8.5
TQ	120	49		6.5			12.8	
YT	140	52		6.4	2.3			5.4
TY	302	80			7.1			1.8
OBO's:								
OP*	67	43						
MG	52	36						
MI	69	40						
MO	108	45	6.2			12.2		
OQ	130	52		2.8			4.9	
NQ	136	53		2.8			3.9	
MT	153	56						
QT	168	62			3.3			9.2
Life-cycle cost			2293	2003	1872	2373	1986	1884

* Able to transit Panama Canal.

than a part-empty larger ship. The latter has a greater area (equivalent to tons per inch immersion), so it carries a greater payload at a given draft. Also, vessel construction costs, and especially operating costs, increase very gradually with size. This is especially true for U.S.-flag operation, where crew wages are expensive. The cost data supplied to us assumed no increase of crew with size. These two effects, in general, produce an optimum (lowest cost per ton-mile) ship size so large that operating partly empty was economically feasible. Of course, if this is carried to an extreme, light ship-weight and similar counterbalancing variables finally produce rising costs. One wonders whether shipowners often take this view into consideration in planning new construction.

Second, partial loading of ships is enhanced because of the availability of backhauls. It may well be that the backhaul ports are shallower than those of the primary haul, but the limited load of backhaul cargo is still feasible and attractive economically.

SUMMARY AND CONCLUSIONS

The work described hereinabove produced results as summarized below.

Guidance as to Ship Designs

1. A modular set of ships (tanker, bulker, OBO) of about 110,000 dwt is distinctly preferred over a similar but somewhat larger set (120,000–130,000). Hence, the technique is able to indicate optimum ship designs fairly precisely.
2. The savings of multiple modular construction more than offset the extra costs of ship standardization.
3. Forcing 15 OBO's into the fleet causes only a very small increase in total cost, which most likely is more than offset by the advantages of the OBO's flexibility.

Sensitivities

1. The composition of the optimum fleet was insensitive to small changes in the mission to be performed, and even fairly large changes had no serious effect. This result is essential for using deterministic linear programming for planning the optimal fleet in 1982.
2. Deletion of half or all the oil from a mission has the expected effect. OBO's disappear, and tankers are reduced in importance.
3. The fleet cost and composition are very sensitive to the opportunities for backhauls. If the latter are excluded arbitrarily, the fleet cost rises by 20 percent.
4. Port constraints have a major influence on optimum ship designs and fleet life-cycle costs. In the Delaware–North Jersey area, an extra 10 ft in the channel, or an oil discharge terminal of unlimited depth, would cause a major decrease in overall fleet life-cycle costs.

Techniques Used

The linear-programming approach used in this study has many important advantages. For one thing, it provides a systematic framework in which a very large number of details can be accommodated. For example, the calculations of the effect of port constraints on shiploads were handled easily within the linear program as part of its input calculations. The system also keeps track of all the various voyages and port calls that are involved. Even if it were only a bookkeeping device, the program would be very useful.

The important thing, however, is that it permits the analyst to view the problem from a systems standpoint. Thus, ships were optimized, not with respect to a few specific voyages or ports, but rather as part of the overall system of ships, ports, voyages, and commodity flows. Ships interact with each other because one can partly, but not completely, substitute for another (as to backhauls, payloads, and so on), and also because of serial production economies. It is important to take these interactions into account.

Finally, the approach provides a ready means for performing sensitivity analysis. The importance of being able to explore variations in the data was demonstrated by our analysis of the future uncertainties in the mission.

In this study we have not dealt with the complete shipping system by any means. However, the approach could be extended easily to include more elements in this system. For example, if we were concerned about new materials-handling methods, a slight extension would include these activities in the linear program and thereby optimize both the ship type and the loading and discharging facilities at the ports. As indicated by our studies of port depths, we could extend it still further to perform a simultaneous optimization of terminals and harbors as well as of the ships themselves.

EXERCISES

1. *Problem Definition*

   Discuss the characteristics of the problem presented by the design of the U.S. Merchant Marine Fleet. What are the basic elements of the problem? What assumptions are being made in order to develop a mathematical model? Do these assumptions reduce the realism and, therefore, the usefulness of the model significantly? What is the primary role of the model? How do you envision that the model will be used in supporting decisions regarding the planning of the fleet? What important considerations are omitted from the model formulation?

2. *Time Horizon and Number of Time Periods*

   The model considers only one time period, corresponding to the year 1982. Why not use a multiperiod planning model? How can one study the dynamic issues associated with developing the proposed fleet for 1982? What would be a reasonable way of exploring the evolutionary development of one fleet from 1970 to 1982?

3. *Decision Variables and Constraints*

   Discuss the selection of the decision variables. Are there alternative formulations of the model based upon other decision variables? Should the variables be constrained to be integers? What are the implications of these additional constraints? Is there any useful information that can be obtained from the shadow prices of the constraints? Can you provide an estimate of the number of variables and the number of constraints involved in a typical fleet-composition problem?

4. *Objective Function*

Do you agree with the objective function used in the model? What assumptions are implicit in the computations of the life-cycle cost? Is it reasonable to use life-cycle cost in a one-period model? What alternatives might you consider? How would you determine the value of the discount rate α to use? What about the inflation rate?

5. *Basic Findings*

Analyze the results provided in Table 7.2. What conclusions can be drawn from these results? Discuss the interpretation of the results given in Table 7.3. Would you favor the introduction of OBO's into the fleet?

6. *Sensitivity Analysis*

List the kinds of sensitivity analyses you would propose to run for this problem. How would you program the runs to make the computational time required to process them as short as possible? Interpret the results given in Tables 7.5, 7.6, and 7.7.

7. *Multiple origins*

In practice, some voyages will not be closed by a return to the port of origin after at most two loaded legs, as assumed in the model. Rather, a ship might embark on a new voyage from another origin or might make several loaded legs prior to a return to the origin. How would these modifications be incorporated into the model? Is this omission serious?

8. *Detailed Scheduling*

The planning model analyzed in this chapter has as a viewpoint the overall optimization of the Merchant Marine Fleet. Contrast this viewpoint with the attitude of the owner of a small subset of this fleet who is interested in optimizing the performance of *his* available ships. Indicate what should be the characteristics of an appropriate model to support his decisions. Particular emphasis should be given to the degree of detail required in this new model, and the timing implications. Suggest a hierarchical procedure that links the planning and the detailed scheduling models.

ACKNOWLEDGMENTS

This chapter is a modification of the article "Optimization of a Fleet of Large Tankers and Bulkers: A Linear Programming Approach," which appeared in *Marine Technology*, October 1972, and was co-authored by J. Everett, A. Hax, V. Lewison, and D. Nutts.

Network Models

8

There are several kinds of linear-programming models that exhibit a special structure that can be exploited in the construction of efficient algorithms for their solution. The motivation for taking advantage of their structure usually has been the need to solve larger problems than otherwise would be possible to solve with existing computer technology. Historically, the first of these special structures to be analyzed was the transportation problem, which is a particular type of network problem. The development of an efficient solution procedure for this problem resulted in the first widespread application of linear programming to problems of industrial logistics. More recently, the development of algorithms to efficiently solve particular large-scale systems has become a major concern in applied mathematical programming.

Network models are possibly still the most important of the special structures in linear programming. In this chapter, we examine the characteristics of network models, formulate some examples of these models, and give one approach to their solution. The approach presented here is simply derived from specializing the rules of the simplex method to take advantage of the structure of network models. The resulting algorithms are extremely efficient and permit the solution of network models so large that they would be impossible to solve by ordinary linear-programming procedures. Their efficiency stems from the fact that a pivot operation for the simplex method can be carried out by simple addition and subtraction without the need for maintaining and updating the usual tableau at each iteration. Further, an added benefit of these alogorithms is that the optimal solutions generated turn out to be *integer* if the relevant constraint data are integer.

8.1 THE GENERAL NETWORK-FLOW PROBLEM

A common scenario of a network-flow problem arising in industrial logistics concerns the distribution of a single homogeneous product from plants (origins) to consumer markets (destinations). The total number of units produced at each plant and the total number of units required at each market are assumed to be known. The product need not be sent directly from source to destination, but may be routed through intermediary points reflecting warehouses or distribution centers. Further, there may be capacity restrictions that limit some of the shipping links. The objective is to minimize the variable cost of producing and shipping the products to meet the consumer demand.

The sources, destinations, and intermediate points are collectively called *nodes* of the network, and the transportation links connecting nodes are termed *arcs*. Although a production/distribution problem has been given as the motivating scenario, there are many other applications of the general model. Table 8.1 indicates a few of the many possible alternatives.

Table 8.1 Examples of Network Flow Problems

	Urban transportation	Communication systems	Water resources
Product	Buses, autos, etc.	Messages	Water
Nodes	Bus stops, street intersections	Communication centers, relay stations	Lakes, reservoirs, pumping stations
Arcs	Streets (lanes)	Communication channels	Pipelines, canals, rivers

A numerical example of a network-flow problem is given in Fig. 8.1. The nodes are represented by numbered circles and the arcs by arrows. The arcs are assumed to be *directed* so that, for instance, material can be sent from node 1 to node 2, but not from node 2 to node 1. Generic arcs will be denoted by *i–j*, so that 4–5 means the arc *from* node 4 *to* node 5. Note that some pairs of nodes, for example 1 and 5, are not connected directly by an arc.

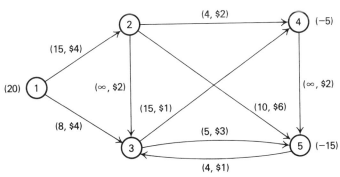

Fig. 8.1 Minimum-cost flow problem.

Figure 8.1 exhibits several additional characteristics of network flow problems. First, a flow capacity is assigned to each arc, and second, a per-unit cost is specified for shipping along each arc. These characteristics are shown next to each arc. Thus, the flow on arc 2–4 must be between 0 and 4 units, and each unit of flow on this arc costs $2.00. The ∞'s in the figure have been used to denote unlimited flow capacity on arcs 2–3 and 4–5. Finally, the numbers in parentheses next to the nodes give the material supplied or demanded at that node. In this case, node 1 is an origin or *source node* supplying 20 units, and nodes 4 and 5 are destinations or *sink nodes* requiring 5 and 15 units, respectively, as indicated by the negative signs. The remaining nodes have no net supply or demand; they are intermediate points, often referred to as *transshipment nodes*.

The objective is to find the minimum-cost flow pattern to fulfill demands from the source nodes. Such problems usually are referred to as *minimum-cost flow* or *capacitated transshipment* problems. To transcribe the problem into a formal linear program, let

$$x_{ij} = \text{Number of units shipped from node } i \text{ to } j \text{ using arc } i\text{–}j.$$

Then the tabular form of the linear-programming formulation associated with the network of Fig. 8.1 is as shown in Table 8.2.

Table 8.2 Tableau for Minimum-Cost Flow Problem

	x_{12}	x_{13}	x_{23}	x_{24}	x_{25}	x_{34}	x_{35}	x_{45}	x_{53}	Righthand side
Node 1	1	1								20
Node 2	−1		1	1	1					0
Node 3		−1	−1			1	1		−1	0
Node 4				−1		−1		1		−5
Node 5					−1		−1	−1	1	−15
Capacities	15	8	∞	4	10	15	5	∞	4	
Objective function	4	4	2	2	6	1	3	2	1	(Min)

The first five equations are flow-balance equations at the nodes. They state the conservation-of-flow law,

$$\begin{pmatrix} \text{Flow out} \\ \text{of a node} \end{pmatrix} - \begin{pmatrix} \text{Flow into} \\ \text{a node} \end{pmatrix} = \begin{pmatrix} \text{Net supply} \\ \text{at a node} \end{pmatrix}.$$

As examples, at nodes 1 and 2 the balance equations are:

$$x_{12} + x_{13} \qquad\qquad = 20$$
$$x_{23} + x_{24} + x_{25} - x_{12} = 0.$$

It is important to recognize the special structure of these balance equations. Note that there is one balance equation for each node in the network. The flow variables x_{ij} have only 0, $+1$, and -1 coefficients in these equations. Further, each variable appears in exactly two balance equations, once with a $+1$ coefficient, corresponding to the node from which the arc emanates; and once with a -1 coefficient, corresponding to the node upon which the arc is incident. This type of tableau is referred to as a *node–arc incidence matrix*; it completely describes the physical layout of the network. It is this particular structure that we shall exploit in developing specialized, efficient algorithms.

The remaining two rows in the table give the upper bounds on the variables and the cost of sending one unit of flow across an arc. For example, x_{12} is constrained by $0 \leqq x_{12} \leqq 15$ and appears in the objective function as $2x_{12}$. In this example the lower bounds on the variables are taken implicitly to be zero, although in general there may also be nonzero lower bounds.

This example is an illustration of the following general *minimum-cost flow* problem with n nodes:

$$\text{Minimize } z = \sum_i \sum_j c_{ij} x_{ij},$$

subject to:

$$\sum_j x_{ij} - \sum_k x_{ki} = b_i \quad (i = 1, 2, \ldots, n), \quad \text{[Flow balance]}$$

$$\ell_{ij} \leqq x_{ij} \leqq u_{ij}. \quad\quad\quad\quad\quad\quad\quad\quad \text{[Flow capacities]}$$

The summations are taken only over the arcs in the network. That is, the first summation in the ith flow-balance equation is over all nodes j such that i–j is an arc of the network, and the second summation is over all nodes k such that k–i is an arc of the network. The objective function summation is over arcs i–j that are contained in the network and represents the total cost of sending flow over the network. The ith balance equation is interpreted as above: it states that the flow out of node i minus the flow into i must equal the net supply (demand if b_i is negative) at the node. u_{ij} is the upper bound on arc flow and may be $+\infty$ if the capacity on arc i–j is unlimited. ℓ_{ij} is the lower bound on arc flow and is often taken to be zero, as in the previous example. In the following sections we shall study variations of this general problem in some detail.

8.2 SPECIAL NETWORK MODELS

There are a number of interesting special cases of the minimum-cost flow model that have received a great deal of attention. This section introduces several of these models, since they have had a significant impact on the development of a general network theory. In particular, algorithms designed for these specific models have motivated solution procedures for the more general minimum-cost flow problem.

The Transportation Problem

The transportation problem is a network-flow model without intermediate locations. To formulate the problem, let us define the following terms:

$$a_i = \text{Number of units available at source } i \ (i = 1, 2, \ldots, m);$$
$$b_j = \text{Number of units required at destination } j \ (j = 1, 2, \ldots, n);$$
$$c_{ij} = \text{Unit transportation cost from source } i \text{ to destination } j$$
$$(i = 1, 2, \ldots, m; j = 1, 2, \ldots, n).$$

For the moment, we assume that the total product availability is equal to the total product requirements; that is,

$$\sum_{i=1}^{m} a_i = \sum_{j=1}^{n} b_j.$$

Later we will return to this point, indicating what to do when this supply–demand balance is not satisfied. If we define the decision variables as:

$$x_{ij} = \text{Number of units to be distributed from source } i \text{ to destination } j$$
$$(i = 1, 2, \ldots, m; j = 1, 2, \ldots, n),$$

we may then formulate the transportation problem as follows:

$$\text{Minimize } z = \sum_{i=1}^{m} \sum_{j=1}^{n} c_{ij} x_{ij}, \tag{1}$$

subject to:

$$\sum_{j=1}^{n} x_{ij} = a_i \qquad (i = 1, 2, \ldots, m), \tag{2}$$

$$\sum_{i=1}^{m} (-x_{ij}) = -b_j \qquad (j = 1, 2, \ldots, n), \tag{3}$$

$$x_{ij} \geq 0 \qquad (i = 1, 2, \ldots, m; j = 1, 2, \ldots, n) \tag{4}$$

Expression (1) represents the minimization of the total distribution cost, assuming a linear cost structure for shipping. Equation (2) states that the amount being shipped from source i to all possible destinations should be equal to the total availability, a_i, at that source. Equation (3) indicates that the amounts being shipped to destination j from all possible sources should be equal to the requirements, b_j, at that destination. Usually Eq. (3) is written with positive coefficients and righthand sides by multiplying through by minus one.

Let us consider a simple example. A compressor company has plants in three locations: Cleveland, Chicago, and Boston. During the past week the total production of a special compressor unit out of each plant has been 35, 50, and 40 units respectively. The company wants to ship 45 units to a distribution center in Dallas, 20 to Atlanta, 30 to San Francisco, and 30 to Philadelphia. The unit production and distribution costs from each plant to each distribution center are given in Table 8.3. What is the best shipping strategy to follow?

Table 8.3 Unit Production and Shipping Costs

	Distribution centers				Availability
Plants	Dallas	Atlanta	San Francisco	Phila.	(units)
Cleveland	8	6	10	9	35
Chicago	9	12	13	7	50
Boston	14	9	16	5	40
Requirements (units)	45	20	30	30	[125]

The linear-programming formulation of the corresponding transportation problem is:

$$\text{Minimize } z = 8x_{11} + 6x_{12} + 10x_{13} + 9x_{14} + 9x_{21} + 12x_{22} + 13x_{23}$$
$$+ 7x_{24} + 14x_{31} + 9x_{32} + 16x_{33} + 5x_{34},$$

subject to:

$$
\begin{array}{lr}
x_{11} + x_{12} + x_{13} + x_{14} & = 35, \\
x_{21} + x_{22} + x_{23} + x_{24} & = 50, \\
x_{31} + x_{32} + x_{33} + x_{34} = & 40, \\
-x_{11} \qquad -x_{21} \qquad -x_{31} & = -45, \\
-x_{12} \qquad -x_{22} \qquad -x_{32} & = -20, \\
-x_{13} \qquad -x_{23} \qquad -x_{33} & = -30, \\
-x_{14} \qquad -x_{24} \qquad -x_{34} = & -30, \\
\end{array}
$$

$$x_{ij} \geq 0 \quad (i = 1, 2, 3; j = 1, 2, 3, 4).$$

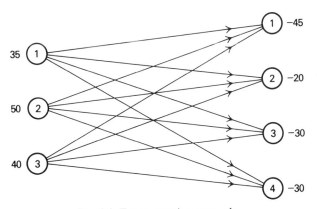

Fig. 8.2 Transportation network.

Because there is no ambiguity in this case, the same numbers normally are used to designate the origins and destinations. For example, x_{11} denotes the flow from source 1 to destination 1, although these are two distinct nodes. The network corresponding to this problem is given in Fig. 8.2.

The Assignment Problem

Though the transportation model has been cast in terms of material flow from sources to destinations, the model has a number of additional applications. Suppose, for example, that n people are to be assigned to n jobs and that c_{ij} measures the performance of person i in job j. If we let

$$x_{ij} = \begin{cases} 1 & \text{if person } i \text{ is assigned to job } j, \\ 0 & \text{otherwise,} \end{cases}$$

we can find the optimal assignment by solving the optimization problem:

$$\text{Maximize } z = \sum_{i=1}^{n} \sum_{j=1}^{n} c_{ij} x_{ij},$$

subject to:

$$\sum_{j=1}^{n} x_{ij} = 1 \qquad (i = 1, 2, \ldots, n),$$

$$\sum_{i=1}^{n} x_{ij} = 1 \qquad (j = 1, 2, \ldots, n),$$

$$x_{ij} = 0 \quad \text{or} \quad 1 \qquad (i = 1, 2, \ldots, n; j = 1, 2, \ldots, n).$$

The first set of constraints shows that each person is to be assigned to exactly one job and the second set of constraints indicates that each job is to be performed by one person. If the second set of constraints were multiplied by minus one, the equations of the model would have the usual network interpretation.

As stated, this assignment problem is formally an integer program, since the decision variables x_{ij} are restricted to be zero or one. However, if these constraints are replaced by $x_{ij} \geq 0$, the model becomes a special case of the transportation problem, with one unit available at each source (person) and one unit required by each destination (job). As we shall see, network-flow problems have integer solutions, and therefore formal specification of integrality constraints is unnecessary. Consequently, application of the simplex method, or most network-flow algorithms, will solve such integer problems directly.

The Maximal Flow Problem

For the maximal flow problem, we wish to send as much material as possible from a specified node s in a network, called the *source*, to another specified node t, called the *sink*. No costs are associated with flow. If v denotes the amount of material sent from node s to node t and x_{ij} denotes the flow from node i to node j over arc i–j, the formulation is:

Maximize v,

subject to:

$$\sum_j x_{ij} - \sum_k x_{ki} = \begin{cases} v & \text{if } i = s \text{ (source)}, \\ -v & \text{if } i = t \text{ (sink)}, \\ 0 & \text{otherwise}, \end{cases}$$

$$0 \leq x_{ij} \leq u_{ij} \qquad (i = 1, 2, \ldots, n; j = 1, 2, \ldots, n).$$

As usual, the summations are taken only over the arcs in the network. Also, the upper bound u_{ij} for the flow on arc i–j is taken to be $+\infty$ if arc i–j has unlimited capacity. The interpretation is that v units are supplied at s and consumed at t.

Let us introduce a fictitious arc t–s with unlimited capacity; that is, $u_{ts} = +\infty$. Now x_{ts} represents the variable v, since x_{ts} simply returns the v units of flow from node t back to node s, and no formal external supply of material occurs. With the introduction of the arc t–s, the problem assumes the following special form of the general network problem:

Maximize x_{ts},

subject to:

$$\sum_j x_{ij} - \sum_k x_{ki} = 0 \qquad (i = 1, 2, \ldots, n),$$

$$0 \leq x_{ij} \leq u_{ij} \qquad (i = 1, 2, \ldots, n; j = 1, 2, \ldots, n).$$

Let us again consider a simple example. A city has constructed a piping system to route water from a lake to the city reservoir. The system is now underutilized and city planners are interested in its overall capacity. The situation is modeled as finding the maximum flow from node 1, the lake, to node 6, the reservoir, in the network shown in Fig. 8.3.

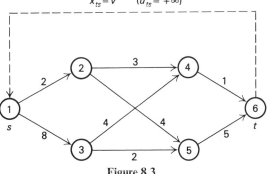

Figure 8.3

The numbers next to the arcs indicate the maximum flow capacity (in 100,000 gallons/day) in that section of the pipeline. For example, at most 300,000 gallons/day can be sent from node 2 to node 4. The city now sends 100,000 gallons/day along each of the paths 1–2–4–6 and 1–3–5–6. What is the maximum capacity of the network for shipping water from node 1 to node 6?

The Shortest-Path Problem

The shortest-path problem is a particular network model that has received a great deal of attention for both practical and theoretical reasons. The essence of the problem can be stated as follows: Given a network with distance c_{ij} (or travel time, or cost, etc.) associated with each arc, find a path through the network from a particular origin (source) to a particular destination (sink) that has the shortest total distance. The simplicity of the statement of the problem is somewhat misleading, because a number of important applications can be formulated as shortest- (or longest-) path problems where this formulation is not obvious at the outset. These include problems of equipment replacement, capital investment, project scheduling, and inventory planning. The theoretical interest in the problem is due to the fact that it has a special structure, in addition to being a network, that results in very efficient solution procedures. (In Chapter 11 on dynamic programming, we illustrate some of these other procedures.) Further, the shortest-path problem often occurs as a subproblem in more complex situations, such as the subproblems in applying decomposition to traffic-assignment problems or the group-theory problems that arise in integer programming.

In general, the formulation of the shortest-path problem is as follows:

$$\text{Minimize } z = \sum_i \sum_j c_{ij}x_{ij},$$

subject to:

$$\sum_j x_{ij} - \sum_k x_{ki} = \begin{cases} 1 & \text{if } i = s \text{ (source)}, \\ 0 & \text{otherwise}, \\ -1 & \text{if } i = t \text{ (sink)}, \end{cases}$$

$$x_{ij} \geqq 0 \qquad \text{for all arcs } i\text{--}j \text{ in the network}.$$

We can interpret the shortest-path problem as a network-flow problem very easily. We simply want to send one unit of flow from the source to the sink at minimum cost. At the source, there is a net supply of one unit; at the sink, there is a net demand of one unit; and at all other nodes there is no net inflow or outflow.

As an elementary illustration, consider the example given in Fig. 8.4, where we wish to find the shortest distance from node 1 to node 8. The numbers next to the arcs are the distance over, or cost of using, that arc. For the network specified in Fig. 8.4, the linear-programming tableau is given in Tableau 1.

Tableau 1 Node–Arc Incidence Tableau for a Shortest-Path Problem

	x_{12}	x_{13}	x_{24}	x_{25}	x_{32}	x_{34}	x_{37}	x_{45}	x_{46}	x_{47}	x_{52}	x_{56}	x_{58}	x_{65}	x_{67}	x_{68}	x_{76}	x_{78}	Relations	RHS
Node 1	1	1																	=	1
Node 2	−1		1	1	−1						−1								=	0
Node 3		−1			1	1	1												=	0
Node 4			−1			−1		1	1	1									=	0
Node 5				−1				−1			1	1	1	−1					=	0
Node 6									−1			−1		1	1	1	−1		=	0
Node 7							−1			−1					−1		1	1	=	0
Node 8													−1			−1		−1	=	−1
Distance	5.1	3.4	0.5	2.0	1.0	1.5	5.0	2.0	3.0	4.2	1.0	3.0	6.0	1.5	0.5	2.2	2.0	2.4	=	z (min)

319

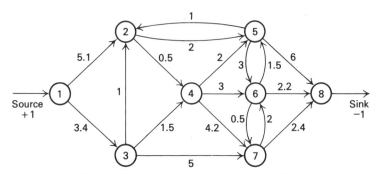

Fig. 8.4 Network for a shortest-path problem.

8.3 THE CRITICAL-PATH METHOD

The Critical-Path Method (CPM) is a project-management technique that is used widely in both government and industry to analyze, plan, and schedule the various tasks of complex projects. CPM is helpful in identifying which tasks are critical for the execution of the overall project, and in scheduling all the tasks in accordance with their prescribed *precedence relationships* so that the total project completion date is minimized, or a target date is met at minimum cost.

Typically, CPM can be applied successfully in large construction projects, like building an airport or a highway; in large maintenance projects, such as those encountered in nuclear plants or oil refineries; and in complex research-and-development efforts, such as the development, testing, and introduction of a new product. All these projects consist of a well specified collection of tasks that should be executed in a certain prescribed sequence. CPM provides a methodology to define the interrelationships among the tasks, and to determine the most effective way of scheduling their completion.

Although the mathematical formulation of the scheduling problem presents a network structure, this is not obvious from the outset. Let us explore this issue by discussing a simple example.

Suppose we consider the scheduling of tasks involved in building a house on a foundation that already exists. We would like to determine in what sequence the tasks should be performed in order to minimize the total time required to execute the project. All we really know is how long it takes to carry out each task and which tasks must be completed before commencing any particular task. In fact, it will be clear that we need only know the tasks that *immediately* precede a particular task, since completion of all *earlier* tasks will be implied by this information. The tasks that need to be performed in building this particular house, their immediate predecessors, and an estimate of their duration are given in Table 8.4.

It is clear that there is no need to indicate that the siding must be put up before the outside painting can begin, since putting up the siding precedes installing the windows, which precedes the outside painting. It is always convenient to identify

Table 8.4 Tasks and Precedence Relationships

No.	Task	Immediate predecessors	Duration	Earliest starting times
0	Start	—	0	—
1	Framing	0	2	t_1
2	Roofing	1	1	t_2
3	Siding	1	3	t_2
4	Windows	3	2.5	t_3
5	Plumbing	3	1.5	t_3
6	Electricity	2, 4	2	t_4
7	Inside Finishing	5, 6	4	t_5
8	Outside Painting	2, 4	3	t_4
9	Finish	7, 8	0	t_6

a "start" task, that is, an immediate predecessor to all tasks, which in itself does not have predecessors; and a "finish" task, which has, as immediate predecessors, *all* tasks that in actuality have no successors.

Although it is by no means required in order to perform the necessary computations associated with the scheduling problem, often it is useful to represent the interrelations among the tasks of a given project by means of a network diagram. In this diagram, nodes represent the corresponding tasks of the project, and arcs represent the precedence relationships among tasks. The network diagram for our example is shown in Fig. 8.5.

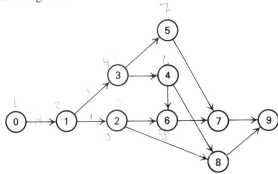

Fig. 8.5 Task-oriented network.

As we can see, there are nine nodes in the network, each representing a given task. For this reason, this network representation is called a task- (or activity-) oriented network.

If we assume that our objective is to minimize the elapsed time of the project, we can formulate a linear-programming problem. First, we define the decision variables t_i for $i = 1, 2, \ldots, 6$, as the earliest starting times for each of the tasks. Table 8.4 gives the earliest starting times where the same earliest starting time is assigned to tasks with the same immediate predecessors. For instance, tasks 4 and 5 have task 3 as their immediate predecessor. Obviously, they cannot start until task 3 is finished; therefore, they should have the *same* earliest starting time. Letting t_6 be the earliest completion time of the entire project, our objective is to minimize the project duration given by

Minimize $t_6 - t_1$,

subject to the precedence constraints among tasks. Consider a particular task, say 6, installing the electricity. The earliest starting time of task 6 is t_4, and its immediate predecessors are tasks 2 and 4. The earliest starting times of tasks 2 and 4 are t_2 and t_3, respectively, while their durations are 1 and 2.5 weeks, respectively. Hence, the earliest starting time of task 6 must satisfy:

$$t_4 \geq t_2 + 1,$$
$$t_4 \geq t_3 + 2.5.$$

In general, if t_j is the earliest starting time of a task, t_i is the earliest starting time of an immediate predecessor, and d_{ij} is the duration of the immediate predecessor, then we have:

$$t_j \geq t_i + d_{ij}.$$

For our example, these precedence relationships define the linear program given in Tableau 2.

Tableau 2

t_1	t_2	t_3	t_4	t_5	t_6	Relation	RHS
-1	1					$\geq$	2
	-1	1				$\geq$	3
	-1		1			$\geq$	1
		-1	1			$\geq$	2.5
		-1		1		$\geq$	1.5
			-1	1		$\geq$	2
			-1		1	$\geq$	3
				-1	1	$\geq$	4
-1					1	$=$	T (min)

(5)

We do not yet have a network flow problem; the constraints of (5) do not satisfy our restriction that each column have only a plus-one and a minus-one coefficient in the constraints. However, this *is* true for the rows, so let us look at the dual of (5). Recognizing that the variables of (5) have not been explicitly restricted to the nonnegative, we will have equality constraints in the dual. If we let x_{ij} be the dual variable associated with the constraint of (5) that has a minus one as a coefficient for t_i and a plus one as a coefficient of t_j, the dual of (5) is then given in Tableau 3.

Tableau 3

x_{12}	x_{23}	x_{24}	x_{34}	x_{35}	x_{45}	x_{46}	x_{56}	*Relation*	RHS
-1								$=$	-1
1	-1	-1						$=$	0
	1		-1	-1				$=$	0
		1	1		-1	-1		$=$	0
				1	1		-1	$=$	0
						1	1	$=$	1
2	3	1	2.5	1.5	2	3	4	$=$	z (max)

(6)

Now we note that each column of (6) has only one plus-one coefficient and one minus-one coefficient, and hence the tableau describes a network. If we multiply each equation through by minus one, we will have the usual sign convention with respect to arcs emanating from or incident to a node. Further, since the righthand side has only a plus one and a minus one, we have flow equations for sending one unit of flow from node 1 to node 6. The network corresponding to these flow equations is given in Fig. 8.6; this network clearly maintains the precedence relationships from Table 8.4. Observe that we have a longest-path problem, since we wish to maximize z (in order to minimize the project completion date T).

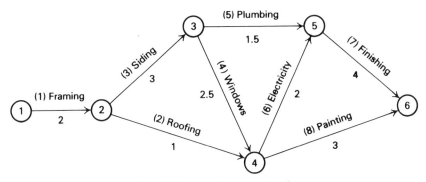

Fig. 8.6 Event-oriented network.

Note that, in this network, the arcs represent the tasks, while the nodes describe the precedence relationships among tasks. This is the opposite of the network representation given in Fig. 8.5. As we can see, the network of Fig. 8.6 contains 6 nodes, which is the number of sequencing constraints prescribed in the task definition of Table 8.4, since only six earliest starting times were required to characterize these constraints. Because the network representation of Fig. 8.6 emphasizes the event associated with the starting of each task, it is commonly referred to as an event-oriented network.

There are several other issues associated with critical-path scheduling that also give rise to network-model formulations. In particular, we can consider allocating funds among the various tasks in order to reduce the total time required to complete the project. The analysis of the cost-*vs.*-time tradeoff for such a change is an important network problem. Broader issues of resource allocation and requirements smoothing can also be interpreted as network models, under appropriate conditions.

8.4 CAPACITATED PRODUCTION—A HIDDEN NETWORK

Network-flow models are more prevalent than one might expect, since many models not cast naturally as networks can be transformed into a network format. Let us illustrate this possibility by recalling the strategic-planning model for aluminum production developed in Chapter 6. In that model, bauxite ore is converted to aluminum products in several smelters, to be shipped to a number of customers. Production and shipment are governed by the following constraints:

$$\sum_a \sum_p Q_{sap} - M_s \quad\quad = 0 \quad\quad (s = 1, 2, \ldots, 11), \quad\quad\quad (7)$$

$$\sum_a Q_{sap} \quad\quad - E_{sp} = 0 \quad\quad (s = 1, 2, \ldots, 11; p = 1, 2, \ldots, 8), \quad\quad (8)$$

$$\sum_s Q_{sap} \quad\quad = d_{ap} \quad\quad (a = 1, 2, \ldots, 40; p = 1, 2, \ldots, 8), \quad\quad (9)$$

$$\underline{m}_s \leqq M_s \leqq \bar{m}_s \quad\quad (s = 1, 2, \ldots, 11),$$

$$\underline{e}_{sp} \leqq E_{sp} \leqq \bar{e}_{sp} \quad\quad (p = 1, 2, \ldots, 40).$$

Variable Q_{sap} is the amount of product p to be produced at smelter s and shipped to customer a. The constraints (7) and (8) merely define the amount M_s produced at smelter s and the amount E_{sp} of product p (ingots) to be "cast" at smelter s. Equations (9) state that the total production from all smelters must satisfy the demand d_{ap} for product p of each customer a. The upper bounds of M_s and E_{sp} reflect smelting and casting capacity, whereas the lower bounds indicate minimum economically attractive production levels.

As presented, the model is not in a network format, since it does not satisfy the property that every variable appear in exactly two constraints, once with a $+1$ coefficient and once with a -1 coefficient. It can be stated as a network, however, by making a number of changes. Suppose, first, that we rearrange all the constraints of (7), as

$$\sum_p \left(\sum_a Q_{sap} \right) - M_s = 0,$$

and substitute, for the term in parenthesis, E_{sp} defined by (8). Let us also multiply the constraints of (9) by (-1). The model is then rewritten as:

$$\sum_p E_{sp} - M_s = 0 \qquad\qquad (s = 1, 2, \ldots, 11), \qquad\qquad (10)$$

$$\sum_a Q_{sap} - E_{sp} \quad\;\; = 0 \qquad\qquad (s = 1, 2, \ldots, 11; p = 1, 2, \ldots, 8), \quad (11)$$

$$\sum_s - Q_{sap} \qquad\quad\;\; = -d_{ap} \qquad (a = 1, 2, \ldots, 40; p = 1, 2, \ldots, 8), \quad (12)$$

$$\underline{m}_s \leqq M_s \leqq \bar{m}_s \qquad\qquad (s = 1, 2, \ldots, 11),$$

$$\underline{e}_{sp} \leqq E_{sp} \leqq \bar{e}_{sp} \qquad\qquad (p = 1, 2, \ldots, 8).$$

Each variable E_{sp} appears once in the equations of (10) with a $+1$ coefficient and once in the equations of (11) with a -1 coefficient; each variable Q_{sap} appears once in the equations of (11) with a $+1$ coefficient and once in the equations of (12) with a -1 coefficient. Consequently, except for the variables M_s, the problem is in the form of a network. Now, suppose that we *add* all the equations to form one additional redundant constraint. As we have just noted, the terms involving the variables Q_{sap} and E_{sp} will all vanish, so that the resulting equation, when multiplied by minus one, is:

$$\sum_s M_s = \sum_a \sum_p d_{ap}. \qquad\qquad (13)$$

Each variable M_s now appears once in the equations of (10) with a $+1$ coefficient and once in the equations of (13) with a -1 coefficient, so appending this constraint to the previous formulation gives the desired network formulation.

The network representation is shown in Fig. 8.7. As usual, each equation in the model defines a node in the network. The topmost node corresponds to the redundant equation just added to the model; it just collects production from the smelters. The other nodes correspond to the smelters, the casting facilities for products at the smelters, and the customer–product demand combinations. The overall supply to the system, $\sum_a \sum_p d_{ap}$, as indicated at the topmost node, is the total production at the smelters, and must equal the demand for all products.

In practice, manipulations like these just performed for re-expressing problems can be used frequently to exhibit network structure that might be hidden in a model formulation. They may not always lead to pure network-flow problems, as in this example, but instead might show that the problem has a substantial network component. The network features might then be useful computationally in conjunction with large-scale systems techniques that exploit the network structure.

Finally, observe that the network in Fig. 8.7 contains only a small percentage of the arcs that could potentially connect the nodes since, for example, the smelters do not connect directly with customer demands. This low density of arcs is common in practice, and aids in both the information storage and the computations for network models.

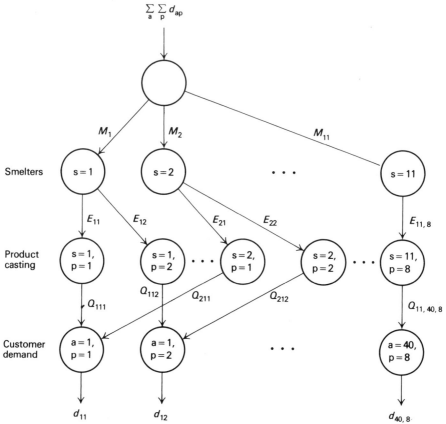

Fig. 8.7 Network formulation of the aluminum production-planning model.

8.5 SOLVING THE TRANSPORTATION PROBLEM

Ultimately in this chapter we want to develop an efficient algorithm for the general minimum-cost flow problem by specializing the rules of the simplex method to take advantage of the problem structure. However, before taking a somewhat formal approach to the general problem, we will indicate the basic ideas by developing a similar algorithm for the transportation problem. The properties of this algorithm for the transportation problem will then carry over to the more general minimum-cost flow problem in a straightforward manner. Historically, the transportation problem was one of the first special structures of linear programming for which an efficient special-purpose algorithm was developed. In fact, special-purpose algorithms have been developed for all of the network structures presented in Section 8.1, but they will not be developed here.

As we have indicated before, many computational algorithms are characterized by three stages:

1. obtaining an initial solution;

2. checking an optimality criterion that indicates whether or not a termination condition has been met (i.e., in the simplex algorithm, whether the problem is infeasible, the objective is unbounded over the feasible region, or an optimal solution has been found);

3. developing a procedure to improve the current solution if a termination condition has not been met.

After an initial solution is found, the algorithm repetitively applies steps 2 and 3 so that, in most cases, after a finite number of steps, a termination condition arises. The effectiveness of an algorithm depends upon its efficiency in attaining the termination condition.

Since the transportation problem is a linear program, each of the above steps can be performed by the simplex method. Initial solutions can be found very easily in this case, however, so phase I of the simplex method need not be performed. Also, when applying the simplex method in this setting, the last two steps become particularly simple.

The transportation problem is a special network problem, and the steps of any algorithm for its solution can be interpreted in terms of network concepts. However, it also is convenient to consider the transportation problem in purely algebraic terms. In this case, the equations are summarized very nicely by a tabular representation like that in Tableau 4.

Tableau 4

Sources	Destinations				Supply
	1	2	$\cdots$	n	
1	c_{11} x_{11}	c_{12} x_{12}	$\cdots$	c_{1n} x_{1n}	a_1
2	c_{21} x_{21}	c_{22} x_{22}	$\cdots$	c_{2n} x_{2n}	a_2
$\vdots$	$\vdots$	$\vdots$		$\vdots$	$\vdots$
m	c_{m1} x_{m1}	c_{m2} x_{m2}	$\cdots$	c_{mn} x_{mn}	a_m
Demand	b_1	b_2	$\cdots$	b_n	Total

Each row in the tableau corresponds to a source node and each column to a destination node. The numbers in the final column are the supplies available at the source nodes and those in the bottom row are the demands required at the destination nodes. The entries in cell i–j in the tableau denote the flow allocation x_{ij} from source i to destination j and the corresponding cost per unit of flow is c_{ij}. The sum of x_{ij} across row i must equal a_i in any feasible solution, and the sum of x_{ij} down column j must equal b_j.

Initial Solutions

In order to apply the simplex method to the transportation problem, we must first determine a basic feasible solution. Since there are $(m + n)$ equations in the constraint set of the transportation problem, one might conclude that, in a nondegenerate situation, a basic solution will have $(m + n)$ strictly positive variables. We should note, however, that, since

$$\sum_{i=1}^{m} a_i = \sum_{j=1}^{n} b_j = \sum_{i=1}^{m}\sum_{j=1}^{n} x_{ij},$$

one of the equations in the constraint set is redundant. In fact, any one of these equations can be eliminated without changing the conditions represented by the original constraints. For instance, in the transportation example in Section 8.2, the last equation can be formed by summing the first three equations and subtracting the next three equations. Thus, the constraint set is composed of $(m + n - 1)$ independent equations, and a corresponding nondegenerate basic solution will have exactly $(m + n - 1)$ basic variables.

There are several procedures used to generate an initial basic feasible solution, but we will consider only a few of these. The simplest procedure imaginable would be one that ignores the costs altogether and rapidly produces a basic feasible solution. In Fig. 8.8, we have illustrated such a procedure for the transportation problem introduced in Section 8.2. We simply send as much as possible from origin 1 to destination 1, i.e., the minimum of the supply and demand, which is 35. Since the supply at origin 1 is then exhausted but the demand at destination 1 is not, we next fulfill the remaining demand at destination 1 from that available at origin 2. At this point destination 1 is completely supplied, so we send as much as possible (20 units) of the remaining supply of 40 at origin 2 to destination 2, exhausting the demand there. Origin 2 still has a supply of 20 and we send as much of this as possible to destination 3, exhausting the supply at origin 2 but leaving a demand of 10 at destination 3. This demand is supplied from origin 3, leaving a supply there of 30, exactly corresponding to the demand of

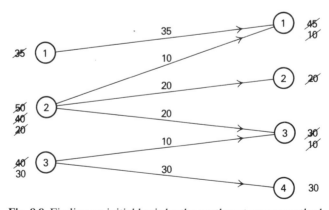

Fig. 8.8 Finding an initial basis by the northwest-corner method.

30 needed at destination 4. The available supply equals the required demand for this final allocation because we have assumed that the total supply equals the total demand, that is,

$$\sum_{i=1}^{m} a_i = \sum_{j=1}^{n} b_j.$$

This procedure is called the *northwest-corner* rule since, when interpreted in terms of the transportation array, it starts with the upper leftmost corner (the northwest corner) and assigns the maximum possible flow allocation to that cell. Then it moves to the right, if there is any remaining supply in the first row, or to the next lower cell, if there is any remaining demand in the first column, and assigns the maximum possible flow allocation to that cell. The procedure repeats itself until one reaches the lowest right corner, at which point we have exhausted all the supply and satisfied all the demand.

Table 8.5 summarizes the steps of the northwest-corner rule, in terms of the transportation tableau, when applied to the transportation example introduced in Section 8.2.

Table 8.5 Finding an Initial Basis by the Northwest Corner Method

	Distribution centers				
Plants	1. Dallas	2. Atlanta	3. San Fran.	4. Phila.	*Supply*
1. Cleveland	35				35
2. Chicago	10	20	20		50 40 20
3. Boston			10	30	40 30
Demand	45 10	20	30 10	30	

The availabilities and requirements at the margin of the table are updated after each allocation assignment. Although the northwest-corner rule is easy to implement, since it does not take into consideration the cost of using a particular arc, it will not, in general, lead to a cost-effective initial solution.

An alternative procedure, which is cognizant of the costs and straightforward to implement, is the *minimum matrix* method. Using this method, we allocate as much as possible to the available arc with the lowest cost. Figure 8.9 illustrates the procedure for the example that we have been considering. The least-cost arc joins origin 3 and destination 4, at a cost of $5/unit, so we allocate the maximum possible, 30 units, to this arc, completely supplying destination 4. Ignoring the arcs entering destination 4, the least-cost remaining arc joins origin 1 and destination 2, at a cost of $6/unit, so we allocate the maximum possible, 20 units, to this arc, completely supplying destination 2. Then, ignoring the arcs entering either destination 2 or 4, the least-cost remaining arc joins origin 1 and destination 1, at a cost of $8/unit, so we allocate the maximum possible, 15 units, to this arc, exhausting the supply at origin 1. Note that this arc is the least-cost *remaining* arc but not the next-lowest-cost

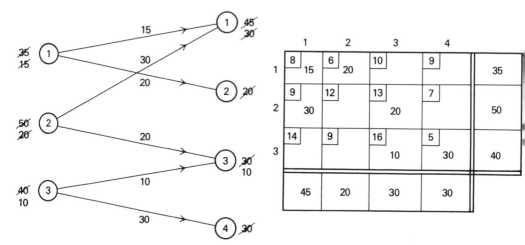

Fig. 8.9 Finding an initial basis by the minimum-matrix method.

unused arc in the entire cost array. Then, ignoring the arcs leaving origin 1 or entering destinations 2 or 4, the procedure continues. It should be pointed out that the last two arcs used by this procedure are relatively expensive, costing \$13/unit and \$16/unit. It is often the case that the minimum matrix method produces a basis that simultaneously employs some of the least expensive arcs and some of the most expensive arcs.

There are many other methods that have been used to find an initial basis for starting the transportation method. An obvious variation of the minimum matrix method is the *minimum row* method, which allocates as much as possible to the available arc with least cost *in each row* until the supply at each successive origin is exhausted. There is clearly the analogous *minimum column* method.

It should be pointed out that any comparison among procedures for finding an initial basis should be made only by comparing *solution times,* including both finding the initial basis *and* determining an optimal solution. For example, the northwest-corner method clearly requires fewer operations to determine an initial basis than does the minimum matrix rule, but the latter generally requires fewer iterations of the simplex method to reach optimality.

The two procedures for finding an initial basic feasible solution resulted in different bases; however, both have a number of similarities. Each basis consists of exactly $(m + n - 1)$ arcs, one less than the number of nodes in the network. Further, each basis is a subnetwork that satisfies the following two properties:

1. Every node in the network is connected to every other node by a sequence of arcs from the subnetwork, where the direction of the arcs is ignored.

2. The subnetwork contains no loops, where a loop is a sequence of arcs connecting a node to itself, where again the direction of the arcs is ignored.

A subnetwork that satisfies these two properties is called a *spanning tree.*

It is the fact that a basis corresponds to a spanning tree that makes the solution of these problems by the simplex method very efficient. Suppose you were told that a feasible basis consists of arcs 1–1, 2–1, 2–2, 2–4, 3–2, and 3–3. Then the allocations to each arc can be determined in a straightforward way without algebraic manipulations of tableaus. Start by selecting any *end* (a node with only 1 arc connecting it to the rest of the network) in the subnetwork. The allocation to the arc joining that end must be the supply or demand available at that end. For example, source 1 is an end node. The allocation on arc 1–1 must then be 35, decreasing the unfulfilled demand at destination 1 from 45 to 10 units. The end node and its connecting arc are then dropped from the subnetwork and the procedure is repeated. In Fig. 8.10, we use these end-node evaluations to determine the values of the basic variables for our example.

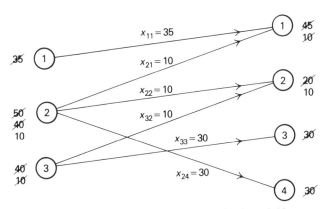

Fig. 8.10 Determining the values of the basic variables.

This example also illustrates that the solution of a transportation problem by the simplex method results in *integer* values for the variables. Since any basis corresponds to a spanning tree, as long as the supplies and demands are integers the amount sent across any arc must be an integer. This is true because, at each stage of evaluating the variables of a basis, as long as the remaining supplies and demands are integer the amount sent to any end must also be an integer. Therefore, if the initial supplies and demands are integers, any basic feasible solution will be an integer solution.

Optimization Criterion

Because it is so common to find the transportation problem written with +1 coefficients for the variables x_{ij} (i.e., with the demand equations of Section 8.2 multiplied by minus one), we will give the optimality conditions for this form. They can be easily stated in terms of the reduced costs of the simplex method. If x_{ij} for $i = 1, 2, \ldots, m$

and $j = 1, 2, \ldots, n$ is a feasible solution to the transportation problem:

$$\text{Minimize } z = \sum_{i=1}^{m} \sum_{j=1}^{n} c_{ij} x_{ij},$$

subject to:

		Shadow prices

$$\sum_{j=1}^{n} x_{ij} = a_i \qquad (i = 1, 2, \ldots, m), \qquad u_i$$

$$\sum_{i=1}^{m} x_{ij} = b_j \qquad (j = 1, 2, \ldots, n), \qquad v_j$$

$$x_{ij} \geq 0,$$

then it is optimal if there exist shadow prices (or simplex multipliers) u_i associated with the origins and v_j associated with the destinations, satisfying:

$$\bar{c}_{ij} = c_{ij} - u_i - v_j \geq 0 \qquad \text{if } x_{ij} = 0, \tag{14}$$

and

$$\bar{c}_{ij} = c_{ij} - u_i - v_j = 0 \qquad \text{if } x_{ij} > 0. \tag{15}$$

The simplex method selects multipliers so that condition (15) holds for all basic variables, even if some basic variable $x_{ij} = 0$ due to degeneracy.

These conditions not only allow us to test whether the optimum solution has been found or not, but provide us with the necessary foundation to reinterpret the characteristics of the simplex algorithm in order to solve the transportation problem efficiently. The algorithm will proceed as follows: after a basic feasible solution has been found (possibly by applying the northwest-corner method or the minimum matrix method), we choose simplex multipliers u_i and v_j ($i = 1, 2, \ldots, m; j = 1, 2, \ldots, n$) that satisfy:

$$u_i + v_j = c_{ij} \tag{16}$$

for basic variables. With these values for the simplex multipliers, we compute the corresponding values of the reduced costs:

$$\bar{c}_{ij} = c_{ij} - u_i - v_j \tag{17}$$

for all nonbasic variables. If every $\bar{c}_{ij}$ is nonnegative, then the optimal solution has been found; otherwise, we attempt to improve the current solution by increasing as much as possible the variable that corresponds to the most negative (since this is a minimization problem) reduced cost.

First, let us indicate the mechanics of determining from (16) the simplex multipliers associated with a particular basis. Conditions (16) consist of $(m + n - 1)$ equations in $(m + n)$ unknowns. However, any one of the simplex multipliers can be given an arbitrary value since, as we have seen, any one of the $(m + n)$ equations of the transportation problem can be considered redundant. Since there are

$m + n - 1$) equations in (16), once one of the simplex multipliers has been specified, the remaining values of u_i and v_j are determined uniquely. For example, Fig. 8.11 gives the initial basic feasible solution produced by the minimum matrix method. The simplex multipliers associated with this basis are easily determined by first arbitrarily setting $u_1 = 0$. Given $u_1 = 0$, $v_1 = 8$ and $v_2 = 6$ are immediate from (16); then $u_2 = 1$ is immediate from v_1, and so on. It should be emphasized that the set of multipliers is not unique, since any multiplier could have been chosen and set to any finite value, positive or negative, to initiate the determination.

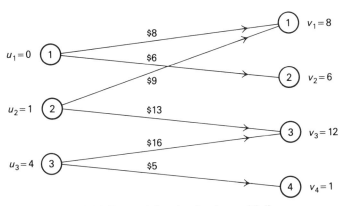

Fig. 8.11 Determining the simplex multipliers.

Once we have determined the simplex multipliers, we can then easily find the reduced costs for all nonbasic variables by applying (17). These reduced costs are given in Tableau 5. The —'s indicate the basic variable, which therefore has a reduced cost of zero. This symbol is used to distinguish between basic variables and nonbasic variables at zero level.

Tableau 5 Reduced Costs

8	6	10	9	
—	—	−2	8	0
9	12	13	7	
—	5	—	5	1
14	9	16	5	
2	−1	—	—	4
8	6	12	1	u_i / v_j

Since, in Tableau 5, $\bar{c}_{13} = -2$ and $\bar{c}_{32} = -1$, the basis construced by the minimum matrix method does not yield an optimal solution. We, therefore, need to find an improved solution.

Improving the Basic Solution

As we indicated, every basic variable has associated with it a value of $\bar{c}_{ij} = 0$. If the current basic solution is not optimal, then there exists at least one nonbasic variable x_{ij} at value zero with $\bar{c}_{ij}$ negative. Let us select, among all those variables, the one with the most negative $\bar{c}_{ij}$ (ties are broken by choosing arbitrarily from those variables that tie); that is,

$$\bar{c}_{st} = \min_{ij} \{\bar{c}_{ij} = c_{ij} - u_i - v_j | \bar{c}_{ij} < 0\}.$$

Thus, we would like to increase the corresponding value of x_{st} as much as possible, and adjust the values of the other basic variables to compensate for that increase. In our illustration, $\bar{c}_{st} = \bar{c}_{13} = -2$, so we want to introduce x_{13} into the basis. If we consider Fig. 8.11 we see that adding the arc 1–3 to the spanning tree corresponding to the current basis creates a unique loop O_1–D_3–O_2–D_1–O_1 where O and D refer to origin and destination, respectively.

It is easy to see that if we make $x_{st} = \theta$, maintaining all other nonbasic variables equal to zero, the flows on this loop must be adjusted by plus or minus θ, to maintain the feasibility of the solution. The limit to which θ can be increased corresponds to the smallest value of a flow on this loop from which θ must be subtracted. In this example, θ may be increased to 15, corresponding to x_{11} being reduced to zero and therefore dropping out of the basis. The basis has x_{13} replacing x_{11}, and the corresponding spanning tree has arc 1–3 replacing arc 1–1 in Fig. 8.12. Given the new basis, the procedure of determining the shadow prices and then the reduced costs, to check the optimality conditions of the simplex method, can be repeated.

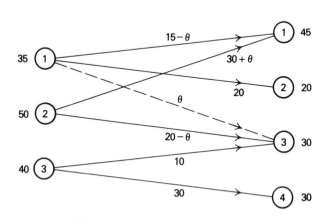

Fig. 8.12 Introducing a new variable.

It should be pointed out that the current solution usually is written out not in network form but in tableau form, for ease of computation. The current solution given in Fig. 8.12 would then be written as in Tableau 6.

Note that the row and column totals are correct for any value of θ satisfying $0 \leq \theta \leq 15$. The tableau form for the current basic solution is convenient to use for computations, but the justification of its use is most easily seen by considering the corresponding spanning tree. In what follows we will use the tableau form to illustrate the computations. After increasing θ to 15, the resulting basic solution is given in Tableau 7 (ignoring θ) and the new multipliers and reduced costs in Tableau 8.

Tableau 6

$15 - \theta$	20	θ		35
$30 + \theta$		$20 - \theta$		50
		10	30	40
45	20	30	30	

Tableau 7 Current Basic Solution

	$20 - \theta$	$15 + \theta$		35
45		5		50
	θ	$10 - \theta$	30	40
45	20	30	30	

Tableau 8 Reduced Costs

8	6	10	9	u_i
2	—	—	10	0
9	12	13	7	
—	3	—	5	3
14	9	16	5	
2	−3	—	—	6
6	6	10	−1	v_j

Since the only negative reduced cost corresponds to $\bar{c}_{32} = -3$, x_{32} is next introduced into the basis. Adding the arc 3–2 to the spanning tree corresponds to increasing the allocation to cell 3–2 in the tableau and creates a unique loop O_3–D_2–O_1–D_3–O_3. The flow θ on this arc may be increased until $\theta = 10$, corresponding to x_{33} dropping from the basis. The resulting basic solution is Tableau 9 and the new multipliers and reduced costs are given in Tableau 10.

Tableau 9 Current Basic Solution

	10	25		35
45		5		50
	10		30	40
45	20	30	30	

Tableau 10 Reduced Costs

8	6	10	9	u_i
2	—	—	7	0
9	12	13	7	
—	3	—	2	3
14	9	16	5	
5	—	3	—	3
6	6	10	2	v_j

Since all of the reduced costs are now nonnegative, we have the optimal solution.

8.6 ADDITIONAL TRANSPORTATION CONSIDERATIONS

In the previous section, we described how the simplex method has been specialized in order to solve transportation problems efficiently. There are three steps to the approach:

1. finding an initial basic feasible solution;
2. checking the optimality conditions; and
3. constructing an improved basic feasible solution, if necessary.

The problem we solved was quite structured in the sense that total supply equaled total demand, shipping between all origins and destinations was permitted, degeneracy did not occur, and so forth. In this section, we address some of the important variations of the approach given in the previous section that will handle these situations.

Supply Not Equal to Demand

We have analyzed the situation where the total availability is equal to the total requirement. In practice, this is usually not the case, since often either demand exceeds supply or supply exceeds demand. Let us see how to transform a problem in which this assumption does not hold to the previously analyzed problem with equality of availability and requirement. Two situations will exhaust all the possibilities:

First, assume that the total availability exceeds the total requirement; that is,

$$\sum_{i=1}^{m} a_i - \sum_{j=1}^{n} b_j = d > 0.$$

In this case, we create an artificial destination $j = n + 1$, with corresponding "requirement" $b_{n+1} = d$, and make the corresponding cost coefficients to this destination equal to zero, that is, $c_{i,\,n+1} = 0$ for $i = 1, 2, \ldots, m$. The variable $x_{i,\,n+1}$ in the optimal solution will show how the excess availability is distributed among the sources.

Second, assume that the total requirement exceeds the total availability, that is,

$$\sum_{j=1}^{n} b_j - \sum_{i=1}^{m} a_i = d > 0.$$

In this case, we create an artificial origin $i = m + 1$, with corresponding "availability" $a_{m+1} = d$, and assign zero cost coefficients to this destination, that is, $c_{m+1,\,j} = 0$ for $j = 1, 2, \ldots, n$. The optimal value for the variable $x_{m+1,\,j}$ will show how the unsatisfied requirements are allocated among the destinations.

In each of the two situations, we have constructed an equivalent transportation problem such that the total "availability" is equal to the total "requirement."

Prohibited Routes

If it is impossible to ship *any* goods from source i to destination j, we assign a very high cost to the corresponding variable x_{ij}, that is, $c_{ij} = M$, where M is a very large number, and use the procedure previously discussed. If these prohibited routes cannot be eliminated from the optimal solution then the problem is infeasible.

Alternatively, we can use the two-phase simplex method. We start with any initial basic feasible solution, which may use prohibited routes. The first phase will ignore the given objective function and minimize the sum of the flow along the prohibited routes. If the flow on the prohibited routes cannot be driven to zero, then no feasible solution exists without permitting flow on at least one of the prohibited routes. If, on the other hand, flow on the prohibited routes *can* be made zero, then an initial basic feasible solution *without* positive flow on prohibited routes has been constructed. It is necessary then simply to prohibit flow on these routes in the subsequent iterations of the algorithm.

Degeneracy

Degeneracy can occur on two different occasions during the computational process described in the previous section. First, during the computation of the initial solution of the transportation problem, we can simultaneously eliminate a row and a column at an intermediate step. This situation gives rise to a basic solution with less than $(m + n - 1)$ strictly positive variables. To rectify this, one simply assigns a zero value to a cell in either the row or column to be simultaneously eliminated, and treats that variable as a basic variable in the remaining computational process.

As an example, let us apply the northwest-corner rule to the case in Tableau 11.

Tableau 11

	D1	D2	D3	D4	Supply
O1	20	5	0		25 5 0
O2			30		30
O3			10		10
O4			10	40	50 40
Demand	20	5	50	40	
	0	0	20		
			10		

In this instance, when making $x_{12} = 5$, we simultaneously satisfy the first row availability and the second column requirement. We thus make $x_{13} = 0$, and treat that variable as a basic in the rest of the computation.

A second situation where degeneracy arises is while improving a current basic solution. A tie might be found when computing the new value to be given to the entering basic variable, $x_{st} = 0$. In this case, more than one of the old basic variables will take the value zero simultaneously, creating a new basic solution with less than $(m + n - 1)$ strictly positive values. Once again, the problem is overcome by choosing the variable to leave the basis arbitrarily from among the basic variables reaching zero at $x_{st} = 0$, and treating the remaining variables reaching zero at $x_{st} = 0$ as basic variables at zero level.

Vogel Approximation

Finally, we should point out that there have been a tremendous number of procedures suggested for finding an initial basic feasible solution to the transportation problem. In the previous section, we mentioned four methods: northwest corner, minimum matrix, minimum column, and minimum row.

The first of these ignores the costs altogether, while the remaining methods allocate costs in such a way that the last few assignments of flows often result in very high costs being incurred. The high costs are due to the lack of choice as to how the final flows are to be assigned to routes, once the initial flows have been established. The initial flows are not chosen with sufficient foresight as to how they might impair later flow choices.

The Vogel approximation method was developed to overcome this difficulty and has proved to be so effective that it is sometimes used to obtain an approximation to the optimal solution of the problem. The method, instead of sequentially using the least-cost remaining arc, bases its selection on the difference between the two lowest-cost arcs leaving an origin or entering a destination. This difference indicates where departure from the lowest-cost allocations will bring the highest increase in cost. Therefore, one assigns the maximum possible amount to that arc that has the lowest cost in that row or column having the greatest cost difference. If this assignment exhausts the demand at that destination, the corresponding column is eliminated from further consideration; similarly, if the assignment exhausts the supply at that origin, the corresponding row is eliminated. In either case, the origin and destination cost differences are recomputed, and the procedure continues in the same way.

The Vogel approximation method is applied to our illustrative example in Tableau 12, and the resulting basic feasible solution is given in Fig. 8.13. It is interesting to note that the approximation finds the optimal solution in this particular case, as can be seen by comparing the initial basis from the Vogel approximation in Fig. 8.13 with the optimal basis in Tableau 7. This, of course, does not mean that the Vogel approximation is the best procedure for determining the initial basic feasible solution. Any such comparison among computational procedures must compare total time required, from preprocessing to final optimal solution.

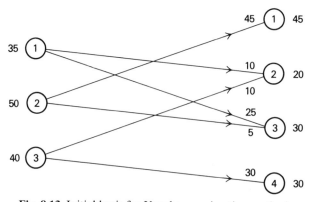

Fig. 8.13 Initial basis for Vogel approximation method.

Tableau 12 Applying the Vogel Approximation Method

First iteration:

	Dallas	Atlanta	San Francisco	Phila-delphia	Supply	Row difference
Cleveland	8	6	10	9	35	2
Chicago	9	12	13	7	50	2
Boston	14	9	16	5 30	40 10	4 ←
Demand	45	20	30	30		
Column difference	1	3	3	2		

Second iteration:

	Dallas	Atlanta	San Francisco	Supply	Row difference
Cleveland	8	6	10	35	2
Chicago	9	12	13	50	3
Boston	14	9 10	16	~~10~~	5 ←
Demand	45	~~20~~ 10	30		
Column difference	1	3	3		

Third iteration:

	Dallas	Atlanta	San Francisco	Supply	Row difference
Cleveland	8	6 10	10	35 25	2
Chicago	9	12	13	50	3
Demand	45	~~10~~	30		
Column difference	1	6 ↑	3		

Fourth iteration:

	Dallas	San Francisco	Supply	Row difference
Cleveland	8	10 25	25	2
Chicago	9 45	13 5	~~50~~ 45	4 ←
Demand	45	30		
Column difference	1	3		

8.7 THE SIMPLEX METHOD FOR NETWORKS

The application of the simplex method to the transportation problem presented in the previous section takes advantage of the network structure of the problem and illustrates a number of properties that extend to the general minimum-cost flow problem. All of the models formulated in Section 8.2 are examples of the general minimum-cost flow problem, although a number, including the transportation problem, exhibit further special structure of their own. Historically, many different algorithms have been developed for each of these models; but, rather than consider each model separately, we will develop the essential step underlying the efficiency of all of the simplex-based algorithms for networks.

We already have seen that, for the transportation problem, a basis corresponds to a spanning tree, and that introducing a new variable into the basis adds an arc to the spanning tree that forms a unique loop. The variable to be dropped from the basis is then determined by finding which variable in the loop drops to zero first when flow is increased on the new arc. It is this property, that bases correspond to spanning trees, that extends to the general minimum-cost flow problem and makes solution by the simplex method very efficient.

In what follows, network interpretations of the steps of the simplex method will be emphasized; therefore, it is convenient to define some of the network concepts that will be used. Though these concepts are quite intuitive, the reader should be cautioned that the terminology of network flow models has not been standardized, and that definitions vary from one author to another.

Formally, a *network* is defined to be any finite collection of points, called nodes, together with a collection of directed arcs that connect particular pairs of these nodes. By convention, we do not allow an arc to connect a node to itself, but we do allow more than one arc to connect the same two nodes. We will be concerned only with *connected* networks in the sense that *every* node can be reached from every other node by following a sequence of arcs, where the direction of the arcs is ignored. In linear programming, if a network is *disconnected*, then the problem it describes can be treated as separate problems, one for each connected subnetwork.

A *loop* is a sequence of arcs, where the direction of the arcs is ignored, connecting a particular node to itself. In Fig. 8.1, the node sequences 3–4–5–3 and 1–2–3–1 are both examples of loops.

A *spanning tree* is a connected subset of a network including all nodes and containing no loops. Figure 8.14 shows two examples of spanning trees for the minimum-cost flow problem of Fig. 8.1.

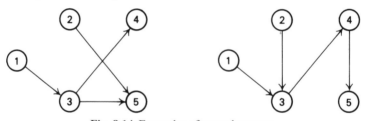

Fig. 8.14 Examples of spanning trees.

It is the concept of a spanning tree, which proved most useful in solving the transportation problem in the previous section, that will be the foundation of the algorithm for the general minimum-cost flow problem.

Finally, an *end* is a node of a network with exactly one arc incident to it. In the first example of Fig. 8.14, nodes 1, 2, and 4 are ends, and in the second examples, nodes 1, 2, and 5 are ends. It is easy to see that every tree must have at least two ends. If you start with any node i in a tree and follow any arc away from it, you eventually come to an end, since the tree contains no loops. If node i is an end, then you have two ends. If node i is not an end, there is another arc from node i that will lead to a second end, since again there are no loops in the tree.

In the transportation problem we saw that there are $(m + n - 1)$ basic variables, since any one of the equations is redundant under the assumption that the sum of the supplies equals the sum of the demands. This implies that the number of arcs in any spanning tree corresponding to a basis in the transportation problem is always *one less than* the number of nodes in the network. Note that the number of arcs is one less than the number of nodes in each of the trees shown in Fig. 8.14, since they each contain 5 nodes and 4 arcs. In fact, this characterization of spanning trees holds for the general minimum-cost flow problem.

> **Spanning-Tree Characterization.** A subnetwork of a network with n nodes is a spanning tree if and only if it is connected and contains $(n - 1)$ arcs.

We can briefly sketch an inductive proof to show the spanning-tree characterization. The result is clearly true for the two node networks containing one arc. First, we show that if a subnetwork of an n-node network is a spanning tree, it contains $(n - 1)$ arcs. Remove any end and incident arc from the n-node network. The reduced network with $(n - 1)$ nodes is still a tree, and by our inductive assumption it must have $(n - 2)$ arcs. Therefore, the original network with n nodes must have had $(n - 1)$ arcs. Next, we show that if an n-node connected subnetwork has $(n - 1)$ arcs and no loops, it is a spanning tree. Again, remove any end and its incident arc from the n-node network. The reduced network is connected, has $(n - 1)$ nodes, $(n - 2)$ arcs, and no loops; and by our inductive assumption, it must be a spanning tree. Therefore, the original network with n nodes and $(n - 1)$ arcs must be a spanning tree.

The importance of the spanning-tree characterization stems from the relationship between a spanning tree and a basis in the simplex method. We have already seen that a basis for the transportation problem corresponds to a spanning tree, and it is this property that carries over to the general network-flow model.

> **Spanning-Tree Property of Network Bases.** In a general minimum-cost flow model, a spanning tree for the network corresponds to a basis for the simplex method.

This is an important property since, together with the spanning-tree characterization, it implies that the number of basic variables is always one less than the number of nodes in a general network-flow problem. Now let us intuitively argue that the spanning-tree property holds, first by showing that the variables corresponding to a spanning tree constitute a basis, and second by showing that a set of basic variables constitutes a spanning tree.

First, assume that we have a network with n nodes, which is a spanning tree. In order to show that the variables corresponding to the arcs in the tree constitute a basis, it is sufficient to show that the $(n - 1)$ tree variables are uniquely determined. In the simplex method, this corresponds to setting the nonbasic variables to specific values and uniquely determining the basic variables. First, set the flows on all arcs not in the tree to either their upper or lower bounds, and update the righthand-side values by their flows. Then choose any node corresponding to an end in the subnetwork, say node k. (There must be at least two ends in the spanning tree since it contains no loops.) Node k corresponds to a row in the linear-programming tableau for the tree with exactly one nonzero coefficient in it. To illustrate this the tree variables of the first example in Fig. 8.14 are given in Tableau 13.

Tableau 13 Tree Variables

x_{13}	x_{25}	x_{34}	x_{35}	Righthand side
1				20
	1			0
-1		1	1	0
		-1		-5
	-1		-1	-15

Since there is only one nonzero coefficient in row k, the corresponding arc incident to node k must have flow across it equal to the righthand-side value for that row. In the example above, $x_{13} = 20$. Now, drop node k from further consideration and bring the determined variable over to the righthand side, so that the righthand side of the third constraint becomes $+20$. Now we have an $(n - 1)$-node subnetwork with $(n - 2)$ arcs and no loops. Hence, we have a tree for the reduced network, and the process may be repeated. At each iteration exactly one flow variable is determined. On the last iteration, there are two equations corresponding to two nodes, and one arc joining them. Since we have assumed that the total net flow into or out of the network is zero, the last tree variable will satisfy the last two equations. Hence we have shown that a spanning tree in a network corresponds to a basis in the simplex method. Further, since we have already shown that a tree for a connected network with n nodes contains $(n - 1)$ arcs, we have shown that the number of basic variables for a connected network-flow problem is $(n - 1)$.

Now assume that we have a network with n nodes and that we know the $(n - 1)$ basic variables. To show that these basic variables correspond to a tree, we need only show that the subnetwork corresponding to the basic variables does not contain any loops. We establish this property by assuming that there is a loop and showing that this leads to a contradiction. In Fig. 8.15, we have a four-arc network containing a loop and its associated tableau.

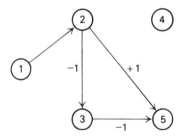

x_{12}	x_{23}	x_{25}	x_{35}	Righthand side
1				20
-1	1	1		0
	-1		1	0
				-5
		-1	-1	-15

Fig. 8.15 Network containing a loop.

If there exists a loop, then choose weights for the columns corresponding to arcs in the loop such that the weight is $+1$ for a forward arc in the loop and -1 for a backward arc in the loop. If we then add the columns corresponding to the loop weighted in this manner, we produce a column containing all zeros. In Fig. 8.15, the loop is 2–5–3–2, and adding the columns for the variables x_{25}, x_{53}, and x_{32} with weights 1, -1, and -1, respectively, produces a zero column. This implies that the columns corresponding to the loop are not independent. Since a basis consists of a set of $(n - 1)$ independent columns, any set of variables containing a loop cannot be a basis. Therefore, $(n - 1)$ variables corresponding to a basis in the simplex method must be a spanning tree for the network.

If a basis in the simplex method corresponds to a tree, what, then, is the interpretation of introducing a new variable into the basis? Introducing a new variable into the basis adds an arc to the tree, and since every node of the tree is connected to every other node by a sequence of arcs, the addition will form a loop in the subnetwork corresponding to the tree. In Fig. 8.16, arc 4–5 is being introduced into the tree, forming the loop 4–5–3–4. It is easy to argue that adding an arc to a tree creates a *unique* loop in the augmented network. *At least one loop* must be created, since adding the arc connects two nodes that were already connected. Further, *no more than one loop* is created, since, if adding the one arc created more than one loop, the entering arc would have to be common to two distinct loops, which, upon removal of their common arcs, would yield a single loop that was part of the original network.

To complete a basis change in the simplex method, one of the variables currently in the basis must be dropped. It is clear that the variable to be dropped must corre-

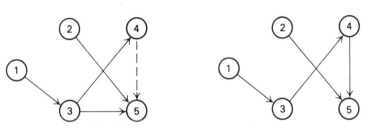

Fig. 8.16 Introducing a new arc in a tree.

spond to one of the arcs in the loop, since dropping any arc not in the loop would leave the loop in the network and hence would not restore the spanning-tree property. We must then be able to determine the arcs that form the loop.

This is accomplished easily by starting with the subnetwork including the loop and eliminating all ends in this network. If the reduced network has no ends, *it* must be the loop. Otherwise, repeat the process of eliminating all the ends in the reduced network, and continue. Since there is a unique loop created by adding an arc to a spanning tree, dropping any arc in that loop will create a new spanning tree, since each node will be connected to every other node by a sequence of arcs, and the resulting network will contain no loops. Figure 8.16 illustrates this process by dropping arc 3–5. Clearly, any arc in the loop could have been dropped, to produce alternative trees.

In the transportation problem, the unique loop was determined easily, although we did not explicitly show how this could be guaranteed. Once the loop is determined, we increased the flow on the incoming arc and adjusted the flows on the other arcs in the loop, until the flow in one of the arcs in the loop was reduced to zero. The variable corresponding to the arc whose flow was reduced to zero was then dropped out of the basis. This is essentially the same procedure that will be employed by the general minimum-cost flow problem, except that the special rules of the simplex method with upper and lower bounds on the variables will be employed. In the next section an example is carried out that applies the simplex method to the general minimum-cost flow problem.

Finally, we should comment on the integrality property of the general minimum-cost flow problem. We saw that, for the transportation problem, since the basis corresponds to a spanning tree, as long as the supplies and demands are integers, the flows on the arcs for a basic solution are integers. This is also true for the general minimum-cost flow problem, so long as the net flows at any node are integers and the upper and lower bounds on the variables are integers.

> *Integrality Property.* In the general minimum-cost flow problem, assuming that the upper and lower bounds on the variables are integers and the righthand-side values for the flow-balance equations are integers, the values of the basic variables are also integers when the nonbasic variables are set to their upper or lower bounds.

In the simplex method with upper and lower bounds on the variables, the non-basic variables are at either their upper or lower bound. If these bounds are integers, then the net flows at all nodes, when the flows on the nonbasic arcs are included, are also integers. Hence, the flows on the arcs corresponding to the basic variables also will be integers, since these flows are determined by first considering all ends in the corresponding spanning tree and assigning a flow to the incident arc equal to the net flow at the node. These assigned flows must clearly be integers. The ends and the arcs incident to them are then eliminated, and the process is repeated, yielding an integer assignment of flows to the arcs in the reduced tree at each stage.

Tableau 14 Basis variables.

x_{13}	x_{25}	x_{34}	x_{35}	Righthand side	Row no.
1				20	1
	1			0	2
-1		1	1	0	3
		-1		-5	4
	-1		-1	-15	5

Tableau 15 A basis is triangular.

x_{25}	x_{34}	x_{35}	x_{13}	Righthand side	Row no.
1				0	2
	-1			-5	4
-1		-1		-15	5
	1	1	-1	0	3

The integrality property of the general minimum-cost flow problem was established easily by using the fact that a basis corresponds to a spanning tree. Essentially,
all ends could be immediately evaluated, then eliminated, and the procedure repeated.
We were able to solve a system of equations by recognizing that at least one variable
in the system could be evaluated by inspection at each stage, since at each stage at
least one equation would have only one basic variable in it. A system of equations
with this property is called *triangular*.

In Tableau 14 we have rewritten the system of equations corresponding to the
tree variables given in Tableau 13. Then we have arbitrarily dropped the first
equation, since a connected network with n nodes has $(n - 1)$ basic variables. We
have rearranged the remaining variables and constraints to exhibit the triangular
form in Tableau 15.

The variables on the diagonal of the triangular system then may be evaluated
sequentially, starting with the first equation. Clearly $x_{25} = 0$. Then, moving the
evaluated variable to the righthand side, we have a new triangular system with one
less equation. Then the next diagonal variable may be evaluated in the same way and
the procedure repeated. It is easy to see that, for our example, the values of the
variables are $x_{25} = 0$, $x_{34} = 5$, $x_{35} = 15$, $x_{13} = 20$. Note that the value of x_{13}
satisfies the first equation that was dropped. It should be pointed out that many other
systems of equations besides network-flow problems can be put in the form of a
triangular system and therefore can be easily solved.

8.8 SOLVING THE MINIMUM-COST FLOW PROBLEM

In this section we apply the simplex method to the general minimum-cost flow
problem, using the network concepts developed in the previous section. Consider the
minimum-cost flow problem given in Section 8.1 and repeated here in Fig. 8.17 for
reference.

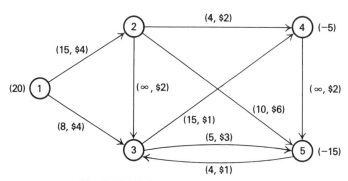

Fig. 8.17 Minimum-cost flow problem.

This problem is more complicated than the transportation problem since it contains intermediate nodes (points of transshipment) and capacities limiting the flow on some of the arcs.

In order to apply the simplex method to this example, we must first determine a basic feasible solution. Whereas, in the case of the transportation problem, an initial basic feasible solution is easy to determine (by the northwest-corner method, the minimum matrix method, or the Vogel approximation method), in the general case an initial basic feasible solution may be difficult to find. The difficulty arises from the fact that the upper and lower bounds on the variables are treated implicitly and, hence, nonbasic variables may be at either bound. We will come back to this question later. For the moment, assume that we have been given the initial basic feasible solution shown in Fig. 8.18.

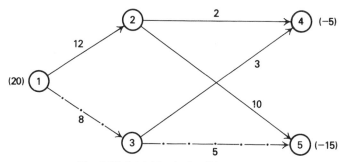

Fig. 8.18 Initial basic feasible solution.

The dash–dot arcs 1–3 and 3–5 indicate nonbasic variables at their upper bounds of 8 and 5, respectively. The arcs not shown are nonbasic at their lower bounds of zero. The solid arcs form a spanning tree for the network and constitute a basis for the problem.

To determine whether this initial basic feasible solution is optimal, we must compute the reduced costs of all nonbasic arcs. To do this, we first determine multipliers y_i ($i = 1, 2, \ldots, n$) and, if these multipliers satisfy:

$$\bar{c}_{ij} = c_{ij} - y_i + y_j \geq 0 \qquad \text{if } x_{ij} = \ell_{ij},$$

$$\bar{c}_{ij} = c_{ij} - y_i + y_j = 0 \qquad \text{if } \ell_{ij} < x_{ij} < u_{ij},$$

$$\bar{c}_{ij} = c_{ij} - y_i + y_j \leq 0 \qquad \text{if } x_{ij} = u_{ij},$$

then we have an optimal solution. Since the network-flow problem contains a redundant constraint, any one multiplier may be chosen artibrarily, as was indicated in previous sections. Suppose $y_2 = 0$ is set arbitrarily; the remaining multipliers are determined from the equations:

$$c_{ij} - y_i + y_j = 0$$

for basic variables. The resulting multipliers for the initial basis are given as node labels in Fig. 8.19. These were determined from the cost data given in Fig. 8.17.

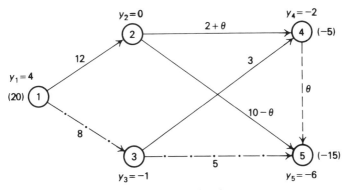

Fig. 8.19 Iteration 1.

Given these multipliers, we can compute the reduced costs of the nonbasis variables by $\bar{c}_{ij} = c_{ij} - y_i + y_j$. The reduced costs are determined by using the given cost data in Fig. 18.17 as:

$$\bar{c}_{13} = 4 - 4 + (-1) \qquad = -1,$$
$$\bar{c}_{23} = 2 - 0 + (-1) \qquad = 1,$$
$$\bar{c}_{35} = 3 - (-1) + (-6) = -2,$$
$$\bar{c}_{45} = 2 - (-2) + (-6) = -2,$$
$$\bar{c}_{53} = 1 - (-6) + (-1) = \quad 6.$$

In the simplex method with bounded variables, the nonbasic variables are at either their upper or lower bounds. An improved solution can be found by either:

1. increasing a variable that has a negative reduced cost and is currently at its lower bound; or

2. decreasing a variable that has a positive reduced cost and is currently at its upper bound.

In this case, the only promising candidate is x_{45}, since the other two negative reduced costs correspond to nonbasic variables at their upper bounds. In Fig. 8.19 we have added the arc 4–5 to the network, forming the unique loop 4–5–2–4 with the basic variables. If the flow on arc 4–5 is increased by θ, the remaining arcs in the loop must be appropriately adjusted. The limit on how far we can increase θ is given by arc 2–4, which has an upper bound of 4. Hence, $\theta = 2$ and x_{24} becomes nonbasic at its upper bound. The corresponding basic feasible solution is given in Fig. 8.20, ignoring the dashed arc 2–3.

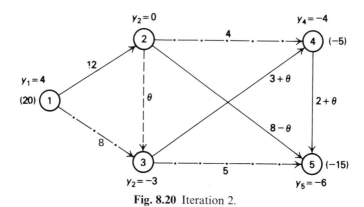

Fig. 8.20 Iteration 2.

The new multipliers are computed as before and are indicated as node labels in Fig. 8.20. Note that not all of the multipliers have to be recalculated. Those multipliers corresponding to nodes that are connected to node 2 by the same sequence of arcs as before will not change labels. The reduced costs for the new basis are then:

$$\bar{c}_{13} = 4 - 4 + (-3) \quad = -3,$$
$$\bar{c}_{23} = 2 - 0 + (-3) \quad = -1,\leftarrow$$
$$\bar{c}_{24} = 2 - 0 + (-4) \quad = -2,$$
$$\bar{c}_{35} = 3 - (-3) + (-6) = \quad 0,$$
$$\bar{c}_{53} = 1 - (-6) + (-3) = \quad 4.$$

Again there is only one promising candidate, x_{23}, since the other two negative reduced costs correspond to nonbasic variables at their upper bounds. In Fig. 8.20 we have added the arc 2–3 to the network, forming the unique loop 2–3–4–5–2 with the basic variables. If we increase the flow on arc 2–3 by θ and adjust the flows on the remaining arcs in the loop to maintain feasibility, the increase in θ is limited by arc 2–5. When $\theta = 8$, the flow on arc 2–5 is reduced to zero and x_{25} becomes nonbasic at its lower bound. Figure 8.21 shows the corresponding basic feasible solution.

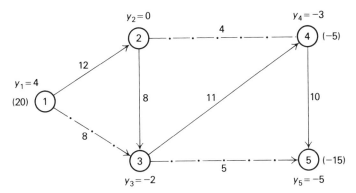

Fig. 8.21 Optimal solution.

The new multipliers are computed as before and are indicated as node labels in Fig. 8.21. The reduced costs for the new basis are then:

$$\bar{c}_{13} = 4 - 4 + (-2) \quad\;\; = -2,$$
$$\bar{c}_{24} = 2 - 0 + (-3) \quad\;\; = -1,$$
$$\bar{c}_{25} = 6 - 0 + (-5) \quad\;\; = \;\;\; 1,$$
$$\bar{c}_{35} = 3 - (-2) + (-5) = \;\;\; 0,$$
$$\bar{c}_{53} = 1 - (-5) + (-2) = \;\;\; 4.$$

This is an optimal solution, since all negative reduced costs correspond to nonbasic variables at their upper bounds and all positive reduced costs correspond to nonbasic variables at their lower bounds.

The reduced cost $\bar{c}_{35} = 0$ indicates that alternative optimal solutions may exist. In fact, it is easily verified that the solution given in Fig. 8.22 is an alternative optimal solution.

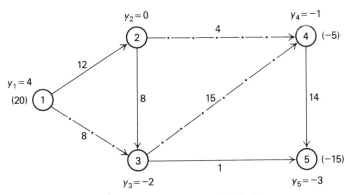

Fig. 8.22 Alternative optimal solution.

Hence, given an initial basic feasible solution, it is straightforward to use the concepts developed in the previous section to apply the simplex method to the general minimum-cost flow problem. There are two essential points to recognize: (1) a basis for the simplex method corresponds to a spanning tree for the network, and (2) introducing a new variable into the basis forms a unique loop in the spanning tree, and the variable that drops from the basis is the limiting variable in this loop.

Finally, we must briefly discuss how to find an initial basic feasible solution if one is not readily available. There are a number of good heuristics for doing this, depending on the particular problem, but almost all of these procedures involve adding some artificial arcs at some point. It is always possible to add uncapacitated arcs from the points of supply to the points of demand in such a fashion that a basis is formed. In our illustrative example, we could have simply added the artificial arcs 1–4 and 1–5 and had the initial basis given in Fig. 8.23.

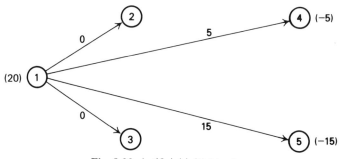

Fig. 8.23 Artificial initial basis.

Then either a phase I procedure is performed, minimizing the sum of the flows on the artificial arcs, or a very high cost is attached to the artificial arcs to drive them out of the basis. Any heuristic procedure for finding an initial basis usually attempts to keep the number of artificial arcs that have to be added as small as possible.

EXERCISES

1. A gas company owns a pipeline network, sections of which are used to pump natural gas from its main field to its distribution center. The network is shown below, where the direction of the arrows indicates the only direction in which the gas can be pumped. The pipeline links of the system are numbered one through six, and the intermediate nodes are large pumping stations. At the present time, the company nets 1200 mcf (million cubic feet) of gas per month from its main field and must transport that entire amount to the distribution center. The following are the maximum usage rates and costs associated with each link:

	1	*2*	*3*	*4*	*5*	*6*
Maximum usage: mcf/month	500	900	700	400	600	1000
Tariff: $/mcf	20	25	10	15	20	40

The gas company wants to find those usage rates that minimize total cost of transportation.

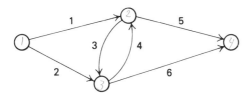

a) What are the decision variables?
b) Formulate the problem as a linear program.
c) For the optimal solution, do you expect the dual variable associated with the maximum usage of link 1 to be positive, zero, or negative and why?
d) Suppose there were maximum usage rates on the pumping stations; how would your formulation change?

2. On a particular day during the tourist season a rent-a-car company must supply cars to four destinations according to the following schedule:

Destination	Cars required
A	2
B	3
C	5
D	7

The company has three branches from which the cars may be supplied. On the day in question, the inventory status of each of the branches was as follows:

Branch	Cars available
1	6
2	1
3	10

The distances between branches and destinations are given by the following table:

	Destination			
Branch	A	B	C	D
1	7	11	3	2
2	1	6	0	1
3	9	15	8	5

Plan the day's activity such that supply requirements are met at a minimum cost (assumed proportional to car-miles travelled).

maximum profit!

3. The National Association of Securities Dealers Automated Quotation Systems (NASDAQ) is a network system in which quotation information in over-the-counter operations is collected. Users of the system can receive, in a matter of seconds, buy and sell prices and the exact bid and ask price of each market maker that deals in a particular security. There are 1700 terminals in 1000 locations in almost 400 cities. The central processing center is in Trumbull, Conn., with concentration facilities in New York, Atlanta, Chicago, and San Francisco. On this particular day, the market is quiet, so there are only a few terminals being used. The information they have has to be sent to one of the main processing facilities. The following table gives terminals (supply centers), processing facilities (demand centers), and the time that it takes to transfer a message.

Terminals	Trumbull	N.Y.	Atlanta	Chicago	San Fran.	Supply
Cleveland	6	6	9	4	10	45
Boston	3	2	7	5	12	90
Houston	8	7	5	6	4	95
Los Angeles	11	12	9	5	2	75
Washington, D.C.	4	3	4	5	11	105
Demand	120	80	50	75	85	

a) Solve, using the minimum matrix method to find an initial feasible solution.
b) Are there alternative optimal solutions?

4. A large retail sporting-goods chain desires to purchase 300, 200, 150, 500, and 400 tennis racquets of five different types. Inquiries are received from four manufacturers who will supply not more than the following quantities (all five types of racquets combined).

$$
\begin{array}{ll}
\text{M1} & 600 \\
\text{M2} & 500 \\
\text{M3} & 300 \\
\text{M4} & 400 \\
\end{array}
$$

The store estimates that its profit per racquet will vary with the manufacturer as shown below:

	Racquets				
Manufacturer	R1	R2	R3	R4	R5
M1	5.50	7.00	8.50	4.50	3.00
M2	6.00	6.50	9.00	3.50	2.00
M3	5.00	7.00	9.50	4.00	2.50
M4	6.50	5.50	8.00	5.00	3.50

How should the orders be placed?

5. A construction project involves 13 tasks; the tasks, their estimated duration, and their immediate predecessors are shown in the table below:

Task	Immediate predecessors	Duration
Task 1	—	1
Task 2	1	2
Task 3	—	3
Task 4	—	4
Task 5	1	2
Task 6	2, 3	1
Task 7	4	2
Task 8	5	6
Task 9	5	10
Task 10	6, 7	5
Task 11	8, 10	3
Task 12	8, 10	3
Task 13	12	2

Our objective is to find the schedule of tasks that minimizes the total elapsed time of the project.

a) Draw the event- and task-oriented networks for this problem and formulate the corresponding linear program.
b) Solve to find the critical path.

6. The Egserk Catering Company manages a moderate-sized luncheon cafeteria featuring prompt service, delectable cuisine, and luxurious surroundings. The desired atmosphere requires fresh linen napkins, which must be available at the start of each day. Normal laundry takes one full day at 1.5 cents per napkin; rapid laundry can be performed overnight but costs 2.5 cents a napkin. Under usual usage rates, the current napkin supply of 350 is adequate to permit complete dependence upon the normal laundry; however, the additional usage resulting from a three-day seminar to begin tomorrow poses a problem. It is known that the napkin requirements for the next three days will be 300, 325, and 275, in that order. It is now midafternoon and there are 175 fresh napkins, and 175 soiled napkins ready to be sent to the laundry. It is against the health code to have dirty napkins linger overnight. The cafeteria will be closed the day after the seminar and, as a result, all soiled napkins on the third day can be sent to normal laundry and be ready for the next business day.

The caterer wants to plan for the napkin laundering so as to minimize total cost, subject to meeting all his fresh napkin requirements and complying with the health code.

a) What are the decision variables?
b) Formulate the problem as a linear program.
c) Interpret the resulting model as a network-flow problem. Draw the corresponding network diagram.
d) For the optimal solution, do you expect the dual variable associated with tomorrow's requirement of 300 to be positive, zero, or negative, and why?
e) Suppose you could hold over dirty napkins at no charge; how would your formulation change?

7. An automobile association is organizing a series of car races that will last for four days. The organizers know that $r_j \geq 0$ special tires in good condition will be required on each of the four successive days, $j = 1, 2, 3, 4$. They can meet these needs either by buying new tires at P dollars apiece or by reshaping used tires (reshaping is a technique by which the grooves on the tire are deepened, using a special profile-shaped tool). Two kinds of service are available for reshaping: normal service, which takes one full day at N dollars a tire, and quick service, which takes overnight at Q dollars a tire. How should the association, which starts out with no special tires, meet the daily requirements at minimal cost?

   a) Formulate a mathematical model for the above problem. Does it exhibit the characteristics of a network problem? Why? (*Hint.* Take into account the fact that, at the end of day j, some used tires may not be sent to reshaping.)
   b) If the answer to (a) is *no*, how can the formulation be manipulated to become a network problem? Draw the associated network. (*Hint.* Add a redundant constraint introducing a fictitious node.)
   c) Assume that a tire may be reshaped only once. How does the above model change? Will it still be a network problem?

8. Conway Tractor Company has three plants located in Chicago, Austin (Texas), and Salem (Oregon). Three customers located respectively in Tucson (Arizona), Sacramento (California), and Charlestown (West Virginia) have placed additional orders with Conway Tractor Company for 10, 8, and 10 tractors, respectively. It is customary for Conway Tractor Company to quote to customers a price on a *delivered* basis, and hence the company absorbs the delivery costs of the tractors. The manufacturing cost does not differ significantly from one plant to another, and the following tableau shows the delivery costs incurred by the firm.

Destination

Plant	Tucson	Sacramento	Charlestown
Chicago	150	200	70
Austin	70	120	80
Salem	80	50	170

   The firm is now facing the problem of assigning the extra orders to its plants to minimize delivery costs and to meet all orders (the Company, over the years, has established a policy of first-class service, and this includes quick and reliable delivery of all goods ordered). In making the assignment, the company has to take into account the limited additional manufacturing capacity at its plants in Austin and Salem, of 8 and 10 tractors, respectively. There are no limits on the additional production capacity at Chicago (as far as these extra orders are concerned).

   a) Formulate as a transportation problem.
   b) Solve completely.

9. A manufacturer of electronic calculators produces its goods in Dallas, Chicago, and Boston, and maintains regional warehousing distribution centers in Philadelphia, Atlanta, Cleveland, and Washington, D.C. The company's staff has determined that shipping costs are directly proportional to the distances from factory to storage center, as listed here.

Warehouses

Mileage from:	Philadelphia	Atlanta	Cleveland	Washington
Boston	300	1000	500	400
Chicago	500	900	300	600
Dallas	1300	1000	1100	1200

The cost per calculator-mile is $0.002 and supplies and demands are:

	Supply		Demand	
Boston	1500	Philadelphia	2000
Chicago	2500	Atlanta	1600
Dallas	4000	Cleveland	1200
		Washington	3200

a) Use the Vogel approximation method to arrive at an initial feasible solution.
b) Show that the feasible solution determined in (a) is optimal.
c) Why does the Vogel approximation method perform so well, compared to other methods of finding an initial feasible solution?

10. Colonel Cutlass, having just taken command of the brigade, has decided to assign men to his staff based on previous experience. His list of major staff positions to be filled is adjutant (personnel officer), intelligence officer, operations officer, supply officer, and training officer. He has five men he feels could occupy these five positions. Below are their years of experience in the several fields.

	Adjutant	Intelligence	Operations	Supply	Training
Major Muddle	3	5	6	2	2
Major Whiteside	2	3	5	3	2
Captain Kid	3	—	4	2	2
Captain Klutch	3	—	3	2	2
Lt. Whiz	—	3	—	1	—

Who, based on experience, should be placed in which positions to give the greatest total years of experience for all jobs? (*Hint.* A basis, even if degenerate, is a spanning tree.)

11. Consider the following linear program:

$$\text{Minimize } z = 3x_{12} + 2x_{13} + 5x_{14} + 2x_{41} + x_{23} + 2x_{24} + 6x_{42} + 4x_{34} + 4x_{43},$$

subject to:

$$
\begin{aligned}
x_{12} + x_{13} + x_{14} - x_{41} &\leq 8, \\
x_{12} - x_{23} - x_{24} + x_{42} &\geq 4, \\
x_{34} - x_{13} - x_{23} - x_{43} &\leq 4, \\
x_{14} + x_{34} + x_{24} - x_{42} - x_{42} - x_{43} &\geq 5, \\
\text{all } x_{ij} &\geq 0.
\end{aligned}
$$

a) Show that this is a network problem, stating it in general minimum-cost flow form. Draw the associated network and give an interpretation to the flow in this network.

b) Find an initial feasible solution. (*Hint.* Exploit the triangular property of the basis.)
c) Show that your initial solution is a spanning tree.
d) Solve completely.

12. A lot of three identical items is to be sequenced through three machines. Each item must be processed first on machine 1, then on machine 2, and finally on machine 3. It takes 20 minutes to process one item on machine 1, 12 minutes on machine 2, and 25 minutes on machine 3. The objective is to minimize the total work span to complete all the items.

a) Write a linear program to achieve our objective. (*Hint.* Let x_{ij} be the starting time of processing item i on machine j. Two items may not occupy the same machine at the same time; also, an item may be processed on machine ($j + 1$) only after it has been completed on machine j.)

b) Cast the model above as a network problem. Draw the associated network and give an interpretation in terms of flow in networks. (*Hint.* Formulate and interpret the dual problem of the linear program obtained in (a).)

c) Find an initial feasible solution; solve completely.

13. A manufacturer of small electronic calculators is working on setting up his production plans for the next six months. One product is particularly puzzling to him. The orders on hand for the coming season are:

Month	Orders
January	100
February	150
March	200
April	100
May	200
June	150

The product will be discontinued after satisfying the June demand. Therefore, there is no need to keep any inventory after June. The production cost, using regular manpower, is $10 per unit. Producing the calculator on overtime costs an additional $2 per unit. The inventory-carrying cost is $0.50 per unit per month. If the regular shift production is limited to 100 units per month and overtime production is limited to an additional 75 units per month, what is the optimal production schedule? (*Hint.* Treat regular and overtime capacities as sources of supply for each month.)

14. Ships are available at three ports of origin and need to be sent to four ports of destination. The number of ships available at each origin, the number required at each destination, and the sailing times are given in the tableau below. Our objective is to minimize the total number of sailing days.

Origin \ Destination	1	2	3	4	Number of ships available
1	5	4	3	2	5
2	10	8	4	7	5
3	9	9	8	4	5
Number of ships required	1	4	4	6	15

a) Find an initial basic feasible solution.

b) Show that your initial basis is a spanning tree.

c) Find an initial basic feasible solution using the other two methods presented in the text. Solve completely, starting from the three initial solutions found in parts (a) and (c). Compare how close these solutions were to the optimal one.

d) Which of the dual variables may be chosen arbitrarily, and why?

e) Give an economic interpretation of the optimal simplex multipliers associated with the origins and destinations.

15. A distributing company has two major customers and three supply sources. The corresponding unit from each supply center to each customer is given in the following table, together with the total customer requirements and supply availabilities.

	Customer		Available
Supply center	1	2	supplies
1	− 1	3	300
2	1	6	400
3	1	5	900
Customer requirements	800	500	

Note that Customer 1 has strong preferences for Supplier 1 and will be willing not only to absorb all the transportation costs but also to pay a premium price of $1 per unit of product coming from Supplier 1.

a) The top management of the distributing company feels it is obvious that Supply Center 1 should send all its available products to Customer 1. Is this necessarily so? (*Hint*. Obtain the least-cost solution to the problem. Explore whether alternative optimal solutions exist where not all the 300 units available in Supply Center 1 are assigned to Customer 1.)

b) Assume Customer 2 is located in an area where all shipments will be subject to taxes defined as a percentage of the unit cost of a product. Will this tax affect the optimal solution of part (a)?

c) Ignore part (b). What will be the optimal solution to the original problem if Supply Center 1 increases its product availability from 300 units to 400 units?

16. After solving a transportation problem with positive shipping costs c_{ij} along all arcs, we increase the supply at one source and the requirement at one destination in a manner that will maintain equality of total supply and total demand.

a) Would you expect the shipping cost in the modified problem with a larger total shipment of goods to be higher than the optimal shipping plan from the original problem?

b) Solve the following transportation problem:

	Unit shipping costs to destinations			
Source	D1	D2	D3	Supplies
S1	4	2	4	15
S2	12	8	4	15
Requirements	10	10	10	

c) Increase the supply at source S1 by 1 unit and the demand at demand center D3 by 1 unit, and re-solve the problem. Has the cost of the optimal shipping plan decreased? Explain this behavior.

17. Consider a very elementary transportation problem with only two origins and two destinations. The supplies, demands, and shipping costs per unit are given in the following tableau.

	D1	D2	Units supplied
S1	5	2	20
S2	8	4	80
Units demanded	50	50	

Since the total number of units supplied equals the total number of units demanded, the problem may be formulated with equality constraints. An optimal solution to the problem is:

$$x_{11} = 20, \quad x_{12} = 0, \quad x_{21} = 30, \quad x_{22} = 50;$$

and a corresponding set of shadow prices on the nodes is:

$$y_{s1} = 4, \quad y_{s2} = 0, \quad y_{d1} = 1, \quad y_{d2} = 4.$$

a) Why is the least expensive route not used?
b) Are the optimal values of the decision variables unique?
c) Are the optimal values of the shadow prices unique?
d) Determine the ranges on the righthand-side values, changed one at a time, for which the basis remains unchanged.
e) What happens when the ranges determined in (d) are exceeded by some small amount?

18. Consider a transportation problem identical to the one given in Exercise 17. One way the model may be formulated is as a linear program with inequality constraints. The formulation and solution are given below.

	x_{11}	x_{12}	x_{21}	x_{22}	Relation	RHS	
Supply 1	1	1	0	0	$\leq$	20	-3
Supply 2	0	0	1	1	$\leq$	80	0 ⎫ Shadow
Demand 1	1	0	1	0	$=$	50	8 ⎬ prices
Demand 2	0	1	0	1	$=$	50	4 ⎭
Costs	5	2	8	4	$=$	z (min)	
	20	0	30	50			

Solution

For this formulation of the model:

a) Are the optimal values of the shadow prices unique?

b) Determine the ranges on the righthand-side values, changed one at a time, for which the basis remains unchanged.

c) Reconcile the results of (b) with those obtained in Exercise 17.

19. Suppose that there are three suppliers S_1, S_2, and S_3 in a distribution system that can supply 5, 5, and 6 units, respectively, of the company's good. The distribution system contains five demand centers, that require 2, 2, 4, 4, and 3 units each of the good. The transportation costs, in dollars per unit, are as follows:

	D_1	D_2	D_3	D_4	D_5
S_1	2	1	2	3	3
S_2	2	2	2	1	-1
S_3	3	3	2	1	2

a) Compute an optimum shipping schedule. Is the optimal solution unique?

b) Find the range over which the cost of transportation from S_1 to D_3 can vary while maintaining the optimal basis found in part (a).

c) To investigate the sensitivity of the solution to this problem, we might consider what happens if the amount supplied from any *one* supplier and the amount demanded by any *one* demand center were both increased. Is it possible for the total shipping costs to decrease by increasing the supply and the demand for any particular choice of supply and demand centers? Establish a limit on these increases as specific pairs of supply and demand centers are selected.

d) A landslide has occurred on the route from S_2 to D_5. If you bribe the state highway crew with $10, they will clear the slide. If not, the route will remain closed. Should you pay the bribe?

20. An oil company has three oil fields and five refineries. The production and transportation costs from each oil field to each refinery, in dollars per barrel, are given in the table below:

Oil field	Refineries					Availability
	R1	R2	R3	R4	R5	
OF1	5	3	3	3	7	4
OF2	5	4	4	2	1	6
OF3	5	4	2	6	2	7
Requirements	2	2	3	4	4	

The corresponding production capacity of each field and requirements of each refinery, in *millions of barrels* per week, are also given.

a) What is the optimum weekly production program?

Parts (b), (c), and (d) are independent, each giving modifications to part (a).

b) Suppose that field OF1 has worked *under* capacity so far, and that its production increases by one unit (i.e., 1 million barrels). What is the new optimal production plan? Has the optimal basis changed? How does the objective function change? What is the range within which the production of field OF1 may vary?

c) Because of pipeline restrictions, it is impossible to send more than 1 million barrels from OF2 to R5. How would you have formulated the problem if it had been stated in this form from the very beginning? (*Hint.* Decompose R5 into two destinations: one with a requirement of one unit (i.e., 1 million barrels), the other with a requirement of three. Prohibit the route from OF2 to the second destination of R5.) Change the optimum solution in (a) to find the new optimum program.

d) Suppose that fields OF1, OF2, and OF3 have additional overtime capacities of 1, 1, and 1.5 units, respectively (that is, 1, 1, and 1.5 million barrels, respectively). This causes an increase in the corresponding production costs of 0.5, 1.5, and 2 dollars per barrel, respectively. Also assume that the refinery requirements are increased by one million barrels at each refinery and that there is no convenient route to ship oil from field OF2 to refinery R3. What is the optimum program?

21. Consider the following transshipment problem where goods are shipped from two plants to either a warehouse or two distribution centers and then on to the two end markets.

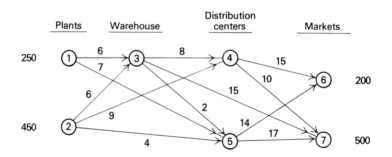

The production rates in units per month are 250 and 450 for plants 1 and 2, respectively. The demands of the two markets occur at rates 200 and 500 units per month. The costs of shipping one unit over a particular route is shown adjacent to the appropriate arc in the network.

a) Redraw the above transshipment network as a network consisting of only origins and destinations, by replacing all intermediate nodes by two nodes, an origin and destination, connected by an arc, from the destination back to the origin, whose cost is zero.

b) The network in (a) is a transportation problem except that a backwards arc, with flow x_{ii}, from newly created destination i to its corresponding newly created source, is required. Convert this to a transportation network by substituting $x'_{ii} = B - x_{ii}$. How do you choose a value for the constant B?

c) Certain arcs are inadmissible in the original transshipment formulation; how can these be handled in the reformulated transportation problem?

d) Interpret the linear-programming tableau of the original transshipment network and that of the reformulated transportation network.

e) Can any transshipment problem be transformed into an equivalent transportation problem?

22.) Consider the following minimum-cost flow model:

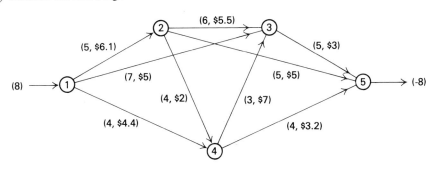

We wish to send eight units from node 1 to node 5 at minimum cost. The numbers next to the arcs indicate upper bounds for the flow on an arc and the cost per unit of flow. The following solution has been proposed, where the numbers next to the arcs are flows.

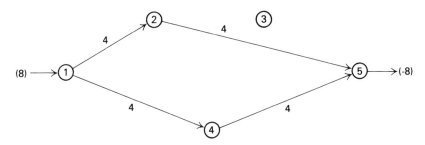

The total cost of the proposed solution is $66.8.

a) Is the proposed solution a feasible solution? Is it a basic feasible solution? Why?
b) How can the proposed solution be modified to constitute a basic feasible solution?
c) Determine multipliers on the nodes associated with the basic feasible solution given in (b). Are these multipliers unique?
d) Show that the basic feasible solution determined in (b) is not optimal.
e) What is the next basis suggested by the reduced costs? What are the *values* of the new basic variables? Nonbasic variables?

23.) For the minimum-cost flow model given in Exercise 22, suppose that the spanning tree indicated by the solid lines in the following network, along with the dash–dot arcs at their upper bounds, has been proposed as a solution:

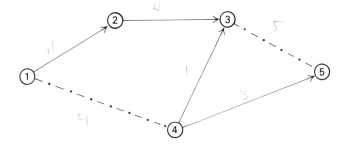

a) What are the flows on the arcs corresponding to this solution?
b) Determine a set of shadow prices for the nodes.
c) Show that this solution is optimal.
d) Is the optimal solution unique?

24. For the minimum-cost flow model given in Exercise 22, with the optimal solution *determined in Exercise 23*, answer the following questions.

   a) For each nonbasic variable, determine the range on its objective-function coefficient so that the current basis remains optimal.
   b) For each basic variable, determine the range on its objective-function coefficient so that the current basis remains optimal.
   c) Determine the range on each righthand-side value so that the basis remains unchanged.
   d) In question (c), the righthand-side ranges are all tight, in the sense that any change in one righthand-side value *by itself* will apparently change the basis. What is happening?

25. The following model represents a simple situation of buying and selling of a seasonal product for profit:

$$\text{Maximize } z = -\sum_{t=1}^{T} p_t x_t + \sum_{t=1}^{T} s_t y_t,$$

subject to:

$$I + \sum_{j=1}^{t} (x_j - y_j) \leq C \quad (t = 1, 2, \ldots, T),$$

$$y_1 \leq I,$$
$$y_2 \leq I + (x_1 - y_1),$$
$$y_3 \leq I + (x_1 - y_1) + (x_2 - y_2),$$
$$y_t \leq I + \sum_{j=1}^{t-1} (x_j - y_j) \quad (t = 4, 5, \ldots, T),$$

$$x_t \geq 0, \ y_t \geq 0 \quad \text{for } t = 1, 2, \ldots, T.$$

In this formulation, the decision variables x_t and y_t denote, respectively, the amount of the product purchased and sold, in time period t. The given data is:

$$p_t = \text{per unit purchase price in period } t,$$
$$s_t = \text{per unit selling price in period } t,$$
$$I = \text{amount of the product on hand initially,}$$
$$C = \text{capacity for storing the product in any period.}$$

The constraints state (i) that the amount of the product on hand at the end of any period cannot exceed the storage capacity C, and (ii) that the amount of the product sold at the beginning of period t cannot exceed its availability from previous periods.

a) Let

$$w_t = C - I - \sum_{j=1}^{t} (x_j - y_j) \quad \text{and} \quad z_t = I + \sum_{j=1}^{t-1} (x_j - y_j) - y_t,$$

denote slack variables for the constraints (with $z_1 = I - y_1$). Show that the given model is equivalent to the formulation on page 363.

$$\text{Maximize } z = -\sum_{t=1}^{T} p_t x_t + \sum_{t=1}^{T} s_t y_t,$$

subject to:

$$
\begin{array}{l}
x_1 \qquad\qquad - y_1 \qquad\qquad\qquad\qquad + w_1 \qquad\qquad\qquad\qquad = C - I, \\
\quad x_2 \qquad\qquad\quad - y_2 \qquad\qquad\quad - w_1 + w_2 \qquad\qquad\qquad = 0, \\
\qquad x_3 \qquad\qquad\quad - y_3 \qquad\qquad\quad - w_2 + w_3 \qquad\qquad = 0, \\
\qquad\qquad \cdots \qquad\qquad \cdots \qquad\qquad\qquad \cdots \qquad\qquad\qquad = 0, \\
\qquad\qquad\quad x_T \qquad\qquad\quad - y_T \qquad\qquad\quad - w_{T-1} + w_T \qquad = 0, \\[4pt]
-x_1 \qquad\qquad\quad y_1 + y_2 \qquad\qquad\qquad\qquad\qquad + z_1 \qquad\qquad = I, \\
\quad - x_2 \qquad\qquad\qquad\quad + y_3 \qquad\qquad\qquad\qquad - z_1 + z_2 \qquad = 0, \\
\qquad \cdots \qquad\qquad\qquad\qquad\qquad\qquad\qquad\qquad - z_2 + z_3 \qquad = 0, \\
\qquad - x_{T-1} \qquad\qquad\qquad + y_T \qquad\qquad\qquad \cdots \qquad\qquad = 0, \\
\qquad\qquad\qquad\qquad\qquad\qquad\qquad\qquad\qquad\qquad - z_{T-1} + z_T = 0,
\end{array}
$$

$$x_t \geqq 0, \qquad y_t \geqq 0, \qquad w_t \geqq 0, \qquad z_t \geqq 0 \qquad \text{for } t = 1, \ldots, T.$$

b) State the dual to the problem formulation in part (a), letting u_i denote the dual variable for the constraint $+x_i$, and v_j denote the dual variable for the constraint containing $+y_i$.

c) Into what class of problems does this model fall? Determine the nature of the solution for $T = 3$.

26. A set of words (for example, ace, bc, dab, dfg, fe) is to be transmitted as messages. We want to investigate the possibility of representing each word by one of the letters *in the word* such that the words will be represented uniquely. If such a representation is possible, we can transmit a single letter instead of a complete word for a message we want to send.

a) Using as few constraints and variables as possible, formulate the possibility of transmitting letters to represent words as the solution to a mathematical program. Is there anything special about the structure of this program that facilitates discovery of a solution?

b) Suppose that you have a computer code that computes the solution to the following transportation problem:

$$\text{Minimize } z = \sum_i \sum_j z_{ij} x_{ij},$$

subject to:

$$\sum_{j=1}^{n} x_{ij} = a_i \qquad (i = 1, 2, \ldots, n),$$

$$\sum_{i=1}^{m} x_{ij} = b_j \qquad (j = 1, 2, \ldots, m),$$

$$x_{ij} \geq 0 \qquad (i = 1, 2, \ldots, m; j = 1, 2, \ldots, n).$$

and requires that:

(i) $\qquad\qquad a_i, \quad b_j, \quad c_{ij} \geq 0 \qquad$ and integer $\qquad (i = 1, 2, \ldots, m; j = 1, 2, \ldots, n)$

(ii) $$\sum_{i=1}^{m} a_i = \sum_{j=1}^{n} b_j.$$

How would you use this code to solve the problem posed in (a)? Answer the question in general, and use the specific problem defined above as an illustration.

27. The Defense Communications Agency is responsible for operating and maintaining a world-wide communications system. It thinks of costs as being proportional to the "message units" transmitted in one direction over a particular link in the system. Hence, under normal operating conditions it faces the following minimum-cost flow problem:

$$\text{Minimize } z = \sum_i \sum_j c_{ij} x_{ij},$$

subject to:

$$\sum_j x_{ij} - \sum_k x_{ki} = b_i \qquad (i = 1, 2, \ldots, n),$$

$$0 \leq x_{ij} \leq u_{ij},$$

where

$$c_{ij} = \text{cost per message unit over link } (i\text{–}j).$$
$$b_i = \text{message units generated (or received) at station } i,$$
$$u_{ij} = \text{upper bound on number of message units that can be transmitted over link } (i\text{–}j).$$

Suppose that the agency has been given a budget of B dollars to spend on increasing the capacity of any link in the system. The price for increasing capacity on link $(i\text{–}j)$ is p_{ij}.

a) Formulate a linear program (not necessarily a network) with only $(n + 1)$ constraints, that will minimize operating costs subject to this budget constraint. (*Hint.* You must allow for investing in additional capacity for each link.)

b) How can the "near" network model formulated in (a) be analyzed by network-programming techniques? (*Hint.* How could parametric programming be used?)

28. The general minimum-cost flow problem is given as follows:

$$\text{Minimize } z = \sum_i \sum_j c_{ij} x_{ij},$$

subject to:

$$\sum_j x_{ij} - \sum_k x_{ki} = b_i \qquad (i = 1, 2, \ldots, n),$$

$$\ell_{ij} \le x_{ij} \le u_{ij}.$$

a) Assuming that the lower bounds on the variables are all finite, this is, $\ell_{ij} > -\infty$, show that any problem of this form can be converted to a transportation problem with lower bounds on the variables of zero and nonnegative, or infinite, upper bounds. (*Hint.* Refer to Exercise 22.)

b) Comment on the efficiency of solving minimum-cost flow problems by a bounded-variables transportation method, versus the simplex method for general networks.

29. One difficulty with solving the general minimum-cost flow problem with upper and lower bounds on the variables lies in determining an initial basic feasible solution. Show that an initial basic feasible solution to this problem can be determined by solving an appropriate maximum-flow problem. (*Hints.* (1) Make a variable substitution to eliminate the nonzero lower bounds on the variables. (2) Form a "super source," connected to all the source nodes, and a "super sink," connected to all the sink nodes, and maximize the flow from super sink to super source.)

ACKNOWLEDGMENTS

Exercises 1 and 6 are due to Sherwood C. Frey, Jr., of the Harvard Business School.
Exercise 6, in turn, is based on "The Caterer Problem," in *Flows in Networks* by L. R. Ford and D. R. Fulkerson, Princeton University Press, 1962. In their 1955 article "Generalizations of the Warehousing Model," from the *Operational Research Quarterly*, A. Charnes and W. W. Cooper introduced the transformation used in Exercise 25.

Integer Programming

9

The linear-programming models that have been discussed thus far all have been *continuous*, in the sense that decision variables are allowed to be fractional. Often this is a realistic assumption. For instance, we might easily produce $102\frac{3}{4}$ gallons of a divisible good such as wine. It also might be reasonable to accept a solution giving an hourly production of automobiles at $58\frac{1}{2}$ if the model were based upon average hourly production, and the production had the interpretation of *production rates*.

At other times, however, fractional solutions are not realistic, and we must consider the optimization problem:

$$\text{Maximize} \sum_{j=1}^{n} c_j x_j,$$

subject to:

$$\sum_{j=1}^{n} a_{ij} x_j = b_i \qquad (i = 1, 2, \ldots, m),$$

$$x_j \geqq 0 \qquad (j = 1, 2, \ldots, n),$$

$$x_j \text{ integer} \qquad (\text{for some or all } j = 1, 2, \ldots, n).$$

This problem is called the (linear) *integer-programming problem*. It is said to be a *mixed* integer program when some, but not all, variables are restricted to be integer, and is called a *pure* integer program when *all* decision variables must be integers. As we saw in the preceding chapter, if the constraints are of a network nature, then an integer solution can be obtained by ignoring the integrality restrictions and solving the resulting linear program. In general, though, variables will be fractional in the linear-programming solution, and further measures must be taken to determine the integer-programming solution.

The purpose of this chapter is twofold. First, we will discuss integer-programming formulations. This should provide insight into the scope of integer-programming applications and give some indication of why many practitioners feel that the integer-programming model is one of the most important models in management science. Second, we consider basic approaches that have been developed for solving integer and mixed-integer programming problems.

9.1 SOME INTEGER-PROGRAMMING MODELS

Integer-programming models arise in practically every area of application of mathematical programming. To develop a preliminary appreciation for the importance of these models, we introduce, in this section, three areas where integer programming has played an important role in supporting managerial decisions. We do not provide the most intricate available formulations in each case, but rather give basic models and suggest possible extensions.

Capital Budgeting In a typical capital-budgeting problem, decisions involve the selection of a number of potential investments. The investment decisions might be to choose among possible plant locations, to select a configuration of capital equipment, or to settle upon a set of research-and-development projects. Often it makes no sense to consider partial investments in these activities, and so the problem becomes a *go–no-go* integer program, where the decision variables are taken to be $x_j = 0$ or 1, indicating that the jth investment is rejected or accepted. Assuming that c_j is the contribution resulting from the jth investment and that a_{ij} is the amount of resource i, such as cash or manpower, used on the jth investment, we can state the problem formally as:

$$\text{Maximize} \sum_{j=1}^{n} c_j x_j,$$

subject to:

$$\sum_{j=1}^{n} a_{ij} x_j \leqq b_i \qquad (i = 1, 2, \ldots, m),$$

$$x_j = 0 \quad \text{or} \quad 1 \qquad (j = 1, 2, \ldots, n).$$

The objective is to maximize total contribution from all investments without exceeding the limited availability b_i of any resource.

One important special scenario for the capital-budgeting problem involves cash-flow constraints. In this case, the constraints

$$\sum_{j=1}^{n} a_{ij} x_i \leqq b_i$$

reflect incremental cash balance in each period. The coefficients a_{ij} represent the net cash flow from investment j in period i. If the investment requires additional cash in period i, then $a_{ij} > 0$, while if the investment *generates* cash in period i, then $a_{ij} < 0$. The righthand-side coefficients b_i represent the incremental exogenous cash flows. If additional funds are made available in period i, then $b_i > 0$, while if funds are withdrawn in period i, then $b_i < 0$. These constraints state that the funds required for investment must be less than or equal to the funds generated from prior investments plus exogenous funds made available (or *minus* exogenous funds withdrawn).

The capital-budgeting model can be made much richer by including logical considerations. Suppose, for example, that investment in a new product line is contingent upon previous investment in a new plant. This *contingency* is modeled simply by the constraint

$$x_j \geqq x_i,$$

which states that if $x_i = 1$ and project i (new product development) is accepted, then necessarily $x_j = 1$ and project j (construction of a new plant) must be accepted. Another example of this nature concerns conflicting projects. The constraint

$$x_1 + x_2 + x_3 + x_4 \leqq 1,$$

for example, states that only one of the first four investments can be accepted. Constraints like this commonly are called *multiple-choice constraints*. By combining these logical constraints, the model can incorporate many complex interactions between projects, in addition to issues of resource allocation.

The simplest of all capital-budgeting models has just one resource constraint, but has attracted much attention in the management-science literature. It is stated as:

$$\text{Maximize} \sum_{j=1}^{n} c_j x_j,$$

subject to:

$$\sum_{j=1}^{n} a_j x_j \leqq b,$$

$$x_j = 0 \quad \text{or} \quad 1 \quad (j = 1, 2, \ldots, n).$$

Usually, this problem is called the 0–1 *knapsack* problem, since it is analogous to a situation in which a hiker must decide which goods to include on his trip. Here c_j is the "value" or utility of including good j, which weighs $a_j > 0$ pounds; the objective is to maximize the "pleasure of the trip," subject to the weight limitation that the hiker can carry no more than b pounds. The model is altered somewhat by allowing more than one unit of any good to be taken, by writing $x_j \geqq 0$ and x_j-integer in place of the 0–1 restrictions on the variables. The knapsack model is important because a number of integer programs can be shown to be equivalent to it, and further, because solution procedures for knapsack models have motivated procedures for solving general integer programs.

Warehouse Location   In modeling distribution systems, decisions must be made about tradeoffs between transportation costs and costs for operating distribution centers. As an example, suppose that a manager must decide which of n warehouses to use for meeting the demands of m customers for a good. The decisions to be made are which warehouses to operate and how much to ship from any warehouse to any customer. Let

$$y_i = \begin{cases} 1 & \text{if warehouse } i \text{ is opened,} \\ 0 & \text{if warehouse } i \text{ is not opened;} \end{cases}$$

$$x_{ij} = \text{Amount to be sent from warehouse } i \text{ to customer } j.$$

The relevant costs are:

$f_i = $ Fixed operating cost for warehouse i, if opened (for example, a cost to lease the warehouse),

$c_{ij} = $ Per-unit operating cost at warehouse i plus the transportation cost for shipping from warehouse i to customer j.

There are two types of constraints for the model:

 i) the demand d_j of each customer must be filled from the warehouses; and

 ii) goods can be shipped from a warehouse only if it is opened.

The model is:

$$\text{Minimize} \quad \sum_{i=1}^{m} \sum_{j=1}^{n} c_{ij} x_{ij} + \sum_{i=1}^{m} f_i y_i, \tag{1}$$

subject to:

$$\sum_{i=1}^{m} x_{ij} = d_j \qquad (j = 1, 2, \ldots, n). \tag{2}$$

$$\sum_{j=1}^{n} x_{ij} - y_i \left(\sum_{j=1}^{n} d_j \right) \leq 0 \qquad (i = 1, 2, \ldots, m), \tag{3}$$

$$x_{ij} \geq 0 \qquad (i = 1, 2, \ldots, m; j = 1, 2, \ldots, n),$$

$$y_i = 0 \quad \text{or} \quad 1 \qquad (i = 1, 2, \ldots, m),$$

The objective function incorporates transportation and variable warehousing costs, in addition to fixed costs for operating warehouses. The constraints (2) indicate that each customer's demand must be met. The summation over the shipment variables x_{ij} in the ith constraint of (3) is the amount of the good shipped from warehouse i. When the warehouse is not opened, $y_i = 0$ and the constraint specifies that nothing can be shipped from the warehouse. On the other hand, when the warehouse is opened and $y_i = 1$, the constraint simply states that the amount to be shipped from warehouse i can be no larger than the total demand, which is always true. Consequently, constraints (3) imply restriction (ii) as proposed above.

Although oversimplified, this model forms the core for sophisticated and realistic distribution models incorporating such features as:

1. multi-echelon distribution systems from plant to warehouse to customer;

2. capacity constraints on both plant production and warehouse throughput;

3. economies of scale in transportation and operating costs;

4. service considerations such as maximum distribution time from warehouses to customers;

5. multiple products; or

6. conditions preventing splitting of orders (in the model above, the demand for any customer can be supplied from several warehouses).

These features can be included in the model by changing it in several ways. For example, warehouse capacities are incorporated by replacing the term involving y_i in constraint (3) with $y_i K_i$, where K_i is the throughput capacity of warehouse i; multi-echelon distribution may require triple-subscripted variables x_{ijk} denoting the amount to be shipped, from plant i to customer k through warehouse j. Further examples of how the simple warehousing model described here can be modified to incorporate the remaining features mentioned in this list are given in the exercises at the end of the chapter.

Scheduling   The entire class of problems referred to as sequencing, scheduling, and routing are inherently integer programs. Consider, for example, the scheduling of students, faculty, and classrooms in such a way that the number of students who cannot take their first choice of classes is minimized. There are constraints on the number and size of classrooms available at any one time, the availability of faculty members at particular times, and the preferences of the students for particular schedules. Clearly, then, the ith student *is* scheduled for the jth class during the nth time period or *not*; hence, such a variable is either zero or one. Other examples of this class of problems include line-balancing, critical-path scheduling with resource constraints, and vehicle dispatching.

As a specific example, consider the scheduling of airline flight personnel. The airline has a number of routing "legs" to be flown, such as 10 A.M. New York to Chicago, or 6 P.M. Chicago to Los Angeles. The airline must schedule its personnel crews on routes to cover these flights. One crew, for example, might be scheduled to fly a route containing the two legs just mentioned. The decision variables, then, specify the scheduling of the crews to routes:

$$x_j = \begin{cases} 1 & \text{if a crew is assigned to route } j, \\ 0 & \text{otherwise.} \end{cases}$$

Let

$$a_{ij} = \begin{cases} 1 & \text{if leg } i \text{ is included on route } j, \\ 0 & \text{otherwise,} \end{cases}$$

and

$$c_j = \text{Cost for assigning a crew to route } j.$$

The coefficients a_{ij} define the acceptable combinations of legs and routes, taking into account such characteristics as sequencing of legs for making connections between flights and for including in the routes ground time for maintenance. The model becomes:

$$\text{Minimize } \sum_{j=1}^{n} c_j x_j,$$

subject to:

$$\sum_{j=1}^{n} a_{ij} x_j = 1 \qquad (i = 1, 2, \ldots, m), \tag{4}$$

$$x_j = 0 \quad \text{or} \quad 1 \qquad (j = 1, 2, \ldots, n).$$

The ith constraint requires that one crew must be assigned on a route to fly leg i. An alternative formulation permits a crew to ride as passengers on a leg. Then the constraints (4) become:

$$\sum_{j=1}^{n} a_{ij} x_j \geqq 1 \qquad (i = 1, 2, \ldots, m). \tag{5}$$

If, for example,

$$\sum_{j=1}^{n} a_{1j} x_j = 3,$$

then two crews fly as passengers on leg 1, possibly to make connections to other legs to which they have been assigned for duty.

These airline-crew scheduling models arise in many other settings, such as vehicle delivery problems, political districting, and computer data processing. Often model (4) is called a *set-partitioning problem*, since the set of legs will be divided, or partitioned, among the various crews. With constraints (5), it is called a *set-covering problem*, since the crews then will cover the set of legs.

Another scheduling example is the so-called *traveling salesman problem*. Starting from his home, a salesman wishes to visit each of $(n - 1)$ other cities and return home at minimal cost. He must visit each city exactly once and it costs c_{ij} to travel from city i to city j. What route should he select? If we let

$$x_{ij} = \begin{cases} 1 & \text{if he goes from city } i \text{ to city } j, \\ 0 & \text{otherwise,} \end{cases}$$

we may be tempted to formulate his problem as the assignment problem:

$$\text{Minimize} \sum_{i=1}^{n} \sum_{j=1}^{n} c_{ij} x_{ij},$$

subject to:

$$\sum_{i=1}^{n} x_{ij} = 1 \qquad (j = 1, 2, \ldots, n),$$

$$\sum_{j=1}^{n} x_{ij} = 1 \qquad (i = 1, 2, \ldots, n),$$

$$x_{ij} \geq 0 \qquad (i = 1, 2, \ldots, n; j = 1, 2, \ldots, n).$$

The constraints require that the salesman must enter and leave each city exactly once. Unfortunately, the assignment model can lead to infeasible solutions. It is possible in a six-city problem, for example, for the assignment solution to route the salesman through two disjoint subtours of the cities instead of on a single trip or tour. (See Fig. 9.1.)

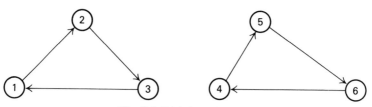

Fig. 9.1 Disjoint subtours.

Consequently, additional constraints must be included in order to eliminate subtour solutions. There are a number of ways to accomplish this. In this example, we can avoid the subtour solution of Fig. 9.1 by including the constraint:

$$x_{14} + x_{15} + x_{16} + x_{24} + x_{25} + x_{26} + x_{34} + x_{35} + x_{36} \geq 1.$$

This inequality ensures that at least one leg of the tour connects cities 1, 2, and 3 with cities 4, 5, and 6. In general, if a constraint of this form is included for each way in which the cities can be divided into two groups, then subtours will be eliminated. The problem with this and related approaches is that, with n cities, $(2^n - 1)$ constraints of this nature must be added, so that the formulation becomes a very large integer-programming problem. For this reason the traveling salesman problem generally is regarded as difficult when there are many cities.

The traveling salesman model is used as a central component of many vehicular routing and scheduling models. It also arises in production scheduling. For example, suppose that we wish to sequence $(n - 1)$ jobs on a single machine, and that c_{ij} is the cost for setting up the machine for job j, given that job i has just been completed. What scheduling sequence for the jobs gives the lowest total setup costs? The problem can be interpreted as a traveling salesman problem, in which the "salesman" corresponds to the machine which must "visit" or perform each of the jobs. "Home" is the initial setup of the machine, and, in some applications, the machine will have to be returned to this initial setup after completing all of the jobs. That is, the "salesman" must return to "home" after visiting the "cities."

9.2 FORMULATING INTEGER PROGRAMS

The illustrations in the previous section not only have indicated specific integer-programming applications, but also have suggested how integer variables can be used to provide broad modeling capabilities beyond those available in linear programming. In many applications, integrality restrictions reflect natural indivisibilities of the problem under study. For example, when deciding how many nuclear aircraft carriers to have in the U.S. Navy, fractional solutions clearly are meaningless, since the optimal number is on the order of one or two. In these situations, the decision variables are inherently integral by the nature of the decision-making problem.

This is not necessarily the case in every integer-programming application, as illustrated by the capital-budgeting and the warehouse-location models from the last section. In these models, integer variables arise from (i) logical conditions, such as *if* a new product is developed, *then* a new plant must be constructed, and from (ii) non-linearities such as fixed costs for opening a warehouse. Considerations of this nature are so important for modeling that we devote this section to analyzing and consolidating specific integer-programming formulation techniques, which can be used as tools for a broad range of applications.

Binary (0–1) Variables

Suppose that we are to determine whether or not to engage in the following activities: (i) to build a new plant, (ii) to undertake an advertising campaign, or (iii) to develop a new product. In each case, we must make a *yes–no* or so-called *go–no–go* decision. These choices are modeled easily by letting $x_j = 1$ if we engage in the jth activity and $x_j = 0$ otherwise. Variables that are restricted to 0 or 1 in this way are termed *binary, bivalent, logical,* or *0–1 variables*. Binary variables are of great importance because they occur regularly in many model formulations, particularly in problems addressing long-range and high-cost strategic decisions associated with capital-investment planning.

If, further, management had decided that *at most one* of the above three activities can be pursued, the following constraint is appropriate:

$$\sum_{j=1}^{3} x_j \leq 1.$$

As we have indicated in the capital-budgeting example in the previous section, this restriction usually is referred to as a *multiple-choice* constraint, since it limits our choice of investments to be *at most one* of the three available alternatives.

Binary variables are useful whenever variables can assume one of two values, as in batch processing. For example, suppose that a drug manufacturer must decide whether or not to use a fermentation tank. If he uses the tank, the processing technology requires that he make B units. Thus, his production y must be 0 or B, and the problem can be modeled with the binary variable $x_j = 0$ or 1 by substituting Bx_j for y everywhere in the model.

Logical Constraints

Frequently, problem settings impose logical constraints on the decision variables (like timing restrictions, contingencies, or conflicting alternatives), which lend themselves to integer-programming formulations. The following discussion reviews the most important instances of these logical relationships.

Constraint Feasibility

Possibly the simplest logical question that can be asked in mathematical programming is whether a given choice of the decision variables satisfies a constraint. More precisely, *when* is the general constraint

$$f(x_1, x_2, \ldots, x_n) \leq b \tag{6}$$

satisfied?

We introduce a binary variable y with the interpretation:

$$y = \begin{cases} 0 & \text{if the constraint is known to be satisfied,} \\ 1 & \text{otherwise,} \end{cases}$$

and write

$$f(x_1, x_2, \ldots, x_n) - By \leq b, \tag{7}$$

where the constant B is chosen to be large enough so that the constraint always is satisfied if $y = 1$; that is,

$$f(x_1, x_2, \ldots, x_n) \leq b + B,$$

for every possible choice of the decision variables $x_1, x_2, \ldots, x_n$ at our disposal. Whenever $y = 0$ gives a feasible solution to constraint (7), we know that constraint (6) must be satisfied. In practice, it is usually very easy to determine a large number to serve as B, although generally it is best to use the smallest possible value of B in order to avoid numerical difficulties during computations.

Alternative Constraints

Consider a situation with the *alternative* constraints:

$$f_1(x_1, x_2, \ldots, x_n) \leq b_1,$$
$$f_2(x_1, x_2, \ldots, x_n) \leq b_2.$$

At least one, but not necessarily both, of these constraints must be satisfied. This restriction can be modeled by combining the technique just introduced with a multiple-choice constraint as follows:

$$f_1(x_1, x_2, \ldots, x_n) - B_1 y_1 \leq b_1,$$
$$f_2(x_1, x_2, \ldots, x_n) - B_2 y_2 \leq b_2,$$
$$y_1 + y_2 \leq 1,$$
$$y_1, y_2 \text{ binary.}$$

The variables y_1 and y_2 and constants B_1 and B_2 are chosen as above to indicate when the constraints are satisfied. The multiple-choice constraint $y_1 + y_2 \leq 1$ implies that at least one variable y_j equals 0, so that, as required, at least one constraint must be satisfied.

We can save one integer variable in this formulation by noting that the multiple-choice constraint can be replaced by $y_1 + y_2 = 1$, or $y_2 = 1 - y_1$, since this constraint also implies that either y_1 or y_2 equals 0. The resulting formulation is given by:

$$f_1(x_1, x_2, \ldots, x_n) - B_1 y_1 \qquad \leq b_1,$$
$$f_2(x_1, x_2, \ldots, x_n) - B_2(1 - y_1) \leq b_2,$$
$$y_1 = 0 \quad \text{or} \quad 1.$$

As an illustration of this technique, consider again the custom-molder example from Chapter 1. That example included the constraint

$$6x_1 + 5x_2 \leq 60, \tag{8}$$

which represented the production capacity for producing x_1 hundred cases of six-ounce glasses and x_2 hundred cases of ten-ounce glasses. Suppose that there were an alternative production process that could be used, having the capacity constraint

$$4x_1 + 5x_2 \leq 50. \tag{9}$$

Then the decision variables x_1 and x_2 must satisfy *either* (8) or (9), depending upon which production process is selected. The integer-programming formulation replaces (8) and (9) with the constraints:

$$6x_1 + 5x_2 - 100y \qquad \leq 60,$$
$$4x_1 + 5x_2 - 100(1 - y) \leq 50,$$
$$y = 0 \quad \text{or} \quad 1.$$

In this case, both B_1 and B_2 are set to 100, which is large enough so that the constraint is not limiting for the production process *not* used.

Conditional Constraints

These constraints have the form:

$$f_1(x_1, x_2, \ldots, x_n) > b_1 \qquad \text{implies that } f_2(x_1, x_2, \ldots, x_n) \leq b_2.$$

Since this implication is not satisfied *only when both* $f_1(x_1, x_2, \ldots, x_n) > b_1$ *and* $f_2(x_1, x_2, \ldots, x_n) > b_2$, the conditional constraint is logically equivalent to the alternative constraints

$$f_1(x_1, x_2, \ldots, x_n) \leq b_1 \qquad \text{and/or} \qquad f_2(x_1, x_2, \ldots, x_n) \leq b_2,$$

where *at least one* must be satisfied. Hence, this situation can be modeled by alternative constraints as indicated above.

k-Fold Alternatives

Suppose that we must satisfy at least k of the constraints:

$$f_j(x_1, x_2, \ldots, x_n) \leq b_j \qquad (j = 1, 2, \ldots, p).$$

For example, these restrictions may correspond to manpower constraints for p potential inspection systems for quality control in a production process. If management has decided to adopt at least k inspection systems, then the k constraints specifying the manpower restrictions for these systems must be satisfied, and the remaining constraints can be ignored. Assuming that B_j for $j = 1, 2, \ldots, p$, are chosen so that the ignored constraints will not be binding, the general problem can be formulated as follows:

$$f_j(x_1, x_2, \ldots, x_n) - B_j(1 - y_j) \leq b_j \qquad (j = 1, 2, \ldots, p),$$

$$\sum_{j=1}^{p} y_j \geq k,$$

$$y_j = 0 \quad \text{or} \quad 1 \qquad (j = 1, 2, \ldots, p).$$

That is, $y_j = 1$ if the jth constraint is to be satisfied, and at least k of the constraints must be satisfied. If we define $y'_j \equiv 1 - y_j$, and substitute for y_j in these constraints, the form of the resulting constraints is analogous to that given previously for modeling alternative constraints.

Compound Alternatives

The feasible region shown in Fig. 9.2 consists of three disjoint regions, each specified by a system of inequalities. The feasible region is defined by alternative sets of constraints, and can be modeled by the system:

$$\left.\begin{array}{l} f_1(x_1, x_2) - B_1 y_1 \leq b_1 \\ f_2(x_1, x_2) - B_2 y_1 \leq b_2 \end{array}\right\} \quad \begin{array}{l}\text{Region 1}\\\text{constraints}\end{array}$$

$$\left.\begin{array}{l} f_3(x_1, x_2) - B_3 y_2 \leq b_3 \\ f_4(x_1, x_2) - B_4 y_2 \leq b_4 \end{array}\right\} \quad \begin{array}{l}\text{Region 2}\\\text{constraints}\end{array}$$

$$\left.\begin{array}{l} f_5(x_1, x_2) - B_5 y_3 \leq b_5 \\ f_6(x_1, x_2) - B_6 y_3 \leq b_6 \\ f_7(x_1, x_2) - B_7 y_3 \leq b_7 \end{array}\right\} \quad \begin{array}{l}\text{Region 3}\\\text{constraints}\end{array}$$

$$y_1 + y_2 + y_3 \leq 2,$$

$$x_1 \geq 0, \qquad x_2 \geq 0,$$

$$y_1, y_2, y_3 \quad \text{binary.}$$

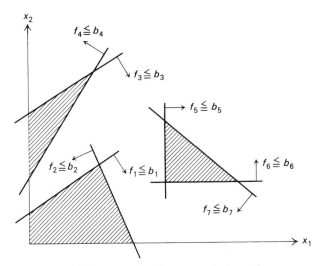

Fig. 9.2 An example of compound alternatives.

Note that we use the same binary variable y_j for each constraint defining one of the regions, and that the constraint $y_1 + y_2 + y_3 \leq 2$ implies that the decision variables x_1 and x_2 lie in *at least one* of the required regions. Thus, for example, if $y_3 = 0$, then each of the constraints

$$f_5(x_1, x_2) \leq b_5, \qquad f_6(x_1, x_2) \leq b_6, \qquad \text{and} \qquad f_7(x_1, x_2) \leq b_7$$

is satisfied.

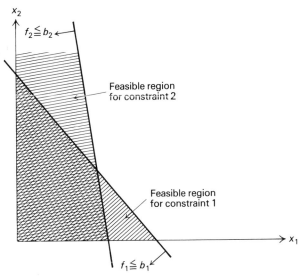

Fig. 9.3 Geometry of alternative constraints.

The regions do not have to be disjoint before we can apply this technique. Even the simple alternative constraint

$$f_1(x_1, x_2) \leq b_1 \qquad \text{or} \qquad f_2(x_1, x_2) \leq b_2$$

shown in Fig. 9.3 contains overlapping regions.

Representing Nonlinear Functions

Nonlinear functions can be represented by integer-programming formulations. Let us analyze the most useful representations of this type.

i) *Fixed Costs*

Frequently, the objective function for a minimization problem contains fixed costs (preliminary design costs, fixed investment costs, fixed contracts, and so forth). For example, the cost of producing x units of a specific product might consist of a fixed cost of setting up the equipment and a variable cost per unit produced on the equipment. An example of this type of cost is given in Fig, 9.4.

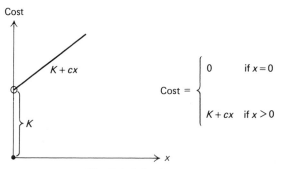

Fig. 9.4 A fixed cost.

Assume that the equipment has a capacity of B units. Define y to be a binary variable that indicates when the fixed cost is incurred, so that $y = 1$ when $x > 0$ and $y = 0$ when $x = 0$. Then the contribution to cost due to x may be written as

$$Ky + cx,$$

with the constraints:

$$x \leq By,$$

$$x \geq 0,$$

$$y = 0 \quad \text{or} \quad 1.$$

As required, these constraints imply that $x = 0$ when the fixed cost is not incurred, i.e., when $y = 0$. The constraints themselves do not imply that $y = 0$ if $x = 0$. But when $x = 0$, the minimization will clearly select $y = 0$, so that the fixed cost is not incurred. Finally, observe that if $y = 1$, then the added constraint becomes $x \leq B$, which reflects the capacity limit on the production equipment.

ii) *Piecewise Linear Representation*
Another type of nonlinear function that can be represented by integer variables is a piecewise linear curve. Figure 9.5 illustrates a cost curve for plant expansion that contains three linear segments with variable costs of 5, 1, and 3 million dollars per 1000 items of expansion.

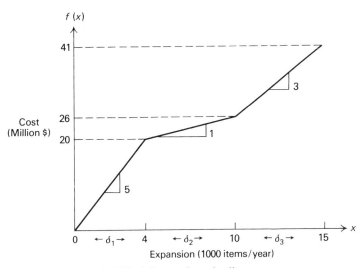

Fig. 9.5 Modeling a piecewise linear curve.

To model the cost curve, we express any value of x as the sum of three variables $\delta_1, \delta_2, \delta_3$, so that the cost for each of these variables is linear. Hence,

$$x = \delta_1 + \delta_2 + \delta_3,$$

where

$$0 \le \delta_1 \le 4,$$

$$0 \le \delta_2 \le 6, \tag{10}$$

$$0 \le \delta_3 \le 5;$$

and the total variable cost is given by:

$$\text{Cost} = 5\delta_1 + \delta_2 + 3\delta_3.$$

Note that we have defined the variables so that:

  δ_1 corresponds to the amount by which x exceeds 0, but is less than or equal to 4;

  δ_2 is the amount by which x exceeds 4, but is less than or equal to 10; and

  δ_3 is the amount by which x exceeds 10, but is less than or equal to 15.

If this interpretation is to be valid, we must also require that $\delta_1 = 4$ whenever $\delta_2 > 0$ and that $\delta_2 = 6$ whenever $\delta_3 > 0$. Otherwise, when $x = 2$, say, the cost would be minimized by selecting $\delta_1 = \delta_3 = 0$ and $\delta_2 = 2$, since the variable δ_2 has the smallest variable cost. However, these restrictions on the variables are simply conditional constraints and can be modeled by introducing binary variables, as before.

If we let

$$w_1 = \begin{cases} 1 & \text{if } \delta_1 \text{ is at its upper bound,} \\ 0 & \text{otherwise,} \end{cases}$$

$$w_2 = \begin{cases} 1 & \text{if } \delta_2 \text{ is at its upper bound,} \\ 0 & \text{otherwise,} \end{cases}$$

then constraints (10) can be replaced by

$$4w_1 \leq \delta_1 \leq 4,$$

$$6w_2 \leq \delta_2 \leq 6w_1,$$

$$0 \leq \delta_3 \leq 5w_2, \tag{11}$$

$$w_1 \quad \text{and} \quad w_2 \text{ binary,}$$

to ensure that the proper conditional constraints hold. Note that if $w_1 = 0$, then $w_2 = 0$, to maintain feasibility for the constraint imposed upon δ_2, and (11) reduces to

$$0 \leq \delta_1 \leq 4, \qquad \delta_2 = 0, \qquad \text{and} \qquad \delta_3 = 0.$$

If $w_1 = 1$ and $w_2 = 0$, then (11) reduces to

$$\delta_1 = 4, \qquad 0 \leq \delta_2 \leq 6, \qquad \text{and} \qquad \delta_3 = 0.$$

Finally, if $w_1 = 1$ and $w_2 = 1$, then (11) reduces to

$$\delta_1 = 4, \qquad \delta_2 = 6, \qquad \text{and} \qquad 0 \leq \delta_3 \leq 5.$$

Hence, we observe that there are three feasible combinations for the values of w_1 and w_2:

$w_1 = 0, \qquad w_2 = 0 \qquad$ corresponding to $0 \leq x \leq 4 \qquad$ since $\delta_2 = \delta_3 = 0$;

$w_1 = 1, \qquad w_2 = 0 \qquad$ corresponding to $4 \leq x \leq 10 \qquad$ since $\delta_1 = 4$ and $\delta_3 = 0$;

and

$w_1 = 1, \qquad w_2 = 1 \qquad$ corresponding to $10 \leq x \leq 15 \quad$ since $\delta_1 = 4$ and $\delta_2 = 6$.

The same general technique can be applied to piecewise linear curves with any number of segments. The general constraint imposed upon the variable δ_j for the jth segment will read:

$$L_j w_j \leq \delta_j \leq L_j w_{j-1},$$

where L_j is the length of the segment.

iii) *Diseconomies of Scale*
An important special case for representing nonlinear functions arises when only diseconomies of scale apply—that is, when marginal costs are increasing for a minimization problem or marginal returns are decreasing for a maximization problem. Suppose that the expansion cost in the previous example now is specified by Fig. 9.6.

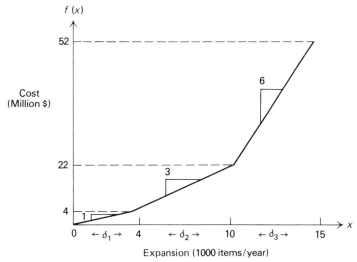

Fig. 9.6 Diseconomies of scale.

In this case, the cost is represented by

$$\text{Cost} = \delta_1 + 3\delta_2 + 6\delta_3,$$

subject only to the linear constraints without integer variables,

$$0 \leq \delta_1 \leq 4,$$

$$0 \leq \delta_2 \leq 6,$$

$$0 \leq \delta_3 \leq 5.$$

The conditional constraints involving binary variables in the previous formulation can be ignored if the cost curve appears in a minimization objective function, since the coefficients of δ_1, δ_2, and δ_3 imply that it is always best to set $\delta_1 = 4$ before taking $\delta_2 > 0$, and to set $\delta_2 = 6$ before taking $\delta_3 > 0$. As a consequence, the integer variables have been avoided completely.

This representation without integer variables is not valid, however, if economies of scale are present; for example, if the function given in Fig. 9.6 appears in a maximization problem. In such cases, it would be best to select the third segment with variable δ_3 before taking the first two segments, since the returns are higher on this segment. In this instance, the model requires the binary-variable formulation of the previous section.

iv) *Approximation of Nonlinear Functions*

One of the most useful applications of the piecewise linear representation is for approximating nonlinear functions. Suppose, for example, that the expansion cost in our illustration is given by the heavy curve in Fig. 9.7.

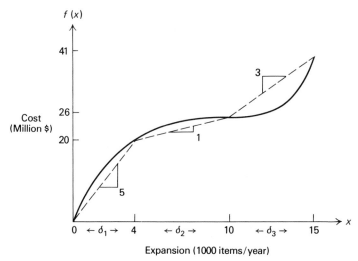

Fig. 9.7 Approximation of a nonlinear curve.

If we draw linear segments joining selected points on the curve, we obtain a *piecewise linear approximation*, which can be used instead of the curve in the model. The piecewise approximation, of course, is represented by introducing integer variables as indicated above. By using more points on the curve, we can make the approximation as close as we desire.

9.3 A SAMPLE FORMULATION[†]

Proper placement of service facilities such as schools, hospitals, and recreational areas is essential to efficient urban design. Here we will present a simplified model for firehouse location. Our purpose is to show formulation devices of the previous section arising together in a meaningful context, rather than to give a comprehensive model for the location problem *per se*. As a consequence, we shall ignore many relevant issues, including uncertainty.

Assume that population is concentrated in I districts within the city and that district i contains p_i people. Preliminary analysis (land surveys, politics, and so forth) has limited the potential location of firehouses to J sites. Let $d_{ij} \geq 0$ be the distance from the center of district i to site j. We are to determine the "best" site selection and

---

[†] This section may be omitted without loss of continuity.

assignment of districts to firehouses. Let

$$y_j = \begin{cases} 1 & \text{if site } j \text{ is selected,} \\ 0 & \text{otherwise;} \end{cases}$$

and

$$x_{ij} = \begin{cases} 1 & \text{if district } i \text{ is assigned to site } j, \\ 0 & \text{otherwise.} \end{cases}$$

The basic constraints are that every district should be assigned to exactly one fire-house, that is,

$$\sum_{j=1}^{J} x_{ij} = 1 \qquad (i = 1, 2, \ldots, I),$$

and that no district should be assigned to an unused site, that is, $y_j = 0$ implies $x_{ij} = 0$ $(i = 1, 2, \ldots, I)$. The latter restriction can be modeled as alternative constraints, or more simply as:

$$\sum_{i=1}^{I} x_{ij} \leq y_j I \qquad (j = 1, 2, \ldots, J).$$

Since x_{ij} are binary variables, their sum never exceeds I, so that if $y_j = 1$, then constraint j is nonbinding. If $y_j = 0$, then $x_{ij} = 0$ for all i.

Next note that d_i, the distance from district i to its assigned firehouse, is given by:

$$d_i = \sum_{j=1}^{J} d_{ij} x_{ij},$$

since one x_{ij} will be 1 and all others 0.

Also, the total population serviced by site j is:

$$s_j = \sum_{i=1}^{I} p_i x_{ij}.$$

Assume that a central district is particularly susceptible to fire and that either sites 1 and 2 or sites 3 and 4 can be used to protect this district. Then one of a number of similar restrictions might be:

$$y_1 + y_2 \geq 2 \qquad \text{or} \qquad y_3 + y_4 \geq 2.$$

We let y be a binary variable; then these alternative constraints become:

$$y_1 + y_2 \geq 2y,$$

$$y_3 + y_4 \geq 2(1 - y).$$

Next assume that it costs $f_j(s_j)$ to build a firehouse at site j to service s_j people and that a total budget of B dollars has been allocated for firehouse construction. Then

$$\sum_{j=1}^{I} f_j(s_j) \leq B.$$

Finally, one possible social-welfare function might be to minimize the distance traveled to the district farthest from its assigned firehouse, that is, to:

Minimize D,

where

$$D = \max d_i;$$

or, equivalently,[‡] to

Minimize D,

subject to:

$$D \geq d_i \qquad (i = 1, 2, \ldots, I).$$

Collecting constraints and substituting above for d_i in terms of its defining relationship

$$d_i = \sum_{j=1}^{J} d_{ij} x_{ij},$$

we set up the full model as:

Minimize D,

subject to:

$$D - \sum_{j=1}^{J} d_{ij} x_{ij} \geq 0 \qquad (i = 1, 2, \ldots, I),$$

$$\sum_{j=1}^{J} x_{ij} = 1 \qquad (i = 1, 2, \ldots, I),$$

$$\sum_{i=1}^{I} x_{ij} \leq y_j I \qquad (j = 1, 2, \ldots, J),$$

$$s_j - \sum_{i=1}^{I} p_i x_{ij} = 0 \qquad (j = 1, 2, \ldots, J),$$

$$\sum_{j=1}^{J} f_j(s_j) \leq B,$$

$$y_1 + y_2 - 2y \geq 0,$$

$$y_3 + y_4 + 2y \geq 2,$$

$$x_{ij}, y_j, y \quad \text{binary} \qquad (i = 1, 2, \ldots, I; j = 1, 2, \ldots, J).$$

[‡] The inequalities $D \geq d_i$ imply that $D \geq \max d_i$. The minimization of D then ensures that it will actually be the maximum of the d_i.

At this point we might replace each function $f_j(s_j)$ by an integer-programming approximation to complete the model. Details are left to the reader. Note that if $f_j(s_j)$ contains a fixed cost, then new fixed-cost variables need not be introduced— the variable y_j serves this purpose.

The last comment, and the way in which the conditional constraint "$y_j = 0$ implies $x_{ij} = 0$ $(i = 1, 2, \ldots, I)$" has been modeled above, indicate that the formulation techniques of Section 9.2 should not be applied without thought. Rather, they provide a common framework for modeling and should be used in conjunction with good modeling "common sense." In general, it is best to introduce as few integer variables as possible.

9.4 SOME CHARACTERISTICS OF INTEGER PROGRAMS—A SAMPLE PROBLEM

Whereas the simplex method is effective for solving linear programs, there is no single technique for solving integer programs. Instead, a number of procedures have been developed, and the performance of any particular technique appears to be highly problem-dependent. Methods to date can be classified broadly as following one of three approaches:

  i) enumeration techniques, including the branch-and-bound procedure;

 ii) cutting-plane techniques; and

iii) group-theoretic techniques.

In addition, several composite procedures have been proposed, which combine techniques using several of these approaches. In fact, there is a trend in computer systems for integer programming to include a number of approaches and possibly utilize them all when analyzing a given problem. In the sections to follow, we shall consider the first two approaches in some detail. At this point, we shall introduce a specific problem and indicate some features of integer programs. Later we will use this example to illustrate and motivate the solution procedures. Many characteristics of this example are shared by the integer version of the custom-molder problem presented in Chapter 1.

The problem is to determine z^* where:

$$z^* = \max z = 5x_1 + 8x_2,$$

subject to:

$$x_1 + x_2 \leqq 6,$$
$$5x_1 + 9x_2 \leqq 45,$$

$$x_1, x_2 \geqq 0 \quad \text{and} \quad \text{integer}.$$

The feasible region is sketched in Fig 9.8. Dots in the shaded region are feasible integer points.

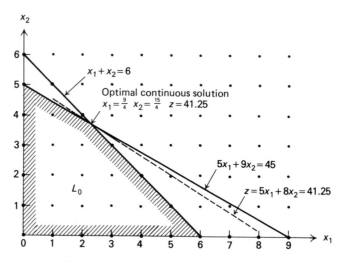

Fig. 9.8 An integer programming example.

If the integrality restrictions on variables are dropped, the resulting problem is a linear program. We will call it the *associated linear program*. We may easily determine its optimal solution graphically. Table 9.1 depicts some of the features of the problem.

Table 9.1 Problem features.

	Continuous optimum	Round off	Nearest feasible point	Integer optimum
x_1	$\frac{9}{4} = 2.25$	2	2	0
x_2	$\frac{15}{4} = 3.75$	4	3	5
z	41.25	Infeasible	34	40

Observe that the optimal integer-programming solution is not obtained by *rounding* the linear-programming solution. The closest point to the optimal linear-program solution is not even feasible. Also, note that the nearest feasible integer point to the linear-program solution is far removed from the optimal integer point. Thus, it is not sufficient simply to round linear-programming solutions. In fact, by scaling the righthand-side and cost coefficients of this example properly, we can construct a problem for which the optimal integer-programming solution lies as far as we like from the rounded linear-programming solution, in either z value or distance on the plane.

In an example as simple as this, almost any solution procedure will be effective. For instance, we could easily enumerate all the integer points with $x_1 \leqq 9$, $x_2 \leqq 6$, and select the best feasible point. In practice, the number of points to be considered is likely to prohibit such an exhaustive enumeration of potentially feasible points, and a more sophisticated procedure will have to be adopted.

9.5 BRANCH-AND-BOUND

Branch-and-bound is essentially a strategy of "divide and conquer." The idea is to partition the feasible region into more manageable subdivisions and then, if required, to further partition the subdivisions. In general, there are a number of ways to divide the feasible region, and as a consequence there are a number of branch-and-bound algorithms. We shall consider one such technique, for problems with only binary variables, in Section 9.7. For historical reasons, the technique that will be described next usually is referred to as *the* branch-and-bound procedure.

Basic Procedure

An integer linear program is a linear program further constrained by the integrality restrictions. Thus, in a maximization problem, the value of the objective function, at the linear-program optimum, will always be an upper bound on the optimal integer-programming objective. In addition, any integer feasible point is always a lower bound on the optimal linear-program objective value.

The idea of branch-and-bound is to utilize these observations to systematically subdivide the linear-programming feasible region and make assessments of the integer-programming problem based upon these subdivisions. The method can be described easily by considering the example from the previous section. At first, the linear-programming region is not subdivided: The integrality restrictions are dropped and the associated linear program is solved, giving an optimal value z^0. From our remark above, this gives the upper bound on z^*, $z^* \leq z^0 = 41\frac{1}{4}$. Since the coefficients in the objective function are integral, z^* must be integral and this implies that $z^* \leq 41$.

Next note that the linear-programming solution has $x_1 = 2\frac{1}{4}$ and $x_2 = 3\frac{3}{4}$. Both of these variables must be integer in the optimal solution, and we can divide the feasible region in an attempt to *make* either integral. We know that, in any integer programming solution, x_2 must be either an integer ≤ 3 or an integer ≥ 4. Thus, our first subdivision is into the regions where $x_2 \leq 3$ and $x_2 \geq 4$ as displayed by the shaded regions L_1 and L_2 in Fig. 9.9. Observe that, by making the subdivisions, we have excluded the old linear-program solution. (If we selected x_1 instead, the region would be subdivided with $x_1 \leq 2$ and $x_1 \geq 3$.)

The results up to this point are pictured conveniently in an *enumeration tree* (Fig. 9.10). Here L_0 represents the associated linear program, whose optimal solution has been included within the L_0 box, and the upper bound on z^* appears to the right of the box. The boxes below correspond to the new subdivisions; the constraints that subdivide L_0 are included next to the lines joining the boxes. Thus, the constraints of L_1 are those of L_0 together with the constraint $x_2 \geq 4$, while the constraints of L_2 are those of L_0 together with the constraint $x_2 \leq 3$.

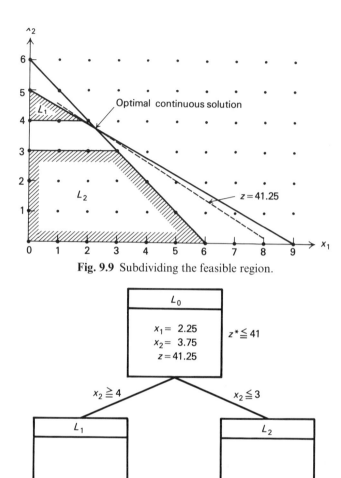

Fig. 9.9 Subdividing the feasible region.

Fig. 9.10 Enumeration tree.

The strategy to be pursued now may be apparent: Simply treat each subdivision as we did the original problem. Consider L_1 first. Graphically, from Fig. 9.9 we see that the optimal linear-programming solution lies on the second constraint with $x_2 = 4$, giving $x_1 = \frac{1}{5}(45 - 9(4)) = \frac{9}{5}$ and an objective value $z = 5(\frac{9}{5}) + 8(4) = 41$. Since x_1 is not integer, we subdivide L_1 further, into the regions L_3 with $x_1 \geq 2$ and L_4 with $x_1 \leq 1$. L_3 is an infeasible problem and so this branch of the enumeration tree no longer needs to be considered.

The enumeration tree now becomes that shown in Fig. 9.12. Note that the constraints of any subdivision are obtained by tracing back to L_0. For example, L_4 contains the original constraints together with $x_2 \geq 4$ and $x_1 \leq 2$. The asterisk (*) below box L_3 indicates that the region need not be subdivided or, equivalently, that the tree will not be extended from this box.

At this point, subdivisions L_2 and L_4 must be considered. We may select one arbitrarily; however, in practice, a number of useful heuristics are applied to make

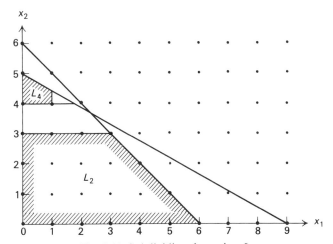

Fig. 9.11 Subdividing the region L_1.

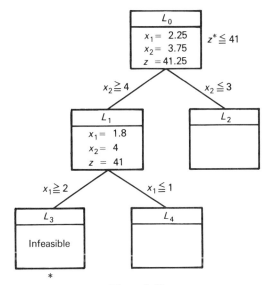

Figure 9.12

this choice. For simplicity, let us select the subdivision most recently generated, here L_4. Analyzing the region, we find that its optimal solution has

$$x_1 = 1, \qquad x_2 = \tfrac{1}{9}(45 - 5) = \tfrac{40}{9}.$$

Since x_2 is not integer, L_4 must be further subdivided into L_5 with $x_2 \leq 4$, and L_6 with $x_2 \geq 5$, leaving L_2, L_5, and L_6 yet to be considered.

   Treating L_5 first (see Fig. 9.13), we see that its optimum has $x_1 = 1$, $x_2 = 4$, and $z = 37$. Since this is the best linear-programming solution for L_5 and the linear program contains every integer solution in L_5, no integer point in that subdivision

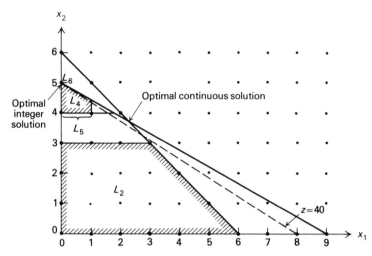

Fig. 9.13 Final subdivisions for the example.

can give a larger objective value than this point. Consequently, other points in L_5 need never be considered and L_5 need not be subdivided further. In fact, since $x_1 = 1$, $x_2 = 4$, $z = 37$, is a feasible solution to the original problem, $z^* \geq 37$ and we now have the bounds $37 \leq z^* \leq 41$. Without further analysis, we could terminate with the integer solution $x_1 = 1$, $x_2 = 4$, knowing that the objective value of this point is within 10 percent of the true optimum. For convenience, the lower bound $z^* \geq 37$ just determined has been appended to the right of the L_5 box in the enumeration tree (Fig. 9.14).

Although $x_1 = 1$, $x_2 = 4$ is the best integer point in L_5, the regions L_2 and L_6 might contain better feasible solutions, and we must continue the procedure by analyzing these regions. In L_6, the only feasible point is $x_1 = 0$, $x_2 = 5$, giving an objective value $z = +40$. This is better than the previous integer point and thus the lower bound on z^* improves, so that $40 \leq z^* \leq 41$. We could terminate with this integer solution knowing that it is within 2.5 percent of the true optimum. However, L_2 *could* contain an even better integer solution.

The linear-programming solution in L_2 has $x_1 = x_2 = 3$ and $z = 39$. This is the best integer point in L_2 but is not as good as $x_1 = 0$, $x_2 = 5$, so the later point (in L_6) must indeed be optimal. It is interesting to note that, even if the solution to L_2 did not give x_1 and x_2 integer, but had $z < 40$, then no feasible (and, in particular, no integer point) in L_2 could be as good as $x_1 = 0$, $x_2 = 5$, with $z = 40$. Thus, again $x_1 = 0$, $x_2 = 5$ would be known to be optimal. This observation has important computational implications, since it is not necessary to drive every branch in the enumeration tree to an integer or infeasible solution, but only to an objective value below the best integer solution.

The problem now is solved and the entire solution procedure can be summarized by the enumeration tree in Fig. 9.15.

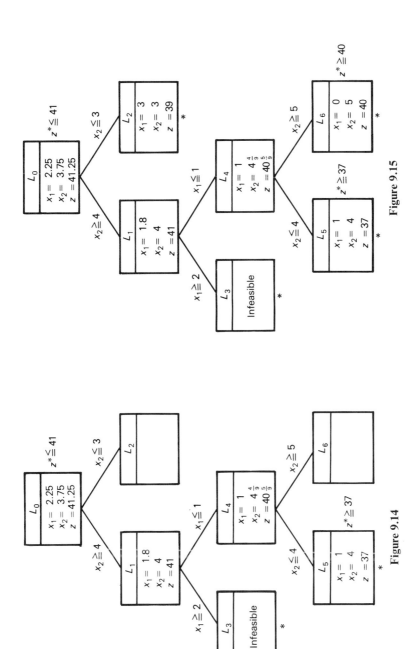

Figure 9.15

Figure 9.14

391

Further Considerations

There are three points that have yet to be considered with respect to the branch-and-bound procedure:

i) Can the linear programs corresponding to the subdivisions be solved efficiently?

ii) What is the best way to subdivide a given region, and which unanalyzed subdivision should be considered next?

iii) Can the upper bound ($z = 41$, in the example) on the optimal value z^* of the integer program be improved while the problem is being solved?

The answer to the first question is an unqualified *yes*. When moving from a region to one of its subdivisions, we add one constraint that is not satisfied by the optimal linear-programming solution over the parent region. Moreover, this was one motivation for the dual simplex algorithm, and it is natural to adopt that algorithm here.

Referring to the sample problem will illustrate the method. The first two subdivisions L_1 and L_2 in that example were generated by adding the following constraints to the original problem:

$$\text{For subdivision 1:}\qquad x_2 \geq 4 \qquad \text{or}\qquad x_2 - s_3 = 4 \qquad (s_3 \geq 0);$$

$$\text{For subdivision 2:}\qquad x_2 \leq 3 \qquad \text{or}\qquad x_2 + s_4 = 3 \qquad (s_4 \geq 0).$$

In either case we add the new constraint to the optimal linear-programming tableau. For subdivision 1, this gives:

$$
\begin{aligned}
(-z) && -\tfrac{5}{4}s_1 - \tfrac{3}{4}s_2 &= -41\tfrac{1}{4} \\
x_1 && +\tfrac{9}{4}s_1 - \tfrac{1}{4}s_2 &= \tfrac{9}{4} \\
\boxed{x_2} && -\tfrac{5}{4}s_1 + \tfrac{1}{4}s_2 &= \tfrac{15}{4} \\
&& -x_2 \qquad + s_3 &= -4,
\end{aligned}
$$

Constraints from the optimal canonical form

Added constraint

$$x_1, x_2, s_1, s_2, s_3 \geq 0,$$

where s_1 and s_2 are slack variables for the two constraints in the original problem formulation. Note that the new constraint has been multiplied by -1, so that the slack variable s_3 can be used as a basic variable. Since the basic variable x_2 appears with a nonzero coefficient in the new constraint, though, we must pivot to isolate this variable in the second constraint to re-express the system as:

$$
\begin{aligned}
(-z) && -\tfrac{5}{4}s_1 - \tfrac{3}{4}s_2 &= -41\tfrac{1}{4}, \\
x_1 && +\tfrac{9}{4}s_1 - \tfrac{1}{4}s_2 &= \tfrac{9}{4}, \\
x_2 && -\tfrac{5}{4}s_1 + \tfrac{1}{4}s_2 &= \tfrac{15}{4}, \\
&& \boxed{-\tfrac{5}{4}}s_1 + \tfrac{1}{4}s_2 + s_3 &= -\tfrac{1}{4}, \\
\end{aligned}
$$

$$x_1, x_2, s_1, s_2, s_3 \geq 0.$$

These constraints are expressed in the proper form for applying the dual simplex algorithm, which will pivot next to make s_1 the basic variable in the third constraint.

The resulting system is given by:

$$
\begin{aligned}
(-z) \qquad\qquad -s_2 - s_3 &= -41, \\
x_1 \qquad\quad + \tfrac{1}{5}s_2 + \tfrac{9}{5}s_3 &= \tfrac{9}{5}, \\
x_2 \qquad\quad - s_3 &= 4, \\
s_1 - \tfrac{1}{5}s_2 - \tfrac{4}{5}s_3 &= \tfrac{1}{5}, \\
x_1, x_2, s_1, s_2, s_3 &\geqq 0.
\end{aligned}
$$

This tableau is optimal and gives the optimal linear-programming solution over the region L_1 as $x_1 = \tfrac{9}{5}$, $x_2 = 4$, and $z = 41$. The same procedure can be used to determine the optimal solution in L_2.

When the linear-programming problem contains many constraints, this approach for recovering an optimal solution is very effective. After adding a new constraint and making the slack variable for that constraint basic, we always have a starting solution for the dual-simplex algorithm with only one basic variable negative. Usually, only a few dual-simplex pivoting operations are required to obtain the optimal solution. Using the primal-simplex algorithm generally would require many more computations.

Issue (ii) raised above is very important since, if we can make our choice of subdivisions in such a way as to rapidly obtain a good (with luck, near-optimal) integer solution $\hat{z}$, then we can eliminate many potential subdivisions immediately. Indeed, if any region has its linear programming value $z \leqq \hat{z}$, then the objective value of no integer point in that region can exceed $\hat{z}$ and the region need not be subdivided. There is no universal method for making the required choice, although several heuristic procedures have been suggested, such as selecting the subdivision with the largest optimal linear-programming value.[†]

Rules for determining which fractional variables to use in constructing subdivisions are more subtle. Recall that any fractional variable can be used to generate a subdivision. One procedure utilized is to look ahead one step in the dual-simplex method for every possible subdivision to see which is most promising. The details are somewhat involved and are omitted here. For expository purposes, we have selected the fractional variable arbitrarily.

Finally, the upper bound $\overline{z}$ on the value z^* of the integer program can be improved as we solve the problem. Suppose for example, that subdivision L_2 was analyzed before subdivisions L_5 or L_6 in our sample problem. The enumeration tree would be as shown in Fig. 9.16.

At this point, the optimal solution must lie in either L_2 or L_4. Since, however, the largest value for any feasible point in either of these regions is $40\tfrac{5}{9}$, the optimal

---

[†] One common method used in practice is to consider subdivisions on a last-generated–first-analyzed basis. We used this rule in our previous example. Note that data to initiate the dual-simplex method mentioned above must be stored for each subdivision that has yet to be analyzed. This data usually is stored in a list, with new information being added to the top of the list. When required, data then is extracted from the *top* of this list, leading to the last-generated–first-analyzed rule. Observe that when we subdivide a region into two subdivisions, one of these subdivisions will be analyzed next. The data required for this analysis already will be in the computer core and need not be extracted from the list.

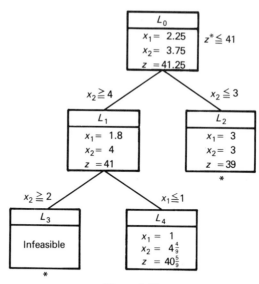

Figure 9.16

value for the problem z^* cannot *exceed* $40\frac{5}{9}$. Because z^* must be integral, this implies that $z^* \leq 40$ and the upper bound has been improved from the value 41 provided by the solution to the linear program on L_0. In general, the upper bound is given in this way as the largest value of any "hanging" box (one that has not been divided) in the enumeration tree.

Summary

The essential idea of branch-and-bound is to subdivide the feasible region to develop *bounds* $\underline{z} < z^* < \bar{z}$ on z^*. For a maximization problem, the lower bound $\underline{z}$ is the highest value of any feasible integer point encountered. The upper bound is given by the optimal value of the associated linear program or by the largest value for the objective function at any "hanging" box. After considering a subdivision, we must *branch* to (move to) another subdivision and analyze it. Also, if *either*

i) the linear program over L_j is infeasible;

ii) the optimal linear-programming solution over L_j is integer; *or*

iii) the value of the linear-programming solution z^j over L_j satisfies $z^j \leq \underline{z}$ (if maximizing),

then L_j need not be subdivided. In these cases, integer-programming terminology says that L_j has been *fathomed*.[†] Case (i) is termed fathoming by infeasibility, (ii) fathoming by integrality, and (iii) fathoming by bounds.

The flow chart in Fig. 9.17 summarizes the general procedure.

[†] To *fathom* is defined as "to get to the bottom of; to understand thoroughly." In this chapter, *fathomed* might be more appropriately defined as "understood enough or already considered."

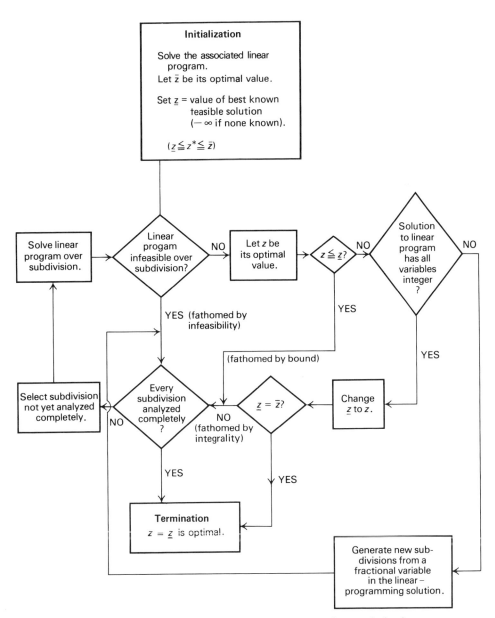

Fig. 9.17 Branch-and-bound for integer-programming maximization.

9.6 BRANCH-AND-BOUND FOR MIXED-INTEGER PROGRAMS

The branch-and-bound approach just described is easily extended to solve problems in which some, but not all, variables are constrained to be integral. Subdivisions then are generated solely by the integral variables. In every other way, the procedure is the same as that specified above. A brief example will illustrate the method.

$$z^* = \max z = -3x_1 - 2x_2 + 10,$$

subject to:

$$x_1 - 2x_2 + x_3 \qquad = \tfrac{5}{2},$$
$$2x_1 + x_2 \qquad + x_4 = \tfrac{3}{2},$$
$$x_j \geq 0 \qquad (j = 1, 2, 3, 4),$$
$$x_2 \quad \text{and} \quad x_3 \quad \text{integer}.$$

The problem, as stated, is in canonical form, with x_3 and x_4 optimal basic variables for the associated linear program.

The continuous variable x_4 cannot be used to generate subdivisions since any value of $x_4 \geq 0$ potentially can be optimal. Consequently, the subdivisions must be defined by $x_3 \leq 2$ and $x_3 \geq 3$. The complete procedure is summarized by the enumeration tree in Fig. 9.18.

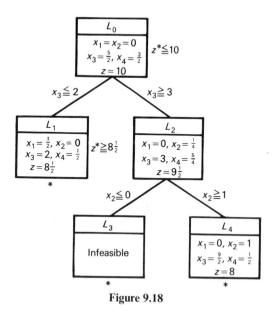

Figure 9.18

The solution in L_1 satisfies the integrality restrictions, so $z^* \geq z = 8\tfrac{1}{2}$. The only integral variable with a fractional value in the optimal solution of L_2 is x_2, so subdivisions L_3 and L_4 are generated from this variable. Finally, the optimal linear-programming value of L_4 is 8, so no feasible mixed-integer solution in that region can be better than the value $8\frac{1}{2}$ already generated. Consequently, that region need not be subdivided and the solution in L_1 is optimal.

The dual-simplex iterations that solve the linear programs in L_1, L_2, L_3, and L_4 are given below in Tableau 1. The variables s_j in the tableaus are the slack variables for the constraints added to generate the subdivisions. The coefficients in the appended constraints are determined as we mentioned in the last section, by eliminating the basic variables x_j from the new constraint that is introduced. To follow the iterations, recall that in the dual-simplex method, pivots are made on negative elements in the generating row; if all elements in this row are *positive*, as in region L_3, then the problem is infeasible.

Tableau 1

L_1

Basic variables	Current values	x_1	x_2	x_3	x_4	s_1	
$(-z)$	-10	-3	-2				
x_3	$\frac{5}{2}$	1	-2	1			
x_4	$\frac{3}{2}$	2	1		1		
s_1	$-\frac{1}{2}$	$\boxed{-1}$	2			1	(i.e., $x_3 \leq 2$)

Basic variables	Current values	x_1	x_2	x_3	x_4	s_1
$(-z)$	$-8\frac{1}{2}$		-8			-3
x_3	2		0	1		1
x_4	$\frac{1}{2}$		5		1	2
x_1	$\frac{1}{2}$	1	-2			-1

L_2

Basic variables	Current values	x_1	x_2	x_3	x_4	s_2	
$(-z)$	-10	-3	-2				
x_3	$\frac{5}{2}$	1	-2	1			
x_4	$\frac{3}{2}$	2	1		1		
s_2	$-\frac{1}{2}$	1	$\boxed{-2}$			1	(i.e., $x_3 \geq 3$)

Basic variables	Current values	x_1	x_2	x_3	x_4	s_2
$(-z)$	$-9\frac{1}{2}$	-4				-1
x_3	3	0		1		-1
x_4	$\frac{5}{4}$	$\frac{5}{2}$			1	$\frac{1}{2}$
x_2	$\frac{1}{4}$	$-\frac{1}{2}$	1			$-\frac{1}{2}$

Tableau 1 (*Continued*)

L_3	Basic variables	Current values	x_1	x_2	x_3	x_4	s_2	s_3	
	$(-z)$	$-9\frac{1}{2}$	-4				-1		
	x_3	3	0		1		-1		
	x_4	$\frac{5}{4}$	$\frac{5}{2}$			1	$\frac{1}{2}$		
	x_2	$\frac{1}{4}$	$-\frac{1}{2}$	1			$-\frac{1}{2}$		
	s_3	$-\frac{1}{4}$	$\frac{1}{2}$					$\frac{1}{2}$	1	(i.e., $x_2 \leqq 0$)

No pivot possible, problem infeasible

L_4	Basic variables	Current values	x_1	x_2	x_3	x_4	s_2	s_4	
	$(-z)$	$-9\frac{1}{2}$	-4				-1		
	x_3	3	0		1		-1		
	x_4	$\frac{5}{4}$	$\frac{5}{2}$			1	$\frac{1}{2}$		
	x_2	$\frac{1}{4}$	$-\frac{1}{2}$	1			$-\frac{1}{2}$		
	s_4	$-\frac{3}{4}$	$-\frac{1}{2}$					$\left(-\frac{1}{2}\right)$	1	(i.e., $x_2 \geqq 1$)

Basic variables	Current values	x_1	x_2	x_3	x_4	s_2	s_4
$(-z)$	-8	-3					-2
x_3	$\frac{9}{2}$	1		1			-2
x_4	$\frac{1}{2}$	2			1		1
x_2	1	0	1				-1
s_2	$\frac{3}{2}$	1				1	-2

9.7 IMPLICIT ENUMERATION

A special branch-and-bound procedure can be given for integer programs with only binary variables. The algorithm has the advantage that it requires no linear-programming solutions. It is illustrated by the following example:

$$z^* = \max z = -8x_1 - 2x_2 - 4x_3 - 7x_4 - 5x_5 + 10,$$

subject to:

$$-3x_1 - 3x_2 + x_3 + 2x_4 + 3x_5 \leqq -2,$$
$$-5x_1 - 3x_2 - 2x_3 - x_4 + x_5 \leqq -4,$$
$$x_j = 0 \quad \text{or} \quad 1 \quad (j = 1, 2, \ldots, 5).$$

One way to solve such problems is complete enumeration. List all possible binary combinations of the variables and select the best such point that is feasible. The approach works very well on a small problem such as this, where there are only a few potential 0–1 combinations for the variables, here 32. In general, though, an n-variable problem contains 2^n 0–1 combinations; for large values of n, the exhaustive approach is prohibitive. Instead, one might implicitly consider every binary combination, just as every integer point was implicitly *considered*, but not necessarily evaluated, for the general problem via branch-and-bound.

Recall that in the ordinary branch-and-bound procedure, subdivisions were analyzed by maintaining the linear constraints and dropping the integrality restrictions. Here, we adopt the opposite tactic of always maintaining the 0–1 restrictions, but ignoring the linear inequalities.

The idea is to utilize a branch-and-bound (or subdivision) process to fix some of the variables at 0 or 1. The variables remaining to be specified are called *free variables*. Note that, if the inequality constraints are ignored, the objective function is maximized by setting the free variables to zero, since their objective-function coefficients are negative. For example, if x_1 and x_4 are fixed at 1 and x_5 at 0, then the free variables are x_2 and x_3. Ignoring the inequality constraints, the resulting problem is:

$$\max \left[-8(1) - 2x_2 - 4x_3 - 7(1) - 5(0) + 10 \right] = \max \left[-2x_2 - 4x_3 - 5 \right],$$

subject to:

$$x_2 \quad \text{and} \quad x_3 \text{ binary.}$$

Since the free variables have negative objective-function coefficients, the maximization sets $x_2 = x_3 = 0$. The simplicity of this trivial optimization, as compared to a more formidable linear program, is what we would like to exploit.

Returning to the example, we start with *no* fixed variables, and consequently every variable is free and set to zero. The solution does not satisfy the inequality constraints, and we must subdivide to search for feasible solutions. One subdivision choice might be:

For subdivision 1: $x_1 = 1$,

For subdivision 2: $x_1 = 0$.

Now variable x_1 is fixed in each subdivision. By our observations above, if the inequalities are ignored, the optimal solution over each subdivision has $x_2 = x_3 = x_4 = x_5 = 0$. The resulting solution in subdivision 1 gives

$$z = -8(1) - 2(0) - 4(0) - 7(0) - 5(0) + 10 = 2,$$

and happens to satisfy the inequalities, so that the optimal solution to the original problem is *at least* 2, $z^* \geq 2$. Also, subdivision 1 has been fathomed: The above solution is best among all 0–1 combinations with $x_1 = 1$; thus it must be best among those satisfying the inequalities. No other feasible 0–1 combination in subdivision 1 needs to be evaluated explicitly. These combinations have been considered implicitly.

The solution with $x_2 = x_3 = x_4 = x_5 = 0$ in subdivision 2 is the same as the original solution with every variable at zero, and is infeasible. Consequently, the region must be subdivided further, say with $x_2 = 1$ or $x_2 = 0$, giving:

For subdivision 3: $x_1 = 0, x_2 = 1$;

For subdivision 4: $x_1 = 0, x_2 = 0$.

The enumeration tree to this point is as given in Fig. 9.19.

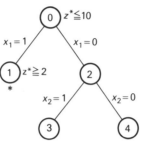

Figure 9.19

Observe that this tree differs from the enumeration trees of the previous sections. For the earlier procedures, the linear-programming solution used to analyze each subdivision was specified explicitly in a box. Here the 0–1 solution (ignoring the inequalities) used to analyze subdivisions is not stated explicitly, since it is known simply by setting free variables to zero. In subdivision ③, for example, $x_1 = 0$ and $x_2 = 1$ are fixed, and the free variables x_3, x_4, and x_5 are set to zero.

Continuing to fix variables and subdivide in this fashion produces the complete tree shown in Fig. 9.20.

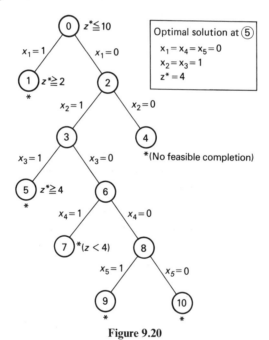

Figure 9.20

The tree is not extended after analyzing subdivisions 4, 5, 7, 9, and 10, for the following reasons.

i) At ⑤, the solution $x_1 = 0$, $x_2 = x_3 = 1$, with free variables $x_4 = x_5 = 0$, is feasible, with $z = 4$, thus providing an improved lower bound on z^*.

ii) At ⑦, the solution $x_1 = x_3 = 0$, $x_2 = x_4 = 1$, and free variable $x_5 = 0$, has $z = 1 < 4$, so that no solution in that subdivision can be as good as that generated at ⑤.

iii) At ⑨ and ⑩, every free variable is fixed. In each case, the subdivisions contain only a single point, which is infeasible, and further subdivision is not possible.

iv) At ④, the second inequality (with fixed variables $x_1 = x_2 = 0$) reads:

$$-2x_3 - x_4 + x_5 \leqq -4.$$

No 0–1 values of x_3, x_4, or x_5 "completing" the fixed variables $x_1 = x_2 = 0$ satisfy this constraint, since the lowest value for the lefthand side of this equation is -3 when $x_3 = x_4 = 1$ and $x_5 = 0$. The subdivision then has no feasible solution and need not be analyzed further.

The last observation is completely general. If, at any point after substituting for the fixed variables, the sum of the remaining negative coefficients in any constraint exceeds the righthand side, then the region defined by these fixed variables has no feasible solution. Due to the special nature of the 0–1 problem, there are a number of other such tests that can be utilized to reduce the number of subdivisions generated. The efficiency of these tests is measured by weighing the time needed to perform them against the time saved by fewer subdivisions.

The techniques used here apply to any integer-programming problem involving only binary variables, so that implicit enumeration is an alternative branch-and-bound procedure for this class of problems. In this case, subdivisions are fathomed if any of three conditions hold:

i) the integer program is known to be infeasible over the subdivision, for example, by the above infeasibility test;

ii) the 0–1 solution obtained by setting free variables to zero satisfies the linear inequalities; or

iii) the objective value obtained by setting free variables to zero is no larger than the best feasible 0–1 solution previously generated.

These conditions correspond to the three stated earlier for fathoming in the usual branch-and-bound procedure. If a region is not fathomed by one of these tests, implicit enumeration subdivides that region by selecting any free variable and fixing its values to 0 or 1.

Our arguments leading to the algorithm were based upon stating the original 0–1 problem in the following standard form:

1. the objective is a maximization with all coefficients negative; and

2. constraints are specified as "less than or equal to" inequalities.

As usual, minimization problems are transformed to maximization by multiplying cost coefficients by -1. If x_j appears in the maximization form with a positive coefficient, then the variable substitution $x_j = 1 - x_j'$ everywhere in the model leaves the binary variable x_j' with a negative objective-function coefficient. Finally, "greater than or equal to" constraints can be multiplied by -1 to become "less than or equal to" constraints; and general equality constraints are converted to inequalities by the special technique discussed in Exercise 17 of Chapter 2.

Like the branch-and-bound procedure for general integer programs, the way we choose to subdivide regions can have a profound effect upon computations. In implicit enumeration, we begin with the zero solution $x_1 = x_2 = \cdots = x_n = 0$ and generate other solutions by setting variables to 1. One natural approach is to subdivide based upon the variable with highest objective contribution. For the sample problem, this would imply subdividing initially with $x_2 = 1$ or $x_2 = 0$.

Another approach often used in practice is to try to drive toward feasibility as soon as possible. For instance, when $x_1 = 0$, $x_2 = 1$, and $x_3 = 0$ are fixed in the example problem, we could subdivide based upon either x_4 or x_5. Setting x_4 or x_5 to 1 and substituting for the fixed variables, we find that the constraints become:

$$x_4 = 1, \quad x_5(\text{free}) = 0: \qquad\qquad\qquad x_5 = 1, \quad x_4(\text{free}) = 0:$$

$$-3(0) - 3(1) + \ (0) + 2(1) + 3(0) \leqq -2, \quad -3(0) - 3(1) + \ (0) + 2(0) + 3(1) \leqq -2,$$

$$-5(0) - 3(1) - 2(0) - 1(1) + \ (0) \leqq -4, \quad -5(0) - 3(1) - 2(0) - 1(0) + \ (1) \leqq -4.$$

For $x_4 = 1$, the first constraint is infeasible by 1 unit and the second constraint is feasible, giving 1 total unit of infeasibility. For $x_5 = 1$, the first constraint is infeasible by 2 units and the second by 2 units, giving 4 total units of infeasibility. Thus $x_4 = 1$ appears more favorable, and we would subdivide based upon that variable. In general, the variable giving the *least total infeasibilities* by this approach would be chosen next. Reviewing the example problem the reader will see that this approach has been used in our solution.

9.8 CUTTING PLANES

The cutting-plane algorithm solves integer programs by modifying linear-programming solutions until the integer solution is obtained. It does not partition the feasible region into subdivisions, as in branch-and-bound approaches, but instead works with a single linear program, which it refines by adding new constraints. The new constraints successively reduce the feasible region until an integer optimal solution is found.

In practice, the branch-and-bound procedures almost always outperform the cutting-plane algorithm. Nevertheless, the algorithm has been important to the evolution of integer programming. Historically, it was the first algorithm developed for integer programming that could be proved to converge in a finite number of steps. In addition, even though the algorithm generally is considered to be very inefficient, it has provided insights into integer programming that have led to other, more efficient, algorithms.

Again, we shall discuss the method by considering the sample problem of the previous sections:

$$z^* = \max 5x_1 + 8x_2,$$

subject to:

$$x_1 + x_2 + s_1 \qquad = 6, \tag{11}$$
$$5x_1 + 9x_2 \qquad + s_2 = 45,$$

$$x_1, x_2, s_1, s_2 \geqq 0.$$

s_1 and s_2 are, respectively, slack variables for the first and second constraints.

Solving the problem by the simplex method produces the following optimal tableau:

$$(-z) \qquad\qquad - \tfrac{5}{4}s_1 - \tfrac{3}{4}s_2 = -41\tfrac{1}{4},$$
$$x_1 \qquad + \tfrac{9}{4}s_1 - \tfrac{1}{4}s_2 = \qquad \tfrac{9}{4},$$
$$x_2 - \tfrac{5}{4}s_1 + \tfrac{1}{4}s_2 = \qquad \tfrac{15}{4},$$

$$x_1, x_2, s_1, s_2 \geqq 0.$$

Let us rewrite these equations in an equivalent but somewhat altered form:

$$(-z) \qquad\qquad - 2s_1 - s_2 + 42 = \tfrac{3}{4} - \tfrac{3}{4}s_1 - \tfrac{1}{4}s_2,$$
$$x_1 \qquad + 2s_1 - s_2 - 2 = \tfrac{1}{4} - \tfrac{1}{4}s_1 - \tfrac{3}{4}s_2,$$
$$x_2 - 2s_1 \qquad - 3 = \tfrac{3}{4} - \tfrac{3}{4}s_1 - \tfrac{1}{4}s_2,$$

$$x_1, x_2, s_1, s_2 \geqq 0.$$

These algebraic manipulations have isolated integer coefficients to one side of the equalities and fractions to the other, in such a way that the constant terms on the righthand side are all nonnegative and the slack variable coefficients on the righthand side are all nonpositive.

In any integer solution, the lefthand side of each equation in the last tableau must be integer. Since s_1 and s_2 are nonnegative and appear to the right with negative coefficients, each righthand side necessarily must be less than or equal to the fractional constant term. Taken together, these two observations show that both sides of every equation must be an integer less than or equal to zero (if an integer is less than or equal to a fraction, it necessarily must be 0 or negative). Thus, from the first equation, we may write:

$$\tfrac{3}{4} - \tfrac{3}{4}s_1 - \tfrac{1}{4}s_2 \leqq 0 \quad \text{and} \quad \text{integer},$$

or, introducing a slack variable s_3,

$$\tfrac{3}{4} - \tfrac{3}{4}s_1 - \tfrac{1}{4}s_2 + s_3 = 0, \qquad s_3 \geqq 0 \quad \text{and} \quad \text{integer}. \tag{C_1}$$

Similarly, other conditions can be generated from the remaining constraints:

$$\tfrac{1}{4} - \tfrac{1}{4}s_1 - \tfrac{3}{4}s_2 + s_4 = 0, \qquad s_4 \geqq 0 \quad \text{and} \quad \text{integer} \tag{C_2}$$
$$\tfrac{3}{4} - \tfrac{3}{4}s_1 - \tfrac{1}{4}s_2 + s_5 = 0, \qquad s_5 \geqq 0 \quad \text{and} \quad \text{integer}. \tag{C_3}$$

Note that, in this case, (C_1) and (C_3) are identical.

The new equations (C_1), (C_2), and (C_3) that have been derived are called *cuts* for the following reason: Their derivation did not exclude any integer solutions to the problem, so that any integer feasible point to the original problem must satisfy the cut constraints. The linear-programming solution had $s_1 = s_2 = 0$; clearly, these do *not* satisfy the cut constraints. In each case, substituting $s_1 = s_2 = 0$ gives either s_3, s_4, or $s_5 < 0$. Thus the net effect of a cut is to cut away the optimal linear-programming solution from the feasible region without excluding any feasible integer points.

The geometry underlying the cuts can be established quite easily. Recall from (11) that slack variables s_1 and s_2 are defined by:

$$s_1 = 6 - x_1 - x_2,$$
$$s_2 = 45 - 5x_1 - 9x_2.$$

Substituting these values in the cut constraints and rearranging, we may rewrite the cuts as:

$$2x_1 + 3x_2 \leqq 15, \qquad\qquad (C_1 \text{ or } C_3)$$
$$4x_1 + 7x_2 \leqq 35. \qquad\qquad (C_2)$$

In this form, the cuts are displayed in Fig. 9.21. Note that they exhibit the features suggested above. In each case, the added cut removes the linear-programming solution $x_1 = \frac{9}{4}$, $x_2 = \frac{15}{4}$, from the feasible region, at the same time including every feasible integer solution.

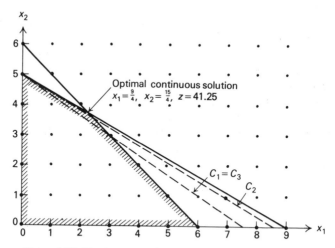

Figure 9.21 Cutting away the linear-programming solution.

The basic strategy of the cutting-plane technique is to add cuts (usually only one) to the constraints defining the feasible region and then to solve the resulting linear program. If the optimal values for the decision variables in the linear program are all integer, they are optimal; otherwise, a new cut is derived from the new optimal linear-programming tableau and appended to the constraints.

Note from Fig. 9.21 that the cut $C_1 = C_3$ leads directly to the optimal solution. Cut C_2 does not, and further iterations will be required if this cut is appended to the problem (without the cut $C_1 = C_3$). Also note that C_1 cuts deeper into the feasible region than does C_2. For problems with many variables, it is generally quite difficult to determine which cuts will be deep in this sense. Consequently, in applications, the algorithm frequently generates cuts that shave very little from the feasible region, and hence the algorithm's poor performance.

A final point to be considered here is the way in which cuts are generated. The linear-programming tableau for the above problem contained the constraint:

$$x_1 + \tfrac{9}{4}s_1 - \tfrac{1}{4}s_2 = \tfrac{9}{4}.$$

Suppose that we round *down* the fractional coefficients to integers, that is, $\tfrac{9}{4}$ to 2, $-\tfrac{1}{4}$ to -1, and $\tfrac{9}{4}$ to 2. Writing these integers to the left of the equality and the remaining fractions to the right, we obtain as before, the equivalent constraint:

$$x_1 + 2s_1 - s_2 - 2 = \tfrac{1}{4} - \tfrac{1}{4}s_1 - \tfrac{3}{4}s_2.$$

By our previous arguments, the cut is:

$$\tfrac{1}{4} - \tfrac{1}{4}s_1 - \tfrac{3}{4}s_2 \le 0 \quad \text{and} \quad \text{integer}.$$

Another example may help to clarify matters. Suppose that the final linear-programming tableau to a problem has the constraint

$$x_1 \qquad + \tfrac{1}{6}x_6 - \tfrac{7}{6}x_7 + 3x_8 \qquad = 4\tfrac{5}{6}.$$

Then the equivalent constraint is:

$$x_1 \qquad\qquad - 2x_7 + 3x_8 - 4 = \tfrac{5}{6} - \tfrac{1}{6}x_6 - \tfrac{5}{6}x_7,$$

and the resulting cut is:

$$\tfrac{5}{6} - \tfrac{1}{6}x_6 - \tfrac{5}{6}x_7 \le 0 \quad \text{and} \quad \text{integer}.$$

Observe the way that fractions are determined for negative coefficients. The fraction in the cut constraint determined by the x_7 coefficient $-\tfrac{7}{6} = -1\tfrac{1}{6}$ is *not* $\tfrac{1}{6}$, but rather it is the fraction generated by rounding down to -2; i.e., the fraction is $-1\tfrac{1}{6} - (-2) = \tfrac{5}{6}$.

Tableau 2 shows the complete solution of the sample problem by the cutting-plane technique. Since cut $C_1 = C_3$ leads directly to the optimal solution, we have chosen to start with cut C_2. Note that, if the slack variable for any newly generated cut is taken as the basic variable in that constraint, then the problem is in the proper form for the dual-simplex algorithm. For instance, the cut in Tableau 2(b) generated from the x_1 constraint

$$x_1 + \tfrac{7}{3}s_1 - \tfrac{1}{3}s_2 = \tfrac{7}{3} \quad \text{or} \quad x_1 + 2s_1 - s_2 - 2 = \tfrac{1}{3} - \tfrac{1}{3}s_1 - \tfrac{2}{3}s_2$$

is given by:

$$\tfrac{1}{3} - \tfrac{1}{3}s_1 - \tfrac{2}{3}s_2 \le 0 \quad \text{and} \quad \text{integer}.$$

Letting s_4 be the slack variable in the constraint, we obtain:

$$-\tfrac{1}{3}s_1 - \tfrac{2}{3}s_2 + s_4 = -\tfrac{1}{3}.$$

Since s_1 and s_2 are nonbasic variables, we may take s_4 to be the basic variable isolated in this constraint (see Tableau 2(b)).

Tableau 2 Dual Simplex Iterations for the Cutting-Plane Algorithm.

(a)

Basic variables	Current values	x_1	x_2	s_1	s_2	s_3	
$(-z)$	$-41\frac{1}{4}$			$-\frac{5}{4}$	$-\frac{3}{4}$		
x_1	$\frac{9}{4}$	1		$\frac{9}{4}$	$-\frac{1}{4}$		
x_2	$\frac{15}{4}$		1	$-\frac{5}{4}$	$\frac{1}{4}$		
s_3	$-\frac{1}{4}$			$-\frac{1}{4}$	$\left(-\frac{3}{4}\right)$	1	(cut generated from x_1 constraint)

(b)

Basic variables	Current values	x_1	x_2	s_1	s_2	s_3	s_4	
$(-z)$	-41			-1		-1		
x_1	$\frac{7}{3}$	1		$\frac{7}{3}$		$-\frac{1}{3}$		
x_2	$\frac{11}{3}$		1	$-\frac{4}{3}$		$\frac{1}{3}$		
s_2	$\frac{1}{3}$			$\frac{1}{3}$	1	$-\frac{4}{3}$		
s_4	$-\frac{1}{3}$			$-\frac{1}{3}$		$\left(-\frac{2}{3}\right)$	1	(cut generated from x_1 constraint)

(c)

Basic variables	Current values	x_1	x_2	s_1	s_2	s_3	s_4	s_5	
$(-z)$	$-40\frac{1}{2}$			$-\frac{1}{2}$			$-\frac{3}{2}$		
x_1	$\frac{5}{2}$	1		$\frac{5}{2}$			$-\frac{1}{2}$		
x_2	$\frac{7}{2}$		1	$-\frac{3}{2}$			$\frac{1}{2}$		
s_2	1			1	1		-2		
s_3	$\frac{1}{2}$			$\frac{1}{2}$		1	$-\frac{3}{2}$		
s_5	$-\frac{1}{2}$			$\left(-\frac{1}{2}\right)$			$-\frac{1}{2}$	1	(cut generated from x_1 constraint)

(d)

Basic variables	Current values	x_1	x_2	s_1	s_2	s_3	s_4	s_5
$(-z)$	-40						-1	-1
x_1	0	1					3	5
x_2	5		1				2	-3
s_2	0				1		-3	2
s_3	0					1	-2	1
s_1	1			1			1	-2

By making slight modifications to the cutting-plane algorithm that has been described, we can show that an optimal solution to the integer-programming problem will be obtained, as in this example, after adding only a finite number of cuts. The proof of this fact by R. Gomory in 1958 was a very important theoretical break-through, since it showed that integer programs can be solved by *some* linear program (the associated linear program plus the added constraints). Unfortunately, the number of cuts to be added, though finite, is usually quite large, so that this result does not have important practical ramifications.

EXERCISES

1. As the leader of an oil-exploration drilling venture, you must determine the least-cost selection of 5 out of 10 possible sites. Label the sites $S_1, S_2, \ldots, S_{10}$, and the exploration costs associated with each as $C_1, C_2, \ldots, C_{10}$.

    Regional development restrictions are such that:

    i) Evaluating sites S_1 *and* S_7 will prevent you from exploring site S_8.
    ii)- Evaluating site S_3 *or* S_4 prevents you from assessing site S_5.
    iii) Of the group S_5, S_6, S_7, S_8, only two sites may be assessed.

    Formulate an integer program to determine the minimum-cost exploration scheme that satisfies these restrictions.

2. A company wishes to put together an academic "package" for an executive training program. There are five area colleges, each offering courses in the six fields that the program is designed to touch upon.

    The package consists of 10 courses; each of the six fields must be covered.

    The tuition (basic charge), assessed when at least one course is taken, at college i is T_i (independent of the number of courses taken). Moreover, each college imposes an additional charge (covering course materials, instructional aids, and so forth) for *each* course, the charge depending on the college and the field of instructions.

    Formulate an integer program that will provide the company with the minimum amount it must spend to meet the requirements of the program.

3. The marketing group of A. J. Pitt Company is considering the options available for its next advertising campaign program. After a great deal of work, the group has identified a selected number of options with the characteristics shown in the accompanying table.

	TV	Trade magazine	Newspaper	Radio	Popular magazine	Promotional campaign	Total resource available
Customers reached	1,000,000	200,000	300,000	400,000	450,000	450,000	—
Cost ($)	500,000	150,000	300,000	250,000	250,000	100,000	1,800,000
Designers needed (man-hours)	700	250	200	200	300	400	1,500
Salesmen needed (man-hours)	200	100	100	100	100	1,000	1,200

The objective of the advertising program is to maximize the number of customers reached, subject to the limitation of resources (money, designers, and salesmen) given in the table above. In addition, the following constraints have to be met:

i) If the promotional campaign is undertaken, it needs either a radio or a popular magazine campaign effort to support it.

ii) The firm cannot advertise in both the trade and popular magazines.

Formulate an integer-programming model that will assist the company to select an appropriate advertising campaign strategy.

4. Three different items are to be routed through three machines. Each item must be processed first on machine 1, then on machine 2, and finally on machine 3. The sequence of items may differ for each machine. Assume that the times t_{ij} required to perform the work on item i by machine j are known and are integers. Our objective is to minimize the total time necessary to process all the items.

a) Formulate the problem as an integer programming problem. [*Hint.* Let x_{ij} be the starting time of processing item i on machine j. Your model must prevent two items from occupying the same machine at the same time; also, an item may not start processing on machine $(j + 1)$ unless it has completed processing on machine j.]

b) Suppose we want the items to be processed in the same sequence on each machine. Change the formulation in part (a) accordingly.

5. Consider the problem:

$$\text{Maximize } z = x_1 + 2x_2,$$

subject to:

$$x_1 + x_2 \leq 8,$$
$$-x_1 + x_2 \leq 2,$$
$$x_1 - x_2 \leq 4,$$
$$x_2 \geq 0 \qquad \text{and integer,}$$
$$x_1 = 0, 1, 4, \text{ or } 6.$$

a) Reformulate the problem as an equivalent integer linear program.

b) How would your answer to part (a) change if the objective function were changed to:

$$\text{Maximize } z = x_1^2 + 2x_2?$$

6. Formulate, but *do not solve*, the following mathematical-programming problems. Also, indicate the type of algorithm used in general to solve each.

a) A courier traveling to Europe can carry up to 50 kilograms of a commodity, all of which can be sold for $40 per kilogram. The round-trip air fare is $450 plus $5 per kilogram of baggage in excess of 20 kilograms (one way). Ignoring any possible profits on the return trip, should the courier travel to Europe and, if so, how much of the commodity should be taken along in order to maximize his profits?

b) A copying service incurs machine operating costs of:

$$\$0.10 \quad \text{for copies 1 to 4,}$$
$$0.05 \quad \text{for copies 5 to 8,}$$
$$0.025 \text{ for copies 9 and over,}$$

and has a capacity of 1000 copies per hour. One hour has been reserved for copying a 10-page article to be sold to MBA students. Assuming that all copies can be sold for $0.50 per article, how many copies of the article should be made?

c) A petrochemical company wants to maximize profit on an item that sells for $0.30 per gallon. Suppose that increased temperature increases output according to the graph in Fig. E9–1. Assuming that production costs are directly proportional to temperature at $7.50 per degree centigrade, how many gallons of the item should be produced?

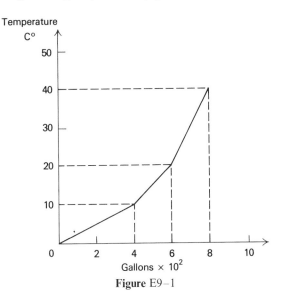

Figure E9–1

7. Suppose that you are a ski buff and an entrepreneur. You own a set of hills, any or all of which can be developed. Figure E9–2 illustrates the nature of the cost for putting ski runs on any hill.

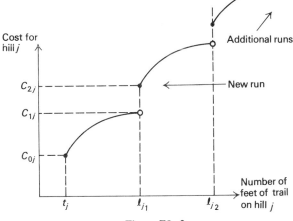

Figure E9–2

The cost includes fixed charges for putting new trails on a hill. For each hill j, there is a limit d_j on the number of trails that can be constructed and a lower limit t_j on the number of feet of trail that must be developed if the hill is used.

Use a piecewise linear approximation to formulate a strategy based on cost minimization for locating the ski runs, given that you desire to have M total feet of developed ski trail in the area.

8. Consider the following word game. You are assigned a number of tiles, each containing a letter $a, b, \ldots,$ or z from the alphabet. For any letter α from the alphabet, your assignment includes N_α (a nonnegative integer) tiles with the letter α. From the letters you are to construct any of the words $w_1, w_2, \ldots, w_n$. This list might, for example, contain all words from a given dictionary.

You may construct any word at most once, and use any tile at most once. You receive $v_j \geq 0$ points for making word w_j and an additional bonus of $b_{ij} \geq 0$ points for making *both* words w_i and w_j $(i = 1, 2, \ldots, n; j = 1, 2, \ldots, n)$.

a) Formulate a *linear* integer program to determine your optimal choice of words.

b) How does the formulation change if you are allowed to select 100 tiles with no restriction on your choice of letters?

9. In Section 9.1 of the text, we presented the following simplified version of the warehouse-location problem:

$$\text{Minimize} \sum_i \sum_j c_{ij} x_{ij} + \sum_i f_i y_i,$$

subject to:

$$\sum_i x_{ij} = d_j \qquad\qquad (j = 1, 2, \ldots, n)$$

$$\sum_j x_{ij} - y_i \left(\sum_j d_j \right) \leq 0 \qquad\qquad (i = 1, 2, \ldots, m)$$

$$x_{ij} \geq 0 \qquad\qquad (i = 1, 2, \ldots, m; j = 1, 2, \ldots, n)$$

$$y_i = 0 \text{ or } 1 \qquad\qquad (i = 1, 2, \ldots, m).$$

a) The above model assumes only one product and two distribution stages (from warehouse to customer). Indicate how the model would be modified if there were three distribution stages (from plant to warehouse to customer) and L different products. [*Hint.* Define a new decision variable $x_{ijk\ell}$ equal to the amount to be sent from plant i, through warehouse j, to customer k, of product ℓ.]

b) Suppose there are maximum capacities for plants and size limits (both upper and lower bounds) for warehouses. What is the relevant model now?

c) Assume that each customer has to be served from a single warehouse; i.e., no splitting of orders is allowed. How should the warehousing model be modified? [*Hint.* Define a new variable z_{jk} = fraction of demand of customer k satisfied from warehouse j.]

d) Assume that each warehouse i experiences economies of scale when shipping above a certain threshold quantity to an individual customer; i.e., the unit distribution cost is c_{ij} whenever the amount shipped is between 0 and d_{ij}, and c'_{ij} (lower than c_{ij}) whenever the amount shipped exceeds d_{ij}. Formulate the warehouse location model under these conditions.

10. Graph the following integer program:

$$\text{Maximize } z = \quad x_1 + 5x_2,$$

subject to:

$$-4x_1 + 3x_2 \leq 6,$$
$$3x_1 + 2x_2 \leq 18,$$
$$x_1, x_2 \geq 0 \text{ and integer.}$$

Apply the branch-and-bound procedure, graphically solving each linear-programming problem encountered. Interpret the branch-and-bound procedure graphically as well.

11. Solve the following integer program using the branch-and-bound technique:

$$\text{Minimize } z = \quad 2x_1 - 3x_2 - 4x_3,$$

subject to:

$$-x_1 + x_2 + 3x_3 \leq 8,$$
$$3x_1 + 2x_2 - x_3 \leq 10,$$
$$x_1, x_2, x_3 \geq 0 \quad \text{and integer.}$$

The optimal tableau to the linear program associated with this problem is:

Basic variables	Current values	x_1	x_2	x_3	x_4	x_5
x_3	$\frac{6}{7}$	$-\frac{5}{7}$		1	$\frac{2}{7}$	$-\frac{1}{7}$
x_2	$\frac{38}{7}$	$\frac{8}{7}$	1		$\frac{1}{7}$	$\frac{3}{7}$
$(-z)$	$\frac{138}{7}$	$\frac{18}{7}$			$\frac{11}{7}$	$\frac{5}{7}$

The variables x_4 and x_5 are slack variables for the two constraints.

12. Use branch-and-bound to solve the mixed-integer program:

$$\text{Maximize } z = -5x_1 - x_2 - 2x_3 + 5,$$

subject to:

$$-2x_1 + x_2 - x_3 + x_4 \qquad = \tfrac{7}{2},$$
$$2x_1 + x_2 + x_3 \qquad + x_5 = 2,$$
$$x_j \geq 0 \quad (j = 1, 2, \dots, 5)$$
$$x_3 \text{ and } x_5 \text{ integer.}$$

13. Solve the mixed-integer programming knapsack problem:

$$\text{Maximize } z = 6x_1 + 4x_2 + 4x_3 + x_4 + \quad x_5,$$

subject to:

$$2x_1 + 2x_2 + 3x_3 + x_4 + 2x_5 = 7,$$
$$x_j \geq 0 \quad (j = 1, 2, \dots, 5),$$
$$x_1 \text{ and } x_2 \text{ integer.}$$

14. Solve the following integer program using implicit enumeration:

$$\text{Maximize } z = 2x_1 - x_2 - x_3 + 10,$$

subject to:

$$2x_1 + 3x_2 - x_3 \leq 9,$$
$$2x_2 + x_3 \geq 4,$$
$$3x_1 + 3x_2 + 3x_3 = 6,$$

$$x_j = 0 \quad \text{or} \quad 1 \quad (j = 1, 2, 3).$$

15. A college intramural four-man basketball team is trying to choose its starting line-up from a six-man roster so as to maximize average height. The roster follows:

Player	Number	Height*	Position
Dave	1	10	Center
John	2	9	Center
Mark	3	6	Forward
Rich	4	6	Forward
Ken	5	4	Guard
Jim	6	−1	Guard

* In inches over 5'6".

The starting line-up must satisfy the following constraints:

  i) At least one guard must start.
  ii) Either John or Ken must be held in reserve.
  iii) Only one center can start.
  iv) If John or Rich starts, then Jim cannot start.

a) Formulate this problem as an integer program.
b) Solve for the optimal starting line-up, using the implicit enumeration algorithm.

16. Suppose that one of the following constraints arises when applying the implicit enumeration algorithm to a 0–1 integer program:

$$-2x_1 - 3x_2 + x_3 + 4x_4 \leq -6, \tag{1}$$
$$-2x_1 - 6x_2 + x_3 + 4x_4 \leq -5, \tag{2}$$
$$-4x_1 - 6x_2 - x_3 + 4x_4 \leq -3. \tag{3}$$

In each case, the variables on the lefthand side of the inequalities are free variables and the righthand-side coefficients include the contributions of the fixed variables.

a) Use the feasibility test to show that constraint (1) contains no feasible completion.
b) Show that $x_2 = 1$ and $x_4 = 0$ in any feasible completion to constraint (2). State a general rule that shows when a variable x_j, like x_2 or x_4 in constraint (2), must be either 0 or 1 in any feasible solution. [*Hint.* Apply the usual feasibility test after setting x_j to 1 or 0.]
c) Suppose that the objective function to minimize is:

$$z = 6x_1 + 4x_2 + 5x_3 + x_4 + 10,$$

and that $z = 17$ is the value of an integer solution obtained previously. Show that $x_3 = 0$ in a feasible completion to constraint (3) that is a potential improvement upon z with $z < \underline{z}$. (Note that either $x_1 = 1$ or $x_2 = 1$ in any feasible solution to constraint (3) having $x_3 = 1$.)

d) How could the tests described in parts (b) and (c) be used to reduce the size of the enumeration tree encountered when applying implicit enumeration?

17. Although the constraint

$$x_1 - x_2 \leqq -1$$

and the constraint

$$-x_1 + 2x_2 \leqq -1$$

to a zero–one integer program both have feasible solutions, the system composed of both constraints is infeasible. One way to exhibit the inconsistency in the system is to add the two constraints, producing the constraint

$$x_2 \leqq -1,$$

which has no feasible solution with $x_2 = 0$ or 1.

More generally, suppose that we multiply the ith constraint of a system

Multipliers

$$a_{11}x_1 + a_{12}x_2 + \cdots + a_{1n}x_n \leqq b_1, \qquad u_1$$
$$\vdots \qquad\qquad \vdots \qquad\qquad \vdots$$
$$a_{i1}x_1 + a_{i2}x_2 + \cdots + a_{in}x_n \leqq b_i, \qquad u_i$$
$$\vdots \qquad\qquad \vdots \qquad\qquad \vdots$$
$$a_{m1}x_1 + a_{m2}x_2 + \cdots + a_{mn}x_n \leqq b_m, \qquad u_m$$
$$x_j = 0 \quad \text{or} \quad 1 \qquad (j = 1, 2, \ldots, n),$$

by nonnegative constraints u_i and add to give the composite, or *surrogate*, constraint:

$$\left(\sum_{i=1}^{m} u_i a_{i1}\right)x_1 + \left(\sum_{i=1}^{m} u_i a_{i2}\right)x_2 + \cdots + \left(\sum_{i=1}^{m} u_i a_{in}\right)x_n \leqq \sum_{i=1}^{m} u_i b_i.$$

a) Show that any feasible solution $x_j = 0$ or 1 $(j = 1, 2, \ldots, n)$ for the system of constraints must also be feasible to the surrogate constraint.

b) How might the fathoming tests discussed in the previous exercise be used in conjunction with surrogate constraints in an implicit enumeration scheme?

c) A "best" surrogate constraint might be defined as one that eliminates as many nonoptimal solutions $x_j = 0$ or 1 as possible. Consider the objective value of the integer program when the system constraints are replaced by the surrogate constraint; that is the problem:

$$v(u) = \text{Min } c_1 x_1 \qquad\qquad + c_2 x_2 \qquad\qquad + \cdots + c_n x_n,$$

subject to:

$$\left(\sum_{i=1}^{m} u_i a_{i1}\right)x_1 + \left(\sum_{i=1}^{m} u_i a_{i2}\right)x_2 + \cdots + \left(\sum_{i=1}^{m} u_i a_{in}\right)x_n \leqq \sum_{i=1}^{m} u_i b_i,$$
$$x_j = 0 \quad \text{or} \quad 1 \qquad\qquad (j = 1, 2, \ldots, n).$$

Let us say that a surrogate constraint with multipliers $u_1, u_2, \ldots, u_m$ is *stronger* than another surrogate constraint with multipliers $\bar{u}_1, \bar{u}_2, \ldots, \bar{u}_m$, if the value $v(u)$ of the surrogate constraint problem with multipliers $u_1, u_2, \ldots, u_m$ is larger than the value of $v(\bar{u})$ with multipliers $\bar{u}_1, \bar{u}_2, \ldots, \bar{u}_m$. (The larger we make $v(u)$, the more nonoptimal solutions we might expect to eliminate.)

Suppose that we estimate the value of $v(u)$ defined above by replacing $x_j = 0$ or 1 by $0 \leq x_j \leq 1$ for $j = 1, 2, \ldots, n$. Then, to estimate the strongest surrogate constraint, we would need to find those values of the multipliers $u_1, u_2, \ldots, u_m$ to maximize $v(u)$, where

$$v(u) = \text{Min } c_1 x_1 + c_2 x_2 + \cdots + c_n x_n,$$

subject to:

$$0 \leq x_j \leq 1 \tag{1}$$

and the surrogate constraint.

Show that the optimal shadow prices to the linear program

$$\text{Maximize } c_1 x_1 + c_2 x_2 + \cdots + c_n x_n,$$

subject to:

$$a_{i1} x_1 + a_{i2} x_2 + \cdots + a_{in} x_n \leq b_i \quad (i = 1, 2, \ldots, m),$$

$$0 \leq x_j \leq 1 \quad (j = 1, 2, \ldots, n),$$

solve the problem of maximizing $v(u)$ in (1).

18. The following tableau specifies the solution to the linear program associated with the integer program presented below:

Basic variables	Current values	x_1	x_2	x_3	x_4
x_1	$\frac{16}{5}$	1		$\frac{2}{5}$	$-\frac{1}{5}$
x_2	$\frac{23}{5}$		1	$\frac{1}{5}$	$\frac{2}{5}$
$(-z)$	$-\frac{133}{5}$			$-\frac{11}{5}$	$-\frac{2}{5}$

$$\text{Maximize } z = 4x_1 + 3x_2,$$

subject to:

$$2x_1 + x_2 + x_3 \qquad = 11,$$

$$-x_1 + 2x_2 \qquad + x_4 \leq 6,$$

$$x_j \geq 0 \quad \text{and integer} \quad (j = 1, 2, 3, 4).$$

a) Derive cuts from each of the rows in the optimal linear-programming tableau, including the objective function.

b) Express the cuts in terms of the variables x_1 and x_2. Graph the feasible region for x_1 and x_2 and illustrate the cuts on the graph.

c) Append the cut derived from the objective function to the linear program, and re-solve. Does the solution to this new linear program solve the integer program? If not, how would you proceed to find the optimal solution to the integer program?

19. Given the following integer linear program:

$$\text{Maximize } z = 5x_1 + 2x_2,$$

subject to:

$$5x_1 + 4x_2 \leq 21,$$
$$x_1, x_2 \geq 0 \quad \text{and integer,}$$

solve, using the cutting-plane algorithm. Illustrate the cuts on a graph of the feasible region.

20. The following knapsack problem:

$$\text{Maximize } \sum_{j=1}^{n} cx_j,$$

subject to:

$$\sum_{j=1}^{n} 2x_j \leq n,$$
$$x_j = 0 \quad \text{or} \quad 1 \quad (j = 1, 2, \ldots, n),$$

which has the same "contribution" for each item under consideration, has proved to be rather difficult to solve for most general-purpose integer-programming codes when n is an *odd* number.

a) What is the optimal solution when n is even? when n is odd?
b) Comment specifically on why this problem might be difficult to solve on general integer-programming codes when n is odd.

21. Suppose that a firm has N large rolls of paper, each W inches wide. It is necessary to cut N_i rolls of width W_i from these rolls of paper. We can formulate this problem by defining variables

$$x_{ij} = \text{Number of smaller rolls of width } W_i \text{ cut from large roll } j.$$

We assume there are m different widths W_i. In order to cut all the required rolls of width W_i, we need constraints of the form:

$$\sum_{j=1}^{n} x_{ij} = N_i \quad (i = 1, 2, \ldots, m).$$

Further, the number of smaller rolls cut from a large roll is limited by the width W of the large roll. Assuming no loss due to cutting, we have constraints of the form:

$$\sum_{i=1}^{m} W_i x_{ij} \leq W \quad (j = 1, 2, \ldots, N).$$

a) Formulate an objective function to minimize the number of large rolls of width W used to produce the smaller rolls.
b) Reformulate the model to minimize the total trim loss resulting from cutting. Trim loss is defined to be that part of a large roll that is unusable due to being smaller than any size needed.
c) Comment on the difficulty of solving each formulation by a branch-and-bound method.
d) Reformulate the problem in terms of the collection of possible patterns that can be cut from a given large roll. (*Hint.* See Exercise 25 in Chapter 1.)
e) Comment on the difficulty of solving the formulation in (c), as opposed to the formulations in (a) or (b), by a branch-and-bound method.

22. The Bradford Wire Company produces wire screening woven on looms in a process essentially identical to that of weaving cloth. Recently, Bradford has been operating at full capacity and is giving serious consideration to a major capital investment in additional looms. They currently have a total of 43 looms, which are in production all of their available hours. In order to justify the investment in additional looms, they must analyze the utilization of their existing capacity.

Of Bradford's 43 looms, 28 are 50 inches wide, and 15 are 80 inches wide. Normally, one or two widths totaling less than the width of the loom are made on a particular loom. With the use of a "center-tucker," up to three widths can be simultaneously produced on one loom; however, in this instance the effective capacities of the 50-inch and 80-inch looms are reduced to 49 inches and 79 inches, respectively. Under no circumstance is it possible to make more than three widths simultaneously on any one loom.

Figure E9–3 gives a typical loom-loading configuration at one point in time. Loom #1, which is 50 inches wide, has two 24-inch widths being produced on it, leaving 2 inches

	50″ Looms			80″ Looms	
Loom number	Widths	Waste	Loom number	Widths	Waste
1	24–24	2	29	48–32	0
2	30	20	30	24–26–28	1
3	40	10	31	48–32	0
4	$30\frac{1}{4}$	$19\frac{3}{4}$	32	48–32	0
5	48	2	33	36–36	8
6	$32–14\frac{1}{2}$	$3\frac{1}{2}$	34	36–36	8
7	28	22	35	60–18	2
8	$26\frac{1}{2}$	$23\frac{1}{2}$	36	36–36	8
9	30	20	37	42–34	4
10	30	20	38	24–26–28	1
11	35	15	39	24–26–28	1
12	31	19	40	24–26–28	1
13	$34\frac{1}{4}$	$15\frac{3}{4}$	41	48–32	0
14	36	14	42	36–36	8
15	32	18	43	42–34	4
16	34–16	0			
17	$29\frac{3}{8}$	$20\frac{5}{8}$	80″ Total		46
18	30	20			$337\frac{3}{4}$
19	36	14	Grand total		$383\frac{3}{4}$
20	$47\frac{1}{2}$	$2\frac{1}{2}$			
21	30	20			
22	20–30	0			
23	$48\frac{7}{8}$	$1\frac{1}{8}$			
24	28–22	0			
25	$30–14\frac{1}{2}$	$5\frac{1}{2}$			
26	42	8			
27	32–18	0			
28	$28\frac{1}{2}$	$21\frac{1}{2}$			
50″ Total		$337\frac{3}{4}$			

Fig. E9–3 Typical loom loading.

of unused or "wasted" capacity. Loom #12 has only a 31-inch width on it, leaving 19 inches of the loom unused. If there were an order for a width of 19 inches or less, then it could be produced at zero marginal machine cost along with the 31-inch width already being produced on this loom. Note that loom #40 has widths of 24, 26, and 28 inches, totaling 78 inches. The "waste" here is considered to be only 1 inch, due to the reduced effective capacity from the use of the center-tucker. Note also that the combination of widths 24, 26, and 30 is not allowed, for similar reasons.

The total of $383\frac{3}{4}$ inches of "wasted" loom capacity represents roughly seven and one-half 50-inch looms; and some members of Bradford's management are sure that there must be a more efficient loom-loading configuration that would save some of this "wasted" capacity. As there are always numerous additional orders to be produced, any additional capacity can immediately be put to good use.

The two types of looms are run at different speeds and have different efficiencies. The 50-inch looms are operated at 240 pics per second, while the 80-inch looms are operated at 214 pics per second. (There are 16 pics to the inch.) Further, the 50-inch looms are more efficient, since their "up time" is about 85% of the total time, as compared to 65% for the 80-inch looms.

The problem of scheduling the various orders on the looms sequentially over time is difficult. However, as a first cut at analyzing how efficiently the looms are currently operating, the company has decided to examine the loom-loading configuration at one point in time as given in Fig. E9–3. If these same orders can be rearranged in such a way as to free up one or two looms, then it would be clear that a closer look at the utilization of existing equipment would be warranted before additional equipment is purchased.

a) Since saving an 80-inch loom is not equivalent to saving a 50-inch loom, what is an appropriate objective function to minimize?

b) Formulate an integer program to minimize the objective function suggested in part (a).

23. In the export division of Lowell Textile Mills, cloth is woven in lengths that are multiples of the piece-length required by the customer. The major demand is for 18-meter piece-lengths, and the cloth is generally woven in 54-meter lengths.

Cloth cannot be sold in lengths greater than the stipulated piece-length. Lengths falling short are classified into four groups. For 18-meter piece-lengths, the categories and the contribution per meter are as follows:

Category	Technical term	Length (Meters)	Contribution/ Meter
A	First sort	18	1.00
B	Seconds	11–17	0.90
C	Short lengths	6–10	0.75
D	Rags	1–5	0.60
J	Joined parts*	18	0.90

* Joined parts consist of lengths obtained by joining two pieces such that the shorter piece is at least 6 meters long.

The current cutting practice is as follows. Each woven length is inspected and defects are flagged prominently. The cloth is cut at these defects and, since the cloth is folded in exact meter lengths, the lengths of each cut piece is known. The cut pieces are then cut again, if necessary, to obtain as many pieces of first sort as possible. The short lengths are joined wherever possible to make joined parts.

Since the process is continuous, it is impossible to pool all the woven lengths and then decide on a cutting pattern. Each woven length has to be classified for cutting before it enters the cutting room.

As an example of the current practice, consider a woven length of 54 meters with two defects, one at 19 meters and one at 44 meters. The woven length is first cut at the defects, giving three pieces of lengths 19, 25, and 10 meters each. Then further cuts are made as follows:

Piece length	Further cuts
25	18 + 7
19	18 + 1
10	10

The resulting contribution is

$$2 \times 18 \times 1.00 + (7 + 10) \times 0.75 + 1 \times 0.60 = 49.35.$$

It is clear that this cutting procedure can be improved upon by the following alternative cutting plan:

Piece length	Further cuts
25	18 + 7
19	8 + 11
10	10

By joining $7 + 11$ and $8 + 10$ to make two joined parts, the resulting contribution is:

$$18 \times 1.00 + 18 \times 2 \times 0.90 = 50.40.$$

Thus with one woven length, contribution can be increased by $1.05 by merely changing the cutting pattern. With a daily output of 1000 such woven lengths (two defects on average), substantial savings could be realized by improved cutting procedures.

a) Formulate an integer program to maximize the contribution from cutting the woven length described above.

b) Formulate an integer program to maximize the contribution from cutting a general woven length with no more than four defects. Assume that the defects occur at integral numbers of meters.

c) How does the formulation in (b) change when the defects are *not* at integral numbers of meters?

24. Custom Pilot Chemical Company is a chemical manufacturer that produces batches of specialty chemicals to order. Principal equipment consists of eight interchangeable reactor vessels, five interchangeable distillation columns, four large interchangeable centrifuges, and a network of switchable piping and storage tanks. Customer demand comes in the form of orders for batches of one or more (usually not more than three) specialty chemicals, normally to be delivered simultaneously for further use by the customer. Custom Pilot Chemical fills these orders by means of a sort of pilot plant for each specialty chemical formed by inter-connecting the appropriate quantities of equipment. Sometimes a specialty chemical requires processing by more than one of these "pilot" plants, in which case one or more batches of intermediate products may be produced for further processing. There is no shortage of piping and holding tanks, but the expensive equipment (reactors, distillation columns, and centrifuges) is frequently inadequate to meet the demand.

The company's schedules are worked out in calendar weeks, with actual production always taking an integer multiple of weeks. Whenever a new order comes in, the scheduler immediately makes up a precedence tree-chart for it. The precedence tree-chart breaks down the order into "jobs." For example, an order for two specialty chemicals, of which one needs an intermediate stage, would be broken down into three jobs (Fig. E9–4). Each job requires certain equipment, and each is assigned a preliminary time-in-process, or "duration." The durations can always be predicted with a high degree of accuracy.

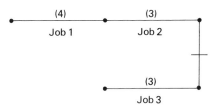

Figure E9–4

Currently, Custom Pilot Chemical has three orders that need to be scheduled (see the accompanying table). Orders can be started at the beginning of their arrival week and should be completed by the end of their *week due*.

							Resource requirements	
Order number	Job number	Precedence relations	Arrival week	Duration in weeks	Week-due	Reactors	Distillation columns	Centri-fuges
AK14	1	None	15	4	22	5	3	2
	2	(1)	15	3	22	0	1	1
	3	None	15	3	22	2	0	2
AK15	1	None	16	3	23	1	1	1
	2	None	16	2	23	2	0	0
	3	(1)	16	2	23	2	2	0
AK16	1	None	17	5	23	2	1	1
	2	None	17	1	23	1	3	0

For example, AK14 consists of three jobs, where Job 2 cannot be started until Job 1 has been completed. Figure E9–4 is an appropriate precedence tree for this order. Generally, the resources required are tied up for the entire duration of a job. Assume that Custom Pilot Chemical is just finishing week 14.

a) Formulate an integer program that minimizes the total completion time for all orders.
b) With a more complicated list of orders, what other objective functions might be appropriate? How would these be formulated as integer programs?

25. A large electronics manufacturing firm that produces a single product is faced with rapid sales growth. Its planning group is developing an overall capacity-expansion strategy, that would balance the cost of building new capacity, the cost of operating the new and existing facilities, and the cost associated with unmet demand (that has to be subcontracted outside at a higher cost).

The following model has been proposed to assist in defining an appropriate strategy for the firm.

Decision variables

Y_{it}	A binary integer variable that will be 1 if a facility exists at site i in period t, and 0 otherwise.
$(IY)_{it}$	A binary integer variable that will be 1 if facility is constructed at site i in period t, and 0 otherwise.
A_{it}	Capacity (sq ft) at site i in period t.
$(IA)_{it}$	The increase in capacity at site i in period t.
$(DA)_{it}$	The decrease in capacity at site i in period t.
P_{it}	Total units produced at site i in period t.
U_t	Total unmet demand (subcontracted) units in period t.

Forecast parameters

D_t	Demand in units in period t.

Cost parameters–Capacity

s_{it}	Cost of operating a facility at site i in period t.
d_{it}	Cost of building a facility at site i in period t.
a_{it}	Cost of occupying 1 sq ft of fully equipped capacity at site i in period t.
b_{it}	Cost of increasing capacity by 1 sq ft at site i in period t.
c_{it}	Cost of decreasing capacity by 1 sq ft at site i in period t.

Cost parameters–Production

o_t	Cost of unmet demand (subcontracting + opportunity cost) for one unit in period t.
v_{it}	Tax-adjusted variable cost (material + labor + taxes) of producing one unit at site i in period t.

Productivity parameters

$(pa)_{it}$	Capacity needed (sq ft years) at site i in period t to produce one unit of the product.

Policy parameters

$\bar{F}_{it}, \underline{F}_{it}$	Maximum, minimum facility size at site i in period t (if the site exists).
G_{it}	Maximum growth (sq ft) of site i in period t.

Objective function

The model's objective function is to minimize total cost:

$$\text{Min} \sum_{t=1}^{T} \sum_{i=1}^{T} s_{it} Y_{it} + d_{it}(IY)_{it}$$

$$+ \sum_{t=1}^{T} \sum_{i=1}^{N} a_{it} A_{it} + b_{it}(IA)_{it} + c_{it}(DA)_{it}$$

$$+ \sum_{t=1}^{T} \sum_{i=1}^{N} v_{it} P_{it} + \sum_{t=1}^{T} o_t U_t$$

Description of constraints

1. *Demand constraint*

$$\sum_{i=1}^{N} P_{it} + U_t = D_t$$

2. *Productivity constraint*

$$(pa)_{it}P_{it} \leq A_{it} \qquad \begin{cases} \text{for } i = 1, 2, \ldots, N, \\ \text{for } t = 1, 2, \ldots, T, \end{cases}$$

3. *Facility-balance constraint*

$$Y_{it-1} + (IY)_{it} = Y_{it} \qquad \begin{cases} \text{for } i = 1, 2, \ldots, N, \\ \text{for } t = 1, 2, \ldots, T, \end{cases}$$

4. *Area-balance constraint*

$$A_{it-1} + (IA)_{it} - (DA)_{it} = A_{it} \qquad \begin{cases} \text{for } i = 1, 2, \ldots, N, \\ \text{for } t = 1, 2, \ldots, T, \end{cases}$$

5. *Facility-size limits*

$$A_{it} \geq Y_{it}\underline{F}_{it}$$
$$A_{it} \leq Y_{it}\overline{F}_{it} \qquad \begin{cases} \text{for } i = 1, 2, \ldots, N, \\ \text{for } t = 1, 2, \ldots, T, \end{cases}$$

6. *Growth constraint*

$$(IA)_{it} - (DA)_{it} \leq G_{it} \qquad \begin{cases} \text{for } i = 1, 2, \ldots, N, \\ \text{for } t = 1, 2, \ldots, T, \end{cases}$$

7. *Additional constraints*

$$0 \leq Y_{it} \leq 1$$
$$Y_{it} \quad \text{integer} \qquad \begin{cases} \text{for } i = 1, 2, \ldots, N, \\ \text{for } t = 1, 2, \ldots, T, \end{cases}$$

All variables nonnegative.

Explain the choice of decision variables, objective function, and constraints. Make a detailed discussion of the model assumptions. Estimate the number of constraints, continuous variables, and integer variables in the model. Is it feasible to solve the model by standard branch-and-bound procedures?

26. An investor is considering a total of I possible investment opportunities ($i = 1, 2, \ldots, I$), during a planning horizon covering T time periods ($t = 1, 2, \ldots, T$). A total of b_{it} dollars is required to initiate investment i in period t. Investment in project i at time period t provides an income stream $a_{i,t+1}, a_{i,t+2}, \ldots, a_{i,T}$ in succeeding time periods. This money is available for reinvestment in other projects. The total amount of money available to the investor at the beginning of the planning period is B dollars.

a) Assume that the investor wants to maximize the net present value of the net stream of cash flows (c_t is the corresponding discount factor for period t). Formulate an integer-programming model to assist in the investor's decision.

b) How should the model be modified to incorporate timing restrictions of the form:

i) Project j cannot be initiated until after project k has been initiated; or

ii) Projects j and k cannot both be initiated in the same period?

c) If the returns to be derived from these projects are uncertain, how would you consider the risk attitudes of the decision-maker? [*Hint.* See Exercises 22 and 23 in Chapter 3.]

27. The advertising manager of a magazine faces the following problem: For week t, $t = 1, 2, \ldots, 13$, she has been allocated a maximum of n_t pages to use for advertising. She has received requests $r_1, r_2, \ldots, r_B$ for advertising, bid r_k indicating:

i) the initial week i_k to run the ad,

ii) the duration d_k of the ad (in weeks),

iii) the page allocation a_k of the ad (half-, quarter-, or full-page),

iv) a price offer p_k.

The manager must determine which bids to accept to maximize revenue, subject to the following restrictions:

i) Any ad that is accepted must be run in consecutive weeks throughout its duration.

ii) The manager cannot accept conflicting ads. Formally, subsets T_j and $\bar{T}_j$ for $j = 1, 2, \ldots, n$ of the bids are given, and she may not select an ad from both T_j and $\bar{T}_j$ ($j = 1, 2, \ldots, n$). For example, if $T_1 = \{r_1, r_2\}$, $\bar{T}_1 = \{r_3, r_4, r_5\}$, and bid r_1 or r_2 is accepted, then bids r_3, r_4, or r_5 must all be rejected; if bid r_3, r_4, or r_5 is accepted, then bids r_1 and r_2 must both be rejected.

iii) The manager must meet the Federal Communication Commission's balanced advertising requirements. Formally, subsets S_j and $\bar{S}_j$ for $j = 1, 2, \ldots, m$ of the bids are given; if she selects a bid from S_j, she must also select a bid from $\bar{S}_j$ ($j = 1, 2, \ldots, m$). For example, if $S_1 = \{r_1, r_3, r_8\}$ and $S_2 = \{r_4, r_6\}$, then either request r_4 or r_6 must be accepted if any of the bids r_1, r_3, or r_8 are accepted.

Formulate as an integer program.

28. The m-traveling-salesman problem is a variant of the traveling-salesman problem in which m salesmen stationed at a home base, city 1, are to make tours to visit other cities, 2, 3, $\ldots$, n. Each salesman must be routed through some, but not all, of the cities and return to his (or her) home base; each city must be visited by one salesman. To avoid redundancy, the sales coordinator requires that no city (other than the home base) be visited by more than one salesman or that any salesman visit any city more than once.

Suppose that it cost c_{ij} dollars for any salesman to travel from city i to city j.

a) Formulate an integer program to determine the minimum-cost routing plan. Be sure that your formulation does not permit subtour solutions for any salesman.

b) The m-traveling-salesman problem can be reformulated as a single-salesman problem as follows: We replace the home base (city 1) by m fictitious cities denoted $1', 2', \ldots, m'$. We link each of these fictitious cities to each other with an arc with high cost $M > \sum_{i=1}^{n} \sum_{j=1}^{n} |c_{ij}|$, and we connect each fictitious city i' to every city j at cost $c_{i'j} = c_{ij}$. Figure E9–5 illustrates this construction for $m = 3$.

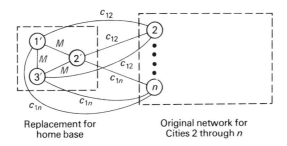

Replacement for
home base

Original network for
Cities 2 through n

Suppose that we solve this reformulation as a single-traveling-salesman problem, but identify each portion of the tour that travels from one of the fictitious cities i' through the original network and back to another fictitious city j' as a tour for one of the m salesmen. Show that these tours solve the m-traveling-salesman problem. [*Hint.* Can the solution to the single-salesman problem ever use any of the arcs with a high cost M? How many times must the single salesman leave from and return to one of the fictitous cities?]

29. Many practical problems, such as fuel-oil delivery, newspaper delivery, or school-bus routing, are special cases of the generic vehicle-routing problem. In this problem, a fleet of vehicles must be routed to deliver (or pick up) goods from a depot, node 0, to n drop-points, $i = 1, 2, \ldots, n$.

Let

$Q_k = $ Loading capacity of the kth vehicle in the fleet $(k = 1, 2, \ldots, m)$;

$d_i = $ Number of items to be left at drop-point i $(i = 1, 2, \ldots, n)$;

$t_i^k = $ Time to unload vehicle k at drop-point i $(i = 1, 2, \ldots, n; k = 1, 2, \ldots, m)$;

$t_{ij}^k = $ Time for vehicle k to travel from drop-point i to drop-point
$\quad j$ $(i = 0, 1, \ldots, n; j = 0, 1, \ldots, n; k = 1, 2, \ldots, m)$;

$c_{ij}^k = $ Cost for vehicle k to travel from node i to node j $(i = 0, 1, \ldots, n;$
$\quad j = 0, 1, \ldots, n; k = 1, 2, \ldots, m)$.

If a vehicle visits drop-point i, then it fulfills the entire demand d_i at that drop-point. As in the traveling or m-traveling salesman problem, only one vehicle can visit any drop-point, and no vehicle can visit the same drop-point more than once. The routing pattern must satisfy the service restriction that vehicle k's route take longer than T_k time units to complete.

Define the decision variables for this problem as:

$$x_{ij}^k = \begin{cases} 1 & \text{if vehicle } k \text{ travels from drop-point } i \text{ to drop-point } j, \\ 0 & \text{otherwise.} \end{cases}$$

Formulate an integer program that determines the minimum-cost routing pattern to fulfill demand at the drop-points that simultaneously meets the maximum routing-time constraints and loads no vehicle beyond its capacity. How many constraints and variables does the model contain when there are 400 drop-points and 20 vehicles?

ACKNOWLEDGMENTS

Exercise 22 is based on the Bradford Wire Company case written by one of the authors. Exercise 23 is extracted from the Muda Cotton Textiles case written by Munakshi Malya and one of the authors.

Exercise 24 is based on the Custom Pilot Chemical Co. case, written by Richard F. Meyer and Burton Rothberg, which in turn is based on "Multiproject Scheduling with Limited Resources: A Zero–One Programming Approach," *Management Science* 1969, by A. A. B. Pritsken, L. J. Watters, and P. M. Wolfe.

In his master's thesis (Sloan School of Management, MIT, 1976), entitled "A Capacity Expansion Planning Model with Bender's Decomposition," M. Lipton presents a more intricate version of the model described in Exercise 25.

The transformation used in Exercise 28 has been proposed by several researchers; it appears in the 1971 book *Distribution Management* by S. Eilson, C. Watson-Gandy, and N. Christofides, Hafner Publishing Co., 1971.

Design of a Naval Tender Job Shop

10

The project described in this chapter deals with the determination of the optimal man/machine configuration of a naval tender job shop. The approach illustrates how to complement the strengths of two important modeling techniques: mathematical programming and simulation. The problem can be characterized as one of capacity planning, where a great deal of uncertainty exists as to the demands on the system.

The approach taken is hierarchical in the sense that an aggregate planning model first suggests a man/machine configuration for the job shop and then a detailed model evaluates the performance of this configuration in a simulated environment. The aggregate model extends over a six-month planning horizon and has a mixed-integer programming structure. Once a proposed configuration for the job shop is generated by the aggregate planning model, the detailed model addresses the uncertainties and the precedence relations that affect the job-shop environment on an hour-by-hour basis. If the detailed evaluation of the configuration is unacceptable, constraints are modified in the aggregate model and the procedure is repeated.

This hierarchical approach, which combines optimization and simulation, provides a viable way of eliminating the weaknesses inherent in the two modeling approaches. The mixed-integer programming model cannot incorporate the detailed precedence relationships among jobs or include uncertainties explicitly, without becoming so large that it would be impossible to solve. On the other hand, the simulation model does not generate alternative man/machine configurations, but merely evaluates those presented to it. By using the two approaches jointly, it is possible both to generate a reasonable set of alternative configurations and to evaluate each one against a set of scheduling environments.

The hierarchical approach also facilitates the decision-maker's interaction with the models, and allows for comprehensive testing of a wide variety of options that can result in robust and efficient solutions.

10.1 THE PROBLEM DESCRIPTION

In order to support its fleet of ships, the U.S. Navy maintains a number of special-purpose ships, called naval tenders, which are dedicated to performing maintenance functions for the fleet. The purpose of this project is to develop an analytic approach for determining the machine configuration and manpower allocation for a naval-tender machine shop. Although this objective might be regarded as quite specific, the naval-tender machine shop can be considered a typical example of an intermittent-production, open job shop wherein general-purpose equipment and trained mechanics are held ready to meet a widely fluctuating demand for repair and manufacturing work. In this specific case, the work is generated by the fleet of ships for which the tender is responsible. The suggested design approach can be extended easily to other job-shop configurations.

The principal functions of the naval-tender machine shop are to repair pumps, valves, and similar mechanical equipment; to manufacture machinery replacement items; to perform grinding and engraving work; and to assist other tender shops. The typical modern naval-tender machine shop contains milling machines, drill presses, grinders, engine lathes, a furnace, a dip tank, bandsaws, shapers, turret lathes, boring mills, a disintegrator, an arbor press, and various other equipment. The shop normally is supervised by three chief petty officers, and the operations personnel include several petty officers (first-, second-, and third-class) as well as a large number of "non-rated" machinery repairmen.

In order to reduce the scope of the study to a more manageable size, we have excluded from our analysis the engraving and grinding sections, since there is virtually no cross-training between these sections and the remaining part of the tender, and their use does not overlap with the remaining activities of the naval tender. It would be straightforward to extend our suggested approach to cover grinding and engraving operations if proper data were available.

The Use of Numerically-Controlled Machines

A primary concern of our study is to examine the applicability of numerically-controlled machine technology to naval tenders. Numerical control provides for the automatic operation of machinery, using as input discrete numerical data and instructions stored on an appropriate medium such as punched or magnetized tape. The motions and operations of numerically-controlled machine tools are controlled primarily, not by an operator, but by an electronic director, which interprets coded instructions and directs a corresponding series of motions on the machine. Numerically-controlled machine tools combine the operations of several conventional machines, such as those used in milling, drilling, boring, and cutting operations.

To evaluate the decision-making problem properly, it is important to examine some of the advantages that numerically-controlled machines offer over conventional machine tools.

First, the combination of many machining activities into one machine may decrease setup losses, transportation times between machine groups, and waiting times in queues. Jobs then tend to spend less time in the shop, and so generally there will be less work-in-process and less need for finished-good inventories.

Second, the programmed instructions provided to the numerically-controlled machines can be transmitted, by conventional data lines or via a satellite system, to tenders in any of the seas and oceans. This creates an opportunity to develop centralized design, engineering, parts-programming, and quality-control organization, which can offer many economical, tactical, and manufacturing advantages.

Third, because numerically-controlled machines can be programmed to perform repetitive tasks very effectively, the jobs that they complete may require less rework and can be expected to result in less scrap. Also, superior quality control can be gained without relying on an operator to obtain close tolerances, and significant savings in inspection time can be realized.

Fourth, numerical control can have major impacts on tooling considerations. Tool wear can be accounted for, at given speed and feed rates, by automatically modifying the tooling operation to compensate for the changes in tool shape. With this compensation, numerous stops to adjust tools are avoided, and the tool is changed only when dull.

Finally, skills that are placed onto tape can be retained even though there are numerous transfers and changes in shop personnel. Despite this characteristic, inferences regarding individual skills and capabilities of operating personnel cannot be drawn so readily. Some advocates of numerical control would say that the personnel skill level could be lower in a numerical-control machine shop. On the other hand, several users of numerical control (e.g., the Naval Air Rework Facility in Quonset Point, R.I.) have indicated no change in overall skill levels when numerically-controlled machines are used, since experienced machinists are needed to ascertain whether the machine is producing the desired part or not.

All of these considerations suggest investigating the possibility of placing numerically-controlled machines on board naval tenders. The justification, or lack of justification, for the replacement of conventional machinery by numerically-controlled machinery is one of the important questions addressed in this study.

The Hierarchical Approach

There are two distinct levels of decision in the design of a naval-tender machine shop. The first level, which involves resource acquisition, encompasses the broad allocation of manpower and equipment for the tender, including the capital investment required for purchasing numerically-controlled machinery. The second level, dealing with the utilization of these resources, is concerned with the detailed scheduling of jobs, equipment, and workmen. Although it is theoretically possible to develop a single model to support these two levels of decisions, that approach seems unacceptable for the following reasons:

First, present computer and methodological capabilities do not permit solution of such a large integrated/detailed production model.

Second, and far more important, a single mathematical model does not provide sufficient cognizance of the distinct characteristics of time horizons, scopes, and information content of the various decisions.

Third, a partitioned, hierarchical model facilitates management interaction at the various levels.

Therefore we approach our problem by means of a hierarchical system in which the two decision levels are represented by two interactive models:

The aggregate model. Utilizing forecast demand as input, it makes decisions regarding machinery purchases and work-force size; and

The detailed model. Utilizing the machinery configuration and work-force schedule derived in the aggregate model as inputs, it simulates scheduling and assignment decisions, and determines shop performance as well as manpower and machinery utilization.

The time horizon of the aggregate model is at least as long as the tender deployment period, typically six months, whereas the detailed model addresses decisions on a daily or hourly basis. The two models are coupled and highly interactive. The aggregate model is oblivious to daily or weekly changes in the demand patterns, and does not consider bottlenecks or queue formations in front of machine groups; the machine configuration and work-force output of the aggregate model does, however, bound the daily and hourly operation of the tender machine shop. On the other hand, in its scheduling of jobs and assignment of machines and work force, the detailed model determines the utilization of manpower (undertime or overtime) and recognizes how demand uncertainties affect measures of shop performance (such as number of tardy jobs, or mean tardiness of jobs). This information can alter the machinery configuration and/or work-force allocations by labor class as determined by the aggregate model. It is proposed that the two models be solved sequentially, the aggregate model first, with iterations between the two models as necessary to address the interactions.

10.2 THE AGGREGATE MODEL

Model Objective and Assumptions

The primary objective of the aggregate model is to provide a preliminary allocation of manpower requirements by skill classes, and to propose a mix of conventional and numerically-controlled machines. These allocations are made by attempting to minimize the relevant costs associated with the recommended manpower and machine configuration, while observing several aggregate constraints on workload requirements, shop-space availability, weight limits, machine substitutability, existing conventional machine configuration, and machine and manpower productivities.

Several assumptions have been made to simplify the model structure, while maintaining an acceptable degree of realism in the problem representation.

First, demand requirements are assumed to be known deterministically. This assumption is relaxed in the detailed model where the impact of uncertainties in workload estimates are evaluated.

The work force (whose size and composition is to be determined by the model) is assumed to be fixed throughout the planning horizon. Hiring and firing options, which are available in *industrial* job-shop operations, are precluded in this application. Moreover, we have not allowed for overtime to be used as a method for absorbing demand fluctuations in the aggregate model; rather, overtime is reserved as an operational device to deal with uncertainties in the detailed model.

Rework due to operator error or machine malfunction is not considered explicitly. Rather, the productivity figures that are used include allocations for a normal amount of rework. Similarly, no provision is made for machine breakdown or preventive maintenance. Field studies indicated that a conventional machine is rarely "down" completely for more than one day; preventive maintenance time can be considered explicitly by the addition of "jobs" requiring manpower and machine time but no throughput material.

Finally, we assume that required raw materials always are available in necessary quantities in inventory on board the tender.

The Aggregation of Information

One of the basic issues to be resolved when designing an aggregate model is the consolidation of the pertinent information to be processed by the model in a meaningful way. Workload requirements are aggregated in terms of labor skill classes, machine types, and time periods. At the detailed level, these requirements are broken down into specific jobs, with precedence relationships, uncertainties in task-performance times, and due dates properly specified.

Now, we will review the major categories of information proposed in this model.

Timing

The planning horizon of the models has a six-month duration, which corresponds to a typical naval-tender deployment period. This planning horizon is divided into six equal time periods of one month each, because much of the data is gathered on a monthly basis and the current planning practices are based on monthly reports. The time periods are designated by t, for $t = 1, 2, \ldots, 6$.

Machines

The machines in the naval tender are grouped into two sections: heavy and light. Due to the nature of the operations performed, the logical candidates for substitution by numerically-controlled machines are the standard lathes in the light section and the universal/plain milling machines in the heavy section. Each of these two classes of machines, then, should be examined as separate machine groups. The remaining machines of the light section can be grouped into one large group, since the tender machine shop is labor-limited in this area; similarly, the remaining machines of the heavy section are grouped together. The machine groups are denoted by i, for $i = 1, 2, 3$, and 4.

Workforce

The workforce is divided into four groups corresponding to the current classes of skill/pay rates. Furthermore, the number of chief petty officers required for shop administration and supervision is assumed to be constant for all machinery configurations; since these costs are fixed, they do not enter into our analysis. The workforce classes are denoted by ℓ, for $\ell = 1, 2, 3$, and 4.

Model Formulation

Prior to presenting the mathematical formulation of the aggregate model, it is useful to introduce the symbolic notation used to describe the decision variables and the parameters of the model.

Decision Variables

Essentially, the decision variables are the number of conventional and numerically-controlled machines, the number of workers of various skill classes needed on the tender, and the allocation of the workers to the machines.

   The following list describes each of the decision variables included in the aggregate model, in terms of conventional machines:

$X_{\ell it}$ Number of hours of conventional machine time used by workers of skill-class ℓ on machine group i in time period t;

N_i Number of conventional machines required in machine group i;

$\underline{N}_{it}$ Number of conventional machines in machine group i required to satisfy the workload demand for that machine group in time period t;

$N_{\ell t}$ Number of skill-class ℓ workers required to meet the workload demand on conventional machinery in time period t;

R_i Number of conventional machines removed from machine group i;

M_ℓ Number of skill-class ℓ workers; and

$M_{\ell t}$ Number of skill-class ℓ workers required to meet the workload demand on *both* conventional and numerically-controlled machinery in time period t;

$X_{\ell it}^*, \underline{N}_{it}^*, N_i^*, N_{\ell t}^*$ Decision variables corresponding to $X_{\ell it}$, $\underline{N}_{it}$, N_i, $N_{\ell t}$, respectively, for the numerically-controlled machines.

Parameters

The parameters of the model reflect cost, productivity, demand, and weight and space limitations. The following list describes each of the parameters included in the aggregate model in terms of conventional machinery:

$C_{\ell t}$ Composite standard military pay rate (salary and benefits equivalent) charged for a worker of skill-class ℓ in time period t;

f_i Productivity factor reflecting an increased throughput rate for jobs that are completed on a numerically-controlled machine rather than on a conventional machine, for a particular machine group i—for example, if in the aggregate a set of jobs requires 100 hours of numerically-controlled lathe time or 300 hours of conventional lathe time, then the productivity factor for a numerically-controlled machine in the lathe-machine group would be 3;

$d_{\ell it}$ Number of hours of conventional machine time to be performed by workers of skill-class ℓ on machine group i in time period t;

h_{it} Number of hours that a conventional machine in machine group i can be productive in time period t;

k_i Constant $(0 \leq k_i \leq 1)$, reflecting the fact that a certain amount of the demand workload cannot be performed on a numerically-controlled machine. In the context of constraint (6), if $k_i = 0$, then *all* of the demand can be accomplished on numerically-controlled machinery, whereas if $k_i = 1.0$, then all of the demand must be met by work on *conventional* machinery;

b_i Original number of conventional machines in machine group i, prior to any substitution by numerically-controlled machinery;

a_i Deck area required for a machine in machine group i;

k' Constant that can be utilized to introduce more (or less) free deck space in the tender machine shop: k' is the ratio of the areas of removed machines to the areas of numerically-controlled machines brought aboard and, as such, reflects the limited deck area available for the mounting of machinery; if $k' = 1$, then the amount of space devoted to machinery *cannot* be altered;

w_i Weight of a conventional machine to be removed from machine group i;

m Maximum permissible machinery weight, reflecting naval architecture (weight-constrained design) or other design constraints on the bringing aboard of *additional* weight;

$h_{\ell t}$ Number of manhours that a worker of skill-class ℓ will be available for productive work on conventional machinery during time period t;

k'' Constant used in smoothing manpower requirements on second and later iterations through the aggregate model; $k'' = 1$ for the first iteration; use of k'' will become clear when constraint (13) is discussed later in this section.

$h_{it}^*, a_i^*, w_i^*, h_{\ell t}^*$ Parameters corresponding to h_{it}, a_i, w_i, $h_{\ell t}$, respectively, for the numerically-controlled machines.

In addition,

C_{it}^* Share of the acquisition, installation and incremental operation, maintenance, and overhead costs attendant to bringing aboard a numerically-controlled machine into machine group i, attributable to time period t.

Mathematical Formulation

The mathematical formulation of the aggregate model is as follows:

$$\text{Minimize} \sum_{t} \sum_{\ell} C_{\ell t} M_{\ell} + \sum_{t} \sum_{i} C_{it}^{*} N_{i} + \sum_{i} (1) R_{i},$$

subject to:

$$X_{\ell it} \quad\quad + f_i X_{\ell it}^{*} \quad\quad = d_{\ell it}, \quad\quad \text{all } \ell, i, t \quad\quad (1)$$

$$\sum_{\ell} X_{\ell it} \quad - h_{it} \underline{N}_{it} \quad\quad = 0, \quad\quad \text{all } i, t \quad\quad (2)$$

$$\sum_{\ell} X_{\ell it}^{*} \quad - h_{it}^{*} \underline{N}_{it}^{*} \quad\quad = 0, \quad\quad \text{all } i, t \quad\quad (3)$$

$$\underline{N}_{it} \quad\quad - N_i \quad\quad \leqq 0, \quad\quad \text{all } i, t \quad\quad (4)$$

$$\underline{N}_{it}^{*} \quad\quad - N_i^{*} \quad\quad \leqq 0, \quad\quad \text{all } i, t \quad\quad (5)$$

$$\sum_{\ell} X_{\ell it} \quad\quad \geqq k_i \sum_{\ell} d_{\ell it}, \quad\quad \text{all } i, t \quad\quad (6)$$

$$R_i \quad\quad + N_i \quad\quad \leqq b_i, \quad\quad \text{all } i \quad\quad (7)$$

$$a_i R_i \quad\quad - k' a_i^{*} N_i^{*} \quad\quad \geqq 0, \quad\quad \text{all } i \quad\quad (8)$$

$$\sum_{i} w_i^{*} N_i^{*} - \sum_{i} w_i R_i \quad \leqq m, \quad\quad (9)$$

$$\sum_{i} X_{\ell it} \quad - h_{\ell t} N_{\ell t} \quad\quad = 0, \quad\quad \text{all } \ell, t \quad\quad (10)$$

$$\sum_{i} X_{\ell it}^{*} \quad - h_{\ell t}^{*} N_{\ell t}^{*} \quad\quad = 0, \quad\quad \text{all } \ell, t \quad\quad (11)$$

$$N_{\ell t} \quad\quad + N_{\ell t}^{*} - M_{\ell t} = 0, \quad\quad \text{all } \ell, t \quad\quad (12)$$

$$k'' M_{\ell t} \quad\quad - M_{\ell} \quad\quad \leqq 0, \quad\quad \text{all } \ell, t \quad\quad (13)$$

$$X_{\ell it}, X_{\ell it}^{*}, N_{it}, N_{it}^{*}, R_i, N_{\ell t}, N_{\ell t}^{*}, M_{\ell t}, \quad\quad \text{all nonnegative}, \quad\quad (14)$$

$$N_i, N_i^{*}, R_i, M_{\ell}, \quad\quad \text{all nonnegative integers.} \quad\quad (15)$$

The resulting model is a mixed-integer program. We will now briefly comment on the model structure.

The objective function attempts to minimize the manpower costs and the acquisition, installation, and incremental costs introduced by the numerically-controlled machines. The third cost component in the objective function discourages the removal of more conventional machines than necessary by assigning a fictitious penalty of one dollar for the removal of one conventional machine. This gives some extra capacity to the naval tender and allows for more flexibility in its operation.

Constraint (1) requires that the demand be satisfied by a proper combination of conventional and numerically-controlled machines. Note that f_i is a factor that represents the increase in throughput for jobs processed in a numerically-controlled machine rather than a conventional machine.

Constraints (2) and (3) convert hours of conventional and numerically-controlled machines required each month to number of machines. Constraints (4) and (5) specify that the numbers of each type of machine required at *any* time cannot exceed the number carried aboard throughout the time horizon.

Constraint (6) requires that a given fraction of the demand must be met by conventional machinery, since *numerical control is not universally applicable.*

Constraint (7) states that the number of machines to be removed from a machine group, R_i, plus the actual number required N_i, *cannot exceed the initial number b_i of machines in the group.* Restrictions (8) and (9) represent constraints on deck area availability, and weight limits.

Constraints (10) and (11) determine the required manpower for conventional and numerically-controlled machinery during each month. Constraint (12) simply computes the sum of the two manpower needs. Finally, constraint (13), when $k'' = 1.0$, requires that a man of skill-class ℓ needed at any time during the planning horizon be ordered aboard at the *start* of the time horizon and kept aboard until the end of the planning horizon (in our case, a six-month deployment). The factor k'' is provided for use on subsequent runs in an iterative process: For example, if the overtime utilization of skill-class ℓ exceeds the desires of the decision-maker, k'' can be set greater than 1.0, thereby requiring additional personnel aboard.

10.3 THE DETAILED MODEL

Model Objective and Assumptions

The objective of the detailed model is to test the preliminary recommendations obtained from the aggregate model regarding machine and manpower mix, against a more realistic environment, which includes the uncertainties present in the daily operation of the job shop, the precedence relationships that exist in scheduling production through the various work centers, and the congestion generated by executing the production tasks.

Demand requirements are being specified in terms of individual jobs. Each job has a given duration, which is defined by means of a probability distribution. Alternative paths through the machine shop, e.g., from a numerically-controlled lathe to a conventional drill press, or from a conventional lathe to a conventional drill press, are specified. Each alternative path includes certain precedence relationships that must be observed; e.g., a shaft must be turned on a lathe and then a keyway has to be cut on a boring mill; the keyway *cannot be cut before* the shaft is turned.

Most machine-shop personnel coming aboard work first in the light section and then move to the heavy section of the tender. Therefore, we will assume that workers in the heavy section can perform work in both sections, whereas those in the light section are not assignable to the heavy section. We also assume that workers of a higher skill class can accomplish work normally assigned to a lower skill class; in other words, there is *downward substitutability* among worker skill classes. Further, a job can be worked on by only one worker of one skill class at a time; this assumption ignores the fact that large bulky items may require more than one man to set up a machine, but the amount of time required for this setup is generally quite small compared to the overall time on one machine with one worker.

It is assumed that each job is broken down into its smallest components. In order to satisfy the constraint stating one man and one machine for each operation, as well as the precedence relationships, we shall permit no overlapping of operations. A job which has two or more parts that can be worked in parallel is decomposed into two or more new jobs, with the appropriate due dates.

No preemption of jobs is allowed. This is not to say that a job leaving its first operation and entering a queue for its second operation cannot be delayed by a higher-priority job, but only that, once a machine and a man *have been committed* to performing an operation on a specific job, that operation on that job will be completed without interruption, irrespective of the higher-priority arrivals at the queue for that machine.

As before, we assumed that required raw materials, or satisfactory substitute materials, are always in inventory on board the tender in necessary quantities.

The Simulation Approach

In Chapter 1 we discussed the basic characteristics of simulation models. The essence of simulation is to provide a realistic and detailed representation of the problem under study, which allows the decision-maker to test various alternatives he might want to consider. The simulation model evaluates each alternative by calculating its corresponding measure of performance. It is important to emphasize that simulation models do not generate an optimum solution, but simply permit the evaluation of alternative solutions supplied externally by the decision-maker.

A simulation model was chosen to represent the detailed characteristics of the job-shop activities. Simulation has proved to be a very effective and flexible modeling tool for dealing with queueing networks such as a job-shop scheduling problem. Basically, the simulation model identifies each machine that is part of the machine shop and each job that has to be processed in the job shop. The dispatching rules that govern the order in which jobs are processed and their sequencing through the shop, the characteristics of the jobs themselves, and the availability of machine and manpower capabilities determine how fast the jobs can be processed and what overall measures of performance will be obtained from the job-shop operation. Common measures of performance are: percentage of jobs to be processed on time; total tardiness in job execution; utilization of manpower and machines, and so forth. The simulation models allow us to incorporate a number of characteristics of the job-shop performance that have not been taken into account in the aggregate model. The most important of these characteristics are uncertainties in job completion times, priority rules associated with job execution (since some jobs are more important than others), precedence relationships associated with the various activities or tasks that are part of an individual job, alternative ways of executing these activities (i.e., using either a conventional or a numerically-controlled machine), and so on.

More specifically, the basic elements of the simulation model are:

Jobs, which flow through a network of machines that perform a variety of operations; where the sequence, machine groups, worker skill levels, and service times at each step are a function of and specified by the job itself.

A job is composed of various activities:

Activities are the basic elements of a job. They utilize multiple resources (machines, manpower, and material), and require time to be performed;

Flow lines connecting a network of activities, defining a sequence of operations, and denoting a direction of flow; and

Boundary elements, that is, points of job origination (sources) and job termination (sinks).

These elements provide the network configuration. In addition, the following input data should be provided by the user:

Service times to perform the various activities; these are random variables specified by their probability distributions, which depend on the individual characteristics of each job;

Job routing through alternative paths within the network;

Queue disciplines, that regulate the order in which jobs waiting at a station are processed. Common queue disciplines to be used are FCFS (first come, first served), shortest processing time (the activity with shortest processing time goes first), due dates (the job with closer due date is processed first), and so on;

Operating schedules for the system, whereby standard workdays can be established, and the system closed or open to arriving jobs according to some predetermined role. The operating schedules keep track of the passage of time, and thus simulate the time dimensions of the problem.

Figures 10.1 and 10.2 present flow charts describing how these elements are integrated into a job-shop simulator model. A job arriving in the shop is characterized by its priority, the minimum skill-class worker required, its preferred and acceptable alternative paths through the shop (where applicable), and the service times needed at each node on the respective paths. For example, consider a job specified by $P/L/R\text{-}ab/S\text{-}cd$: The first digit P refers to priority (1 through 9), which is determined exogenous to the simulator by combining the requesting ship's assigned priority (1, 2, 3, 4) and the initial slack (due date minus arrival date minus expected operating time); the second digit L refers to the minimum skill-class worker (1 through 4) required to accomplish the job; the next group $R\text{-}ab$ refers to the preferred path

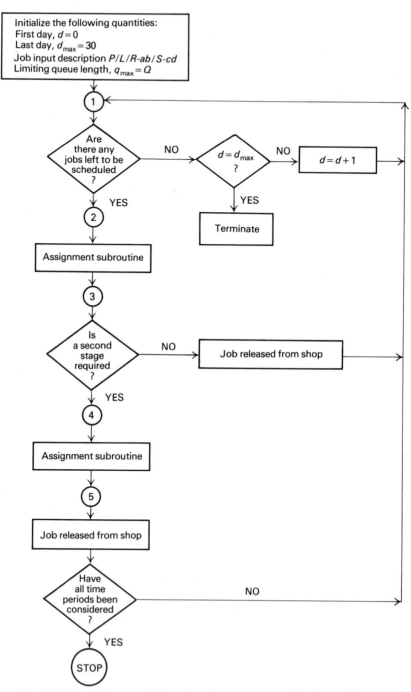

Fig. 10.1 Simulation-model flow chart for a two-stage job-shop problem.

(in this case R) and the probability distribution characterizing the duration of the two-stage production activities (i.e., we are assuming that each job is composed of at most two activities, with processing times a and b, which have to be executed one after the other); the last group S-cd refers to an acceptable alternative path (in this case S) and the operating-time distributions at each stage on the path.

A specific example that provides detailed characterization of the input data is given in Section 10.6. Figure 10.1 provides an overall representation of the simulation model for a two-stage job-shop problem. It is simple, conceptually, to expand the simulation to cover more complex job-shop situations, where jobs are allowed to have any number of activities in parallel and in series.

The assignment subroutine is described in Fig. 10.2. This subroutine assigns each individual job to a specific machine in accordance with the job specification, the machine availabilities, and the queue discipline adopted (in our example, we use FCFS = first come, first served). The flow-chart description is presented in very broad terms, explaining the major transactions that take place in the simulation, but avoiding unnecessary detailed information.

10.4 INTERACTION BETWEEN THE AGGREGATE AND DETAILED MODELS

We have indicated in the previous two sections how the resource acquisition and resource utilization decisions associated with the job-shop tender problem have been partitioned into two manageable models. We now analyze the way in which the two models are linked and the iterative nature of their interaction. Figure 10.3 illustrates this integrative scheme.

First, the aggregate model is solved, obtaining an initial recommendation for machine and manpower requirements. Then, these requirements are examined by the decision-maker to check their consistency with existing managerial policies that have not been included explicitly in the initial model formulation. New constraints or changes in the cost structure might be used to eliminate potential inconsistencies.

For example, the manpower requirements might have violated a desired pyramid-like manning organization, which can be preserved by adding the following constraints:

$$M_1 \leqq M_2, \qquad M_2 \leqq M_3, \qquad \text{and} \qquad M_3 \leqq M_4.$$

These constraints might not be included initially, to give an idea of an optimum manpower composition without these additional requirements.

In order to prevent excessive undertime, the following constraint might be used:

$$\sum_{t=1}^{T} M_{\ell t} - T(0.75)M_\ell \geqq 0,$$

on appropriate labor-class ℓ over T aggregate time periods. This constraint would require the average utilization of the workers of class ℓ to be at least 75 percent over the T time periods.

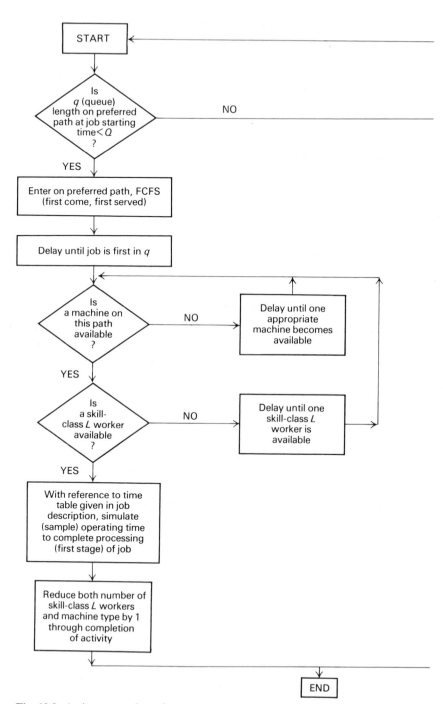

Fig. 10.2 Assignment subroutines.

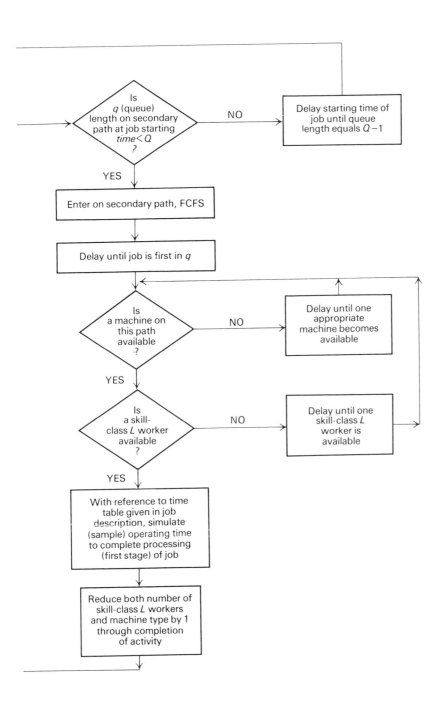

Fig. 10.2 (*continued*)

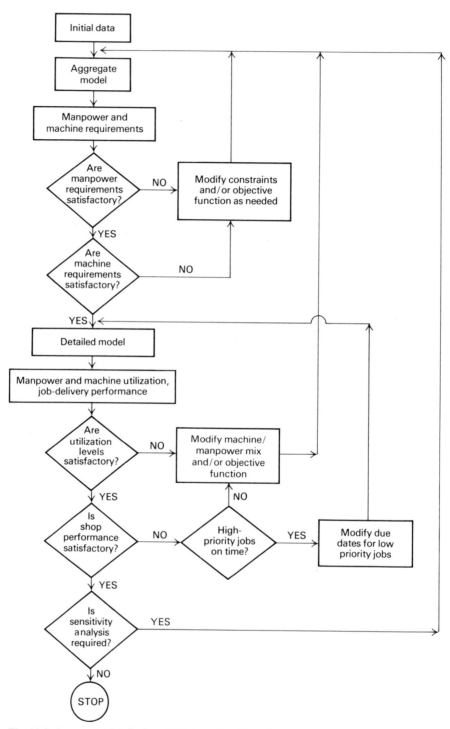

Fig. 10.3 Aggregate-detailed model interaction flow chart.

Alternatively, if the manpower utilization rate seems to be excessive at the aggregate level, leaving little or no room for absorbing demand uncertainties, the following constraint could be added for the appropriate skill class ℓ:

$$\sum_{t=1}^{T} M_{\ell t} - T(0.90)M_{\ell} \leq 0,$$

forcing the average utilization of workers belonging to skill-class ℓ to be less than 90 percent over T time periods.

Similar types of constraints may be utilized for the machinery mix. For example, if the decision-maker desires to impose an upper bound on the ratio of numerically-controlled lathes to conventional lathes, the following constraint could be introduced:

$$N_1^* - 3N_1 \leq 0,$$

which would require the machinery mix to provide at least three conventional lathes for each numerically-controlled lathe. Direct upper bounds also can be imposed on the number of machines to be used (for example, $N_1^* \leq 2$), thus permitting better machine utilization

Adding new constraints to the problem will allow the decision-maker to explore the cost sensitivity to the proposed changes. In a linear-programming model, most of this information is provided directly by the shadow prices associated with the original model constraints. Our aggregate model, however, is of a mixed-integer programming type, which does not generate similar shadow-price information. This is the reason for the more elaborate sensitivity analysis.

Changes also can be performed in the cost structure of the initial model, by including hiring and firing costs (although this is not applicable in the naval-tender case), overtime penalties, backordering costs, and so forth.

Once a satisfactory combination of manpower and machinery requirements have been obtained, a simulation is conducted with these data as input parameters to the detailed model. The manpower and machine utilization levels obtained from the simulation then are examined. If these levels are not considered acceptable, new changes in the manpower and machine composition and/or in the cost structure may be indicated. These changes then will modify the aggregate model formulation, which will force a new iteration to take place. If the utilization levels are satisfactory, the shop performance (in terms of delivery dates versus due dates) can be checked. In the actual tender environment, jobs with lower priorities could be "slipped" for completion at a later time. Once both acceptable utilization and shop performance levels are obtained, a sensitivity analysis can be conducted to test how robust the performed manpower and machine configurations are to changes in the problem parameters. The results of the sensitivity analysis may indicate that some of the parameters, constraints, or demand characteristics should be modified, and the problem is run again starting with the aggregate model.

The proposed hierarchical approach provides the decision-maker with an effective tool to test the performance of the job-shop operations under a wide range of anticipated conditions, and thus permits a satisfactory solution to the tender design problem, which performs well against a wide collection of possible job characteristics.

10.5 IMPLEMENTING THE MODEL

The Experimental Environment

The naval-tender machine-shop configuration used is similar to that found on the latest generation of destroyer tenders, exemplified by USS PUGET SOUND (AD–38). A field study was conducted, in which the historical workload over several months was examined, and the month of May 1973 was chosen as typical. Several days were spent with the leading petty officers of both the light and heavy sections. Each job was analyzed in detail regarding work accomplished and problems encountered, and the following data were collected for each job: description; skill class required; machine(s) required; prescribed sequence of operations; time distributions of each node in the sequence (to determine favorable, most likely, and pessimistic estimates); setup times; job release date to shop; job due date from shop; lot-size or number of items to be manufactured; and job priority as assigned by the customer ship. These jobs (approximately 178) comprised the workload for the first month ($t = 1$). For the other five months, various perturbations about this benchmark month were permitted; changes also were made in terms of both skill-class and machine-group requirements for each month.

Once these data were determined, a field study was continued at the Naval Air Rework Facility, Quonset Point, Rhode Island. Each job from the month of May 1973 was explained in detail by a leading petty officer from USS PUGET SOUND to one or more numerically-controlled machine specialists. For those jobs (or portions of jobs) that could be accomplished readily on numerically-controlled lathes or machining centers, data was gathered similar to that specified above for the conventional machinery. We assumed that the same skill-class worker could perform the job on either conventional or numerically-controlled machinery. It was found that numerically-controlled machinery could be utilized for approximately half of the jobs (representing approximately half of the required conventional man-hours).

A description of the data input for the aggregate and detailed models is provided in Section 10.6.

Results of the Model Experimentation

Several tests were conducted with the aggregate model in order to assess the sensitivity of the results to varying conditions of demand and productivity improvements introduced by the numerically-controlled machines. Three different demand levels were analyzed, corresponding to 100, 110, and 120 percent of the May 1973 demand; and two productivity factors ($f_i = 3$ and 5) were tried. A summary of the sensitivity analysis results is provided in Fig. 10.4.

| | $f_i = 3.0$ | | | | | | $f_i = 5.0$ | | | | | |
| | $1.00d_{tit}$ | | $1.10d_{tit}$ | | $1.20d_{tit}$ | | $1.00d_{tit}$ | | $1.10d_{tit}$ | | $1.20d_{tit}$ | |
	Integer solution	Overall utilization	Integer solution	Overall utilization	Integer solution	Overall utilization	Integer solution	Overall utilization	Integer solution	Overall utilization	Integer solution	Overall utilization
N_1	6	0.81	6	0.88	7	0.80	6	0.83	8	0.70	8	0.78
N_2	4	0.65	4	0.72	2	0.83	4	0.67	4	0.72	4	0.50
N_1^*	2	0.81	2	0.89	2	0.92	1	0.93	1	0.91	1	0.97
N_2^*	0	—	0	—	1	0.59	0	—	0	—	1	0.44
M_1	3	0.93	3	0.90	3	0.92	3	0.74	3	0.75	3	0.91
M_2	3	0.58	3	0.63	2	0.52	2	0.80	3	0.61	2	0.77
M_3	4	0.95	4	0.97	4	0.91	3	0.87	4	0.77	3	0.96
M_4	8	0.86	9	0.86	10	0.84	9	0.84	9	0.90	11	0.83
Cost	$78,838		$81,966		$90,416		$68,276		$75,972		$83,981	

Fig. 10.4 Summary of sensitivity-analysis results with the aggregate model.

443

Detailed model results for the base case ($1.00d_{\ell it}$ and $f_i = 3$) are given in Fig. 10.5. These results were obtained by using the IBM Mathematical Programming System Expanded (MPSX), which provides a mixed-integer programming capability. It is interesting to note that, in the trial cases, when the productivity factors are 3.0 and 5.0, for the $1.00d_{\ell it}$ (i.e., original demand data) and $1.10d_{\ell it}$ (i.e., 10 percent increase in the original demand data) cases, the purchase of numerically-controlled machining centers was not recommended; purchase of a numerically-controlled machining center was recommended only when the demand was increased by twenty percent of the base case. Additionally, it is worth observing the results regarding numerically-controlled lathes: in the $f_i = 3.0$ case, utilization of two numerically-controlled lathes was recommended, whereas for the $f_i = 5.0$ case, only one such lathe was recommended. Upon first examination, the latter result may appear counterintuitive—if the machines were more efficient, it may be reasoned, more of them should have been introduced. Alternatively, however, since the machines were more efficient ($f_i = 5.0$ versus $f_i = 3.0$), and since the fractions of the total work that could be performed on them was constrained, only one numerically-controlled lathe was required to accomplish its share of the workload.

	N_{i1}	N_{i2}	N_{i3}	N_{i4}	N_{i5}	N_{i6}	Integer solution	Overall utilization
N_1	4.464	6.00	4.571	4.850	4.164	5.164	6	0.811
N_2	2.457	2.750	2.471	2.371	2.578	3.064	4	0.654
N_1^*	1.843	1.048	1.526	1.617	1.886	1.828	2	0.812
N_2^*	0	0	0	0	0	0	0	—

	M_{11}	M_{12}	M_{13}	M_{14}	M_{15}	M_{16}	Integer solution	Overall utilization
M_1	3.00	2.625	3.00	2.083	3.00	3.00	3	0.928
M_2	1.439	1.542	1.458	2.008	1.480	2.432	3	0.575
M_3	4.00	4.00	4.00	3.450	4.00	3.402	4	0.952
M_4	5.795	7.119	6.309	7.071	6.812	7.997	8	0.856

Total cost = $78,838

Fig. 10.5 Aggregate-model results for the case $1.00d_{\ell it}$ and $f_i = 3.0$.

With respect to the manpower/machinery costs shown in the aggregate-model results, direct comparison was possible between $1.00d_{\ell it}$ ($f_i = 3.0$ and $f_i = 5.0$) cases and a $1.00d_{\ell it}$ ($f_i = 0$, i.e., numerically-controlled machinery not introduced) case. When $f_i = 0$, the required manpower can be obtained by dividing the total conventional workload for skill-class ℓ by the available number of hours $h_{\ell t}$ for the appropriate skill class; since no numerically-controlled machines are introduced, there is no acquisition cost, and therefore no penalty cost was incurred for removing existing machinery. Then, the required manpower would be 3 of skill-class 1, 3 of skill-class 2, 7 of skill-class 3, and 9 of skill-class 4, for a total cost of $80,675. Comparing this cost figure with the two earlier cited cases shows that, in the aggregate, the incorporation of numerically-controlled machine technology can indeed result in manpower reductions and a lower total cost. Using the manpower and machinery configuration recommended by the aggregate model, the detailed simulation tested what would actually occur on an hour-by-hour basis. The output of the detailed simulation reflects the performance of the configuration recommended by the aggregate model—measures of effectiveness presented here include the number of jobs completed, mean flow times of completed jobs, and manpower and machinery utilization. The first simulation was based on the manpower/machinery configuration determined by the aggregate model with $1.10d_{lit}$ and $f_i = 3$: 6 conventional lathes, 4 conventional milling machines, 2 numerically-controlled lathes, 0 numerically-controlled machining centers, the existing configuration of "other lights and heavies," 3 workers of skill-class 1, 3 workers of skill-class 2, 4 workers of skill-class 3, and 9 workers of skill-class 4.

The simulation run with the above configuration showed that 34 jobs (of 178 total jobs) were not completed; the average elapsed time to perform a job (including delays) was 25 hours and 55 minutes, and the average delay for a job was 9 hours and 38 minutes. The utilization data for the various manpower levels indicated that the eighth and ninth members of skill-class 4 were needed only 11.2 percent of the time; however, all four of the skill-class 3 workers were needed 88.4 percent of the time. The other manpower and machinery utilization appeared to be satisfactory.

A second simulation, with skill-class 3 augmented by one worker and skill-class 4 reduced by two workers (all other manpower and machinery pools unchanged) was carried out. For this case, 20 jobs were not performed; although the average elapsed time per job (including delays) increased slightly to 27 hours and 41 minutes, the average delay time for a job was reduced to 8 hours and 36 minutes, indicating that more jobs that required increased machining time were actually completed. The manpower utilization data shifted in such a way that all 7 members of skill-class 4 were needed 20 percent of the time, while all 5 members of skill-class 3 were needed 67.1 percent of the time.

Finally, a run was made with 6 skill-class 3 workers (the other resources unchanged). Marked improvement in the machine shop performance resulted: only 4 jobs were not completed; the average elapsed time (including delays) was relatively unchanged at 27 hours and 50 minutes, while the average delay was reduced significantly to 6 hours and 42 minutes. The utilization data for machines and manpower groups of interest are presented, for this last simulation, in Fig. 10.6. Utilization data for manpower skill-class 1 is not presented because this skill-class was assigned only 2 jobs; although 3 members of this skill-class were indicated (since $h_{lt} = 10$ hour/week), only 1 member (at 35 hours/week availability) was utilized in the simulation. Additionally, data for conventional milling machines is not reported: The aggregate model

Conventional lathes	
Number	%
0	0
1	0
2	2.6
3	7.3
4	10.4
5	16.8
6	62.9

Numerically-controlled lathe	
Number	%
0	0
1	61.2
2	38.8

Skill-class 2 manpower	
Number	%
0	0
1	1.8
2	20.2
3	78.0

Skill-class 3 manpower	
Number	%
0	0
1	0
2	4.5
3	17.2
4	19.7
5	12.9
6	45.8

Skill-class 4 manpower	
Number	%
0	1.4
1	18.3
2	11.5
3	19.0
4	17.1
5	10.3
6	1.6
7	20.6

Fig. 10.6 Utilization data for selected machinery types and manpower skill classes, as determined by the detailed model.

indicated that 5 milling machines were required (since they did not need to be removed), while the maximum simultaneous usage for these machines in the simulation was 3. The six conventional lathes recommended were utilized simultaneously in the detailed simulation 62.9 percent of the time, whereas the aggregate model indicated 4.464 were needed in the first time period. The higher utilization in the simulation reflected the congestion that occurred in the machine shop, which the aggregate model was designed to ignore. This last run seemed to offer a satisfactory solution.

The manpower and machinery configurations suggested by the model are as follows:

a) Remove two conventional lathes, and replace them with two numerically-controlled lathes;

b) Do not replace any of the existing conventional milling machines with numerically-controlled machining centers; and

c) Assign three machinery repairmen first class, three machinery repairmen second class, six machinery repairmen third class, and seven machinery repairmen "strikers" (a skill-class 4 worker in the model formulation).

10.6 DESCRIPTION OF THE DATA

This section is essentially an appendix describing the data used for implementing the model.

Data Input for Aggregate Model

1. $d_{\ell it}$ = Demand in conventional hours, representing workload in May 1973 ($t = 1$):

		Machine i			
Skill-class ℓ	1	2	3	4	Total
1	0	0	0	70	70
2	53	116	4	35	208
3	523	118	7	98	746
4	673	110	50	173	1006
Total	1249	344	61	376	2030

Small perturbations on these data generated demand for time period $t = 2, 3, 4, 5,$ and 6.

Source: Fieldwork on board of USS PUGET SOUND (AD–38)

2. $C = \sum_{t=1}^{6} C_{\ell t}$: Composite standard military pay rate (salary and benefits) for worker in skill-class ℓ, for a six-month period:

$$C_1 = \$5060, \qquad C_2 = \$4130, \qquad C_3 = \$3566, \qquad C_4 = \$3127.$$

Source: Navy Composite Standard Military Rate Table

3. C_i^* = Discounted acquisition, installation, and incremental operation, maintenance and overhead costs for numerically-controlled machine i, attributable to a six-month period:

$$C_1^* = \$5994 \qquad \text{(NC lathe)}$$

$$C_2^* = \$9450 \qquad \text{(NC machining center)}.$$

Source: Naval Ship Research and Development Center, Carderock, Md.

Assumption: Incremental expense of $1000 per machine. Economic life of 4 years, with salvage value assumed to be one-half of initial acquisition cost. Discount rate of 10 percent.

4. f_i = Factor that reflects the increase in productivity for numerically-controlled machine i with respect to corresponding conventional machine.

In the first set of runs $f_1 = f_2 = 3$. A second set of runs was conducted with $f_1 = f_2 = 5$. These values represent reasonable expected performance.

The remaining factors f_3 and f_4 have been set at $f_3 = f_4 = 0$, since machine groups 3 and 4 are not candidates for numerically-controlled replacement.

5. h_{it}, h_{it}^* = Number of hours during month t a machine must be available for the accomplishment of productive work. Set at:

$$\left(4\frac{\text{week}}{\text{month}}\right) \times \left(35\frac{\text{hours}}{\text{week}}\right) = 140\frac{\text{hours}}{\text{month}} \qquad \text{for all groups.}$$

6. k_i = Proportionality constant relating the minimum fraction of work that must be accomplished on conventional machinery:

$$k_1 = 0.5, \qquad k_2 = 0.5, \qquad k_3 = 1.0, \qquad k_4 = 1.0.$$

7. b_i = Original number of conventional machines aboard of USS PUGET SOUND:

$$b_1 = 9, \qquad b_2 = 5, \qquad b_3 = 12, \qquad b_4 = 15.$$

8. a_i, a_i^* = Deck area required for conventional and numerically-controlled machines, respectively:

$$a_1 = \ 96 \text{ sq ft}, \qquad a_2 = 225 \text{ sq ft},$$
$$a_1^* = 105 \text{ sq ft}, \qquad a_2^* = 225 \text{ sq ft}.$$

Since there is no substitution allowed for machine groups 3 and 4, a_3, a_4, a_3^* and a_4^* were set to zero.

The factor k', used to introduce more (or less) free deck space, was set at unity.

Source: Naval Ship Research and Development Center, Carderock, Md.

9. $h_{\ell t} = h_{\ell t}^*$ = Number of man-hours that a worker of skill-class ℓ must be available for productive work. Set at the following values:

$$h_{1t} = \left(10\frac{\text{hours}}{\text{week}}\right) \times \left(4\frac{\text{weeks}}{\text{month}}\right) = \ 40\frac{\text{hours}}{\text{month}},$$

$$h_{2t} = \left(30\frac{\text{hours}}{\text{week}}\right) \times \left(4\frac{\text{weeks}}{\text{month}}\right) = 120\frac{\text{hours}}{\text{month}},$$

$$h_{3t} = h_{4t} = \left(35\frac{\text{hours}}{\text{week}}\right) \times \left(4\frac{\text{weeks}}{\text{month}}\right) = 140\frac{\text{hours}}{\text{month}}.$$

Assumption: The figures of allowable productive hours worked per week by the first- and second-class petty officers ($\ell = 1$, $\ell = 2$) were chosen arbitrarily to permit their participation in various shop administration, supervision, and training functions. Since the basic shop work-week for planning purposes is 35 hours, this figure was chosen for the lower rated personnel.

Data Input for Detailed Model

1. Partial listing of jobs input to the detailed model:

Arrival date	Priority	Labor class	Preferred path	Timetables 1	Timetables 2	Alternative path	Timetables 1	Timetables 2
0	4	4	C	F	A	K	H	H
	5	3	B	L	H			
	5	3	B	L	H			
	5	3	B	L	H			
	9	4	D	A	—	A	H	—
	3	3	B	F	G			
	4	4	A	K	—			
	4	2	A	G	—			
	7	2	C	A	C	B	G	E
	8	4	A	I	—			
	4	4	D	G	—	A	M	—
	2	4	D	A	—	B	G	B
	4	4	T	H	B			
	2	3	D	C	—	A	A	—
	4	4	A	G	—			
	4	4	N	E	—			
	5	3	A	J	—			
	4	4	A	B	—			
1	3	4	A	C	—			
	6	3	H	J	C			
	5	4	A	D	—			
	5	4	A	C	—			
	1	4	X	K	—			
	1	4	A	E	—			
	8	2	M	D	I			
	5	4	Y	F	—	U	A	S
	7	4	A	G	—			
2	9	4	A	D	—			
	6	3	I	A	E	H	D	E
	5	2	B	K	K			
	1	3	A	G	—			
	7	4	A	K	—			
3	9	3	E	F	A	B	M	I
	9	4	C	A	A	B	E	B
	9	3	Y	H	—	B	P	E
	9	3	D	F	—	A	F	—
	4	3	D	A	—	A	D	—
	9	4	F	B	C	B	N	C
	7	2	C	A	C	B	G	E
	7	2	B	A	M			

Note. The complete listing included 30 days. Descriptions of machine paths and timetables used are given in the following pages.

2. Paths in simulation network:

Path	Node 1	Node 2	Node 3	Node 4
A	QA1	Conv. Lathe		
B	QB1	Conv. Lathe	QB2	Conv. Mill
C	QC1	NC Mill	QC2	NC Lathe
D	QD1	NC Lathe		
E	QE1	NC Lathe	QE2	NC Mill
F	QF1	NC Lathe	QF2	Conv. Mill
G	QG1	Vertical Mill		
H	QH1	Conv. Lathe	QH2	Cleerman Drill
I	QI1	NC Lathe	QI2	Cleerman Drill
J	QJ1	Band Saw	QJ2	Conv. Lathe
K	QK1	Cleerman Drill	QK2	Conv. Lathe
L	QL1	Gap Lathe	QL2	Conv. Mill
M	QM1	Monarch Lathe	QM2	Conv. Mill
N	QN1	Horiz. Bar Mill		
P	QP1	Band Saw		
Q	QQ1	Horiz. Tur. Lathe		
R	QR1	Radial Drill		
T	QT1	Conv. Mill	QT2	Cleerman Drill
U	QU1	Conv. Lathe	QU2	Wells Index
V	QV1	Vert. Tur. Lathe		
W	QW1	Conv. Mill	QW2	Vertical Mill
X	QX1	Drill Press		
Y	QY1	NC Mill		
Z	QZ1	Bullard		

3. Processing timetables utilized in detailed model:

Table	Cum. prob.	Time	Cum. prob.	Time	Cum. prob.	Time	Cum. prob.	Time	Cum. prob.	Time
A	0.0	0.25	0.25	0.33	0.5	0.5	0.75	0.75	1.0	1.0
B	0.0	0.5	0.25	0.6	0.5	1.0	0.75	1.3	1.0	1.5
C	0.0	1.5	0.25	1.6	0.5	2.0	0.75	2.3	1.0	2.5
D	0.0	2.5	0.25	2.6	0.5	3.0	0.75	3.3	1.0	3.5
E	0.0	3.5	0.25	3.6	0.5	4.0	0.75	4.3	1.0	4.5
F	0.0	4.5	0.25	4.6	0.5	5.0	0.75	5.3	1.0	5.5
G	0.0	5.0	0.25	5.4	0.5	7.0	0.75	8.6	1.0	9.0
H	0.0	6.0	0.25	6.4	0.5	8.0	0.75	9.6	1.0	10.0
I	0.0	7.5	0.25	8.0	0.5	10.0	0.75	12.5	1.0	13.0
J	0.0	9.0	0.25	9.5	0.5	12.0	0.75	15.0	1.0	15.5
K	0.0	11.0	0.25	11.5	0.5	14.0	0.75	17.0	1.0	17.5
L	0.0	13.0	0.25	13.5	0.5	16.0	0.75	19.0	1.0	20.0

Table	Cum. prob.	Time	Cum. prob.	Time	Cum. prob.	Time	Cum. prob.	Time	Cum. prob.	Time
M	0.0	18.0	0.25	14.0	0.5	21.0	0.75	23.0	1.0	25.0
N	0.0	23.0	0.25	23.7	0.5	25.0	0.75	26.5	1.0	27.0
P	0.0	26.0	0.25	26.7	0.5	28.0	0.75	30.0	1.0	31.0
Q	0.0	28.2	0.25	28.7	0.5	30.0	0.75	33.5	1.0	35.0
R	0.0	30.0	0.25	31.5	0.5	35.0	0.75	40.0	1.0	42.0
S	0.0	35.0	0.25	36.0	0.5	40.0	0.75	45.0	1.0	49.0
T	0.0	55.0	0.25	56.0	0.5	60.0	0.75	64.0	1.0	65.0

EXERCISES

Problem Description

1. Try to structure the overall nature of the naval-tender job-shop design problem. What is the relevant planning horizon for the manpower and machine configuration decisions? What are the appropriate decision variables, parameters, constraints, and objective function? Are the uncertainties of the problem very significant? Can the problem be formulated as a single model? What are the difficulties in approaching the problem via a single model? What is the essence of the proposed hierarchical approach? What are the advantages and disadvantages of the hierarchical approach versus a single-model approach?

The Aggregate Model

2. Discuss the stages of model formulation with respect to the aggregate model. In particular, interpret the objective function and constraints given by expressions (1) to (15). How many decision variables and constraints are there? How many of those decision variables are required to assume only integer values? Discuss the interpretation of the shadow prices associated with every constraint type. What important elements of the problem have been left out of this model formulation? Why?

The Detailed Model

3. Contrast the characteristics of optimization models and simulation models. Why has a simulation model been suggested as the detailed model of the naval-tender job-shop design problem? Would it have been possible to formulate the detailed model as an optimization model? Review and discuss the model description provided in the test. How would you change the flow chart of Figs. 10.1 and 10.2 if every job consisted of several activities in series and/or in parallel? What measures of performance do you propose to use to evaluate the job-shop efficiency? What alternative can be evaluated by means of the simulation model? How are these alternatives generated?

Interaction between the Aggregate and Detailed Models

4. Discuss the nature of the proposed interaction between the aggregate and detailed models represented in Fig. 10.3. What outputs of the aggregate model become inputs to the detailed model? How does the detailed model modify the aggregate-model recommendations? What mechanisms would you propose to enhance the interaction of the models?

Implementation of the Models

5. Discuss the implementation approach and the results of the model experimentation. Analyze the summaries provided in Figs. 10.4 and 10.5. What kind of experimental design would you have suggested? What conclusions can you draw from the existing results?

ACKNOWLEDGMENTS

The material in this chapter is based on the paper by Robert J. Armstrong and Arnoldo C. Hax, "A Hierarchical Approach for a Naval Tender Job-Shop Design," M.I.T. Operations Research Center, *Technical Report No. 101*, August 1974.

Dynamic Programming

11

Dynamic programming is an optimization approach that transforms a complex problem into a sequence of simpler problems; its essential characteristic is the multi-stage nature of the optimization procedure. More so than the optimization techniques described previously, dynamic programming provides a general framework for analyzing many problem types. Within this framework a variety of optimization techniques can be employed to solve particular aspects of a more general formulation. Usually creativity is required before we can recognize that a particular problem can be cast effectively as a dynamic program; and often subtle insights are necessary to restructure the formulation so that it can be solved effectively.

We begin by providing a general insight into the dynamic programming approach by treating a simple example in some detail. We then give a formal characterization of dynamic programming under certainty, followed by an in-depth example dealing with optimal capacity expansion. Other topics covered in the chapter include the discounting of future returns, the relationship between dynamic-programming problems and shortest paths in networks, an example of a continuous-state-space problem, and an introduction to dynamic programming under uncertainty.

11.1 AN ELEMENTARY EXAMPLE

In order to introduce the dynamic-programming approach to solving multistage problems, in this section we analyze a simple example. Figure 11.1 represents a street map connecting homes and downtown parking lots for a group of commuters in a model city. The arcs correspond to streets and the nodes correspond to intersections. The network has been designed in a diamond pattern so that every commuter must traverse five streets in driving from home to downtown. The design characteristics and traffic pattern are such that the total time spent by any commuter between intersections is independent of the route taken. However, substantial delays are

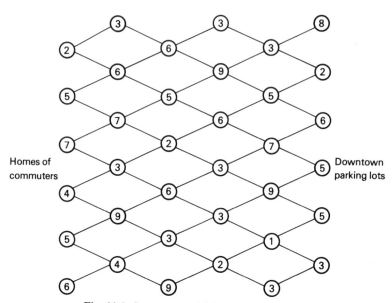

Fig. 11.1 Street map with intersection delays.

experienced by the commuters in the intersections. The lengths of these delays, in minutes, are indicated by the numbers within the nodes. We would like to minimize the total delay any commuter can incur in the intersections while driving from his home to downtown.

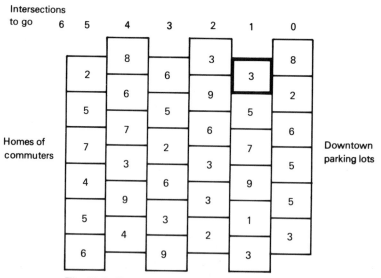

Fig. 11.2 Compact representation of the network.

Figure 11.2 provides a compact tabular representation for the problem that is convenient for discussing its solution by dynamic programming. In this figure, boxes correspond to intersections in the network. In going from home to downtown, any commuter must move from left to right through this diagram, moving at each stage only to an adjacent box in the next column to the right. We will refer to the "stages to go," meaning the number of intersections left to traverse, not counting the intersection that the commuter is currently in.

The most naive approach to solving the problem would be to enumerate all 150 paths through the diagram, selecting the path that gives the smallest delay. Dynamic programming reduces the number of computations by moving systematically from one side to the other, building the best solution as it goes.

Suppose that we move backward through the diagram from right to left. If we are in any intersection (box) with no further intersections to go, we have no decision to make and simply incur the delay corresponding to that intersection. The last column in Fig. 11.2 summarizes the delays with no (zero) intersections to go.

Our first decision (from right to left) occurs with one stage, or intersection, left to go. If for example, we are in the intersection corresponding to the highlighted box in Fig. 11.2, we incur a delay of three minutes in this intersection and a delay of either *eight* or *two* minutes in the last intersection, depending upon whether we move up or down. Therefore, the smallest possible delay, or optimal solution, in this intersection is $3 + 2 = 5$ minutes. Similarly, we can consider each intersection (box) in this column in turn and compute the smallest total delay as a result of being in each intersection. The solution is given by the bold-faced numbers in Fig. 11.3. The arrows indicate the optimal decision, up or down, in any intersection with one stage, or one intersection, to go.

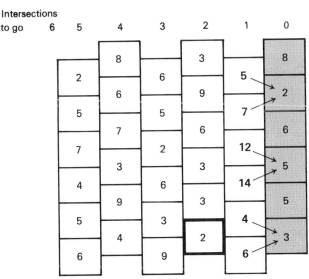

Fig. 11.3 Decisions and delays with one intersection to go.

Note that the numbers in bold-faced type in Fig. 11.3 completely summarize, for decision-making purposes, the total delays over the last two columns. Although the original numbers in the last two columns have been used to determine the bold-faced numbers, whenever we are making decisions to the left of these columns we need only know the bold-faced numbers. In an intersection, say the topmost with one stage to go, we know that our (optimal) remaining delay, including the delay in this intersection, is five minutes. The bold-faced numbers summarize all delays from this point on. For decision-making to the left of the bold-faced numbers, the last column can be ignored.

With this in mind, let us back up one more column, or stage, and compute the optimal solution in each intersection with two intersections to go. For example, in the bottom-most intersection, which is highlighted in Fig. 11.3, we incur a delay of two minutes in the intersection, plus *four* or *six* additional minutes, depending upon whether we move up or down. To minimize delay, we move *up* and incur a total delay in this intersection and *all remaining intersections* of $2 + 4 = 6$ minutes. The remaining computations in this column are summarized in Fig. 11.4, where the bold-faced numbers reflect the optimal total delays in each intersection with two stages, or two intersections, to go.

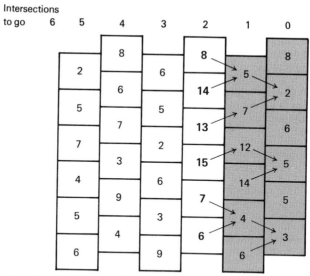

Fig. 11.4 Decisions and delays with two intersections to go.

Once we have computed the optimal delays in each intersection with two stages to go, we can again move back one column and determine the optimal delays and the optimal decisions with three intersections to go. In the same way, we can continue to move back one stage at a time, and compute the optimal delays and decisions

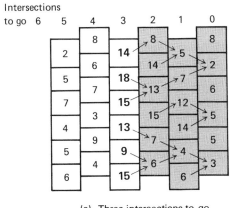

(a) Three intersections to go

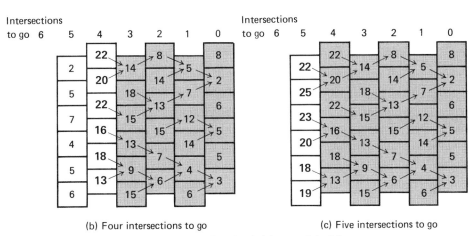

(b) Four intersections to go   (c) Five intersections to go

Fig. 11.5 Charts of optimal delays and decisions.

with four and five intersections to go, respectively. Figure 11.5 summarizes these calculations.

Figure 11.5(c) shows the optimal solution to the problem. The least possible delay through the network is 18 minutes. To follow the least-cost route, a commuter has to start at the second intersection from the bottom. According to the optimal decisions, or arrows, in the diagram, we see that he should next move down to the bottom-most intersection in column 4. His following decisions should be up, down, up, down, arriving finally at the bottom-most intersection in the last column.

However, the commuters are probably not free to arbitrarily choose the intersection they wish to start from. We can assume that their homes are adjacent to only one of the leftmost intersections, and therefore each commuter's starting point is fixed. This assumption does not cause any difficulty since we have, in fact, determined the routes of minimum delay from the downtown parking lots to *all* the commuter's homes. Note that this assumes that commuters do not care in which downtown lot they park. Instead of solving the minimum-delay problem for only a particular commuter, we have *embedded* the problem of the particular commuter in the more general problem of finding the minimum-delay paths from all homes to the group of downtown parking lots. For example, Fig. 11.5 also indicates that the commuter starting at the topmost intersection incurs a delay of 22 minutes if he follows his optimal policy of down, up, up, down, and then down. He presumably parks in a lot close to the second intersection from the top in the last column. Finally, note that three of the intersections in the last column are not entered by any commuter. The analysis has determined the minimum-delay paths from each of the commuter's homes to the group of downtown parking lots, not to each particular parking lot.

Using dynamic programming, we have solved this minimum-delay problem sequentially by keeping track of how many intersections, or stages, there were to go. In dynamic-programming terminology, each point where decisions are made is usually called a *stage* of the decision-making process. At any stage, we need only know which intersection we are in to be able to make subsequent decisions. Our subsequent decisions do not depend upon how we arrived at the particular intersection. Information that summarizes the knowledge required about the problem in order to make the current decisions, such as the intersection we are in at a particular stage, is called a *state* of the decision-making process.

In terms of these notions, our solution to the minimum-delay problem involved the following intuitive idea, usually referred to as the *principle of optimality*.

> *Any optimal policy has the property that, whatever the current state and decision, the remaining decisions must constitute an optimal policy with regard to the state resulting from the current decision.*

To make this principle more concrete, we can define the *optimal-value function* in the context of the minimum-delay problem.

> $v_n(s_n)$ = Optimal value (minimum delay) over the current and subsequent stages (intersections), given that we are in state s_n (in a particular intersection) with n stages (intersections) to go.

The optimal-value function at each stage in the decision-making process is given by the appropriate column of Fig. 11.5(c). We can write down a *recursive* relationship for computing the optimal-value function by recognizing that, at each stage, the decision in a particular state is determined simply by choosing the minimum total delay. If we number the states at each stage as $s_n = 1$ (bottom intersection) up to

$s_n = 6$ (top intersection), then

$$v_n(s_n) = \text{Min } \{t_n(s_n) + v_{n-1}(s_{n-1})\}, \tag{1}$$

subject to:

$$s_{n-1} = \begin{cases} s_n + 1 & \text{if we choose up and } n \text{ even,} \\ s_n - 1 & \text{if we choose down and } n \text{ odd,} \\ s_n & \text{otherwise,} \end{cases}$$

where $t_n(s_n)$ is the delay time in intersection s_n at stage n.

The columns of Fig. 11.5(c) are then determined by starting at the right with

$$v_0(s_0) = t_0(s_0) \qquad (s_0 = 1, 2, \ldots, 6), \tag{2}$$

and successively applying Eq. (1). Corresponding to this optimal-value function is an *optimal-decision function*, which is simply a list giving the optimal decision for each state at every stage. For this example, the optimal decisions are given by the arrows leaving each box in every column of Fig. 11.5(c).

The method of computation illustrated above is called *backward induction*, since it starts at the right and moves back one stage at a time. Its analog, *forward induction*, which is also possible, starts at the left and moves forward one stage at a time. The spirit of the calculations is identical but the interpretation is somewhat different. The optimal-value function for forward induction is defined by:

$u_n(s_n) = $ Optimal value (minimum delay) over the current and completed stages (intersections), given that we are in state s_n (in a particular intersection) with n stages (intersections) to go.

The recursive relationship for forward induction on the minimum-delay problem is

$$u_{n-1}(s_{n-1}) = \text{Min } \{u_n(s_n) + t_{n-1}(s_{n-1})\}, \tag{3}$$

subject to:

$$s_{n-1} = \begin{cases} s_n + 1 & \text{if we choose up and } n \text{ even,} \\ s_n - 1 & \text{if we choose down and } n \text{ odd,} \\ s_n & \text{otherwise,} \end{cases}$$

where the stages are numbered in terms of intersections to go. The computations are carried out by setting

$$u_5(s_5) = t_5(s_5) \qquad (s_5 = 1, 2, \ldots, 6), \tag{4}$$

and successively applying (3). The calculations for forward induction are given in Fig. 11.6. When performing forward induction, the stages are usually numbered in terms of the number of stages *completed* (rather than the number of stages to go). However, in order to make a comparison between the two approaches easier, we have avoided using the "stages completed" numbering.

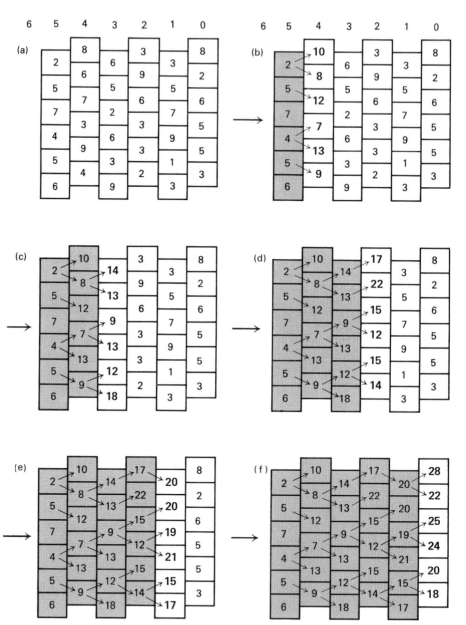

Fig. 11.6 Solution by forward induction.

The columns of Fig. 11.6(f) give the optimal-value function at each stage for the minimum-delay problem, computed by forward induction. This figure gives the minimum delays from each particular downtown parking lot to the *group* of homes of the commuters. Therefore, this approach will only guarantee finding the minimum delay path from the downtown parking lots to *one* of the commuters' homes. The method, in fact, finds the minimum-delay path to a particular origin only if that origin may be reached from a downtown parking lot by a backward sequence of arrows in Fig. 11.6(f).

If we select the minimum-delay path in Fig. 11.6(f), lasting 18 minutes, and follow the arrows backward, we discover that this path leads to the intersection second from the bottom in the first column. This is the same minimum-delay path determined by backward induction in Fig. 11.5(c).

Forward induction determined the minimum-delay paths from each individual parking lot to the *group of homes*, while backward induction determined the minimum-delay paths from each individual home to the *group of downtown parking lots*. The minimum-delay path between the two groups is guaranteed to be the same in each case but, in general, the remaining paths determined may be different. Therefore, when using dynamic programming, it is necessary to think about whether forward or backward induction is best suited to the problem you want to solve.

11.2 FORMALIZING THE DYNAMIC-PROGRAMMING APPROACH

The elementary example presented in the previous section illustrates the three most important characteristics of dynamic-programming problems:

Stages

The essential feature of the dynamic-programming approach is the structuring of optimization problems into multiple *stages*, which are solved sequentially one stage at a time. Although each one-stage problem is solved as an ordinary optimization problem, its solution helps to define the characteristics of the next one-stage problem in the sequence.

Often, the stages represent different time periods in the problem's planning horizon. For example, the problem of determining the level of inventory of a single commodity can be stated as a dynamic program. The decision variable is the amount to order at the beginning of each month; the objective is to minimize the total ordering and inventory-carrying costs; the basic constraint requires that the demand for the product be satisfied. If we can order only at the beginning of each month and we want an optimum ordering policy for the coming year, we could decompose the problem into 12 stages, each representing the ordering decision at the beginning of the corresponding month.

Sometimes the stages do not have time implications. For example, in the simple situation presented in the preceding section, the problem of determining the routes of minimum delay from the homes of the commuters to the downtown parking lots was formulated as a dynamic program. The decision variable was whether to choose *up* or *down* in any intersection, and the stages of the process were defined to be the number of intersections to go. Problems that can be formulated as dynamic programs with stages that do not have time implications are often difficult to recognize.

States

Associated with each stage of the optimization problem are the *states* of the process. The states reflect the information required to fully assess the consequences that the current decision has upon future actions. In the inventory problem given in this section, each stage has only one variable describing the state: the inventory level on hand of the single commodity. The minimum-delay problem also has one state variable: the intersection a commuter is in at a particular stage.

The specification of the states of the system is perhaps the most critical design parameter of the dynamic-programming model. There are no set rules for doing this. In fact, for the most part, this is an art often requiring creativity and subtle insight about the problem being studied. The essential properties that should motivate the selection of states are:

i) The states should convey enough information to make future decisions without regard to how the process reached the current state; and

ii) The number of state variables should be small, since the computational effort associated with the dynamic-programming approach is prohibitively expensive when there are more than two, or possibly three, state variables involved in the model formulation.

This last feature considerably limits the applicability of dynamic programming in practice.

Recursive Optimization

The final general characteristic of the dynamic-programming approach is the development of a *recursive optimization* procedure, which builds to a solution of the overall N-stage problem by first solving a one-stage problem and sequentially including one stage at a time and solving one-stage problems until the overall optimum has been found. This procedure can be based on a *backward induction* process, where the first stage to be analyzed is the final stage of the problem and problems are solved moving back one stage at a time until all stages are included. Alternatively, the recursive procedure can be based on a *forward induction* process, where the first stage to be solved is the initial stage of the problem and problems are solved moving forward one stage at a time, until all stages are included. In certain problem settings, only one of these induction processes can be applied (e.g., only backward induction is allowed in most problems involving uncertainties).

The basis of the recursive optimization procedure is the so-called *principle of optimality*, which has already been stated: an optimal policy has the property that, whatever the current state and decision, the remaining decisions must constitute an optimal policy with regard to the state resulting from the current decision.

General Discussion

In what follows, we will formalize the ideas presented thus far. Suppose we have a multistage decision process where the *return* (or cost) for a particular *stage* is:

$$f_n(d_n, s_n), \tag{5}$$

where d_n is a permissible *decision* that may be chosen from the set D_n, and s_n is the *state* of the process with n stages to go. Normally, the set of feasible decisions, D_n, available at a given stage depends upon the state of the process at that stage, s_n, and could be written formally as $D_n(s_n)$. To simplify our presentation, we will denote the set of feasible decisions simply as D_n. Now, suppose that there are a total of N stages in the process and we continue to think of n as the number of stages *remaining* in the process. Necessarily, this view implies a finite number of stages in the decision process and therefore a specific horizon for a problem involving time. Further, we assume that the state s_n of the system with n stages to go is a full description of the system for decision-making purposes and that knowledge of prior states is unnecessary. The next state of the process depends entirely on the current state of the process and the current decision taken. That is, we can define a *transition function* such that, given s_n, the state of the process with n stages to go, the subsequent state of the process with $(n - 1)$ stages to go is given by

$$s_{n-1} = t_n(d_n, s_n), \tag{6}$$

where d_n is the decision chosen for the current stage and state. Note that there is no uncertainty as to what the next state will be, once the current state and current decision are known. In Section 11.7, we will extend these concepts to include uncertainty in the formulation.

Our multistage decision process can be described by the diagram given in Fig. 11.7. Given the current state s_n which is a complete description of the system for decision-making purposes with n stages to go, we want to choose the decision d_n that will maximize the total return over the remaining stages. The decision d_n, which must be chosen from a set D_n of permissible decisions, produces a return at this stage of $f_n(d_n, s_n)$ and results in a new state s_{n-1} with $(n - 1)$ stages to go. The new state at the beginning of the next stage is determined by the transition function $s_{n-1} = t_n(d_n, s_n)$, and the new state is a complete description of the system for decision-making purposes with $(n - 1)$ stages to go. Note that the stage returns are independent of one another.

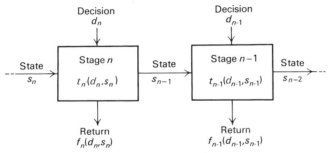

Fig. 11.7 Multistage decision process.

In order to illustrate these rather abstract notions, consider a simple inventory example. In this case, the state s_n of the system is the inventory level I_n with n months to go in the planning horizon. The decision d_n is the amount O_n to order this month. The resulting inventory level I_{n-1} with $(n - 1)$ months to go is given by the usual inventory-balance relationship:

$$I_{n-1} = I_n + O_n - R_n,$$

where R_n is the demand requirement this month. Thus, formally, the transition function with n stages to go is defined to be:

$$I_{n-1} = t_n(I_n, O_n) = I_n + O_n - R_n.$$

The objective to be minimized is the total ordering and inventory-carrying costs, which is the sum of the one-stage costs $C_n(I_n, O_n)$.

For the general problem, our objective is to maximize the sum of the return functions (or minimize the sum of cost functions) over all stages of the decision process; and our only constraints on this optimization are that the decisions chosen for each stage belong to some set D_n of permissible decisions and that the transitions from state to state be governed by Eq. (6). Hence, given that we are in state s_n with n stages to go, our optimization problem is to choose the decision variables $d_n, d_{n-1}, \ldots, d_0$ to solve the following problems:

$$v_n(s_n) = \text{Max} \left[f_n(d_n, s_n) + f_{n-1}(d_{n-1}, s_{n-1}) + \cdots + f_0(d_0, s_0) \right],$$

subject to:

$$
\begin{aligned}
s_{m-1} &= t_m(d_m, s_m) & (m = 1, 2, \ldots, n), \\
d_m &\in D_m & (m = 0, 1, \ldots, n).
\end{aligned}
\tag{7}
$$

We call $v_n(s_n)$ the *optimal-value function*, since it represents the maximum return possible over the n stages to go. Formally, we define:

$v_n(s_n) =$ Optimal value of all subsequent decisions, given that we are in state s_n with n stages to go.

Now since $f_n(d_n, s_n)$ involves only the decision variable d_n and not the decision variables $d_{n-1}, \ldots, d_0$, we could first maximize over this latter group for every possible d_n and then choose d_n so as to maximize the entire expression. Therefore, we can rewrite Eq. (7) as follows:

$$v_n(s_n) = \text{Max} \left\{ f_n(d_n, s_n) + \text{Max} \left[f_{n-1}(d_{n-1}, s_{n-1}) + \cdots + f_0(d_0, s_0) \right] \right\},$$

subject to: subject to:

$$
\begin{aligned}
s_{n-1} &= t_n(d_n, s_n) & s_{m-1} &= t_m(d_m, s_m) & (m = 1, 2, \ldots, n-1), \\
d_n &\in D_n, & d_m &\in D_m, & (m = 0, 1, \ldots, n-1).
\end{aligned}
\tag{8}
$$

Note that the second part of Eq. (8) is simply the optimal-value function for the $(n-1)$-stage dynamic-programming problem defined by replacing n with $(n-1)$ in (7). We can therefore rewrite Eq. (8) as the following recursive relationship:

$$v_n(s_n) = \text{Max} \left[f_n(d_n, s_n) + v_{n-1}(s_{n-1}) \right],$$

subject to:

$$s_{n-1} = t_n(d_n, s_n), \tag{9}$$

$$d_n \in D_n.$$

To emphasize that this is an optimization over d_n, we can rewrite Eq. (9) equivalently as:

$$v_n(s_n) = \text{Max} \left\{ f_n(d_n, s_n) + v_{n-1}[t_n(d_n, s_n)] \right\}, \tag{10}$$

subject to:

$$d_n \in D_n.$$

The relationship in either Eq. (9) or (10) is a formal statement of the *principle of optimality*. As we have indicated, this principle says that an optimal sequence of decisions for a multistage problem has the property that, regardless of the current decision d_n and current state s_n, all subsequent decisions must be optimal, given the state s_{n-1} resulting from the current decision.

Since $v_n(s_n)$ is defined recursively in terms of $v_{n-1}(s_{n-1})$, in order to solve Eqs. (9) or (10) it is necessary to initiate the computation by solving the "stage-zero" problem. The stage-zero problem is not defined recursively, since there are no more stages after the final stage of the decision process. The stage-zero problem is then the following:

$$v_0(s_0) = \text{Max} \ f_0(d_0, s_0), \tag{11}$$

subject to:

$$d_0 \in D_0.$$

Often there *is* no stage-zero problem, as $v_0(s_0)$ is identically zero for all final stages. In the simple example of the previous section, where we were choosing the path of minimum delay through a sequence of intersections, the stage-zero problem consisted of accepting the delay for the intersection corresponding to each final state.

In this discussion, we have derived the optimal-value function for *backward induction*. We could easily have derived the optimal-value function for *forward induction*, as illustrated in the previous section. However, rather than develop the analogous result, we will only state it here. Assuming that we continue to number the states "backwards," we can define the optimal-value function for forward induction as follows:

$u_n(s_n) = $ Optimal value of all prior decisions, given that we are in state s_n with n stages to go.

The optimal-value function is then given by:

$$u_{n-1}(s_{n-1}) = \text{Max}\ [u_n(s_n) + f_n(d_n, s_n)], \tag{12}$$

subject to:

$$s_{n-1} = t_n(d_n, s_n),$$

$$d_n \in D_n,$$

where the computations are usually initialized by setting

$$u_n(s_n) = 0,$$

or by solving some problem, external to the recursive relationship, that gives a value to being in a particular initial state. Note that, for forward induction, you need to think of the problem as one of examining all the combinations of current states and actions that produce a specific state at the next stage, and then choose optimally among these combinations.

It should be pointed out that nothing has been said about the specific form of the stage-return functions or the set of permissible decisions at each stage. Hence, what we have said so far holds regardless of whether the decisions are discrete, continuous, or mixtures of the two. All that is necessary is that the recursive relationship be solvable for the optimal solution at each stage, and then a *global* optimal solution to the overall problem is determined. The optimization problem that is defined at each stage could lead to the application of a wide variety of techniques, i.e., linear programming, network theory, integer programming, and so forth, depending on the nature of the transition function, the constraint set D_n, and the form of the function to be optimized.

It should also be pointed out that nowhere in developing the fundamental recursive relationship of dynamic programming was any use made of the fact that there were a finite number of states at each stage. In fact, Eqs. (9), (10), and (12) hold independent of the number of states. The recursive relationship merely needs to be solved for all possible states of the system at each stage. If the state space, i.e., the set of possible states, is continuous, and therefore an infinite number of states are possible at each stage, then the number of states is usually made finite by making a discrete approximation of the set of possible states, and the same procedures are used. An example of a dynamic-programming problem with a continuous state space is given in Section 11.6.

Finally, we have assumed certainty throughout our discussion so far; this assumption will be relaxed in Section 11.7, and a very similar formal structure will be shown to hold.

11.3 OPTIMAL CAPACITY EXPANSION

In this section, we further illustrate the dynamic-programming approach by solving a problem of optimal capacity expansion in the electric power industry.

Table 11.1 Demand and cost per plant ($ $\times$ 1000)

Year	Cumulative demand (in number of plants)	Cost per plant ($ $\times$ 1000)
1981	1	5400
1982	2	5600
1983	4	5800
1984	6	5700
1985	7	5500
1986	8	5200

A regional electric power company is planning a large investment in nuclear power plants over the next few years. A total of eight nuclear power plants must be built over the next six years because of both increasing demand in the region and the energy crisis, which has forced the closing of certain of their antiquated fossil-fuel plants. Suppose that, for a first approximation, we assume that demand for electric power in the region is known with certainty and that we must satisfy the minimum levels of cumulative demand indicated in Table 11.1. The demand here has been converted into equivalent numbers of nuclear power plants required by the end of each year. Due to the extremely adverse public reaction and subsequent difficulties with the public utilities commission, the power company has decided at least to meet this minimum-demand schedule.

The building of nuclear power plants takes approximately one year. In addition to a cost directly associated with the construction of a plant, there is a common cost of $1.5 million incurred when any plants are constructed in any year, independent of the number of plants constructed. This common cost results from contract preparation and certification of the impact statement for the Environmental Protection Agency. In any given year, at most three plants can be constructed. The cost of construction per plant is given in Table 11.1 for each year in the planning horizon. These costs are currently increasing due to the elimination of an investment tax credit designed to speed investment in nuclear power. However, new technology should be available by 1984, which will tend to bring the costs down, even given the elimination of the investment tax credit.

We can structure this problem as a dynamic program by defining the state of the system in terms of the cumulative capacity attained by the end of a particular year. Currently, we have no plants under construction, and by the end of each year in the planning horizon we must have completed a number of plants equal to or greater than the cumulative demand. Further, it is assumed that there is no need ever to construct more than eight plants. Figure 11.8 provides a graph depicting the allowable capacity (states) over time. Any node of this graph is completely described by the corresponding year number and level of cumulative capacity, say the node (n, p). Note that we have chosen to measure time in terms of *years to go* in the planning horizon. The cost of traversing any upward-sloping arc is the common cost of $1.5 million plus the plant costs, which depend upon the year of construction and whether 1, 2, or 3 plants are completed. Measured in thousands of dollars, these

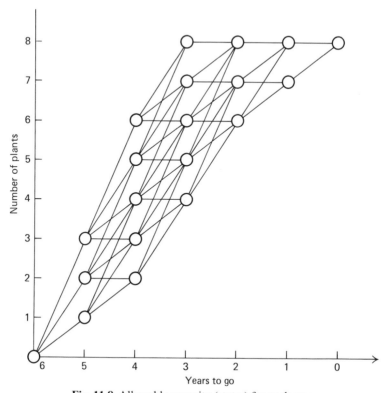

Fig. 11.8 Allowable capacity (states) for each stage.

costs are

$$1500 + c_n x_n,$$

where c_n is the cost per plant in the year n and x_n is the number of plants constructed. The cost for traversing any horizontal arc is zero, since these arcs correspond to a situation in which no plant is constructed in the current year.

Rather than simply developing the optimal-value function in equation form, as we have done previously, we will perform the identical calculations in tableau form to highlight the dynamic-programming methodology. To begin, we label the final state zero or, equivalently define the "stage-zero" optimal-value function to be zero for all possible states at stage zero. We will define a state as the cumulative total number of plants completed. Since the only permissible final state is to construct the entire cumulative demand of eight plants, we have $s_0 = 8$ and

$$v_0(8) = 0.$$

Now we can proceed recursively to determine the optimal-value function with one stage remaining. Since the demand data requires 7 plants by 1985, with one year to go the only permissible states are to have completed 7 or 8 plants. We can describe the situation by Tableau 1.

Tableau 1

Possible
new plants

s_1 \ d_1	0	1	$v_1(s_1)$	$d_1^*(s_1)$
Plants completed { 8	0	—	0	0
7	—	6,700	6,700	1

$c_1(s_1, d_1)$

The dashes indicate that the particular combination of current state and decision results in a state that is not permissible. In this table there are no choices, since, if we have not already completed eight plants, we will construct one more to meet the demand. The cost of constructing the one additional plant is the $1500 common cost plus the $5200 cost per plant, for a total of $6700. (All costs are measured in thousands of dollars.) The column headed $d_1^*(s_1)$ gives the optimal decision function, which specifies the optimal number of plants to construct, given the current state of the system.

Now let us consider what action we should take with two years (stages) to go. Tableau 2 indicates the possible costs of each state:

Tableau 2

s_2 \ d_2	0	1	2	$v_2(s_2)$	$d_2^*(s_2)$
8	0	—	—	0	0
7	6,700	7,000	—	6,700	0
6	—	13,700	12,500	12,500	2

$c_2(s_2, d_2) + v_1(s_1)$

If we have already completed eight plants with two years to go, then clearly we will not construct any more. If we have already completed seven plants with two years to go, then we can either construct the one plant we need this year or postpone its construction. Constructing the plant now costs $1500 in common costs plus $5500 in variable costs, and results in state 8 with one year to go. Since the cost of state 8 with one year to go is zero, the total cost over the last two years is $7000. On the other hand, delaying construction costs zero this year and results in state 7 with one year to go. Since the cost of state 7 with one year to go is $6700, the total cost over the last two years is $6700. If we arrive at the point where we have two years to go and have completed seven plants, it pays to delay the production of the last plant needed. In a similar way, we can determine that the optimal decision when in state 6 with two years to go is to construct two plants during the next year.

To make sure that these ideas are firmly understood, we will determine the optimal-value function and optimal decision with three years to go. Consider

Tableau 3 for three years to go:

Tableau 3

s_3 \ d_3	0	1	2	3	$v_3(s_3)$	$d_3^*(s_3)$
8	0	—	—	—	0	0
7	6,700	7,200	—	—	6,700	0
6	12,500	13,900	12,900	—	12,500	0
5	—	19,700	19,600	18,600	18,600	3
4	—	—	25,400	25,300	25,300	3

$$c_3(s_3, d_3) + v_2(s_2)$$

Now suppose that, with three years to go, we have completed five plants. We need to construct at least one plant this year in order to meet demand. In fact, we can construct either 1, 2, or 3 plants. If we construct one plant, it costs $1500 in common costs plus $5700 in plant costs, and results in state 6 with two years to go. Since the minimum cost following the optimal policy for the remaining two years is then $12,500, our total cost for three years would be $19,700. If we construct two plants, it costs the $1500 in common costs plus $11,400 in plant costs and results in state 7 with two years to go. Since the minimum cost following the optimal policy for the remaining two years is then $6700, our total cost for three years would be $19,600. Finally, if we construct three plants, it costs the $1500 in common costs plus $17,100 in plant costs and results in state 8 with two years to go. Since the minimum cost following the optimal policy for the remaining two years is then zero, our total cost for three years would be $18,600. Hence, the optimal decision, having completed five plants (being in state 5) with three years (stages) to go, is to construct three plants this year. The remaining tableaus for the entire dynamic-programming solution are determined in a similar manner (see Fig. 11.9).

Since we start the construction process with no plants (i.e., in state 0) with six years (stages) to go, we can proceed to determine the optimal sequence of decisions by considering the tableaus in the reverse order. With six years to go it is optimal to construct three plants, resulting in state 3 with five years to go. It is then optimal to construct three plants, resulting in state 6 with four years to go, and so forth. The optimal policy is then shown in the tabulation below:

Years to go	Construct	Resulting state
6	3	3
5	3	6
4	0	6
3	0	6
2	2	8
1	0	8

Hence, from Tableau 6, the total cost of the policy is $48.8 million.

Tableau 4

s_4 \ d_4	0	1	2	3	$v_4(s_4)$	$d_4^*(s_4)$
6	12,500	14,000	13,100	—	12,500	0
5	18,600	19,800	19,800	18,900	18,600	0
4	25,300	25,900	25,600	25,600	25,300	0
3	—	32,600	31,700	31,400	31,400	3
2	—	—	38,400	37,500	37,500	3

$$c_4(s_4, d_4) + v_3(s_3)$$

Tableau 5

s_5 \ d_5	0	1	2	3	$v_5(s_5)$	$d_5^*(s_5)$
3	31,400	32,400	31,300	30,800	30,800	3
2	37,500	38,500	38,000	36,900	36,900	3
1	—	44,600	44,100	43,600	43,600	3

$$c_5(s_5, d_5) + v_4(s_4)$$

Tableau 6

s_6 \ d_6	0	1	2	3	$v_6(s_6)$	$d_6^*(s_6)$
0	—	50,500	49,200	48,800	48,800	3

$$c_6(s_6, d_6) + v_5(s_5)$$

Fig. 11.9 Tableaus to complete power-plant example.

11.4 DISCOUNTING FUTURE RETURNS

In the example on optimal capacity expansion presented in the previous section, a very legitimate objection might be raised that the *present value of money* should have been taken into account in finding the optimal construction schedule. The issue here is simply that a dollar received today is clearly worth more than a dollar received one year from now, since the dollar received today could be invested to yield some additional return over the intervening year. It turns out that dynamic programming is extremely well suited to take this into account.

We will define, in the usual way, the one-period *discount factor* β as the present value of one dollar received *one period from now*. In terms of interest rates, if the interest rate for the period were i, then one dollar invested now would accumulate to $(1 + i)$ at the end of one period. To see the relationship between the discount factor β and the interest rate i, we ask the question "How much must be invested now to yield one dollar one period from now?" This amount is clearly the present value of a dollar received one period from now, so that $\beta(1 + i) = 1$ determines the relationship between β and i, namely, $\beta = 1/(1 + i)$. If we invest one dollar now for n periods at an interest rate per period of i, then the accumulated value at the end of n periods, assuming the interest is compounded, is $(1 + i)^n$. Therefore, the present value of one dollar received n periods from now is $1/(1 + i)^n$ or, equivalently, β^n.

The concept of discounting can be incorporated into the dynamic-programming framework very easily since we often have a return per period (stage) that we may wish to discount by the per-period discount factor. If we have an n-stage dynamic-programming problem, the optimal-value function, including the appropriate discounting of future returns, is given by

$$v_n(s_n) = \text{Max} \left[f_n(d_n, s_n) + \beta f_{n-1}(d_{n-1}, s_{n-1}) + \beta^2 f_{n-2}(d_{n-2}, s_{n-2}) \right. $$
$$\left. + \cdots + \beta^n f_0(d_0, s_0) \right],$$

subject to:

$$s_{m-1} = t_m(d_m, s_m) \qquad (m = 1, 2, \ldots, n),$$
$$d_m \in D_m, \qquad\qquad (m = 0, 1, \ldots, n), \tag{13}$$

where the stages (periods) are numbered in terms of *stages to go*. Making the same argument as in Section 11.3 and factoring out the β, we can rewrite Eq. (13) as:

$$v_n(s_n) = \text{Max} \left\{ f_n(d_n, s_n) \quad + \quad \beta \, \text{Max} \left[f_{n-1}(d_{n-1}, s_{n-1}) + \beta f_{n-2}(d_{n-2}, s_{n-2}) \right. \right. $$
$$\left. + \cdots + \beta^{n-1} f_0(d_0, s_0) \right],$$

subject to: subject to:
$$s_{n-1} = t_n(d_n, s_n) \qquad s_{m-1} = t_m(d_m, s_m) \quad (m = 1, 2, \ldots, n-1),$$
$$d_n \in D_n \qquad\qquad d_m \in D_m \qquad\qquad (m = 0, 1, \ldots, n-1). \tag{14}$$

Since the second part of Eq. (14) is simply the optimal-value function for the $(n-1)$-stage problem multiplied by β, we can rewrite Eq. (14) as

$$v_n(s_n) = \text{Max} \left[f_n(d_n, s_n) + \beta v_{n-1}(s_{n-1}) \right],$$

subject to:

$$s_{n-1} = t_n(d_n, s_n), \tag{15}$$
$$d_n \in D_n,$$

which is simply the recursive statement of the optimal-value function for backward induction with discounting. If $\beta = 1$, we have the case of *no discounting* and Eq. (15) is identical to Eq. (9). Finally, if the discount rate depends on the period, β can be replaced by β_n and (15) still holds.

We can look at the impact of discounting future-stage returns by considering again the optimal capacity expansion problem presented in the previous section. Suppose that the alternative uses of funds by the electric power company result in a 15 percent return on investment. This corresponds to a yearly discount factor of approximately 0.87. If we merely apply backward induction to the capacity expansion problem according to Eq. (15), using $\beta = 0.87$, we obtain the optimal-value function for each stage as given in Fig. 11.10.

s_1 \ d_1	0	1	$v_1(s_1)$	$d_1^*(s_1)$
8	0	—	0	0
7	—	6,700	6,700	1

$$c_1(s_1, d_1)$$

s_2 \ d_2	0	1	2	$v_2(s_2)$	$d_2^*(s_2)$
8	0	—	—	0	0
7	5,829	7,000	—	5,829	0
6	—	12,825	12,500	12,500	2

$$c_2(s_2, d_2) + \beta v_1(s_1)$$

s_3 \ d_3	0	1	2	3	$v_3(s_3)$	$d_3^*(s_3)$
8	0	—	—	—	0	0
7	5,071	7,200	—	—	5,071	0
6	10,875	12,271	12,900	—	10,875	0
5	—	18,075	17,971	18,600	17,971	2
4	—	—	23,775	23,671	23,671	3

$$c_3(s_3, d_3) + \beta v_2(s_2)$$

s_4 \ d_4	0	1	2	3	$v_4(s_4)$	$d_4^*(s_4)$
6	9,461	11,712	13,100	—	9,461	0
5	15,635	16,761	17,512	18,900	15,635	0
4	20,594	22,935	22,561	23,312	20,594	0
3	—	27,894	28,735	28,361	27,894	1
2	—	—	33,694	34,535	33,694	2

$$c_4(s_4, d_4) + \beta v_3(s_3)$$

s_5 \ d_5	0	1	2	3	$v_5(s_5)$	$d_5^*(s_5)$
3	24,268	25,017	26,302	26,531	24,268	0
2	29,314	31,368	30,617	31,902	29,314	0
1	—	36,414	36,968	36,217	36,217	3

$$c_5(s_5, d_5) + \beta v_4(s_4)$$

s_6 \ d_6	0	1	2	3	$v_6(s_6)$	$d_6^*(s_6)$
0	—	38,409	37,803	38,813	37,803	2

$$c_6(s_6, d_6) + \beta v_5(s_5)$$

Fig. 11.10 Optimal-value and decision functions with discounting.

Given that the system is in state zero with six stages to go, we determine the optimal construction strategy by considering the optimal decision function $d_n^*(s_n)$ from stage 6 to stage 0. The optimal construction sequence is then shown in the following tabulation:

Stages to go	Construct	Resulting state
6	2	2
5	0	2
4	2	4
3	3	7
2	0	7
1	1	8

and the optimal value of the criterion function, *present value of total future costs*, is $37.8 million for this strategy. Note that this optimal strategy is significantly different from that computed in the previous section without discounting. The effect of the discounting of future costs is to delay construction in general, which is what we would expect.

*11.5 SHORTEST PATHS IN A NETWORK

Although we have not emphasized this fact, dynamic-programming and shortest-path problems are very similar. In fact, as illustrated by Figs. 11.1 and 11.8, our previous examples of dynamic programming can both be interpreted as shortest-path problems.

In Fig. 11.8, we wish to move through the network from the starting node (initial state) at stage 6, with no plants yet constructed, to the end node (final state) at stage 0 with eight plants constructed. Every path in the network specifies a strategy indicating how many new plants to construct each year. Since the cost of a strategy sums the cost at each stage, the total cost corresponds to the "length" of a path from the starting to ending nodes. The minimum-cost strategy then is just the shortest path.

Figure 11.11 illustrates a shortest-path network for the minimum-delay problem presented in Section 11.1. The numbers next to the arcs are delay times. An end node representing the group of downtown parking lots has been added. This emphasizes the fact that we have assumed that the commuters do not care in which lot they park. A start node has also been added to illustrate that the dynamic-programming solution by *backward* induction finds the shortest path from the end node to the start node. In fact, it finds the shortest paths from the end node to *all* nodes in the network, thereby solving the minimum-delay problem for each commuter. On the other hand, the dynamic-programming solution by *forward* induction finds the shortest path from the start node to the end node. Although the *shortest path* will be the same for both methods, forward induction will *not* solve the minimum-delay problem for *all* commuters, since the commuters are not indifferent to which home they arrive.

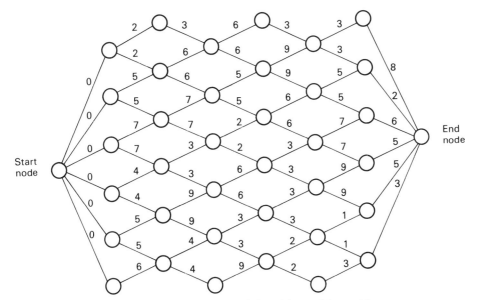

Fig. 11.11 Shortest-path network for minimum-delay problem.

To complete the equivalence that we have suggested between dynamic program-ming and shortest paths, we next show how shortest-path problems can be solved by dynamic programming. Actually, several different dynamic-programming solutions can be given, depending upon the structure of the network under study. As a general rule, the more *structured* the network, the more efficient the algorithm that can be developed. To illustrate this point we give two separate algorithms applicable to the following types of networks:

i) *Acyclic networks.* These networks contain no directed cycles. That is, we cannot start from any node and follow the arcs in their given directions to return to the same node.

ii) *Networks without negative cycles.* These networks may contain cycles, but the distance around any cycle (i.e., the sum of the lengths of its arcs) must be nonnegative.

In the first case, to take advantage of the acyclic structure of the network, we order the nodes so that, if the network contains the arc $i-j$, then $i > j$. To obtain such an ordering, begin with the terminal node, which can be thought of as having only entering arcs, and number it "one." Then ignore that node and the incident arcs, and number any node that has only incoming arcs as the next node. Since the network is acyclic, there must be such a node. (Otherwise, from any node, we can move along an arc to another node. Starting from any node and continuing to move away from any node encountered, we eventually would revisit a node, determining a cycle, con-tradicting the acyclic assumption.) By ignoring the numbered nodes and their incident arcs, the procedure is continued until all nodes are numbered.

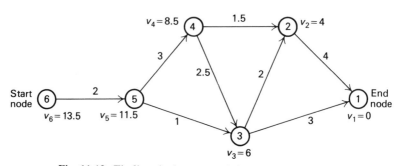

Fig. 11.12 Finding the longest path in an acyclic network.

This procedure is applied, in Fig. 11.12, to the longest-path problem introduced as a critical-path scheduling example in Section 8.1.

We can apply the dynamic-programming approach by viewing each node as a stage, using either backward induction to consider the nodes in ascending order, or forward induction to consider the nodes in reverse order. For backward induction, v_n will be interpreted as the longest distance from node n to the end node. Setting $v_1 = 0$, dynamic programming determines $v_2, v_3, \ldots, v_N$ in order, by the recursion

$$v_n = \text{Max} \left[d_{nj} + v_j \right], \qquad j < n,$$

where d_{nj} is the given distance on arc n–j. The results of this procedure are given as node labels in Fig. 11.13 for the critical-path example.

For a shortest-path problem, we use minimization instead of maximization in this recursion. Note that the algorithm finds the longest (shortest) paths from every node to the end node. If we want only the longest path to the start node, we can terminate the procedure once the start node has been labeled. Finally, we could have found the longest distances from the start node to all other nodes by labeling the nodes in the reverse order, beginning with the start node.

A more complicated algorithm must be given for the more general problem of finding the shortest path between two nodes, say nodes 1 and N, in a network without negative cycles. In this case, we can devise a dynamic-programming algorithm based upon a value function defined as follows:

$v_n(j) =$ Shortest distance from node 1 to node j along paths using at most n intermediate nodes.

By definition, then,

$$v_0(j) = d_{1j} \qquad \text{for } j = 2, 3, \ldots, N,$$

the length d_{1j} of arc 1–j since no intermediate nodes are used. The dynamic-programming recursion is

$$v_n(j) = \text{Min} \{d_{ij} + v_{n-1}(i)\}, \qquad 1 \leq j \leq N, \tag{16}$$

which uses the principle of optimality: that any path from node 1 to node j, using at most n intermediate nodes, arrives at node j from node i along arc $i\text{--}j$ after using the shortest path with at most $(n - 1)$ intermediate nodes from node j to node i. We allow $i = j$ in the recursion and take $d_{jj} = 0$, since the optimal path using at most n intermediate nodes may coincide with the optimal path with length $v_{n-1}(j)$ using at most $(n - 1)$ intermediate nodes.

The algorithm computes the shortest path from node 1 to every other node in the network. It terminates when $v_n(j) = v_{n-1}(j)$ for every node j, since computations in Eq. (16) will be repeated at every stage from n on. Because no path (without cycles) uses any more than $(N - 1)$ intermediate nodes, where N is the total number of nodes, the algorithm terminates after at most $(N - 1)$ steps.

As an application of the method, we solve the shortest-path problem introduced in Chapter 8 and given in Fig. 11.13.

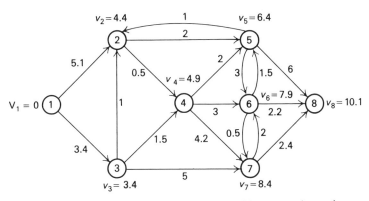

Fig. 11.13 Shortest paths in a network without negative cycles.

Initially the values $v_0(j)$ are given by

$$v_0(1) = 0, \qquad v_0(2) = d_{12} = 5.1, \qquad v_0(3) = d_{13} = 3.4,$$

and

$$v_0(j) = \infty \quad \text{for } j = 4, 5, 6, 7, 8,$$

since these nodes are not connected to node 1 by an arc. The remaining steps are specified in Tableaus 7, 8, and 9. The computations are performed conveniently by maintaining a table of distances d_{ij}. If the list $v_0(i)$ is placed to the left of this table, then recursion Eq. (14) states that $v_1(j)$ is given by the smallest of the comparisons:

$$v_0(i) + d_{ij} \qquad \text{for } i = 1, 2, \ldots, 8.$$

That is, place the column $v_0(i)$ next to the jth column of the d_{ij} table, add the corresponding elements, and take $v_1(j)$ as the smallest of the values. If $v_1(j)$ is recorded below the jth column, the next iteration to find $v_2(j)$ is initiated by replacing the column $v_0(i)$ with the elements $v_1(j)$ from below the distance table.

Tableau 7[†]

Node i	$v_0(i)$	\multicolumn{8}{c}{Node j}
		1	2	3	4	5	6	7	8
1	0	0	5.1	3.4					
2	5.1		0		.5	2			
3	3.4		1	0	1.5			5	
4	$+\infty$				0	2	3	4.2	
5	$+\infty$					0	3		6
6	$+\infty$					2	0	.5	2.2
7	$+\infty$						2	0	2.4
8	$+\infty$								0
$v_1(j) = \min\{d_{ij} + v_0(i)\}$		0	4.4	3.4	4.9	7.1	$+\infty$	8.4	$+\infty$

Tableau 8[†]

Node i	$v_1(i)$	\multicolumn{8}{c}{Node j}
		1	2	3	4	5	6	7	8
1	0	0	5.1	3.4					
2	4.4		0		.5	2			
3	3.4		1	0	1.5			5	
4	4.9				0	2	3	4.2	
5	7.1					0	3		6
6	$+\infty$					2	0	.5	2.2
7	8.4						2	0	2.4
8	$+\infty$								0
$v_2(j) = \min\{d_{ij} + v_1(i)\}$		0	4.4	3.4	4.9	6.4	7.9	8.4	10.8

Tableau 9[†]

Node i	$v_2(i)$	\multicolumn{8}{c}{Node j}
		1	2	3	4	5	6	7	8
1	0	0	5.1	3.4					
2	4.4		0		.5	2			
3	3.4		1	0	1.5			5	
4	4.9				0	2	3	4.2	
5	6.4					0	3		6
6	7.9					2	0	.5	2.2
7	8.4						2	0	2.4
8	10.8								0
$v_3(j) = \min\{d_{ij} + v_2(i)\}$		0	4.4	3.4	4.9	6.4	7.9	8.4	10.1

[†] $d_{ij} = +\infty$, if blank.

As the reader can verify, the next iteration gives $v_4(j) = v_3(j)$ for all j. Consequently, the values $v_3(j)$ recorded in Tableau 9 are the shortest distances from node 1 to each of the nodes $j = 2, 3, \ldots, 8$.

11.6 CONTINUOUS STATE-SPACE PROBLEMS

Until now we have dealt only with problems that have had a finite number of states associated with each stage. Since we also have assumed a finite number of stages, these problems have been identical to finding the shortest path through a network with special structure. Since the development, in Section 11.3, of the fundamental recursive relationship of dynamic programming did not depend on having a finite number of states at each stage, here we introduce an example that has a continuous state space and show that the same procedures still apply.

Suppose that some governmental agency is attempting to perform cost/benefit analysis on its programs in order to determine which programs should receive funding for the next fiscal year. The agency has managed to put together the information in Table 11.2. The benefits of each program have been converted into equivalent tax savings to the public, and the programs have been listed by decreasing benefit-to-cost ratio. The agency has taken the position that there will be no partial funding of programs. Either a program *will* be funded at the indicated level or it will *not* be considered for this budget cycle. Suppose that the agency is fairly sure of receiving a budget of $34 million from the state legislature if it makes a good case that the money is being used effectively. Further, suppose that there is some possibility that the budget will be as high as $42 million. How can the agency make the most effective use of its funds at *either* possible budget level?

Table 11.2 Cost/benefit information by program.

Program	Expected benefit	Expected cost	Benefit/Cost
A	$ 59.2 M	$ 2.8 M	21.1
B	31.4	1.7	18.4
C	15.7	1.0	15.7
D	30.0	3.2	9.4
E	105.1	15.2	6.9
F	11.6	2.4	4.8
G	67.3	16.0	4.2
H	2.3	.7	3.3
I	23.2	9.4	2.5
J	18.4	10.1	1.8
	$364.2 M	$62.5 M	

We should point out that mathematically this problem is an integer program. If b_j is the benefit of the jth program and c_j is the cost of that program, then an integer-programming formulation of the agency's budgeting problem is determined easily.

Letting

$$x_j = \begin{cases} 1 & \text{if program } j \text{ is funded,} \\ 0 & \text{if program } j \text{ is } not \text{ funded,} \end{cases}$$

the integer-programming formulation is:

$$\text{Maximize} \sum_{j=1}^{n} b_j x_j,$$

subject to:

$$\sum_{j=1}^{n} c_j x_j \leq B,$$

$$x_j = 0 \quad \text{or} \quad 1 \quad (j = 1, 2, \ldots, n),$$

where B is the total budget allocated. This cost/benefit example is merely a variation of the well-known knapsack problem that was introduced in Chapter 9. We will ignore, for the moment, this integer-programming formulation and proceed to develop a highly efficient solution procedure using dynamic programming.

In order to approach this problem via dynamic programming, we need to define the stages of the system, the state space for each stage, and the optimal-value function. Let

$v_k(B) =$ Maximum total benefit obtainable, choosing from the first k programs, with budget limitation B.

With this definition of the optimal-value function, we are letting the first k programs included be the number of "stages to go" and the available budget at each stage be the state space. Since the possible budget might take on any value, we are allowing for a continuous state space for each stage. In what follows the order of the projects is immaterial although the order given in Table 11.2 may have some computational advantages.

Let us apply the dynamic-programming reasoning as before. It is clear that with $k = 0$ programs, the total benefit must be zero regardless of the budget limitation. Therefore

$$v_0(B_0) = 0 \quad \text{for } B_0 \geq 0.$$

If we now let $k = 1$, it is again clear that the optimal-value function can be determined easily since the budget is either large enough to fund the first project, or *not*. (See Tableau 10.)

Tableau 10

B_1 \ d_1	$x_1 = 0$	$x_1 = 1$	$v_1(B_1)$	$d_1^*(B_1)$
$2.8 \leq B$	0	59.2	59.2	1
$0 \leq B < 2.8$	0	—	0	0

$$\underbrace{\qquad\qquad\qquad\qquad}_{c_1(B_1, d_1)}$$

Now consider which programs to fund when the first two programs are available. The optimal-value function $v_2(B_2)$ and optimal decision function $d_2^*(B_2)$ are developed in Tableau 11.

Tableau 11

B_2 \ d_2	$x_2 = 0$	$x_2 = 1$	$v_2(B_2)$	$d_2^*(B_2)$
$4.5 \leq B_2$	59.2	59.2 + 31.4	90.6	1
$2.8 \leq B_2 < 4.5$	59.2	31.4	59.2	0
$1.7 \leq B_2 < 2.8$	0	31.4	31.4	1
$0 \leq B_2 < 1.7$	0	—	0	0

$$c_2(B_2, d_2) + v_1(B_1)$$

Here again the dash means that the current state and decision combination will result in a state that is not permissible. Since this tableau is fairly simple, we will go on and develop the optimal-value function $v_3(B_3)$ and optimal decision function $d_3^*(B_3)$ when the first three programs are available (see Tableau 12).

Tableau 12

B_3 \ d_3	$x_3 = 0$	$x_3 = 1$	$v_3(B_3)$	$d_3^*(B_3)$
$5.5 \leq B_3$	90.6	90.6 + 15.7	106.3	1
$4.5 \leq B_3 < 5.5$	90.6	59.2 + 15.7	90.6	0
$3.8 \leq B_3 < 4.5$	59.2	59.2 + 15.7	74.9	1
$2.8 \leq B_3 < 3.8$	59.2	31.4 + 15.7	59.2	0
$2.7 \leq B_3 < 2.8$	31.4	31.4 + 15.7	47.1	1
$1.7 \leq B_3 < 2.7$	31.4	0 + 15.7	31.4	0
$1.0 \leq B_3 < 1.7$	0	0 + 15.7	15.7	1
$0 \leq B_3 < 1.0$	0	—	0	0

$$c_3(B_3, d_3) + v_2(B_2)$$

For any budget level, for example, 4.0 M, we merely consider the two possible decisions: either funding program C ($x_3 = 1$) or not ($x_3 = 0$). If we fund program C, then we obtain a benefit of $15.7 M while consuming $1.0 M of our own budget. The remaining $3.0 M of our budget is then optimally allocated to the remaining programs, producing a benefit of $59.2 M, which we obtain from the optimal-value function with the first two programs included (Tableau 11). If we do not fund program C, then the entire amount of $4.0 M is optimally allocated to the remaining two programs (Tableau 11), producing a benefit of $59.2. Hence, we should clearly fund program C if our budget allocation is $4.0 M. Optimal decisions taken for other budget levels are determined in a similar manner.

Although it is straightforward to continue the recursive calculation of the optimal-value function for succeeding stages, we will not do so since the number of ranges that need to be reported rapidly becomes rather large. The general recursive relationship that determines the optimal-value function at each stage is given by:

$$v_n(B_n) = \text{Max}\left[c_n x_n + v_{n-1}(B_n - c_n x_n)\right],$$

subject to:

$$x_n = 0 \quad \text{or} \quad 1.$$

The calculation is initialized by observing that

$$v_0(B_0) = 0$$

for all possible values of B_0. Note that the state transition function is simply

$$B_{n-1} = t_n(x_n, B_n) = B_n - c_n x_n.$$

We can again illustrate the usual principle of optimality: Given budget B_n at stage n, whatever decision is made with regard to funding the nth program, the remaining budget must be allocated optimally among the first $(n - 1)$ programs. If these calculations were carried to completion, resulting in $v_{10}(B_{10})$ and $d^*_{10}(B_{10})$, then the problem would be solved for all possible budget levels, not just \$3.4 M and \$4.2 M.

Although this example has a continuous state space, a finite number of ranges can be constructed because of the zero–one nature of the decision variables. In fact, all breaks in the range of the state space either are the breaks from the previous stage, or they result from adding the cost of the new program to the breaks in the previous range. This is not a general property of continuous state space problems, and in most cases such ranges cannot be determined. Usually, what is done for continuous state space problems is that they are converted into discrete state problems by defining an appropriate grid on the continuous state space. The optimal-value function is then computed only for the points on the grid. For our cost/benefit example, the total budget must be between zero and \$62.5 M, which provides a range on the state space, although at any stage a tighter upper limit on this range is determined by the sum of the budgets of the first n programs. An appropriate grid would consist of increments of \$0.1 M over the limits of the range at each stage, since this is the accuracy with which the program costs have been estimated. The difference between problems with continuous state spaces and those with discrete state spaces essentially then disappears for computational purposes.

11.7 DYNAMIC PROGRAMMING UNDER UNCERTAINTY

Up to this point we have considered exclusively problems with deterministic behavior. In a deterministic dynamic-programming process, if the system is in state s_n with n stages to go and decision d_n is selected from the set of permissible decisions for this stage and state, then the stage return $f_n(d_n, s_n)$ and the state of the system at the next stage, given by $s_{n-1} = t_n(d_n, s_n)$, are both known with certainty. This deterministic process can be represented by means of the decision tree in Fig. 11.14. As one can observe, given the current state, a specific decision leads with complete certainty to a particular state at the next stage. The stage returns are also known with certainty and are associated with the branches of the tree.

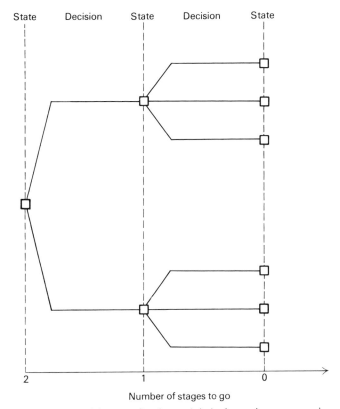

Fig. 11.14 Decision tree for deterministic dynamic programming.

When uncertainty is present in a dynamic-programming problem, a specific decision for a given state and stage of the process does not, by itself, determine the state of the system at the next stage; this decision may not even determine the return for the current stage. Rather, in dynamic programming under uncertainty, given the state of the system s_n with n stages to go and the current decision d_n, an uncertain event occurs which is determined by a random variable $\tilde{e}_n$ whose outcome e_n is *not* under the control of the decision maker. The stage return function may depend on this random variable, that is,

$$f_n(d_n, s_n, \tilde{e}_n),$$

while the state of the system s_{n-1} with $(n-1)$ stages to go invariably will depend on the random variable by

$$\tilde{s}_{n-1} = t_n(d_n, s_n, \tilde{e}_n).$$

The outcomes of the random variable are governed by a probability distribution, $p_n(e_n|d_n, s_n)$, which may be the same for every stage or may be conditional on the stage, the state at the current stage, and even the decision at the current stage.

Figure 11.15 depicts dynamic programming under uncertainty as a *decision tree,* where squares represent states where decisions have to be made and circles represent uncertain events whose outcomes are not under the control of the decision maker. These diagrams can be quite useful in analyzing decisions under uncertainty if the number of possible states is not too large. The decision tree provides a pictorial representation of the sequence of decisions, outcomes, and resulting states, *in the order in which* the decisions must be made and the outcomes become known to the

STATE DECISION EVENT OUTCOME STATE DECISION EVENT OUTCOME STATE

(s_2) (d_2) $(\tilde{e}_2)$ (e_2) (s_1) (d_1) $(\tilde{e}_1)$ (e_1) (s_0)

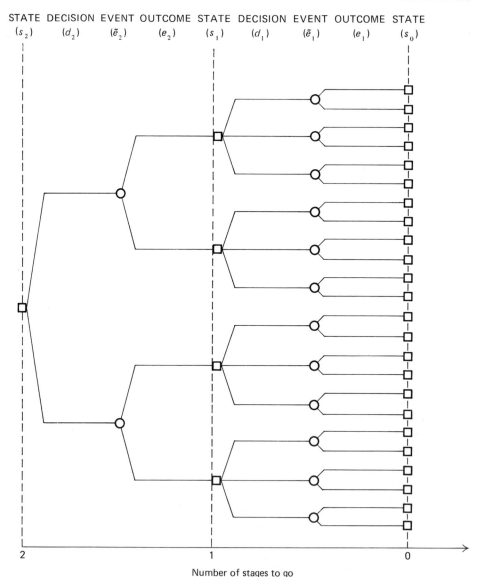

Number of stages to go

Fig. 11.15 Decision tree for dynamic programming under uncertainty.

decision maker. Unlike deterministic dynamic programming wherein the optimal decisions at each stage can be specified at the outset, in dynamic programming under uncertainty, the optimal decision at each stage can be selected only after we know the outcome of the uncertain event at the previous stage. At the outset, all that can be specified is a set of decisions that would be made *contingent* on the outcome of a sequence of uncertain events.

In dynamic programming under uncertainty, since the stage returns and resulting state may both be uncertain at each stage, we cannot simply optimize the sum of the stage-return functions. Rather, we must optimize the *expected return* over the stages of the problem, taking into account the sequence in which decisions can be made and the outcomes of uncertain events become known to the decision maker. In this situation, backward induction can be applied to determine the optimal strategy, but forward induction cannot. The difficulty with forward induction is that it is impossible to assign values to states at the next stage that are independent of the uncertain evolution of the process from that future state on. With backward induction, on the other hand, no such difficulties arise since the states with zero stages to go are evaluated first, and then the states with one stage to go are evaluated by computing the expected value of any decision and choosing optimally.

We start the backward induction process by computing the optimal-value function at stage 0. This amounts to determining the value of ending in each possible state with 0 stages to go. This determination may involve an optimization problem or the value of the assets held at the horizon. Next, we compute the optimal-value function at the previous stage. To do this, we first compute the expected value of each uncertain event, weighting the stage return plus the value of the resulting state for each outcome by the probability of each outcome. Then, for each state at the previous stage, we select the decision that has the maximum (or minimum) expected value. Once the optimal-value function for stage 1 has been determined, we continue in a similar manner to determine the optimal-value functions at prior stages by backward induction.

The optimal-value function for dynamic programming under uncertainty is then defined in the following recursive form:

$$v_n(s_n) = \text{Max } E[f_n(d_n, s_n, \tilde{e}_n) + v_{n-1}(\tilde{s}_{n-1})], \tag{17}$$

subject to:

$$\tilde{s}_{n-1} = t_n(d_n, s_n, \tilde{e}_n),$$

$$d_n \in D_n,$$

where $E[\cdot]$ denotes the expected value of the quantity in brackets. To initiate the recursive calculations we need to determine the optimal-value function with zero stages to go, which is given by:

$$v_0(s_0) = \text{Max } E[f_0(d_0, s_0, \tilde{e}_0)],$$

subject to:

$$d_0 \in D_0.$$

The optimization problems that determine the optimal-value function with zero stages to go are not determined recursively, and therefore may be solved in a straight-forward manner. If the objective function is to maximize the expected discounted costs, then Eq. (17) is modified as in Section 11.4 by multiplying the term $v_{n-1}(\mathfrak{T}_{n-1})$ by β_n, the discount factor for period n.

We can make these ideas more concrete by considering a simple example. A manager is in charge of the replenishment decisions during the next two months for the inventory of a fairly expensive item. The production cost of the item is \$1000/unit, and its selling price is \$2000/unit. There is an inventory-carrying cost of \$100/unit per month on each unit left over at the end of the month. We assume there is no setup cost associated with running a production order, and further that the production process has a short lead time; therefore any amount produced during a given month is available to satisfy the demand during that month. At the present time, there is no inventory on hand. Any inventory left at the end of the next two months has to be disposed of at a salvage value of \$500/unit.

The demand for the item is uncertain, but its probability distribution is identical for each of the coming two months. The probability distribution of the demand is as follows:

Demand	Probability
0	0.25
1	0.40
2	0.20
3	0.15

The issue to be resolved is how many units to produce during the first month and, *depending on the actual demand in the first month,* how many units to produce during the second month. Since demand is uncertain, the inventory at the end of each month is also uncertain. In fact, demand could exceed the available units on hand in any month, in which case all excess demand results in lost sales. Consequently, our pro-duction decision must find the proper balance between production costs, lost sales, and final inventory salvage value.

The states for this type of problem are usually represented by the inventory level I_n at the beginning of each month. Moreover, the problem is characterized as a two-stage problem, since there are two months involved in the inventory-replenishment decision. To determine the optimal-value function, let

$v_n(I_n)$ = Maximum contribution, given that we have I_n units of inventory with n stages to go.

We initiate the backward induction procedure by determining the optimum-value function with 0 stages to go. Since the salvage value is \$500/unit, we have:

I_0	$v_0(I_0)$
0	0
1	500
2	1000
3	1500

To compute the optimal-value function with one stage to go, we need to determine, for each inventory level (state), the corresponding contribution associated with each possible production amount (decision) and level of sales (outcome). For each inventory level, we select the production amount that maximizes the expected contribution.

Table 11.3 provides all the necessary detailed computations to determine the optimal-value function with one stage to go. Column 1 gives the state (inventory level) of the process with one stage to go. Column 2 gives the possible decisions (amount to produce) for each state, and, since demand cannot be greater than three, the amount produced is at most three. Column 3 gives the possible outcomes for the uncertain level of sales for each decision and current state, and column 4 gives the

Table 11.3 Computation of optimal-value function with one stage to go.

(1) State I_1	(2) Pro-duce d_1	(3) Sell S_1	(4) Proba-bility $(\tilde{S}_1 = s_1)$	(5) Resulting state $\tilde{I}_0$	(6) Produc-tion cost	(7) Sales rev-enue	(8) Inven-tory cost	(9) $v_0(I_0)$	(10) Proba-bility × $	(11) Expected contri-bution
0	0	0	1.	0	0	0	0	0	0	0
	1	0	.25	1	−1000	0	−100	500	−150	} 600*
		1	.75	0	−1000	2000	0	0	750	
	2	0	.25	2	−2000	0	−200	1000	−300	
		1	.40	1	−2000	2000	−100	500	160	} 560
		2	.35	0	−2000	4000	0	0	700	
	3	0	.25	3	−3000	0	−300	1500	−450	
		1	.40	2	−3000	2000	−200	1000	−80	} 200
		2	.20	1	−3000	4000	−100	500	280	
		3	.15	0	−3000	6000	0	0	450	
1	0	0	.25	1	0	0	−100	500	100	} 1600*
		1	.75	0	0	2000	0	0	1500	
	1	0	.25	2	−1000	0	−200	1000	−50	
		1	.40	1	−1000	2000	−100	500	560	} 1560
		2	.35	0	−1000	4000	0	0	1050	
	2	0	.25	3	−2000	0	−300	1500	−200	
		1	.40	2	−2000	2000	−200	1000	320	} 1200
		2	.20	1	−2000	4000	−100	500	480	
		3	.15	0	−2000	6000	0	0	600	
2	0	0	.25	2	0	0	−200	1000	200	} 2560*
		1	.40	1	0	2000	−100	500	960	
		2	.35	0	0	4000	0	0	1400	
	1	0	.25	3	−1000	0	−300	1500	50	
		1	.40	2	−1000	2000	−200	1000	720	} 2200
		2	.20	1	−1000	4000	−100	500	680	
		3	.15	0	−1000	6000	0	0	750	
3	0	0	.25	3	0	0	−300	1500	300	
		1	.40	2	0	2000	−200	1000	1120	} 3200*
		2	.20	1	0	4000	−100	500	880	
		3	.15	0	0	6000	0	0	900	

probability of each of these possible outcomes. Note that, in any period, it is impossible to sell more than the supply, which is the sum of the inventory currently on hand plus the amount produced. Hence, the probability distribution of sales differs from that of demand since, whenever demand exceeds supply, the entire supply is sold and the excess demand is lost. Column 5 is the resulting state, given that we currently have I_1 on hand, produce d_1, and sell s_1. The transition function in general is just:

$$\tilde{I}_{n-1} = I_n + d_n - \tilde{s}_n,$$

where the tildes ($\sim$) indicate that the level of sales is uncertain and, hence, the resulting state is also uncertain. Columns 6, 7, and 8 reflect the revenue and costs for each state, decision, and sales level, and column 9 reflects the value of being in the resulting state at the next stage. Column 10 merely weights the sum of columns 6 through 9 by the probability of their occurring, which is an intermediate calculation in determining the expected value of making a particular decision, given the current state. Column 11 is then just this expected value; and the asterisk indicates the optimal decision for each possible state.

The resulting optimal-value function and the corresponding optimum-decision function are determined directly from Table 11.3 and are the following:

I_1	$v_1(I_1)$	$d_1^*(I_1)$
0	600	1
1	1600	0
2	2560	0
3	3200	0

Next we need to compute the optimum-value function with two stages to go. However, since we have assumed that there is no initial inventory on hand, it is not necessary to describe the optimal-value function for every possible state, but only for $I_2 = 0$. Table 11.4 is similar to Table 11.3 and gives the detailed computations required to evaluate the optimal-value function for this case.

Table 11.4 Computation of optimal-value function with two stages to go, $I_2 = 0$ only.

( ) State I_2	(2) Produce d_2	(3) Sell s_2	(4) Probability $(\tilde{s}_2 = s_2)$	(5) Resulting state $\tilde{I}_1$	(6) Production cost	(7) Sales revenue	(8) Inventory cost	(9) $v_1(I_1)$	(10) Probability $\times$ $	(11) Expected contribution
0	0	0	1.	0	0	0	0	650.	650	650
	1	0	.25	1	-1000	0	0	1600.	150	} 1350
		1	.75	0	-1000	2000	0	600.	1200	
	2	0	.25	2	-2000	0	-200	2560.	90	
		1	.40	1	-2000	2000	-100	1600.	600	} 1600.*
		2	.35	0	-2000	4000	0	600.	910	
	3	0	.25	3	-3000	0	-300	3200.	-25	
		1	.40	2	-3000	2000	-200	2560.	544	} 1559
		2	.20	1	-3000	4000	-100	1600.	500	
		3	.15	0	-3000	6000	0	600.	540	

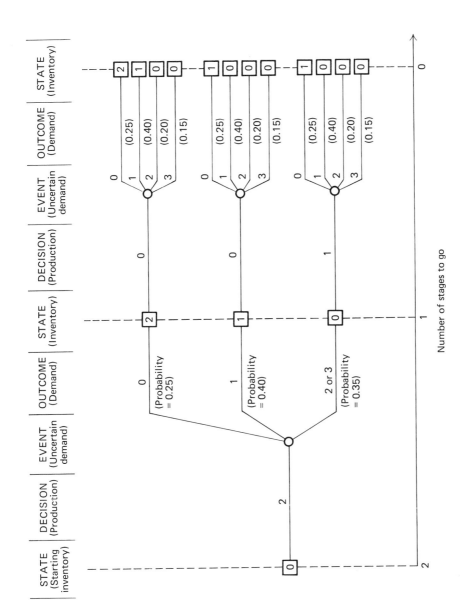

489

The optimal-value function and the corresponding decision function for $I_2 = 0$ are taken directly from Table 11.4 and are the the following:

I_2	$v_2(I_2)$	$d_2^*(I_2)$
0	1600	2

The optimum strategy can be summarized by the decision tree given in Fig. 11.16. The expected contribution determined by the dynamic-programming solution corresponds to weighting the contribution of every path in this tree by the *probability* that this path occurs. The decision tree in Fig. 11.16 emphasizes the contingent nature of the optimal strategy determined by dynamic programming under uncertainty.

EXERCISES

Solutions to exercises marked with an asterisk (*) involve extensive computations. Formulate these problems as dynamic programs and provide representative computations to indicate the nature of the dynamic programming recursions; solve to completion only if a computer system is available.

1. In solving the minimum-delay routing problem in Section 11.1, we assumed the same delay along each street (arc) in the network. Suppose, instead, that the delay when moving along any arc upward in the network is 2 units greater than the delay when moving along any arc downward. The delay at the intersections is still given by the data in Fig. 11.1. Solve for the minimum-delay route by both forward and backward induction.

2. Decatron Mills has contracted to deliver 20 tons of a special coarsely ground wheat flour at the end of the current month, and 140 tons at the end of the next month. The production cost, based on which the Sales Department has bargained with prospective customers, is $c_1(x_1) = 7500 + (x_1 - 50)^2$ per ton for the first month, and $c_2(x_2) = 7500 + (x_2 - 40)^2$ per ton for the second month; x_1 and x_2 are the number of tons of the flour produced in the first and second months, respectively. If the company chooses to produce more than 20 tons in the first month, any excess production can be carried to the second month at a storage cost of $3 per ton.

   Assuming that there is no initial inventory and that the contracted demands must be satisfied in each month (that is, no back-ordering is allowed), derive the production plan that minimizes total cost. Solve by both backward and forward induction. Consider x_1 and x_2 as continuous variables, since any fraction of a ton may be produced in either month.

3. A construction company has four projects in progress. According to the current allocation of manpower, equipment, and materials, the four projects can be completed in 15, 20, 18, and 25 weeks. Management wants to reduce the completion times and has decided to allocate an additional $35,000 to all four projects. The new completion times as functions of the additional funds allocated to each project are given in Table E11.1.

   How should the $35,000 be allocated among the projects to achieve the largest total reduction in completion times? Assume that the additional funds can be allocated only in blocks of $5000.

Table E11.1 Completion times (in weeks)

Additional funds (× 1000 dollars)	Project 1	Project 2	Project 3	Project 4
0	15	20	18	25
5	12	16	15	21
10	10	13	12	18
15	8	11	10	16
20	7	9	9	14
25	6	8	8	12
30	5	7	7	11
35	4	7	6	10

4. The following table specifies the unit weights and values of five products held in storage. The quantity of each item is unlimited.

Product	Weight (W_i)	Value (V_i)
1	7	9
2	5	4
3	4	3
4	3	2
5	1	$\frac{1}{2}$

   A plane with a capacity of 13 weight units is to be used to transport the products. How should the plane be loaded to maximize the value of goods shipped? (Formulate the problem as an integer program and solve by dynamic programming.)

5. Any linear-programming problem with n decision variables and m constraints can be converted into an n-stage dynamic-programming problem with m state parameters.
   Set up a dynamic-programming formulation for the following linear program:

$$\text{Minimize} \sum_{j=1}^{n} c_j x_j,$$

subject to:

$$\sum_{j=1}^{n} a_{ij} x_j \leq b_i \qquad (i = 1, 2, \ldots, m),$$

$$x_j \geq 0 \qquad (j = 1, 2, \ldots, n).$$

   Why is it generally true that the simplex method rather than dynamic programming is recommended for solving linear programs?

6. Rambling Roger, a veteran of the hitchhiking corps, has decided to leave the cold of a Boston winter and head for the sunshine of Miami. His vast experience has given him an indication of the expected time in hours it takes to hitchhike over certain segments of the highways. Knowing he will be breaking the law in several states and wishing to reach the warm weather quickly, Roger wants to know the least-time route to take. He summarized his expected travel times on the map in Fig. E11.1. Find his shortest time route.

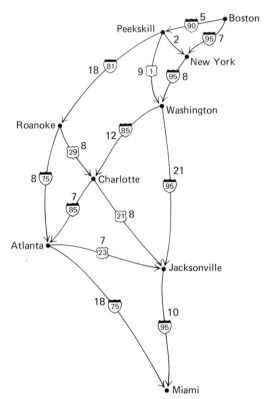

Fig. E11.1 Travel times on highways.

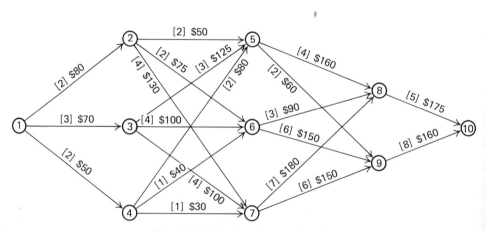

Fig. E11.2 Routing times and costs.

7. J. J. Jefferson has decided to move from the West Coast, where he lives, to a mid-western town, where he intends to buy a small farm and lead a quiet life. Since J. J. is single and has accumulated little furniture, he decides to rent a small truck for $200 a week or fraction of a week (one-way, no mileage charge) and move his belongings by himself. Studying the map, he figures that his trip will require four stages, regardless of the particular routing. Each node shown in Fig. E11.2 corresponds to a town where J. J. has either friends or relatives and where he plans to spend one day resting and visiting if he travels through the town. The numbers in brackets in Fig. E11.2 specify the travel time in days between nodes. (The position of each node in the network is not necessarily related to its geographical position on the map.) As he will travel through different states, motel rates, tolls, and gas prices vary significantly; Fig. E11.2 also shows the cost in dollars for traveling (excluding truck rental charges) between every two nodes. Find J. J.'s cheapest route between towns 1 and 10, including the truck rental charges.

8. At THE CASINO in Las Vegas, a customer can bet only in dollar increments. Betting a certain amount is called "playing a round." Associated with each dollar bet on a round, the customer has a 40% chance to win another dollar and a 60% chance to lose his, or her, dollar. If the customer starts with $4 and wants to maximize the chances of finishing with at least $7 after two rounds, how much should be bet on each round? [*Hint.* Consider the number of dollars available at the beginning of each round as the state variable.]

* 9. In a youth contest, Joe will shoot a total of ten shots at four different targets. The contest has been designed so that Joe will not know whether or not he hits any target until after he has made all ten shots. He obtains 6 points if any shot hits target 1, 4 points for hitting target 2, 10 points for hitting target 3, and 7 points for hitting target 4. At each shot there is an 80% chance that he will miss target 1, a 60% chance of missing target 2, a 90% chance of missing target 3, and a 50% chance of missing target 4, given that he aims at the appropriate target.

    If Joe wants to maximize his expected number of points, how many shots should he aim at each target?

10. A monitoring device is assembled from five different components. Proper functioning of the device depends upon its total weight q so that, among other tests, the device is weighed; it is accepted only if $r_1 \leq q \leq r_2$, where the two limits r_1 and r_2 have been prespecified.

    The weight q_j ($j = 1, 2, \ldots, 5$) of each component varies somewhat from unit to unit in accordance with a normal distribution with mean μ_j and variance σ_j^2. As $q_1, q_2, \ldots, q_5$ are independent, the total weight q will also be a normal variable with mean $\mu = \sum_{j=1}^{5} \mu_j$ and variance $\sigma^2 = \sum_{j=1}^{5} \sigma_j^2$.

    Clearly, even if μ can be adjusted to fall within the interval $[r_1, r_2]$, the rejection rate will depend upon σ^2; in this case, the rejection rate can be made as small as desired by making the variance σ^2 sufficiently small. The design department has decided that $\sigma^2 = 5$ is the largest variance that would make the rejection rate of the monitoring device acceptable. The cost of manufacturing component j is $c_j = 1/\sigma_j^2$.

    Determine values for the design parameters σ_j^2 for $j = 1, 2, \ldots, 5$ that would minimize the manufacturing cost of the components while ensuring an acceptable rejection rate. [*Hint.* Each component is a stage; the state variable is that portion of the total variance σ^2 not yet distributed. Consider σ_j^2's as continuous variables.]

*11. A scientific expedition to Death Valley is being organized. In addition to the scientific equipment, the expedition also has to carry a stock of spare parts, which are likely to fail under the extreme heat conditions prevailing in that area. The estimated number of times that the six critical parts, those sensitive to the heat conditions, will fail during the expedition are shown below in the form of probability distributions.

Part 1

# of Failures	Probability
0	0.5
1	0.3
2	0.2

Part 2

# of Failures	Probability
0	0.4
1	0.3
2	0.2
3	0.1

Part 3

# of Failures	Probability
0	0.7
1	0.2
2	0.1

Part 4

# of Failures	Probability
0	0.9
1	0.1

Part 5

# of Failures	Probability
0	0.8
1	0.1
2	0.1

Part 6

# of Failures	Probability
0	0.8
1	0.2

The spare-part kit should not weigh more than 30 pounds. If one part is needed and it is not available in the spare-part kit, it may be ordered by radio and shipped by helicopter at unit costs as specified in Table E11.2, which also gives the weight of each part.

Table E11.2 Spare-Part Data

Part	Weight (pounds/unit)	Shipping cost ($/unit)
1	4	100
2	3	70
3	2	90
4	5	80
5	3	60
6	2	50

Determine the composition of the spare-part kit to minimize total expected ordering costs.

*12. After a hard day at work I frequently wish to return home as quickly as possible. I must choose from several alternate routes (see Fig. E11.3); the travel time on any road is uncertain and depends upon the congestion at the nearest major intersection preceding that route. Using the data in Table E11.3, determine my best route, given that the congestion at my starting point is heavy.

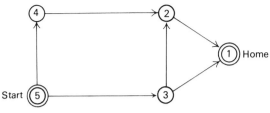

Figure E11.3A

Table E11.3 Travel time on the road

Road $i-j$	Congestion at initial intersection (i)	Travel-time distribution	
		Travel time (minutes)	Probability
5–4	Heavy	4	$\frac{1}{4}$
		6	$\frac{1}{2}$
		10	$\frac{1}{4}$
	Light	2	$\frac{1}{3}$
		3	$\frac{1}{3}$
		5	$\frac{1}{3}$
5–3	Heavy	5	$\frac{1}{2}$
		12	$\frac{1}{2}$
	Light	3	$\frac{1}{2}$
		6	$\frac{1}{2}$
4–2	Heavy	7	$\frac{1}{3}$
		14	$\frac{2}{3}$
	Light	4	$\frac{1}{2}$
		6	$\frac{1}{2}$
3–2	Heavy	5	$\frac{1}{4}$
		11	$\frac{3}{4}$
	Light	3	$\frac{1}{3}$
		5	$\frac{1}{3}$
		7	$\frac{1}{3}$
3–1	Heavy	3	$\frac{1}{2}$
		5	$\frac{1}{2}$
	Light	2	$\frac{1}{2}$
		3	$\frac{1}{2}$
2–1	Heavy	2	$\frac{1}{2}$
		4	$\frac{1}{2}$
	Light	1	$\frac{1}{2}$
		2	$\frac{1}{2}$

Assume that if I am at intersection i with heavy congestion and I take road i–j, then

$$\text{Prob (intersection } j \text{ is heavy)} = 0.8.$$

If the congestion is light at intersection i and I take road i–j, then

$$\text{Prob (intersection } j \text{ is heavy)} = 0.3.$$

13. Find the shortest path from node 1 to every other node in the network given in Fig. E11.4, using the shortest-route algorithm for acyclic networks. The number next to each arc is the "length" of that arc.

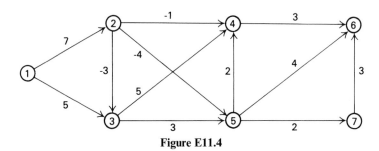

Figure E11.4

14. Find the shortest path from node 1 to every other node in the network given in Fig. E11.5.

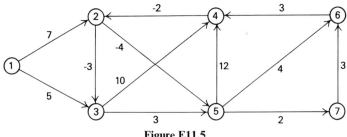

Figure E11.5

15. a) Give a dynamic-programming recursion for finding the shortest path from every node to a particular node, node k, in a network without negative cycles.
    b) Apply the recursion from part (a) to find the shortest path from every node to node 6 in the network specified in the previous exercise.

16. A state's legislature has R representatives. The state is sectioned into s districts, where District j has a population p_j and $s < R$. Under strictly proportional representation, District j would receive $Rp_j/(\sum_{j=1}^{s} p_j) = r_j$ representatives; this allocation is not feasible, however, because r_j may not be integer-valued. The objective is to allocate y_j representatives to District j for $j = 1, 2, \ldots, s$, so as to minimize, over all the districts, the maximum difference between y_j and r_j; that is, minimize $[\text{maximum } (|y_1 - r_1|, |y_2 - r_2|, \ldots, |y_s - r_s|)]$.
    a) Formulate the model in terms of a dynamic-programming recursion.
    b) Apply your method to the data $R = 4$, $s = 3$, and $r_1 = 0.4$, $r_2 = 2.4$, and $r_3 = 1.2$.
    c) Discuss whether the solution seems reasonable, given the context of the problem.

17. In a textile plant, cloth is manufactured in rolls of length L. Defects sometimes occur along the length of the cloth. Consider a specific roll with ($N - 1$) defects appearing at distances $y_1, y_2, \ldots, y_{N-1}$ from the start of the roll ($y_{i+1} > y_i$ for all i). Denote the start of the roll by y_0, the end by y_N.

The roll is cut into pieces for sale. The value of a piece depends on its length and the number of defects. Let

$$v(x, m) = \text{Value of a piece of length } x \text{ having } m \text{ defects.}$$

Assume that all cuts are made *through* defects and that such cutting removes the defect. Specify how to determine where to cut the cloth to maximize total value.

18. A manufacturing company, receiving an order for a special product, has worked out a production plan for the next 5 months. All components will be manufactured internally except for one electronic part that must be purchased. The purchasing manager in charge of buying the electronic part must meet the requirements schedule established by the production department. After negotiating with several suppliers, the purchasing manager has determined the best possible price for the electronic part for each of the five months in the planning horizon. Table E11.4 summarizes the requirement schedule and purchase price information.

Table E11.4 Requirements schedule and purchasing prices.

Month	Requirements (thousands)	Purchasing price ($/thousand pieces)
1	5	10
2	10	11
3	6	13
4	9	10
5	4	12

The storage capacity for this item is limited to 12,000 units; there is no initial stock, and after the five-month period the item will no longer be needed. Assume that the orders for the electronic part are placed once every month (at the beginning of each month) and that the delivery lead time is very short (delivery is made practically instantaneously). No back-ordering is permitted.

a) Derive the monthly purchasing schedule if total purchasing cost is to be minimized.
b) Assume that a storage charge of $250 is incurred for each 1000 units found in inventory at the end of a month. What purchasing schedule would minimize the purchasing and storage costs?

19. Rebron, Inc., a cosmetics manufacturer, markets five different skin lotions and creams: A, B, C, D, E. The company has decided to increase the advertising budget allocated to this group of products by 1 million dollars for next year. The marketing department has conducted a research program to establish how advertising affects the sales levels of these products. Table E11.5 shows the increase in each product's contribution to net profits as a function of the additional advertisement expenditures.

Given that maximization of net profits is sought, what is the optimal allocation of the additional advertising budget among the five products? Assume, for simplicity, that advertising funds must be allocated in blocks of $100,000.

Table E11.5 Profits in response to advertising

Additional investment in advertising (in $100,000)	Profits (in $100,000)				
	Product A	Product B	Product C	Product D	Product E
0	0	0	0	0	0
1	0.20	0.18	0.23	0.15	0.25
2	0.35	0.30	0.43	0.30	0.45
3	0.50	0.42	0.60	0.45	0.65
4	0.63	0.54	0.75	0.58	0.80
5	0.75	0.64	0.80	0.70	0.90
6	0.83	0.74	0.92	0.81	0.95
7	0.90	0.84	0.98	0.91	0.98
8	0.95	0.92	1.02	1.00	1.01
9	0.98	1.00	1.05	1.04	1.02
10	0.98	1.05	1.06	1.07	1.03

*20. A machine tool manufacturer is planning an expansion program. Up to 10 workers can be hired and assigned to the five divisions of the company. Since the manufacturer is currently operating with idle machine capacity, no new equipment has to be purchased.

Hiring new workers adds $250/day to the indirect costs of the company. On the other hand, new workers add value to the company's output (i.e., sales revenues in excess of direct costs) as indicated in Table E11.6. Note that the value added depends upon both the number of workers hired and the division to which they are assigned.

Table E11.6 Value added by new workers

New workers (x_n)	Increase in contribution to overhead ($/day)				
	Division 1	Division 2	Division 3	Division 4	Division 5
0	0	0	0	0	0
1	30	25	35	32	28
2	55	50	65	60	53
3	78	72	90	88	73
4	97	90	110	113	91
5	115	108	120	133	109
6	131	124	128	146	127
7	144	138	135	153	145
8	154	140	140	158	160
9	160	150	144	161	170
10	163	154	145	162	172

The company wishes to hire workers so that the value that they add exceeds the $250/day in indirect costs. What is the minimum number of workers the company should hire and how should they be allocated among the five divisions?

21. A retailer wishes to plan the purchase of a certain item for the next five months. Suppose that the demand in these months is known and given by:

Month	Demand (units)
1	10
2	20
3	30
4	30
5	20

The retailer orders at the beginning of each month. Initially he has no units of the item. Any units left at the end of a month will be transferred to the next month, but at a cost of 10¢ per unit. It costs $20 to place an order. Assume that the retailer can order only in lots of 10, 20, ... units and that the maximum amount he can order each month is 60 units. Further assume that he receives the order immediately (no lead time) and that the demand occurs just after he receives the order. He attempts to stock whatever remains but cannot stock more than 40 units—units in excess of 40 are discarded at no additional cost and with no salvage value. How many units should the retailer order each month?

* 22. Suppose that the retailer of the previous exercise does not know demand with certainty. All assumptions are as in Exercise 21 except as noted below. The demand for the item is the same for each month and is given by the following distribution:

Demand	Probability
10	0.2
30	0.5
30	0.3

Each unit costs $1. Each unit demanded in excess of the units on hand is lost, with a penalty of $2 per unit. How many units should be ordered each month to minimize total expected costs over the planning horizon? Outline a dynamic-programming formulation and complete the calculations for the last two stages only.

* 23. The owner of a hardware store is surprised to find that he is completely out of stock of "Safe-t-lock," an extremely popular hardened-steel safety lock for bicycles. Fortunately, he became aware of this situation before anybody asked for one of the locks; otherwise he would have lost $2 in profits for each unit demanded but not available. He decides to use his pickup truck and immediately obtain some of the locks from a nearby warehouse.

Although the demand for locks is uncertain, the probability distribution for demand is known; it is the same in each month and is given by:

Demand	Probability
0	0.1
100	0.3
200	0.4
300	0.2

The storage capacity is 400 units, and the carrying cost is $1 per unit per month, charged to the month's *average inventory* [i.e., (initial + ending)/2]. Assume that the withdrawal rate is uniform over the month. The lock is replenished monthly, at the beginning of the month, in lots of one hundred.

What is the replenishment strategy that minimizes the expected costs (storage and shortage costs) over a planning horizon of four months? No specific inventory level is required for the end of the planning horizon.

* 24. In anticipation of the Olympic games, Julius, a famous Danish pastry cook, has opened a coffee-and-pastry shop not far from the Olympic Village. He has signed a contract to sell the shop for $50,000 after operating it for 5 months.

Julius has several secret recipes that have proved very popular with consumers during the last Olympic season, but now that the games are to be held on another continent, variations in tastes and habits cast a note of uncertainty over his chances of renewed success.

The pastry cook plans to sell all types of common pastries and to use his specialties to attract large crowds to his shop. He realizes that the popularity of his shop will depend upon how well his specialties are received; consequently, he may alter the offerings of these pastries from month to month when he feels that he can improve business. When his shop is not popular, he may determine what type of specialties to offer by running two-day market surveys. Additionally, Julius can advertise in the *Olympic Herald Daily* and other local newspapers to attract new customers.

The shop's popularity may change from month to month. These transitions are uncertain and depend upon advertising and market-survey strategies. Table E11.7 summarizes the various possibilities. The profit figures in this table include advertising expenditures and market-survey costs.

Table E11.7 Profit possibilities (p = probability; E = expected profit)

	Popular next month		Not popular next month	
Popular this month, no advertising	$p = \frac{6}{10}$,	$E = \$6000$	$p = \frac{4}{10}$,	$E = \$2000$
Popular this month, advertising	$p = \frac{3}{4}$,	$E = \$4000$	$p = \frac{1}{10}$,	$E = \$3000$
Not popular this month, market survey	$p = \frac{1}{3}$,	$E = \$3000$	$p = \frac{2}{3}$,	$E = -\$2000$
Not popular this month, advertising	$p = \frac{6}{10}$,	$E = \$1000$	$p = \frac{4}{10}$,	$E = -\$5000$

Note that Julius has decided either to advertise or to run the market survey whenever the shop is not popular.

Assume that, during his first month of operation, Julius passively waits to see how popular his shop will be. What is the optimal strategy for him to follow in succeeding months to maximize his expected profits?

25. A particular city contains six significant points of interest. Figure E11.6 depicts the network of major two-way avenues connecting the points; the figure also shows travel time (in both

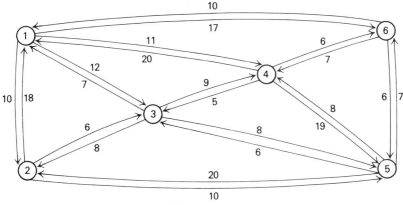

Figure E11.6

directions) along each avenue. Other streets, having travel times exceeding those along the major avenues, link the points but have been dropped from this analysis.

In an attempt to reduce congestion of traffic in the city, the city council is considering converting some two-way avenues to one-way avenues.

The city council is considering two alternative planning objectives:

a) Given that point (1) is the tourist information center for the city, from which most visitors depart, which avenues should be made one-way so as to minimize the travel times from point (1) to every other point?

b) If the travel times from each point to every other point were to be minimized, which avenues would be converted to one-way?

In both cases, assume that the travel times shown in Fig. E11.6 would not be affected by the conversion. If the total conversion cost is proportional to the number of avenues converted to one-way, which of the above solutions has the lowest cost?

* 26. Consider the following one-period problem: a certain item is produced centrally in a factory and distributed to four warehouses. The factory can produce up to 12 thousand pieces of the item. The transportation cost from the factory to warehouse n is t_n dollars per thousand pieces.

From historical data, it is known that the demand per period from warehouse n for the item is governed by a Poisson distribution[†] with mean λ_n (in thousands of pieces). If demand exceeds available stock a penalty of π_n dollars per thousand units out of stock is charged at warehouse n.

The current inventory on hand at warehouse n is q_n thousand units.

a) Formulate a dynamic program for determining the amount to be produced and the optimal allocation to each warehouse, in order to minimize transportation and expected stockout costs.

b) Solve the problem for a four-warehouse system with the data given in Table E11.8.

[†] The Poisson distribution is given by Prob. $(\tilde{k} = k) = \dfrac{\lambda^k e^{-\lambda}}{k!}$.

Table E11.8

	Demand	Inventory	Transportation cost	Stockout penalty
Warehouse (n)	λ_n (thousand units)	q_n (thousand units)	t_n ($ per 1000 units)	π_n ($ per 1000 units)
1	3	1	100	1500
2	4	2	300	2000
3	5	1	250	1700
4	2	0	200	2200

* 27. Precision Equipment, Inc., has won a government contract to supply 4 pieces of a high-precision part that is used in the fuel throttle-valve of an orbital module. The factory has three different machines capable of producing the item. They differ in terms of setup cost, variable production cost, and the chance that every single item will meet the high-quality standards (see Table E11.9).

Table E11.9

Machine	Setup cost ($)	Variable cost ($/unit)	Probability of meeting standards
A	100	20	0.50
B	300	40	0.80
C	500	60	0.90

After the parts are produced, they are sent to the engine assembly plant where they are tested. There is no way to recondition a rejected item. Any parts in excess of four, even if good, must be scrapped. If less than 4 parts are good, the manufacturer has to pay a penalty of $200 for each undelivered item.

How many items should be produced on each machine in order to minimize total expected cost? [*Hint.* Consider each machine as a stage and define the state variable as the number of acceptable parts still to be produced.]

* 28. One of the systems of a communications satellite consists of five electronic devices connected in series; the system as a whole would fail if any one of these devices were to fail. A common engineering design to increase the reliability of the system is to connect several devices of the same type in parallel, as shown in Fig. E11.7. The parallel devices in each group are

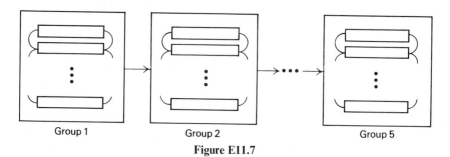

Group 1   Group 2   Group 5

Figure E11.7

controlled by a monitoring system, so that, if one device fails, another one immediately becomes operative.

The total weight of the system may not exceed 20 pounds. Table E11.10 shows the weight in pounds and the probability of failure for each device in group j of the system design.

Table E11.10

Group	Weight (lbs./device)	Probability of failure for each device
1	1	0.20
2	2	0.10
3	1	0.30
4	2	0.15
5	3	0.05

How many devices should be connected in parallel in each group so as to maximize the reliability of the overall system?

* 29. The production manager of a manufacturing company has to devise a production plan for item AK102 for the next four months. The item is to be produced at most once monthly; because of capacity limitations the monthly production may not exceed 10 units. The cost of one setup for any positive level of production in any month is $10.

The demand for this item is uncertain and varies from month to month; from past experience, however, the manager concludes that the demand in each month can be approximated by a Poisson distribution with parameter λ_n (n shows the month to which the distribution refers).

Inventory is counted at the end of each month and a holding cost of $10 is charged for each unit; if there are stockouts, a penalty of $20 is charged for every unit out of stock. There is no initial inventory and no outstanding back-orders; no inventory is required at the end of the planning period. Assume that the production lead time is short so that the amount released for production in one month can be used to satisfy demand within the same month.

What is the optimal production plan, assuming that the optimality criterion is the minimum expected cost? Assume that $\lambda_1 = 3$, $\lambda_2 = 5$, $\lambda_3 = 2$, $\lambda_4 = 4$ units.

* 30. Just before the Christmas season, Bribham of New England, Inc., has signed a large contract to buy four varieties of Swiss chocolate from a local importer. As it was already late, the distributor could arrange for only a limited transportation of 20 tons of Swiss chocolate to be delivered in time for Christmas.

Chocolate is transported in containers; the weight and the transportation cost per container are given in Table E11.11.

Table E11.11

Variety	Weight (tons/container)	Transportation ($/container)	Shortage cost ($/container)	λ_n (tons)
1	2	50	500	3
2	3	100	300	4
3	4	150	800	2
4	4	200	1000	1

A marketing consulting firm has conducted a study and has estimated the demand for the upcoming holiday season as a Poisson distribution with parameter λ_n ($n = 1, 2, 3, 4$ indicates the variety of the chocolate). Bribham loses contribution (i.e., shortage cost) for each container that can be sold (i.e., is demanded) but is not available.

How many containers of each variety should the company make available for Christmas in order to minimize total expected cost (transportation and shortage costs)?

ACKNOWLEDGMENTS

The example in Section 11.6 is taken from the State Department of Public Health case by Richard F. Meyer.

Exercise 10 is based on Section 10.6 of *Nonlinear and Dynamic Programming*, Addison-Wesley, 1962, by George Hadley.

Exercise 24 is inspired by *Dynamic Programming and Markov Processes*, John Wiley & Sons, 1960, by Ronald A. Howard.

Large-Scale Systems

12

As mathematical-programming techniques and computer capabilities evolve, the spectrum of potential applications also broadens. Problems that previously were considered intractable, from a computational point of view, now become amenable to practical mathematical-programming solutions. Today, commercial linear-programming codes can solve general linear programs of about 4000 to 6000 constraints. Although this is an impressive accomplishment, many applied problems lead to formulations that greatly exceed this existing computational limit. Two approaches are available to deal with these types of problems.

One alternative, that we have discussed in Chapter 5, leads to the partitioning of the overall problem into manageable subproblems, which are linked by means of a hierarchical integrative system. An application of this approach was presented in Chapter 6, where two interactive linear-programming models were designed to support strategic and tactical decisions in the aluminum industry, including resource acquisition, swapping contracts, inventory plans, transportation routes, production schedules and market-penetration strategies. This hierarchical method of attacking large problems is particularly effective when the underlying managerial process involves various decision makers, whose areas of concern can be represented by a specific part of the overall problem and whose decisions have to be coordinated within the framework of a hierarchical organization.

Some large-scale problems are not easily partitioned in this way. They present a monolithic structure that makes the interaction among the decision variables very hard to separate, and lead to situations wherein there is a single decision maker responsible for the actions to be taken, and where the optimum solution is very sensitive to the overall variable interactions. Fortunately, these large-scale problems invariably contain special structure. The large-scale system approach is to treat the problem as a unit, devising specialized algorithms to exploit the structure of the problem. This alternative will be explored in this chapter, where two of the most

important large-scale programming procedures—decomposition and colunm generation—will be examined.

The idea of taking computational advantage of the special structure of a specific
problem to develop an efficient algorithm is not new. The upper-bounding technique
introduced in Chapter 2, the revised simplex method presented in Appendix B, and
the network-solution procedures discussed in Chapter 8 all illustrate this point. This
chapter further extends these ideas.

12.1 LARGE-SCALE PROBLEMS

Certain structural forms of large-scale problems reappear frequently in applications,
and large-scale systems theory concentrates on the analysis of these problems. In
this context, structure means the pattern of zero and nonzero coefficients in the
constraints; the most important such patterns are depicted in Fig. 12.1. The first
illustration represents a problem composed of independent subsystems. It can be

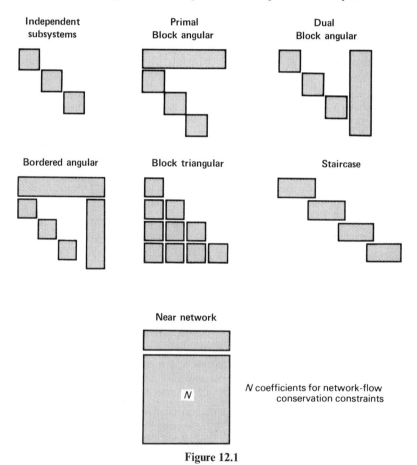

Figure 12.1

written as:

$$\text{Minimize} \sum_{j=1}^{r} c_j x_j + \sum_{j=r+1}^{s} c_j x_j + \sum_{j=s+1}^{n} c_j x_j,$$

subject to:

$$\sum_{j=1}^{r} a_{ij} x_j = b_i \quad (i = 1, 2, \ldots, t),$$

$$\sum_{j=r+1}^{s} a_{ij} x_j = b_i \quad (i = t + 1, t + 2, \ldots, u),$$

$$\sum_{j=s+1}^{n} a_{ij} x_j = b_i \quad (i = u + 1, u + 2, \ldots, m),$$

$$x_j \geq 0 \quad (j = 1, 2, \ldots, n).$$

Observe that the variables $x_1, x_2, \ldots, x_r$, the variables $x_{r+1}, x_{r+2}, \ldots, x_s$, and the variables $x_{s+1}, x_{s+2}, \ldots, x_n$ do not appear in common constraints. Consequently, these variables are independent, and the problem can be approached by solving one problem in the variables $x_1, x_2, \ldots, x_r$, another in the variables $x_{r+1}, x_{r+2}, \ldots, x_s$ and a third in the variables $x_{s+1}, x_{s+2}, \ldots, x_n$. This separation into smaller and independent subproblems has several important implications.

First, it provides significant computational savings, since the computations for linear programs are quite sensitive to m, the number of constraints, in practice growing proportionally to m^3. If each subproblem above contains $\frac{1}{3}$ of the constraints, then the solution to each subproblem requires on the order of $(m/3)^3 = m^3/27$ computations. All three subproblems then require about $3(m^3/27) = m^3/9$ computations, or approximately $\frac{1}{9}$ the amount for an m-constraint problem without structure. If the number of subsystems were k, the calculations would be only $1/k^2$ times those required for an unstructured problem of comparable size.

Second, each of the independent subproblems can be treated separately. Data can be gathered, analyzed, and stored separately. The problem also can be solved separately and, in fact, simultaneously. Each of these features suggests problems composed solely with independent subsystems as the most appealing of the structured problems.

The most natural extensions of this model are to problems with *nearly* independent subsystems, as illustrated by the next three structures in Fig. 12.1. In the primal block angular structure, the subsystem variables appear together, sharing common resources in the uppermost "coupling" constraints. For example, the subsystems might interact via a corporate budgetary constraint specifying that *total* capital expenditures of all subsystems cannot exceed available corporate resources.

The dual block angular structure introduces complicating "coupling" variables. In this case, the subsystems interact only by engaging in some common activities. For example, a number of otherwise independent subsidiaries of a company might join together in pollution-abatement activities that utilize some resources from each of the subsidiaries.

The bordered angular system generalizes these models by including complications from both coupling variables and coupling constraints. To solve any of these problems, we would like to decompose the system, removing the complicating variables or constraints, to reduce the problem to one with independent subsystems. Several of the techniques in large-scale system theory can be given this interpretation.

Dynamic models, in the sense of multistage optimization, provide another major source of large-scale problems. In dynamic settings, decisions must be made at several points in time, e.g., weekly or monthly. Usually decisions made in any time period have an impact upon other time periods, so that, even when every instantaneous problem is small, timing effects compound the decision process and produce large, frequently *extremely* large, optimization problems. The staircase and block triangular structures of Fig. 12.1 are common forms for these problems. In the staircase system, some activities, such as holding of inventory, couple succeeding time periods. In the block triangular case, decisions in each time period can directly affect resource allocation in any future time period.

The last structure in Fig. 12.1 concerns problems with large network subsystems. In these situations, we would like to exploit the special characteristics of network problems.

It should be emphasized that the special structures introduced here do not exhaust all possibilities. Other special structures, like Leontief systems arising in economic planning, could be added. Rather, the examples given are simply types of problems that arise frequently in applications. To develop a feeling for potential applications, let us consider a few examples.

Multi-Item Production Scheduling

Many industries must schedule production and inventory for a large number of products over several periods of time, subject to constraints imposed by limited resources. These problems can be cast as large-scale programs as follows. Let

$$\theta_{jk} = \begin{cases} 1 & \text{if the } k\text{th production schedule is used for item } j, \\ 0 & \text{otherwise}; \end{cases}$$

$$K_j = \text{Number of possible schedules for item } j.$$

Each production schedule specifies how item j is to be produced in each time period $t = 1, 2, \ldots, T$; for example, ten items in period 1 on machine 2, fifteen items in period 2 on machine 4, and so forth. The schedules must be designed so that production plus available inventory in each period is sufficient to satisfy the (known) demand for the items in that period. Usually it is not mandatory to consider every potential production schedule; under common hypotheses, it is known that at most 2^{T-1} schedules must be considered for each item in a T-period problem. Of course, this number can lead to enormous problems; for example, with $J = 100$ items and $T = 12$ time periods, the total number of schedules (θ_{jk} variables) will be $100 \times 2^{11} = 204,800$.

Next, we let

c_{jk} = Cost of the kth schedule for item j (inventory plus production cost, including machine setup costs for production),

b_i = Availability of resource i ($i = 1, 2, \ldots, m$),

a^i_{jk} = Consumption of resource i in the kth production plan for item j.

The resources might include available machine hours or labor skills, as well as cash-flow availability. We also can distinguish among resource availabilities in each time period; e.g., b_1 and b_2 might be the supply of a certain labor skill in the first and second time periods, respectively.

The formulation becomes:

$$\text{Minimize} \sum_{k=1}^{K_1} c_{1k}\theta_{1k} + \sum_{k=1}^{K_2} c_{2k}\theta_{2k} + \cdots + \sum_{k=1}^{K_J} c_{Jk}\theta_{Jk},$$

subject to:

$$\sum_{k=1}^{K_1} a^i_{1k}\theta_{1k} + \sum_{k=1}^{K_2} a^i_{2k}\theta_{2k} + \cdots + \sum_{k=1}^{K_J} a^i_{Jk}\theta_{Jk} \leqq b_i \quad (i = 1, 2, \ldots, m),$$

$$\left.\begin{aligned} \sum_{k=1}^{K_1} \theta_{1k} &= 1\\[1em] \sum_{k=1}^{K_2} \theta_{2k} &= 1\\[1em] &\;\;\ddots\\[1em] \sum_{k=1}^{K_J} \theta_{Jk} &= 1 \end{aligned}\right\} J \text{ constraints}$$

$$\theta_{jk} \geqq 0 \quad \text{and integer}, \quad \text{all } j \text{ and } k.$$

The J equality restrictions, which usually comprise most of the constraints, state that exactly one schedule must be selected for each item. Note that any basis for this problem contains ($m + J$) variables, and at least one basic variable must appear in each of the last J constraints. The remaining m basic variables can appear in no more than m of these constraints. When m is much smaller than J, this implies that most (at least $J - m$) of the last constraints contain one basic variable whose value must be 1. Therefore, most variables will be integral in any linear-programming basis. In practice, then, the integrality restrictions on the variables will be dropped to obtain an approximate solution by linear programming. Observe that the problem has a block angular structure. It is approached conveniently by either the decomposition procedure discussed in this chapter or a technique referred to as generalized upper bounding (GUB), which is available on many commercial mathematical-programming systems.

Exercises 11–13 at the end of this chapter discuss this multi-item production scheduling model in more detail.

Multicommodity Flow

Communication systems such as telephone systems or national computer networks must schedule message transmission over communication links with limited capacity. Let us assume that there are K types of messages to be transmitted and that each type is to be transmitted from its source to a certain destination. For example, a particular computer program might have to be sent to a computer with certain running or storage capabilities.

The communication network includes sending stations, receiving stations, and relay stations. For notation, let:

$$x_{ij}^k = \text{Number of messages of type } k \text{ transmitted along the communication link from station } i \text{ to station } j,$$

$$u_{ij} = \text{Message capacity for link } i, j,$$

$$c_{ij}^k = \text{Per-unit cost for sending a type-}k \text{ message along link } i, j,$$

$$b_i^k = \text{Net messages of type } k \text{ generated at station } i.$$

In this context, $b_i^k < 0$ indicates that i is a receiving station for type-k messages; $(-b_i^k) > 0$ then is the number of type-k messages that this station will process; $b_i^k = 0$ for relay stations. The formulation is:

$$\text{Minimize} \sum_{i,j} c_{ij}^1 x_{ij}^1 + \sum_{i,j} c_{ij}^2 x_{ij}^2 + \cdots + \sum_{i,j} c_{ij}^K x_{ij}^K,$$

subject to:

$$x_{ij}^1 + x_{ij}^2 + \cdots + x_{ij}^K \leqq u_{ij}, \quad \text{all } i, j$$

$$\left(\sum_j x_{ij}^1 - \sum_r x_{ri}^1 \right) = b_i^1, \quad \text{all } i$$

$$\left(\sum_j x_{ij}^2 - \sum_r x_{ri}^2 \right) = b_i^2, \quad \text{all } i$$

$$\ddots \qquad \vdots \qquad \vdots$$

$$\left(\sum_j x_{ij}^K - \sum_r x_{ri}^K \right) = b_i^K, \quad \text{all } i$$

$$x_{ij}^k \geq 0, \quad \text{all } i, j, k.$$

The summations in each term are made only over indices that correspond to arcs in the underlying network. The first constraints specify the transmission capacities u_{ij} for the links. The remaining constraints give flow balances at the communication stations i. For each fixed k, they state that the total messages of type k sent from station i must equal the number received at that station plus the number generated there. Since these are network-flow constraints, the model combines both the block angular and the near-network structures.

There are a number of other applications for this multicommodity-flow model. For example, the messages can be replaced by goods in an import–export model.

Stations then correspond to cities and, in particular, include airport facilities and ship ports. In a traffic-assignment model, as another example, vehicles replace messages and roadways replace communication links. A numerical example of the multi-commodity-flow problem is solved in Section 12.5.

Economic Development

Economic systems convert resources in the form of goods and services into output resources, which are other goods and services. Assume that we wish to plan the economy to consume b_i^t units of resource i at time t $(i = 1, 2, \ldots, m; t = 1, 2, \ldots, T)$. The b_i^t's specify a desired consumption schedule. We also assume that there are n production (service) activities for resource conversion, which are to be produced to meet the consumption schedule. Let

$$x_j = \text{Level of activity } j,$$
$$a_{ij}^t = \text{Number of units of resource } i \text{ that activity } j \text{ ``produces'' at time } t$$
$$\text{per unit of the activity level.}$$

By convention, $a_{ij}^t < 0$ means that activity j consumes resource i at time t in its production of another resource; this consumption is internal to the production process and does not count toward the b_i^t desired by the ultimate consumers. For example, if $a_{1j}^1 = -2, a_{2j}^1 = -3$, and $a_{3j}^1 = 1$, it takes 2 and 3 units of goods one and two, respectively, to produce 1 unit of good three in the first time period.

  It is common to assume that activities are defined so that each produces exactly one output; that is, for each j, $a_{it}^t > 0$ for one combination of i and t. An activity that produces output in period t is assumed to utilize input resources only from the current or previous periods (for example, to produce at time t we may have to train workers at time $t - 1$, "consuming" a particular skill from the labor market during the previous period). If J_t are the activities that produce an output in period t and j is an activity from J_t, then the last assumption states that $a_{ij}^\tau = 0$ whenever $\tau > t$. The feasible region is specified by the following linear constraints:

$$\sum_{j \text{ in } J_1} a_{ij}^1 x_j + \sum_{j \text{ in } J_2} a_{ij}^1 x_j + \sum_{j \text{ in } J_3} a_{ij}^1 x_j + \cdots + \sum_{j \text{ in } J_T} a_{ij}^1 x_j = b_i^1 \qquad (i = 1, 2, \ldots, m),$$

$$\sum_{j \text{ in } J_2} a_{ij}^2 x_j + \sum_{j \text{ in } J_3} a_{ij}^2 x_j + \cdots + \sum_{j \text{ in } J_T} a_{ij}^2 x_j = b_i^2 \qquad (i = 1, 2, \ldots, m),$$

$$\sum_{j \text{ in } J_3} a_{ij}^3 x_j + \cdots + \sum_{j \text{ in } J_T} a_{ij}^3 x_j = b_i^3 \qquad (i = 1, 2, \ldots, m),$$

$$\ddots \qquad \qquad \vdots \qquad\qquad \vdots$$

$$\sum_{j \text{ in } J_T} a_{ij}^T x_j = b_i^T \qquad (i = 1, 2, \ldots, m),$$

$$x_j \geq 0 \qquad (j = 1, 2, \ldots, n).$$

  One problem in this context is to see if a feasible plan exists and to find it by linear programming. Another possibility is to specify one important resource, such as labor, and to

$$\text{Minimize} \sum_{j=1}^{n} c_j x_j,$$

where c_j is the per-unit consumption of labor for activity j.

In either case, the problem is a large-scale linear program with triangular structure. The additional feature that each variable x_j has a positive coefficient in exactly one constraint (i.e., it produces exactly one output) can be used to devise a special algorithm that solves the problem as several small linear programs, one at each point in time $t = 1, 2, \ldots, T$.

12.2 DECOMPOSITION METHOD—A PREVIEW

Several large-scale problems including any with block angular or near-network structure become much easier to solve when some of their constraints are removed. The decomposition method is one way to approach these problems. It essentially considers the problem in two parts, one with the "easy" constraints and one with the "complicating" constraints. It uses the shadow prices of the second problem to specify resource prices to be used in the first problem. This leads to interesting economic interpretations, and the method has had an important influence upon mathematical economics. It also has provided a theoretical basis for discussing the coordination of decentralized organization units, and for addressing the issue of transfer prices among such units.

This section will introduce the algorithm and motivate its use by solving a small problem. Following sections will discuss the algorithm formally, introduce both geometric and economic interpretations, and develop the underlying theory.

Consider a problem with bounded variables and a single resource constraint:

$$\text{Maximize } z = 4x_1 + x_2 + 6x_3,$$

subject to:

$$3x_1 + 2x_2 + 4x_3 \leq 17 \quad \text{(Resource constraint)},$$

$$
\begin{aligned}
x_1 & \leq 2, \\
x_2 & \leq 2, \\
x_3 & \leq 2, \\
x_1 & \geq 1, \\
x_2 & \geq 1, \\
x_3 & \geq 1.
\end{aligned}
$$

We will use the problem in this section to illustrate the decomposition procedure, though in practice it would be solved by bounded-variable techniques.

First, note that the resource constraint complicates the problem. Without it, the problem is solved trivially as the objective function is maximized by choosing

x_1, x_2, and x_3 as large as possible, so that the solution $x_1 = 2$, $x_2 = 2$, and $x_3 = 2$ is optimal.

In general, given any objective function, the problem

Maximize $c_1x_1 + c_2x_2 + c_3x_3$,

subject to:

$$
\begin{aligned}
x_1 &\leqq 2, \\
x_2 &\leqq 2, \\
x_3 &\leqq 2, \\
x_1 &\geqq 1, \\
x_2 &\geqq 1, \\
x_3 &\geqq 1,
\end{aligned}
\tag{1}
$$

is also trivial to solve: One solution is to set x_1, x_2, or x_3 to 2 if its objective coefficient is positive, or to 1 if its objective coefficient is nonpositive.

Problem (1) contains some but not all of the original constraints and is referred to as a *subproblem* of the original problem. Any feasible solution to the subproblem potentially can be a solution to the original problem, and accordingly may be called a subproblem *proposal*. Suppose that we are given two subproblem proposals and that we combine them with weights as in Table 12.1.

Table 12.1 Weighting subproblem proposals.

	Activity levels			Resource	Objective	
	x_1	x_2	x_3	usage	value	Weights
Proposal 1	2	2	2	18	22	λ_1
Proposal 2	1	1	2	13	17	λ_2
Weighted proposal	$2\lambda_1 + \lambda_2$	$2\lambda_1 + \lambda_2$	$2\lambda_1 + 2\lambda_2$	$18\lambda_1 + 13\lambda_2$	$22\lambda_1 + 17\lambda_2$	

Observe that if the weights are nonnegative and sum to 1, then the weighted proposal also satisfies the subproblem constraints and is also a proposal. We can ask for those weights that make this composite proposal best for the overall problem, by solving the optimization problem:

Maximize $z = 22\lambda_1 + 17\lambda_2$,

subject to:

$$
\begin{aligned}
18\lambda_1 + 13\lambda_2 &\leqq 17, \\
\lambda_1 + \lambda_2 &= 1, \\
\lambda_1 \geqq 0, \quad \lambda_2 &\geqq 0.
\end{aligned}
$$

Optimal
shadow
prices
———
1
4

The first constraint states that the composite proposal should satisfy the resource limitation. The remaining constraints define λ_1 and λ_2 as weights. The linear-programming solution to this problem has $\lambda_1 = \frac{4}{5}$, $\lambda_2 = \frac{1}{5}$, and $z = 21$.

We next consider the effect of introducing any new proposal to be weighted with the two above. Assuming that each unit of this proposal contributes p_1 to the objective function and uses r_1 units of the resource, we have the modified problem:

$$\text{Maximize } 22\lambda_1 + 17\lambda_2 + p_1\lambda_3,$$

subject to:

$$18\lambda_1 + 13\lambda_2 + r_1\lambda_3 \leq 17,$$
$$\lambda_1 + \lambda_2 + \lambda_3 = 1,$$
$$\lambda_1 \geq 0, \quad \lambda_2 \geq 0, \quad \lambda_3 \geq 0.$$

To discover whether any new proposal would aid the maximization, we price out the general new activity to determine its reduced cost coefficient $\bar{p}_1$. In this case, applying the shadow prices gives:

$$\bar{p}_1 = p_1 - (1)r_1 - (4)1. \tag{2}$$

Note that we must specify the new proposal by giving numerical values to p_1 and r_1 before $\bar{p}_1$ can be determined.

By the simplex optimality criterion, the weighting problem cannot be improved if $\bar{p}_1 \leq 0$ for every new proposal that the subproblem can submit. Moreover, if $\bar{p}_1 > 0$, then the proposal that gives $\bar{p}_1$ improves the objective value. We can check both conditions by solving max $\bar{p}_1$ over all potential proposals.

Recall, from the original problem statement, that, for any proposal x_1, x_2, x_3,

$$p_1 \text{ is given by } 4x_1 + x_2 + 6x_3,$$

and

$$r_1 \text{ is given by } 3x_1 + 2x_2 + 4x_3. \tag{3}$$

Substituting in (2), we obtain:

$$\bar{p}_1 = (4x_1 + x_2 + 6x_3) - 1(3x_1 + 2x_2 + 4x_3) - 4(1),$$
$$= x_1 - x_2 + 2x_3 - 4,$$

and

$$\text{Max } \bar{p}_1 = \text{Max } (x_1 - x_2 + 2x_3 - 4).$$

Checking potential proposals by using this objective in the subproblem, we find that the solution is $x_1 = 2$, $x_2 = 1$, $x_3 = 2$, and $\bar{p}_1 = (2) - (1) + 2(2) - 4 = 1 > 0$. Equation (3) gives $p_1 = 21$ and $r_1 = 16$.

Consequently, the new proposal is useful, and the weighting problem becomes:

$$\text{Maximize } z = 22\lambda_1 + 17\lambda_2 + 21\lambda_3,$$

subject to:

	Optimal shadow prices	
$18\lambda_1 + 13\lambda_2 + 16\lambda_3 \leq 17,$	$\frac{1}{2}$	(4)
$\lambda_1 + \lambda_2 + \lambda_3 = 1,$	13	
$\lambda_1 \geq 0, \quad \lambda_2 \geq 0, \quad \lambda_3 \geq 0.$		

The solution is $\lambda_1 = \lambda_3 = \frac{1}{2}$, and z has increased from 21 to $21\frac{1}{2}$. Introducing a new proposal with contribution p_2, resource usage r_2, and weight λ_4, we now may repeat the same procedure. Using the new shadow prices and pricing out this proposal to determine its reduced cost, we find that:

$$\begin{aligned}
\bar{p}_2 &= p_2 - \tfrac{1}{2}r_2 - 13(1), \\
&= (4x_1 + x_2 + 6x_3) - \tfrac{1}{2}(3x_1 + 2x_2 + 4x_3) - 13, \\
&= \tfrac{5}{2}x_1 + 4x_3 - 13.
\end{aligned} \tag{5}$$

Solving the subproblem again, but now with expression (5) as an objective function, gives $x_1 = 2$, $x_2 = 1$, $x_3 = 2$, and

$$\bar{p}_2 = \tfrac{5}{2}(2) + 4(2) - 13 = 0.$$

Consequently, no new proposal improves the current solution to the weighting problem (4). The optimal solution to the overall problem is given by weighting the first and third proposals each by $\frac{1}{2}$; see Table 12.2.

Table 12.2 Optimal weighting of proposals.

	Activity levels			Resource usage	Objective value	Weights
	x_1	x_2	x_3			
Proposal 1	2	2	2	18	22	$\frac{1}{2}$
Proposal 3	2	1	2	16	21	$\frac{1}{2}$
Optimal solution	2	$\frac{3}{2}$	2	17	$21\frac{1}{2}$	

The algorithm determines an optimal solution by successively generating a new proposal from the subproblem at each iteration, and then finding weights that maximize the objective function among all combinations of the proposals generated thus far. Each proposal is an extreme point of the subproblem feasible region; because this region contains a finite number of extreme points, at most a finite number of subproblem solutions will be required.

The following sections discuss the algorithm more fully in terms of its geometry, formal theory, and economic interpretation.

12.3 GEOMETRICAL INTERPRETATION OF DECOMPOSITION

The geometry of the decomposition procedure can be illustrated by the problem solved in the previous section:

Maximize $4x_1 + x_2 + 6x_3$,

subject to:

$$3x_1 + 2x_2 + 4x_3 \leq 17, \qquad \text{(Complicating resource constraint)}$$

$$1 \leq x_j \leq 2 \ (j = 1, 2, 3). \qquad \text{(Subproblem)}$$

The feasible region is plotted in Fig. 12.2.

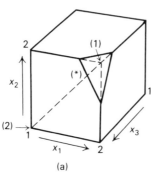 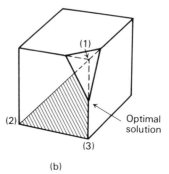

(a)   (b)

Fig. 12.2 Geometry of the decomposition method. (a) First approximation to feasible region; (b) final approximation to feasible region.

The feasible region to the subproblem is the cube $1 \leq x_j \leq 2$, and the resource constraint

$$3x_1 + 2x_2 + 4x_3 \leq 17$$

cuts away one corner from this cube.

The decomposition solution in Section 12.2 started with proposals (1) and (2) indicated in Fig. 12.2(a). Note that proposal (1) is not feasible since it violates the resource constraint. The initial weighting problem considers all combinations of proposals (1) and (2); these combinations correspond to the line segment joining points (1) and (2). The solution lies at (∗) on the intersection of this line segment and the resource constraint.

Using the shadow prices from the weighting problem, the subproblem next generates the proposal (3). The new weighting problem considers all weighted combinations of (1), (2) and (3). These combinations correspond to the triangle determined by these points, as depicted in Fig. 12.2(b). The optimal solution lies on the midpoint of the line segment joining (1) and (3), or at the point $x_1 = 2$, $x_2 = \frac{3}{2}$, and $x_3 = 2$. Solving the subproblem indicates that no proposal can improve upon this point and so it is optimal.

Note that the first solution to the weighting problem at (∗) is not an extreme point of the feasible region. This is a general characteristic of the decomposition algorithm that distinguishes it from the simplex method. In most applications, the method will consider many nonextreme points while progressing toward the optimal solution.

Also observe that the weighting problem approximates the feasible region of the overall problem. As more subproblem proposals are added, the approximation improves by including more of the feasible region. The efficiency of the method is predicated on solving the problem before the approximation becomes too fine and many proposals are generated. In practice, the algorithm usually develops a fairly good approximation quickly, but then expends considerable effort refining it. Consequently, when decomposition is applied, the objective value usually increases rapidly and then "tails off" by approaching the optimal objective value very slowly. This phenomenon is illustrated in Fig. 12.3 which plots the progress of the objective function for a typical application of the decomposition method.

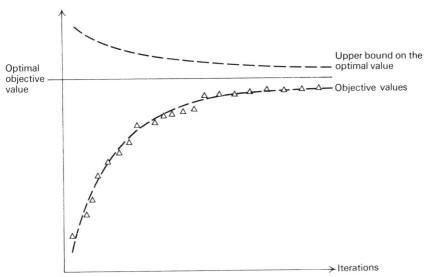

Fig. 12.3 Objective progress for a typical application of decomposition.

Fortunately, as discussed in Section 12.7, one feature of the decomposition algorithm is that it provides an upper bound on the value of the objective function at each iteration (see Fig. 12.3). As a result, the procedure can be terminated, prior to finding an optimal solution, with a conservative estimate of how far the current value of the objective function can be from its optimal value. In practice, since the convergence of the algorithm has proved to be slow in the final stages, such a termination procedure is employed fairly often.

12.4 THE DECOMPOSITION ALGORITHM

This section formalizes the decomposition algorithm, discusses implementation issues, and introduces a variation of the method applicable to primal block angular problems.

Formal Algorithm

Decomposition is applied to problems with the following structure.

Maximize $z = c_1 x_1 + c_2 x_2 + \cdots + c_n x_n$,

subject to:

$$
\left.
\begin{aligned}
a_{11} x_1 + a_{12} x_2 + \cdots + a_{1n} x_n &= b_1 \\
&\vdots \\
a_{m1} x_1 + a_{m2} x_2 + \cdots + a_{mn} x_n &= b_m
\end{aligned}
\right\}
\begin{aligned}
&\text{Complicating} \\
&\text{resource} \\
&\text{constraints}
\end{aligned}
$$

$$
\left.
\begin{aligned}
e_{11} x_1 + e_{12} x_2 + \cdots + e_{1n} x_n &= d_1 \\
&\vdots \\
e_{q1} x_1 + e_{q2} x_2 + \cdots + e_{qn} x_n &= d_q \\
x_j \geq 0 \quad (j = 1, 2, \ldots, n).
\end{aligned}
\right\}
\begin{aligned}
&\text{Subproblem} \\
&\text{constraints}
\end{aligned}
$$

The constraints are divided into two groups. Usually the problem is much easier to solve if the complicating a_{ij} constraints are omitted, leaving only the "easy" e_{ij} constraints.

Given any subproblem proposal $x_1, x_2, \ldots, x_n$ (i.e., a feasible solution to the subproblem constraints), we may compute:

$$r_i = a_{i1}x_1 + a_{i2}x_2 + \cdots + a_{in}x_n \qquad (i = 1, 2, \ldots, m),$$

and

$$p = c_1 x_1 + c_2 x_2 + \cdots + c_n x_n,$$

(6)

which are, respectively, the amount of resource r_i used in the ith complicating constraint and the profit p associated with the proposal.

When k proposals to the subproblem are known, the procedure acts to weight these proposals optimally. Let superscripts distinguish between proposals so that r_i^j is the use of resource i by the jth proposal and p^j is the profit for the jth proposal. Then the weighting problem is written as:

$$\text{Maximize } [p^1 \lambda_1 + p^2 \lambda_2 + \cdots + p^k \lambda_k],$$

Optimal shadow prices

subject to:

$$
\begin{aligned}
r_1^1 \lambda_1 + r_1^2 \lambda_2 + \cdots + r_1^k \lambda_k &= b_1, & \pi_1 \\
r_2^1 \lambda_1 + r_2^2 \lambda_2 + \cdots + r_2^k \lambda_k &= b_2, & \pi_2 \\
\vdots \qquad\qquad\qquad &\ \ \vdots & \vdots \\
r_m^1 \lambda_1 + r_m^2 \lambda_2 + \cdots + r_m^k \lambda_k &= b_m, & \pi_m \\
\lambda_1 + \lambda_2 + \cdots + \lambda_k &= 1, & \sigma
\end{aligned}
$$

(7)

$$\lambda_j \geq 0 \qquad (j = 1, 2, \ldots, k).$$

The weights $\lambda_1, \lambda_2, \ldots, \lambda_k$ are variables and r_i^j and p^j are known data in this problem.

Having solved the weighting problem and determined optimal shadow prices, we next consider adding new proposals. As we saw in Chapters 3 and 4, the reduced cost for a new proposal in the weighting linear program is given by:

$$\bar{p} = p - \sum_{i=1}^{m} \pi_i r_i - \sigma.$$

Substituting from the expressions in (6) for p and the r_i, we have

$$\bar{p} = \sum_{j=1}^{n} c_j x_j - \sum_{i=1}^{m} \pi_i \left(\sum_{j=1}^{n} a_{ij} x_j \right) - \sigma$$

or, rearranging,

$$\bar{p} = \sum_{j=1}^{n} \left(c_j - \sum_{i=1}^{m} \pi_i a_{ij} \right) x_j - \sigma.$$

Observe that the coefficient for x_j is the same reduced cost that was used in normal linear programming when applied to the complicating constraints. The additional term σ, introduced for the weighting constraint in problem (7), is added because of the subproblem constraints.

To determine whether any new proposal will improve the weighting linear program, we seek max $\bar{p}$ by solving the subproblem

$$v^k = \text{Max} \sum_{j=1}^{n} \left(c_j - \sum_{i=1}^{m} \pi_i a_{ij} \right) x_j, \tag{8}$$

subject to the subproblem constraints. There are two possible outcomes:

i) If $v^k \leq \sigma$, then max $\bar{p} \leq 0$. No new proposal improves the weighting linear program, and the procedure terminates. The solution is specified by weighting the subproblem proposals by the optimal weights $\lambda_1^*, \lambda_2^*, \ldots, \lambda_k^*$ to problem (7).

ii) If $v^k > \sigma$, then the optimal solution $x_1^*, x_2^*, \ldots, x_n^*$ to the subproblem is used in the weighting problem, by calculating the resource usages $r_1^{k+1}, r_2^{k+1}, \ldots, r_m^{k+1}$ and profit p^{k+1} for this proposal from the expressions in (6), and adding these coefficients with weight λ_{k+1}. The weighting problem is solved with this additional proposal and the procedure is repeated.

Section 12.7 develops the theory of this method and shows that it solves the original problem after a *finite* number of steps. This property uses the fact that the subproblem is a linear program, so that the simplex method for its solution determines each new proposal as an extreme point of the subproblem feasible region. Finite convergence then results, since there are only a finite number of potential proposals (i.e., extreme points) for the subproblem.

Computation Considerations

Initial Solutions

When solving linear programs, initial feasible solutions are determined by Phase I of the simplex method. Since the weighting problem is a linear program, the same technique can be used to find an initial solution for the decomposition method. Assuming that each righthand-side coefficient b_i is nonnegative, we introduce artificial variables $a_1, a_2, \ldots, a_m$ and solve the Phase I problem:

$$w = \text{Maximize} \, (-a_1 - a_2 - \cdots - a_m),$$

subject to:

$$a_i + r_i^1 \lambda_1 + r_i^2 \lambda_2 + \cdots + r_i^k \lambda_k = b_i \quad (i = 1, 2, \ldots, m),$$
$$\lambda_1 + \lambda_2 + \cdots + \lambda_k = 1,$$
$$\lambda_j \geq 0 \quad (j = 1, 2, \ldots, k),$$
$$a_i \geq 0 \quad (i = 1, 2, \ldots, m).$$

This problem weights subproblem proposals as in the original problem and decomposition can be used in its solution. To initiate the procedure, we might include only the artificial variables $a_1, a_2, \ldots, a_m$ and any known subproblem proposals.

If no subproblem proposals are known, one can be found by ignoring the complicating constraints and solving a linear program with only the subproblem constraints. New subproblem proposals are generated by the usual decomposition procedure. In this case, though, the profit contribution of every proposal is zero for the Phase I objective; i.e., the pricing calculation is:

$$\bar{p} = 0 - \sum \pi_i r_i - \sigma.$$

Otherwise, the details are the same as described previously.

If the optimal objective value $w* < 0$, then the original constraints are infeasible and the procedure terminates. If $w* = 0$, the final solution to the phase I problem identifies proposals and weights that are feasible in the weighting problem. We continue by applying decomposition with the phase II objective function, starting with these proposals.

The next section illustrates this phase I procedure in a numerical example.

Resolving Problems

Solving the weighting problem determines an optimal basis. After a new column (proposal) is added from the subproblem, this basis can be used as a starting point to solve the new weighting problem by the revised simplex method (see Appendix B). Usually, the old basis is near-optimal and few iterations are required for the new problem. Similarly, the optimal basis for the last subproblem can be used to initiate the solution to that problem when it is considered next.

Dropping Nonbasic Columns

After many iterations the number of columns in the weighting problem may become large. Any nonbasic proposal to that problem can be dropped to save storage. If it is required, it is generated again by the subproblem.

Variation of the Method

The decomposition approach can be modified slightly for treating primal block-angular structures. For notational convenience, let us consider the problem with only two subsystems:

Maximize $z = c_1 x_1 + c_2 x_2 + \cdots + c_t x_t + c_{t+1} x_{t+1} + \cdots + c_n x_n$,

subject to:

$$a_{i1} x_1 + a_{i2} x_2 \cdots + a_{it} x_t + a_{i,t+1} x_{t+1} + \cdots + a_{in} x_n = b_i \quad (i = 1, 2, \ldots, m),$$
$$e_{s1} x_1 + e_{s2} x_2 \cdots + e_{st} x_t \qquad\qquad = d_s \quad (s = 1, 2, \ldots, \bar{q}),$$
$$e_{s,t+1} x_{t+1} + \cdots + e_{sn} x_n = d_s \quad (s = \bar{q} + 1, \bar{q} + 2, \ldots, q)$$
$$x_j \geq 0, \quad (j = 1, 2, \ldots, n).$$

The easy e_{ij} constraints in this case are composed of two independent subsystems, one containing the variables $x_1, x_2, \ldots, x_t$ and the other containing the variables $x_{t+1}, x_{t+2}, \ldots, x_n$.

Decomposition may be applied by viewing the e_{ij} constraints as a single subproblem. Alternately, each subsystem may be viewed as a separate subproblem. Each will submit its own proposals and the weighting problem will act to coordinate these proposals in the following way. For any proposal $x_1, x_2, \ldots, x_t$ from subproblem 1, let

$$r_{i1} = a_{i1}x_1 + a_{i2}x_2 + \cdots + a_{it}x_t \qquad (i = 1, 2, \ldots, m)$$

and

$$p_1 = c_1 x_1 + c_2 x_2 + \cdots + c_t x_t$$

denote the resource usage r_{i1} in the ith a_{ij} constraint and profit contribution p_1 for this proposal. Similarly, for any proposal $x_{t+1}, x_{t+2}, \ldots, x_n$ from subproblem 2, let

$$r_{i2} = a_{i, t+1}x_{t+1} + a_{i, t+2}x_{t+2} + \cdots + a_{in}x_n \qquad (i = 1, 2, \ldots, m),$$

and

$$p_2 = c_{t+1}x_{t+1} + c_{t+2}x_{t+2} + \cdots + c_n x_n$$

denote its corresponding resource usage and profit contribution.

Suppose that, at any stage in the algorithm, k proposals are available from subproblem 1, and ℓ proposals are available from subproblem 2. Again, letting superscripts distinguish between proposals, we have the weighting problem:

Max $p_1^1 \lambda_1 + p_1^2 \lambda_2 + \cdots + p_1^k \lambda_k + p_2^1 \mu_1 + p_2^2 \mu_2 + \cdots + p_2^\ell \mu_\ell,$ *Optimal shadow prices*

subject to:

$$r_{i1}^1 \lambda_1 + r_{i1}^2 \lambda_2 + \cdots + r_{i1}^k \lambda_k + r_{i2}^1 \mu_1 + r_{i2}^2 \mu_2 + \cdots + r_{i2}^\ell \mu_\ell = b_i$$

$$\hspace{6cm} (i = 1, \ldots, m) \qquad \pi_i$$

$$\lambda_1 + \lambda_2 + \cdots + \lambda_k \hspace{3.5cm} = 1, \qquad \sigma_1$$

$$\mu_1 + \mu_2 + \cdots + \mu_\ell = 1, \qquad \sigma_2$$

$$\lambda_j \geqq 0, \qquad \mu_s \geqq 0 \qquad (j = 1, 2, \ldots, k; s = 1, 2, \ldots, \ell).$$

The variables $\lambda_1, \lambda_2, \ldots, \lambda_k$ weight subproblem 1 proposals and the variables $\mu_1, \mu_2, \ldots, \mu_\ell$ weight subproblem 2 proposals. The objective function adds the contribution from both subproblems and the ith constraint states that the total resource usage from both subsystems should equal the resource availability b_i of the ith resource.

After solving this linear program and determining the optimal shadow prices $\pi_1, \pi_2, \ldots, \pi_m$ and σ_1, σ_2, optimality is assessed by pricing out potential proposals from each subproblem:

$$\bar{p}_1 = p_1 - \sum_{i=1}^m \pi_i r_{i1} - \sigma_1 \qquad \text{for subproblem 1,}$$

$$\bar{p}_2 = p_2 - \sum_{i=1}^m \pi_i r_{i2} - \sigma_2 \qquad \text{for subproblem 2.}$$

Substituting for p_1, p_2, r_{i1}, and r_{i2} in terms of the variables x_j, we make these assessments, as in the usual decomposition procedure, by solving the subproblems:

Subproblem 1

$$v_1 = \text{Max} \sum_{j=1}^{t} \left(c_j - \sum_{i=1}^{m} \pi_i a_{ij} \right) x_j,$$

subject to:

$$e_{s1}x_1 + e_{s2}x_2 + \cdots + e_{st}x_t = d_s \qquad (s = 1, 2, \ldots, \bar{q}),$$
$$x_j \geq 0, \qquad (j = 1, 2, \ldots, t).$$

Subproblem 2

$$v_2 = \text{Max} \sum_{j=t+1}^{n} \left(c_j - \sum_{i=1}^{m} \pi_i a_{ij} \right) x_j,$$

subject to:

$$e_{s,t+1}x_{t+1} + e_{s,t+2}x_{t+2} + \cdots + e_{sn}x_n = d_s \qquad (s = \bar{q} + 1, \bar{q} + 2, \ldots, q),$$
$$x_j \geq 0, \qquad (j = t + 1, t + 2, \ldots, n).$$

If $v_i \leq \sigma_i$ for $i = 1$ and 2, then $\bar{p}_1 \leq 0$ for every proposal from subproblem 1 and $\bar{p}_2 \leq 0$ for every proposal from subproblem 2; the optimal solution has been obtained. If $v_1 > \sigma_1$, the optimal proposal to the first subproblem is added to the weighting problem; if $v_2 > \sigma_2$, the optimal proposal to the second subproblem is added to the weighting problem. The procedure then is repeated.

This modified algorithm easily generalizes when the primal block-angular system contains more than two subsystems. There then will be one weighting constraint for each subsystem. We should point out that it is not necessary to determine a new proposal from each subproblem at every iteration. Consequently, it is not necessary to solve each subproblem at every iteration, but rather subproblems must be solved until the condition $v_i > \sigma_i$ (that is, $\bar{p}_i > 0$) is achieved for one solution, so that at least one new proposal is added to the weighting problem at each iteration.

12.5 AN EXAMPLE OF THE DECOMPOSITION PROCEDURE

To illustrate the decomposition procedure with an example that indicates some of its computational advantages, we consider a special case of the multicommodity-flow problem introduced as an example in Section 12.1.

An automobile company produces luxury and compact cars at two of its regional plants, for distribution to three local markets. Tables 12.3 and 12.4 specify the transportation characteristics of the problem on a per-month basis, including the transportation solution. The problem is formulated in terms of profit maximization.

One complicating factor is introduced by the company's delivery system. The company has contracted to ship from plants to destinations with a trucking company.

Table 12.3 Luxury cars

Plant	Market 1	Market 2	Market 3	Supply
1	100 15	120 10	90	25
2	80 5	70	140 10	15
Demand	20	10	10	

Profit = 4500

Table 12.4 Compact cars

Plant	Market 1	Market 2	Market 3	Supply
1	40 20	20 10	30 20	50
2	20 20	40 30	10	30
Demand	20	40	20	

Profit = 2800

The routes from plant 1 to both markets 1 and 3 are hazardous, however; for this reason, the trucking contract specifies that no more than 30 cars in total should be sent along either of these routes in any single month. The above solutions sending 35 cars (15 luxury and 20 compact) from plant 1 to market 1 does not satisfy this restriction, and must be modified.

Let superscript 1 denote luxury cars, superscript 2 denote compact cars, and let x_{ij}^k be the number of cars of type k sent from plant i to market j. The model is formulated as a primal block-angular problem with objective function

$$\text{Maximize } [100x_{11}^1 + 120x_{12}^1 + 90x_{13}^1 + 80x_{21}^1 + 70x_{22}^1 + 140x_{23}^1$$
$$+ 40x_{11}^2 + 20x_{12}^2 + 30x_{13}^2 + 20x_{21}^2 + 40x_{22}^2 + 10x_{23}^2].$$

The five supply and demand constraints of each transportation table and the following two trucking restrictions must be satisfied.

$$x_{11}^1 + x_{11}^2 \leq 30 \quad \text{(Resource 1)},$$
$$x_{13}^1 + x_{13}^2 \leq 30 \quad \text{(Resource 2)}.$$

This linear program is easy to solve without the last two constraints, since it then reduces to two separate transportation problems. Consequently, it is attractive to use decomposition, with the transportation problems as two separate subproblems.

The initial weighting problem considers the transportation solutions as one proposal from each subproblem. Since these proposals are infeasible, a Phase I version of the weighting problem with artificial variable a_1 must be solved first:

Maximize $(-a_1)$,

subject to:

$$-a_1 + 15\lambda_1 + 20\mu_1 + \quad s_1 \quad\quad = 30, \qquad \pi_1 = 1$$
$$0\lambda_1 + 20\mu_1 \quad\quad\quad + s_2 = 30, \qquad \pi_2 = 0$$
$$\lambda_1 \quad\quad\quad\quad\quad\quad = 1, \qquad \sigma_1 = -15$$
$$\mu_1 \quad\quad\quad\quad\quad = 1, \qquad \sigma_2 = -20$$

$$\lambda_1 \geq 0, \quad \mu_1 \geq 0, \quad s_1 \geq 0, \quad s_2 \geq 0.$$

Optimal shadow prices

In this problem, s_1 and s_2 are slack variables for the complicating resource constraints. Since the two initial proposals ship $15 + 20 = 35$ cars on route 1–1, the first constraint is infeasible, and we must introduce an artificial variable in this constraint. Only 20 cars are shipped on route 1–3, so the second constraint is feasible, and the slack variable s_2 can serve as an initial basic variable in this constraint. No artificial variable is required.

The solution to this problem is $a_1 = 5$, $\lambda_1 = 1$, $\mu_1 = 1$, $s_1 = 0$, $s_2 = 10$, with the optimal shadow prices indicated above. Potential new luxury-car proposals are assessed by using the Phase I objective function and pricing out:

$$\bar{p}_1 = 0 - \pi_1 r_{11} + \pi_2 r_{21} - \sigma_1$$
$$= 0 - (1)r_{11} - (0)r_{21} + 15.$$

Since the two resources for the problem are the shipping capacities from plant 1 to markets 1 and 3, $r_{11} = x^1_{11}$ and $r_{21} = x^1_{13}$, and this expression reduces to:

$$\bar{p}_1 = -x^1_{11} + 15.$$

The subproblem becomes the transportation problem for luxury cars with objective coefficients as shown in Table 12.5. Note that this problem imposes a penalty of $1 for sending a car along route 1–1.

Table 12.5

Plant	Market 1	Market 2	Market 3	Supply
	-1	0	0	
1	5	10	10	25
	0	0	0	
2	15			15
Demand	20	10	10	

Phase II profit contribution $= 3800$

The solution indicated in the transportation tableau has an optimal objective value $v_1 = -5$. Since $\bar{p}_1 = v_1 + 15 > 0$, this proposal, using 5 units of resource 1 and 10 units of resource 2, is added to the weighting problem. The inclusion of this proposal causes a_1 to leave the basis, so that Phase I is completed.

Using the proposals now available, we may formulate the Phase II weighting problem as:

Maximize $4500\lambda_1 + 3800\lambda_2 + 2800\mu_1$,

subject to:

$$
\begin{array}{llll}
15\lambda_1 + & 5\lambda_2 + & 20\mu_1 + s_1 & = 30, \\
0\lambda_1 + & 10\lambda_2 + & 20\mu_1 & + s_2 = 30, \\
\lambda_1 + & \lambda_2 & & = 1, \\
& & \mu_1 & = 1,
\end{array}
$$

$$\lambda_1 \geq 0, \quad \lambda_2 \geq 0, \quad \mu_1 \geq 0, \quad s_1 \geq 0, \quad s_2 \geq 0.$$

Optimal shadow prices

$$
\begin{aligned}
\pi_1 &= 70 \\
\pi_2 &= 0 \\
\sigma_1 &= 3450 \\
\sigma_2 &= 1400
\end{aligned}
$$

The optimal solution is given by $\lambda_1 = \lambda_2 = \frac{1}{2}$, $\mu_1 = 1$, $s_1 = 0$, and $s_2 = 5$, with an objective value of \$6950. Using the shadow prices to price out potential proposals gives:

$$\bar{p}^j = p^j - \pi_1 r_{1j} - \pi_2 r_{2j} - \sigma_j$$
$$= p^j - \pi_1(x_{11}^j) - \pi_2(x_{13}^j) - \sigma_j,$$

or

$$\bar{p}^1 = p^1 - 70(x_{11}^1) - 0(x_{13}^1) - 3450 = p^1 - 70x_{11}^1 - 3450,$$

$$\bar{p}^2 = p^2 - 70(x_{11}^2) - 0(x_{13}^2) - 1400 = p^2 - 70x_{11}^2 - 1400.$$

In each case, the per-unit profit for producing in plant 1 for market 1 has decreased by \$70. The decomposition algorithm has imposed a penalty of \$70 on route 1–1 shipments, in order to divert shipments to an alternative route. The solution for luxury cars is given in Table 12.6.

Table 12.6 Luxury cars

Plant	Market 1	2	3	Supply
1	30 120 90 15 10			25
2	80 70 140 5 10			15
Demand	20	10	10	

$$v_1 = 30(15) + 120(10) + 80(5)$$
$$+ 140(10) = 3450$$

Since $v_1 - \sigma_1 = 3450 = 0$, no new luxury-car proposal is profitable, and we must consider compact cars, as in Table 12.7.

Table 12.7 Compact cars

Plant	Market 1	2	3	Supply
1	−30 20 30 30 20			50
2	20 40 10 20 10			30
Demand	20	40	20	

$$v_2 = 20(30) + 30(20) + 20(20)$$
$$+ 40(10) = 2000$$

Here $v_2 - \sigma_2 = 2000 - 1400 > 0$, so that the given proposal improves the weighting problem. It uses no units of resource 1, 20 units of resource 2, and its profit contribution is 2000, which in this case happens to equal v_2. Inserting this proposal in the weighting problem, we have:

Maximize $4500\lambda_1 + 3800\lambda_2 + 2800\mu_1 + 2000\mu_2$,

subject to:

Optimal shadow prices

$$
\begin{array}{llllll}
15\lambda_1 + & 5\lambda_2 + & 20\mu_1 + & 0\mu_2 + & s_1 & = 30, & \pi_1 = & 40 \\
0\lambda_1 + & 10\lambda_2 + & 20\mu_1 + & 20\mu_2 & + s_2 = 30, & \pi_2 = & 0 \\
\lambda_1 + & \lambda_2 & & & = 1, & \sigma_1 = & 3900 \\
& & \mu_1 + & \mu_2 & = 1, & \sigma_2 = & 2000
\end{array}
$$

$$\lambda_1 \geq 0, \quad \lambda_2 \geq 0, \quad \mu_1 \geq 0, \quad \mu_2 \geq 0, s_1 \geq 0, s_2 \geq 0.$$

The optimal basic variables are $\lambda_1 = 1$, $\mu_1 = \frac{3}{4}$, $\mu_2 = \frac{1}{4}$, and $s_2 = 10$, with objective value $7100, and the pricing-out operations become:

$$\bar{p}^1 = p^1 - 40(x_{11}^1) - 0(x_{13}^1) - 3900 \quad \text{for luxury cars,}$$

and

$$\bar{p}^2 = p^2 - 40(x_{11}^2) - 0(x_{13}^2) - 2000 \quad \text{for compact cars.}$$

The profit contribution for producing in plant 1 for market 1 now is penalized by $40 per unit for both types of cars. The transportation solutions are given by Tables 12.8 and 12.9.

Table 12.8 Luxury cars

Plant	Market			Supply
	1	2	3	
1	60 / 15	120 / 10	90	25
2	80 / 5	70	140 / 10	15
Demand	20	10	10	

$v_1 = 60(15) + 120(10) + 80(5)$
$\quad + 140(10) = 3900$

Table 12.9 Compact cars

Plant	Market			Supply
	1	2	3	
1	0 / 20	20 / 10	30 / 20	50
2	20 / 20	40 / 30	10	30
Demand	20	40	20	

$v_2 = 0(20) + 20(10) + 30(20)$
$\quad + 40(30) = 2000$

Since $v_1 - \sigma_1 = 0$ and $v_2 - \sigma_2 = 0$, neither subproblem can submit proposals to improve the last weighting problem and the optimal solution uses the first luxury car proposal, since $\lambda_1 = 1$, and weights the two compact car proposals with $\mu_1 = \frac{3}{4}$, $\mu_2 = \frac{1}{4}$, giving the composite proposal shown in Table 12.10.

Table 12.10 Compact cars

Plant	Market				Market				Market		
	1	2	3		1	2	3		1	2	3
1	15	15	20	$= \frac{3}{4}$	20	10	20	$+ \frac{1}{4}$	0	30	20
2	5	25	0		0	30	0		20	10	0

Observe that, although both of the transportation proposals shown on the righthand side of this expression solve the final transportation subproblem for compact cars with value $v_2 = 2000$, neither is an optimal solution to the overall problem. The unique solution to the overall problem is the composite compact-car proposal shown on the left together with the first luxury-car proposal.

12.6 ECONOMIC INTERPRETATION OF DECOMPOSITION

The connection between prices and resource allocation has been a dominant theme in economics for some time. The analytic basis for pricing systems is rather new, however, and owes much to the development of mathematical programming. Chapter 4 established a first connection between pricing theory and mathematical programming by introducing an economic interpretation of linear-programming duality theory. Decomposition extends this interpretation to decentralized decision making. It provides a mechanism by which prices can be used to coordinate the activities of several decision makers.

For convenience, let us adopt the notation of the previous section and discuss primal block-angular systems with two subsystems. We interpret the problem as a profit maximization for a firm with two divisions. There are two levels of decision making—corporate and subdivision. Subsystem constraints reflect the divisions' allocation of their own resources, assuming that these resources are not shared. The complicating constraints limit corporate resources, which are shared and used in any proposal from either division.

Frequently, it is very expensive to gather detailed information about the divisions in a form usable by either corporate headquarters or other divisions, or to gather detailed corporate information for the divisions. Furthermore, each level of decision making usually requires its own managerial skills with separate responsibilities. For these reasons, it is often best for each division and corporate headquarters to operate somewhat in isolation, passing on only that information required to coordinate the firm's activities properly.

As indicated in Fig. 12.4, in a decomposition approach the information passed on are *prices*, from corporate headquarters to the divisions, and *proposals*, from the divisions to the corporate coordinator. Only the coordinator knows the full corporate constraints and each division knows its own operating constraints.

The corporate coordinator acts to weight subproblem proposals by linear programming, to maximize profits. From the interpretation of duality given in Chapter 4, the optimal shadow prices from its solution establish a per-unit value for each resource. These prices are an internal evaluation of resources by the firm, indicating how profit will be affected by changes in the resource levels.

To ensure that the divisions are cognizant of the firm's evaluation of resources, the coordinator "charges" the divisions for their use of corporate resources. That is, whatever decisions $x_1, x_2, \ldots, x_t$ the first division makes, its gross revenue is

$$p_1 = c_1 x_1 + c_2 x_2 + \cdots + c_t x_t,$$

and its use of resource i is given by

$$r_{i1} = a_{i1} x_1 + a_{i2} x_2 + \cdots + a_{it} x_t.$$

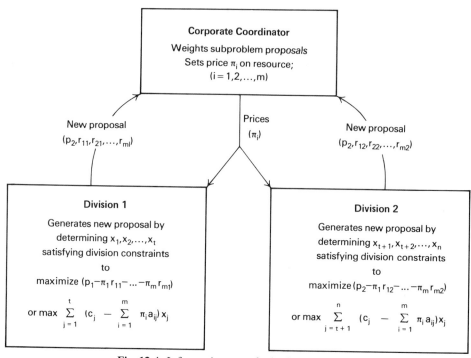

Fig. 12.4 Information transfer in decomposition.

Consequently, its net profit is computed by:

$$(\text{Net profit}) = (\text{Gross revenue}) - (\text{Resource cost})$$

$$= p_1 - \pi_1 r_{11} - \pi_2 r_{21} - \cdots - \pi_m r_{m1},$$

or, substituting in terms of the x_j's and rearranging,

$$(\text{Net profit}) = \sum_{j=1}^{t} \left(c_j - \sum_{i=1}^{m} \pi_i a_{ij} \right) x_j.$$

In this objective, c_j is the per-unit gross profit for activity x_j. The shadow price π_i is the value of the ith corporate resource, $\pi_i a_{ij}$ is the cost of resource i for activity j, and $\sum_{i=1}^{m} \pi_i a_{ij}$ is the total corporate resource cost, or opportunity cost, to produce each unit of this activity.

The cost $\sum_{i=1}^{m} \pi_i a_{ij}$ imposes a penalty upon activity j that reflects impacts resulting from this activity that are external to the division. That is, by engaging in this activity, the firm uses additional units a_{ij} of the corporate resources. Because the firm has limited resources, the activities of other divisions must be modified to compensate for this resource usage. The term $\sum_{i=1}^{m} \pi_i a_{ij}$ is the revenue lost by the firm as a result of the modifications that the other divisions must make in their use of resources.

Once each division has determined its optimal policy with respect to its net profit objective, it conveys this information in the form of a proposal to the co-ordinator. If the coordinator finds that no *new* proposals are better than those currently in use in the weighting problem, then prices π_i have stabilized (since the former linear-programming solution remains optimal), and the procedure terminates. Otherwise, new prices are determined, they are transmitted to the divisions, and the process continues.

Finally, the coordinator assesses optimality by pricing out the newly generated proposal in the weighting problem. For example, for a new proposal from division 1, the calculation is:

$$\bar{p}_1 = (p_1 - \pi_1 r_{11} - \pi_2 r_{21} - \cdots - \pi_m r_{m1}) - \sigma_1,$$

where σ_1 is the shadow price of the weighting constraint for division 1 proposals. The first term is the net profit of the new proposal as just calculated by the division. The term σ_1 is interpreted as the value (gross profit – resource cost) of the optimal composite or weighted proposal from the previous weighting problem. If $\bar{p}_1 > 0$, the new proposal's profit exceeds that of the composite proposal, and the coordinator alters the plan. The termination condition is that $\bar{p}_1 \leq 0$ and $\bar{p}_2 \leq 0$, when no new proposal is better than the current composite proposals of the weighting problem.

Example: The final weighting problem to the automobile example of the previous section was:

Maximize $4500\lambda_1 + 3800\lambda_2 + 2800\mu_1 + 2000\mu_2$,

subject to:

					Optimal shadow prices
$15\lambda_1 +$	$5\lambda_2 +$	$20\mu_1 +$	$0\mu_2 + s_1$	$= 30,$	$\pi_1 = \quad 40$
$0\lambda_1 +$	$10\lambda_2 +$	$20\mu_1 +$	$20\mu_2 \quad + s_2$	$= 30,$	$\pi_2 = \quad 0$
$\lambda_1 +$	λ_2			$= 1,$	$\sigma_1 = 3900$
		$\mu_1 +$	μ_2	$= 1,$	$\sigma_2 = 2000$

$$\lambda_j, \mu_j, s_j \geq 0 \qquad (j = 1, 2),$$

with optimal basic variables $\lambda_1 = 1$, $\mu_1 = \frac{3}{4}$, $\mu_2 = \frac{1}{4}$, and $s_2 = 10$. The first truck route from plant 1 to market 1 (constraint 1) is used to capacity, and the firm evaluates sending another car along this route at \$40. The second truck route from plant 1 to market 3 is not used to capacity, and accordingly its internal evaluation is $\pi_2 = \$0$.

The composite proposal for subproblem 1 is simply its first proposal with $\lambda_1 = 1$. Since this proposal sends 15 cars on the first route at \$40 each, its net profit is given by:

$$\sigma_1 = \text{(Gross profit)} - \text{(Resource cost)}$$
$$= \$4500 - \$40(15) = \$3900.$$

Similarly, the composite proposal for compact cars sends

$$20(\tfrac{3}{4}) + 0(\tfrac{1}{4}) = 15$$

cars along the first route. Its net profit is given by weighting its gross profit coefficients with $\mu_1 = \frac{3}{4}$ and $\mu_2 = \frac{1}{4}$ and subtracting resource costs, that is, as

$$\sigma_2 = [\$2800(\tfrac{3}{4}) + \$2000(\tfrac{1}{4})] - \$40(15) = \$2000.$$

By evaluating its resources, the corporate weighting problem places a cost of $40 on each car sent along route 1–1. Consequently, when solving the subproblems, the gross revenue in the transportation array must be decreased by $40 along the route 1–1; the profit of luxury cars along route 1–1 changes from $100 to $60(= \$100 - \$40)$, and the profit of compact cars changes from $40 to $0(= \$40 - \$40)$.

To exhibit the effect of externalities between the luxury and compact divisions, suppose that the firm ships 1 additional luxury car along route 1–1. Then the capacity along this route decreases from 30 to 29 cars. Since $\lambda_1 = 1$ is fixed in the optimal basic solution, μ_1 must decrease by $\frac{1}{20}$ to preserve equality in the first resource constraint of the basic solution. Since $\mu_1 + \mu_2 = 1$, this means that μ_2 must increase by $\frac{1}{20}$. Profit from the compact-car proposals then changes by $\$2800(-\frac{1}{20}) + \$2000(+\frac{1}{20}) = -\$40$, as required. Note that the 1-unit change in luxury-car operations induces a change in the composite proposal of compact cars for subproblem 2. The decomposition algorithm allows the luxury-car managers to be aware of this external impact through the price information passed on from the coordinator.

Finally, note that, although the price concept introduced in this economic interpretation provides an internal evaluation of resources that permits the firm to coordinate subdivision activities, the prices by themselves do not determine the optimal production plan at each subdivision. As we observed in the last section, the compact-car subdivision for this example has several optimal solutions to its subproblem transportation problem with respect to the optimal resource prices of $\pi_1 = \$40$ and $\pi_2 = \$0$. Only one of these solutions however, is optimal for the *overall* corporate plan problem. Consequently, the coordinator must negotiate the final solution used by the subdivisions; merely passing on the optimal resources will not suffice.

*12.7 DECOMPOSITION THEORY

In this section, we assume that decomposition is applied to a problem with only one subproblem:

Maximize $z = c_1x_1 + c_2x_2 + \cdots + c_nx_n$,

subject to:

$$a_{i1}x_1 + a_{i2}x_2 + \cdots + a_{in}x_n = b_i$$
$$(i = 1, 2, \ldots, m), \qquad \text{(Complicating constants)}$$

$$e_{s1}x_1 + e_{s2}x_2 + \cdots + e_{sn}x_n = d_s$$
$$(s = 1, 2, \ldots, q), \qquad \text{(Subproblem)}$$

$$x_j \geq 0 \qquad (j = 1, 2, \ldots, n).$$

The discussion extends to primal block-angular problems in a straightforward manner.

The theoretical justification for decomposition depends upon a fundamental result from convex analysis. This result is illustrated in Fig. 12.5 for a feasible region determined by linear inequalities. The region is bounded and has five extreme points denoted $x^1, x^2, \ldots, x^5$. Note that any point y in the feasible region can be expressed as a weighted (convex) combination of extreme points. For example, the weighted combination of the extreme points x^1, x^2, and x^5 given by

$$y = \lambda_1 x^1 + \lambda_2 x^2 + \lambda_3 x^5,$$

for some selection of

$$\lambda_1 \geq 0, \qquad \lambda_2 \geq 0, \qquad \text{and} \qquad \lambda_3 \geq 0,$$

with

$$\lambda_1 + \lambda_2 + \lambda_3 = 1,$$

determines the shaded triangle in Fig. 12.5. Note that the representation of y as an extreme point is not unique; y can also be expressed as a weighted combination of x^1, x^4, and x^5, or x^1, x^3, and x^5.

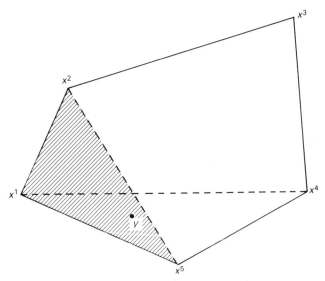

Fig. 12.5 Extreme point representation.

The general result that we wish to apply is stated as the Representation Property, defined as:

Representation Property. Let $x^1, x^2, \ldots, x^K$ be the extreme points [each x^k specifies values for each variable x_j as $(x_1^k, x_2^k, \ldots, x_n^k)$] of a feasible region determined by the constraints

$$e_{s1}x_1 + e_{s2}x_2 + \cdots + e_{sn}x_n = d_s \qquad (s = 1, 2, \ldots, q),$$

$$x_j \geq 0 \qquad (j = 1, 2, \ldots, n),$$

and assume that the points in this feasible region are bounded. Then any feasible point $x = (x_1, x_2, \ldots, x_n)$ can be expressed as a convex (weighted) combination of the points $x^1, x^2, \ldots, x^K$, as

$$x_j = \lambda_1 x_j^1 + \lambda_2 x_j^2 + \cdots + \lambda_K x_j^K \qquad (j = 1, 2, \ldots, n)$$

with

$$\lambda_1 \quad + \lambda_2 \quad + \cdots + \lambda_K \quad = 1,$$

$$\lambda_k \quad \geq 0 \qquad (k = 1, 2, \ldots, K).$$

By applying this result, we can express the overall problem in terms of the extreme points $x^1, x^2, \ldots, x^K$ of the subproblem. Since *every* feasible point to the subproblem is generated as the coefficients λ_k vary, the original problem can be re-expressed as a linear program in the variables λ_k:

Max $z = c_1(\lambda_1 x_1^1 + \lambda_2 x_1^2 + \cdots + \lambda_K x_1^K) + \cdots + c_n(\lambda_1 x_n^1 + \lambda_2 x_n^2 + \cdots + \lambda_K x_n^K),$

subject to:

$$a_{i1}(\lambda_1 x_1^1 + \lambda_2 x_1^2 + \cdots + \lambda_K x_1^K) + \cdots + a_{in}(\lambda_1 x_n^1 + \lambda_2 x_n^2 + \cdots + \lambda_K x_n^K)$$
$$= b_i \qquad (i = 1, 2, \ldots, m),$$

$$\lambda_1 \quad + \lambda_2 \quad + \cdots + \lambda_K \quad = 1,$$

$$\lambda_k \geq 0 \qquad (k = 1, 2, \ldots, K).$$

or, equivalently, by collecting coefficients for the λ_k:

Max $z = p_1\lambda_1 + p_2\lambda_2 + \cdots + p_K\lambda_K,$

subject to:

$$r_i^1\lambda_1 + r_i^2\lambda_2 + \cdots + r_i^K\lambda_K = b_i \qquad \text{(Resource constraint)}$$
$$(i = 1, 2, \ldots, m)$$

$$\lambda_1 + \lambda_2 + \cdots + \lambda_K = 1, \qquad \text{(Weighting constraint)}$$

$$\lambda_k \geq 0 \qquad (k = 1, 2, \ldots, K),$$

where

$$p_k = c_1 x_1^k + c_2 x_2^k + \cdots + c_n x_n^k,$$

and

$$r_i^k = a_{i1} x_1^k + a_{i2} x_2^k + \cdots + a_{in} x_n^k \qquad (i = 1, 2, \ldots, m)$$

indicate, respectively, the profit and resource usage for the kth extreme point $x^k = (x_1^k, x_2^k, \ldots, x_n^k)$. Observe that this notation corresponds to that used in Section 12.4. Extreme points here play the role of proposals in that discussion.

It is important to recognize that the new problem is equivalent to the original problem. The weights ensure that the solution $x_j = \lambda_1 x_j^1 + \lambda_2 x_j^2 + \cdots + \lambda_K x_j^K$ satisfies the subproblem constraints, and the resource constraints for b_i are equivalent to the original complicating constraints. The new form of the problem includes all the characteristics of the original formulation and is often referred to as the *master problem*.

Note that the reformulation has reduced the number of constraints by replacing the subproblem constraints with the single weighting constraint. At the same time, the new version of the problem usually has many more variables, since the number of extreme points in the subproblem may be enormous (hundreds of thousands). For this reason, it seldom would be tractable to generate all the subproblem extreme points in order to solve the master problem directly by linear programming.

Decomposition avoids solving the full master problem. Instead, it starts with a subset of the subproblem extreme points and generates the remaining extreme points only as needed. That is, it starts with the *restricted master problem*

$$z^J = \text{Max } z = p_1\lambda_1 + p_2\lambda_2 + \cdots + p_J\lambda_J,$$

Optimal shadow prices

subject to:

$$r_i^1\lambda_1 + r_i^2\lambda_2 + \cdots + r_i^J\lambda_J = b_i \qquad (i = 1, 2, \ldots, m), \qquad \pi_i$$
$$\lambda_1 + \lambda_2 + \cdots + \lambda_J = 1, \qquad\qquad \sigma$$
$$\lambda_k \geq 0 \qquad (k = 1, 2, \ldots, J),$$

where J is usually so much less than K that the simplex method can be employed for its solution.

Any feasible solution to the restricted master problem is feasible for the master problem by taking $\lambda_{J+1} = \lambda_{J+2} = \cdots = \lambda_K = 0$. The theory of the simplex method shows that the solution to the restricted master problem is optimal for the overall problem if every column in the master problem prices out to be nonnegative; that is, if

$$p_k - \pi_1 r_1^k - \pi_2 r_2^k - \cdots - \pi_m r_m^k - \sigma \leq 0 \qquad (k = 1, 2, \ldots, K)$$

or, equivalently, in terms of the variables x_j^k generating p_k and the r_i^k's,

$$\sum_{j=1}^{n} \left[c_j - \sum_{i=1}^{m} \pi_i a_{ij} \right] x_j^k - \sigma \leq 0 \qquad (k = 1, 2, \ldots, K). \qquad (9)$$

This condition can be checked easily without enumerating every extreme point. We must solve only the linear-programming subproblem

$$v^J = \text{Max } \sum_{j=1}^{n} \left[c_j - \sum_{i=1}^{m} \pi_i a_{ij} \right] x_j,$$

subject to:

$$e_{s1}x_1 + e_{s2}x_2 + \cdots + e_{sn}x_n = d_s \qquad (s = 1, 2, \ldots, q),$$
$$x_j \geq 0 \qquad (j = 1, 2, \ldots, n).$$

If $v^J - \sigma \leq 0$, then the optimality condition (9) is satisfied, and the problem has been solved. The optimal solution $x_1^*, x_2^*, \ldots, x_n^*$ is given by weighting the extreme points $x_j^1, x_j^2, \ldots, x_j^J$ used in the restricted master problem by the optimal weights $\lambda_1^*, \lambda_2^*, \ldots, \lambda_J^*$ to that problem, that is,

$$x_j^* = \lambda_1^* x_j^1 + \lambda_2^* x_j^2 + \cdots + \lambda_J^* x_j^J \qquad (j = 1, 2, \ldots, n).$$

If $v^J - \sigma > 0$, then the optimal extreme point solution to the subproblem $x_1^{J+1}, x_2^{J+1}, \ldots, x_n^{J+1}$ is used at the $(J + 1)$st extreme point to improve the restricted master. A new weight λ_{J+1} is added to the restricted master problem with coefficients

$$p_{J+1} = c_1 x_1^{J+1} + c_2 x_2^{J+1} + \cdots + c_n x_n^{J+1},$$
$$r_i^{J+1} = a_{i1} x_1^{J+1} + a_{i2} x_2^{J+1} + \cdots + a_{in} x_n^{J+1} \qquad (i = 1, 2, \ldots, m),$$

and the process is then repeated.

Convergence Property. The representation property has shown that decomposition is solving the master problem by generating coefficient data as needed. Since the master problem is a linear program, the decomposition algorithm inherits finite convergence from the simplex method. Recall that the simplex method solves linear programs in a finite number of steps. For decomposition, the subproblem calculation ensures that the variable introduced into the basis has a positive reduced cost, just as in applying the simplex method to the master problem. Consequently, from the linear-programming theory, the master problem is solved in a finite number of steps; the procedure thus determines an optimal solution by solving the restricted master problem and subproblem alternately a *finite* number of times.

Bounds on the Objective Value

We previously observed that the value z^J to the restricted master problem tends to tail off and approach z^*, the optimal value to the overall problem, very slowly. As a result, we may wish to terminate the algorithm before an optimal solution has been obtained, rather than paying the added computational expense to improve the current solution only slightly. An important feature of the decomposition approach is that it permits us to assess the effect of terminating with a suboptimal solution by indicating how far z^J is removed from z^*.

For notation let $\pi_1^J, \pi_2^J, \ldots, \pi_m^J$, and σ^J denote the optimal shadow prices for the m resource constraints and the weighting constraint in the current restricted master problem. The current subproblem is:

$$v^J = \text{Max} \sum_{j=1}^{n} \left(c_j - \sum_{i=1}^{m} \pi_i^J a_{ij} \right) x_j,$$

Optimal shadow prices

subject to:

$$\sum_{j=1}^{n} e_{sj} x_j = d_s \qquad (s = 1, 2, \ldots, q), \qquad \alpha_s$$

$$x_j \geq 0 \qquad (j = 1, 2, \ldots, n),$$

with optimal shadow prices $\alpha_1, \alpha_2, \ldots, \alpha_q$. By linear programming duality theory these shadow prices solve the dual to the subproblem, so that

$$c_j - \sum_{i=1}^{m} \pi_i^J a_{ij} - \sum_{s=1}^{q} \alpha_s e_{sj} \leq 0 \qquad (j = 1, 2, \ldots, n)$$

and

$$\sum_{s=1}^{q} \alpha_s d_s = v^J. \tag{10}$$

But these inequalities are precisely the dual feasibility conditions of the original problem, and so the solution to every subproblem provides a dual feasible solution to that problem. The weak duality property of linear programming, though, shows that every feasible solution to the dual gives an upper bound to the primal objective value z^*. Thus

$$\sum_{i=1}^{m} \pi_i^J b_i + \sum_{s=1}^{q} \alpha_s d_s \geq z^*. \tag{11}$$

Since the solution to every restricted master problem determines a feasible solution to the original problem (via the master problem), we also know that

$$z^* \geq z^J.$$

As the algorithm proceeds, the lower bounds z^J increase and approach z^*. There is, however, no guarantee that the dual feasible solutions are improving. Consequently, the upper bound generated at any step may be worse than those generated at previous steps, and we always record the *best* upper bound generated thus far.

The upper bound can be expressed in an alternative form. Since the variables π_i^J and σ^J are optimal shadow prices to the restricted master problem, they solve its dual problem, so that:

$$z^J = \sum_{i=1}^{m} \pi_i^J b_i + \sigma^J = \text{Dual objective value}.$$

Substituting this value together with the equality (10), in expression (11), gives the alternative form for the bounds:

$$z^J \leq z^* \leq z^J - \sigma^J + v^J.$$

This form is convenient since it specifies the bounds in terms of the objective values of the subproblem and restricted master problem. The only dual variable used corresponds to the weighting constraint in the restricted master.

To illustrate these bounds, reconsider the preview problem introduced in Section 12.2. The first restricted master problem used two subproblem extreme points and was given by:

$$z^2 = \text{Max } z = 22\lambda_1 + 17\lambda_2,$$

subject to:

	Optimal shadow prices
$18\lambda_1 + 13\lambda_2 \leq 17,$	$\pi_1^2 = 1$
$\lambda_1 + \lambda_2 = 1,$	$\sigma^2 = 4$
$\lambda_1 \geq 0, \qquad \lambda_2 \geq 0.$	

Here $z^2 = 21$ and the subproblem

$$v^2 = \text{Max} \,(x_1 - x_2 + 2x_3),$$

subject to:

$$1 \leqq x_j \leqq 2 \qquad (j = 1, 2, 3),$$

has the solution $v^2 = 5$. Thus,

$$21 \leqq z^* \leqq 21 - 4 + 5 = 22.$$

At this point, computations could have been terminated, with the assurance that the solution of the current restricted master problem is within 5 percent of the optimal objective value, which in this case is $z^* = 21\frac{1}{2}$.

Unbounded Solution to the Subproblem

For expositional purposes, we have assumed that every subproblem encountered has an optimal solution, even though its objective value might be unbounded. First, we should note that an unbounded objective value to a subproblem does not necessarily imply that the overall problem is unbounded, since the constraints that the subproblem ignores may prohibit the activity levels leading to an unbounded subproblem solution from being feasible to the full problem. Therefore, we cannot simply terminate computations when the subproblem is unbounded; a more extensive procedure is required; reconsidering the representation property underlying the theory will suggest the appropriate procedure.

When the subproblem is unbounded, the representation property becomes more delicate. For example, the feasible region in Fig. 12.6 contains three extreme points. Taking the weighted combination of these points generates only the shaded portion of the feasible region. Observe, though, that by moving from the shaded region in a *direction parallel to either of the unbounded edges*, every feasible point can be generated.

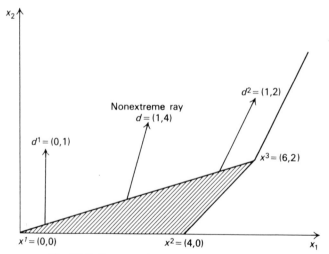

Fig. 12.6 Representing an unbounded region.

This suggests that the general representation property should include *directions* as well as extreme points. Actually, we do not require all possible movement directions, but only those that are analogous to extreme points.

Before exploring this idea, let us introduce a definition.

Definition.

i) A direction $d = (d_1, d_2, \ldots, d_n)$ is called a *ray* for the subproblem if, whenever $x_1, x_2, \ldots, x_n$ is a feasible solution, then the point

$$x_1 + \theta d_1, x_2 + \theta d_2, \ldots, x_n + \theta d_n$$

also is feasible for any choice of $\theta \geq 0$.

ii) A ray $d = (d_1, d_2, \ldots, d_n)$ is called an *extreme ray* if it cannot be expressed as a weighted combination of two other rays; that is, if there are no two rays $d' = (d'_1, d'_2, \ldots, d'_n)$ and $d'' = (d''_1, d''_2, \ldots, d''_n)$ and weight $0 < \lambda < 1$ such that

$$d_j = \lambda d'_j + (1 - \lambda)d''_j \qquad (j = 1, 2, \ldots, n).$$

A ray is a direction that points from any feasible point only toward other feasible points. An extreme ray is an unbounded edge of the feasible region. In Fig. 12.6, d^1 and d^2 are the only two extreme rays. Any other ray such as d can be expressed as a weighted combination of these two rays; for example,

$$d = (1, 4) = 2d^1 + d^2 = 2(0, 1) + (1, 2).$$

The extended representation result states that there are only a finite number of extreme rays $d^1, d^2, \ldots, d^L$ to the subproblem and that any feasible solution $x = (x_1, x_2, \ldots, x_n)$ to the subproblem can be expressed as a weighted combination of extreme points plus a nonnegative combination of extreme rays, as:

$$x_j = \lambda_1 x_j^1 + \lambda_2 x_j^2 + \cdots + \lambda_k x_j^k + \theta_1 d_j^1 + \theta_2 d_j^2 + \cdots + \theta_L d_j^L \qquad (j = 1, 2, \ldots, n),$$
$$\lambda_1 \quad + \lambda_2 \quad + \cdots + \lambda_k \qquad\qquad\qquad\qquad\qquad\quad = 1,$$
$$\lambda_k \geq 0, \qquad \theta_\ell \geq 0 \qquad (k = 1, 2, \ldots, K; \ell = 1, 2, \ldots, L).$$

Observe that the θ_j need not sum to one.

Let $\hat{p}_k$ and $\hat{r}_i^k$ denote, respectively, the per-unit profit and the ith resource usage of the kth extreme ray; that is,

$$\hat{p}_k = c_1 d_1^k + c_2 d_2^k + \cdots + c_n d_n^k,$$
$$\hat{r}_i^k = a_{i1} d_1^k + a_{i2} d_2^k + \cdots + a_{in} d_n^k \qquad (i = 1, 2, \ldots, m).$$

Substituting as before for x_j in the complicating constraints in terms of extreme points and extreme rays gives the master problem:

$$\text{Max } z = p_1 \lambda_1 + p_2 \lambda_2 + \cdots + p_K \lambda_K + \hat{p}_1 \theta_1 + \hat{p}_2 \theta_2 + \cdots + \hat{p}_L \theta_L,$$

subject to:

$$r_i^1 \lambda_1 + r_i^2 \lambda_2 + \cdots + r_i^K \lambda_K + \hat{r}_i^1 \theta_1 + \hat{r}_i^2 \theta_2 + \cdots + \hat{r}_i^L \theta_L = b_i \quad (i = 1, 2, \ldots, m),$$
$$\lambda_1 + \lambda_2 + \cdots + \lambda_K \qquad\qquad\qquad\qquad\qquad\qquad\quad = 1,$$
$$\lambda_k \geq 0 \qquad \theta_\ell \geq 0 \qquad (k = 1, 2, \ldots, K; \ell = 1, 2, \ldots, L).$$

The solution strategy parallels that given previously. At each step, we solve a restricted master problem containing only a subset of the extreme points and extreme rays, and use the optimal shadow prices to define a subproblem. If the subproblem has an optimal solution, a new extreme point is added to the restricted master problem and it is solved again. When the subproblem is unbounded, though, an extreme ray is added to the restricted master problem. To be precise, we must specify how an extreme ray is identified. It turns out that an extreme ray is determined easily as a byproduct of the simplex method, as illustrated by the following example.

Example

Maximize $z = 5x_1 - x_2$,

subject to:

$$x_1 \qquad\qquad\qquad \leqq\ 8, \qquad \text{(Complicating constraint)}$$

$$
\left.
\begin{aligned}
x_1 - x_2 + x_3 \qquad\ &=\ 4 \\
2x_1 - x_2 \qquad + x_4 &= 10 \\
x_j \geqq 0 \qquad (j = 1, 2, &\ 3, 4).
\end{aligned}
\right\} \qquad \text{(Subproblem)}
$$

The subproblem has been identified as above solely for purposes of illustration. The feasible region to the subproblem was given in terms of x_1 and x_2 in Fig. 12.6 by viewing x_3 and x_4 as slack variables.

As an initial restricted master problem, let us use the extreme points $(x_1, x_2, x_3, x_4) = (4, 0, 0, 2), (x_1, x_2, x_3, x_4) = (6, 2, 0, 0)$, and no extreme rays. These extreme points, respectively, use 4 and 6 units of the complicating resource and contribute 20 and 28 units to the objective function. The restricted master problem is given by:

$$z^2 = \text{Max } 20\lambda_1 + 28\lambda_2,$$

Optimal shadow prices

subject to:

$$4\lambda_1 + 6\lambda_2 \leqq 8, \qquad\qquad 0$$

$$\lambda_1 + \lambda_2 = 1, \qquad\qquad 28$$

$$\lambda_1 \geqq 0, \qquad \lambda_2 \geqq 0.$$

The solution is $\lambda_1 = 0$, $\lambda_2 = 1$, $z^2 = 28$, with a price of 0 on the complicating constraint.

The subproblem is

$$v^2 = \text{Max } 5x_1 - x_2,$$

subject to:

$$x_1 - x_2 + x_3 \qquad\ = 4,$$

$$2x_1 - x_2 \qquad + x_4 = 10,$$

$$x_j \geqq 0 \qquad (j = 1, 2, 3, 4).$$

Solving by the simplex method leads to the canonical form:

Maximize $z = \qquad 3x_3 - 4x_4 + 28,$

subject to:

$$x_1 \qquad - \quad x_3 + \quad x_4 = 6,$$
$$x_2 - 2x_3 + \quad x_4 = 2,$$
$$x_j \geq 0 \qquad (j = 1, 2, 3, 4).$$

Since the objective coefficient for x_3 is positive and x_3 does not appear in any constraint with a positive coefficient, the solution is unbounded. In fact, as we observed when developing the simplex method, by taking $x_3 = 0$, the solution approaches $+\infty$ by increasing θ and setting

$$z \qquad\qquad = 28 + 3\theta,$$
$$x_1 \qquad = 6 + \quad \theta,$$
$$x_2 = 2 + 2\theta.$$

This serves to alter x_1, x_2, x_3, x_4, from $x_1 = 6, x_2 = 2, x_3 = 0, x_4 = 0,$ to $x_1 = 6 + \theta, x_2 = 2 + 2\theta, x_3 = 0, x_4 = 0,$ so that we move in the direction $d = (1, 2, 1, 0)$ by a multiple of θ. This direction has a per-unit profit of 3 and uses 1 unit of the complicating resource. It is the extreme ray added to the restricted master problem, which becomes:

$$z^3 = \text{Max } 20\lambda_1 + 28\lambda_2 + 30_1,$$

subject to:

	Optimal shadow prices
$4\lambda_1 + 6\lambda_2 + 0_1 \leq 8,$	3
$\lambda_1 + \lambda_2 \qquad = 1,$	10

$$\lambda_1 \geq 0, \qquad \lambda_2 \geq 0, \qquad 0_1 \geq 0,$$

and has optimal solution $\lambda_1 = 0, \lambda_2 = 1, 0_1 = 2,$ and $z^3 = 34.$

Since the price of the complicating resource is 3, the new subproblem objective function becomes:

$$v^3 = \text{Max } 5x_1 - x_2 - 3x_1 = \text{Max } 2x_1 - x_2.$$

Graphically we see from Fig. 12.6, that an optimal solution is $x_1 = 6, x_2 = 2, x_3 = x_4 = 0, v^3 = 10.$ Since $v^3 \leq \sigma^3 = 10,$ the last solution solves the full master problem and the procedure terminates. The optimal solution uses the extreme point $(x_1, x_2, x_3, x_4) = (6, 2, 0, 0),$ plus two times the extreme ray $d = (1, 2, 1, 0);$ that is,

$$x_1 = 6 + 2(1) = 8, \qquad x_2 = 2 + 2(2) = 6,$$
$$x_3 = 0 + 2(1) = 2, \qquad x_4 = 0 + 2(0) = 0.$$

In general, whenever the subproblem is unbounded, the simplex method determines a canonical form with $\bar{c}_j > 0$ and $\bar{a}_{ij} \leq 0$ for each coefficient of some nonbasic variable x_j. As above, the extreme ray $d = (d_1, d_2, \ldots, d_n)$ to be submitted to the restricted master problem has a profit coefficient $\bar{c}_j$ and coefficients d_k given by

$$
d_k = \begin{cases}
1 & \text{if } k = s \text{ (increasing nonbasic } x_s); \\
-\bar{a}_{ij} & \text{if } x_k \text{ is the } i\text{th basic variable (changing the basis to compensate} \\
& \text{for } x_s); \\
0 & \text{if } x_k \text{ is nonbasic and } k \neq s \text{ (hold other nonbasics at 0).}
\end{cases}
$$

The coefficients of this extreme ray simply specify how the values of the basic variables change per unit change in the nonbasic variable x_s being increased.

12.8 COLUMN GENERATION

Large-scale systems frequently result in linear programs with enormous numbers of variables, that is, linear programs such as:

$$
z^* = \text{Max } z = c_1 x_1 + c_2 x_2 + \cdots + c_n x_n,
$$

subject to:

$$
a_{i1} x_1 + a_{i2} x_2 + \cdots + a_{in} x_n = b_i \qquad (i = 1, 2, \ldots, m), \qquad (12)
$$

$$
x_j \geq 0 \qquad (j = 1, 2, \ldots, n),
$$

where n is very large. These problems arise directly from applications such as the multi-item production scheduling example from Section 12.1, or the cutting-stock problem to be introduced below. They may arise in other situations as well. For example, the master problem in decomposition has this form; in this case, problem variables are the weights associated with extreme points and extreme rays.

Because of the large number of variables, direct solution by the simplex method may be inappropriate. Simply generating all the coefficient data a_{ij} usually will prohibit this approach. Column generation extends the technique introduced in the decomposition algorithm, of using the simplex method, but generating the coefficient data only as needed. The method is applicable when the data has inherent structural properties that allow numerical values to be specified easily. In decomposition, for example, we exploited the fact that the data for any variable corresponds to an extreme point or extreme ray of another linear program. Consequently, new data could be generated by solving this linear program with an appropriate objective function.

The column-generation procedure very closely parallels the mechanics of the decomposition algorithm. The added wrinkle concerns the subproblem, which now need not be a linear program, but can be any type of optimization problem, including nonlinear, dynamic, or integer programming problems. As in decomposition, we assume *a priori* that certain variables, say $x_{J+1}, x_{J+2}, \ldots, x_n$ are nonbasic and restrict their values to zero. The resulting problem is:

$$z^J = \text{Max } c_1 x_1 + c_2 x_2 + \cdots + c_J x_J,$$

Optimal shadow prices

subject to:

$$a_{i1}x_1 + a_{i2}x_2 + \cdots + a_{iJ}x_J = b_i \quad (i = 1, 2, \ldots, m), \quad \pi_i^J \qquad (13)$$

$$x_j \geq 0 \quad (j = 1, 2, \ldots, J);$$

this is now small enough so that the simplex method can be employed for its solution. The original problem (12) includes all of the problem characteristics and again is called a *master problem*, whereas problem (13) is called the *restricted master problem*.

Suppose that the restricted master problem has been solved by the simplex method and that $\pi_1^J, \pi_2^J, \ldots, \pi_m^J$ are the optimal shadow prices. The optimal solution together with $x_{J+1} = x_{J+2} = \cdots = x_n = 0$ is feasible and so potentially optimal for the master problem (12). It is optimal if the simplex optimality condition holds, that is, if $\bar{c}_j = c_j - \sum_{i=1}^{m} \pi_i^J a_{ij} \leq 0$ for every variable x_j. Stated in another way, the solution to the restricted master problem is optimal if $v^J \leq 0$ where:

$$v^J = \underset{1 \leq j \leq n}{\text{Max}} \left[c_j - \sum_{i=1}^{m} \pi_i^J a_{ij} \right]. \qquad (14)$$

If this condition is satisfied, the original problem has been solved without specifying all of the a_{ij} data or solving the full master problem.

If $v^J = c_s - \sum_{i=1}^{m} \pi_i^J a_{is} > 0$, then the simplex method, when applied to the master problem, would introduce variable x_s into the basis. Column generation accounts for this possibility by adding variable x_s as a new variable to the restricted master problem. The new restricted master can be solved by the simplex method and the entire procedure can be repeated.

This procedure avoids solving the full master problem; instead it alternately solves a restricted master problem and makes the computations (14) to generate data $a_{1s}, a_{2s}, \ldots, a_{ms}$ for a new variable x_s. Observe that (14) is itself an optimization problem, with variables $j = 1, 2, \ldots, n$. It is usually referred to as a *subproblem*.

The method is specified in flow-chart form in Fig. 12.7. Its efficiency is predicated upon:

i) Obtaining an optimal solution before many columns have been added to the restricted master problem. Otherwise the problems inherent in the original formulation are encountered.

ii) Being able to solve the subproblem effectively.

Details concerning the subproblem depend upon the structural characteristics of the problem being studied. By considering a specific example, we can illustrate how the subproblem can be an optimization problem other than a linear program.

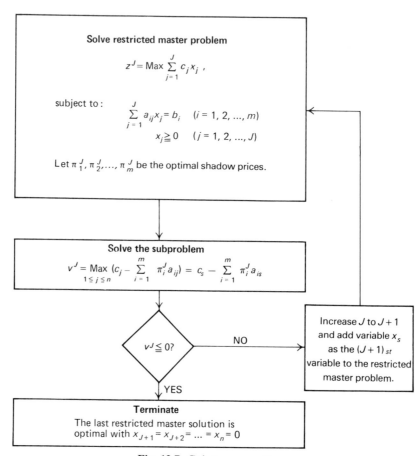

Fig. 12.7 Column generation.

Example. (Cutting-stock problem) A paper (textile) company must produce various sizes of its paper products to meet demand. For most grades of paper, the production technology makes it much easier to first produce the paper on large rolls, which are then cut into smaller rolls of the required sizes. Invariably, the cutting process involves some waste. The company would like to minimize waste or, equivalently, to meet demand using the fewest number of rolls.

For notational purposes, assume that we are interested in one grade of paper and that this paper is produced only in rolls of length ℓ for cutting. Assume, further, that the demand requires d_i rolls of size ℓ_i $(i = 1, 2, \ldots, m)$ to be cut. In order for a feasible solution to be possible, we of course need $\ell_i \leq \ell$.

One approach to the problem is to use the possible cutting patterns for the rolls as decision variables. Consider, for example, $\ell = 200$ inches and rolls required for 40 different lengths ℓ_i ranging from 20 to 80 inches. One possible cutting pattern produces lengths of

$$35'', \qquad 40'', \qquad 40'', \qquad 70'',$$

with a waste of 15 inches. Another is

$$20'', \quad 25'', \quad 30'', \quad 50'', \quad 70'',$$

with a waste of 5 inches. In general, let

$n =$ Number of possible cutting patterns,

$x_j =$ Number of times cutting pattern j is used,

$a_{ij} =$ Number of rolls of size ℓ_i used on the jth cutting pattern.

Then $a_{ij}x_j$ is the number of rolls of size ℓ_i cut using pattern j, and the problem of minimizing total rolls used to fulfill demand becomes:

Minimize $x_1 + \quad x_2 + \cdots + \quad x_n,$

subject to:

$$a_{i1}x_1 + a_{i2}x_2 + \cdots + a_{in}x_n \geq d_i \qquad\qquad (i = 1, 2, \ldots, m),$$

$$x_j \geq 0 \quad \text{and integer} \qquad (j = 1, 2, \ldots, n).$$

For the above illustration, the number of possible cutting patterns n exceeds 10 million, and this problem is a large-scale integer-programming problem. Fortunately, the demands d_i are usually high, so that rounding optimal linear-programming solutions to integers leads to good solutions.

If we drop the integer restrictions, the problem becomes a linear program suited for the column-generation algorithm. The subproblem becomes:

$$v^J = \operatorname*{Min}_{1 \leq j \leq n} \left[1 - \sum_{i=1}^{m} \pi_i^J a_{ij} \right], \tag{15}$$

since each objective coefficient is equal to one. Note that the subproblem is a minimization, since the restricted master problem is a minimization problem seeking variables x_j with $\bar{c}_j < 0$ (as opposed to seeking $\bar{c}_j > 0$ for maximization).

The subproblem considers all potential cutting plans. Since a cutting plan j is feasible whenever

$$\sum_{i=1}^{m} \ell_i a_{ij} \leq \ell, \tag{16}$$

$$a_{ij} \geq 0 \quad \text{and integer},$$

the subproblem must determine the coefficients a_{ij} of a new plan to minimize (15). For example, if the roll length is given by $\ell = 100''$ and the various lengths ℓ_i to be cut are 25, 30, 35, 40, 45, and 50'', then the subproblem constraints become:

$$25a_{1j} + 30a_{2j} + 35a_{3j} + 40a_{4j} + 45a_{5j} + 50a_{6j} \leq 100,$$

$$a_{ij} \geq 0 \quad \text{and integer} \qquad (i = 1, 2, \ldots, 6).$$

The optimal values for a_{ij} indicate how many of each length ℓ_i should be included in the new cutting pattern j. Because subproblem (15) and (16) is a one-constraint integer-programming problem (called a *knapsack* problem), efficient special-purpose dynamic-programming algorithms can be used for its solution.

As this example illustrates, column generation is a flexible approach for solving linear programs with many columns. To be effective, the algorithm requires that the subproblem can be solved efficiently, as in decomposition or the cutting-stock problem, to generate a new column or to show that the current restricted master problem is optimal. In the next chapter, we discuss another important application by using column generation to solve *nonlinear* programs.

EXERCISES

1. Consider the following linear program:

$$\text{Maximize } 9x_1 + \quad x_2 - 15x_3 - 5x_4,$$

subject to:

$$-3x_1 + 2x_2 + 9x_3 + x_4 \leqq 7,$$
$$6x_1 + 16x_2 - 12x_3 - 2x_4 \leqq 10,$$
$$0 \leqq x_j \leqq 1 \quad (j = 1, 2, 3, 4).$$

Assuming no bounded-variable algorithm is available, solve by the decomposition procedure, using $0 \leqq x_j \leqq 1$ $(j = 1, 2, 3, 4)$ as the subproblem constraints.

Initiate the algorithm with two proposals: the optimum solution to the subproblem and the proposal $x_1 = 1, x_2 = x_3 = x_4 = 0$.

2. Consider the following linear-programming problem with special structure:

$$\text{Maximize } z = 15x_1 + 7x_2 + 15x_3 + 20y_1 + 12y_2,$$

subject to:

$$\begin{array}{l}
\left.\begin{array}{rrrrrl}
x_1 + & x_2 + & x_3 + & y_1 + & y_2 \leqq & 5 \\
3x_1 + & 2x_2 + & 4x_3 + & 5y_1 + & 2y_2 \leqq & 16
\end{array}\right\} \text{Master problem} \\[2ex]
\left.\begin{array}{rrrr}
4x_1 + & 4x_2 + & 5x_3 & \leqq 20 \\
2x_1 + & x_2 & & \leqq 4 \\
x_1 \geqq 0, x_2 \geqq 0, x_3 \geqq 0
\end{array}\right\} \text{Subproblem I} \\[3ex]
\left.\begin{array}{r}
y_1 + \tfrac{1}{2}y_2 \leqq 3 \\
\tfrac{1}{2}y_1 + \tfrac{1}{2}y_2 \leqq 2 \\
y_1 \geqq 0, y_2 \geqq 0.
\end{array}\right\} \text{Subproblem II}
\end{array}$$

Tableau 1 represents the solution of this problem by the decomposition algorithm in the midst of the calculations. The variables s_1 and s_2 are slack variables for the first two constraints of the master problem; the variables a_1 and a_2 are artificial variables for the weighting constraints of the master problem.

Tableau 1 Data at an iteration of the decomposition method

Basic variables	Current values	Subproblem I		Subproblem II		Slacks		Artificials	
		λ_1	λ_2	μ_1	μ_2	s_1	s_2	a_1	a_2
λ_1	$\frac{1}{2}$	1				$\frac{1}{3}$	$-\frac{1}{6}$	$\frac{4}{3}$	
λ_2	$\frac{2}{3}$		1			$-\frac{1}{3}$	$\frac{1}{6}$	$-\frac{1}{3}$	
μ_1	$\frac{5}{12}$			1		$\frac{5}{12}$	$-\frac{1}{12}$	$-\frac{1}{3}$	
μ_2	$\frac{7}{12}$				1	$-\frac{5}{12}$	$\frac{1}{12}$	$\frac{1}{3}$	1
$(-z)$	$-\frac{80}{3}$					-10	-1	-4	

The extreme points generated thus far are:

x_1	x_2	x_3	Weights
2	0	0	λ_1
0	0	4	λ_2

for subproblem I and

y_1	y_2	Weights
0	4	μ_1
0	0	μ_2

for subproblem II.

a) What are the shadow prices associated with each constraint of the restricted master?
b) Formulate the two subproblems that need to be solved at this stage using the shadow prices determined in part (a).
c) Solve each of the subproblems graphically.
d) Add any newly generated extreme points of the subproblems to the restricted master.
e) Solve the new restricted master by the simplex method continuing from the previous solution. (See Exercise 29 in Chapter 4.)
f) How do we know whether the current solution is optimal?

3. Consider a transportation problem for profit maximization:

$$\text{Maximize } z = c_{11}x_{11} + c_{12}x_{12} + c_{13}x_{13} + c_{21}x_{21} + c_{22}x_{22} + c_{23}x_{23},$$

subject to:

$$
\begin{aligned}
x_{11} + x_{12} + x_{13} &= a_1, \\
x_{21} + x_{22} + x_{23} &= a_2, \\
x_{11} + x_{21} &= b_1, \\
x_{12} + x_{22} &= b_2, \\
x_{13} + x_{23} &= b_3, \\
x_{ij} \geq 0 \quad (i = 1, 2; \ j = 1, 2, 3).
\end{aligned}
$$

a) Suppose that we solve this problem by decomposition, letting the requirement b_i constraints and nonnegativity $x_{ij} \geq 0$ constraints compose the subproblem. Is it easy to solve the subproblem at each iteration? Does the restricted master problem inherit the network structure of the problem, or is the network structure "lost" at the master-problem level?

b) Use the decomposition procedure to solve for the data specified in the following table:

	Distribution profits (c_{ij})			Availabilities a_i
	100	150	200	20
	50	50	75	40
Requirements b_i	10	30	20	

Initiate the algorithm with two proposals:

	Activity levels						
	x_{11}	x_{12}	x_{13}	x_{21}	x_{22}	x_{23}	Profit
Proposal 1	0	0	0	10	30	20	9500
Proposal 2	10	30	20	0	0	0	3500

To simplify calculations, you may wish to use the fact that a transportation problem contains a redundant equation and remove the second supply equation from the problem.

4. A small city in the northeast must racially balance its 10 elementary schools or sacrifice all federal aid being issued to the school system. Since the recent census indicated that approximately 28% of the city's population is composed of minorities, it has been determined that each school in the city must have a minority student population of 25% to 30% to satisfy the federal definition of "racial balance." The decision has been made to bus children in order to meet this goal. The parents of the children in the 10 schools are very concerned about the additional travel time for the children who will be transferred to new schools. The School Committee has promised these parents that the busing plan will minimize the total time that the children of the city have to travel. Each school district is divided into 2 zones, one which is close to the local school and one which is far from the school, as shown in Fig. E12.1.

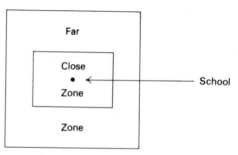

Figure E12.1

The School Committee has also promised the parents of children who live in a "close zone" that they will attempt to discourage the busing of this group of children (minority and nonminority) away from their present neighborhood school. The School Committee members are intent on keeping their promises to this group of parents.

An additional problem plaguing the Committee is that any school whose enrollment drops below 200 students must be closed; this situation would be unacceptable to the Mayor and to the taxpayers who would still be supporting a "closed school" serving no one.

The available data include the following:

For each district $i = 1, 2, \ldots, 10$, we have

N_i^{NONc} = Number of nonminority children in the close zone of school district i.
N_i^{MINc} = Number of minority children in the close zone of school district i.
N_i^{NONf} = Number of nonminority children in the far zone of school district i.
N_i^{MINf} = Number of minority children in the far zone of school district i.

For each pair (i, j) of school districts, we have the travel time t_{ij}.

For each school i, the capacity D_i is known (all $D_i > 200$ and there is enough school capacity to accommodate all children).

a) Formulate the problem as a linear program. [*Hint.* To discourage the busing of students who live close to their neighborhood school, you may add a penalty, p, to the travel time of any student who lives in the close zone of school district i and is assigned to school district j $(i \neq j)$. Assume that a student who lives in the close zone of school i and is assigned to school i does not have to be bused.]

b) There is only a small-capacity minicomputer in the city, which cannot solve the linear program in its entirety. Hence, the decomposition procedure could be applied to solve the problem. If you were a mathematical programming specialist hired by the School Committee, how would you decompose the program formulated in part (a)? Identify the subproblem, the weighting program, and the proposal-generating program. Do not attempt to solve the problem.

5. A food company blames seasonality in production for difficulties that it has encountered in scheduling its activities efficiently. The company has to cope with three major difficulties:

   I) Its food products are perishable. On the average, one unit spoils for every seven units kept in inventory from one month to another.

   II) It is costly to change the level of the work force to coincide with requirements imposed by seasonal demands. It costs \$750 to hire and train a new worker, and \$500 to fire a worker.

   III) On the average, one out of eight workers left idle in any month decides to leave the firm.

   Because of the ever-increasing price of raw materials, the company feels that it should design a better scheduling plan to reduce production costs, rather than lose customers by increasing prices of its products.

The task of the team hired to study this problem is made easier by the following operating characteristics of the firm:

i) Practically, the firm has no problems procuring any raw materials that it requires;
ii) Storage capacity is practically unlimited at the current demand level; and
iii) The products are rather homogeneous, so that all output can be expressed in standard units (by using certain equivalence coefficients).

The pertinent information for decision-making purposes is:

iv) The planning horizon has $T = 12$ months (one period = one month);
v) Demand D_i is known for each period ($i = 1, 2, \ldots, 12$);
vi) Average productivity is 1100 units per worker per month;
vii) The level of the work force at the start of period 1 is L_1; S_0 units of the product are available in stock at the start of period 1;
viii) An employed worker is paid W_t as wages per month in period t;
ix) An idle worker is paid a minimum wage of M_t in month t, to be motivated not to leave;
x) It costs I dollars to keep one unit of the product in inventory for one month.

With the above information, the company has decided to construct a pilot linear program to determine work-force level, hirings, firings, inventory levels, and idle workers.

a) Formulate the linear program based on the data above. Show that the model has a staircase structure.
b) Restate the constraints in terms of cumulative demand and work force; show that the model now has block triangular structure.

6. A firm wishing to operate with as decentralized an organizational structure as possible has two separate operating divisions. The divisions can operate independently except that they compete for the firm's two scarce resources—working capital and a particular raw material. Corporate headquarters would like to set prices for the scarce resources that would be paid by the divisions, in order to ration the scarce resources. The goal of the program is to let each division operate independently with as little interference from corporate headquarters as possible.

Division #1 produces 3 products and faces capacity constraints as follows:

$$4x_1 + 4x_2 + 5x_3 \leq 20,$$
$$4x_1 + 2x_2 \qquad \leq 8,$$

$$x_1 \geq 0, \quad x_2 \geq 0, \quad x_3 \geq 0.$$

The contribution to the firm per unit from this division's products are 2.50, 1.75, and 0.75, respectively. Division #2 produces 2 different products and faces its own capacity constraints as follows:

$$2y_1 + y_2 \leq 6,$$
$$y_1 + y_2 \leq 4,$$

$$y_1 \geq 0, \quad y_2 \geq 0.$$

The contribution to the firm per unit from this division's products are 3 and 2, respectively. The joint constraints that require coordination among the divisions involve working capital and one raw material. The constraint on working capital is

$$x_1 + x_2 + x_3 + y_1 + y_2 \le 7,$$

and the constraint on the raw material is

$$3x_1 + 2x_2 + 4x_3 + 5y_1 + 2y_2 \le 16.$$

Corporate headquarters has decided to use decomposition to set the prices for the scarce resources. The optimal solution using the decomposition algorithm indicated that division #1 should produce $x_1 = 1$, $x_2 = 2$, and $x_3 = 0$, while division #2 should produce $y_1 = \frac{1}{2}$ and $y_2 = 3\frac{1}{2}$. The shadow prices turned out to be $\frac{2}{3}$ and $\frac{1}{3}$ for working capital and raw material, respectively. Corporate headquarters congratulated itself for a fine piece of analysis. They then announced these prices to the divisions and told the divisions to optimize their own operations independently. Division #1 solved its subproblem and reported an operating schedule of $x_1 = 0$, $x_2 = 4$, $x_3 = 0$. Similarly, division #2 solved its subproblem and reported an operating schedule of $y_1 = 2$, $y_2 = 2$.

Corporate headquarters was aghast—together the divisions requested more of both working capital and the raw material than the firm had available!

a) Did the divisions cheat on the instructions given them by corporate headquarters?

b) Were the shadow prices calculated correctly?

c) Explain the paradox.

d) What can the corporate headquarters do with the output of the decomposition algorithm to produce overall optimal operations?

7. For the firm described in Exercise 6, analyze the decomposition approach in detail.

a) Graph the constraints of each subproblem, division #1 in three dimensions and division #2 in two dimensions.

b) List *all* the extreme points for each set of constraints.

c) Write out the full master problem, including all the extreme points.

d) The optimal shadow prices are $\frac{2}{3}$ and $\frac{1}{3}$ on the working capital and raw material, respectively. The shadow prices on the weighting constraints are $\frac{5}{8}$ and $\frac{8}{3}$ for divisions #1 and #2, respectively. Calculate the reduced costs of all variables.

e) Identify the basic variables and determine the weights on each extreme point that form the optimal solution.

f) Solve the subproblems using the above shadow prices. How do you know that the solution is optimal after solving the subproblems?

g) Show graphically that the optimal solution to the overall problem is not an extreme solution to either subproblem.

8. To plan for long-range energy needs, a federal agency is modeling electrical-power investments as a linear program. The agency has divided its time horizon of 30 years into six periods $t = 1, 2, \ldots, 6$, of five years each. By the end of each of these intervals, the government can construct a number of plants (hydro, fossil, gas turbine, nuclear, and so forth). Let x_{ij} denote the capacity of plant j when initiated at the end of interval i, with per-unit construction cost of c_{ij}. Quantities x_{0j} denote capacities of plants currently in use.

Given the decisions x_{ij} on plant capacity, the agency must decide how to operate the plants to meet energy needs. Since these decisions require more detailed information to account for seasonal variations in energy demand, the agency has further divided each of the time intervals $t = 1, 2, \ldots, 6$ into 20 subintervals $s = 1, 2, \ldots, 20$. The agency has estimated the electrical demand in each (interval t, subinterval s) combination as d_{ts}. Let o_{ijts} denote the operating level during the time period ts of plant j that has been constructed in interval i. The plants must be used to meet demand requirements and incur per-unit operating costs of v_{ijts}. Because of operating limitations and aging, the plants cannot always operate at full construction capacity. Let a_{ijt} denote the availability during time period t of plant j that was constructed in time interval i. Typically, the coefficient a_{ijt} will be about 0.9. Note that $a_{ijt} = 0$ for $t \leq i$, since the plant is not available until after the end of its construction interval i.

To model uncertainties in its demand forecasts, the agency will further constrain its construction decisions by introducing a margin m of reserve capacity; in each period the total operating capacity from all plants must be at least as large as $d_{ts}(1 + m)$.

Finally, the total output of hydroelectric power in any time interval t cannot exceed the capacity H_{it} imposed by availability of water sources. (In a more elaborate model, we might incorporate H_{it} as a decision variable.)

The linear-programming model developed by the agency is:

$$\text{Minimize} \sum_{j=1}^{20} \sum_{i=1}^{6} c_{ij} x_{ij} + \sum_{j=1}^{20} \sum_{t=1}^{6} \sum_{i=0}^{6} \sum_{s=1}^{20} v_{ijts} o_{ijts} \theta_s,$$

subject to:

$$\sum_{j=1}^{20} \sum_{i=0}^{6} o_{ijts} \geq d_{ts} \qquad\qquad (t = 1, 2, \ldots, 6; \quad s = 1, 2, \ldots, 20),$$

$$o_{ijts} \leq a_{ijt} x_{ij} \qquad\qquad \begin{aligned} &(i = 0, 1, \ldots, 6; \quad t = 1, 2, \ldots, 6; \\ &\ j = 1, 2, \ldots, 20; \quad s = 1, 2, \ldots, 20), \end{aligned}$$

$$\sum_{s=1}^{20} o_{ihts} \theta_s \leq H_{it} \qquad\qquad (t = 1, 2, \ldots, 6; \quad i = 0, 1, \ldots, 6),$$

$$\sum_{j=1}^{20} \sum_{i=0}^{6} x_{ij} \geq d_{ts}(1 + m) \qquad\qquad (t = 1, 2, \ldots, 6; \quad s = 1, 2, \ldots, 20),$$

$$x_{ij} \geq 0, \quad o_{ijts} \geq 0 \qquad\qquad \begin{aligned} &(i = 0, 1, \ldots, 6; \quad t = 1, 2, \ldots, 6; \\ &\ j = 1, 2, \ldots, 20; \quad s = 1, 2, \ldots, 20). \end{aligned}$$

In this formulation, θ_s denotes the length of time period s; the values of x_{ij} are given. The subscript h denotes hydroelectric.

a) Interpret the objective function and each of the constraints in this model. How large is the model?

b) What is the structure of the constraint coefficients for this problem?

c) Suppose that we apply the decomposition algorithm to solve this problem; for each plant j and time period t, let the constraints

$$o_{ijts} \leq a_{ijt} x_{ij} \qquad\qquad (i = 0, 1, \ldots, 6; \quad s = 1, 2, \ldots, 20),$$

$$o_{ijts} \geq 0, \quad x_{ij} \geq 0 \qquad (i = 0, 1, \ldots, 6; \quad s = 1, 2, \ldots, 20),$$

form a subproblem. What is the objective function to the subproblem at each step? Show that each subproblem either solves at $o_{ijts} = 0$ and $x_{ij} = 0$ for all i and s, or is unbounded. Specify the steps for applying the decomposition algorithm with this choice of subproblems.

d) How would the application of decomposition discussed in part (c) change if the constraints

$$\sum_{s=1}^{20} o_{ihts}\theta_s \leq H_{it}, \qquad\qquad (i = 0, 1, \ldots, 6),$$

are added to each subproblem in which $j = h$ denotes a hydroelectric plant?

9. The decomposition method can be interpreted as a "cutting-plane" algorithm. To illustrate this viewpoint, consider the example:

Maximize $z = 3x_1 + 8x_2$,

subject to:

$$2x_1 + 4x_2 \leq 3,$$

$$0 \leq x_1 \leq 1,$$

$$0 \leq x_2 \leq 1.$$

Applying decomposition with the constraints $0 \leq x_1 \leq 1$ and $0 \leq x_2 \leq 1$ as the subproblem, we have four extreme points to the subproblem:

		Weights
Extreme point 1:	$x_1 = 0$, $x_2 = 0$	λ_1
Extreme point 2:	$x_1 = 0$, $x_2 = 1$	λ_2
Extreme point 3:	$x_1 = 1$, $x_2 = 0$	λ_3
Extreme point 4:	$x_1 = 1$, $x_2 = 1$	λ_4

Evaluating the objective function $3x_1 + 8x_2$ and resource usage $2x_1 + 4x_2$ at these extreme-point solutions gives the following master problem:

Maximize $z = 0\lambda_1 + 8\lambda_2 + 3\lambda_3 + 11\lambda_4$,   *Dual*
                                        variables

subject to:

$$0\lambda_1 + 4\lambda_2 + 2\lambda_3 + 6\lambda_4 \leq 3, \qquad \pi$$

$$\lambda_1 + \lambda_2 + \lambda_3 + \lambda_4 = 1, \qquad \sigma$$

$$\lambda_j \geq 0 \qquad (j = 1, 2, 3, 4).$$

a) Let the variable w be defined in terms of the dual variables π and σ as $w = \sigma + 3\pi$. Show that the dual to the master problem in terms of w and π is:

Minimize w,

subject to:

$$w - 3\pi \geq 0,$$

$$w + \pi \geq 8,$$

$$w - \pi \geq 3,$$

$$w + 3\pi \geq 11.$$

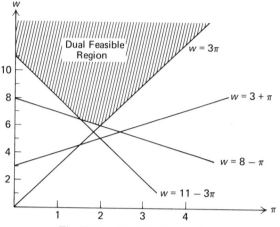

Fig. E12.2 Dual feasible region.

b) Figure E12.2 depicts the feasible region for the dual problem. Identify the optimal solution to the dual problem in this figure. What is the value of z^*, the optimal objective value of the original problem?

c) Suppose that we initiate the decomposition with a restricted master problem containing only the third extreme point $x_1 = 1$ and $x_2 = 0$. Illustrate the feasible region to the dual of this restricted master in terms of w and π, and identify its optimal solution w^* and π^*. Does this dual feasible region contain the feasible region to the full dual problem formulated in part (a)?

d) Show that the next step of the decomposition algorithm adds a new constraint to the dual of the restricted master problem. Indicate which constraint in Fig. E12.2 is added next. Interpret the added constraint as "cutting away" the optimal solution w^* and π^* found in part (c) from the feasible region. What are the optimal values of the dual variables after the new constraint has been added?

e) Note that the added constraint is found by determining which constraint is most violated at $\pi = \pi^*$; that is, by moving vertically in Fig. E12.2 at $\pi = \pi^*$, crossing all violated constraints until we reach the dual feasible region at $w = \hat{w}$. Note that the optimal objective value z^* to the original problem satisfies the inequalities:

$$w^* \leqq z^* \leqq \hat{w}.$$

Relate this bound to the bounds discussed in this chapter.

f) Solve this problem to completion, using the decomposition algorithm. Interpret the solution in Fig. E12.2, indicating at each step the cut and the bounds on z^*.

g) How do extreme rays in the master problem alter the formulation of the dual problem? How would the cutting-plane interpretation discussed in this problem be modified when the subproblem is unbounded?

10. In this exercise we consider a two-dimensional version of the cutting stock problem.

a) Suppose that we have a W-by-L piece of cloth. The material can be cut into a number of smaller pieces and sold. Let π_{ij} denote the revenue for a smaller piece with dimensions w_i by ℓ_j $(i = 1, 2, \ldots, m; j = 1, 2, \ldots, n)$.

Operating policies dictate that we first cut the piece along its width into strips of size w_i. The strips are then cut into lengths of size ℓ_j. Any waste is scrapped, with no additional revenue.

For example, a possible cutting pattern for a 9-by-10 piece might be that shown in Fig. E12.3. The shaded regions correspond to trim losses.

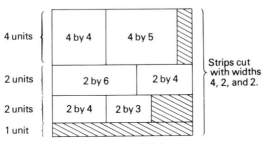

Figure E12.3

Formulate a (nonlinear) integer program for finding the maximum-revenue cutting pattern. Can we solve this integer program by solving several knapsack problems? [*Hint.* Can we use the same-length cuts in any strips with the same width? What is the optimal revenue v_i obtained from a strip of width w_i? What is the best way to choose the widths w_i to maximize the total value of the v_i's?]

b) A firm has unlimited availabilities of W-by-L pieces to cut in the manner described in part (a). It must cut these pieces into smaller pieces in order to meet its demand of d_{ij} units for a piece with width w_i and length ℓ_j $(i = 1, 2, \ldots, m; j = 1, 2, \ldots, n)$. The firm wishes to use as few W-by-L pieces as possible to meet its sales commitments.

Formulate the firm's decision-making problem in terms of cutting patterns. How can column generation be used to solve the linear-programming approximation to the cutting-pattern formulation?

11. In Section 12.1 we provided a formulation for a large-scale multi-item production-scheduling problem. The purpose of this exercise (and of Exercises 12 and 13) is to explore the implications of the suggested formulation, as well as techniques that can be developed to solve the problem.

The more classical formulation of the multi-item scheduling problem can be stated as follows:

$$\text{Minimize } z = \sum_{j=1}^{J} \sum_{t=1}^{T} [s_{jt}\delta(x_{jt}) + v_{jt}x_{jt} + h_{jt}I_{jt}],$$

subject to:

$$x_{jt} + I_{j,t-1} - I_{jt} = d_{jt} \qquad\qquad (t = 1, 2, \ldots, T; \ \ j = 1, 2, \ldots, J),$$

$$\sum_{j=1}^{J} [\ell_j\delta(x_{jt}) + k_jx_{jt}] \leq b_t \qquad\qquad (t = 1, 2, \ldots, T),$$

$$x_{jt} \geq 0, \quad I_{jt} \geq 0 \qquad\qquad (t = 1, 2, \ldots, T; \ j = 1, 2, \ldots, J),$$

where

$$\delta(x_{jt}) = \begin{cases} 0 & \text{if } x_{jt} = 0, \\ 1 & \text{if } x_{jt} > 0, \end{cases}$$

and $\quad x_{jt} =$ Units of item j to be produced in period t,

$\qquad I_{jt} =$ Units of inventory of item j left over at the end of period t,

$\qquad s_{jt} =$ Setup cost of item j in period t,

$\qquad v_{jt} =$ Unit production cost of item j in period t,

$\qquad h_{jt} =$ Inventory holding cost for item j in period t,

$\qquad d_{jt} =$ Demand for item j in period t,

$\qquad \ell_{j} =$ Down time consumed in performing a setup for item j,

$\qquad k_{j} =$ Man-hours required to produce one unit of item j,

$\qquad b_{t} =$ Total man-hours available for period t.

a) Interpret the model formulation. What are the basic assumptions of the model? Is there any special structure to the model?

b) Formulate an equivalent (linear) mixed-integer program for the prescribed model. If $T = 12$ (that is, we are planning for twelve time periods) and $J = 10,000$ (that is, there are 10,000 items to schedule), how many integer variables, continuous variables, and constraints does the model have? Is it feasible to solve a mixed-integer programming model of this size?

12. Given the computational difficulties associated with solving the model presented in Exercise 11, A. S. Manne conceived of a way to approximate the mixed-integer programming model as a linear program. This transformation is based on defining for each item j a series of production sequences over the planning horizon T. Each sequence is a set of T nonnegative integers that identify the amount to be produced of item j at each time period t during the planning horizon, in such a way that demand requirements for the item are met. It is enough to consider production sequences such that, at a given time period, the production is either zero or the sum of consecutive demands for some number of periods into the future. This limits the number of production sequences to a total of 2^{T-1} for each item. Let

$$x_{jkt} = \text{amount to be produced of item } j \text{ in period } t \text{ by means of production sequence } k.$$

To illustrate how the production sequences are constructed, assume that $T = 3$. Then the total number of production sequences for item j is $2^{3-1} = 4$. The corresponding sequences are given in Table E12.1.

Table E12.1

| Sequence | Time period | | |
number	$t = 1$	$t = 2$	$t = 3$
$k = 1$	$x_{j11} = d_{j1} + d_{j2} + d_{j3}$	$x_{j12} = 0$	$x_{j13} = 0$
$k = 2$	$x_{j21} = d_{j1} + d_{j2}$	$x_{j22} = 0$	$x_{j23} = d_{j3}$
$k = 3$	$x_{j31} = d_{j1}$	$x_{j32} = d_{j2} + d_{j3}$	$x_{j33} = 0$
$k = 4$	$x_{j41} = d_{j1}$	$x_{j42} = d_{j2}$	$x_{j43} = d_{j3}$

The total cost associated with sequence k for the production of item j is given by

$$c_{jk} = \sum_{t=1}^{T} [s_{jt}\delta(x_{jkt}) + v_{jt}x_{jkt} + h_{jt}I_{jt}],$$

and the corresponding man-hours required for this sequence in period t is

$$a_{jkt} = \ell_j \delta(x_{jkt}) + k_j x_{jkt}.$$

a) Verify that, if the model presented in Exercise 11 is restricted to producing each item in production sequences, then it can be formulated as follows:

$$\text{Minimize } z = \sum_{j=1}^{J} \sum_{k=1}^{K} c_{jk}\theta_{jk},$$

subject to:

$$\sum_{j=1}^{J} \sum_{k=1}^{K} a_{jkt}\theta_{jk} \leqq b_t \qquad\qquad (t = 1, 2, \ldots, T),$$

$$\sum_{j=1}^{J} \theta_{jk} = 1 \qquad\qquad (k = 1, 2, \ldots, K),$$

$$\theta_{jk} \geqq 0 \quad \text{and integer} \qquad (j = 1, 2, \ldots, J; \quad k = 1, 2, \ldots, K).$$

b) Study the structure of the resulting model. How could you define the structure? For $T = 12$ and $J = 10{,}000$, how many rows and columns does the model have?

c) Under what conditions can we eliminate the integrality constraints imposed on variables θ_{jk} without significantly affecting the validity of the model? [*Hint.* Read the comment made on the multi-term scheduling problem in Section 12.1 of the text.]

d) Propose a decomposition approach to solve the resulting large-scale linear-programming model. What advantages and disadvantages are offered by this approach? (Assume that at this point the resulting subproblems are easy to solve. See Exercise 13 for details.)

13. Reconsider the large-scale linear program proposed in the previous exercise:

$$\text{Minimize } z = \sum_{j=1}^{J} \sum_{k=1}^{K} c_{jk}\theta_{jk},$$

subject to:

$$\sum_{j=1}^{J} \sum_{k=1}^{K} a_{jkt}\theta_{jk} \leqq b_t \qquad\qquad (t = 1, 2, \ldots, T), \qquad (1)$$

$$\sum_{j=1}^{J} \theta_{jk} = 1 \qquad\qquad (k = 1, 2, \ldots, K), \qquad (2)$$

$$\theta_{jk} \geqq 0 \qquad\qquad (j = 1, 2, \ldots, J; \quad k = 1, 2, \ldots, K), \qquad (3)$$

a) Let us apply the column-generation algorithm to solve this problem. At some stage of the process, let π_t for $t = 1, 2, \ldots, T$ be the shadow prices associated with constraints (1), and let π_{T+k} for $k = 1, 2, \ldots, K$ be the shadow prices associated with constraints (2), in the restricted master problem. The reduced cost $\bar{c}_{jk}$ for variable θ_{jk} is given by the following expression:

$$\bar{c}_{jk} = c_{jk} - \sum_{t=1}^{T} \pi_t a_{jkt} - \pi_{T+k}.$$

Show, in terms of the original model formulation described in Exercise 11, that $\bar{c}_{jk}$ is defined as:

$$\bar{c}_{jk} = \sum_{t=1}^{T} \left[(s_{jt} - \pi_t \ell_j)\delta(x_{jkt}) + (v_{jt} - \pi_t k_j)x_{jkt} + h_{jt}I_{jt} \right] - \pi_{T+k}.$$

b) The subproblem has the form:

Minimize $\left[\underset{k}{\text{minimize}} \, \bar{c}_{jk}\right]$.
   j

The inner minimization can be interpreted as finding the minimum-cost production sequence for a specific item j. This problem can be interpreted as an uncapacitated single-item production problem under fluctuating demand requirements d_{jt} throughout the planning horizon $t = 1, 2, \ldots, T$. Suggest an effective dynamic-programming approach to determine the optimum production sequence under this condition.

c) How does the above approach eliminate the need to generate all the possible production sequences for a given item j? Explain the interactions between the master problem and the subproblem.

14. The "traffic-assignment" model concerns minimizing travel time over a large network, where traffic enters the network at a number of origins and must flow to a number of different destinations. We can consider this model as a multicommodity-flow problem by defining a commodity as the traffic that must flow between a particular origin–destination pair. As an alternative to the usual node–arc formulation, consider chain flows. A *chain* is merely a *directed* path through a network from an origin to a destination. In particular, let

$$a_{ij}^k = \begin{cases} 1 & \text{if arc } i \text{ is in chain } j, \text{ which connects origin–destination pair } k, \\ 0 & \text{otherwise.} \end{cases}$$

In addition, define

$$z_j^k = \text{Flow over chain } j \text{ between origin–destination pair } k.$$

For example, the network in Fig. E12.4, shows the arc flows of one of the commodities, those vehicles entering node 1 and flowing to node 5. The chains connecting the origin–destination pair 1–5 can be used to express the flow in this network as:

	Chain 1	Chain 2	Chain 3	Chain 4	Chain 5
Chain j	1–2–5	1–2–4–5	1–4–5	1–3–5	1–3–4–5
Flow value z_j	3	1	2	3	2

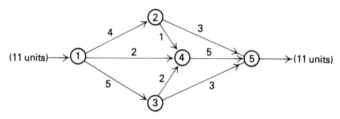

Figure E12.4

Frequently an upper bound u_i is imposed upon the total flow on each arc i. These restrictions are modeled as:

$$\sum_{j_1} a_{ij}^1 z_j^1 + \sum_{j_2} a_{ij}^2 z_j^2 + \cdots + \sum_{j_K} a_{ij}^K z_j^K \leqq u_i \qquad (i = 1, 2, \ldots, I).$$

The summation indices j_k correspond to chains joining the kth origin–destination pair. The total number of arcs is I and the total number of origin–destination pairs is K. The requirement that certain levels of traffic must flow between origin–destination pairs can be formulated as follows:

$$\sum_{j_k} z_j^k = v_k \qquad (k = 1, 2, \ldots, K),$$

where

$v_k = $ Required flow between origin–destination pair (commodity) k.

Finally, suppose that the travel time over any arc is t_i, so that, if x_i units travel over arc i, the total travel time on arc i is $t_i x_i$.

a) Complete the "arc–chain" formulation of the traffic-assignment problem by specifying an objective function that minimizes total travel time on the network. [*Hint.* Define the travel time over a chain, using the a_{ij} data.]
b) In reality, generating all the chains of a network is very difficult computationally. Suppose enough chains have been generated to determine a basic feasible solution to the linear program formulated in part (a). Show how to compute the reduced cost of the next chain to enter the basis from those generated thus far.
c) Now consider the chains not yet generated. In order for the current solution to be optimal, the minimum reduced costs of these chains must be nonnegative. How would you find the chain with the minimum reduced cost for each "commodity"? [*Hint.* The reduced costs are, in general,

$$\bar{c}_j^k = \sum_i a_{ij}(t_i - \pi_i) - u_k,$$

where π_i and u_k are the shadow prices associated with the capacity restriction on the ith constraint and flow requirement between the kth origin–destination pair. What is the sign of π_i?]
d) Give an economic interpretation of π_i. In the reduced cost of part (c), do the values of π_i depend on which commodity flows over arc i?

15. Consider the node–arc formulation of the "traffic-assignment" model. Define a "commodity" as the flow from an origin to a destination. Let

$$x_{ij}^k = \text{Flow over arc } i\text{–}j \text{ of commodity } k.$$

The conservation-of-flow equations for each commodity are:

$$\sum_i x_{in}^k - \sum_j x_{nj}^k = \begin{cases} v_k & \text{if } n = \text{origin for commodity } k, \\ -v_k & \text{if } n = \text{destination for commodity } k, \\ 0 & \text{otherwise.} \end{cases}$$

The capacity restrictions on the arcs can be formulated as follows:

$$\sum_{k=1}^{K} x_{ij}^k \leqq u_{ij} \qquad \text{for all arcs } i\text{--}j,$$

assuming that t_{ij} is the travel time on arc i–j. To minimize total travel time on the network, we have the following objective function:

$$\text{Minimize} \sum_k \sum_i \sum_j t_{ij} x_{ij}^k.$$

a) Let the conservation-of-flow constraints for a commodity correspond to a subproblem, and the capacity restrictions on the arcs correspond to the master constraints in a decomposition approach. Formulate the restricted master, and the subproblem for the kth commodity. What is the objective function of this subproblem?

b) What is the relationship between solving the subproblems of the node–arc formulation and finding the minimum reduced cost for each commodity in the arc–chain formulation discussed in the previous exercise?

c) Show that the solution of the node–arc formulation by decomposition is identical to solving the arc–chain formulation discussed in the previous exercise. [*Hint*. In the arc–chain formulation, define new variables

$$\lambda_j^k = \frac{x_j^k}{v_k}.\Bigg]$$

16. In the node–arc formulation of the "traffic-assignment" problem given in Exercise 15, the subproblems correspond to finding the shortest path between the kth origin–destination pair. In general, there may be a large number of origin–destination pairs and hence a large number of such subproblems. However, in Chapter 11 on dynamic programming, we saw that we can solve simultaneously for the shortest paths from a particular origin to all destinations. We can then consolidate the subproblems by defining one subproblem for each node where traffic originates. The conservation-of-flow constraints become:

$$\sum_i y_{in}^s - \sum_j y_{nj}^s = \begin{cases} \sum v_k & \text{if } n = \text{origin node } s, \\ -v_k & \text{if } n = \text{a destination node in the origin–destination pair } k = (s, n), \\ 0 & \text{otherwise,} \end{cases}$$

where the summation $\sum v_k$ is the total flow emanating from origin s for all destination nodes. In this formulation, $y_{ij}^s = \sum x_{ij}^k$ denotes the total flow on arc i–j that emanates from origin s; that is, the summation is carried over all origin–destination pairs $k = (s, t)$ whose origin is node s.

a) How does the decomposition formulation developed in Exercise 15 change with this change in definition of a subproblem? Specify the new formulation precisely.

b) Which formulation has more constraints in its restricted master?

c) Which restricted master is more restricted? [*Hint*. Which set of constraints implies the other?]

d) How does the choice of which subproblems to employ affect the decomposition algorithm? Which choice would you expect to be more efficient? Why?

17. Consider a "nested decomposition" as applied to the problem

$$\text{Maximize} \sum_{j=1}^{n} c_j x_j,$$

subject to:

$$\sum_{j=1}^{n} a_{ij} x_j = b_i \qquad\qquad (i = 1, 2, \ldots, k), \qquad (1)$$

$$\sum_{j=1}^{n} d_{ij} x_j = d_i \qquad\qquad (i = k+1, k+2, \ldots, \ell), \qquad (2) \qquad (\text{P})$$

$$\sum_{j=1}^{n} g_{ij} x_j = g_i \qquad\qquad (i = \ell+1, \ell+2, \ldots, m), \qquad (3)$$

$$x_j \geq 0 \qquad\qquad (j = 1, 2, \ldots, n).$$

Let (1) be the constraints of the (first) restricted master problem. If π_i $(i = 1, 2, \ldots, k)$ are shadow prices for the constraints (1) in the weighting problem, then

$$\text{Maximize} \sum_{j=1}^{n} \left(c_j - \sum_{i=1}^{k} \pi_i a_{ij} \right) x_j,$$

subject to:

$$\sum_{j=1}^{n} d_{ij} x_j = d_i \qquad (i = k+1, k+2, \ldots, \ell), \qquad (2')$$

$$\sum_{j=1}^{n} g_{ij} x_j = g_i \qquad (i = \ell+1, \ell+2, \ldots, m), \qquad (3') \qquad (\text{Subproblem 1})$$

$$x_j \geq 0, \qquad\qquad (j = 1, 2, \ldots, n),$$

constitutes subproblem 1 (the proposal-generating problem).

Suppose, though, that the constraints (3') complicate this problem and make it difficult to solve. Therefore, to solve the subproblem we further apply decomposition on subproblem 1. Constraints (2') will be the constraints of the "second" restricted master. Given any shadow prices α_i $(i = k+1, k+2, \ldots, \ell)$ for constraints (2') in the weighting problem, the subproblem 2 will be:

$$\text{Maximize} \sum_{j=1}^{n} \left(c_j - \sum_{i=1}^{k} \pi_i a_{ij} - \sum_{i=k+1}^{\ell} \alpha_i d_{ij} \right) x_j,$$

subject to:

$$\sum_{j=1}^{n} g_{ij} x_j = g_i \qquad\qquad (i = \ell+1, \ell+2, \ldots, m), \qquad (\text{Subproblem 2})$$

$$x_j \geq 0 \qquad\qquad (j = 1, 2, \ldots, n).$$

a) Consider the following decomposition approach: Given shadow prices π_i, solve subproblem (1) to completion by applying decomposition with subproblem (2). Use the solution to this problem to generate a new weighting variable to the first restricted master problem, or show that the original problem (P) [containing all constraints (1), (2), (3)] has been solved. Specify details of this approach.

b) Show finite convergence and bounds on the objective function to (P).

c) Now consider another approach: Subproblem 1 need not be solved to completion, but merely until a solution x_j ($j = 1, 2, \ldots, n$) is found, so that

$$\sum_{j=1}^{n} \left(c_j - \sum_{i=1}^{k} \pi_i a_{ij} \right) x_j > \gamma,$$

where γ is the shadow price for the weighting constraint to the first restricted master. Indicate how to identify such a solution x_j ($j = 1, 2, \ldots, n$) while solving the second restricted master problem; justify this approach.

d) Discuss convergence and objective bounds for the algorithm proposed in part (c).

ACKNOWLEDGMENTS

A number of the exercises in this chapter are based on or inspired by articles in the literature.
Exercise 8: D. Anderson, "Models for Determining Least-Cost Investments in Electricity Supply," *The Bell Journal of Economics and Management Science*, **3**, No. 1, Spring 1972.
Exercise 10: P. E. Gilmore and R. E. Gomory, "A Linear Programming Approach to the Cutting Stock Problem-II," *Operations Research*, **11**, No. 6, November–December 1963.
Exercise 12: A. S. Manne, "Programming of Economic Lot Sizes," *Management Science*, **4**, No. 2, January 1958.
Exercise 13: B. P. Dzielinski and R. E. Gomory, "Optimal Programming of Lot Sizes, Inventory, and Labor Allocations," *Management Science*, **11**, No. 9, July 1965; and L. S. Lasdon and R. C. Terjung, "An Efficient Algorithm for Multi-Item Scheduling," *Operations Research*, **19**, No. 4, July–August 1971.
Exercises 14 through 16: S. P. Bradley, "Solution Techniques for the Traffic Assignment Problem," Operations Research Center Report ORC 65–35, University of California, Berkeley.
Exercise 17: R. Glassey, "Nested Decomposition and Multi-Stage Linear Programs," *Management Science*, **20**, No. 3, 1973.

Nonlinear Programming

13

Numerous mathematical-programming applications, including many introduced in previous chapters, are cast naturally as linear programs. Linear programming assumptions or approximations may also lead to appropriate problem representations over the range of decision variables being considered. At other times, though, nonlinearities in the form of either nonlinear objective functions or nonlinear constraints are crucial for representing an application properly as a mathematical program. This chapter provides an initial step toward coping with such nonlinearities, first by introducing several characteristics of nonlinear programs and then by treating problems that can be solved using simplex-like pivoting procedures. As a consequence, the techniques to be discussed are primarily algebra-based. The final two sections comment on some techniques that do not involve pivoting.

As our discussion of nonlinear programming unfolds, the reader is urged to reflect upon the linear-programming theory that we have developed previously, contrasting the two theories to understand why the nonlinear problems are intrinsically more difficult to solve. At the same time, we should try to understand the similarities between the two theories, particularly since the nonlinear results often are motivated by, and are direct extensions of, their linear analogs. The similarities will be particularly visible for the material of this chapter where simplex-like techniques predominate.

13.1 NONLINEAR PROGRAMMING PROBLEMS

A general optimization problem is to select n decision variables $x_1, x_2, \ldots, x_n$ from a given feasible region in such a way as to optimize (minimize or maximize) a given objective function

$$f(x_1, x_2, \ldots, x_n)$$

of the decision variables. The problem is called a *nonlinear programming problem* (NLP) if the objective function is nonlinear and/or the feasible region is determined by nonlinear constraints. Thus, in maximization form, the general nonlinear program is stated as:

$$\text{Maximize } f(x_1, x_2, \ldots, x_n),$$

subject to:

$$g_1(x_1, x_2, \ldots, x_n) \leqq b_1,$$
$$\vdots \qquad\qquad \vdots$$
$$g_m(x_1, x_2, \ldots, x_n) \leqq b_m,$$

where each of the constraint functions g_1 through g_m is given. A special case is the linear program that has been treated previously. The obvious association for this case is

$$f(x_1, x_2, \ldots, x_n) = \sum_{j=1}^{n} c_j x_j,$$

and

$$g_i(x_1, x_2, \ldots, x_n) = \sum_{j=1}^{n} a_{ij} x_j \qquad (i = 1, 2, \ldots, m).$$

Note that nonnegativity restrictions on variables can be included simply by appending the additional constraints:

$$g_{m+i}(x_1, x_2, \ldots, x_n) = -x_i \leqq 0 \qquad (i = 1, 2, \ldots, n).$$

Sometimes these constraints will be treated explicitly, just like any other problem constraints. At other times, it will be convenient to consider them implicitly in the same way that nonnegativity constraints are handled implicitly in the simplex method.

For notational convenience, we usually let x denote the vector of n decision variables $x_1, x_2, \ldots, x_n$ — that is, $x = (x_1, x_2, \ldots, x_n)$ — and write the problem more concisely as

$$\text{Maximize } f(x),$$

subject to:

$$g_i(x) \leqq b_i \qquad (i = 1, 2, \ldots, m).$$

As in linear programming, we are not restricted to this formulation. To minimize $f(x)$, we can of course maximize $-f(x)$. Equality constraints $h(x) = b$ can be written as two inequality constraints $h(x) \leqq b$ and $-h(x) \leqq -b$. In addition, if we introduce a slack variable, each inequality constraint is transformed to an equality constraint.

Thus sometimes we will consider an alternative equality form:

Maximize $f(x)$,

subject to:

$$h_i(x) = b_i \qquad (i = 1, 2, \ldots, m)$$

$$x_j \geq 0 \qquad (j = 1, 2, \ldots, n).$$

Usually the problem context suggests either an equality or inequality formulation (or a formulation with both types of constraints), and we will not wish to force the problem into either form.

The following three simplified examples illustrate how nonlinear programs can arise in practice.

Portfolio Selection An investor has \$5000 and two potential investments. Let x_j for $j = 1$ and $j = 2$ denote his allocation to investment j in thousands of dollars. From historical data, investments 1 and 2 have an expected annual return of 20 and 16 percent, respectively. Also, the total risk involved with investments 1 and 2, as measured by the variance of total return, is given by $2x_1^2 + x_2^2 + (x_1 + x_2)^2$, so that risk increases with total investment and with the amount of each individual investment. The investor would like to maximize his expected return and at the same time minimize his risk. Clearly, both of these objectives cannot, in general, be satisfied simultaneously. There are several possible approaches. For example, he can minimize risk subject to a constraint imposing a lower bound on expected return. Alternatively, expected return and risk can be combined in an objective function, to give the model:

$$\text{Maximize } f(x) = 20x_1 + 16x_2 - \theta[2x_1^2 + x_2^2 + (x_1 + x_2)^2],$$

subject to:

$$g_1(x) = x_1 + x_2 \leq 5,$$

$$x_1 \geq 0, \quad x_2 \geq 0, \qquad (\text{that is, } g_2(x) = -x_1, \quad g_3(x) = -x_2).$$

The nonnegative constant θ reflects his tradeoff between risk and return. If $\theta = 0$, the model is a linear program, and he will invest completely in the investment with greatest expected return. For very large θ, the objective contribution due to expected return becomes negligible and he is essentially minimizing his risk.

Water Resources Planning In regional water planning, sources emitting pollutants might be required to remove waste from the water system. Let x_j be the pounds of Biological Oxygen Demand (an often-used measure of pollution) to be removed at source j.

One model might be to minimize total costs to the region to meet specified pollution standards:

$$\text{Minimize} \sum_{j=1}^{n} f_j(x_j),$$

subject to:

$$\sum_{j=1}^{n} a_{ij}x_j \geqq b_i \qquad (i = 1, 2, \ldots, m)$$

$$0 \leqq x_j \leqq u_j \qquad (j = 1, 2, \ldots, n),$$

where

$\quad f_j(x_j) =$ Cost of removing x_j pounds of Biological Oxygen Demand at source j,

$\quad\quad b_i =$ Minimum desired improvement in water quality at point i in the system,

$\quad\quad a_{ij} =$ Quality response, at point i in the water system, caused by removing one pound of Biological Oxygen Demand at source j,

$\quad\quad u_j =$ Maximum pounds of Biological Oxygen Demand that can be removed at source j.

Constrained Regression A university wishes to assess the job placements of its graduates. For simplicity, it assumes that each graduate accepts either a government, industrial, or academic position. Let

$$N_j = \text{Number of graduates in year } j \qquad (j = 1, 2, \ldots, n),$$

and let G_j, I_j, and A_j denote the number entering government, industry, and academia, respectively, in year j ($G_j + I_j + A_j = N_j$).

One model being considered assumes that a given fraction of the student population joins each job category each year. If these fractions are denoted as λ_1, λ_2, and λ_3, then the predicted number entering the job categories in year j is given by the expressions

$$\hat{G}_j = \lambda_1 N_j,$$

$$\hat{I}_j = \lambda_2 N_j,$$

$$\hat{A}_j = \lambda_3 N_j.$$

A reasonable performance measure of the model's validity might be the difference between the actual number of graduates G_j, I_j, and A_j entering the three job categories and the predicted numbers $\hat{G}_j$, $\hat{I}_j$, and $\hat{A}_j$, as in the least-squares estimate:

$$\text{Minimize} \sum_{j=1}^{n} [(G_j - \hat{G}_j)^2 + (I_j - \hat{I}_j)^2 + (A_j - \hat{A}_j)^2],$$

subject to the constraint that all graduates are employed in one of the professions. In terms of the fractions entering each profession, the model can be written as:

$$\text{Minimize} \sum_{j=1}^{n} [(G_j - \lambda_1 N_j)^2 + (I_j - \lambda_2 N_j)^2 + (A_j - \lambda_3 N_j)^2],$$

subject to:

$$\lambda_1 + \lambda_2 + \lambda_3 = 1,$$

$$\lambda_1 \geq 0, \quad \lambda_2 \geq 0, \quad \lambda_3 \geq 0.$$

This is a nonlinear program in three variables λ_1, λ_2, and λ_3.

There are alternative ways to approach this problem. For example, the objective function can be changed to:

$$\text{Minimize} \sum_{j=1}^{n} [|G_j - \hat{G}_j| + |I_j - \hat{I}_j| + |A_j - \hat{A}_j|].^{\dagger}$$

This formulation is appealing since the problem now can be transformed into a linear program. Exercise 28 (see also Exercise 20) from Chapter 1 illustrates this transformation.

The range of nonlinear-programming applications is practically unlimited. For example, it is usually simple to give a nonlinear extension to any linear program. Moreover, the constraint $x = 0$ or 1 can be modeled as $x(1 - x) = 0$ and the constraint x integer as $\sin(\pi x) = 0$. Consequently, *in theory* any application of integer programming can be modeled as a nonlinear program. We should not be overly optimistic about these formulations, however; later we shall explain why nonlinear programming is not attractive for solving these problems.

13.2 LOCAL vs. GLOBAL OPTIMUM

Geometrically, nonlinear programs can behave much differently from linear programs, even for problems with linear constraints. In Fig. 13.1, the portfolio-selection example from the last section has been plotted for several values of the tradeoff parameter θ. For each fixed value of θ, contours of constant objective values are concentric ellipses. As Fig. 13.1 shows, the optimal solution can occur:

a) at an interior point of the feasible region;

b) on the boundary of the feasible region, which is not an extreme point; or

c) at an extreme point of the feasible region.

As a consequence, procedures, such as the simplex method, that search only extreme points may not determine an optimal solution.

$^{\dagger}|\quad|$ denotes absolute value; that is, $|x| = x$ if $x \geq 0$ and $|x| = -x$ if $x < 0$.

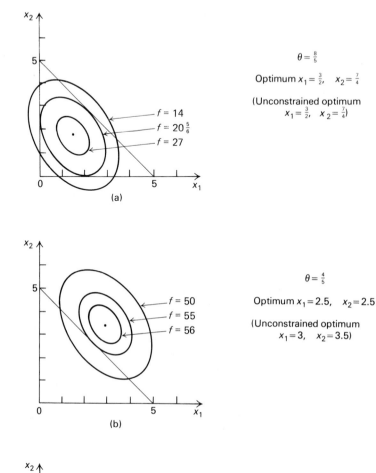

$\theta = \frac{8}{5}$

Optimum $x_1 = \frac{3}{2}$, $x_2 = \frac{7}{4}$

(Unconstrained optimum
$x_1 = \frac{3}{2}$, $x_2 = \frac{7}{4}$)

$f = 14$

$f = 20\frac{5}{6}$

$f = 27$

(a)

$\theta = \frac{4}{5}$

Optimum $x_1 = 2.5$, $x_2 = 2.5$

(Unconstrained optimum
$x_1 = 3$, $x_2 = 3.5$)

$f = 50$

$f = 55$

$f = 56$

(b)

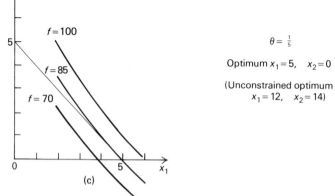

$f = 100$

$f = 85$

$f = 70$

$\theta = \frac{1}{5}$

Optimum $x_1 = 5$, $x_2 = 0$

(Unconstrained optimum
$x_1 = 12$, $x_2 = 14$)

(c)

Fig. 13.1 Portfolio-selection example for various values of θ. (Lines are contours of constant objective values.)

Figure 13.2 illustrates another feature of nonlinear-programming problems. Suppose that we are to minimize $f(x)$ in this example, with $0 \leq x \leq 10$. The point $x = 7$ is optimal. Note, however, that in the indicated dashed interval, the point $x = 0$ is the best feasible point; i.e., it is an optimal feasible point in the local vicinity of $x = 0$ specified by the dashed interval.

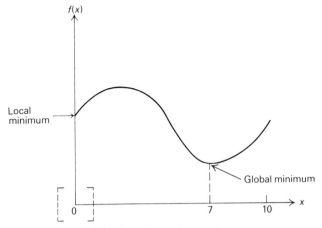

Fig. 13.2 Local and global minima.

The latter example illustrates that a solution optimal in a local sense need not be optimal for the overall problem. Two types of solution must be distinguished. A global optimum is a solution to the overall optimization problem. Its objective value is as good as any other point in the feasible region. A local optimum, on the other hand, is optimum only with respect to feasible solutions close to that point. Points far removed from a local optimum play no role in its definition and may actually be preferred to the local optimum. Stated more formally,

> **Definition.** Let $x = (x_1, x_2, \ldots, x_n)$ be a feasible solution to a maximization problem with objective function $f(x)$. We call x

1. A *global maximum* if $f(x) \geq f(y)$ for every feasible point $y = (y_1, y_2, \ldots, y_n)$;
2. A *local maximum* if $f(x) \geq f(y)$ for every feasible point $y = (y_1, y_2, \ldots, y_n)$ sufficiently close to x. That is, if there is a number $\epsilon > 0$ (possibly quite small) so that, whenever each variable y_j is within ϵ of x_j — that is, $x_j - \epsilon \leq y_j \leq x_j + \epsilon$ — and y is feasible, then $f(x) \geq f(y)$.

Global and local minima are defined analogously. The definition of local maximum simply says that if we place an n-dimensional box (e.g., a cube in three dimensions) about x, whose side has length 2ϵ, then $f(x)$ is as small as $f(y)$ for every feasible point y lying within the box. (Equivalently, we can use n-dimensional spheres in this definition.) For instance, if $\epsilon = 1$ in the above example, the one-dimensional box, or interval, is pictured about the local minimum $x = 0$ in Fig. 13.2.

The concept of a local maximum is extremely important. As we shall see, most general-purpose nonlinear-programming procedures are near-sighted and can do no better than determine local maxima. We should point out that, since every global maximum is also a local maximum, the overall optimization problem can be viewed as seeking the best local maxima.

Under certain circumstances, local maxima and minima are known to be global. Whenever a function "curves upward" as in Fig. 13.3(a), local minima will be global. These functions are called *convex*. Whenever a function "curves downward" as in Fig. 13.3(b) a local maximum will be a global maximum. These functions are called *concave.*[†] For this reason we usually wish to minimize convex functions and maximize concave functions. These observations are formalized below.

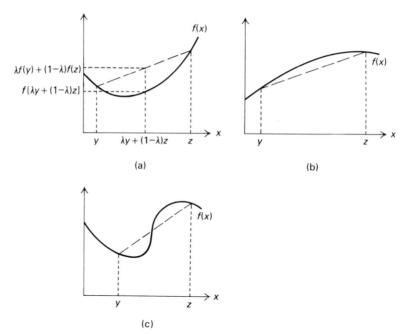

(a)

(b)

(c)

Fig. 13.3 a) Convex function; b) concave function; c) nonconvex, nonconcave function.

13.3 CONVEX AND CONCAVE FUNCTIONS

Because of both their pivotal role in model formulation and their convenient mathematical properties, certain functional forms predominate in mathematical programming. Linear functions are by far the most important. Next in importance are functions which are convex or concave. These functions are so central to the theory that we take some time here to introduce a few of their basic properties.

[†] As a mnemonic, the "A" in concAve reflects the shape of these functions.

An essential assumption in a linear-programming model for profit maximization is constant returns to scale for each activity. This assumption implies that if the level of one activity doubles, then that activity's profit contribution also doubles; if the first activity level changes from x_1 to $2x_1$, then profit increases proportionally from say \$20 to \$40 [i.e., from $c_1 x_1$ to $c_1(2x_1)$]. In many instances, it is realistic to assume constant returns to scale over the range of the data. At other times, though, due to economies of scale, profit might increase disproportionately, to say \$45; or, due to diseconomies of scale (saturation effects), profit may be only \$35. In the former case, marginal returns are increasing with the activity level, and we say that the profit function is *convex* (Fig. 13.3(a)). In the second case, marginal returns are decreasing with the activity level and we say that the profit function is *concave* (Fig. 13.3(b)). Of course, marginal returns may increase over parts of the data range and decrease elsewhere, giving functions that are neither convex nor concave (Fig. 13.3(c)).

An alternative way to view a convex function is to note that linear interpolation overestimates its values. That is, for any points y and z, the line segment joining $f(y)$ and $f(z)$ lies above the function (see Fig. 13.3). More intuitively, convex functions are "bathtub like" and hold water. Algebraically,

Definition. A function $f(x)$ is called *convex* if, for every y and z and every $0 \leq \lambda \leq 1$,

$$f[\lambda y + (1 - \lambda)z] \leq \lambda f(y) + (1 - \lambda)f(z).$$

It is called *strictly convex* if, for every two distinct points y and z and every $0 < \lambda < 1$,

$$f[\lambda y + (1 - \lambda)z] < \lambda f(y) + (1 - \lambda)f(z).$$

The lefthand side in this definition is the function evaluation on the line joining x and y; the righthand side is the linear interpolation. Strict convexity corresponds to profit functions whose marginal returns are strictly increasing.

Note that although we have pictured f above to be a function of one decision variable, this is not a restriction. If $y = (y_1, y_2, \ldots, y_n)$ and $z = (z_1, z_2, \ldots, z_n)$, we must interpret $\lambda y + (1 - \lambda)z$ only as weighting the decision variables one at a time, i.e., as the decision vector $(\lambda y_1 + (1 - \lambda)z_1, \ldots, \lambda y_n + (1 - \lambda)z_n)$.

Concave functions are simply the negative of convex functions. In this case, linear interpolation underestimates the function. The definition above is altered by reversing the direction of the inequality. Strict concavity is defined analogously. Formally,

Definition. A function $f(x)$ is called *concave* if, for every y and z and every $0 \leq \lambda \leq 1$,

$$f[\lambda y + (1 - \lambda)z] \geq \lambda f(y) + (1 - \lambda)f(z).$$

It is called *strictly concave* if, for every y and z and every $0 < \lambda < 1$,

$$f[\lambda y + (1 - \lambda)z] > \lambda f(y) + (1 - \lambda)f(z).$$

We can easily show that a linear function is both convex and concave. Consider the linear function:

$$f(x) = \sum_{j=1}^{n} c_j x_j,$$

and let $0 \leq \lambda \leq 1$. Then

$$f(\lambda y + (1 - \lambda)z) = \sum_{j=1}^{n} c_j(\lambda y_j + (1 - \lambda)z_j)$$

$$= \lambda \left[\sum_{j=1}^{n} c_j y_j \right] + (1 - \lambda)\left[\sum_{j=1}^{n} c_j z_j \right]$$

$$= \lambda f(y) + (1 - \lambda)f(z).$$

These manipulations state, quite naturally, that linear interpolation gives exact values for f and consequently, from the definitions, that a linear function is both convex and concave. This property is essential, permitting us to either maximize or minimize linear functions by computationally attractive methods such as the simplex method for linear programming.

Other examples of convex functions are x^2, x^4, e^x, e^{-x} or $-\log x$. Multiplying each example by minus one gives a concave function. The definition of convexity implies that the sum of convex functions is convex and that any nonnegative multiple of a convex function also is convex. Utilizing this fact, we can obtain a large number of convex functions met frequently in practice by combining these simple examples, giving, for instance,

$$2x^2 + e^x, \qquad e^x + 4x,$$

or

$$-3 \log x + x^4.$$

Similarly, we can easily write several concave functions by multiplying these examples by minus one.

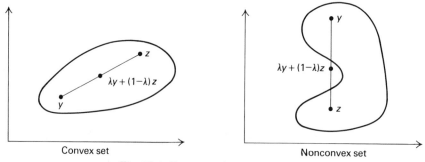

Convex set   Nonconvex set

Fig. 13.4 Convex and nonconvex sets.

A notion intimately related to convex and concave functions is that of a *convex set*. These sets are "fat," in the sense that, whenever y and z are contained in the set, every point on the line segment joining these points is also in the set (see Fig. 13.4). Formally,

> **Definition.** A set of points C is called *convex* if, for all λ in the interval $0 \leq \lambda \leq 1$, $\lambda y + (1 - \lambda)z$ is contained in C whenever x and y are contained in C.

Again we emphasize that y and z in this definition are decision vectors; in the example, each of these vectors has two components.

We have encountered convex sets frequently before, since the feasible region for a linear program is convex. In fact, the feasible region for a nonlinear program is convex if it is specified by less-than-or-equal-to equalities with convex functions. That is, if $f_i(x)$ for $i = 1, 2, \ldots, m$, are convex functions and if the points $x = y$ and $x = z$ satisfy the inequalities

$$f_i(x) \leq b_i \qquad (i = 1, 2, \ldots, m),$$

then, for any $0 \leq \lambda \leq 1$, $\quad \lambda y + (1 - \lambda)z$ is feasible also, since the inequalities

$$f_i(\lambda y + (1 - \lambda)z) \leq \lambda f_i(y) + (1 - \lambda)f_i(z) \leq \lambda b_i + (1 - \lambda)b_i = b_i$$
$$\qquad\qquad \uparrow \qquad\qquad\qquad\qquad\qquad \uparrow$$
$$\qquad \text{Convexity} \qquad\qquad\qquad \text{Feasibility of } y \text{ and } z$$

hold for every constraint. Similarly, if the constraints are specified by greater-than-or-equal-to inequalities and the functions are concave, then the feasible region is convex. In sum, for convex feasible regions we want convex functions for less-than-or-equal-to constraints and concave functions for greater-than-or-equal-to constraints. Since linear functions are both convex and concave, they may be treated as equalities.

An elegant mathematical theory, which is beyond the scope of this chapter, has been developed for convex and concave functions and convex sets. Possibly the most important property for the purposes of nonlinear programming was previewed in the previous section. Under appropriate assumptions, a local optimal can be shown to be a global optimum.

Local Minimum and Local Maximum Property

1. A local $\begin{Bmatrix} \text{minimum} \\ \text{maximum} \end{Bmatrix}$ of a $\begin{Bmatrix} \text{convex} \\ \text{concave} \end{Bmatrix}$ function on a convex feasible region is

   also a global $\begin{Bmatrix} \text{minimum} \\ \text{maximum} \end{Bmatrix}$.

2. A local $\begin{Bmatrix} \text{minimum} \\ \text{maximum} \end{Bmatrix}$ of a strictly $\begin{Bmatrix} \text{convex} \\ \text{concave} \end{Bmatrix}$ function on a convex feasible

   region is the unique global $\begin{Bmatrix} \text{minimum} \\ \text{maximum} \end{Bmatrix}$.

We can establish this property easily by reference to Fig. 13.5. The argument is for convex functions; the concave case is handled similarly. Suppose that y is a local minimum. If y is not a global minimum, then, by definition, there is a feasible point z with $f(z) < f(y)$. But then if f is convex, the function must lie on or below the dashed linear interpolation line. Thus, in any box about y, there must be an x on the line segment joining y and z, with $f(x) < f(y)$. Since the feasible region is convex, this x is feasible and we have contradicted the hypothesis that y is a local minimum. Consequently, no such point z can exist and any local minimum such as y must be a global minimum.

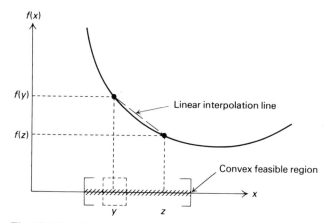

Fig. 13.5 Local minima are global minima for convex functions.

To see the second assertion, suppose that y is a local minimum. By Property 1 it is also a global minimum. If there is another global minimum z (so that $f(z) = f(y)$), then $\frac{1}{2}x + \frac{1}{2}z$ is feasible and, by the definition of strict convexity,

$$f(\tfrac{1}{2}x + \tfrac{1}{2}z) < \tfrac{1}{2}f(y) + \tfrac{1}{2}f(z) = f(y).$$

But this states that $\frac{1}{2}x + \frac{1}{2}z$ is preferred to y, contradicting our premise that y is a global minimum. Consequently, no other global minimum such as z can possibly exist; that is, y must be the unique global minimum.

13.4 PROBLEM CLASSIFICATION

Many of the nonlinear-programming solution procedures that have been developed do not solve the general problem

Maximize $f(x)$,

subject to:

$$g_i(x) \leqq b_i \qquad (i = 1, 2, \ldots, m),$$

but rather some special case. For reference, let us list some of these special cases:

1. Unconstrained optimization:

f arbitrary, $m = 0$ (no constraints).

2. Linear programming:

$$f(x) = \sum_{j=1}^{n} c_j x_j, \qquad g_i(x) = \sum_{j=1}^{n} a_{ij} x_j \qquad (i = 1, 2, \ldots, m),$$

$$g_{m+i}(x) = -x_i \qquad (i = 1, 2, \ldots, n).$$

3. Quadratic programming:

$$f(x) = \sum_{j=1}^{n} c_j x_j + \tfrac{1}{2} \sum_{i=1}^{n} \sum_{j=1}^{n} q_{ij} x_i x_j \qquad \text{(Constraints of case 2),}$$

$$(q_{ij} \text{ are given constants).}$$

4. Linear constrained problem:

$$f(x) \text{ general,} \qquad g_i(x) = \sum_{j=1}^{n} a_{ij} x_j \qquad (i = 1, 2, \ldots, m),$$

(Possibly $x_j \geqq 0$ will be included as well).

5. Separable programming:

$$f(x) = \sum_{j=1}^{n} f_j(x_j), \qquad g_i(x) = \sum_{j=1}^{n} g_{ij}(x_j) \qquad (i = 1, 2, \ldots, m);$$

i.e., the problem "separates" into functions of single variables. The functions f_j and g_{ij} are given.

6. Convex programming:

f is a concave function.   The functions g_i $(i = 1, 2, \ldots, m)$
(In a minimization problem,   are all convex.
f would be a convex function.)

Note that cases 2, 3, and 4 are successive generalizations. In fact linear programming is a special case of every other problem type except for case 1.

13.5 SEPARABLE PROGRAMMING

Our first solution procedure is for separable programs, which are optimization problems of the form:

$$\text{Maximize} \sum_{j=1}^{n} f_j(x_j),$$

subject to:

$$\sum_{j=1}^{n} g_{ij}(x_j) \leq 0 \qquad (i = 1, 2, \ldots, m),$$

where each of the functions f_j and g_{ij} is known. These problems are called separable because the decision variables appear separately, one in each function g_{ij} in the constraints and one in each function f_j in the objective function.

Separable problems arise frequently in practice, particularly for time-dependent optimization. In this case, the variable x_j usually corresponds to an activity level for time period j and the separable model assumes that the decisions affecting resource utilization and profit (or cost) are additive over time. The model also arises when optimizing over distinct geographical regions, an example being the water-resources planning formulation given in Section 13.1.

Actually, instead of solving the problem directly, we make an appropriate approximation so that linear programming can be utilized. In practice, two types of approximations, called the δ-*method* and the λ-*method*, are often used. Since we have introduced the δ-method when discussing integer programming, we consider the λ-method in this section.

The general technique is motivated easily by solving a specific example. Consider the portfolio-selection problem introduced in Section 13.1. Taking $\theta = 1$, that problem becomes:

$$\text{Maximize } f(x) = 20x_1 + 16x_2 - 2x_1^2 - x_2^2 - (x_1 + x_2)^2,$$

subject to:

$$x_1 + x_2 \leq 5,$$

$$x_1 \geq 0, \qquad x_2 \geq 0.$$

As stated, the problem is not separable, because of the term $(x_1 + x_2)^2$ in the objective function. Letting $x_3 = x_1 + x_2$, though, we can re-express it in separable form as:

$$\text{Maximize } f(x) = 20x_1 + 16x_2 - 2x_1^2 - x_2^2 - x_3^2,$$

subject to:

$$x_1 + x_2 \leq 5,$$

$$x_1 + x_2 - x_3 = 0,$$

$$x_1 \geq 0, \qquad x_2 \geq 0, \qquad x_3 \geq 0.$$

The objective function is now written as $f(x) = f_1(x_1) + f_2(x_2) + f_3(x_3)$, where

$$f_1(x_1) = 20x_1 - 2x_1^2,$$
$$f_2(x_2) = 16x_2 - x_2^2,$$

and

$$f_3(x_3) = -x_3^2.$$

Thus it is separable. Clearly, the linear constraints are also separable.

To form the approximation problem, we approximate each nonlinear term by a piecewise-linear curve, as pictured in Fig. 13.6. We have used three segments to approximate the function f_1 and two segments to approximate the functions f_2 and f_3. Note that the constraints imply that

$$x_1 \leq 5, \qquad x_2 \leq 5, \qquad \text{and} \qquad x_3 \leq 5,$$

so that we need not extend the approximation beyond these bounds on the variables.

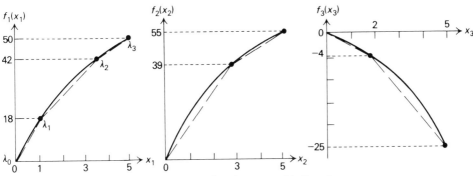

Fig. 13.6 Approximating separable functions.

The dashed approximation curves for $f_1(x_1)$, $f_2(x_2)$, and $f_3(x_3)$ are determined by linear approximation between breakpoints. For example, if $1 \leq x_1 \leq 3$, then the approximation f_1^a for f_1 is given by weighting the function's values at $x_1 = 1$ and $x_1 = 3$; that is, as

$$f_1^a(x_1) = 18\lambda_1 + 42\lambda_2,$$

where the nonnegative variables λ_1 and λ_2 express x_1 as a weighted combination of 1 and 3; thus,

$$x_1 = 1\lambda_1 + 3\lambda_2, \qquad \lambda_1 + \lambda_2 = 1.$$

For instance, evaluating the approximation at $x_1 = 1.5$ gives

$$f_1^a(1.5) = 18(\tfrac{3}{4}) + 42(\tfrac{1}{4}) = 24,$$

since

$$1.5 = 1(\tfrac{3}{4}) + 3(\tfrac{1}{4}).$$

The overall approximation curve $f_1^a(x_1)$ for $f_1(x_1)$ is expressed as:

$$f_1^a(x_1) = 0\lambda_0 + 18\lambda_1 + 42\lambda_2 + 50\lambda_3,$$

where

$$x_1 = 0\lambda_0 + 1\lambda_1 + 3\lambda_2 + 5\lambda_3, \tag{1}$$

$$\lambda_0 + \lambda_1 + \lambda_2 + \lambda_3 = 1,$$

$$\lambda_j \geq 0 \qquad (j = 1, 2, 3, 4),$$

with the provision that the λ_j variables satisfy the following restriction:

Adjacency Condition. At most two λ_j weights are positive. If two weights are positive, then they are adjacent, i.e., of the form λ_j and λ_{j+1}. A similar restriction applies to each approximation.

Figure 13.7 illustrates the need for the adjacency condition. If the weights $\lambda_0 = \frac{1}{3}$ and $\lambda_2 = \frac{2}{3}$, then the approximation (1) gives

$$f_1^a(x_1) = 0(\tfrac{1}{3}) + 42(\tfrac{2}{3}) = 28,$$

$$x_1 = 0(\tfrac{1}{3}) + 3(\tfrac{2}{3}) = 2,$$

as shown in Fig. 13.7(a) by the light curve joining λ_0 and λ_2. In contrast, at $x_1 = 2$ the approximation curve gives $f_1^a(2) = 30$.

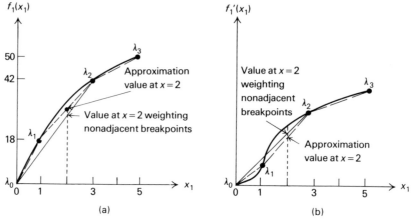

Fig. 13.7 Need for the adjacency condition.

An essential point to note here is that, for *concave* objective functions, the adjacency condition will always be enforced by the maximization and can be ignored. This property is easy to see geometrically by considering Fig 13.7(a). Suppose that the weights λ_0 and λ_2 are positive. By concavity, the function value of 18 at $x_1 = 1$ associated with the intermediate weight λ_1 lies above the line segment joining λ_0 and λ_2 in the figure. Consequently, the approximation curve must also lie above this line segment. The maximization, therefore, will select the dashed approximation curve with only the adjacent weights λ_0 and λ_1, or λ_1 and λ_2, positive, rather than any solution with both λ_0 and λ_2 positive. A similar argument applies if three or more weights are positive. For example, if λ_0, λ_2, and λ_3 are all positive, then the additional weight λ_3 can be viewed as weighting a point on the line segment joining λ_0 and λ_2 with the point at λ_3. Again, concavity implies that this point lies below the approximation curve and will not be accepted by the maximization. Note, however, that for the nonconcave function of Fig. 13.7(b), nonadjacent weights λ_0 and λ_2 are actually preferred to the approximation curve. Consequently, for nonconcave functions some effort must be expended to ensure that the adjacency condition is satisfied.

Returning to the portfolio-selection example, we can write the approximation problem:

Maximize $z = f_1^a(x_1) + f_2^a(x_2) + f_3^a(x_3)$,

subject to:

$$
\begin{aligned}
x_1 + x_2 &\leq 5, \\
x_1 + x_2 - x_3 &= 0, \\
x_1 \geq 0, \quad x_2 \geq 0, \quad x_3 &\geq 0,
\end{aligned}
$$

in terms of weighting variables λ_{ij}. Here we use the first subscript i to denote the weights to attach to variable i. The weights $\lambda_0, \lambda_1, \lambda_2$, and λ_3 used above for variable x_1 thus become $\lambda_{10}, \lambda_{11}, \lambda_{12}$, and λ_{13}. The formulation is:

·   Maximize $z =$

$$0\lambda_{10} + 18\lambda_{11} + 42\lambda_{12} + 50\lambda_{13} + 0\lambda_{20} + 39\lambda_{21} + 55\lambda_{22} - 0\lambda_{30} - 4\lambda_{31} - 25\lambda_{32},$$

subject to:

$$
\begin{aligned}
0\lambda_{10} + 1\lambda_{11} + 3\lambda_{12} + 5\lambda_{13} + 0\lambda_{20} + 3\lambda_{21} + 5\lambda_{22} &\leq 5, \\
0\lambda_{10} + 1\lambda_{11} + 3\lambda_{12} + 5\lambda_{13} + 0\lambda_{20} + 3\lambda_{21} + 5\lambda_{22} - 0\lambda_{30} - 2\lambda_{31} - 5\lambda_{32} &= 0, \\
\lambda_{10} + \lambda_{11} + \lambda_{12} + \lambda_{13} &= 1 \\
\lambda_{20} + \lambda_{21} + \lambda_{22} &= 1 \\
\lambda_{30} + \lambda_{31} + \lambda_{32} &= 1
\end{aligned}
\right\} \quad (2)
$$

$$\lambda_{ij} \geq 0, \qquad \text{for all } i \text{ and } j.$$

Since each of the functions $f_1(x_1), f_2(x_2)$, and $f_3(x_3)$ is concave, the adjacency condition can be ignored and the problem can be solved as a *linear program*. Solving by the simplex method gives an optimal objective value of 44 with $\lambda_{11} = \lambda_{12} = 0.5$, $\lambda_{21} = 1$, and $\lambda_{32} = 1$ as the positive variables in the optimal solution. The corresponding values for the original problem variables are:

$$x_1 = (0.5)(1) + (0.5)(3) = 2, \qquad x_2 = 3, \qquad \text{and} \qquad x_3 = 5.$$

This solution should be contrasted with the true solution

$$x_1 = \tfrac{7}{3}, \qquad x_2 = \tfrac{8}{3}, \qquad x_3 = 5, \qquad \text{and} \qquad f(x_1, x_2, x_3) = 46\tfrac{1}{3},$$

which we derive in Section 13.7.

Note that the approximation problem has added several λ variables and that one weighting constraint in (2) is associated with each x_j variable. Fortunately, these weighting constraints are of a special generalized upper-bounding type, which add little to computational effort and keep the number of effective constraints essentially unchanged. Thus, the technique can be applied to fairly large nonlinear programs, depending of course upon the capabilities of available linear-programming codes.

Once the approximation problem has been solved, we can obtain a better solution by introducing more breakpoints. Usually more breakpoints will be added near the optimal solution given by the original approximation.

Adding a single new breakpoint at $x_1 = 2$ leads to an improved approximation for this problem with a linear-programming objective value of 46 and

$$x_1 = 2, \qquad x_2 = 3, \qquad \text{and} \qquad x_3 = 5.$$

In this way, an approximate solution can be found as close as desired to the actual solution.

General Procedure

The general problem must be approached more carefully, since linear programming can give nonadjacent weights. The procedure is to express each variable* in terms of breakpoints, e.g., as above

$$x_1 = 0\lambda_{10} + 1\lambda_{11} + 3\lambda_{12} + 5\lambda_{13},$$

and then use these breakpoints to approximate the objective function and each constraint, giving the approximation problem:

$$\text{Maximize} \sum_{j=1}^{n} f_j^a(x_j),$$

subject to: (3)

$$\sum_{j=1}^{n} g_{ij}^a(x_j) \leq b_i \qquad (i = 1, 2, \ldots, m).$$

If each original function $f_j(x_j)$ is concave and each $g_{ij}(x_j)$ convex,[†] then the λ_{ij} version is solved as a linear program. Otherwise, the simplex method is modified to enforce the adjacency condition. A natural approach is to apply the simplex method as usual, except for a modified rule that maintains the adjacency condition at each step. The alteration is:

> **Restricted-Entry Criterion.** Use the simplex criterion, but do not introduce a λ_{ik} variable into the basis unless there is only one λ_{ij} variable currently in the basis and it is of the form $\lambda_{i, k-1}$ or $\lambda_{i, k+1}$, i.e., is adjacent to λ_{ik}.

Note that, when we use this rule, the optimal solution may contain a nonbasic variable λ_{ik} that would ordinarily be introduced into the basis by the simplex method (since its objective cost in the canonical form is positive), but is not introduced because of the restricted-entry criterion. If the simplex method would choose a variable to enter the basis that is unacceptable by the restricted-entry rule, then we choose the next best variable according to the greatest positive reduced cost.

---

* Variables that appear in the model in only a linear fashion should not be approximated and remain as x_j variables.
[†] Because the constraints are written as ($\leq$), the constraints should be convex; they should be concave for ($\geq$) inequalities. Similarly, for a minimization problem, the objective functions $f_j(x_j)$ should be convex.

An attractive feature of this procedure is that it can be obtained by making very minor modifications to any existing linear-programming computer code. As a consequence, most commercial linear-programming packages contain a variation of this separable-programming algorithm. However, the solution determined by this method in the general case can only be shown to be a local optimum to the approximation problem (3).

Inducing Separability

Nonseparable problems frequently can be reduced to a separable form by a variety of formulation tricks. A number of such transformations are summarized in Table 13.1.

Table 13.1 Representative transformations

Term	Substitution	Additional constraints	Restriction
$x_1 x_2$	$x_1 x_2 = y_1^2 - y_2^2$	$y_1 = \frac{1}{2}(x_1 + x_2)$ $y_2 = \frac{1}{2}(x_1 - x_2)$	None
$x_1 x_2$	$x_1 x_2 = y_1$	$\log y_1 = \log x_1 + \log x_2$	$x_1 > 0, \quad x_2 > 0$
$x_1^{x_2}$	$x_1^{x_2} = y_1$	$y_1 = 10^{y_2 x_2 *}$ $x_1 = 10^{y_2}$	$x_1 > 0$
$2^{x_1 + x_2^2}$	$2^{x_1 + x_2^2} = y_1$	$\log y_1 = (\log 2)(x_1 + x_2^2)$	None

* The term $y_2 x_2$ should now be separated by the first transformation, followed by an application of the last transformation to separate the resulting power-of-10 term.

To see the versatility of these transformations, suppose that the nonseparable term $x_1 x_2^2/(1 + x_3)$ appears either in the objective function or in a constraint of a problem. Letting $y_1 = 1/(1 + x_3)$, the term becomes $x_1 x_2^2 y_1$. Now if $x_1 > 0, x_1^2 > 0$, and $y_1 > 0$ over the feasible region, then letting $y_2 = x_1 x_2^2 y_1$, the original term is replaced by y_2 and the separable constraints

$$y_1 = \frac{1}{1 + x_3}$$

and

$$\log y_2 = \log x_1 + \log x_2^2 + \log y_1$$

are introduced. If the restrictions $x_1 > 0, x_1^2 > 0, y_1 > 0$ are not met, we may let $y_2 = x_1 x_2^2$, substitute $y_1 y_2$ for the original term, and append the constraints:

$$y_1 = \frac{1}{1 + x_3}, \qquad y_2 = x_1 x_2^2.$$

The nonseparable terms $y_1 y_2$ and $x_1 x_2^2$ can now be separated using the first transformation in Table 13.1 (for the last expression, let x_2^2 replace x_2 in the table).

Using the techniques illustrated by this simple example, we may, in theory, state almost any optimization problem as a separable program. Computationally, though, the approach is limited since the number of added variables and constraints may make the resulting separable program too large to be manageable.

13.6 LINEAR APPROXIMATIONS OF NONLINEAR PROGRAMS

Algebraic procedures such as pivoting are so powerful for manipulating linear equalities and inequalities that many nonlinear-programming algorithms replace the given
problem by an approximating linear problem. Separable programming is a prime
example, and also one of the most useful, of these procedures. As in separable programming, these nonlinear algorithms usually solve several linear approximations
by letting the solution of the last approximation suggest a new one.

By using different approximation schemes, this strategy can be implemented in
several ways. This section introduces three of these methods, all structured to exploit
the extraordinary computational capabilities of the simplex method and the widespread availability of its computer implementation.

There are two general schemes for approximating nonlinear programs. The last
section used linear approximation for separable problems by weighting selected values
of each function. This method is frequently referred to as *inner linearization* since,
as shown in Fig. 13.8 when applied to a convex programming problem (i.e., constraints
$g_i(x) \geq 0$ with g_i concave, or $g_i(x) \leq 0$ with g_i convex), the feasible region for the
approximating problem lies inside that of the original problem. In contrast, other
approximation schemes use slopes to approximate each function. These methods are
commonly referred to as *outer linearizations* since, for convex-programming problems,
the feasible region for the approximating problem encompasses that of the original
problem. Both approaches are illustrated further in this section.

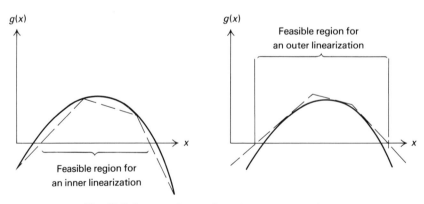

Fig. 13.8 Inner and outer linearizations of $g(x) \geq 0$.

Frank–Wolfe Algorithm

Let $x^0 = (x_1^0, x_2^0, \ldots, x_n^0)$ be any feasible solution to a nonlinear program with
linear constraints:

$$\text{Maximize } f(x_1, x_2, \ldots, x_n),$$

subject to:

$$\sum_{j=1}^{n} a_{ij}x_j \quad b_i \quad (i = 1, 2, \ldots, m),$$

$$x_j \geq 0 \quad (j = 1, 2, \ldots, n).$$

Here x^0 might be determined by phase I of the simplex method. This algorithm forms a linear approximation at the point x^0 by replacing the objective function with its current value plus a linear correction term; that is, by the linear objective

$$f(x^0) + \sum_{j=1}^{n} c_j(x_j - x_j^0),$$

where c_j is the slope, or partial derivative, of f with respect to x_j, evaluated at the point x^0. Since $f(x^0)$, c_j, and x_j^0 are fixed, maximizing this objective function is equivalent to maximizing:

$$z = \sum_{j=1}^{n} c_j x_j.$$

The linear approximation problem is solved, giving an optimal solution $y = (y_1, y_2, \ldots, y_n)$. At this point the algorithm recognizes that, although the linear approximation problem indicates that the objective improves steadily from x^0 to y, the *nonlinear* objective might not continue to improve from x^0 to y. Therefore, the algorithm uses a procedure to determine the maximum value for $f(x_1, x_2, \ldots, x_n)$ along the line segment joining x^0 to y. Special methods for performing the optimization along the line segment are discussed in Section 13.9. For now, let us assume that there is a method to accomplish this line-segment maximization for us.

Letting $x^1 = (x_1^1, x_2^1, \ldots, x_n^1)$ denote the optimal solution of the line-segment optimization, we repeat the procedure by determining a new linear approximation to the objective function with slopes c_j evaluated at x^1. Continuing in this way, we determine a sequence of points $x^1, x^2, \ldots, x^n, \ldots$; any point $x^* = (x_1^*, x_2^*, \ldots, x_n^*)$ that these points approach in the limit is an optimal solution to the original problem.

Let us illustrate the method with the portfolio-selection example from Section 13.1:

Maximize $f(x) = 20x_1 + 16x_2 - 2x_1^2 - x_2^2 - (x_1 + x_2)^2,$

subject to:

$$x_1 + x_2 \leq 5,$$

$$x_1 \geq 0, \qquad x_2 \geq 0.$$

The partial derivatives of the objective function at any point $x = (x_1, x_2)$ are given by:

$$c_1 = 20 - 4x_1 - 2(x_1 + x_2) = 20 - 6x_1 - 2x_2,$$
$$c_2 = 16 - 2x_2 - 2(x_1 + x_2) = 16 - 2x_1 - 4x_2.$$

Suppose that the initial point is $x^0 = (0, 0)$. At this point $c_1 = 20$ and $c_2 = 16$ and the linear approximation uses the objective function $20x_1 + 16x_2$. The optimal solution to this linear program is $y_1 = 5$, $y_2 = 0$, and the line-segment optimization is made on the line joining $x^0 = (0, 0)$ to $y = (5, 0)$; that is, with $0 \leq x_1 \leq 5$, $x_2 = 0$. The optimal solution can be determined to be $x_1 = 3\frac{1}{3}$, $x_2 = 0$, so that the procedure is repeated from $x^1 = (3\frac{1}{3}, 0)$.

The partial derivatives are now $c_1 = 0$, $c_2 = 9\frac{1}{3}$, and the solution to the resulting linear program of maximizing $0x_1 + 9\frac{1}{3}x_2$ is $y_1 = 0$, $y_2 = 5$. The line segment joining x^1 and y is given by

$$x_1 = \theta y_1 + (1 - \theta)x_1^1 = 3\tfrac{1}{3}(1 - \theta),$$
$$x_2 = \theta y_2 + (1 - \theta)x_2^1 = 5\theta,$$

as θ varies between 0 and 1. The optimal value of θ over the line segment is $\theta = \frac{7}{15}$, so that the next point is:

$$x_1^2 = (\tfrac{10}{3})(\tfrac{8}{15}) = 1\tfrac{7}{9} \qquad \text{and} \qquad x_2^2 = 2\tfrac{1}{3}.$$

Figure 13.9 illustrates these two steps of the algorithm and indicates the next few points x^4, x^5, and x^6 that it generates.

The Frank–Wolfe algorithm is convenient computationally because it solves linear programs with the same constraints as the original problem. Consequently, any structural properties of these constraints are available when solving the linear

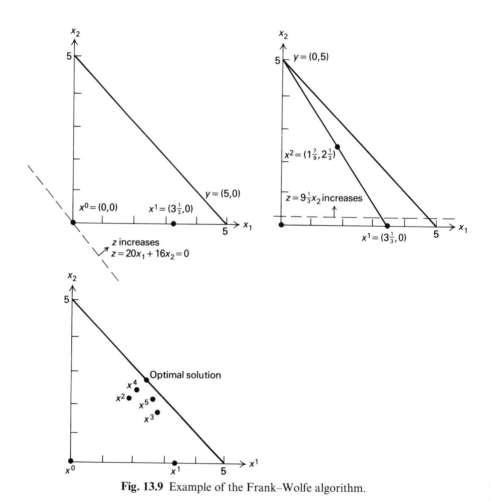

Fig. 13.9 Example of the Frank–Wolfe algorithm.

programs. In particular, network-flow techniques or large-scale system methods can be used whenever the constraints have the appropriate structure. Also, note from Fig. 13.9 that the points $x^1, x^2, x^3, \ldots$ oscillate. The even-numbered points $x^2, x^4, x^6, \ldots$ lie on one line directed toward the optimal solution, and the odd-numbered points $x^1, x^3, x^5, \ldots$, lie on another such line. This general tendency of the algorithm slows its convergence. One approach for exploiting this property to speed convergence is to make periodic optimizations along the line generated by every other point x^{k+2} and x^k for some values of k.

MAP (Method of Approximation Programming)

The Frank–Wolfe algorithm can be extended to general nonlinear programs by making linear approximations to the constraints as well as the objective function. When the constraints are highly nonlinear, however, the solution to the approximation problem can become far removed from the feasible region since the algorithm permits large moves from any candidate solution. The Method of Approximation Programming (MAP) is a simple modification of this approach that limits the size of any move. As a result, it is sometimes referred to as a *small-step procedure*.

Let $x^0 = (x^0_1, x^0_2, \ldots, x^0_n)$ be any candidate solution to the optimization problem:

Maximize $f(x_1, x_2, \ldots, x_n)$,

subject to:

$$g_i(x_1, x_2, \ldots, x_n) \leq 0 \qquad (i = 1, 2, \ldots, m).$$

Each constraint can be linearized, using its current value $g_i(x^0)$ plus a linear correction term, as:

$$\hat{g}_i(x) = g_i(x^0) + \sum_{j=1}^{n} a_{ij}(x_j - x^0_j) \leq 0,$$

where a_{ij} is the partial derivative of constraint g_i with respect to variable x_j evaluated at the point x^0. This approximation is a linear inequality, which can be written as:

$$\sum_{j=1}^{n} a_{ij}x_j \leq b^0_i \equiv \sum_{j=1}^{n} a_{ij}x^0_j - g_i(x^0),$$

since the terms on the righthand side are all constants.

The MAP algorithm uses these approximations, together with the linear objective-function approximation, and solves the linear-programming problem:

Maximize $z = \sum_{j=1}^{n} c_j x_j$,

subject to: (4)

$$\sum_{j=1}^{n} a_{ij}x_j \leq b^0_i \qquad (i = 1, 2, \ldots, m),$$

$$x^0_j - \delta_j \leq x_j \leq x^0_j + \delta_j \qquad (j = 1, 2, \ldots, n).$$

The last constraints restrict the step size; they specify that the value for x_j can vary from x_j^0 by no more than the user-specified constant δ_j.

When the parameters δ_j are selected to be small, the solution to this linear program is not far removed from x^0. We might expect then that the additional work required by the line-segment optimization of the Frank–Wolfe algorithm is not worth the slightly improved solution that it provides. MAP operates on this premise, taking the solution to the linear program (4) as the new point x^1. The partial-derivative data a_{ij}, b_i, and c_j is recalculated at x^1, and the procedure is repeated. Continuing in this manner determines points $x^1, x^2, \ldots, x^k, \ldots$ and as in the Frank–Wolfe procedure, any point $x^* = (x_1^*, x_2^*, \ldots, x_n^*)$ that these points approach in the limit is considered a solution.

Steps of the MAP Algorithm

STEP (0): Let $x^0 = (x_1^0, x_2^0, \ldots, x_n^0)$ be any candidate solution, usually selected to be feasible or near-feasible. Set $k = 0$.

STEP (1): Calculate c_j and a_{ij} $(i = 1, 2, \ldots, m)$, the partial derivatives of the objective function and constraints evaluated at $x^k = (x_1^k, x_2^k, \ldots, x_n^k)$. Let $b_i^k = a_{ij}x^k - g_i(x^k)$.

STEP (3): Solve the linear-approximation problem (4) with b_i^k and x_j^k replacing b_i^0 and x_j^0, respectively. Let $x^{k+1} = (x_1^{k+1}, x_2^{k+1}, \ldots, x_n^{k+1})$ be its optimal solution. Increment k to $k + 1$ and return to STEP 1.

Since many of the constraints in the linear approximation merely specify upper and lower bounds on the decision variables x_j, the bounded-variable version of the simplex method is employed in its solution. Also, usually the constants δ_j are reduced as the algorithm proceeds. There are many ways to implement this idea. One method used frequently in practice, is to reduce each δ_j by between 30 and 50 percent at each iteration.

To illustrate the MAP algorithm, consider the problem:

Maximize $f(x) = [(x_1 - 1)^2 + x_2^2]$,

subject to:

$$g_1(x) = \quad x_1^2 + 6x_2 - 36 \ \leqq \ 0,$$
$$g_2(x) = -4x_1 + \quad x_2^2 - 2x_2 \leqq 0,$$
$$x_1 \geqq 0, \qquad x_2 \geqq 0.$$

The partial derivatives evaluated at the point $x = (x_1, x_2)$ are given by:

$$c_1 = 2x_1 - 2, \qquad c_2 = 2x_2,$$
$$a_{11} = 2x_1, \qquad a_{12} = 6,$$
$$a_{21} = -4, \qquad a_{22} = 2x_2 - 2.$$

Since linear approximations of any linear function gives that function again, no data needs to be calculated for the linear constraints $x_1 \geqq 0$ and $x_2 \geqq 0$.

Using these relationships and initiating the procedure at $x^0 = (0, 2)$ with $\delta_1 = \delta_2 = 2$ gives the linear-approximation problem:

Maximize $z = -2x_1 + 4x_2$,

subject to:

$$0x_1 + 6x_2 \leq \quad 0(0) + 6(2) - (-24) = 36,$$
$$-4x_1 + 2x_2 \leq -4(0) + 2(2) - \quad 0 \quad = \quad 4.$$
$$-2 \leq x_1 \leq 2, \qquad 0 \leq x_2 \leq 4.$$

The righthand sides are determined as above by evaluating $a_{i1}x_1^0 + a_{i2}x_2^0 - g_i(x^0)$.

The feasible region and this linear approximation are depicted in Fig. 13.10. Geometrically, we see that the optimal solution occurs at $x_1^1 = 1$, $x_2^1 = 4$. Using this

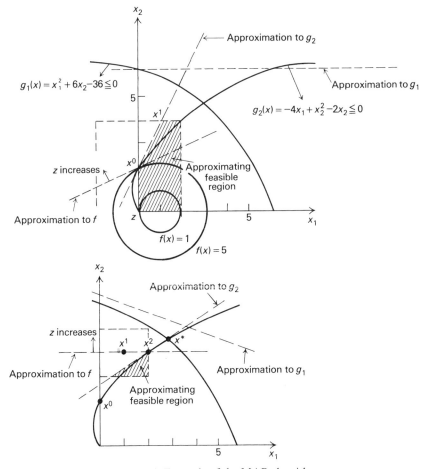

Fig. 13.10 Example of the MAP algorithm.

point and reducing both δ_1 and δ_2 to 1 generates the new approximation:

Maximize $z = \quad 0x_1 + 8x_2,$

subject to:

$$2x_1 + 6x_2 \leq \quad 2(1) + 6(4) - (-11) = 37,$$
$$-4x_1 + 6x_2 \leq -4(1) + 6(4) - \quad (4) \quad = 16,$$

$$0 \leq x_1 \leq 2, \qquad 3 \leq x_2 \leq 5.$$

The solution indicated in Fig. 13.10 occurs at $x_1^2 = 2$, $x_2^2 = 4$.

If the procedure is continued, the points $x^3, x^4, \ldots$ that it generates approach the optimal solution x^* shown in Fig. 13.10. As a final note, let us observe that the solution x^1 is not feasible for the linear program that was constructed by making linear approximations at x^1. Thus, in general, both Phase I and Phase II of the simplex method may be required to solve each linear-programming approximation.

Generalized Programming

This algorithm is applied to the optimization problem of selecting decision variables $x_1, x_2, \ldots, x_n$ from a region C to:

Maximize $f(x_1, x_2, \ldots, x_n),$

subject to:

$$g_i(x_1, x_2, \ldots, x_n) \leq 0 \qquad (i = 1, 2, \ldots, m).$$

The procedure uses inner linearization and extends the decomposition and column-generation algorithms introduced in Chapter 12. When the objective function and constraints g_i are linear and C consists of the feasible points for a linear-programming subproblem, generalized programming reduces to the decomposition method.

As in decomposition, the algorithm starts with k candidate solutions $x^j = (x_1^j, x_2^j, \ldots, x_n^j)$ for $j = 1, 2, \ldots, k$, all lying in the region C. Weighting these points by $\lambda_1, \lambda_2, \ldots, \lambda_k$ generates the candidate solution:

$$x_i = \lambda_1 x_i^1 + \lambda_2 x_i^2 + \cdots + \lambda_k x_i^k \qquad (i = 1, 2, \ldots, n). \tag{5}$$

Any choices can be made for the weights as long as they are nonnegative and sum to one. The "best" choice is determined by solving the linear-approximation problem in the variables $\lambda_1, \lambda_2, \ldots, \lambda_k$:

Maximize $\lambda_1 f(x^1) + \lambda_2 f(x^2) + \cdots + \lambda_k f(x^k),$

subject to:

$$\lambda_1 g_1(x^1) + \lambda_2 g_1(x^2) + \cdots + \lambda_k g_1(x^k) \leq 0,$$

Optimal shadow prices

$$y_1$$

$$\vdots \qquad\qquad\qquad\qquad \vdots \qquad \vdots \qquad \vdots$$

$$\lambda_1 g_m(x^1) + \lambda_2 g_m(x^2) + \cdots + \lambda_k g_m(x^k) \leq 0,$$

$$y_m$$

$$\lambda_1 \qquad + \lambda_2 \qquad\qquad + \lambda_k \qquad = 1.$$

$$\sigma$$

$$\tag{6}$$

$$\lambda_j \geq 0 \qquad (j = 1, 2, \ldots, k).$$

The coefficients $f(x^j)$ and $g_i(x^j)$ of the weights λ_j in this linear program are fixed by our choice of the candidate solutions $x^1, x^2, \ldots, x^k$. In this problem, the original objective function and constraints have been replaced by linear approximations. When x is determined from expression (5) by weighting the points $x^1, x^2, \ldots, x^k$ by the solution of problem (6), the linear approximations at x are given by applying the same weights to the objective function and constraints evaluated at these points; that is,

$$f^a(x) = \lambda_1 f(x^1) + \lambda_2 f(x^2) + \cdots + \lambda_k f(x^k)$$

and

$$g_i^a(x) = \lambda_1 g_i(x^1) + \lambda_2 g_i(x^2) + \cdots + \lambda_k g_i(x^k).$$

The approximation is refined by applying column generation to this linear program. In this case, a new column with coefficients $f(x^{k+1}), g_1(x^{k+1}), \ldots, g_m(x^{k+1})$ is determined by the pricing-out operation:

$$v = \text{Max} \left[f(x) - y_1 g_1(x) - y_2 g_2(x) - \cdots - y_m g_m(x) \right],$$

subject to:

$$x \in C.$$

This itself is a nonlinear programming problem but without the g_i constraints. If $v - \sigma \leq 0$, then no new column improves the linear program and the procedure terminates. If $v - \sigma > 0$, then the solution x^{k+1} giving v determines a new column to be added to the linear program (6) and the procedure continues.

The optimal weights $\lambda_1^*, \lambda_2^*, \ldots, \lambda_k^*$ to the linear program provide a candidate solution

$$x_i^* = \lambda_1^* x_i^1 + \lambda_2^* x_i^2 + \cdots + \lambda_k^* x_i^k$$

to the original optimization problem. This candidate solution is most useful when C is a convex set and each constraint is convex, for then x_i^* is a weighted combination of points $x^1, x^2, \ldots, x^k$ in C and thus belongs to C; and, since linear interpolation overestimates convex functions,

$$g_i(x^*) \leq \lambda_1^* g_i(x^1) + \lambda_2^* g_i(x^2) + \cdots + \lambda_k^* g_i(x^k).$$

The righthand side is less than or equal to zero by the linear-programming constraints; hence, $g_i(x^*) \leq 0$ and x^* is a feasible solution to the original problem.

As an illustration of the method, consider the nonlinear program:

$$\text{Maximize } f(x) = x_1 - (x_2 - 5)^2 + 9,$$

subject to:

$$g(x) = x_1^2 + x_2^2 - 16 \leq 0,$$

$$x_1 \geq 0, \qquad x_2 \geq 0.$$

We let C be the region with $x_1 \geq 0$ and $x_2 \geq 0$, and start with the three points $x^1 = (0, 0)$, $x^2 = (5, 0)$, and $x^3 = (0, 5)$ from C. The resulting linear approximation problem:

$$\text{Maximize } z = -16\lambda_1 - 11\lambda_2 + 9\lambda_3,$$

subject to:

$$-16\lambda_1 + 9\lambda_2 + 9\lambda_3 \leqq 0,$$
$$\lambda_1 + \lambda_2 + \lambda_3 = 1,$$
$$\lambda_1 \geqq 0, \qquad \lambda_2 \geqq 0, \qquad \lambda_3 \geqq 0,$$

Optimal
shadow
prices

$$y = 1$$
$$\sigma = 0$$

is sketched in Fig. 13.11, together with the original problem. The feasible region for the approximation problem is given by plotting the points defined by (5) that correspond to feasible weights in this linear program.

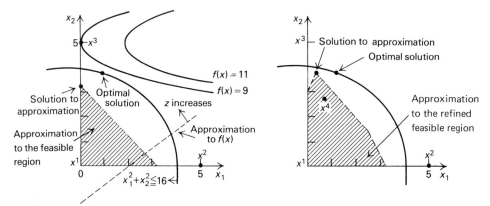

Fig. 13.11 An example of generalized programming.

The solution to the linear program is:

$$\lambda_1^* = 0.36, \qquad \lambda_2^* = 0, \qquad \text{and} \qquad \lambda_3^* = 0.64,$$

or

$$x^* = 0.36(0, 0) + 0.64(0, 5) = (0, 3.2).$$

The nonlinear programming subproblem is:

$$\text{Maximize } [f(x) - yg(x)] = x_1 - (x_2 - 5)^2 - yx_1^2 - yx_2^2 + 16y + 9,$$

subject to:

$$x_1 \geqq 0, \qquad x_2 \geqq 0.$$

For $y > 0$ the solution to this problem can be shown to be $x_1 = 1/(2y)$ and $x_2 = 5/(1 + y)$, by setting the partial derivatives of $f(x) - yg(x)$, with respect to x_1 and x_2, equal to zero.

In particular, the optimal shadow price $y = 1$ gives the new point $x^4 = (\frac{1}{2}, 2\frac{1}{2})$. Since $f(x^4) = 3\frac{1}{4}$ and $g(x^4) = -9\frac{1}{2}$, the updated linear programming approximation is:

$$\text{Maximize } z = -16\lambda_1 - 11\lambda_2 + 9\lambda_3 + 3\tfrac{1}{4}\lambda_4,$$

subject to:

$$-16\lambda_1 + 9\lambda_2 + 9\lambda_3 - 9\tfrac{1}{2}\lambda_4 \leqq 0, \qquad y = \tfrac{23}{74}$$
$$\lambda_1 + \lambda_2 + \lambda_3 + \lambda_4 = 1, \qquad \sigma = \tfrac{459}{74}$$
$$\lambda_j \geqq 0 \qquad (j = 1, 2, 3, 4).$$

The optimal solution has λ_3 and λ_4 basic with $\lambda_3^* = \tfrac{19}{37}$, $\lambda_4^* = \tfrac{18}{37}$. The corresponding value for x is:

$$x^* = \lambda_3^*(0, 5) + \lambda_4^*(\tfrac{1}{2}, 2\tfrac{1}{2})$$
$$= \tfrac{19}{37}(0, 5) + \tfrac{18}{37}(\tfrac{1}{2}, 2\tfrac{1}{2}) = (\tfrac{9}{37}, 3\tfrac{29}{37}).$$

As we continue in this way, the approximate solutions will approach the optimal solution $x_1 = 1.460$ and $x_2 = 3.724$ to the original problem.

We should emphasize that the generalized programming is unlike decomposition for linear programs in that it does not necessarily determine the optimal solution in a finite number of steps. This is true because nonlinearity does not permit a finite number of extreme points to completely characterize solutions to the subproblem.

13.7 QUADRATIC PROGRAMMING

Quadratic programming concerns the maximization of a quadratic objective function subject to linear constraints, i.e., the problem:

$$\text{Maximize } f(x) = \sum_{j=1}^{n} c_j x_j + \frac{1}{2} \sum_{j=1}^{n} \sum_{k=1}^{n} q_{jk} x_j x_k,$$

subject to:

$$\sum_{j=1}^{n} a_{ij} x_j \leqq b_i \qquad (i = 1, 2, \ldots, m),$$
$$x_j \geqq 0 \qquad (j = 1, 2, \ldots, n).$$

The data c_j, a_{ij}, and b_i are assumed to be known, as are the additional q_{jk} data. We also assume the symmetry condition $q_{jk} = q_{kj}$. This condition is really no restriction, since q_{jk} can be replaced by $\tfrac{1}{2}(q_{jk} + q_{kj})$. The symmetry condition is then met, and a straightforward calculation shows that the old and new q_{jk} coefficients give the same quadratic contribution to the objective function. The factor $\tfrac{1}{2}$ has been included to simplify later calculations; if desired, it can be incorporated into the q_{jk} coefficients. Note that, if every $q_{jk} = 0$, then the problem reduces to a linear program.

The first and third examples of Section 13.1 show that the quadratic-programming model arises in constrained regression and portfolio selection. Additionally, the model is frequently applied to approximate problems of maximizing general functions subject to linear constraints.

In Chapter 4 we developed the optimality conditions for linear programming. Now we will indicate the analogous results for quadratic programming. The motivating idea is to note that, for a linear objective function

$$f(x) = \sum_{j=1}^{n} c_j x_j,$$

the derivative (i.e., the slope or marginal return) of f with respect to x_j is given by c_j, whereas, for a quadratic program, the slope at a point is given by

$$c_j + \sum_{k=1}^{n} q_{jk} x_k.$$

The quadratic optimality conditions are then stated by replacing every c_j in the linear-programming optimality conditions by $c_j + \sum_{k=1}^{n} q_{jk} x_k$. (See Tableau 1.)

Tableau 1 Optimality conditions

	Linear program	Quadratic program
Primal feasibility	$\sum_{j=1}^{n} a_{ij} x_j \leqq b_i,$ $x_j \geqq 0$	$\sum_{j=1}^{n} a_{ij} x_j \leqq b_i,$ $x_j \geqq 0$
Dual feasibility	$\sum_{i=1}^{m} y_i a_{ij} \geqq c_j,$ $y_i \geqq 0$	$\sum_{i=1}^{m} y_i a_{ij} \geqq c_j + \sum_{k=1}^{n} q_{jk} x_k,$ $y_i \geqq 0$
Complementary slackness	$y_i \left[b_i - \sum_{j=1}^{n} a_{ij} x_j \right] = 0,$ $\left[\sum_{i=1}^{m} y_i a_{ij} - c_j \right] x_j = 0$	$y_i \left[b_i - \sum_{j=1}^{n} a_{ij} x_j \right] = 0,$ $\left[\sum_{i=1}^{m} y_i a_{ij} - c_j - \sum_{k=1}^{n} q_{jk} x_k \right] x_j = 0$

Note that the primal and dual feasibility conditions for the quadratic program are linear inequalities in nonnegative variables x_j and y_i. As such, they can be solved by the Phase I simplex method. A simple modification to that method will permit the complementary-slackness conditions to be maintained as well. To discuss the modification, let us introduce slack variables s_i for the primal constraints and surplus variables v_j for the dual constraints; that is,

$$s_i = b_i - \sum_{j=1}^{n} a_{ij} x_j,$$

$$v_j = \sum_{i=1}^{m} y_i a_{ij} - c_j - \sum_{k=1}^{n} q_{jk} x_k.$$

Then the complementary slackness conditions become:

$$y_i s_i = 0 \qquad (i = 1, 2, \ldots, m),$$

and

$$v_j x_j = 0 \qquad (j = 1, 2, \ldots, n).$$

The variables y_i and s_i are called *complementary*, as are the variables v_j and x_j. With this notation, the technique is to solve the primal and dual conditions by the Phase I simplex method, but not to allow any complementary pair of variables to appear in the basis both at the same time. More formally, the Phase I simplex method is modified by the:

Restricted-Entry Rule. Never introduce a variable into the basis if its complementary variable is already a member of the basis, even if the usual simplex criterion says to introduce the variable.

Otherwise, the Phase I procedure is applied as usual. If the Phase I procedure would choose a variable to enter the basis that is unacceptable by the restricted-entry rule, then we choose the next best variable according to the greatest positive reduced cost in the Phase I objective function.

An example should illustrate the technique. Again, the portfolio-selection problem of Section 13.1 will be solved with $\theta = 1$; that is,

$$\text{Maximize } f(x) = 20x_1 + 16x_2 - 2x_1^2 - x_2^2 - (x_1 + x_2)^2,$$

subject to:

$$x_1 + x_2 \leq 5,$$

$$x_1 \geq 0, \qquad x_2 \geq 0.$$

Expanding $(x_1 + x_2)^2$ as $x_1^2 + 2x_1 x_2 + x_2^2$ and incorporating the factor $\frac{1}{2}$, we rewrite the objective function as:

$$\text{Maximize } f(x) = 20x_1 + 16x_2 + \tfrac{1}{2}(-6x_1 x_1 - 4x_2 x_2 - 2x_1 x_2 - 2x_2 x_1),$$

so that $q_{11} = -6, q_{12} = -2, q_{21} = -2, q_{22} = -4$, and the optimality conditions are:

$$s_1 = 5 - x_1 - x_2, \qquad\qquad\qquad\qquad \text{(Primal constraint)}$$

$$\left.\begin{aligned} v_1 &= y_1 - 20 + 6x_1 + 2x_2 \\ v_2 &= y_1 - 16 + 2x_1 + 4x_2 \end{aligned}\right\} \qquad\qquad \text{(Dual constraints)}$$

$$x_1, \quad x_2, \quad s_1, \quad v_1, \quad v_2, \quad y_1 \geq 0, \qquad\qquad \text{(Nonnegativity)}$$

$$y_1 s_1 = 0, \qquad v_1 x_1 = 0, \qquad v_2 x_2 = 0. \qquad\qquad \text{(Complementary slackness)}$$

Letting s_1 be the basic variable isolated in the first constraint and adding artificial variables a_1 and a_2 in the second and third constraints, the Phase I problem is solved in Table 13.2.

Table 13.2 Solving a quadratic program.

Basic variables	Current values	x_1	x_2	s_1	$y_1{}^*$	v_1	v_2	a_1	a_2
s_1	5	1	1	1	0	0	0		
a_1	20	⑥	2		1	-1	0	1	
a_2	16	2	4		1	0	-1		1
$(-w)$	36	8	6		2	-1	-1		

Basic variables	Current values	x_1	x_2	s_1	$y_1{}^*$	$v_1{}^*$	v_2	a_1	a_2
s_1	$\frac{5}{3}$		⬭$\frac{2}{3}$	1	$-\frac{1}{6}$	$\frac{1}{6}$	0	$-\frac{1}{6}$	
x_1	$\frac{10}{3}$	1	$\frac{1}{3}$		$\frac{1}{6}$	$-\frac{1}{6}$	0	$\frac{1}{6}$	
a_2	$\frac{28}{3}$		$\frac{10}{3}$		$\frac{2}{3}$	$\frac{1}{3}$	-1	$-\frac{1}{3}$	1
$(-w)$	$\frac{28}{3}$		$\frac{10}{3}$		$\frac{2}{3}$	$\frac{1}{3}$	0	$-\frac{4}{3}$	

Basic variables	Current values	x_1	x_2	s_1	y_1	$v_1{}^*$	$v_2{}^*$	a_1	a_2
x_2	$\frac{5}{2}$		1	$\frac{3}{2}$	$-\frac{1}{4}$	$\frac{1}{4}$	0	$-\frac{1}{4}$	
x_1	$\frac{5}{2}$	1		$-\frac{1}{2}$	$\frac{1}{4}$	$-\frac{1}{4}$	0	$\frac{1}{4}$	
a_2	1			-5	⬭$\frac{3}{2}$	$-\frac{1}{2}$	-1	$\frac{1}{2}$	1
$(-w)$	1			-5	$\frac{3}{2}$	$-\frac{1}{2}$	0	$-\frac{1}{2}$	

Basic variables	Current values	x_1	x_2	$s_1{}^*$	y_1	$v_1{}^*$	$v_2{}^*$	a_1	a_2
x_2	$\frac{8}{3}$		1	$\frac{2}{3}$		$\frac{1}{6}$	$-\frac{1}{6}$	$-\frac{1}{6}$	$\frac{1}{6}$
x_1	$\frac{7}{3}$	1		$\frac{1}{3}$		$-\frac{1}{6}$	$\frac{1}{6}$	$\frac{1}{6}$	$-\frac{1}{6}$
y_1	$\frac{2}{3}$			$\frac{10}{3}$	1	$\frac{1}{3}$	$-\frac{2}{3}$	$\frac{1}{3}$	$\frac{2}{3}$
$(-w)$	0			0		0	1	-1	-1

* Starred variables cannot be introduced into the basis since their complementary variable is in the basis.

For this problem, the restricted-entry variant of the simplex Phase I procedure has provided the optimal solution. It should be noted, however, that the algorithm will *not* solve every quadratic program. As an example, consider the problem:

$$\text{Maximize } f(x_1, x_2) = \tfrac{1}{4}(x_1 - x_2)^2,$$

subject to :

$$x_1 + x_2 \leq 5,$$

$$x_1 \geq 0, \qquad x_2 \geq 0.$$

As the reader can verify, the algorithm gives the solution $x_1 = x_2 = 0$. But this solution is not even a local optimum, since increasing either x_1 or x_2 increases the objective value.

It can be shown, however, that the algorithm does determine an optimal solution if $f(x)$ is strictly concave for a maximization problem or strictly convex for a minimization problem. For the quadratic problem, $f(x)$ is strictly concave whenever

$$\sum_{i=1}^{n} \sum_{j=1}^{n} \alpha_i q_{ij} \alpha_j < 0 \qquad \text{for every choice of } \alpha_1, \alpha_2, \ldots, \alpha_n,$$

such that some $\alpha_j \neq 0$. In this case, the matrix of coefficients (q_{ij}) is called *negative definite*. Thus the algorithm will always work for a number of important applications including the least-square regression and portfolio-selection problems introduced previously.

13.8* UNCONSTRAINED MINIMIZATION AND SUMT

Conceptually, the simplest type of optimization is unconstrained. Powerful solution techniques have been developed for solving these problems, which are based primarily upon calculus, rather than upon algebra and pivoting, as in the simplex method. Because the linear-programming methods and unconstrained-optimization techniques are so efficient, both have been used as the point of departure for constructing more general-purpose nonlinear-programming algorithms. The previous sections have indicated some of the algorithms using the linear-programming-based approach. This section briefly indicates the nature of the unconstrained-optimization approaches by introducing algorithms for unconstrained maximization and showing how they might be used for problems with constraints.

Unconstrained Minimization

Suppose that we want to maximize the function $f(x)$ of n decision variables $x = (x_1, x_2, \ldots, x_n)$ and that this function is differentiable. Let $\partial f/\partial x_j$ denote the partial derivative of f with respect to x_j, defined by

$$\frac{\partial f}{\partial x_j} = \lim_{\theta \to 0} \frac{f(x + \theta u_j) - f(x)}{\theta}, \qquad (7)$$

where u_j is a decision vector $u_j = (0, 0, \ldots, 0, 1, 0, \ldots, 0)$, with all zeros except for the jth component, which is 1. Thus, $x + \theta u_j$ corresponds to the decisions $(x_1, x_2, \ldots, x_{j-1}, x_j + \theta, x_{j+1}, \ldots, x_n)$, in which only the jth decision variable is changed from the decision vector x.

---

* This section requires some knowledge of differential calculus.

Note that if $\partial f/\partial x_j > 0$, then

$$\frac{f(x + \theta u_j) - f(x)}{\theta} > 0 \qquad \text{for } \theta > 0 \text{ small enough,}$$

and therefore $x + \theta u_j$ is preferred to x for a maximization problem. Similarly, if $\partial f/\partial x_j < 0$, then

$$\frac{f(x + \theta u_j) - f(x)}{\theta} < 0 \qquad \text{for } \theta < 0 \text{ small enough,}$$

and $x + \theta u_j$ is again preferred to x. Therefore, at any given point with at least one partial derivative $\partial f/\partial x_j > 0$, we can improve the value of f by making a one-dimensional search

$$\underset{\theta \geq 0}{\text{Maximize}} \, f(x_1, x_2, \ldots, x_{j-1}, x_j + \theta, x_{j+1}, \ldots, x_n),$$

in which only the value of x_j is varied. Similarly, if $\partial f/\partial x_j < 0$, we can alter x_j to $x_j - \theta$ and search on $\theta \geq 0$ to improve the function's value. The one-dimensional search can be made by using any of a number of procedures that are discussed in the next section. One algorithm using this idea, called *cyclic coordinate ascent*, searches by optimizing in turn with respect to x_1, then x_2, then x_3 and so on, holding all other variables constant at stage j when optimizing with respect to variable x_j. After optimizing with x_n, the method starts over again with x_1.

In fact, there is a large class of algorithms known as *ascent algorithms* that implement the "uphill movement" philosophy of cyclic coordinate ascent. Suppose that we consider moving in a general direction $d = (d_1, d_2, \ldots, d_n)$ instead of in a coordinate, or unit, direction. Then, instead of considering the partial derivative, as in (7), we consider the *directional* derivative. The directional derivative, which indicates how the function varies as we move away from x in the direction d, is defined by:

$$\lim_{\theta \to 0} \frac{f(x + \theta d) - f(x)}{\theta} = \frac{\partial f}{\partial x_1} d_1 + \frac{\partial f}{\partial x_2} d_2 + \cdots + \frac{\partial f}{\partial x_n} d_n. \tag{8}$$

The directional derivative is just the slope of the function $f(x)$ in the direction d and reduces to the definition of the partial derivative $\partial f/\partial x_j$ in Eq. (7) when the direction is taken to be $d = u_j$. Just as in the case of partial derivatives, if the directional derivative in Eq. (8) is positive, then f increases in the direction d; that is,

$$f(x_1 + \theta d_1, x_2 + \theta d_2, \ldots, x_n + \theta d_n) > f(x_1, x_2, \ldots, x_n)$$

for $\theta > 0$ small enough. At any given point x, the ascent algorithms choose an increasing direction d (i.e., such that Eq. (8) is positive), and then select the next point $\bar{x}_i = x_i + \bar{\theta} d_i$ as the solution $\theta = \bar{\theta}$ to the one-dimensional problem

$$\underset{\theta \geq 0}{\text{Maximize}} \, f(x_1 + \theta d_1, x_2 + \theta d_2, \ldots, x_n + \theta d_n).$$

From $\bar{x}$, the ascent algorithms select a new direction and solve another one-dimensional problem, and then continue by repeating this procedure. Cyclic coordinate ascent is a special case in which, at each step, all but one d_j, say d_i, is set to zero; d_i is set to $d_i = +1$ if $\partial f/\partial x_i > 0$ and $d_i = -1$ if $\partial f/\partial x_i < 0$.

One natural choice for the direction d for this general class of algorithms is:

$$d_1 = \frac{\partial f}{\partial x_1}, \qquad d_2 = \frac{\partial f}{\partial x_2}, \qquad \dots, \qquad d_n = \frac{\partial f}{\partial x_n},$$

since then Eq. (8) is positive as long as $\partial f/\partial x_j \neq 0$ for some variable x_j. This choice for d is known as *Cauchy's method*, or *steepest ascent*.

To illustrate the ascent algorithms, consider the objective function

$$f(x_1, x_2) = 20x_1 + 16x_2 - 2x_1^2 - x_2^2 - (x_1 + x_2)^2$$

introduced earlier in this chapter for a portfolio-selection problem. The partial derivatives of this function are:

$$\frac{\partial f}{\partial x_1} = 20 - 6x_1 - 2x_2 \qquad \text{and} \qquad \frac{\partial f}{\partial x_2} = 16 - 2x_1 - 4x_2.$$

Figure 13.12 shows the result of applying the first few iterations of both cyclic coordinate ascent and steepest ascent for maximizing this function, starting at the point $x_1 = x_2 = 0$.

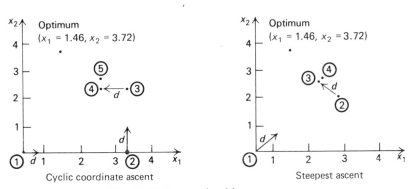

Fig. 13.12 Ascent algorithms.

Cyclic coordinate ascent first increases x_1 from the starting point $x_1 = x_2 = 0$ since $\partial f/\partial x_1 = 20 > 0$. Holding $x_2 = 0$ and setting $\partial f/\partial x_1 = 0$ gives the solution to the first one-dimensional problem at ② as $x_1 = 3\frac{1}{3}$ and $x_2 = 0$. At this point $\partial f/\partial x_2 = 9\frac{1}{3} > 0$ and we increase x_2, holding $x_1 = 3\frac{1}{3}$. The optimum to this one-dimensional problem occurs at point ③ with $\partial f/\partial x_2 = 0$ and $x_1 = 3\frac{1}{3}$, $x_2 = 2\frac{1}{3}$. We now decrease x_1 since $\partial f/\partial x_1 = -4\frac{2}{3}$, generating the next point $x_1 = 2.56$ and $x_2 = 2\frac{1}{3}$. The last point ⑤ shown has $x_1 = 2.56$ and $x_2 = 2.72$. Continuing in this way from this point, the algorithm approaches the optimal solution $x_1 = 2.4$ and $x_2 = 2.8$.

Since the partial derivatives at $x_1 = x_2 = 0$ are $\partial f/\partial x_1 = 20$ and $\partial f/\partial x_2 = 16$, the first direction for steepest ascent is $d_1 = 20$ and $d_2 = 16$, giving the one-dimensional problem:

$$\underset{\theta \geq 0}{\text{Maximize }} f(0 + 20\theta, 0 + 16\theta) = 20(20\theta) + 16(16\theta) - 2(20\theta)^2 - (16\theta)^2$$
$$- (20\theta + 16\theta)^2 = 656\theta - 2352\theta^2.$$

Setting the derivative with respect to θ equal to zero gives the optimal solution

$$\bar{\theta} = \frac{656}{2(2352)} = 0.1395$$

and the next point

$$\bar{x}_1 = 0 + 20\bar{\theta} = 2.79 \qquad \text{and} \qquad x_2 = 0 + 16\bar{\theta} = 2.23.$$

At this point, the partial derivatives of f are

$$\frac{\partial f}{\partial x_1} = 20 - 6x_1 - 2x_2 = 20 - 6(2.79) - 2(2.23) = -1.2,$$

and

$$\frac{\partial f}{\partial x_2} = 16 - 2x_1 - 4x_2 = 16 - 2(2.79) - 4(2.23) = 1.5.$$

Therefore, the next one-dimensional optimization is in the direction $d_1 = -1.2$ and $d_2 = 1.5$ from the last point; that is,

$$\underset{\theta \geq 0}{\text{Maximize }} f(2.79 - 1.2\theta, 2.23 + 1.5\theta).$$

The optimal choice for θ is $\bar{\theta} = 0.354$, giving

$$x_1 = 2.79 - 1.2(0.354) = 2.37 \qquad \text{and} \qquad x_2 = 2.23 + 1.5(0.354) = 2.76$$

as the next point. Evaluating derivatives gives the direction $d_1 = 0.26$ and $d_2 = 0.22$ and the point ④ with $x_1 = 2.40$ and $x_2 = 2.79$, which is practically the optimal solution $x_1 = 2.4$ and $x_2 = 2.8$.

As we have noted, if any partial derivative is nonzero, then the current solution can be improved by an ascent algorithm. Therefore any optimal solution must satisfy the (*first order*) *optimality conditions*

$$\frac{\partial f}{\partial x_j} = 0 \qquad \text{for } j = 1, 2, \ldots, n. \tag{9}$$

As Fig. 13.13 illustrates, these conditions can be satisfied for nonoptimal points as well. Nevertheless, solving the system (9) permits us to generate potential solutions. In fact, we have used this observation above by setting $\partial f/\partial \theta$ to zero to find the solution to the one-dimensional problems. Usually, though, we cannot easily solve for a point that gives zero derivatives, and must rely on numerical methods such as those given in the next section.

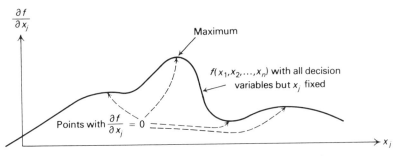

Fig. 13.13 Partial derivatives are zero at maximum points.

Since the partial derivatives of the objective function are zero at an optimal solution, "first-order" methods like those described in this section, which rely only upon first-derivative information, may encounter numerical difficulties near an optimal solution. Also, in general, first-order methods do not converge to an optimal solution particularly fast. Other methods that use curvature, or second-derivative, information (or more often, approximations to the inverse of the matrix of second partial derivatives) overcome these difficulties, at the expense of more computational work at each step. The *Newton–Raphson* algorithm, which is described in the next section for the special case of one-dimensional optimization, is one popular example of these methods. An entire class of methods known as *conjugate direction* algorithms also use second-derivative information. Instead of reviewing these more advanced methods here, we show how unconstrained algorithms can be used to solve constrained optimization problems.

SUMT (Sequential Unrestrained Maximization Technique)

In principle, any optimization problem can be converted into an unconstrained optimization, as illustrated by the example

    Minimize x^2,

subject to:
$$x \le 4,$$
$$x \ge 1. \tag{10}$$

Suppose that we let $P(x)$ denote a penalty for being infeasible, given by:

$$P(x) = \begin{cases} +\infty & \text{if } x \text{ is infeasible (that is, } x > 4 \quad \text{or} \quad x > 1\text{),} \\ 0 & \text{if } x \text{ is feasible (that is, } 1 \le x \le 4\text{).} \end{cases}$$

Then the constrained optimization problem (10) can be restated in unconstrained form as

    Minimize $\{x^2 + P(x)\}$,

since the objective function with the penalty term agrees with the original objective function for any feasible point and is $+\infty$ for every infeasible point.

Although this conceptualization in terms of penalties is useful, the method cannot be implemented easily because of numerical difficulties caused by the $+\infty$ terms. In fact, even if a large penalty $M > 0$ is used to replace the $+\infty$ penalty, the method is difficult to implement, since the objective function is discontinuous (i.e., jumps) at the boundary of the feasible region; for example, with $M = 1000$, the term $x^2 + P(x)$ would equal $x^2 + 1000$ to the left of $x = 1$ and only x^2 at $x = 1$. Consequently, we cannot use the algorithms for unconstrained optimization presented in this section, which require differentiability.

We can overcome this difficulty by approximating the penalty term by a smooth function and refining the approximation sequentially by a method known as the *Sequential Unconstrained Maximization Technique*, or SUMT. In this method, instead of giving every infeasible point the same penalty, such as $+\infty$ or a large constant M, we impose a penalty that increases the more a point becomes infeasible.

The curve with $r = 1$ in Fig. 13.14 shows a penalty term used frequently in practice, in which the penalty $P(x)$ grows quadratically as points move farther from the feasible region. In this case, the penalty is zero for feasible points $1 \leq x \leq 4$. To the left of $x = 1$, the penalty is the quadratic term $(1 - x)^2$, and to the right of $x = 4$, the penalty is the quadratic term $(x - 4)^2$; that is,

$$P(x) = \begin{cases} (1 - x)^2 & \text{if } x \leq 1, \\ 0 & \text{if } 1 \leq x \leq 4, \\ (x - 4)^2 & \text{if } x \geq 4, \end{cases} \tag{11}$$

which is stated compactly as:

$$P(x) = \text{Max } (1 - x, 0)^2 + \text{Max } (x - 4, 0)^2.$$

Note that when $1 \leq x \leq 4$, both maximum terms in this expression are zero and no penalty is incurred; when $x \leq 1$ or $x \geq 4$, the last expression reduces to the appropriate penalty term in (11).

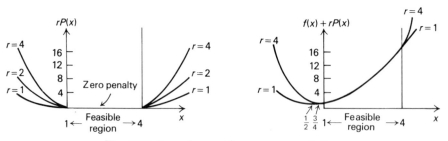

Fig. 13.14 Imposing penalty terms sequentially.

As Fig. 13.14 illustrates, the infeasible point $x = \frac{1}{2}$ solves the problem

$$\text{Minimize } \{f(x) + P(x)\} = \text{Minimize } \{x^2 + P(x)\},$$

since the quadratic penalty term does not impose a stiff enough penalty for near-feasible points. Note, however, that if the penalty term $P(x)$ is replaced by $2P(x)$,

$3P(x)$, $4P(x)$, or, in general, $rP(x)$ for $r > 1$, the penalty increases for any infeasible point. As the penalty scale-factor r becomes very large, the penalty associated with any particular infeasible point also becomes large, so that the solution to the modified penalty problem

Minimize $\{f(x) + rP(x)\}$

is driven closer and closer to the feasible region. The example problem illustrates this behavior. From Fig. 13.14 we see that, for any $r > 1$, the solution to the penalty problem occurs to the left of $x = 1$, where the penalty term is $r(1 - x)^2$. In this region, the penalty problem of minimizing $x^2 + rP(x)$ reduces to:

Minimize $\{x^2 + r(1 - x)^2\}$.

Setting the first derivative of this objective function to zero gives

$$2x - 2r(1 - x) = 0$$

or

$$x = \frac{r}{r + 1}$$

as the optimal solution. At this point, the objective value of the penalty problem is:

$$x^2 + r\,\text{Max}\,(1 - x, 0)^2 + r\,\text{Max}\,(4 - x, 0)^2 = \left(\frac{r}{r + 1}\right)^2 + r\left(1 - \frac{r}{r + 1}\right)^2 + r(0)^2$$

$$= \frac{r}{r + 1}.$$

Consequently, as r approaches $+\infty$, both the optimal solution $r/(r + 1)$ and the optimal value $r/(r + 1)$ to the penalty problem approach the optimal values $x^* = 1$ and $f(x^*) = 1$ to the original problem.

    Although the penalty function cannot always be visualized as nicely as in this simple example, and the computations may become considerably more involved for more realistic problems, the convergence properties exhibited by this example are valid for any constrained problem. The general problem

$$z^* = \text{Min } f(x),$$

subject to:
$$g_i(x) \leq b_i \quad \text{for } i = 1, 2, \ldots, m,$$

is converted into a sequence of unconstrained penalty problems

Minimize $\{f(x) + rP(x)\}$ (12)

as r increases to $+\infty$. The penalty term

$$P(x) = \sum_{i=1}^{m} \text{Max}\,(g_i(x) - b_i, 0)^2$$

introduces a quadratic penalty $[g_i(x) - b_i]^2$ for any constraint i that is violated; that is, $g_i(x) > b_i$. If x^r denotes the solution to the penalty problem (12) when the penalty scale factor is r, then any point x^* that these points approach in the limit solves the original constrained problem. Moreover, the optimal objective values $f(x^r) + rP(x^r)$ to the penalty problems approach the optimal value z^* to the constrained optimization problem.

The general theory underlying penalty-function approaches to constrained optimization permits many variations to the methodology that we have presented, but most are beyond the scope of our introduction to the subject. For example, the absolute value, or any of several other functions of the terms Max $(g_i(x) - b_i, 0)$, can be used in place of the quadratic terms. When equality constraints $h_i(x) = b_i$ appear in the problem formulation, the term $r(h_i(x) - b_i)^2$ is used in the penalty function.

In addition, *barrier methods* can be applied, in which the penalty terms are replaced by barrier terms:

$$\frac{1}{r} \sum_{i=1}^{m} \frac{1}{(g_i(x) - b_i)^2}. \tag{13}$$

In contrast to the SUMT procedure, these methods always remain within the feasible region. Since the barrier term $1/(g_i(x) - b_i)^2$ becomes infinite as x approaches the boundary of the constraint $g_i(x) \leqq b_i$, where $g_i(x) = b_i$, if the method starts with an initial feasible point, the minimization will not let it cross the boundary and become infeasible. As r becomes large, the barrier term decreases near the boundary and the terms (13) begin to resemble the penalty function with $P(x) = 0$ when x is feasible and $P(x) = +\infty$ when x is infeasible. Figure 13.15 illustrates this behavior when the barrier method is applied to our example problem with $r = 1$ and $r = 4$.

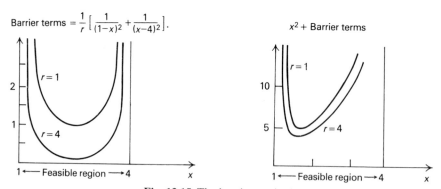

Fig. 13.15 The barrier method.

To conclude this section, let us see how the standard penalty procedure, without any of these variations, performs on the portfolio-selection problem that has been solved by several other methods in this chapter. The problem formulation

$$\text{Maximize } \{20x_1 + 16x_2 - 2x_1^2 - x_2^2 - (x_1 + x_2)^2\},$$

subject to:

$$x_1 + x_2 \leq 5,$$
$$x_1 \qquad\qquad \geq 0 \qquad (\text{or} - x_1 \leq 0)$$
$$x_2 \geq 0 \qquad (\text{or} - x_2 \leq 0)$$

leads to the penalty problem*:

$$\text{Max } [20x_1 + 16x_2 - 2x_1^2 - x_2^2 - (x_1 + x_2)^2 - r \text{ Max } (x_1 + x_2 - 5, 0)^2$$
$$- r \text{ Max } (-x_1, 0)^2 - r \text{ Max } (-x_2, 0)^2].$$

We can find the solution to this problem for any value of r by setting to zero the partial derivatives of the terms in braces with respect to both x_1 and x_2, that is, by solving the equations.†

$$20 - 6x_1 - 2x_2 - 2r \text{ Max } (x_1 + x_2 - 5, 0) - 2r \text{ Max } (-x_1, 0) = 0,$$
$$16 - 2x_1 - 4x_2 - 2r \text{ Max } (x_1 + x_2 - 5, 0) - 2r \text{ Max } (-x_2, 0) = 0.$$

The solution to these equations is given by

$$x_1 = \frac{7r^2 + 33r + 36}{3r^2 + 14r + 15}$$

and

$$x_2 = \frac{8r + 14}{3r + 5},$$

as can be verified by substitution in the equations. Figure 13.16 shows how the solutions approach the optimal solution $x_1^* = 2\frac{1}{3}$, $x_2^* = 2\frac{2}{3}$, and the optimal objective value $46\frac{1}{3}$ as r increases.

r	x_1	x_2	Optimal value for the penalty problem
1	2.375	2.75	46.39
2	2.36	2.73	46.38
5	2.35	2.70	46.36
10	2.34	2.69	46.35
100	2.334	2.669	46.34
$+\infty$	2.333...	2.666...	46.33

Fig. 13.16. Penalty method for portfolio selection.

---

* Note that we subtract the penalty term from the objective function in this example because we are maximizing rather than minimizing.
† The partial derivative of Max $(x_1 + x_2 - 5, 0)^2$ with respect to x_1 or x_2 equals $2(x_1 + x_2 - 5)$ if $x_1 + x_2 \geq 5$ and equals zero if $x_1 + x_2 \leq 5$. Therefore it equals $2 \text{ Max } (x_1 + x_2 - 5, 0)$. Generally, the partial derivative of Max $(g_i(x) - b_i, 0)^2$ with respect to x_j equals

$$2 \text{ Max } (g_i(x) - b_i, 0) \frac{\partial g_i(x)}{\partial x_j}.$$

13.9* ONE-DIMENSIONAL OPTIMIZATION

Many algorithms in optimization are variations on the following general approach: given a current feasible solution $x = (x_1, x_2, \ldots, x_n)$, find a direction of improvement $d = (d_1, d_2, \ldots, d_n)$ and determine the next feasible solution $\bar{x} = (\bar{x}_1, \bar{x}_2, \ldots, \bar{x}_n)$ as the point that optimizes the objective function along the line segment $\bar{x}_i = x_i + \theta d_i$ pointing away from x in the direction d. For a maximization problem, the new point is the solution to the problem

$$\underset{\theta \geq 0}{\text{Maximize }} f(x_1 + \theta d_1, x_2 + \theta d_2, \ldots, x_n + \theta d_n).$$

Since the current point $(x_1, x_2, \ldots, x_n)$ and direction $(d_1, d_2, \ldots, d_n)$ are fixed, this problem is a one-dimensional optimization in the variable θ. The direction d used in this method is chosen so that the solution to this one-dimensional problem improves upon the previous point; that is,

$$f(\bar{x}_1, \bar{x}_2, \ldots, \bar{x}_n) > f(x_1, x_2, \ldots, x_n).$$

Choosing the direction-finding procedure used to implement this algorithm in various ways gives rise to many different algorithms. The Frank-Wolfe algorithm discussed in Section 13.6, for example, determines the direction from a linear program related to the given optimization problem; algorithms for unconstrained optimization presented in the last section select the direction based upon the partial derivatives of the objective function evaluated at the current point x. This section briefly discusses procedures for solving the one-dimensional problem common to this general algorithmic approach; i.e., it considers maximizing a function $g(\theta)$ of a single variable θ.

Search Techniques

Whenever $g(\theta)$ is concave and differentiable, we can eliminate certain points as nonoptimal by evaluating the slope $g'(\theta)$ of the function as shown in Fig. 13.17; if $g'(\theta_1) > 0$, then no point $\theta \leq \theta_1$ can maximize g, and if $g'(\theta_2) < 0$, then no point $\theta \geq \theta_2$ can maximize g. This observation is the basis for a method known as *bisection* or *Bolzano search*.

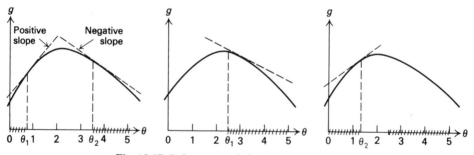

Fig. 13.17 Bolzano search for a concave function.

---

* This section requires differential calculus.

Suppose that the maximum of $g(\theta)$ is known to occur in some interval such as $0 \leq \theta \leq 5$ in Fig. 13.17. Then, evaluating the slope at the midpoint $\theta = 2\frac{1}{2}$ of the interval permits us to eliminate half of the interval from further consideration—here, $2\frac{1}{2} \leq \theta \leq 5$. By evaluating the slope again at the midpoint $\theta = 1\frac{1}{4}$ of the remaining interval, we can eliminate half of *this* interval. Continuing, we can halve the search interval in which the optimum is known to lie at each step, until we obtain a very small interval. We can then find a point as close as we like to a point that optimizes the given function.

The same type of procedure can be used without evaluating slopes. This extension is important, since derivatives must be evaluated numerically in many applications. Such numerical calculations usually require several function calculations. Instead of evaluating the slope at any point $\theta = \bar{\theta}$, we make function evaluations at two points $\theta = \theta_1$ and $\theta = \theta_2$ close to $\theta = \bar{\theta}$, separated from each other only enough so that we can distinguish between the values $g(\theta_1)$ and $g(\theta_2)$. If $g(\theta_1) < g(\theta_2)$, then every point with $\theta \leq \theta_1$ can be eliminated from further consideration; if $g(\theta_1) > g(\theta_2)$, then points with $\theta \geq \theta_2$ can be eliminated. With this modification, Bolzano search can be implemented by making two function evaluations near the midpoint of any interval, in place of the derivative calculation.

Bolzano's method does not require concavity of the function being maximized. All that it needs is that, if $\theta_1 \leq \theta_2$, then (i) $g(\theta_1) \leq g(\theta_2)$ implies that $g(\theta) \leq g(\theta_1)$ for all $\theta \leq \theta_1$, and (ii) $g(\theta_1) \geq g(\theta_2)$ implies that $g(\theta) \leq g(\theta_2)$ for all $\theta \geq \theta_2$. We call such functions *unimodal*, or single-peaked, since they cannot contain any more than a single local maximum. Any concave function, of course, is unimodal. Figure 13.18 illustrates Bolzano's method, without derivative evaluations, for a unimodal function.

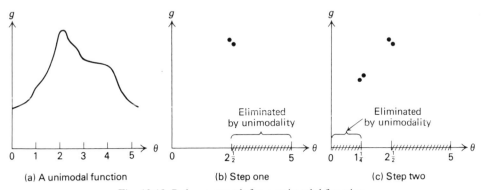

(a) A unimodal function   (b) Step one   (c) Step two

Fig. 13.18 Bolzano search for a unimodal function.

The Bolzano search procedure for unimodal functions can be modified to be more efficient by a method known as *Fibonacci search*. This method is designed to reduce the size of the search interval at each step by making only a single function evaluation instead of two evaluations or a derivative calculation, as in Bolzano search.

Figure 13.19 illustrates the method for the previous example. The first two points selected at $\theta_1 = 2$ and $\theta_2 = 3$ eliminate the interval $\theta \leq 2$ from further

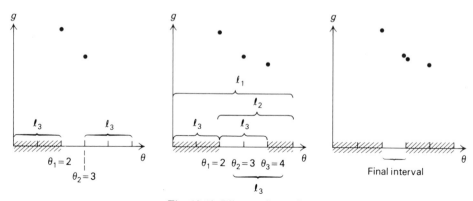

Fig. 13.19 Fibonacci search.

consideration, by the unimodel property. Note that $\theta_2 = 3$ stays within the search interval that remains. By selecting the next point at $\theta_3 = 4$, we eliminate $\theta \geq 4$ from further consideration. Finally, by selecting the last point close to θ_2 at $\theta = 3$, we eliminate $\theta \geq 3$. Consequently, by making four function evaluations, Fibonacci search has reduced the length of the final search interval to 1 unit, whereas four function evaluations (or two, usually more expensive, derivative calculations) in Bolzano search reduced the length of the final search interval to only $1\frac{1}{4}$.

In general, Fibonacci search chooses the function evaluations so that each new point is placed symmetrically in the remaining search interval with respect to the point already in that interval. As Fig. 13.19 illustrates for $k = 1$, the symmetry in placing the function evaluations implies that the length ℓ_k of successive search intervals is given by:

$$\ell_k = \ell_{k+1} + \ell_{k+2}. \tag{14}$$

This expression can be used to determine how many function evaluations are required in order to reduce the final interval to length ℓ_n. By scaling our units of measurement, let us assume that $\ell_n = 1$ (for example, if we want the final interval to have length 0.001, we measure in units of 0.001). Since the final function evaluation just splits the last interval in two, we know that the second-to-last interval has length $\ell_{n-1} = 2$. Then, from Eq. (14), the length of succeeding intervals for function evaluations numbered $3, 4, 5, 6, 7, \ldots$, is given by the "Fibonacci numbers":*

$$3 = 1 + 2, \qquad 5 = 2 + 3, \qquad 8 = 3 + 5,$$
$$13 = 5 + 8, \qquad 21 = 8 + 13, \qquad \ldots \tag{15}$$

Consequently, if the initial interval has length 21 (i.e., 21 times larger than the desired length of the final interval), then seven function evaluations are required, with the initial two evaluations being placed at $\theta = 8$ and $\theta = 13$, and each succeeding evaluation being placed symmetrically with respect to the point remaining in the search interval to be investigated further. Figure 13.20 shows a possibility for the first three steps of this application.

---

* The method is called Fibonacci search because the numbers in expression (15), which arise in many other applications, are called the Fibonacci numbers.

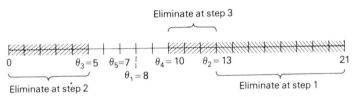

Fig. 13.20 A seven-point Fibonacci search.

Fibonacci search is known to be optimal in the sense that, of all search methods guaranteed to reduce the length of the final interval to ℓ_n, it uses the fewest function evaluations. Unfortunately, the length of the final interval must be known in advance, before the location of the initial two function evaluations can be determined. This *a priori* determination of ℓ_n may be inconvenient in practice. Therefore, an approximation to Fibonacci search, called the *method of golden sections*, is used frequently. As the number of function evaluations becomes large, the ratio between succeeding Fibonacci numbers approaches $1/\gamma \approx 0.618$, where $\gamma = (1 + \sqrt{5})/2$ is known as the *golden ratio*. This observation suggests that the first two evaluations be placed symmetrically in the initial interval, 61.8 percent of the distance between the two endpoints, as in golden-section search. Each succeeding point then is placed, as in Fibonacci search, symmetrically with respect to the point remaining in the search interval. Note that this approximation is very close to Fibonacci search even for a procedure with as few as seven points, as in the example shown in Fig. 13.20, since the initial points are at $\frac{13}{21} = 0.619$, or 61.9 percent, of the distance between the two endpoints when applying Fibonacci search. When applied to this example, the first five golden-section points are $\theta_1 = 8.022$, $\theta_2 = 12.978$, $\theta_3 = 4.956$, $\theta_4 = 9.912$, and $\theta_5 = 6.846$, as compared to $\theta_1 = 8$, $\theta_2 = 13$, $\theta_3 = 5$, $\theta_4 = 10$, and $\theta_5 = 7$ from Fibonacci search.

Curve Fitting and Newton's Method

Another technique for one-dimensional optimization replaces the given function to be maximized by an approximation that can be optimized easily, such as a quadratic or cubic function. Figure 13.21, for instance, illustrates a quadratic approximation

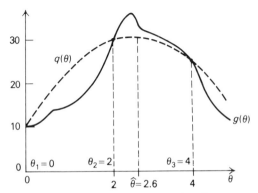

Fig. 13.21 Quadratic approximation.

to our previous example. By evaluating the given function $g(\theta)$ at three points $\theta = \theta_1, \theta_2$, and θ_3, we can determine a quadratic function

$$q(\theta) = a\theta^2 + b\theta + c,$$

which agrees with $g(\theta)$ at θ_1, θ_2, and θ_3. In this case, the resulting quadratic approximation is:

$$q(\theta) = -\frac{25}{8}\theta^2 + \frac{65}{4}\theta + 10.$$

By setting the first derivative $q'(\theta)$ of the approximation to 0, we can solve for an approximate optimal solution $\hat{\theta}$. Here,

$$q'(\theta) = -\frac{50}{8}\theta + \frac{65}{4},$$

and

$$\hat{\theta} = \frac{(65/4)}{(50/8)} = 2.6 \quad \text{with } q(\hat{\theta}) = 31.1.$$

After determining this approximation, we can use the new point $\hat{\theta}$, together with two of the points θ_1, θ_2, and θ_3, to define a new approximation. We select the three points, $\hat{\theta}$ and two others, so that the middle point has the highest value for the objective function, since then the approximating quadratic function has to be concave. Since $g(\hat{\theta}) = g(2.6) > 30$ in this case, we take $\theta_2, \hat{\theta}$, and θ_3 as the new points. If $g(\hat{\theta}) \leq 30$, we would take θ_1, θ_2, and $\hat{\theta}$ as the three points to make the next approximation. By continuing to make quadratic approximations in this way, the points $\hat{\theta}_1, \hat{\theta}_2, \ldots$ determined at each step converge to an optimal solution θ^*.[†] In practice, the method may not find an optimal solution θ^* and stop in a finite number of steps; therefore, some finite termination rule is used. For example, we might terminate when the function values $g(\hat{\theta}_j)$ and $g(\hat{\theta}_{j+1})$ for two succeeding points become sufficiently small—say, within $\epsilon = 0.00001$ of each other.

Similar types of approximation methods can be devised by fitting with cubic functions, or by using other information such as derivatives of the function for the approximation. Newton's method for example, which is possibly the most famous of all the approximating schemes, uses a quadratic approximation based upon first- and second-derivative information at the current point θ_j, instead of function evaluations at three points as in the method just described. The approximation is given by the quadratic function:

$$q(\theta) = g(\theta_j) + g'(\theta_j)(\theta - \theta_j) + \tfrac{1}{2}g''(\theta_j)(\theta - \theta_j)^2. \tag{16}$$

Note that the $q(\theta_j) = g(\theta_j)$ and that the first and second derivatives of $q(\theta)$ agree with the first derivative $g'(\theta_j)$ and the second derivative $g''(\theta_j)$ of $g(\theta)$ at $\theta = \theta_j$.

[†] In the sense that for any given $\epsilon > 0$, infinitely many of the $\hat{\theta}_j$ are within ϵ of θ^*; that is,

$$\theta^* - \epsilon \leq \hat{\theta}_j \leq \theta^* + \epsilon.$$

The next point θ_{j+1} is chosen as the point maximizing $q(\theta)$. Setting the first derivative of $q(\theta)$ in Eq. (16) to zero and solving for θ_{j+1} gives the general formula

$$\theta_{j+1} = \theta_j - \frac{g'(\theta_j)}{g''(\theta_j)}, \tag{17}$$

for generating successive points by the algorithm.

Letting $h(\theta) = g'(\theta)$ denote the first derivative of $g(\theta)$ suggests an interesting interpretation of expression (17). Since $g'(\hat\theta) = 0$ at any point $\hat\theta$ maximizing $g(\theta)$, $\hat\theta$ must be a solution to the equation $h(\hat\theta) = 0$. Newton's method for solving this equation approximates $h(\theta)$ at any potential solution θ_j by a line tangent to the function at $\theta = \theta_j$ given by:

$$\text{Approximation: } y = h(\theta_j) + h'(\theta_j)(\theta - \theta_j).$$

Since $h(\theta_j) = g'(\theta_j)$ and $h'(\theta_j) = g''(\theta_j)$, the solution $y = 0$ to this approximation is the point θ_{j+1} specified in (17). It is because of this association with Newton's well-known method for solving equations $h(\theta) = 0$ that the quadratic-approximation scheme considered here is called Newton's method.

To illustrate Newton's method, let $g(\theta) = (\theta - 3)^4$. Then $h(\theta) = g'(\theta) = 4(\theta - 3)^3$ and $h'(\theta) = g''(\theta) = 12(\theta - 3)^2$. Starting with $\theta_0 = 0$, Newton's method gives:

$$\theta_{j+1} = \theta_j - \frac{h(\theta_j)}{h'(\theta_j)} = \theta_j - \frac{4(\theta_j - 3)^3}{12(\theta_j - 3)^2} = \theta_j - \frac{\theta_j - 3}{3} = \tfrac{2}{3}\theta_j + 1.$$

The points $\theta_1 = 1$, $\theta_2 = 1\tfrac{2}{3}$, $\theta_3 = 2\tfrac{1}{9}$, $\theta_3 = 2\tfrac{11}{27}$, $\theta_4 = 2\tfrac{49}{81}, \ldots$ that the method generates converge to the optimal solution $\theta^* = 3$. Figure 13.22 shows the first two approximations to $h(\theta)$ for this example.

Newton's method is known to converge to an optimal solution θ^* that maximizes $g(\theta)$ as long as $g''(\theta^*) \neq 0$ and the initial point θ_0 is close enough to θ^*. The precise definition of what is meant by "close enough to θ^*" depends upon the problem data

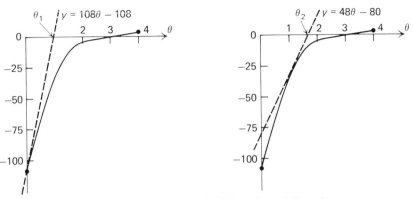

Fig. 13.22 Newton's method for solving $h(\theta) = 0$.

and will not be developed here. We simply note that, because the rate of convergence to an optimal solution has been shown to be good for Newton's method, it often is used in conjunction with other procedures for one-dimensional optimizations that are used first to find a point θ_0 close to an optimal solution.

EXERCISES

1. a) Over what region is the function $f(x) = x^3$ convex? Over what region is f concave?
   b) Over what region is the function $f(x) = -12x + x^3$ convex (concave)?
   c) Plot the function $f(x) = -12x + x^3$ over the region $-3 \le x \le 3$. Identify local minima and local maxima of f over this region. What is the global minimum and what is the global maximum?

2. a) Which of the following functions are convex, which are concave, and which are neither convex nor concave?

   i) $f(x) = |x|$.

   ii) $f(x) = \dfrac{1}{x}$ over the region $x > 0$.

   iii) $f(x) = \log(x)$ over the region $x > 0$.
   iv) $f(x) = e^{-x^2}$.
   v) $f(x_1, x_2) = x_1 x_2$.
   vi) $f(x_1, x_2) = x_1^2 + x_2^2$.

   b) Graph the feasible solution region for each of the following constraints:

   i) $x_1^2 + x_2^2 \le 4$.
   ii) $x_1^2 + x_2^2 = 4$.
   iii) $x_1^2 + x_2^2 \ge 4$.
   iv) $x_2 - |x_1| \le 0$.
   v) $x_2 - |x_1| \ge 0$.

   Which of these regions is convex?

   c) Is the function $f(x_1, x_2) = x_2 - |x_1|$ concave?

3. Because of the significance of convexity in nonlinear programming, it is useful to be able to identify convex sets and convex (or concave) functions easily. Apply the definitions given in this chapter to establish the following properties:

   a) If C_1 and C_2 are two convex sets, then the intersection of C_1 and C_2 (that is, points lying in both) is a convex set. For example, since the set C_1 of solutions to the inequality

   $$x_1^2 + x_2^2 \le 4$$

   is convex and the set C_2 of points satisfying

   $$x_1 \ge 0 \quad \text{and} \quad x_2 \ge 0$$

   is convex, then the feasible solution to the system

   $$x_1^2 + x_2^2 \le 4,$$

   $$x_1 \ge 0, \quad x_2 \ge 0,$$

   which is the intersection of C_1 and C_2, is also convex. Is the intersection of more than two convex sets a convex set?

b) Let $f_1(x)$, $f_2(x)$, ..., $f_m(x)$ be convex functions and let α_1, α_2, ..., α_m be nonnegative numbers; then the function

$$f(x) = \sum_{j=1}^{m} \alpha_j f_j(x)$$

is convex. For example, since $f_1(x_1, x_2) = x_1^2$ and $f_2(x_1, x_2) = |x_2|$ are both convex functions of x_1 and x_2, the function

$$f(x_1, x_2) = 2x_1^2 + |x_1|$$

is convex.

c) Let $f_1(x)$ and $f_2(x)$ be convex functions; then the function $f(x) = f_1(x)f_2(x)$ need not be convex. [Hint. See part (a(v)) of the previous exercise.]

d) Let $g(y)$ be a convex and nondecreasing function [that is, $y_1 \le y_2$ implies that $g(y_1) \le g(y_2)$] and let $f(x)$ be a convex function; then the composite function $h(x) = g[f(x)]$ is convex. For example, since the function e^y is convex and nondecreasing, and the function $f(x_1, x_2) = x_1^2 + x_2^2$ is convex, the function

$$h(x_1, x_2) = e^{x_1^2 + x_2^2}$$

is convex.

4. One of the equivalent definitions for convexity of a differentiable function f of single variable x given in this chapter is that the slopes of the function be nondecreasing with respect to the variable x; that is,

$$\left.\frac{df}{dx}\right|_{x=y} \ge \left.\frac{df}{dx}\right|_{x=z} \qquad \text{whenever } y \ge z.*$$

For a strictly convex function, the condition becomes:

$$\left.\frac{df}{dx}\right|_{x=y} > \left.\frac{df}{dx}\right|_{x=z} \qquad \text{whenever } y > z.$$

Similar conditions can be given for concave functions. If the function f has second derivatives, these conditions are equivalent to the very useful second-order criteria shown in Table E13.1.

Table E13.1

Function property	Condition on the second derivative $\dfrac{d^2f}{dx^2}$ implying the function property
Convex	$\dfrac{d^2f}{dx^2} \ge 0$ for all values of x
Concave	$\dfrac{d^2f}{dx^2} \le 0$ for all values of x
Strictly convex	$\dfrac{d^2f}{dx^2} > 0$ for all values of x
Strictly concave	$\dfrac{d^2f}{dx^2} < 0$ for all values of x

* $\left.\dfrac{df}{dx}\right|_{x=y}$ means the derivative of f with respect to x evaluated at $x = y$.

For example, if $f(x) = x^2$, then

$$\frac{d^2f}{dx^2} = 2,$$

implying that x^2 is strictly convex.

Not only do the second-order conditions imply convexity or concavity, but any convex or concave function must also satisfy the second-order condition. Consequently, since

$$\frac{d^2f}{dx^2} = 6x$$

can be both positive and negative for the function $f(x) = x^3$, we know that x^3 is neither convex nor concave.

Use these criteria to show which of the following functions are convex, concave, strictly convex, or strictly concave.

a) $4x + 2$

b) e^x

c) $\dfrac{1}{x}$ for $x > 0$

d) $\dfrac{1}{x}$ for $x < 0$

e) x^5

f) $\log(x)$ for $x > 0$

g) $\log(x^2)$ for $x > 0$

h) $f(x) = \begin{cases} 0 & \text{if } x \leq 0, \\ x^2 & \text{if } x > 0. \end{cases}$

i) x^4 [*Hint.* Can a function be strictly convex even if its second derivative is zero at some point?]

5. Brite-lite Electric Company produces two types of electric lightbulbs; a 100-watt bulb and a 3-way (50–100–150) bulb. The machine that produces the bulbs is fairly old. Due to higher maintenance costs and down time, the contribution per unit for each additional unit produced decreases as the machine is used more intensely. Brite-lite has fit the following functions to per-unit contribution (in $) for the two types of lightbulbs it produces:

$$f_1(x_1) = 50 - 2x_1, \quad \text{for } x_1 \text{ units of 100-watt bulbs};$$

$$f_2(x_2) = 70 - 3x_2, \quad \text{for } x_2 \text{ units of 3-way bulbs.}$$

The Brite-lite Company is in the unusual position of being able to determine the number of units demanded of a particular type of lightbulb from the amount (in $) it spends on advertisements for that particular bulb.

The equations are:

$$150 - \frac{1000}{10 + y_1} = \text{Number of units of 100-watt bulbs demanded,}$$

$$250 - \frac{2000}{10 + y_2} = \text{Number of units of 3-way bulbs demanded,}$$

where y_1 is the amount (in $) spent on advertisements for 100-watt bulbs, and y_2 is the amount (in $) spent on advertisements for 3-way bulbs.

Brite-lite has an advertising budget of $1500. Its management is committed to meeting demand. Present production capacity is 125 and 175 units of 100-watt and 3-way bulbs, respectively.

Finally, the production of a unit of lightbulb j consumes a_{ij} units of resource i for $i = 1, 2, \ldots, m$, of which only b_i units are available.

Brite-lite wishes to determine optimal production and advertising levels for 100-watt bulbs and 3-way bulbs. Formulate as a nonlinear program.

6. Besides its main activity of providing diaper service, Duke-Dee Diapers, a small workshop owned by a large fabrics firm, produces two types of diapers. The diapers are constructed from a special thin cloth and an absorbent material. Although both materials are manufactured by the fabrics company, only limited monthly amounts, b_1 and b_2, respectively, are available because this activity always has been considered as secondary to production of other fabrics. Management has decided to analyze the activity of the workshop in terms of profits, to decide whether or not to continue its diaper-service operation. A limited market survey showed that the following linear functions give good approximations of the demand curves for the two types of diapers:

$$p_j = \beta_j - \alpha_{j1} x_1 - \alpha_{j2} x_2 \qquad (j = 1, 2),$$

where

p_j = Unit price for the diapers,

β_j = Constant,

α_{jk} = Coefficient related to the substitutability of the two types of diapers (if $\alpha_{11} = 1$, $\alpha_{12} = 0$, $\alpha_{22} = 1$, $\alpha_{21} = 0$, there is no substitutability), and

x_k = Units of diapers of type k sold.

Each unit of type j diaper costs c_j $(j = 1, 2)$ to produce and uses a_{ij} $(i = 1, 2; j = 1, 2)$ units of resource i.

a) Formulate a model to determine the profits from diaper production.
b) Show how you would transform the model into a separable program.

7. A connoisseur has m bottles of rare wine in his cellar. Bottle j is currently t_j years old. Starting next year, he wishes to drink one bottle each year on his birthday, until his supply is exhausted. Suppose that his utility for drinking bottle j is given by $U_j(t)$ (t is its age when consumed) and that utility is additive. The connoisseur wishes to know in what order to drink the wine to maximize his total utility of consumption.

a) Formulate his decision problem of which bottle to select each year as an integer linear program. What is the special structure of the formulation? How can this structure help in solving the problem?
b) Suppose that the utility for consuming each bottle of wine is the same, given by a convex function $U(t)$. Show that it is optimal to select the youngest available bottle each year.

8. Consider a directed communications network, where each arc represents the communication link between two geographically distant points. Each communication link between points i and j has a probability of failure of p_{ij}. Then, for any path through the network, the probability of that particular path being operative is the product of the probabilities $(1 - p_{ij})(1 - p_{jk})(1 - p_{kl}) \ldots$ of the communication links connecting the two points of interest.

a) Formulate a network model to determine the path from s to t having the highest probability of being operative.
b) In what class does the problem formulated in (a) fall?
c) How can this formulation be transformed into a network problem?

9. In Section 4 of Chapter 1, we formulated the following nonlinear-programming version of the custom-molder example:

$$\text{Maximize } 60x_1 - 5x_1^2 + 80x_2 - 4x_2^2,$$

subject to:

$$6x_1 + 5x_2 \leq 60,$$
$$10x_1 + 12x_2 \leq 150,$$
$$x_1 \leq 8,$$
$$x_1 \geq 0, \quad x_2 \geq 0.$$

Using the following breakpoints, indicate a separable-programming approximation to this problem:

$$x_1 = 2, \qquad x_1 = 4, \qquad x_1 = 6,$$

and

$$x_2 = 3, \qquad x_2 = 6, \qquad x_2 = 10.$$

Do *not* solve the separable program.

10. In the network of Fig. E13.1, we wish to ship 10 units from node 1 to node 4 as cheaply as possible.

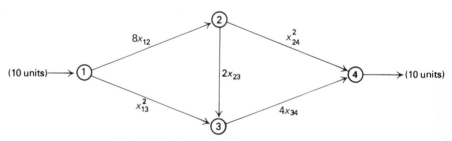

Fig. E13.1 Network with arc costs.

The flow on each arc is uncapacitated. The costs on arcs 1–2, 2–3, and 3–4 are linear, with costs per unit of 8, 2, and 4, respectively; the costs on arcs 1–3 and 2–4 are quadratic and are given by x_{13}^2 and x_{24}^2.

a) Suppose that we apply the separable-programming δ-technique discussed in Chapter 9, using the breakpoints

$$x_{13} = 0, \qquad x_{13} = 2, \qquad x_{13} = 6, \qquad x_{13} = 10$$

and

$$x_{24} = 0, \qquad x_{24} = 2, \qquad x_{24} = 6, \qquad x_{24} = 10.$$

Interpret the resulting linear-programming approximation as a network-flow problem with parallel arcs joining nodes 1 and 3 and nodes 2 and 4, as in Fig. E13.2.

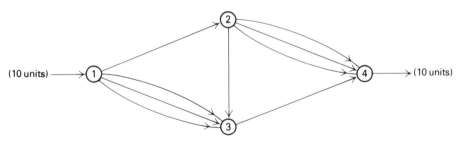

Fig. E13.2 Linear approximation with parallel arcs.

      Specify the per-unit cost and the arc capacity for each arc in the linear approximation.

b) Solve the separable-programming approximation from part (a), comparing the solution with the optimal solution

$$x_{12} = 5, \qquad x_{13} = 5, \qquad x_{23} = 3.5, \qquad x_{24} = 1.5, \qquad x_{34} = 8.5,$$

and minimum cost $= 110.25$, to the original nonlinear problem formulation.

11. a) What is the special form of the linear approximation problem when the Frank-Wolfe algorithm is applied to the network example in the previous exercise?

   b) Complete Table E13.2 for applying the first two steps of the Frank-Wolfe algorithm to this example.

Table E13.2

Arc	Initial solution	Solution to linear approximation	Second solution	Solution to linear approximation	Next solution
1–2	10				
1–3	0				
2–3	10				
2–4	0				
3–4	10				
Total cost	140	/////////		/////////	

12. Solve the following nonlinear program using the λ-method of separable programming described in this chapter.

$$\text{Maximize } 2x_1 - x_1^2 + x_2,$$

subject to:
$$x_1^2 + x_2^2 \leqq 4,$$
$$x_2 \leqq 1.8,$$
$$x_1, \quad x_2 \geqq 0.$$

Carry out calculations using two decimal places. For each decision variable, use a grid of 5 points (including the extreme values).

13. Two chemicals can be produced at one facility, and each takes one hour per ton to produce. Both exhibit diseconomies of scale: If x_1 tons of the first are produced, the contribution is $6x_1 - (x_1^2/2)$; for the second chemical, the contribution is $\sqrt{50x_2}$ for x_2 tons produced. The facility can work 23 hours per day (one hour is required for maintenance) and, because of raw materials availability, no more than 10 tons of the first chemical and 18 tons of the second can be produced daily. What is the optimal daily production schedule, to maximize contribution?

    Solve by separable programming, using the δ-method described in Chapter 9. For x_1 use a grid: 0, 3, 6, 10, and for x_2 use a grid: 0, 2, 5, 18.

14. A young R & D engineer at Carron Chemical Company has synthesized a sensational new fertilizer made of just two interchangeable basic raw materials. The company wants to take advantage of this opportunity and produce as much as possible of the new fertilizer. The company currently has $40,000 to buy raw materials at a unit price of $8000 and $5000 per unit, respectively. When amounts x_1 and x_2 of the basic raw materials are combined, a quantity q of fertilizer results given by:

$$q = 4x_1 + 2x_2 - 0.5x_1^2 - 0.25x_2^2.$$

   a) Formulate as a nonlinear program.
   b) Apply four iterations of the Frank-Wolfe algorithm; graphically identify the optimal point, using the property that even- and odd-numbered points lie on lines directly toward the optimum. Start from $x^0 = (0, 0)$ and use three decimal places in your computations.
   c) Solve the problem using the algorithm for quadratic programming discussed in Section 13.7.

15. A balloon carrying an x-ray telescope and other scientific equipment must be designed and launched. A rough measure of performance can be expressed in terms of the height reached by the balloon and the weight of the equipment lifted. Clearly, the height itself is a function of the balloon's volume.

    From past experience, it has been concluded that a satisfactory performance function to be maximized is $P = f(V, W) = 100V - 0.3V^2 + 80W - 0.2W^2$ where V is the volume, and W the equipment weight.

    The project to be undertaken has a budget constraint of $1040. The cost associated with the volume V is $2V$, and the cost of the equipment is $4W$. In order to ensure that a reasonable balance is obtained between performance due to the height and that due to the scientific equipment, the designer has to meet the constraint $80W \geq 100V$.

    Find the optimal design in terms of volume and equipment weight, solving by the Frank-Wolfe algorithm.

16. Consider the nonlinear-programming problem:

$$\text{Maximize } 2x_1 - x_1^2 + x_2,$$

   subject to:
$$x_1^2 + x_2^2 \leq 4,$$
$$x_2 \leq 1.8,$$
$$x_1 \geq 0, \quad x_2 \geq 0.$$

   a) Carry out two iterations of the generalized programming algorithm on this problem. Let

$$C = \{(x_1, x_2) \mid x_1 \geq 0, \quad 0 \leq x_2 \leq 1.8\},$$

leaving only one constraint to handle explicitly. Start with the following two initial candidate solutions:

$$x_1 = 0, \qquad x_2 = 1.8, \qquad \text{and} \qquad x_1 = 2, \qquad x_2 = 0.$$

Is the solution optimal after these two iterations?

b) Carry out two iterations of the MAP algorithm on the same problem. Start with $x_1 = 0$, $x_2 = 0$, as the initial solution, and set the parameters $\delta_1 = \delta_2 = 2$ at iteration 1 and $\delta_1 = \delta_2 = 1$ at iteration 2. Is the solution optimal after these two iterations?

17. An orbital manned scientific lab is placed on an eccentric elliptical orbit described by $x^2 + 5y^2 + x + 3y = 10$, the reference system (x, y) being the center of the earth. All radio communications with the ground stations are going to be monitored via a satellite that will be fixed with respect to the reference system. The power required to communicate between the satellite and the lab is proportional to the square of the distance between the two.

Assuming that the satellite will be positioned in the plane of the ellipse, what should be the position of the satellite so that the maximum power required for transmissions between the lab and the satellite can be minimized? Formulate as a nonlinear program.

18. An investor has $2 million to invest. He has 5 opportunities for investment, with the following characteristics:

i) The yield on the first investment is given by a linear function:

$$r_1 = 3 + 0.000012x_1,$$

where $r_1 = $ yield per year (%), and $x_1 = $ amount invested ($).

Minimum required: $ 100,000
Maximum allowed: $1,000,000
Years to maturity: 6

ii) The second investment yields:

$$r_2 = 2 + 0.000018x_2,$$

where $r_2 = $ yield per year (%), and $x_2 = $ amount invested ($).

Minimum required: $ 200,000
Maximum allowed: $1,000,000
Years to maturity: 10

iii) An investment at 5% per year with interest continuously compounded. (An amount A invested at 5% per year with continuously compounded interest becomes $Ae^{0.05}$ after one year.)

Years to maturity: 1

iv) Category 1 of government bonds that yield 6% per year.

Years to maturity: 4

v) Category 2 of government bonds that yield 5.5% per year.

Years to maturity: 3

The average years to maturity of the entire portfolio must not exceed 5 years.

a) The objective of the investor is to maximize accumulated earnings at the end of the first year. Formulate a model to determine the amounts to invest in each of the alternatives. Assume all investments are held to maturity.

b) Identify the special type of nonlinear program obtained in part (a).

19. Since its last production diversification, New Home Appliances, Inc. (NHA), a kitchen equipment manufacturer, has encountered unforeseen difficulty with the production scheduling and pricing of its product line. The linear model they have used for a number of years now no longer seems to be a valid representation of the firm's operations.

   In the last decade, the output of the firm has more than tripled and they have become the major supplier of kitchen equipment in the southeastern United States market. After conducting a market survey as well as a production-cost analysis, the consulting firm hired by NHA has concluded that the old linear model failed to optimize the activity of the firm because it did not incorporate the following new characteristics of NHA as a major supplier in the southeastern market;

I. NHA was no longer in a perfectly competitive situation, so that it had to take into account the market demand curve for its products. The consulting firm found that for NHA's 15 principal products, the price elasticity of demand was roughly 1.2; hence, the market demand curve can be expressed by:

$$x_j p_j^{1.2} = 1,600,000 \qquad (j = 1, 2, \ldots, 15), \tag{1}$$

where x_j = units of product j sold; and p_j = unit price of product j. For the remaining 25 products, the price is known and can be considered constant for all levels of sales.

II. Since output has increased, the firm has reached the stage where economies of scale prevail; thus, the per-unit production cost c_j decreases according to:

$$c_j = \gamma_j - \delta_j x_j \qquad (j = 1, 2, \ldots, 15), \tag{2}$$

where γ_j and δ_j are coefficients determined individually for the 15 principal products. For the remaining 25 products, constant returns to scale is a reasonable assumption (i.e., a linear relationship exists between the amount produced and the cost); consider the production costs per unit as known.

III. Production of each unit of product j consumes a_{ij} units of resource i. Resource utilization is limited by a budget of B dollars, which is available for the purchase of raw materials and labor. The suppliers of sheet steel and aluminum are offering discount prices for larger quantities. Linear regression leads to the following equations that show the relationship between the amount ordered and the unit price of each:

$$\mu_s = \alpha_s - \beta_s b_s,$$

$$\mu_a = \alpha_a - \beta_a b_a,$$

where μ_s, μ_a are the unit prices for steel and aluminum, respectively; b_s, b_a are the amounts contracted, and α_s, α_a, β_s, β_a are coefficients. No discounts are available for other resources, because NHA's consumption falls below the level at which such discounts are offered. Unit prices for all other resources are constant and known. Besides steel and aluminum, 51 resources are used.

Formulate a mathematical program that incorporates the above information; the objective is the maximization of contribution (revenues − costs). What type of nonlinear program have you obtained?

20. A rent-a-car company operating in New York City serves the three major airports—Kennedy, La Guardia, and Newark—and has two downtown distribution centers. On Sunday evenings most of the cars are returned to the downtown locations by city residents returning from weekend travels. On Monday morning, most of the cars are needed at the airports and must be "deadheaded" by company drivers.

The two downtown distribution centers have a_i ($i = 1, 2$) excess cars available. The three airports must be supplied with cars at a transportation cost of c_{ij}($i = 1, 2$; $j = 1, 2, 3$) for deadheading from distribution center i to airport j. The Monday morning demand r_j for cars at airport j is uncertain and is described by the probability distribution $p_j(r_j)$. If the demand exceeds supply at airport j, the unsatisfied demand is lost, with an average lost contribution per car of u_j.

a) Formulate a mathematical program to minimize total deadheading transportation cost plus expected lost contribution.

b) Suppose now that the manager of fleet allocation is also concerned with supply over demand. All cars in excess of 50 cars above demand must be parked in an overflow lot at a cost of s_j per car. Reformulate the program to include these expected overage costs.

21. After the admissions decisions have been made for the graduate engineering school, it is the Scholarship Committee's job to award financial aid. There is never enough money to offer as much scholarship aid as each needy applicant requires.

Each admitted applicant's financial need is determined by comparing an estimate of his sources of revenue with a reasonable school and personal expense budget for the normal academic year. An admittee's need, if any, is the difference between the standard budget and the expected contribution from him and his family. Scholarship offers provide an amount of aid equal to some fraction of each applicant's need. In cases where need is not met in full, the school is able to supply low-cost loans to cover the difference between scholarships and need.

Besides receiving funds from the university, a needy admittee might receive a scholarship from nonuniversity funds. In this case the admittee, if he decides to matriculate, is expected to accept the outside funds. His total scholarship award is then the greater of the university offer or the outside offer, because the university supplements any outside offer up to the level awarded by the scholarship committee. Prior to the deadline for determining a school scholarship-offer policy, the committee has a good estimate of the amount of outside aid that each needy admittee will be offered.

The most important function of the scholarship policy is to enroll the *highest-quality* needy admittees possible. The admissions committee's rank list of all needy admittees is used as the measurement of quality for the potential students.

In using this list, the top $100\alpha\%$ of the needy group ordered by quality is expected to yield at least βT enrollees, where T is the total desired number of enrollees from the needy group. In addition to satisfying the above criteria, the dean wants to minimize the total expected cost of the scholarship program to the university.

As a last point, P_i, the probability that needy admittee i enrolls, is an increasing function of y_i, the fraction of the standard budget B covered by the total scholarship offer. An estimate of this function is given in Fig. E13.3. Here, x_iB is the dollar amount of aid offered admittee i and n_iB is the dollar amount of need for admittee i.

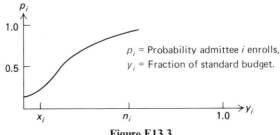

Figure E13.3

a) Formulate a nonlinear-programming model that will have an expected number T of enrolling needy admittees, and minimize the scholarship budget. (Assume that P_i can be approximated by a linear function.)

b) Suggest two different ways of solving the model formulated in (a). [*Hint.* What special form does the objective function have?]

c) Reformulate the model so as to maximize the expected number of enrolling needy students from the top $100\alpha\%$ of the need group, subject to a fixed total scholarship budget. Comment on how to solve this variation of the model.

d) Suppose that, in the formulation proposed in (a), the probability P_i that admittee i enrolls is approximated by $P_i = a_i + b_i y_i^{1/2}$. Comment on how to solve this variation of the model.

22. A well-known model to solve the aggregate production-planning problem that permits quadratic cost functions in the model's objective was developed by Holt, Modigliani, Muth, and Simon.* The model allocates manpower, production, and inventories during a prescribed planning horizon, divided into t time periods denoted by $t = 1, 2, \ldots, T$. The decision variables of the model are:

P_t = Production rate for period t;

W_t = Work-force size used in period t;

I_t = Ending inventory at period t.

If d_t is the demand to be satisfied during period t, the constraints of the model are:

$$P_t + I_{t-1} - I_t = d_t \qquad (t = 1, 2, \ldots, T).$$

The objective function is to minimize the sum of the cost elements involved in the production process. Holt, Modigliani, Muth, and Simon identified the following cost elements for each time period:

i) Regular payroll cost $= c_1 W_t + c_{13}$;

ii) Hiring and firing cost $= c_2(W_t - W_{t-1} - c_{11})^2$;

iii) Overtime and idle cost $= c_3(P_t - c_4 W_t)^2 + c_5 P_t - c_6 W_t + c_{12} P_t W_t$;

iv) Inventory and back-order cost $= c_7[I_t - (c_8 + c_9 d_t)]^2$.

a) Discuss and interpret the assumptions made on the behavior of the cost components. Is it reasonable to assume quadratic functions to characterize costs (ii), (iii), and (iv)?

b) Formulate the overall objective function and the constraints of the model.

c) Suggest a procedure to obtain the optimum solution to the model.

---

* Holt, C. C., F. Modigliani, J. F. Muth, H. A. Simon, *Planning Production, Inventories, and Work-Force*, Prentice-Hall, Inc., Englewood Cliffs, N.J., 1960.

23. In an application of the Holt, Modigliani, Muth, and Simon model (see Exercise 22) to a paint factory, the following decision rules were derived for optimum values of P_t and W_t for a twelve-month period starting with the forthcoming month, t.

$$P_t = \begin{Bmatrix} +0.458d_t \\ +0.233d_{t+1} \\ +0.111d_{t+2} \\ +0.046d_{t+3} \\ +0.014d_{t+4} \\ -0.001d_{t+5} \\ -0.007d_{t+6} \\ -0.008d_{t+7} \\ -0.008d_{t+8} \\ -0.007d_{t+9} \\ -0.005d_{t+10} \\ -0.004d_{t+11} \end{Bmatrix} + 1.005W_{t-1} + 153 - 0.464I_{t-1}$$

$$W_t = 0.742W_{t-1} + 2.00 - 0.010I_{t-1} + \begin{Bmatrix} +0.0101d_t \\ +0.0088d_{t+1} \\ +0.0071d_{t+2} \\ +0.0055d_{t+3} \\ +0.0042d_{t+4} \\ +0.0031d_{t+5} \\ +0.0022d_{t+6} \\ +0.0016d_{t+7} \\ +0.0011d_{t+8} \\ +0.0008d_{t+9} \\ +0.0005d_{t+10} \\ +0.0004d_{t+11} \end{Bmatrix}$$

a) Study the structure of the decision rules. How would you apply them? Are you surprised that the decision rules are linear (as opposed to quadratic)?

b) Note the weights that are given to the demand forecast d_t ($t = 1, 2, \ldots, T$). Comment on the implication of these weights.

c) How would you obtain the resulting optimum inventory levels, I_t?

24. An important problem in production management is the allocation of a given production quantity (determined by an aggregate model or by subjective managerial inputs) among a group of items. For example, let us assume that we have decided to produce $P = 6000$ units of a given product line consisting of three individual items. The allocation of the total quantity among the three items will be decided by the following mathematical model:

$$\text{Minimize } c = \sum_{i=1}^{3} \left(h_i \frac{Q_i}{2} + S_i \frac{d_i}{Q_i} \right),$$

subject to:

$$\sum_{i=1}^{3} Q_i = P,$$

where

Q_i = Production quantity for item i (in units),

h_i = Inventory holding cost for item i (in $/month $\times$ unit),

S_i = Setup cost for item i (in $),

d_i = Demand for item i (in units/month),

P = Total amount to be produced (in units).

a) Interpret the suggested model. What is the meaning of the objective function? What implicit assumption is the model making?

b) The model can be proved to be equivalent to the following unconstrained minimization problem (by means of Lagrange multiplier theory):

$$\text{Minimize } L = \sum_{i=1}^{3} \left(h_i \frac{Q_i}{2} + S_i \frac{d_i}{Q_i} \right) + \lambda \left(\sum_{i=1}^{3} Q_i - P \right).$$

State the optimality conditions for this unconstrained problem (see Section 13.8). Note that the unknowns are Q_i, $i = 1, 2, 3$, and λ. What is the interpretation of λ?

c) Given the following values for the parameters of the problem, establish a procedure to obtain the optimum values of Q_i.

		Items	
Parameter	1	2	3
h_i	1	1	2
S_i	100	50	400
d_i	20,000	40,000	40,000
$Q = 6000$			

[*Hint.* Perform a search on λ; plot the resulting values of Q and λ. Select the optimum value of λ from the graph corresponding to $Q = 6000$.]

d) Apply the SUMT method to the original model. How do you compare the SUMT procedure with the approach followed in part (c)?

25. When applying the Frank-Wolfe algorithm to the optimization problem

$$\text{Maximize } f(x_1, x_2, \ldots, x_n),$$

subject to:

$$\sum_{j=1}^{n} a_{ij} x_j \le b_i \qquad (i = 1, 2, \ldots, m),$$

$$x_j \ge 0 \qquad (j = 1, 2, \ldots, n),$$

(1)

we replace the given problem by a linear approximation at the current solution $x^* = (x_1^*, x_2^*, \ldots, x_n^*)$ given by

$$\text{Maximize } \left[f(x^*) + \frac{\partial f}{\partial x_1} x_1 + \frac{\partial f}{\partial x_2} x_2 + \cdots + \frac{\partial f}{\partial x_n} x_n \right],$$

subject to:

$$\sum_{j=1}^{n} a_{ij} x_j \le b_i \qquad (i = 1, 2, \ldots, m)$$

$$x_j \ge 0 \qquad (j = 1, 2, \ldots, n).$$

(2)

The partial derivatives $(\partial f / \partial x_j)$ are evaluated at the point x^*. We then perform a one-dimensional optimization along the line segment joining x^* with the solution to the linear approximation.

a) Suppose that x^* solves the linear approximation problem so that the solution does not change after solving the linear approximation problem. Show that there are "Lagrange multipliers" $\lambda_1, \lambda_2, \ldots, \lambda_m$ satisfying the *Kuhn-Tucker Optimality Conditions* for linearly-constrained nonlinear programs:

Primal feasibility
$$\begin{cases} \displaystyle\sum_{j=1}^{n} a_{ij}x_j^* \leq b_i & (i = 1, 2, \ldots, m), \\[2ex] x_j^* \geq 0 & (j = 1, 2, \ldots, n), \end{cases}$$

Dual feasibility
$$\begin{cases} \displaystyle\frac{\partial f}{\partial x_j} - \sum_{i=1}^{m} \lambda_i a_{ij} \leq 0 & (j = 1, 2, \ldots, n) \\[2ex] \lambda_i \geq 0 & (i = 1, 2, \ldots, m), \end{cases}$$

Complementary slackness
$$\begin{cases} \displaystyle\left[\frac{\partial f}{\partial x_j} - \sum_{i=1}^{m} \lambda_i a_{ij}\right] x_j^* = 0 & (j = 1, 2, \ldots, n), \\[2ex] \lambda_i\left[\displaystyle\sum_{j=1}^{n} a_{ij}x_j^* - b_i\right] = 0 & (i = 1, 2, \ldots, m). \end{cases}$$

b) What is the form of these Kuhn-Tucker conditions when (1) is a linear program or a quadratic program?

c) Suppose that $x^* = (x_1^*, x_2^*, \ldots, x_n^*)$ solves the original optimization problem (1). Show that x^* also solves the linear approximation problem (2) and therefore satisfies the Kuhn-Tucker conditions. [*Hint.* Recall that

$$\lim_{\theta \to 0} \frac{f(x^* + \theta x_j) - f(x^*)}{\theta} = \frac{\partial f}{\partial x_1} x_1 + \frac{\partial f}{\partial x_2} x_2 + \cdots + \frac{\partial f}{\partial x_n} x_n.\Bigg]$$

d) Suppose that $f(x_1, x_2, \ldots, x_n)$ is a convex function. Show that if $x^* = (x_1^*, x_2^*, \ldots, x_n^*)$ solves the Kuhn-Tucker conditions, then x^* is a global minimum to the original non-linear program (1).

26. When discussing sensitivity analysis of linear programs in Chapter 3, we indicated that the optimal objective value of a linear program is a concave function of the righthand-side values. More generally, consider the optimization problem

$$v(b_1, b_2, \ldots, b_m) = \text{Maximize } f(x),$$

subject to:
$$g_i(x) \leq b_i \qquad (i = 1, 2, \ldots, m)$$

where $x = (x_1, x_2, \ldots, x_n)$ are the problem variables; for linear programs

$$f(x) = \sum_{j=1}^{n} c_j x_j \quad \text{and} \quad g_i(x) = \sum_{j=1}^{m} a_{ij} x_j \quad \text{for } i = 1, 2, \ldots, m.$$

a) Show that if each $g_i(x)$ is a convex function, then the values of $b_1, b_2, \ldots, b_m$ for which the problem has a feasible solution form a convex set C.

b) Show that if, in addition, $f(x)$ is a concave function, then the optimal objective value $v(b_1, b_2, \ldots, b_m)$ is a concave function of the righthand-side values $b_1, b_2, \ldots, b_m$ on the set C.

27. In many applications, the ratio of two linear functions is to be maximized subject to linear constraints. The linear fractional programming problem is:

$$\text{Maximize } \frac{\sum_{j=1}^{n} c_j x_j + c_0}{\sum_{j=1}^{n} d_j x_j + d_0},$$

subject to:

$$\sum_{j=1}^{n} a_{ij} x_j \leq b_i \qquad (i = 1, 2, \ldots, m),$$

$$x_j \geq 0 \qquad (j = 1, 2, \ldots, n). \tag{1}$$

A related linear program is

$$\text{Maximize } \sum_{j=1}^{n} c_j y_j + c_0 y_0,$$

subject to:

$$\sum_{j=1}^{n} a_{ij} y_j - b_i y_0 \leq 0 \qquad (i = 1, 2, \ldots, m),$$

$$\sum_{j=1}^{n} d_j y_j + d_0 y_0 = 1, \tag{2}$$

$$y_j \geq 0 \qquad (j = 0, 1, 2, \ldots, n).$$

a) Assuming that the optimal solution to the linear fractional program occurs in a region where the denominator of the objective function is strictly positive, show that:

   i) If y_j^* $(j = 0, 1, 2, \ldots, n)$ is a finite optimal solution to (2) with $y_0^* > 0$, then $x_j^* = y_j^*/y_0^*$ is a finite optimal solution to (1).

   ii) If λy_j^* $(j = 0, 1, 2, \ldots, n)$ is an unbounded solution of (2) as $x \to \infty$ and $y_0^* > 0$, then $\lambda x_j^* = \lambda y_j^*/y_0^*$ $(j = 1, 2, \ldots, n)$ is an unbounded solution of (1) as $\lambda \to \infty$.

b) Assuming that it is not known whether the optimal solution to the linear fractional program occurs in a region where the denominator is positive or in a region where it is negative, generalize the approach of (a) to solve this problem.

28. A *primal* optimization problem is:

$$\text{Maximize } f(x),$$

subject to:

$$g_i(x) \leq 0 \qquad (i = 1, 2, \ldots, m),$$

$$x \in X$$

where x is understood to mean $(x_1, x_2, \ldots, x_n)$ and the set X usually means $x_j \geq 0$ $(j = 1, 2, \ldots, n)$ but allows other possibilities as well. The related Lagrangian form of the problem is:

$$L(y) = \underset{x \in X}{\text{Maximize}} \left[f(x) - \sum_{i=1}^{m} y_i g_i(x) \right].$$

The *dual* problem is defined to be:

Minimize $L(y)$,

subject to:

$$y_i \geq 0 \qquad (i = 1, 2, \ldots, m).$$

Without making any assumptions on the functions $f(x)$ and $g_i(x)$, show the following:

a) (*Weak duality*) If $\bar{x}_j$ ($j = 1, 2, \ldots, n$) is feasible to the primal and $\bar{y}_i$ ($i = 1, 2, \ldots, m$) is feasible to the dual, then $f(\bar{x}) \leq L(\bar{y})$.

b) (*Unboundedness*) If the primal (dual) is unbounded, then the dual (primal) is infeasible.

c) (*Optimality*) If $\hat{x}_j$ ($j = 1, 2, \ldots, n$) is feasible to the primal and $\hat{y}_i$ ($i = 1, 2, \ldots, m$) is feasible to the dual, and, further, if $f(\hat{x}) = L(\hat{y})$, then $\hat{x}_j$ ($j = 1, 2, \ldots, n$) solves the primal and $\hat{y}_i$ ($i = 1, 2, \ldots, m$) solves the dual.

29. Suppose $\hat{x} = (\hat{x}_1, \hat{x}_2, \ldots, \hat{x}_n)$ and $\hat{y} = (\hat{y}_1, \hat{y}_2, \ldots, \hat{y}_m)$ satisfy the following saddlepoint condition:

$$f(x) - \sum_{i=1}^{m} \hat{y}_i g_i(x) \underset{\uparrow}{\leq} f(\hat{x}) - \sum_{i=1}^{m} \hat{y}_i g_i(\hat{x}) \underset{\uparrow}{\leq} f(\hat{x}) - \sum_{i=1}^{m} y_i g_i(\hat{x}),$$

$$\text{All } (x_1, x_2, \ldots, x_n) \in X \qquad\qquad \text{All } y_i \geq 0 \, (i = 1, 2, \ldots, m)$$

a) Show that $\hat{x}$ solves the nonlinear program

Maximize $f(x)$,

subject to:

$$g_i(x) \leq 0 \qquad (i = 1, 2, \ldots, m),$$

$$x \in X$$

where x refers to $(x_1, x_2, \ldots, x_n)$. [*Hint.* (1) Show that complementary slackness holds; i.e., $\sum_{i=1}^{m} \hat{y}_i g_i(\hat{x}) = 0$, using the righthand inequality with $y_i = 0 (i = 1, 2, \ldots, m)$ to show "$\geq$", and the signs of y_i and $g_i(x)$ to show "$\leq$". (2) Use the lefthand inequality and the sign of $g_i(x)$ to complete the proof.]

b) Show that the saddlepoint condition implies a strong duality of the form

Maximize $f(x)$	$=$	Minimize $L(y)$
subject to:		subject to:
$g_i(x) \leq 0$		$y_i \geq 0$
$x \in X$		

where $L(y)$ is defined in Exercise 28.

30. In linear programming, the duality property states that, whenever the primal (dual) has a finite optimal solution, then so does the dual (primal), and the values of their objective functions are equal. In nonlinear programming, this is not necessarily the case.

a) (*Duality gap*) Consider the nonlinear program:

Maximize $\{\text{Min } |(x_1 x_2)^{1/2}; 1|\}$,

subject to:
$$x_1 = 0,$$
$$x_1 \geqq 0, \quad x_2 \geqq 0.$$

The optimal objective-function value for the primal problem is clearly 0. The Lagrangian problem is:

$$L(y) = \text{Maximize } \{\text{Min } |(x_1 x_2)^{1/2}; 1| + y x_1\},$$

subject to:
$$x_1 \geqq 0, \quad x_2 \geqq 0.$$

Show that the optimal objective value of the dual problem

$$\text{Minimize } L(y)$$
$$y$$

is 1. (Note that y is unrestricted in the dual, since it is associated with an equality constraint in the primal.)

b) (*No finite shadow prices*) Consider the nonlinear program:

$$\text{Maximize } x_1,$$

subject to:
$$x_1^2 \leqq 0,$$
$$x_1 \text{ unrestricted.}$$

The optimal solution to this problem is clearly $x_1 = 0$. The Lagrangian problem is:

$$L(y) = \text{Maximize } \{x_1 - y x_1^2\},$$
$$x_1$$

where x_1 is unrestricted. Show that the optimal solution to the dual does not exist but that $L(y) \to 0$ as $y \to \infty$.

ACKNOWLEDGMENTS

A number of the exercises in this chapter are based on or inspired by articles in the literature.

Exercise 21: L. S. White, "Mathematical Programming Models for Determining Freshman Scholarship Offers," M.I.T. Sloan School Working Paper No. 379–9g.

Exercises 22 and 23: C. C. Holt, F. Modigliani, J. F. Muth, and H. A. Simon, *Planning Production, Inventories, and Work-Force*, Prentice-Hall, Inc., Englewood Cliffs, New Jersery, 1960.

Exercise 24: P. R. Winters, "Constrained Inventory Rules for Production Smoothing," *Management Science*, **8**, No. 4, July 1962.

Exercise 27: A. Charnes and W. W. Cooper, "Programming with Linear Fractional Functionals," *Naval Research Logistics Quarterly*, **9**, 1962; and S. P. Bradley and S. C. Frey, Jr. "Fractional Programming with Homogeneous Functions," *Operations Research*, **22**, No. 2, March–April 1974.

Exercise 30: M. Slater, "Lagrange Multipliers Revisited: A Contribution to Nonlinear Programming," Cowles Commission Paper, Mathematics 403, November 1950; and R. M. Van Slyke and R. J. Wets, "A Duality Theory for Abstract Mathematical Programs with Applications to Optimal Control Theory," *Journal of Mathematical Analysis and Applications*, **22**, No. 3, June 1968.

A System for Bank Portfolio Planning

14

Commercial banks and, to a lesser degree, other financial institutions have substantial holdings of various types of federal, state, and local government bonds. At the beginning of 1974, approximately twenty-five percent of the assets of commercial banks were held in these types of securities. Banks hold bonds for a variety of reasons. Basically, bonds provide banks with a liquidity buffer against fluctuations in demand for funds in the rest of the bank, generate needed taxable income, satisfy certain legal requirements tied to specific types of deposits, and make up a substantial part of the bank's investments that are low-risk in the eyes of the bank examiners.

In this chapter, we present a stochastic programming model to aid the investment-portfolio manager in his planning. The model does not focus on the day-to-day operational decisions of bond trading but rather on the strategic and tactical questions underlying a successful management policy over time. In the hierarchical framework presented in Chapter 5, the model is generally used for tactical planning, with certain of its constraints specified outside the model by general bank policy; the output of the model then provides guidelines for the operational aspects of daily bond trading.

The model presented here is a large-scale linear program under uncertainty. The solution procedure employs the decomposition approach presented in Chapter 12, while the solution of the resulting subproblems can be carried out by dynamic programming, as developed in Chapter 11. The presentation does not require knowledge of stochastic programming in general but illustrates one particular aspect of this discipline, that of "scenario planning." The model is tested by managing a hypothetical portfolio of municipal bonds within the environment of historical interest rates.

14.1 OVERVIEW OF PORTFOLIO PLANNING

The bond-portfolio management problem can be viewed as a multiperiod decision problem under uncertainty, in which portfolio decisions are periodically reviewed and revised. At each decision point, the portfolio manager has an inventory of securities and funds on hand. Based on present credit-market conditions and his assessment of future interest-rate movements and demand for funds, the manager must decide which bonds to hold in the portfolio over the next time period, which bonds to sell, and which bonds to purchase from the marketplace. These decisions are made subject to constraints on total portfolio size, exposure to risk in the sense of realized and unrealized capital losses,* and other policy limitations on the makeup of the portfolio. At the next decision point, the portfolio manager faces a new set of interest rates and bond prices, and possibly new levels for the constraints, and he must then make another set of portfolio decisions that take the new information into account.

Before describing the details of the portfolio-planning problem, it is useful to point out some of the properties of bonds. A bond is a security with a known fixed life, called its maturity, and known fixed payment schedule, usually a semiannual coupon rate plus cash value at maturity. Bonds are bought and sold in the marketplace, sometimes above their face value, or par value, and sometimes below this value. If we think of a bond as having a current price, coupon schedule, and cash value at maturity, there is an internal rate of return that makes the price equal to the present value of the subsequent cash flows, including both the interest income from the coupon schedule and the cash value at maturity. This rate of return is known as the "yield to maturity" of a bond.

Given the attributes of a bond, knowing the price of a bond is equivalent to knowing the yield to maturity of that bond. Since the payment schedule is fixed when the bond is first issued, as bond prices rise the yield to maturity falls, and as bond prices fall the yield to maturity rises. Bond prices are a function of general market conditions and thus rise and fall with the tightening and easing of credit. Usually the fluctuations in bond prices are described in terms of yields to maturity, since these can be thought of as interest rates in the economy. Hence, bond prices are often presented in the form of yield curves. Figure 14.1 gives a typical yield curve for "good-grade" municipal bonds. Usually the yield curve for a particular class of securities rises with increasing maturity, reflecting higher perceived market risk associated with the longer maturities.

One final point concerns the transaction costs associated with bond trading. Bonds are purchased at the "asked" price and, if held to maturity, have no transaction cost. However, if bonds are sold before their maturity, they are sold at the "bid" price, which is lower than the "asked" price. The spread between these prices can be thought of as the *transaction cost* paid at the time the securities are sold.

---

* *Realized* capital losses refer to actual losses incurred on bonds sold, while *unrealized* capital losses refer to losses that would be incurred if bonds currently held had to be sold.

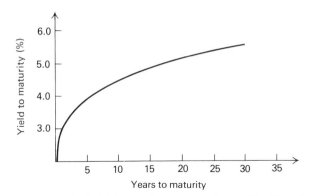

Fig. 14.1 Typical yield curve for good-grade municipal bonds.

At the heart of the portfolio-planning problem is the question of what distribution of maturities to hold during the next period and over the planning horizon in general. The difficulty of managing an investment portfolio stems not only from the uncertainty in future interest-rate movements but from the conflicting uses made of the portfolio. On the one hand, the portfolio is used to generate income, which argues for investing in the highest-yielding securities. On the other hand, the portfolio acts as a liquidity buffer, providing or absorbing funds for the rest of the bank, depending upon other demand for funds. Since this demand on the portfolio is often high when interest rates are high, a conflict occurs, since this is exactly when bond prices are low and the selling of securities could produce capital losses that a bank is unwilling to take. Since potential capital losses on longer maturities are generally higher than on shorter maturities, this argues for investing in relatively shorter maturities.

Even without using the portfolio as a liquidity buffer, there is a conflict over what distribution of maturities to hold. When interest rates are low, the bank often has a need for additional income from the portfolio; this fact argues for investing in longer maturities with their correspondingly higher yields. However, since interest rates are generally cyclical, if interest rates are expected to rise, the investment in longer maturities could build up substantial capital losses in the future, thus arguing for investing in shorter maturities. The opposite is also true. When interest rates are high, the short-term rates approach (and sometimes exceed) the long-term rates; this fact argues for investing in shorter maturities with their associated lower risk. However, if interest rates are expected to fall, this is exactly the time to invest in longer maturities with their potential for substantial capital gains in a period of falling interest rates.

Many commercial banks manage their investment portfolio using a "laddered" maturity structure, in which the amount invested in each maturity is the same for all maturities up to some appropriate length, say 15 years. Generally, the longer the ladder, the more risky the portfolio is considered. Figure 14.2(a) illustrates a 15-year ladder. Each year one fifteenth (i.e., $6\frac{2}{3}$ percent) of the portfolio matures and needs to

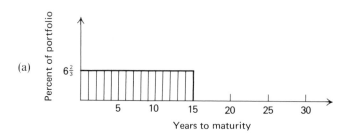

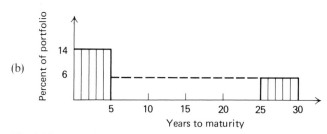

Fig. 14.2 (a) Typical laddered portfolio. (b) Typical barbell portfolio.

be reinvested, along with the usual interest income. In a laddered portfolio, the cash from maturing securities is reinvested in fifteen-year bonds while the interest income is reinvested equally in all maturities to maintain the laddered structure. The advantages of a laddered portfolio are: no transaction costs or realized losses, since bonds are always held to maturity rather than sold; generally high interest income, since the yield curve is usually rising with increasing maturity; and ease of implementation, since theoretically no forecasting is needed and a relatively small percentage of the portfolio needs to be reinvested each year.

Some banks, on the other hand, manage their portfolio using a "barbell" maturity structure, in which the maturities held are clustered at the short and long ends of the maturity spectrum, say 1 to 5 years and 26 to 30 years, with little if any investment in intermediate maturities. Figure 14.2(b) illustrates a typical barbell portfolio structure with 70 percent short- and 30 percent long-term maturities. The riskiness of the portfolio is judged by the percentage of the portfolio that is invested in the long maturities. Each end of the barbell portfolio is managed similarly to a ladder. On the short end, the maturing securities are reinvested in 5-year bonds, while on the long end the 25-year securities are sold and the proceeds reinvested in 30-year securities. The interest income is then used to keep the percentages of the portfolio in each maturity roughly unchanged.

The advantages of a barbell portfolio are usually stated in terms of being more "efficient" than a laddered portfolio. The securities on the long end provide relatively high interest income, as well as potential for capital gains in the event of falling interest rates, while the securities on the short end provide liquid assets to meet various demands for cash from the portfolio for other bank needs. In the barbell portfolio illustrated in Fig. 14.2(b), 20 percent of the portfolio is reinvested each year, since 14 percent matures on the short end and roughly 6 percent is sold on the long end. Comparing this with the $6\frac{2}{3}$ percent maturing in the 15-year ladder, it is argued that a barbell portfolio is more flexible than a laddered portfolio for meeting liquidity needs or anticipating movements in interest rates.

However, effectively managing a barbell portfolio over time presents a number of difficulties. First, significant transaction costs are associated with maintaining a barbell structure since, as time passes, the long-term securities become shorter and must be sold and reinvested in new long-term securities. Second, the short-term securities are not risk-free, since the income and capital received at maturity must be reinvested in new securities at rates that are currently uncertain. To what extent is a barbell portfolio optimal to maintain over time? One might conjecture that often it would not be advantageous to sell the long-term securities of the barbell structure and, hence, that over time the barbell would eventually evolve into a laddered structure.

In order to systematically address the question of what distribution of maturities should be held over time, a stochastic programming model was developed. The basic approach of this model, referred to as the **BONDS** model, is one of "scenario planning." The essential idea of scenario planning is that a limited number of possible evolutions of the economy, or scenarios, is postulated, and probabilities are assigned to each. All the uncertainty in the planning process is then reduced to the question of which scenario will occur. For each scenario, a fairly complex set of attributes might have to be determined; but, given a particular scenario, these attributes are known with certainty.

We can illustrate this process by considering the tree of yield curves given in Fig. 14.3. We can define a collection of scenarios in terms of the yield curves assumed to be possible. Actually, a continuum of yield curves can occur in each of the future planning periods; however, we approximate our uncertainty as to what will occur by selecting a few representative yield curves. Suppose we say that in a three-period example, interest rates can rise, remain unchanged, or fall, in each period with equal probability. Further, although the *levels* of interest rates are serially correlated, there is satistical evidence that the distributions of *changes* in interest rates from one period to the next are independent. If we make this assumption, then there are three possible yield curves by the end of the first period, nine at the end of the second, and twenty-seven by the end of the third. (The yield curves at the end of the third period have not been shown in Fig. 14.3.) A scenario refers to *one specific sequence* of yield curves that might occur; for example, rates might rise, remain unchanged, and then fall over the three periods. The total number of scenarios in this example is $3 \times 9 \times 27$, or 729. Of course, the large number of scenarios results from our independence assumption, and it might be reasonable to eliminate some of these alternatives to reduce the problem size.

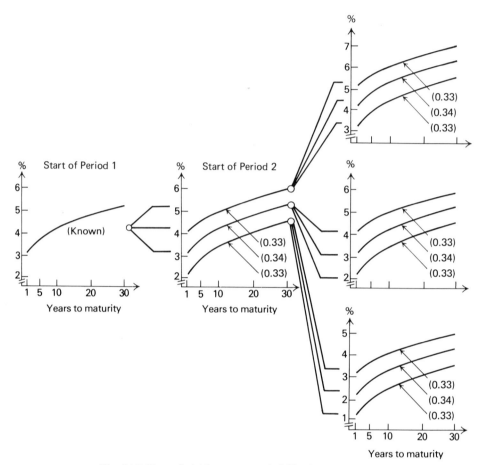

Fig. 14.3 Tree of yield curves; probability in parentheses.

A scenario, defined by a sequence of yield curves, will have additional characteristics that place constraints on the portfolio strategy for that scenario. Since rising interest rates mean a tightening of credit, often funds are withdrawn from the portfolio, under such scenarios, to meet the demands for funds in the rest of the bank. When interest rates are falling, funds are usually plentiful, and additional funds are often made available to the portfolio. Further, a limitation on the investment strategy is imposed by the level of risk the bank is willing to tolerate. This can be expressed for each scenario by limiting the losses that may be realized within a tax year, as well as by limiting the unrealized capital losses that are allowed to build up in the portfolio over the planning horizon. Another limitation on investment strategy results from the bank's "pledging" requirements. The holdings of government securities, as well as the holdings of some state and local bonds, are affected by the levels of certain types of deposits. The fluctuations of these deposits are then forecast for each planning scenario, to indicate the minimum holdings of the securities that will satisfy the pledging requirements. The minimum holdings of government securities may also be affected by the bank's need for taxable income, although this taxable-income requirement also could be a characteristic of each scenario directly specified by the portfolio manager.

Scenario planning is the key to being able to concentrate on the investment portfolio. The interface between the investment portfolio and the rest of the bank is accounted for by using consistent definitions of scenarios for planning throughout the bank. For planning the investment portfolio, this interface is characterized by the demand on the portfolio for funds, the allowable levels of realized and un-realized losses in the portfolio, the limits on the holdings of certain broad categories of securities, as well as any other element of a scenario that the portfolio manager deems important for the planning problem being addressed. These characteristics of the scenarios are then tied to interest-rate movements by using the same definitions of scenarios for assessing them as for forecasting yield-curve movements. The scenario-planning process is illustrated in Section 14.4 where we discuss managing a hypothetical portfolio.

14.2 FORMULATION OF THE BONDS MODEL

The most important assumption in the formulation of the model is that the planning is being carried out with a limited number of economic scenarios. The scenarios are usually keyed to the movement of some appropriate short-term interest rate, such as the 90-day treasury bill rate. The possible movements of the short-term rate generate a collection of scenarios each of which consists of a particular sequence of yield curves and exogenous cash flows, as well as other characteristics for each period in the planning horizon. The assumption of a finite number of scenarios is equivalent to making a discrete approximation of the continuous distribution of *changes* in the short-term rate, and this in turn, along with the finite number of planning periods, permits the formulation of an ordinary linear program that ex-plicitly takes uncertainty into account. Associated with any particular scenario is its probability of occurrence, which is used to structure the objective function of the linear program so as to maximize the expected horizon value of the portfolio.

The remaining characteristics of the economic scenarios are policy considerations involving the interface between the investment portfolio and the rest of the bank. For each tax year in the planning horizon, a maximum level of losses that may be realized is usually specified for each scenario. Further, the maximum level of un-realized losses that potentially could build up in the portfolio over the planning horizon is often specified. In the situation where more than one broad category of securities is being analyzed, either maximum or minimum levels of the holdings of a particular category might be specified. For example, a minimum level of U.S. Treasury holdings typically is specified, to cover the pledging of specific securities to secure certain types of state and municipal deposits.

For any particular analysis that the portfolio manager is considering, he must first group the securities to be included in the planning by broad categories, and then aggregate the securities available for purchase into a number of security classes within each category. The broad categories usually refer to securities described by the same yield curve, such as U.S. Treasury bonds or a particular grade of municipal bonds. The aggregation of securities within these broad categories is by *time to maturity*, such as 3 months, 6 months, 1 year, 2 years, . . . , 30 years. These security classes will usually not include all maturities that are available but some appropriate aggregation of these maturities.

The remainder of this section specifies the details of the mathematical formulation of the BONDS model. The discussion is divided into three parts: the decision variables, the constraints, and the objective function.

Decision Variables

At the beginning of each planning period, a particular portfolio of securities is currently held, and funds are either available for investment or required from the portfolio. The portfolio manager must decide how much of each security class k to *buy*, $b_n^k(e_n)$, and how much of each security class currently held to *sell* $s_{m,n}^k(e_n)$ or continue to *hold* $h_{m,n}^k(e_n)$. The subscript n identifies the current period and m indicates the period when the security class was purchased. Since the amount of capital gain or loss when a security class is sold will depend on the difference between its purchase price and sales price, the portfolio manager must keep track of the amount of each security class held, by its period of purchase. Further, since the model computes the optimal decisions at the beginning of every period for each scenario, the variables that represent decisions at the start of period n must be conditional on the scenario evolution e_n up to the start of period n. An example of a scenario evolution up to the start of period 3 would be "interest rates rise in period 1 and remain unchanged in period 2." More precisely, the decision variables are defined as follows:

$b_n^k(e_n)$ = Amount of security class k *purchased* at the beginning of period n, conditional on scenario evolution e_n; in dollars of initial purchase price.

$s_{m,n}^k(e_n)$ = Amount of security class k, which had been purchased at the beginning of period m, *sold* at the beginning of period n, conditional on scenario evolution e_n; in dollars of initial purchase price.

$h_{m,n}^k(e_n)$ = Amount of security class k, which had been purchased at the beginning of period m, *held* (as opposed to sold) at the beginning of period n, conditional on scenario evolution e_n; in dollars of initial purchase price.

It should be pointed out that liabilities, as well as assets, can be included in the model at the discretion of the planner. Banks regularly borrow funds by participating in various markets open to them, such as the CD (negotiable certificate of deposit) or Eurodollar markets. The portfolio manager can then use these "purchased funds" for either financing a withdrawal of funds from the portfolio or increasing the size of the portfolio. However, since the use of these funds is usually a policy decision external to the investment portfolio, an elaborate collection of liabilities is not needed. The portfolio planner may include in the model a short-term liability available in each period with maturity equal to the length of that period and cost somewhat above the price of a short-term asset with the same maturity.

Constraints

The model maximizes the expected value of the portfolio at the end of the planning horizon subject to five types of constraints on the decision variables as well as non-negativity of these variables. The types of constraints, each of which will be discussed below, include the following: funds flow, inventory balance, current holdings, net capital loss (realized and unrealized), and broad category limits. In general, there are separate constraints for every time period in each of the planning scenarios. The mathematical formulation is given in Table 14.1, where e_n is a particular scenario

Table 14.1 Formulation of the BONDS model

Objective function	Maximize $\displaystyle\sum_{e_N \in E_N} p(e_N) \sum_{k=1}^{K}\left[\sum_{m=0}^{N-1}(y_m^k(e_m) + v_{m,N}^k(e_N))h_{m,N}^k(e_N)\right.$ $\left. + (y_N^k(e_N) + v_{N,N}^k(e_N))b_N^k(e_N)\right]$
Funds flow	$\displaystyle\sum_{k=1}^{K} b_n^k(e_n) - \sum_{k=1}^{K}\left[\sum_{m=0}^{n-2} y_m^k(e_m)h_{m,n-1}^k(e_{n-1}) + y_{n-1}^k(e_{n-1})b_{n-1}^k(e_{n-1})\right]$ $\displaystyle - \sum_{k=1}^{K}\sum_{m=0}^{n-1}(1 + g_{m,n}^k(e_n))s_{m,n}^k(e_n) = f_n(e_n)$ $\forall e_n \in E_n \qquad (n = 1, 2, \ldots, N)$
Inventory balance	$-h_{m,n-1}^k(e_{n-1}) + s_{m,n}^k(e_n) + h_{m,n}^k(e_n) = 0 \qquad (m = 0, 1, \ldots, n-2)$ $-b_{n-1}^k(e_{n-1}) + s_{n-1,n}^k(e_n) + h_{n-1,n}^k(e_n) = 0$ $\forall e_n \in E_n \qquad (n = 1, 2, \ldots, N; k = 1, 2, \ldots, K)$
Initial holdings	$h_{0,0}^k(e_0) = h_0^k \qquad (k = 1, 2, \ldots, K)$
Capital losses	$\displaystyle -\sum_{k=1}^{K}\sum_{m=n'}^{n} g_{m,n}^k(e_n)s_{m,n}^k(e_n) \leq L_n(e_n) \qquad \forall e_n \in E_n, \qquad \forall n \in N'$
Category limits	$\displaystyle \sum_{k \in K^i}\left[b_n^k(e_n) + \sum_{m=0}^{n-1} h_{m,n}^k(e_n)\right] \begin{matrix}\geq \\ (\leq)\end{matrix} C_n^i(e_n) \qquad \forall e_n \in E_n$ $(n = 1, 2, \ldots, N; i = 1, 2, \ldots, I)$
Nonnegativity	$b_n^k(e_n) \geq 0, \quad s_{m,n}^k(e_n) \geq 0, \quad h_{m,n}^k(e_n) \geq 0 \qquad \forall e_n \in E_n,$ $(m = 1, 2, \ldots, n-1; n = 1, 2, \ldots, N; k = 1, 2, \ldots, K)$

evolution prior to period n and E_n is the set of *all possible scenario evolutions* prior to period n.

Funds Flow

The funds-flow constraints require that the funds used for purchasing securities be equal to the sum of the funds generated from the coupon income on holdings during the previous period, funds generated from sales of securities, and exogenous funds flow. We need to assess coefficients reflecting the income *yield* stemming from the semiannual coupon interest from holding a security and the capital *gain or loss* from selling a security, where each is expressed as a percent of initial purchase price. It is assumed that taxes are paid when income and/or gains are received, so that these coefficients are defined as after-tax. Transaction costs are taken into account by adjusting the gain coefficient for the broker's commission; i.e., bonds are purchased at the "asked" price and sold at the "bid" price. We also need to assess the exogenous funds flow, reflecting changes in the level of funds made available to the portfolio. The exogenous funds flow may be either positive or negative, depending on whether funds are being made available to or withdrawn from the portfolio, respectively.

The income yield from coupon interest, the capital gain or loss from selling a security, and the exogenous funds flow can be defined as follows:

$g_{m,n}^k(e_n)$ = Capital gain or loss on security class k, which had been purchased at the beginning of period m and was sold at the beginning of period n conditional on scenario evolution e_n; per dollar of initial purchase price.

$y_m^k(e_n)$ = Income yield from interest coupons on security class k, which was purchased at the beginning of period m, conditional on scenario evolution e_n; per dollar of initial purchase price.

$f_n(e_n)$ = Incremental amount of funds either made available to or withdrawn from the portfolio at the beginning of period n, conditional on scenario evolution e_n; in dollars.

Since it is always possible to purchase a one-period security that has no transaction cost, the funds-flow constraints hold with equality implying that the portfolio is at all times fully invested. Finally, if short-term liabilities are included in the model, then these funds-flow constraints would also reflect the possibility of generating additional funds by selling a one-period liability.

Inventory Balance

The current holdings of each security class purchased in a particular period need to be accounted for in order to compute capital gains and losses. The inventory-balance constraints state that the amount of these holdings sold, plus the remaining amount held at the beginning of a period, must equal the amount on hand at the end of the previous period. The amount on hand at the end of the previous period is either the amount purchased at the beginning of the previous period or the amount held from an earlier purchase.

It is important to point out that this formulation of the problem includes security classes that mature before the time horizon of the model. This is accomplished by setting the hold variable for a matured security to zero (actually dropping the variable from the model). This has the effect, through the inventory-balance constraints, of forcing the "sale" of the security at the time the security matures. In this case, the gain coefficient reflects the fact that there are no transaction costs when securities mature.

Current Holdings

The inventory-balance constraints also allow us to take into account the securities held in the initial portfolio. If the amounts of these holdings are:

h_0^k = Amount of security class k held in the initial portfolio; in
dollars of initial purchase price,

the values of the variables that refer to the holdings of securities in the initial portfolio, $h_{0,0}^k(e_0)$, are set to these amounts.

Capital Losses

Theoretically, we might like to maximize the bank's expected utility for coupon income and capital gains over time. However, such a function would be difficult for a portfolio manager to specify; and further, management would be unlikely to have much confidence in recommendations based on such a theoretical construct. Therefore, in lieu of management's utility function, a set of constraints is added that limit the net *realized* capital loss during any year, as well as the net *unrealized* capital loss that is allowed to build up over the planning horizon.

Loss constraints are particularly appropriate for banks, in part because of a general aversion to capital losses, but also because of capital adequacy and tax considerations. Measures of adequate bank capital, such as that of the Federal Reserve Board of Governors, relate the amount of capital required to the amount of "risk" in the bank's assets. Thus, a bank's capital position affects its willingness to hold assets with capital-loss potential. Further, capital losses can be offset against taxable income to reduce the size of the after-tax loss by roughly 50 percent. As a result, the amount of taxable income, which is sometimes relatively small in commercial banks, imposes an upper limit on the level of losses a bank is willing to absorb.

The loss constraints sum over the periods contained in a particular year the gains or losses from sales of securities in that year, and limit this value to:

$L_n(e_n)$ = Upper bound on the realized net capital loss (after taxes)
from sales during the year ending with period n, conditional
on scenario evolution e_n; in dollars.

In Table 14.1, N' is the set of indices of periods that correspond to the end of fiscal years, and n' is the index of the first period in a year defined by an element of N'. Thus the loss constraints sum the losses incurred in all periods that make up a fiscal year. Since the model forces the "sale" of all securities at the horizon without transaction costs, the unrealized loss constraints have the same form as the realized loss constraints.

Category Limits

It may be of considerable interest to segment the portfolio into a number of broad asset categories each of which is described by different yield curves, transaction costs, income-tax rates, and distribution of maturities. There is no conceptual difficulty with this; however, some computational difficulties may arise due to problem size. Typically the investment portfolio might be segmented into U.S. Treasury and tax-exempt securities. In addition, the tax-exempt securities might be further segmented by quality such as prime-, good-, and medium-grade municipals. In making such a segmentation, we often impose upper- or lower-bound constraints on the total holdings of some of the asset categories. The example cited earlier involved placing lower limits on the amount of U.S. Treasury bonds held for pledging purposes. Defining

$C_n^i(e_i)$ = Lower (upper) bound on the level of holdings of asset
category i, at the beginning of period n, conditional on
scenario evolution e_n; in dollars of initial purchase price,

and letting K^i be the index set of the ith asset category, the constraints in Table 14.1 merely place an upper or lower bound on the holdings of a particular broad asset category.

Objective Function

The objective of the model is to maximize the expected value of the portfolio at the end of the final period. It should be pointed out that this assumes that the portfolio manager is indifferent between revenues received from interest and revenues received from capital gains, since each add equivalent dollars to the value at the horizon. If desired, it would be possible to include interest-income constraints to ensure that sufficient income would be achieved by the portfolio during each period.

The final value of the portfolio consists of the interest income received in the final period, plus the value of securities held at the horizon. It is not obvious, though, how the value of the portfolio holdings should be measured, since they are likely to contain some unrealized gains and losses. Should these gains or losses be calculated before or after taxes? Before or after transaction costs? At one extreme, it would be possible to assume that the portfolio would be sold at the horizon, so that its final value would be after taxes and transaction costs. This approach would tend to artificially encourage short maturities in the portfolio, since they have low transaction costs. The alternative approach of valuing the portfolio before taxes and transaction costs is equally unrealistic. For simplicity, it is usually assumed that the value of the securities at the horizon is after taxes but before transaction costs.

The objective function can be defined in terms of the decisions made at the start of the final period, which are conditional on the evolution of each scenario up to that point. For each scenario, the value of any holdings at the start of the final period should reflect the expected capital gain or loss over the final period and the coupon income to be received in the final period. The objective function can be formalized by defining the *probability* of scenario evolution and the *value* of the noncash holdings at the start of the final period as follows:

$v_{m,N}(e_N)$ = Expected (over period N) cash value per dollar of initial
purchase price of security class k, which had been purchased
at the beginning of period m, and held at the start of
period N, conditional on scenario evolution e_n;

$p(e_N)$ = Probability that scenario evolution e_N occurs prior to
period N.

The expected horizon value of the portfolio given in Table 14.1 is then determined by weighting the value of the holdings at the start of the final period by the probability of each scenario.

14.3 PROBLEM SIZE AND STRUCTURE

In order to have a feeling for the potential size of the model formulated, the number of constraints implied under various model assumptions can be computed. Assume for the moment that the number of events in each time period is the same, and equal to D. Thus there are D scenario evolutions for the first period, D^2 for the second, and so forth. Further, let there be a total of n time periods with n_i periods in year i. Then if K is the total number of different security classes in all categories, and I is the number of broad asset categories, the number of equations can be calculated as follows:

Cash flow: $1 + D + D^2 + \cdots + D^{n-1}$

Net capital loss: $D^{n_1} + D^{n_1 + n_2} + \cdots$

Category limits: $(I - 1)[1 + D + D^2 + \cdots + D^{n-1}]$

Inventory balance: $K[D + 2D^2 + \cdots + nD^{n-1}]$

Table 14.2 indicates the number of each type of constraint under a variety of assumptions. It is clear that, for even a relatively small number of events and time periods, the problem size rapidly becomes completely unmanageable. However, it is also clear that the main difficulty lies with the number of inventory-balance constraints. Hence, an efficient solution procedure is likely to treat these constraints implicitly instead of explicitly.

Figure 14.4 illustrates the structure of the linear-programming tableau for a model involving three time periods and three securities. Note that the inventory-balance constraints exhibit a block diagonal structure. Given this structure, the inventory-balance constraints can be treated implicitly rather than explicitly by a number of techniques of large-scale mathematical programming. For this particular application, the decomposition approach, which was introduced in Chapter 12, was used, since the resulting special structure of the subproblems could be exploited readily, both in the solution of the subproblems and in structuring the restricted master. In this approach, there is one subproblem for each diagonal block of inventory-balance constraints, while the restricted master linear program that must be solved at each iteration has constraints corresponding to the funds flow, net capital loss, and category limit constraints.

Table 14.2 Number of constraints in various models

Period/year	$I = 2, \; K = 8$					
	$D = 3$			$D = 5$		
	$n_1 = 1, \; n_2 = 1$ $n_3 = 1$	$n_1 = 2, \; n_2 = 1$ $n_3 = 1$	$n_1 = 3, \; n_2 = 1$ $n_3 = 1$	$n_1 = 1, \; n_2 = 1$ $n_3 = 1$	$n_1 = 2, \; n_2 = 1$ $n_3 = 1$	$n_1 = 3, \; n_2 = 1$ $n_3 = 1$
Cash flow	13	40	121	31	160	781
Net capital loss	39	117	350	155	775	3,875
Category limit	13	40	121	31	160	781
Inventory balance	168	816	3408	440	3440	23,440
Total constraints	233	1613	4000	657	4535	28,877

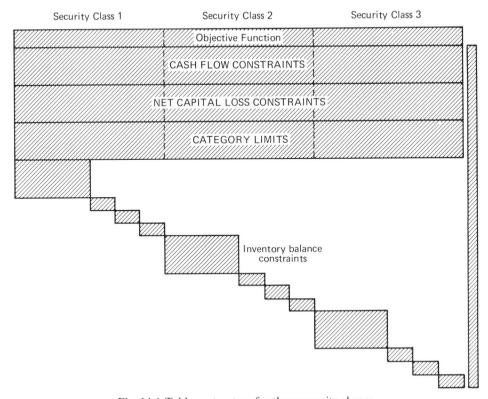

Fig. 14.4 Tableau structure for three security classes.

The subproblems correspond to purchasing each security class at the start of a given period, conditional on every scenario evolution up to that period. For example, purchasing security class k at the start of period one defines one subproblem and purchasing the same security class at the start of period two defines one additional subproblem for each scenario evolution in period one. This is indicated in the tableau structure given in Fig. 14.4 by the three small rectangles following one large one on the diagonal. The decision variables of a subproblem involve selling and holding in subsequent periods the security class purchased. This allows capital gains and losses on sales of securities to be determined.

To illustrate the subproblems further, consider first the period 1 subproblems. Security class k is available for purchase at the start of period 1. If a decision to purchase is made, this security class is then available for sale at the start of period 2, or it may be held during the second period. The amount that is held is then available for sale or holding at the start of the third period. This multistage problem, involving a sequence of sell and hold decisions after a purchase, can be solved by a recursive procedure. The problem has a dynamic-programming structure (see Chapter 11) where the state of the system at time n is defined by the amount of the initial purchase still held. This amount is constrained by the inventory-balance equations, which limit the amount sold in any period to be less than or equal to the amount on hand.

Note that if security class k is purchased at the start of period 2, its purchase price and income yield are conditional on the scenario evolution which occurred during period 1. Thus, one subproblem is defined for each security class that can be purchased and every possible scenario evolution that precedes the purchase period of that class. As would be expected, the subproblems have no decision variables in common with one another, since each set of inventory-balance constraints simply keeps track of the remaining holdings of a particular purchase. This approach leads to a relatively large number of subproblems; however, the rationale is that the subproblems should be efficient to solve, since the state variable of each is one-dimensional.

Another interesting point to note about the subproblem constraints is that they are homogeneous systems of equations (i.e., zero righthand sides). As we saw in Chapter 12, the fundamental theorem employed in decomposition is that the feasible region defined by a system of linear equations may be represented by a convex combination of its extreme points, plus a nonnegative combination of its extreme rays. The solutions of any subproblem have only one extreme point, all decision variables equal to zero. For any nonzero point satisfying the subproblem constraints, a scalar times that point also satisfies the constraints; and hence, with a linear objective function, there exists an associated unbounded solution. As a result we need consider only the extreme rays of the subproblems. These extreme rays may be constructed in an efficient manner either by dynamic programming or by exploiting the triangular structure of the dual of a related "ray-finding" linear program. The ray-finding problem is defined by setting the "buy" variable of any subproblem to one, and then determining the optimal sequence of sell and hold decisions for this one unit.

The restricted master for this decomposition scheme reflects the fact that only extreme rays of the subproblems need to be considered. The usual constraints that require that the solution be a convex combination of extreme points of the subproblems are not necessary, since the only restriction on the weights on the extreme rays is that they be nonnegative. When the model is solved, all profitable rays found at an iteration are added as columns to the restricted master. The restricted master is then solved by continuing the simplex method from the previous solution, which yields new shadow prices, or dual variables. The shadow prices are then used to modify the objective functions of the subproblems and the process is repeated. If no profitable ray is found for any subproblem, then the algorithm terminates, and we have an optimal solution.

The value of the optimal solution is merely given by the nonnegative weighted combination of the subproblem solutions when the weights are determined by the values of the variables in the final restricted master. In general, we need *not* add any unprofitable ray to the restricted master. However, a ray that is unprofitable at one iteration may become profitable at a future iteration, as the objective functions of the subproblems are modified by the shadow prices. Hence, if one profitable ray is generated for any subproblem at an iteration, all new rays generated, profitable or not, are in fact added to the restricted master as columns.

As the restricted master is augmented by more and more columns, those columns not in the current basis are retained, provided storage limitations permit. As storage limitations become binding, those columns that price out most negatively are dropped. Any dropped column will be regenerated automatically if needed.

Further details in the computational procedure are included in the exercises.

14.4 MANAGING A HYPOTHETICAL PORTFOLIO

In the remainder of this chapter, the use of the BONDS model is illustrated by addressing the question of what portfolio strategy should be adopted over time. The model is used to "manage" a hypothetical portfolio of municipal securities over a 10-year historical period. The important advantage derived from using such a model to aid in planning the portfolio maturity structure is that it gives the portfolio manager the opportunity to take explicitly into account the characteristics of the current portfolio, as well as expected interest-rate swings, liquidity needs, programs for realized losses, and exposure to unrealized losses.

Ideally, the performance of the model should be evaluated using Monte Carlo simulation. However, such an experiment would involve a significant number of simulation trials, where portfolio revisions would have to be made by the optimization model at the beginning of each year of the simulation. Updating and running the BONDS model such a large number of times would be prohibitively expensive, from the standpoints of both computer and analyst time.

As an alternative to the Monte Carlo simulation, it is possible to perform a *historical* simulation, which considers how interest rates actually behaved over a particular period of time, and then attempts to plan a portfolio strategy that could have been followed over this period. To implement this approach, the ten-year historical period starting with January 1, 1964 was chosen. In order to keep the simulation simple, portfolio decisions were allowed to be made once a year and the resulting portfolio was held for the entire year. It is not suggested that any bank would have followed the strategy proposed by the model for the entire year; however, it can be considered a rough approximation of such a strategy over the ten-year period.

It should be strongly emphasized that it is difficult to draw firm conclusions from the results of a historical simulation against one particular realization of interest rates. A strategy that performed well against that particular sequence of rates might have performed poorly against some other sequence of rates that had a relatively high likelihood of occurring. The opposite is, of course, also true. However, it does allow us to make comparisons between strategies for a particular sequence of interest rates that indeed did occur.

The historical simulation covered January 1964 through January 1974, a period of ten years. However, a number of years of history prior to the beginning of the simulation period were included, since it was necessary to assess interest-rate expectations for a portfolio manager at the beginning of 1964. Figure 14.5 gives the yields on one-, ten-, and thirty-year maturities for good-grade municipal bonds covering

Fig. 14.5 Yields of good-grade municipals.

the appropriate period. Although some years exhibited a great deal of variation within a year, with a potential for improving performance through an active trading policy, this variation was not included, since in the historical simulation the portfolio was revised only at the beginning of each year.

The basic approach of the historical simulation was to use the BONDS model to make portfolio decisions at the beginning of each year, given the current actual yield curve and the portfolio manager's "reasonable expectations" of future interest-rate movements. These decisions were then implemented, the actual performance of the portfolio in that year was revealed, and the process was repeated. There were two steps in modeling the yield curves needed for the simulation. First, eleven actual yield curves were of interest—one for the beginning of each year when portfolio decisions were made, and one for the final performance evaluation. These yield curves were developed by fitting a functional form to historical data. Second, the portfolio manager's reasonable expectations about future interest-rate fluctuations were modeled by constructing a tree of yield curves, similar to those given in Fig. 14.3, for each year in the simulation.

Modeling the eleven yield curves that occurred was relatively straightforward. Data were available from the actual yield curves at each point in time covering the 1-, 2-, 5-, 10-, 20-, and 30-year maturities. Since yield curves are generally considered to be smooth curves, the data for each curve were fitted with the following functional form:

$$R_m = am^b e^{cm},$$

where R_m is the yield to maturity on securities with m years to maturity, and a, b, and c are constants to be determined from the data. In the simulation, the yield for any maturity was then taken from the derived yield curve for the appropriate year.

Modeling the tree of yield curves reflecting the portfolio manager's reasonable expectations of future interest-rate movements was more complicated. In the historical simulation, portfolio decisions were made as if it were January 1964; it was important not to use any information that was *not available* to a portfolio manager at that time. It is difficult to imagine or reconstruct exactly what a portfolio manager would have forecast for future interest-rate changes in January 1964. Therefore, for the purpose of the simulation, the portfolio manager's interest-rate assessments were mechanically based on the previous seven years of data at each stage. The simulation started with the interest-rate data for the years 1957 through and including 1963; and, based on these data, a tree of yield curves reflecting the portfolio manager's expectations of interest rates was constructed. This tree of yield curves was then used in making portfolio decisions for the year 1964. For each subsequent year, the data corresponding to the year just past was added to the data base and the data more than seven years old were dropped; a new tree of yield curves was then constructed.

Two separate analyses were performed to estimate the tree of yield curves representing the portfolio manager's reasonable expectations of interest rates at the beginning of each of the ten years of the simulation.

First, the distributions of one-year changes in the one-year rate were estimated for each year in the simulation. A monthly time series covering the prior seven years was used to determine the actual distribution of the one-year changes in the one-year rate. Table 14.3 gives the means and standard deviations of these distributions.

Table 14.3 Distribution of one-year changes in one-year rate

Year	Mean (Basis points)*	Standard deviation (Basis points)
'64	−0.2	74.6
'65	−3.3	72.4
'66	12.4	59.8
'67	15.2	61.9
'68	11.6	59.0
'69	24.4	52.3
'70	41.9	68.6
'71	37.7	83.5
'72	11.2	109.5
'73	6.7	110.5

* 100 basis-point change equals 1 percentage-point change.

Second, the changes in two other rates, the twenty- and thirty-year rates, were forecast, *conditional* on the changes in the one-year rate. Then, given a forecast change in the one-year rate, three points on the corresponding forecast future-yield curve could be determined, by adding the current levels for these rates to the forecast changes in these rates. The new levels for the one-, twenty-, and thirty-year rates then determined the constants *a*, *b*, and *c* for the functional form of the yield curve given above.

To forecast the changes in the twenty- and thirty-year rates, conditional on the changes in the one-year rate, two regressions were performed for each year in the simulation. The changes in each of these two rates were separately regressed against changes in the one-year rate. The two regression equations were then used to compute the changes in these rates as deterministic functions of the changes in the one-year rate. Table 14.4 shows the means and standard deviations of the regression coefficients as well as a goodness-of-fit measure. The mean of the regression coefficient gives the change in the twenty-year rate or the change in the thirty-year rate as a fraction of the change in the one-year rate. Given a forecast change in the one-year rate, forecasts of the changes in the twenty- and thirty-year rates were determined by multiplying the change in the one-year rate by the appropriate fraction.

Table 14.4 Changes in the 20- and 30-year rates as a fraction of the changes in the one-year rate

Year	20-year fraction			30-year fraction		
	Mean	*Standard deviation*	*R-square*	*Mean*	*Standard deviation*	*R-square*
'64	.3615	.0416	85.6	.3289	.0398	86.8
'65	.2820	.0312	91.9	.2473	.0274	93.7
'66	.3010	.0354	89.1	.2762	.0298	92.3
'67	.3862	.0331	90.4	.3222	.0261	94.0
'68	.3578	.0399	86.3	.2792	.0358	88.9
'69	.4413	.0501	74.7	.3609	.0504	74.3
'70	.6753	.0421	79.8	.6187	.0478	74.0
'71	.6777	.0507	74.3	.6176	.0549	69.8
'72	.6585	.0431	84.4	.6004	.0468	81.6
'73	.6702	.0437	84.1	.6119	.0478	81.0

An important question to address regarding the assessment of future scenarios for each year of the simulation concerned the handling of the time trend in the interest-rate data. In Table 14.3, the mean of the distribution of one-year changes in the one-year rate is almost always positive, indicating increasing rates, on the average. Would a portfolio manager have assumed that interest rates would continue to increase according to these historical means, when he was forecasting future rates at the beginning of each year of the simulation? It was assumed that a portfolio manager would have initially forecast the upward drift in rates prior to 1970, corresponding to the mean based on the previous seven years of data. In 1969, interest rates dropped precipitously and uncertainty increased about the direction of future changes. Hence, from 1970 until the end of the simulation, it was assumed that the portfolio manager would have forecast no net drift in interest rates, but would have had a variance of that forecast corresponding to that observed in the previous seven years.

Finally, a tree of scenarios for bank planning purposes was determined at each stage by making a discrete approximation of the distribution of changes in the one-year rate. For ease of computation, it was assumed that there were only three possible changes that could occur during each period in the planning horizon. Further, the distribution of one-year changes in the one-year rate was approximately normal; this allowed these changes to be approximated by the mean and the mean plus-or-minus one standard deviation.

By approximating the distribution of one-year changes in the one-year rate with three points, the number of branches on the scenario-planning tree at each stage is 3 in the first period, 9 in the second period, and 27 in the third period. The method just described generates a yield curve for each branch of the planning tree similar to those given in Fig. 14.3 Normally, the portfolio manager would also assess the exogenous cash flow either to or from the portfolio on each branch of the scenario. However, since the main interest was in evaluating the performance of the portfolio, the assumption was made that all cash generated by the portfolio was reinvested, and that no net cash was either made available to or withdrawn from the portfolio after the initial investment.

The only securities under consideration in the stimulation were good-grade municipal bonds. The purchase of nine different maturities was allowed—1-, 2-, 3-, 4-, 5-, 10-, 15-, 20-, and 30-years. This is a robust enough collection to show how the model behaves, although at times some of the maturities that were not included might have been slightly preferable. Trading these securities involves a cost, which is paid at the point of sale of the securities and amounts to the spread between the "bid" and "asked" prices of the market. For bond prices quoted in terms of $100 per bond, the bid–asked spread ranged from $\frac{1}{8}$ for bonds with two years or less to maturity, to $\frac{3}{4}$ for bonds with 11 years or more to maturity.

The planning horizon for the bank was taken to be three years. Since this was a yearly simulation, three one-year periods were then used in structuring the model. Assuming a three-year horizon for planning purposes indicates that the bank is willing to say that its objective is to maximize the value of the portfolio at the end of three years. Therefore, at any point in time, the bank is planning as if it wants to maximize its portfolio value at the end of the next three years, but in fact it is always rolling over the horizon, so the end is never reached.

Throughout the simulation, the limit on realized losses within any one year on any scenario was held to 0.5 percent of the initial book value of the portfolio. That is, at the beginning of each year in the simulation, the current book value of the portfolio was computed and the losses that were allowed to be realized that year and planned for over each of the years of the planning horizon were limited to 0.5 percent of this value.

The unrealized losses were handled differently. Since the actual interest rates rose over the course of the simulation, fairly large amounts of unrealized losses tended to build up in the portfolio. The potential unrealized losses were constrained to be the same for all scenarios, and the level of these unrealized losses, under moderately adverse circumstances, was kept as small as possible. With this view, the limits on potential unrealized losses were as low as 1 percent and as high as 5 percent, depending on how much the interest rates had risen to date. In the first year of the simulation, the allowable unrealized losses on all scenarios at the end of the three years in the planning horizon were limited to 1 percent of the initial book value.

14.5 EVALUATING THE PERFORMANCE OF THE MODEL

Rather than examine the details of the transactions each time the portfolio was revised, let us consider only the general structure of the portfolio over time. Figure 14.6 illustrates the holdings of the portfolio over time in broad maturity categories. It is interesting to note the extent to which a roughly barbell portfolio strategy was maintained. Initially, some intermediate maturities (15- and 20-year) were purchased. However, the 15-year maturity was sold off as soon as the program for realized losses would allow, and the 20-year maturity was gradually sold off. No other investments in intermediate securities were made during the simulation. The final portfolio was essentially a barbell structure with 1-, 2-, 3-, and 5-year maturities on the short end; and 26-, 28-, and 30-year maturities on the long end.

We can get an idea of how the value of the portfolio increased over time from Table 14.5. It should be emphasized that the period of the simulation shows a very

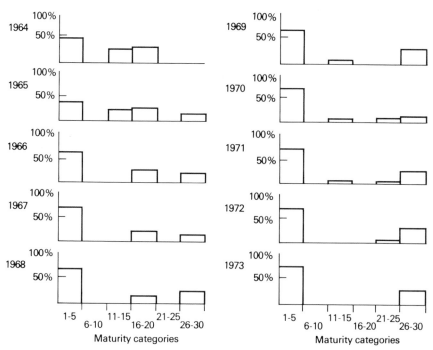

Fig. 14.6 Maturity distributions of managed portfolios (beginning of year).

Table 14.5 Portfolio characteristics for each year

Year	Beginning book value	Interest income	Income return on book value	Realized net losses	Potential unrealized losses at horizon
1964	$100,000.	$2962.	2.96%	$ 0	1.0%
1965	102,959.	3150.	3.06	3	1.0
1966	105,578.	3296.	3.12	531.*	1.5
1967	108,327.	3808.	3.52	547.*	1.5
1968	111,574.	4305.	3.86	561.*	2.0
1969	115,300.	4601.	3.99	579.*	2.5
1970	119,301.	5935.	4.97	600.*	5.0
1971	124,610.	5358.	4.30	626.*	2.5
1972	129,631.	5635.	4.35	337.	2.5
1973	134,729.	6017.	4.47	537.	2.5
1974	141,269.	—	—	−523.	—

* Maximum allowable realized losses taken.

large increase in interest rates in general, and therefore a potential for large capital losses for programs involving active trading of securities. The generally high interest income exhibited by the portfolio is not completely reflected in the increased book value of the portfolio the next year, because losses resulting from trading have reduced the book value of the portfolio. However, the year-to-year increase in the book value of the portfolio generally follows the interest-rate pattern. The final value of $141,269 for the portfolio in January 1974 corresponds to a compounded rate of return on the initial investment of 3.52 percent per year.

Table 14.5 also indicates the level of the losses realized by selling securities, and the limit placed on the level of unrealized losses that could potentially build up in the portfolio. The realized losses were constrained in each year on every scenario to be less than or equal to one-half percent of the current book value of the portfolio. In the years 1966 through 1971, the maximum level of realized losses was attained. In each year the potential unrealized losses at the horizon three years hence were approximately equal to the level of losses already built up in the portfolio.

The performance of the BONDS model can be compared with the results of applying various laddered and barbell strategies to the same historical interest-rate sequence. Alternative portfolio strategies were implemented essentially the same way as in the BONDS model, except that no forecasting procedure was needed since the strategies were applied in a mechanical manner. For the laddered portfolios, at the beginning of each year of the simulation, the funds from maturing securities were reinvested in the longest security allowed in the ladder, and the coupon income was distributed equally among all maturities to keep the proportions in each fixed. For the barbell portfolios, at the beginning of each year of the simulation, the shortest securities in the long end of the barbell were sold and reinvested in the longest security allowed. Similarly, the maturing securities were reinvested in the longest security allowed on the short end. The coupon income was allocated between the long or short ends of the barbell to maintain the stated proportions, and within either end it was distributed equally among the maturities included.

Table 14.6 summarizes the results of using various laddered portfolio strategies in the same historical environment. Any relatively short ladder would seem to be a good strategy against such a rising interest-rate sequence. The five-year ladder gives the highest final value, which is slightly less than that obtained by the BONDS model. However, had the interest-rate sequence been decreasing or had it shown more fluctuation, the BONDS model would have performed significantly better than the five-year ladder, by being able to balance capital losses against gains in each period while retaining a relatively large proportion of the portfolio in longer maturities with higher interest income.

Table 14.6 Laddered portfolio strategies

Strategy	Total interest	Total realized losses (A/T)	Final unrealized gain (A/T)	Final value	Compound rate of return
1–10	$39,775.	$0	$ 642.	$140,416	3.45%
1–7	40,553.	0	345.	140,898	3.49
1–5	41,355.	0	− 175.	141,180	3.51
1–3	41,043.	0	− 409.	140,634	3.47
1	37,596.	0	0	137,596	3.24

Table 14.7 summarizes the performance of using various barbell portfolio strategies in the same historical environment. The column labeled "Strategy" gives the maturities held in the portfolio at all times and the percentage of the portfolio that was reinvested in the long end of the barbell. For example, 1–7, 24–30, 20% means a barbell portfolio structure with seven bonds on the short end, and seven bonds on the long end, but with only twenty percent of the total value of the portfolio invested in the long end.

Table 14.7 Barbell portfolio strategies

Strategy			Total interest	Total realized losses (A/T)	Final unrealized gain (A/T)	Final value	Compound rate of return
1–10,	21–30,	20%	$40,949.	$1723.	$ 189.	$139,415	3.38%
		40	42,135.	3456.	− 266.	138,413	3.30
1–7,	24–30,	20	41,879.	2579.	635.	139,935	3.42
		40	43,220.	5178.	931.	138,973	3.35
1–5,	26–30,	20	42,835.	2953.	458.	140,340	3.45
		40	44,334.	5933.	1102.	139,503	3.39
1–3,	28–30,	20	42,896.	2422.	− 415.	140,059	3.43
		40	44,775.	4865.	− 420.	139,490	3.38
1,	30,	20	40,242.	3020.	0	137,222	3.21
		40	42,941.	6066.	0	136,875	3.19

It is useful to compare laddered portfolios with barbell portfolios having the same number of bonds on the short end. It is clear that, for this particular realization of interest rates, the laddered portfolios did better than the comparable barbell portfolios. This is due to the fact that, although the barbell portfolios had higher interest incomes, these were offset by having to realize losses from selling a relatively long maturity during a period of increasing interest rates. If interest rates had fallen, the barbell portfolios would still have had higher average interest incomes than the ladders but would have realized capital gains rather than losses. Hence, for this historical simulation in a period of rising rates, the laddered portfolios outperformed the barbell portfolios but this will not generally be the case.

Within any barbell structure, as the percent of the portfolio invested in the long end was increased, the performance of the portfolio deteriorated. This is again due to the fact that the realized losses were so large over this period of rising interest rates. Larger amounts invested in the long end produced larger total interest income, since the yield curves were generally increasing with longer maturities. However, the increased interest income was not sufficient to offset the capital losses incurred in trading.

Finally, the riskiness of the portfolio strategies can be analyzed by looking at the amount of unrealized losses that had built up under each strategy in the year of peak interest rates, 1970. Table 14.8 gives the book value, market value, and un-realized after-tax losses as a percent of book value, for the strategy followed by the BONDS model and for various laddered and barbell strategies. In a period of rising rates, strategies that keep most of their assets in short-term securities would be expected to have lower unrealized losses. In the extreme case, a ladder consisting of only the one-year maturity would have *zero* unrealized losses at the beginning of each year, since the entire portfolio would mature at the end of each period.

Table 14.8 Unrealized losses, January 1, 1970

Type of portfolio				Book value	Market value	Unrealized after-tax losses
BONDS				$119,299	$107,536	4.93%
Ladder	1–5			119,198	115,252	1.66
	1–10			119,403	113,958	4.56
Barbell	1–5,	26–30,	20%	118,051	113,688	3.70
	1–5,	26–30,	40%	116,889	110,107	5.80
Barbell	1–10,	21–30,	20%	119,222	111,200	6.73
	1–10,	21–30,	40%	119,039	108,415	8.92

On this dimension, the strategy followed by the BONDS model builds up significant losses, roughly comparable to a ten-year ladder, or a barbell with five maturities on each end and, say, thirty percent of the portfolio value invested in the long end. However, this general level of unrealized losses in the credit crunch of 1969 might have been considered very reasonable. Any strategy that places a significant proportion of its assets in relatively short-term securities will not develop unrealized losses. However, just as there is potential for losses with these strategies, there is also potential for similar gains.

We can sum up the performance of the BONDS model by comparing it to the best of the mechanically-managed laddered or barbell strategies. The performance of the five-year laddered portfolio was almost as good as that of the BONDS model against the particular sequence of interest rates that occurred. However, it should be noted that a portfolio laddered out only to five years is a very conservative strategy, which is rather unlikely to be followed in practice. It happened to turn out, after the fact, that this was a good strategy to have adopted; but it is not clear that any portfolio manager would have been motivated to adopt such a conservative strategy consistently over the period of the simulation.

Finally, had interest rates been level or falling, the BONDS model would certainly have outperformed the five-year ladder. In these instances, some other laddered or barbell portfolio might perform competitively with the BONDS model. However, the important characteristic of the BONDS model is that it produces a strategy that is adaptive to the environment over time. The BONDS model should perform well against *any* interest-rate sequence, while a particular laddered or barbell portfolio will perform well for some realizations of interest rates and poorly for others. In actual practice, the portfolio manager would be *actively forecasting* interest rates, and the BONDS model provides a method of systematically taking advantage of these forecasts.

EXERCISES

1. A bank considering using the BONDS model to aid in portfolio planning divides its portfolio into two pools of funds—U.S. Governments and all grades of municipals. The bank managers forecast a yield curve for their group of municipals and are willing to treat them as one category. The investment portfolio decisions are revised monthly, but no securities with less than three months to maturity are purchased.

   a) Assuming that the bank employs a two-year planning horizon with four planning periods of 3 months, 3 months, 6 months, and one year, how many constraints of each type will their model have, using 5-point approximations to the uncertainties in the first two periods and 3-point approximations in the last two periods? (Assume no initial holdings, but include the remaining constraints of Table 14.1.)

   b) The bank feels that it can aggregate the purchase decisions on individual securities into the following maturities: 3 months, 6 months, 1, 2, 3, 5, 10, and 20 years for U.S. Governments and 1, 2, 3, 4, 5, 10, 20, and 30 years for the municipal group. How many decision variables will the model sketched in (a) have? (Again, assume no initial holdings.)

   c) In fact, the bank does have initial holdings of both Governments and municipals. The bank is willing to aggregate these holdings in the same maturities as used for new purchases. How many additional constraints and variables need to be added to account for the initial holdings?

   d) How is a subproblem defined for the model described in (a), (b), and (c)? How many such subproblems are there? Why is it impossible to combine the subproblem from initial holdings with those of subsequent purchases?

   e) How many constraints does the restricted master have? How would you find an initial basic feasible solution to the restricted master?

2. For the model described in Exercise 1, the demands for information from the portfolio manager are extensive.

   a) How many yield curves need to be assessed to use the proposed planning model?
   b) How would you think about the problem of assessing the exogenous cash flows in such a way that they are consistent with the yield curves assessed?
   c) What is the interpretation of using the same level of realized and unrealized losses on all scenarios? Which scenarios are likely to produce the binding realized and unrealized loss constraints? Can the loss constraints on the remaining scenarios be dropped?
   d) Suppose that the lower-bound constraint on the holdings of government securities could be dropped from the model. How might this change in the model formulation affect the optimal solution? Could the value of the objective function decrease?

3. The objective function of the BONDS model is given in terms of the horizon value of the portfolio. Suppose that we wish to reformulate the model in terms of the interest income and capital gains in each period. The decision variables remain the same; all constraints are unchanged, but the objective function changes.

   a) Formulate a new objective function that maximizes the expected *increase* in the size of the portfolio by summing, over all time periods, the interest income plus capital gains (or losses) in each time period.
   b) Show that the two formulations are equivalent, in the sense that the optimal values of the decision variable using each objective function are identical.
   c) Show that the difference between the optimal values of the two objective functions is the expected exogenous cash flow.

4. Upon seeing your formulation in Exercise 2, the portfolio manager argues that the cash flows in the objective function should be discounted.

   a) Will discounting the cash flows change the optimal values of the decision variables?
   b) What are the arguments for and against discounting in this situation? How could a single discount rate be chosen?
   c) What is the relationship between the shadow prices on the funds-flow constraints of the undiscounted formulation and the discount rate? (See exercises in Chapter 4.)
   d) Suppose you win the argument with the portfolio manager and do not use a discounted objective function. Other departments in the bank place demands on the portfolio for funds in various time periods. Can you suggest an approach to choosing among various requests for funds? How does the approach relate to discounting the cash flows?

5. Consider the subproblems generated by the decomposition approach. Formulate the subproblem corresponding to buying a 20-year U.S. Government security at the beginning of the first period of a model consisting of 3 one-year periods. The generic decision variables to use are as follows:

$$b_1, \quad S_{21}(e_2), \quad h_{21}(e_2), \quad S_{31}(e_3), \quad h_{31}(e_3).$$

(Do not include buying a similar security at the beginning of the second or third periods.)

   a) How many constraints and decision variables does the subproblem have?
   b) The constraints of the subproblems are homogeneous (i.e., zero righthand sides). Suppose that purchasing 1 unit of this security, $b_1 = 1$, gives a positive rate of return. What can be said about purchasing λb_1 units of this security?
   c) Formulate a dynamic-programming model to solve this subproblem, assuming that $b_1 = 1$. Show that this solution determines a *ray* of the subproblem.

6. Suppose that, for the subproblems formulated in Exercise 5, we define a *ray-finding* sub-problem as follows: b_1 is set equal to 1 and moved to the righthand side; the resulting subproblem is solved by linear programming.

   a) Formulate the ray-finding problem.

   b) Find the dual of the ray-finding problem.

   c) Show that a basis for the dual problem is triangular.

   d) Write down a recursive method for calculating the optimal solution of the dual of the ray-finding problem. [*Hint.* Exploit the triangular property of the basis to solve for the dual variables by back-substitution.]

   e) How is the solution of the primal ray-finding problem determined from the solution of the dual?

7. As a follow-up to Exercises 5 and 6, once the subproblems have been solved, the rays generated are added to the restricted master:

   a) Describe how the columns of the restricted master are computed.

   b) Why are weighting constraints *not* needed for the restricted master in this particular application?

   c) What is the stopping rule for this variation of the decomposition algorithm?

   d) Suppose that storage limitations force you to drop some of the nonbasic columns from the restricted master at some iteration. Is it possible that the algorithm will be unable to find the overall optimal solution as a result?

8. There are at least two ways to define subproblems for this model. First, a subproblem can be defined for each maturity and each year it can be purchased. In Fig. 14.4 there are a total of 12 subproblems using this definition. Second, a subproblem can be defined for each maturity regardless of when it is purchased. In Fig. 14.4 there are a total of 3 subproblems using this definition.

   a) Explain why the choice of subproblem definitions need have no impact on the solution procedure adopted for the subproblems.

   b) Explain how the restricted master will differ under each definition of the subproblems.

   c) Which choice of subproblem definitions will make the restricted master more efficient to solve? Why?

   d) If there was a weighting constraint for each subproblem, how would your answer to (c) be affected? [*Hint.* Which definition would add more weighting constraints?]

9. Suppose that a new objective function is to be considered that is a nonlinear function of the holdings entering the final period. More precisely, assume that we wish to maximize the *expected* utility of these holdings, with the utility function given by:

$$u(x) = ax^{1-c},$$

where $a > 0$ and $0 < c < 1$. The objective function is then of the form:

$$\sum_{e_N \in E_N} p(e_N) u\left\{ \sum_{k=1}^{K} \left[\sum_{m=0}^{N-1} \left(y_m^k(e_m) + v_{m,N}^k(e_N) \right) h_{m,N}^k(e_N) + \left(y_N^k(e_N) + v_{N,N}^k(e_N) \right) b_N^k(e_N) \right] \right\}.$$

   a) Show that this problem cannot be handled directly by decomposition. [*Hint.* Is this objective function separable in the appropriate way?]

   b) If the nonlinear-programming problem were solved by the Frank-Wolfe algorithm, a sequence of linear programs would be solved. How can the decomposition approach presented in this chapter be used to solve one of these linear programs?

c) How can the Frank-Wolfe algorithm be efficiently combined with the decomposition approach presented in this chapter, to find the optimal solution to the nonlinear program defined by maximizing the expected utility given above?

d) Does your proposed method generalize to other nonlinear problems?

10. Describe the method of producing scenarios used in the historical simulation over the first three decisions (two years).

a) How many yield curves need to be assessed at the beginning of '64 for the years '64, '65, and '66?

b) Draw a decision tree depicting the probability that each of the yield curves indicated in (a) occurs. What use is made of the uncertainty in the estimates of the change in the twenty-year rate (thirty-year rate) as a fraction of the change in the one-year rate? How could this be modified?

c) Show how to compute the parameters of the functional form for each of the yield curves forecast for the start of '65.

d) Explain how you might assess the required information in actual practice, using a model similar to that described in Exercise 1. Draw a distinction between short-term forecasting (3 months) and long-term forecasting (2 years).

11. The performance of the laddered portfolio structure with five equally spaced maturities proved to be the best among the ladders investigated. Similarly, the performance of the 1–5, 26–30 barbell, with 20% invested in the long end, proved the best among the barbells. The BONDS model outperformed these strategies in the historical simulation over a period of rising rates.

a) How would you expect these strategies to compare with the BONDS model in a period of falling rates?

b) How would you choose an appropriate ladder or barbell strategy without the benefit of 20/20 hindsight?

c) What benefits does the BONDS model have over either a ladder or barbell approach to portfolio management?

ACKNOWLEDGMENTS

This chapter is based primarily on material drawn from "Managing a Bank Bond Portfolio Over Time," by S. P. Bradley and D. B. Crane, to appear in *Stochastic Programming* (ed. M. A. H. Dempster), Academic Press, in press. A broad coverage of the area is contained in *Management of Bank Portfolios* by S. P. Bradley and D. B. Crane, John Wiley & Sons, New York, 1975.

Vectors and
Matrices

Appendix A

Vectors and matrices are notational conveniences for dealing with systems of linear equations and inequalities. In particular, they are useful for compactly representing and discussing the linear programming problem:

$$\text{Maximize} \sum_{j=1}^{n} c_j x_j,$$

subject to:

$$\sum_{j=1}^{n} a_{ij} x_j = b_i \qquad (i = 1, 2, \ldots, m),$$

$$x_j \geq 0 \qquad (j = 1, 2, \ldots, n).$$

This appendix reviews several properties of vectors and matrices that are especially relevant to this problem. We should note, however, that the material contained here is more technical than is required for understanding the rest of this book. It is included for completeness rather than for background.

A.1 VECTORS

We begin by defining vectors, relations among vectors, and elementary vector operations.

Definition. A *k-dimensional vector* y is an ordered collection of k real numbers $y_1, y_2, \ldots, y_k$, and is written as $y = (y_1, y_2, \ldots, y_k)$. The numbers y_j $(j = 1, 2, \ldots, k)$ are called the *components* of the vector y.

Each of the following are examples of vectors:

i) $(1, -3, 0, 5)$ is a four-dimensional vector. Its first component is 1, its second component is -3, and its third and fourth components are 0 and 5, respectively.

ii) The coefficients $c_1, c_2, \ldots, c_n$ of the linear-programming objective function determine the n-dimensional vector $c = (c_1, c_2, \ldots, c_n)$.

iii) The activity levels $x_1, x_2, \ldots, x_n$ of a linear program define the n-dimensional vector $x = (x_1, x_2, \ldots, x_n)$.

iv) The coefficients $a_{i1}, a_{i2}, \ldots, a_{in}$ of the decision variables in the ith equation of a linear program determine an n-dimensional vector $A^i = (a_{i1}, a_{i2}, \ldots, a_{in})$.

v) The coefficients $a_{1j}, a_{2j}, \ldots, a_{nj}$ of the decision variable x_j in constraints 1 through m of a linear program define an m-dimensional vector which we denote as $A_j = (a_{1j}, a_{2j}, \ldots, a_{mj})$.

Equality and ordering of vectors are defined by comparing the vectors' individual components. Formally, let $y = (y_1, y_2, \ldots, y_k)$ and $z = (z_1, z_2, \ldots, z_k)$ be two k-dimensional vectors. We write:

$$y = z \qquad\qquad \text{when } y_j = z_j \qquad (j = 1, 2, \ldots, k),$$

$$y \geqq z \quad \text{or} \quad z \leqq y \qquad \text{when } y_j \geqq z_j \qquad (j = 1, 2, \ldots, k),$$

$$y > z \quad \text{or} \quad z < y \qquad \text{when } y_j > z_j \qquad (j = 1, 2, \ldots, k),$$

and say, respectively, that y equals z, y is *greater than or equal to* z and that y is *greater than* z. In the last two cases, we also say that z is *less than or equal to* y and *less than* y. It should be emphasized that *not all vectors* are ordered. For example, if $y = (3, 1, -2)$ and $x = (1, 1, 1)$, then the first two components of y are greater than or equal to the first two components of x but the third component of y is *less* than the corresponding component of x.

A final note: 0 is used to denote the *null vector* $(0, 0, \ldots, 0)$, where the dimension of the vector is understood from context. Thus, if x is a k-dimensional vector, $x \geqq 0$ means that each component x_j of the vector x is nonnegative.

We also define scalar multiplication and addition in terms of the components of the vectors.

Definition. *Scalar multiplication* of a vector $y = (y_1, y_2, \ldots, y_k)$ and a scalar α is defined to be a new vector $z = (z_1, z_2, \ldots, z_k)$, written $z = \alpha y$ or $z = y\alpha$, whose components are given by $z_j = \alpha y_j$.

Definition. *Vector addition* of two k-dimensional vectors $x = (x_1, x_2, \ldots, x_k)$ and $y = (y_1, y_2, \ldots, y_k)$ is defined as a new vector $z = (z_1, z_2, \ldots, z_k)$, denoted $z = x + y$, with components given by $z_j = x_j + y_j$.

As an example of scalar multiplication, consider

$$4(3, 0, -1, 8) = (12, 0, -4, 32),$$

and for vector addition,

$$(3, 4, 1, -3) + (1, 3, -2, 5) = (4, 7, -1, 2).$$

Using both operations, we can make the following type of calculation:

$$\begin{aligned}(1, 0)x_1 + (0, 1)x_2 + (-3, -8)x_3 &= (x_1, 0) + (0, x_2) + (-3x_3, -8x_3)\\ &= (x_1 - 3x_3, x_2 - 8x_3).\end{aligned}$$

It is important to note that y and z must have the same dimensions for vector addition and vector comparisons. Thus $(6, 2, -1) + (4, 0)$ is *not* defined, and $(4, 0, -1) = (4, 0)$ makes *no* sense at all.

A.2 MATRICES

We can now extend these ideas to any rectangular array of numbers, which we call a *matrix*.

Definition.   A *matrix* is defined to be a rectangular array of numbers

$$A = \begin{bmatrix} a_{11} & a_{12} & \cdots & a_{1n} \\ a_{21} & a_{22} & \cdots & a_{2n} \\ \vdots & & & \vdots \\ a_{m1} & a_{m2} & \cdots & a_{mn} \end{bmatrix},$$

whose *dimension* is m by n. A is called *square* if $m = n$. The numbers a_{ij} are referred to as the *elements* of A.

The tableau of a linear programming problem is an example of a matrix.

We define equality of two matrices in terms of their elements just as in the case of vectors.

Definition.   Two matrices A and B are said to be *equal*, written $A = B$, if they have the same dimension and their corresponding elements are equal, i.e., $a_{ij} = b_{ij}$ for all i and j.

In some instances it is convenient to think of vectors as merely being *special cases* of matrices. However, we will later prove a number of properties of vectors that do not have straightforward generalizations to matrices.

Definition.   A k-by-1 matrix is called a *column vector* and a 1-by-k matrix is called a *row vector*.

The coefficients in row i of the matrix A determine a row vector $A^i = (a_{i1}, a_{i2}, \ldots, a_{in})$, and the coefficients of column j of A determine a column vector $A_j = \langle a_{1j}, a_{2j}, \ldots, a_{mj} \rangle$. For notational convenience, column vectors are frequently written horizontally in angular brackets.

We can define scalar multiplication of a matrix, and addition of two matrices, by the obvious analogs of these definitions for vectors.

Definition. *Scalar multiplication* of a matrix A and a real number α is defined to be a new matrix B, written $B = \alpha A$ or $B = A\alpha$, whose elements b_{ij} are are given by $b_{ij} = \alpha a_{ij}$.

For example,

$$3 \begin{bmatrix} 1 & 2 \\ 0 & -3 \end{bmatrix} = \begin{bmatrix} 3 & 6 \\ 0 & -9 \end{bmatrix}.$$

Definition. *Addition* of two matrices A and B, both with dimension m by n, is defined as a new matrix C, written $C = A + B$, whose elements c_{ij} are given by $c_{ij} = a_{ij} + b_{ij}$.

For example,

$$\underset{(A)}{\begin{bmatrix} 1 & 2 & 4 \\ 0 & -3 & 1 \end{bmatrix}} + \underset{(B)}{\begin{bmatrix} 2 & 6 & -3 \\ -1 & 4 & 0 \end{bmatrix}} = \underset{(C)}{\begin{bmatrix} 3 & 8 & 1 \\ -1 & 1 & 1 \end{bmatrix}}$$

If two matrices A and B do not have the same dimension, then $A + B$ is undefined.

The product of two matrices can also be defined if the two matrices have appropriate dimensions.

Definition. The *product* of an m-by-p matrix A and a p-by-n matrix B is defined to be a new m-by-n matrix C, written $C = AB$, whose elements c_{ij} are given by:

$$c_{ij} = \sum_{k=1}^{p} a_{ik} b_{kj}.$$

For example,

$$\begin{bmatrix} 1 & 2 \\ 0 & -3 \\ 3 & 1 \end{bmatrix} \begin{bmatrix} 2 & 6 & -3 \\ 1 & 4 & 0 \end{bmatrix} = \begin{bmatrix} 4 & 14 & -3 \\ -3 & -12 & 0 \\ 7 & 22 & -9 \end{bmatrix}$$

and

$$\begin{bmatrix} 2 & 6 & -3 \\ 1 & 4 & 0 \end{bmatrix} \begin{bmatrix} 1 & 2 \\ 0 & -3 \\ 3 & 1 \end{bmatrix} = \begin{bmatrix} -7 & -17 \\ 1 & -10 \end{bmatrix}.$$

If the number of columns of A does not equal the number of rows of B, then AB is undefined. Further, from these examples, observe that matrix multiplication is *not commutative*; that is, $AB \neq BA$, in general.

If $\pi = (\pi_1, \pi_2, \ldots, \pi_m)$ is a row vector and $q = \langle q_1, q_2, \ldots, q_m \rangle$ a column vector, then the special case

$$\pi q = \sum_{i=1}^{m} \pi_i q_i$$

of matrix multiplication is sometimes referred to as an *inner product*. It can be visualized by placing the elements of π next to those of q and *adding*, as follows:

$$\pi_1 \times q_1 = \pi_1 q_1,$$
$$\pi_2 \times q_2 = \pi_2 q_2,$$
$$\vdots \qquad \vdots$$
$$\pi_m \times q_m = \pi_m q_m.$$

$$\pi q = \sum_{i=1}^{m} \pi_i q_i.$$

In these terms, the elements c_{ij} of matrix $C = AB$ are found by taking the inner product of A^i (the ith row of A) with B_j (the jth column of B); that is, $c_{ij} = A^i B_j$.

The following properties of matrices can be seen easily by writing out the appropriate expressions in each instance and rearranging the terms:

$$A + B = B + A \qquad \text{(Commutative law)}$$
$$A + (B + C) = (A + B) + C \qquad \text{(Associative law)}$$
$$A(BC) = (AB)C \qquad \text{(Associative law)}$$
$$A(B + C) = AB + AC \qquad \text{(Distributive law)}$$

As a result, $A + B + C$ or ABC is well defined, since the evaluations can be performed in any order.

There are a few special matrices that will be useful in our discussion, so we define them here.

Definition. The *identity matrix* of order m, written I_m (or simply I, when no confusion arises) is a square m-by-m matrix with *ones* along the diagonal and *zeros* elsewhere.

For example,

$$I_3 = \begin{bmatrix} 1 & 0 & 0 \\ 0 & 1 & 0 \\ 0 & 0 & 1 \end{bmatrix}.$$

It is important to note that for any m-by-m matrix B, $BI_m = I_m B = B$. In particular, $I_m I_m = I_m$ or $II = I$.

Definition. The *transpose* of a matrix A, denoted A^t, is formed by *interchanging* the rows and columns of A; that is, $a_{ij}^t = a_{ji}$.

If

$$A = \begin{bmatrix} 2 & 4 & -1 \\ -3 & 0 & 4 \end{bmatrix},$$

then the transpose of A is given by:

$$A^t = \begin{bmatrix} 2 & -3 \\ 4 & 0 \\ -1 & 4 \end{bmatrix}.$$

We can show that $(AB)^t = B^t A^t$ since the ijth element of both sides of the equality is $\sum_k a_{jk} b_{ki}$.

Definition. An *elementary matrix* is a square matrix with one arbitrary column, but otherwise *ones* along the diagonal and *zeros* elsewhere (i.e., an identity matrix with the exception of one column).

For example,

$$E = \begin{bmatrix} 1 & 0 & -1 & 0 \\ 0 & 1 & 3 & 0 \\ 0 & 0 & 2 & 0 \\ 0 & 0 & 4 & 1 \end{bmatrix}$$

is an elementary matrix.

A.3 LINEAR PROGRAMMING IN MATRIX FORM

The linear-programming problem

$$\text{Maximize } c_1 x_1 + c_2 x_2 + \cdots + c_n x_n,$$

subject to:

$$a_{11} x_1 + a_{12} x_2 + \cdots + a_{1n} x_n \leqq b_1,$$
$$a_{12} x_1 + a_{22} x_2 + \cdots + a_{2n} x_n \leqq b_2,$$
$$\vdots \qquad\qquad\qquad \vdots \qquad \vdots$$
$$a_{1m} x_1 + a_{2m} x_2 + \cdots + a_{mn} x_n \leqq b_m,$$
$$x_1 \geqq 0, \quad x_2 \geqq 0, \quad \ldots, \quad x_n \geqq 0,$$

can now be written in matrix form in a straightforward manner. If we let:

$$x = \begin{bmatrix} x_1 \\ x_2 \\ \vdots \\ x_n \end{bmatrix} \quad \text{and} \quad b = \begin{bmatrix} b_1 \\ b_2 \\ \vdots \\ b_m \end{bmatrix}$$

be column vectors, the linear system of inequalities is written in matrix form as $Ax \leqq b$. Letting $c = (c_1, c_2, \ldots, c_n)$ be a row vector, the objective function is written as cx. Hence, the linear program assumes the following compact form:

Maximize cx,

subject to:

$$Ax \leq b, \qquad x \geq 0.$$

The same problem can also be written in terms of the column vectors A_j of the matrix A as:

Maximize $c_1 x_1 + c_2 x_2 + \cdots + c_n x_n$,

subject to:

$$A_1 x_1 + A_2 x_2 + \cdots + A_n x_n \leq b,$$

$$x_j \geq 0 \qquad (j = 1, 2, \ldots, n).$$

At various times it is convenient to use either of these forms.

The appropriate *dual* linear program is given by:

Minimize $b_1 y_1 + b_2 y_2 + \cdots + b_m y_m$,

subject to:

$$a_{11} y_1 + a_{21} y_2 + \cdots + a_{m1} y_m \geq c_1,$$

$$a_{12} y_1 + a_{22} y_2 + \cdots + a_{m2} y_m \geq c_2,$$

$$\vdots \qquad\qquad \vdots \qquad \vdots$$

$$a_{1n} y_1 + a_{2n} y_2 + \cdots + a_{mn} y_m \geq c_n,$$

$$y_1 \geq 0, \quad y_2 \geq 0, \quad \ldots, \quad y_m \geq 0.$$

Letting

$$y^t = \begin{bmatrix} y_1 \\ y_2 \\ \vdots \\ y_n \end{bmatrix}$$

be a column vector, since the dual variables are associated with the constraints of the primal problem, we can write the dual linear program in compact form as follows:

Minimize $b^t y^t$,

subject to:

$$A^t y^t \geq c^t, \qquad y^t \geq 0.$$

We can also write the dual in terms of the untransposed vectors as follows:

Minimize yb,

subject to:

$$yA \geq c, \qquad y \geq 0.$$

In this form it is easy to write the problem in terms of the row vectors A^i of the matrix A, as:

Minimize $y_1 b_1 + y_2 b_2 + \cdots + y_m b_m,$

subject to:

$$y_1 A^1 + y_2 A^2 + \cdots + y_m A^m \geq c,$$

$$y_i \geq 0 \quad (i = 1, 2, \ldots, m).$$

Finally, we can write the primal and dual problems in equality form. In the primal, we merely define an m-dimensional column vector s measuring the amount of slack in each constraint, and write:

Maximize $cx,$

subject to:

$$Ax + Is = b,$$

$$x \geq 0, \qquad s \geq 0.$$

In the dual, we define an n-dimensional row vector u measuring the amount of surplus in each dual constraint and write:

Minimize $yb,$

subject to:

$$yA - uI = c,$$

$$y \geq 0, \qquad u \geq 0.$$

A.4 THE INVERSE OF A MATRIX

Definition. Given a square m-by-m matrix B, if there is an m-by-m matrix D such that

$$DB = BD = I,$$

then D is called the *inverse* of B and is denoted B^{-1}.

Note that B^{-1} does not mean $1/B$ or I/B, since division is *not* defined for matrices. The symbol B^{-1} is just a convenient way to emphasize the relationship between the inverse matrix D and the original matrix B.

There are a number of simple properties of inverses that are sometimes helpful to know.

i) The inverse of a matrix B is unique if it exists.

Proof. Suppose that B^{-1} and A are both inverses of B. Then

$$B^{-1} = IB^{-1} = (AB)B^{-1} = A(BB^{-1}) = A.$$

ii) $I^{-1} = I$ since $II = I$.

iii) If the inverse of A and B exist, then the inverse of AB exists and is given by $(AB)^{-1} = B^{-1}A^{-1}$.

  Proof. $(AB)(B^{-1}A^{-1}) = A(BB^{-1})A^{-1} = AIA^{-1} = AA^{-1} = I$.

iv) If the inverse of B exists, then the inverse of B^{-1} exists and is given by $(B^{-1})^{-1} = B$.

  Proof. $I = I^{-1} = (B^{-1}B)^{-1} = B^{-1}(B^{-1})^{-1}$.

v) If the inverse of B exists, then the inverse of B^t exists and is given by $(B^t)^{-1} = (B^{-1})^t$.

  Proof. $I = I^t = (B^{-1}B)^t = B^t(B^{-1})^t$.

The natural question that arises is: Under what circumstances does the inverse of a matrix exist? Consider the square system of equations given by:

$$Bx = Iy = y.$$

If B has an inverse, then multiplying on the left by B^{-1} yields

$$Ix = B^{-1}y,$$

which "solves" the original square system of equations for any choice of y. The second system of equations has a unique solution in terms of x for any choice of y, since one variable x_j is isolated in each equation. The first system of equations can be derived from the second by multiplying on the left by B; hence, the two systems are identical in the sense that any $\bar{x}$, $\bar{y}$ that satisfies one system will also satisfy the other. We can now show that a square matrix B has an inverse if the square system of equations $Bx = y$ has a unique solution x for an arbitrary choice of y.

The solution to this system of equations can be obtained by successively isolating one variable in each equation by a procedure known as *Gauss–Jordan elimination,* which is just the method for solving square systems of equations learned in high-school algebra. Assuming $b_{11} \ne 0$, we can use the first equation to eliminate x_1 from the other equations, giving:

$$x_1 \qquad + \frac{b_{12}}{b_{11}} x_2 + \cdots + \qquad \frac{b_{1m}}{b_{11}} x_m = \frac{1}{b_{11}} y_1,$$

$$\left(b_{22} - b_{21}\frac{b_{12}}{b_{11}}\right)x_2 + \cdots + \left(b_{2m} - b_{21}\frac{b_{1m}}{b_{11}}\right)x_m = -\frac{b_{21}}{b_{11}} y_1 + y_2,$$

$$\vdots$$

$$\left(b_{m2} - b_{m1}\frac{b_{12}}{b_{11}}\right)x_2 + \cdots + \left(b_{mm} - b_{m1}\frac{b_{1m}}{b_{11}}\right)x_m = -\frac{b_{m1}}{b_{11}} y_1 \qquad + y_m.$$

If $b_{11} = 0$, we merely choose some other variable to isolate in the first equation. In matrix form, the new matrices of the x and y coefficients are given respectively by $E_1 B$ and $E_1 I$, where E_1 is an elementary matrix of the form:

$$E_1 = \begin{bmatrix} k_1 & 0 & 0 & \cdots & 0 \\ k_2 & 1 & 0 & \cdots & 0 \\ k_3 & 0 & 1 & \cdots & 0 \\ \vdots & & & \ddots & \\ k_m & 0 & 0 & \cdots & 1 \end{bmatrix}, \qquad \begin{aligned} k_1 &= \frac{1}{b_{11}}, \\ &\vdots \\ k_i &= -\frac{b_{i1}}{b_{11}} \qquad (i = 2, 3, \ldots, m). \end{aligned}$$

Further, since b_{11} is chosen to be nonzero, E_1 has an inverse given by:

$$E_1^{-1} = \begin{bmatrix} 1/k_1 & 0 & 0 & \cdots & 0 \\ -k_2 & 1 & 0 & \cdots & 0 \\ -k_3 & 0 & 1 & \cdots & 0 \\ \vdots & & & \ddots & \\ -k_m & 0 & 0 & \cdots & 1 \end{bmatrix}.$$

Thus by property (iii) above, if B has an inverse, then $E_1 B$ has an inverse and the procedure may be repeated. Some x_j coefficient in the second row of the updated system *must be nonzero*, or no variable can be isolated in the second row, implying that the inverse does not exist. The procedure may be repeated by eliminating this x_j from the other equations. Thus, a new elementary matrix E_2 is defined, and the new system

$$(E_2 E_1 B)x = (E_2 E_1)y$$

has x_1 isolated in equation 1 and x_2 in equation 2.

Repeating the procedure finally gives:

$$(E_m E_{m-1} \cdots E_2 E_1 B)x = (E_m E_{m-1} \cdots E_2 E_1)y$$

with one variable isolated in each equation. If variable x_j is isolated in equation j, the final system reads:

$$\begin{aligned} x_1 \qquad &= \beta_{11} y_1 + \beta_{12} y_2 + \cdots + \beta_{1m} y_m, \\ x_2 \qquad &= \beta_{21} y_1 + \beta_{22} y_2 + \cdots + \beta_{2m} y_m, \\ &\ \ \vdots \\ x_m &= \beta_{m1} y_1 + \beta_{m2} y_2 + \cdots + \beta_{mm} y_m, \end{aligned}$$

and

$$B^{-1} = \begin{bmatrix} \beta_{11} & \beta_{12} & \cdots & \beta_{1m} \\ \beta_{21} & \beta_{22} & \cdots & \beta_{2m} \\ \vdots & & & \vdots \\ \beta_{m1} & \beta_{m2} & \cdots & \beta_{mm} \end{bmatrix}.$$

Equivalently, $B^{-1} = E_m E_{m-1} \cdots E_2 E_1$ is expressed in *product form* as the matrix product of elementary matrices. If, at any stage in the procedure, it is not possible to isolate a variable in the row under consideration, then the inverse of the original matrix does not exist.

If x_j has not been isolated in the jth equation, the equations may have to be permuted to determine B^{-1}. This point is illustrated by the following example:

$$B = \begin{array}{c} \begin{array}{ccc} x_1 & x_2 & x_3 \end{array} \\ \begin{bmatrix} 0 & 2 & 4 \\ 2 & 2 & 0 \\ 4 & 6 & 0 \end{bmatrix} \end{array}. \qquad \text{Isolate } x_2 \text{ in equation 1.}$$

$$E_1 = \begin{bmatrix} \frac{1}{2} & 0 & 0 \\ -1 & 1 & 0 \\ -3 & 0 & 1 \end{bmatrix} \qquad \overset{E_1 B}{\begin{bmatrix} 0 & 1 & 2 \\ 2 & 0 & -4 \\ 4 & 0 & -12 \end{bmatrix}} \quad \middle| \quad \overset{E_1 I}{\begin{bmatrix} \frac{1}{2} & 0 & 0 \\ -1 & 1 & 0 \\ -3 & 0 & 1 \end{bmatrix}}$$

Isolate x_1 in Eq. 2.

$$E_2 = \begin{bmatrix} 1 & 0 & 0 \\ 0 & \frac{1}{2} & 0 \\ 0 & -2 & 1 \end{bmatrix} \qquad \overset{E_2 E_1 B}{\begin{bmatrix} 0 & 1 & 2 \\ 1 & 0 & -2 \\ 0 & 0 & -4 \end{bmatrix}} \quad \middle| \quad \overset{E_2 E_1 I}{\begin{bmatrix} \frac{1}{2} & 0 & 0 \\ -\frac{1}{2} & \frac{1}{2} & 0 \\ -1 & -2 & 1 \end{bmatrix}}$$

Isolate x_3 in Eq. 3.

$$E_3 = \begin{bmatrix} 1 & 0 & \frac{1}{2} \\ 0 & 1 & -\frac{1}{2} \\ 0 & 0 & -\frac{1}{4} \end{bmatrix} \qquad \overset{E_3 E_2 E_1 B}{\begin{bmatrix} 0 & 1 & 0 \\ 1 & 0 & 0 \\ 0 & 0 & 1 \end{bmatrix}} \quad \middle| \quad \overset{E_3 E_2 E_1 I}{\begin{bmatrix} 0 & -1 & \frac{1}{2} \\ 0 & \frac{3}{2} & -\frac{1}{2} \\ \frac{1}{4} & \frac{1}{2} & -\frac{1}{4} \end{bmatrix}}.$$

$$\qquad\qquad\quad \underset{\text{x coefficients}}{\uparrow} \qquad\qquad\qquad \underset{\text{y coefficients}}{\uparrow}$$

Rearranging the first and second rows of the last table gives the desired transformation of B into the identity matrix, and shows that:

$$B^{-1} = \begin{bmatrix} 0 & \frac{3}{2} & -\frac{1}{2} \\ 0 & -1 & \frac{1}{2} \\ \frac{1}{4} & \frac{1}{2} & -\frac{1}{4} \end{bmatrix}.$$

Alternately, if the first and second columns of the last table are interchanged, an identity matrix is produced. Interchanging the first and second columns of B, and performing the same operations as above, has this same effect. Consequently,

$$E_3 E_2 E_1 = \begin{bmatrix} 0 & -1 & \frac{1}{2} \\ 0 & \frac{3}{2} & -\frac{1}{2} \\ \frac{1}{4} & \frac{1}{2} & -\frac{1}{4} \end{bmatrix} \quad \text{is the inverse of} \quad \begin{bmatrix} 2 & 0 & 4 \\ 2 & 2 & 0 \\ 6 & 4 & 0 \end{bmatrix}.$$

In many applications the column order, i.e., the indexing of the variables x_j, is arbitrary, and this last procedure is utilized. That is, one variable is isolated in each row and the variable isolated in row j is considered the jth basic variable (above, the second basic variable would be x_1). Then the product form gives the inverse of the columns to B, reindexed to agree with the ordering of the basic variables.

In computing the inverse of a matrix, it is often helpful to take advantage of any special structure that the matrix may have. To take advantage of this structure, we may *partition* a matrix into a number of smaller matrices, by subdividing its rows and columns.

For example, the matrix A below is partitioned into four submatrices A_{11}, A_{12}, A_{21}, and A_{22}:

$$A = \begin{bmatrix} a_{11} & a_{12} & a_{13} \\ a_{21} & a_{22} & a_{23} \\ a_{31} & a_{32} & a_{33} \\ a_{41} & a_{42} & a_{43} \end{bmatrix} = \begin{bmatrix} A_{11} & A_{12} \\ A_{21} & A_{22} \end{bmatrix}.$$

The important point to note is that partitioned matrices obey the usual rules of matrix algebra. For example, multiplication of two partitioned matrices

$$A = \begin{bmatrix} A_{11} & A_{12} \\ A_{21} & A_{22} \\ A_{31} & A_{32} \end{bmatrix}, \quad \text{and} \quad B = \begin{bmatrix} B_{11} & B_{12} \\ B_{21} & B_{22} \end{bmatrix}$$

results in

$$AB = \begin{bmatrix} A_{11}B_{11} + A_{12}B_{21} & A_{11}B_{12} + A_{12}B_{22} \\ A_{21}B_{11} + A_{22}B_{21} & A_{21}B_{12} + A_{22}B_{22} \\ A_{31}B_{11} + A_{32}B_{21} & A_{31}B_{12} + A_{32}B_{22} \end{bmatrix},$$

assuming the indicated products are defined; i.e., the matrices A_{ij} and B_{jk} have the appropriate dimensions.

To illustrate that partitioned matrices may be helpful in computing inverses, consider the following example. Let

$$M = \begin{bmatrix} I & Q \\ 0 & R \end{bmatrix},$$

where 0 denotes a matrix with all zero entries. Then

$$M^{-1} = \begin{bmatrix} A & B \\ C & D \end{bmatrix}$$

satisfies

$$MM^{-1} = I \quad \text{or} \quad \begin{bmatrix} I & Q \\ 0 & R \end{bmatrix} \begin{bmatrix} A & B \\ C & D \end{bmatrix} = \begin{bmatrix} I & 0 \\ 0 & I \end{bmatrix},$$

which implies the following matrix equations:

$$A + QC = I, \quad B + QD = 0,$$
$$RC = 0, \quad RD = I.$$

Solving these simultaneous equations gives

$$C = 0, \quad A = I, \quad D = R^{-1}, \quad \text{and} \quad B = -QR^{-1},$$

or, equivalently,

$$M^{-1} = \begin{bmatrix} I & -QR^{-1} \\ 0 & R^{-1} \end{bmatrix}.$$

Note that we need only compute R^{-1} in order to determine M^{-1} easily. This type of use of partitioned matrices is the essence of many schemes for handling large-scale linear programs with special structures.

A.5 BASES AND REPRESENTATIONS

In Chapters 2, 3, and 4, the concept of a basis plays an important role in developing the computational procedures and fundamental properties of linear programming. In this section, we present the algebraic foundations of this concept.

Definition. *m-dimensional real space* R^m *is defined as the collection of all m-dimensional vectors* $y = (y_1, y_2, \ldots, y_m)$.

Definition. A set of *m*-dimensional vectors $A_1, A_2, \ldots, A_k$ is *linearly dependent* if there exist real numbers $\alpha_1, \alpha_2, \ldots, \alpha_k$, *not all zero*, such that

$$\alpha_1 A_1 + \alpha_2 A_2 + \cdots + \alpha_k A_k = 0. \tag{1}$$

If the only set of α_j's for which (1) holds is $\alpha_1 = \alpha_2 = \cdots = \alpha_k = 0$, then the *m*-vectors $A_1, A_2, \ldots, A_k$ are said to be *linearly independent*.

For example, the vectors $(4, 1, 0, -1)$, $(3, 1, 1, -2)$, and $(1, 1, 3, -4)$ are linearly dependent, since

$$2(4, 1, 0, -1) - 3(3, 1, 1, -2) + 1(1, 1, 3, -4) = 0.$$

Further, the *unit m-dimensional vectors* $u_j = (0, \ldots, 0, 1, 0, \ldots, 0)$ for $j = 1, 2, \ldots, m$, with a *plus one* in the *j*th component and zeros elsewhere, are linearly independent, since

$$\sum_{j=1}^{m} \alpha_j u_j = 0$$

implies that

$$\alpha_1 = \alpha_2 = \cdots = \alpha_m = 0.$$

If any of the vectors $A_1, A_2, \ldots, A_k$, say A_r, is the 0 vector (i.e., has all zero components), then, taking $\alpha_r = 1$ and all other $\alpha_j = 0$ shows that the vectors are linearly dependent. Hence, the null vector is linearly dependent on any set of vectors.

Definition. An *m*-dimensional vector Q is said to be *dependent* on the set of *m*-dimensional vectors $A_1, A_2, \ldots, A_k$ if Q can be written as a linear combination of these vectors; that is,

$$Q = \lambda_1 A_1 + \lambda_2 A_2 + \cdots + \lambda_k A_k$$

for some real numbers $\lambda_1, \lambda_2, \ldots, \lambda_k$. The *k*-dimensional vector $(\lambda_1, \lambda_2, \ldots, \lambda_k)$ is said to be the *representation* of Q in terms of $A_1, A_2, \ldots, A_k$.

Note that $(1, 1, 0)$ is not dependent upon $(0, 4, 2)$ and $(0, -1, 3)$, since $\lambda_1(0, 4, 2) + \lambda_2(0, -1, 3) = (0, 4\lambda_1 - \lambda_2, 2\lambda_1 + 3\lambda_2)$ and can never have 1 as its first component.

The m-dimensional vector $(\lambda_1, \lambda_2, \ldots, \lambda_m)$ is dependent upon the m-dimensional unit vectors $u_1, u_2, \ldots, u_m$, since

$$(\lambda_1, \lambda_2, \ldots, \lambda_m) = \sum_{j=1}^{m} \lambda_j u_j.$$

Thus, any m-dimensional vector is dependent on the m-dimensional unit vectors. This suggests the following important definition.

Definition. A *basis* of R^m is a set of linearly independent m-dimensional vectors with the property that every vector of R^m is dependent upon these vectors.

Note that the m-dimensional unit vectors $u_1, u_2, \ldots, u_m$ are a basis for R^m, since they are linearly independent and any m-dimensional vector is dependent on them.

We now sketch the proofs of a number of important properties relating bases of real spaces, representations of vectors in terms of bases, changes of bases, and inverses of basis matrices.

Property 1. A set of m-dimensional vectors $A_1, A_2, \ldots, A_r$ is linearly dependent if and only if one of these vectors is dependent upon the others.

Proof. First, suppose that

$$A_r = \sum_{j=1}^{r-1} \lambda_j A_j,$$

so that A_r is dependent upon $A_1, A_2, \ldots, A_{r-1}$. Then, setting $\lambda_r = -1$, we have

$$\sum_{j=1}^{r-1} \lambda_j A_j - \lambda_r A_r = 0,$$

which shows that $A_1, A_2, \ldots, A_r$ are linearly dependent.

Next, if the set of vectors is dependent, then

$$\sum_{j=1}^{r} \alpha_j A_j = 0,$$

with at least one $\alpha_j \neq 0$, say $\alpha_r \neq 0$. Then,

$$A_r = \sum_{j=1}^{r-1} \lambda_j A_j,$$

where

$$\lambda_j = -\left(\frac{\alpha_j}{\alpha_r}\right),$$

and A_r depends upon $A_1, A_2, \ldots, A_{r-1}$. ∎

Property 2. The representation of any vector Q in terms of basis vectors $A_1, A_2, \ldots, A_m$ is unique.

Proof. Suppose that Q is represented as both

$$Q = \sum_{j=1}^{m} \lambda_j A_j \qquad \text{and} \qquad Q = \sum_{j=1}^{m} \lambda'_j A_j.$$

Eliminating Q gives $0 = \sum_{j=1}^{m} (\lambda_j - \lambda'_j) A_j$. Since $A_1, A_2, \ldots, A_m$ constitute a basis, they are linearly independent and each $(\lambda_j - \lambda'_j) = 0$. That is, $\lambda_j = \lambda'_j$, so that the representation must be unique. ∎

This proposition actually shows that if Q can be represented in terms of the linearly independent vectors $A_1, A_2, \ldots, A_m$, whether a basis or not, then the representation is unique. If $A_1, A_2, \ldots, A_m$ is a basis, then the representation is always possible because of the definition of a basis.

Several mathematical-programming algorithms, including the simplex method for linear programming, move from one basis to another by introducing a vector into the basis in place of one already there.

Property 3. Let $A_1, A_2, \ldots, A_m$ be a basis for R^m; let $Q \neq 0$ be any m-dimensional vector; and let $(\lambda_1, \lambda_2, \ldots, \lambda_m)$ be the representation of Q in terms of this basis; that is,

$$Q = \sum_{j=1}^{m} \lambda_j A_j. \tag{2}$$

Then, if Q replaces any vector A_r in the basis with $\lambda_r \neq 0$, the new set of vectors is a basis for R^m.

Proof. Suppose that $\lambda_m \neq 0$. First, we show that the vectors $A_1, A_2, \ldots, A_{m-1}$, Q are linearly independent. Let α_j for $j = 1, 2, \ldots, m$ and α_Q be any real numbers satisfying:

$$\sum_{j=1}^{m-1} \alpha_j A_j + \alpha_Q Q = 0. \tag{3}$$

If $\alpha_Q \neq 0$, then

$$Q = \sum_{j=1}^{m-1} \left(-\frac{\alpha_j}{\alpha_Q}\right) A_j,$$

which with (2) gives two representations of Q in terms of the basis $A_1, A_2, \ldots, A_m$. By Property 2, this is impossible, so $\alpha_Q = 0$. But then, $\alpha_1 = \alpha_2 = \cdots = \alpha_{m-1} = 0$, since $A_1, A_2, \ldots, A_{m-1}$ are linearly independent. Thus, as required, $\alpha_1 = \alpha_2 = \cdots = \alpha_{m-1} = \alpha_Q = 0$ is the only solution to (3).

Second, we show that any m-dimensional vector P can be represented in terms of the vectors $A_1, A_2, \ldots, A_{m-1}, Q$. Since $A_1, A_2, \ldots, A_m$ is a basis, there are constants $\alpha_1, \alpha_2, \ldots, \alpha_m$ such that

$$P = \sum_{j=1}^{m} \alpha_j A_j.$$

Using expression (2) to eliminate A_m, we find that

$$P = \sum_{j=1}^{m-1} \left[\alpha_j - \alpha_m \left(\frac{\lambda_j}{\lambda_m} \right) A_j \right] + \frac{\alpha_m}{\lambda_m} Q,$$

which by definition shows that $A_1, A_2, \ldots, A_{m-1}, Q$ is a basis. ∎

Property 4. Let $Q_1, Q_2, \ldots, Q_k$ be a collection of linearly independent m-dimensional vectors, and let $A_1, A_2, \ldots, A_r$ be a basis for R^m. Then $Q_1, Q_2, \ldots, Q_k$ can replace k vectors from $A_1, A_2, \ldots, A_r$ to form a new basis.

Proof. First recall that the 0 vector is not one of the vectors Q_j, since 0 vector is dependent on any set of vectors. For $k = 1$, the result is a consequence of Property 3. The proof is by induction. Suppose, by reindexing if necessary, that $Q_1, Q_2, \ldots, Q_j, A_{j+1}, A_{j+2}, \ldots, A_r$ is a basis. By definition of basis, there are real numbers $\lambda_1, \lambda_2, \ldots, \lambda_r$ such that

$$Q_{j+1} = \lambda_1 Q_1 + \lambda_2 Q_2 + \cdots + \lambda_j Q_j + \lambda_{j+1} A_{j+1} + \lambda_{j+2} A_{j+2} + \cdots + \lambda_r A_r.$$

If $\lambda_i = 0$ for $i = j + 1, j + 2, \ldots, r$, then Q is represented in terms of $Q_1, Q_2, \ldots, Q_j$, which, by Property 1, contradicts the linear independence of $Q_1, Q_2, \ldots, Q_k$. Thus some $\lambda_i \neq 0$ for $i = j + 1, j + 2, \ldots, r$, say, $\lambda_{j+1} \neq 0$. By Property 3, then, $Q_1, Q_2, \ldots, Q_{j+1}, A_{j+1}, A_{j+2}, \ldots, A_r$ is also a basis. Consequently, whenever $j < k$ of the vectors Q_i can replace j vectors from $A_1, A_2, \ldots, A_r$ to form a basis, $(j + 1)$ of them can be used as well, and eventually $Q_1, Q_2, \ldots, Q_k$ can replace k vectors from $A_1, A_2, \ldots, A_r$ to form a basis. ∎

Property 5. Every basis for R^m contains m vectors.

Proof. If $Q_1, Q_2, \ldots, Q_k$ and $A_1, A_2, \ldots, A_r$ are two bases, then Property 4 implies that $k \leqq r$. By reversing the roles of the Q_j and A_i, we also have $r \leqq k$ and thus $k = r$, and every two bases contain the same number of vectors. But the unit m-dimensional vectors $u_1, u_2, \ldots, u_m$ constitute a basis with m-dimensional vectors, and consequently, every basis of R^m must contain m vectors. ∎

Property 6. Every collection $Q_1, Q_2, \ldots, Q_k$ of linearly independent m-dimensional vectors is contained in a basis.

Proof. Apply Property 4 with $A_1, A_2, \ldots, A_m$ the unit m-dimensional vectors. ∎

Property 7. Every m linearly-independent vectors of R^m form a basis. Every collection of $(m + 1)$ or more vectors in R^m are linearly dependent.

Proof. Immediate, from Properties 5 and 6. ∎

If a matrix B is constructed with m linearly-independent column vectors B_1, $B_2, \ldots, B_m$, the properties just developed for vectors are directly related to the concept of a *basis inverse* introduced previously. We will show the relationships by defining the concept of a nonsingular matrix in terms of the independence of its vectors. The usual definition of a nonsingular matrix is that the determinant of the matrix is nonzero. However, this definition stems historically from calculating inverses by the method of cofactors, which is of little computational interest for our purposes and will not be pursued.

Definition. An m-by-m matrix B is said to be *nonsingular* if both its column vectors $B_1, B_2, \ldots, B_m$ and row vectors $B^1, B^2, \ldots, B^m$ are linearly independent.

Although we will not establish the property here, defining nonsingularity of B merely in terms of linear independence of *either* its column vectors or row vectors is equivalent to this definition. That is, linear independence of either its column or row vectors automatically implies linear independence of the other vectors.

Property 8. An m-by-m matrix B has an inverse if and only if it is nonsingular.

Proof. First, suppose that B has an inverse and that

$$B_1 \alpha_1 + B_2 \alpha_2 + \cdots + B_m \alpha_m = 0.$$

Letting $\alpha = \langle \alpha_1, \alpha_2, \ldots, \alpha_m \rangle$, in matrix form, this expression says that

$$B\alpha = 0.$$

Thus $(B^{-1})(B\alpha) = B^{-1}(0) = 0$ or $(B^{-1}B)\alpha = I\alpha = \alpha = 0$. That is, $\alpha_1 = \alpha_2 = \cdots = \alpha_m = 0$, so that vectors $B_1, B_2, \ldots, B_m$ are linearly independent. Similarly, $\alpha B = 0$ implies that

$$\alpha = \alpha(BB^{-1}) = (\alpha B)B^{-1} = 0B^{-1} = 0,$$

so that the rows $B^1, B^2, \ldots, B^m$ are linearly independent.

Next, suppose that $B_1, B_2, \ldots, B_m$ are linearly independent. Then, by Property 7, these vectors are a basis for R^m, so that each unit m-dimensional vector u_j is dependent upon them. That is, for each j,

$$B_1 \lambda_1^j + B_2 \lambda_2^j + \cdots + B_m \lambda_m^j = u_j \tag{4}$$

for some real numbers $\lambda_1^j, \lambda_2^j, \ldots, \lambda_m^j$. Letting D_j be the column vector $D_j = \langle \lambda_1^j, \lambda_2^j, \ldots, \lambda_m^j \rangle$, Eq. (4) says that

$$BD_j = u_j \quad \text{or} \quad BD = I,$$

where D is a matrix with columns $D_1, D_2, \ldots, D_m$. The same argument applied to the row vectors $B^1, B^2, \ldots, B^m$ shows that there is a matrix D' with $D'B = I$. But $D = ID = (D'B)D = D'(BD) = D'I = D'$, so that $D = D'$ is the inverse of B. ∎

Property 8 shows that the rows and columns of a nonsingular matrix inherit properties of bases for R^m and suggests the following definition.

Definition. Let A be an m-by-n matrix and B be any m-by-m submatrix of A. If B is nonsingular, it is called a *basis* for A.

Let B be a basis for A and let A_j be any column of A. Then there is a unique solution $\bar{A}_j = \langle \bar{a}_{1j}, \bar{a}_{2j}, \ldots, \bar{a}_{nj} \rangle$ to the system of equations $B\bar{A}_j = A_j$ given by multiplying both sides of the equality by B^{-1}; that is, $\bar{A} = B^{-1}A_j$. Since

$$B\bar{A}_j = B_1\bar{a}_{1j} + B_2\bar{a}_{2j} + \cdots + B_m\bar{a}_{nj} = A_j,$$

the vector $\bar{A}_j$ is the representation of the column A_j in terms of the basis. Applying Property 3, we see that A_j can replace column B_k to form a new basis if $\bar{a}_{kj} \neq 0$. This result is essential for several mathematical-programming algorithms, including the simplex method for solving linear programs.

A.6 EXTREME POINTS OF LINEAR PROGRAMS

In our discussion of linear programs in the text, we have alluded to the connection between extreme points, or corner points, of feasible regions and basic solutions to linear programs. The material in this section delineates this connection precisely, using concepts of vectors and matrices. In pursuing this objective, this section also indicates why a linear program can always be solved at a basic solution, an insight which adds to our seemingly ad hoc choice of basic feasible solutions in the text as the central focus for the simplex method.

Definition. Let S be a set of points in R^n. A point y in S is called an *extreme point* of S if y cannot be written as $y = \lambda w + (1 - \lambda)x$ for two distinct points w and x in S and $0 < \lambda < 1$. That is, y does not lie on the line segment joining any two points of S.

For example, if S is the set of feasible points to the system

$$x_1 + x_2 \leq 6,$$

$$x_2 \leq 3, \quad x_1 \geq 0, \quad x_2 \geq 0,$$

then the extreme points are (0, 0), (0, 3), (3, 3), and (6, 0) (see Fig. A.1).

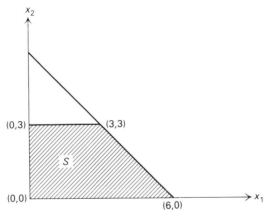

Figure A.1

The next result interprets the geometric notion of an extreme point for linear programs algebraically in terms of linear independence.

Feasible Extreme Point Theorem.   Let S be the set of feasible solutions to the linear program $Ax = b$, $x \geq 0$. Then the feasible point $y = (y_1, y_2, \ldots, y_n)$ is an extreme point of S if and only if the columns of A with $y_i > 0$ are linearly independent.

Proof.   By reindexing if necessary, we may assume that only the first r components of y are positive; that is,

$$y_1 > 0, \qquad y_2 > 0, \qquad \ldots, \qquad y_r > 0, \qquad y_{r+1} = y_{r+2} = \cdots = y_n = 0.$$

We must show that any vector y solving $Ay = b$, $y \geq 0$, is an extreme point if and only if the first r columns $A_1, A_2, \ldots, A_r$ of A are linearly independent. First, suppose that these columns are not linearly independent, so that

$$A_1\alpha_1 + A_2\alpha_2 + \cdots + A_r\alpha_r = 0 \tag{5}$$

for some real numbers $\alpha_1, \alpha_2, \ldots, \alpha_r$ not all zero. If we let x denote the vector $x = (\alpha_1, \alpha_2, \ldots, \alpha_r, 0, \ldots, 0)$, then expression (5) can be written as $Ax = 0$. Now let $w = y + \lambda x$ and $\bar{w} = y - \lambda x$. Then, as long as λ is chosen small enough to satisfy $\lambda|\alpha_j| \leq y_j$ for each component $j = 1, 2, \ldots, r$, both $w \geq 0$ and $\bar{w} \geq 0$. But then, both w and $\bar{w}$ are contained in S, since

$$A(y + \lambda x) = Ay + \lambda Ax = Ay + \lambda(0) = b,$$

and, similarly, $A(y - \lambda x) = b$. However, since $y = \frac{1}{2}(w + \bar{w})$, we see that y is not an extreme point of S in this case. Consequently, every extreme point of S satisfies the linear independence requirement.
Conversely, suppose that $A_1, A_2, \ldots, A_r$ are linearly independent. If $y = \lambda w + (1 - \lambda)x$ for some points w and x of S and some $0 < \lambda < 1$, then $y_j = \lambda w_j + (1 - \lambda)x_j$. Since $y_j = 0$ for $j \geq r + 1$ and $w_j \geq 0$, $x_j \geq 0$, then necessarily $w_j = x_j = 0$ for $j \geq r + 1$. Therefore,

$$A_1y_1 + A_2y_2 + \cdots + A_ry_r = A_1w_1 + A_2w_2 + \cdots + A_rw_r$$
$$= A_1x_1 + A_2x_2 + \cdots + A_rx_r = b.$$

Since, by Property 2 in Section A.5, the representation of the vector b in terms of the linearly independent vectors $A_1, A_2, \ldots, A_r$ is unique, then $y_j = z_j = x_j$. Thus the two points w and x cannot be distinct and therefore y is an extreme point of S. ∎

If A contains a basis (i.e., the rows of A are linearly independent), then, by Property 6, any collection $A_1, A_2, \ldots, A_r$ of linearly independent vectors can be extended to a basis $A_1, A_2, \ldots, A_m$. The extreme-point theorem shows, in this case, that every extreme point y can be associated with a basic feasible solution, i.e., with a solution satisfying $y_j = 0$ for nonbasic variables y_j, for $j = m + 1, m + 2, \ldots, n$.

Chapter 2 shows that optimal solutions to linear programs can be found at basic feasible solutions or equivalently, now, at extreme points of the feasible region. At this point, let us use the linear-algebra tools of this appendix to derive this result independently. This will motivate the simplex method for solving linear programs algebraically.

Suppose that y is a feasible solution to the linear program

Maximize cx,

subject to:

$$Ax = b, \qquad x \geq 0, \tag{6}$$

and, by reindexing variables if necessary, that $y_1 > 0$, $y_2 > 0$, ..., $y_{r+1} > 0$ and $y_{r+2} = y_{r+3} = \cdots = y_n = 0$. If the column A_{r+1} is linearly dependent upon columns $A_1, A_2, \ldots, A_r$, then

$$A_{r+1} = A_1\alpha_1 + A_2\alpha_2 + \cdots + A_r\alpha_r, \tag{7}$$

with at least one of the constants α_j nonzero for $j = 1, 2, \ldots, r$. Multiplying both sides of this expression by θ gives

$$A_{r+1}\theta = A_1(\alpha_1\theta) + A_2(\alpha_2\theta) + \cdots + A_r(\alpha_r\theta), \tag{8}$$

which states that we may simulate the effect of setting $x_{r+1} = \theta$ in (6) by setting $x_1, x_2, \ldots, x_r$, respectively, to $(\alpha_1\theta), (\alpha_2\theta), \ldots, (\alpha_r\theta)$. Taking $\theta = 1$ gives:

$$\tilde{c}_{r+1} = \alpha_1 c_1 + \alpha_2 c_2 + \cdots + \alpha_r c_r$$

as the per-unit profit from the simulated activity of using α_1 units of x_1, α_2 units of x_2, through α_r units of x_r, in place of 1 unit of x_{r+1}.

Letting $\bar{x} = (-\alpha_1, -\alpha_2, \ldots, -\alpha_r, +1, 0, \ldots, 0)$, Eq. (8) is rewritten as $A(\theta x) = \theta A\bar{x} = 0$. Here $\bar{x}$ is interpreted as setting x_{r+1} to 1 and decreasing the simulated activity to compensate. Thus,

$$A(y + \theta\bar{x}) = Ay + \theta A\bar{x} = Ay + 0 = b,$$

so that $y + \theta\bar{x}$ is feasible as long as $y + \theta\bar{x} \geq 0$ (this condition is satisfied if θ is chosen so that $|\theta\alpha_j| \leq y_j$ for every component $j = 1, 2, \ldots, r$). The return from $y + \theta\bar{x}$ is given by:

$$c(y + \theta\bar{x}) = cy + \theta c\bar{x} = cy + \theta(c_{r+1} - \tilde{c}_{r+1}).$$

Consequently, if $\tilde{c}_{r+1} < c_{r+1}$, the simulated activity is less profitable than the $(r + 1)$st activity itself, and return improves by increasing θ. If $\tilde{c}_{r+1} > c_{r+1}$, return increases by decreasing θ (i.e., decreasing y_{r+1} and increasing the simulated activity). If $\tilde{c}_{r+1} = c_{r+1}$, return is unaffected by θ.

These observations imply that, if the objective function is bounded from above over the feasible region, then by increasing the simulated activity and decreasing activity y_{r+1}, or vice versa, we can find a new feasible solution whose objective value is at least as large as cy but which contains at least one more zero component than y.

For, suppose that $\tilde{c}_{r+1} \geq c_{r+1}$. Then by decreasing θ from $\theta = 0$, $c(y + \theta\bar{x}) \geq cy$; eventually $y_j + \theta\bar{x}_j = 0$ for some component $j = 1, 2, \ldots, r + 1$ (possibly $y_{r+1} + \theta\bar{x}_{r+1} = y_{r+1} + \theta = 0$). On the other hand, if $\tilde{c}_{r+1} < c_{r+1}$, then $c(y + \theta\bar{x}) > cy$ as θ increases from $\theta = 0$; if some component of α_j from (7) is positive, then eventually $y_j + \theta\bar{x}_j = y_j - \theta\alpha_j$ reaches 0 as θ increases. (If every $\alpha_j \leq 0$, then we may increase θ indefinitely, $c(y + \theta\bar{x}) \to +\infty$, and the objective value is unbounded over the constraints, contrary to our assumption.) Therefore, if

$$\text{either } \tilde{c}_{r+1} \geq c_{r+1} \qquad \text{or} \qquad \tilde{c}_{r+1} < c_{r+1},$$

we can find a value for θ such that at least one component of $y_j + \theta\bar{x}_j$ becomes zero for $j = 1, 2, \ldots, r + 1$. Since $y_j = 0$ and $\bar{x}_j = 0$ for $j > r + 1$, $y_j + \theta\bar{x}_j$ remains at 0 for $j > r + 1$. Thus, the entire vector $y + \theta\bar{x}$ contains at least one more positive component than y and $c(y + \theta\bar{x}) \geq cy$.

With a little more argument, we can use this result to show that there must be an optimal extreme-point solution to a linear program.

Optimal Extreme-Point Theorem. If the objective function for a feasible linear program is bounded from above over the feasible region, then there is an optimal solution at an extreme point of the feasible region.

Proof. If y is any feasible solution and the columns A_j of A, with $y_j > 0$, are linearly dependent, then one of these columns depends upon the others (Property 1).

From above, there is a feasible solution x to the linear program with both $cx \geq cy$ and x having one less positive component than y. Either the columns of A with $x_j > 0$ are linearly independent, or the argument may be repeated to find another feasible solution with *one less* positive component. Continuing, we eventually find a feasible solution w with $cw \geq cy$, and the columns of A with $w_j > 0$ are linearly independent. By the *feasible extreme-point theorem*, w is an extreme point of the feasible region.

Consequently, given any feasible point, there is always an extreme point whose objective value is at least as good. Since the number of extreme points is finite (the number of collections of linear independent vectors of A is finite), the extreme point giving the maximum objective value solves the problem. ∎

Linear Programming
in Matrix Form

Appendix B

We first introduce matrix concepts in linear programming by developing a variation of the simplex method called the *revised simplex method*. This algorithm, which has become the basis of all commercial computer codes for linear programming, simply recognizes that much of the information calculated by the simplex method at each iteration, as described in Chapter 2, is not needed. Thus, efficiencies can be gained by computing only what is absolutely required.

Then, having introduced the ideas of matrices, some of the material from Chapters 2, 3, and 4 is recast in matrix terminology. Since matrices are basically a notational convenience, this reformulation provides essentially nothing new to the simplex method, the sensitivity analysis, or the duality theory. However, the economy of the matrix notation provides added insight by streamlining the previous material and, in the process, highlighting the fundamental ideas. Further, the notational convenience is such that extending some of the results of the previous chapters becomes more straightforward.

B.1 A PREVIEW OF THE REVISED SIMPLEX METHOD

The revised simplex method, or the simplex method *with multipliers*, as it is often referred to, is a modification of the simplex method that significantly reduces the total number of calculations that must be performed at each iteration of the algorithm. Essentially, the revised simplex method, rather than updating the entire tableau at each iteration, computes *only* those coefficients that are needed to identify the pivot element. Clearly, the reduced costs must be determined so that the entering variable can be chosen. However, the variable that leaves the basis is determined by the minimum-ratio rule, so that only the updated coefficients of the entering variable and the current righthand-side values are needed for this purpose. The revised simplex method then keeps track of only enough information to compute the reduced costs and the minimum-ratio rule at each iteration.

The motivation for the revised simplex method is closely related to our discussion of simple sensitivity analysis in Section 3.1, and we will re-emphasize some of that here. In that discussion of sensitivity analysis, we used the shadow prices to help evaluate whether or not the contribution from engaging in a new activity was sufficient to justify diverting resources from the current optimal group of activities. The procedure was essentially to "price out" the new activity by determining the opportunity cost associated with introducing one unit of the new activity, and then comparing this value to the contribution generated by engaging in one unit of the activity. The opportunity cost was determined by valuing each resource consumed, by introducing one unit of the new activity, at the shadow price associated with that resource.

Tableau B.1

Basic variables	Current values	x_1	x_2	x_3	x_4	x_5	x_6	$-z$
x_4	60	6	5	8	1			0
x_5	150	10	20	10		1		0
x_6	8	1	0	0			1	0
$(-z)$	0	5	4.5	6				1

The custom-molder example used in Chapter 3 to illustrate this point is reproduced in Tableau B.1. Activity 3, producing one hundred cases of champagne glasses, consumes 8 hours of production capacity and 10 hundred cubic feet of storage space. The shadow prices, determined in Chapter 3, are $\$\frac{11}{14}$ per hour of production time and $\$\frac{1}{35}$ per hundred cubic feet of storage capacity, measured in hundreds of dollars. The resulting opportunity cost of diverting resources to produce champagne glasses is then:

$$(\tfrac{11}{14})8 + (\tfrac{1}{35})10 = \tfrac{46}{7} = 6\tfrac{4}{7}.$$

Comparing this opportunity cost with the $6 contribution results in a net loss of $\$\frac{4}{7}$ per case, or a loss of $\$57\frac{1}{7}$ per one hundred cases. It would clearly not be advantageous to divert resources from the current basic solution to the new activity. If, on the other hand, the activity had priced out positively, then bringing the new activity into the current basic solution would appear worthwhile. The essential point is that, by using the shadow prices and the original data, it is possible to decide, without elaborate calculations, whether or not a new activity is a promising candidate to enter the basis.

It should be quite clear that this procedure of pricing out a new activity is not restricted to knowing in advance whether the activity will be promising or not. The pricing-out mechanism, therefore, could in fact be used at each iteration of the simplex method to compute the reduced costs and choose the variable to enter the basis. To do this, we need to define shadow prices at each iteration.

In transforming the initial system of equations into another system of equations at some iteration, we maintain the canonical form at all times. As a result, the objective-function coefficients of the variables that are currently basic are zero at each iteration. We can therefore define simplex multipliers, which are essentially the shadow prices associated with a particular basic solution, as follows:

Definition.   The simplex multipliers $(y_1, y_2, \ldots, y_m)$ associated with a particular basic solution are the multiples of the initial system of equations such that, when all of these equations are multiplied by their respective simplex multipliers and subtracted from the initial objective function, the coefficients of the basic variables are zero.

Thus the basic variables must satisfy the following system of equations:

$$y_1 a_{1j} + y_2 a_{2j} + \cdots + y_m a_{mj} = c_j \qquad \text{for } j \text{ basic.}$$

The implication is that the reduced costs associated with the nonbasic variables are then given by:

$$\overline{c}_j = c_j - (y_1 a_{1j} + y_2 a_{2j} + \cdots + y_m a_{mj}) \qquad \text{for } j \text{ nonbasic.}$$

If we then performed the simplex method in such a way that we knew the simplex multipliers, it would be straightforward to find the largest reduced cost by pricing out all of the nonbasic variables and comparing them.

In the discussion in Chapter 3 we showed that the shadow prices were readily available from the final system of equations. In essence, since varying the righthand-side value of a particular constraint is similar to adjusting the slack variable, it was argued that the shadow prices are the negative of the objective-function coefficients of the slack (or artificial) variables in the final system of equations. Similarly, the simplex multipliers at each intermediate iteration are the negative of the objective-function coefficients of these variables. In transforming the initial system of equations into any intermediate system of equations, a number of iterations of the simplex method are performed, involving subtracting multiples of a row from the objective function at each iteration. The simplex multipliers, or shadow prices in the final tableau, then reflect a summary of all of the operations that were performed on the objective function during this process. If we then keep track of the coefficients of the slack (or artificial) variables in the objective function at each iteration, we immediately have the necessary simplex multipliers to determine the reduced costs $\overline{c}_j$ of the nonbasic variables as indicated above. Finding the variable to introduce into the basis, say x_s, is then easily accomplished by choosing the maximum of these reduced costs.

The next step in the simplex method is to determine the variable x_r to drop from the basis by applying the minimum-ratio rule as follows:

$$\frac{\overline{b}_r}{\overline{a}_{rs}} = \underset{i}{\text{Min}} \left\{ \frac{\overline{b}_i}{\overline{a}_{is}} \,\middle|\, \overline{a}_{is} > 0 \right\}.$$

Hence, in order to determine which variable to drop from the basis, we need both the current righthand-side values and the current coefficients, in each equation, of the variable we are considering introducing into the basis. It would, of course, be easy to keep track of the current righthand-side values for each iteration, since this comprises only a single column. However, if we are to *significantly reduce* the number of computations performed in carrying out the simplex method, we cannot keep track of the coefficients of each variable in each equation on every iteration. In fact, the only coefficients we need to carry out the simplex method are those of the variable to be introduced into the basis at the current iteration. If we could find a way to generate these coefficients after we knew which variable would enter the basis, we would have a genuine economy of computation over the standard simplex method.

It turns out, of course, that there is a way to do exactly this. In determining the new reduced costs for an iteration, we used only the initial data and the simplex multipliers, and further, the simplex multipliers were the negative of the coefficients of the slack (or artificial) variables in the objective function at that iteration. In essence, the coefficients of these variables summarize all of the operations performed on the objective function. Since we began our calculations with the problem in canonical form with respect to the slack (or artificial) variables and $-z$ in the objective function, it would seem intuitively appealing that the coefficients of these variables in any equation summarize all of the operations performed on that equation.

To illustrate this observation, suppose that we are given only the part of the final tableau that corresponds to the slack variables for our custom-molder example. This is reproduced from Chapter 3 in Tableau B.2. In performing the simplex method, multiples of the equations in the initial Tableau B.1 have been added to and subtracted from one another to produce the final Tableau B.2. What multiples of Eq. 1 have been added to Eq. 2? Since x_4 is isolated in Eq. 1 in the initial tableau, any multiples of Eq. 1 that have been added to Eq. 2 must appear as the coefficient of x_4 in Eq. 2 of the final tableau. Thus, without knowing the actual sequence of pivot operations, we know their *net* effect has been to subtract $\frac{2}{7}$ times Eq. 1 from Eq. 2 in the initial tableau to produce the final tableau. Similarly, we can see that $\frac{2}{7}$ times Eq. 1 has been added to Eq. 3. Finally, we see that Eq. 1 (in the initial tableau) has been scaled by multiplying it by $-\frac{1}{7}$ to produce the final tableau. The coefficients of x_5 and x_6 in the final tableau can be similarly interpreted as the multiples of Eqs. 2 and 3, respectively, in the initial tableau that have been added to each equation of the initial tableau to produce the final tableau.

Tableau B.2

Basic variables	Current values	x_4	x_5	x_6
x_2	$4\frac{2}{7}$	$-\frac{1}{7}$	$\frac{3}{35}$	
x_6	$1\frac{4}{7}$	$-\frac{2}{7}$	$\frac{1}{14}$	1
x_1	$6\frac{3}{7}$	$\frac{2}{7}$	$-\frac{1}{14}$	
$(-z)$	$-51\frac{3}{7}$	$-\frac{11}{14}$	$-\frac{1}{35}$	

We can summarize these observations by remarking that the equations of the final tableau must be given in terms of multiples of the equations of the initial tableau as follows:

$$\text{Eq. 1:} \quad (-\tfrac{1}{7})(\text{Eq. 1}) + (\tfrac{3}{35})(\text{Eq. 2}) + 0(\text{Eq. 3});$$
$$\text{Eq. 2:} \quad (-\tfrac{2}{7})(\text{Eq. 1}) + (\tfrac{1}{14})(\text{Eq. 2}) + 1(\text{Eq. 3});$$
$$\text{Eq. 3:} \quad (\tfrac{2}{7})(\text{Eq. 1}) + (-\tfrac{1}{14})(\text{Eq. 2}) + 0(\text{Eq. 3}).$$

The coefficients of the slack variables in the final tableau thus summarize the operations performed on the equations of the initial tableau to produce the final tableau.

We can now use this information to determine the coefficients in the final tableau of x_3, the production of champagne glasses, from the initial tableau and the coefficients of the slack variables in the final tableau. From the formulas developed above we have:

$$\text{Eq. 1:} \quad (-\tfrac{1}{7})(8) + (\tfrac{3}{35})(10) + (0)(0) = -\tfrac{2}{7};$$
$$\text{Eq. 2:} \quad (-\tfrac{2}{7})(8) + (\tfrac{1}{14})(10) + (1)(0) = -\tfrac{11}{7};$$
$$\text{Eq. 3:} \quad (\tfrac{2}{7})(8) + (-\tfrac{1}{14})(10) + (0)(0) = \tfrac{11}{7}.$$

The resulting values are, in fact, the appropriate coefficients of x_3 for each of the equations in the final tableau, as determined in Chapter 3. Hence, we have found that it is only necessary to keep track of the coefficients of the slack variables at each iteration.

The coefficients of the slack variables are what is known as the "inverse" of the current basis. To see this relationship more precisely, let us multiply the matrix* corresponding to the slack variables by the matrix of columns from the initial tableau corresponding to the current basis, arranged in the order in which the variables are basic. In matrix notation, we can write these multiplications as follows:

$$\begin{bmatrix} -\tfrac{1}{7} & \tfrac{3}{35} & 0 \\ -\tfrac{2}{7} & \tfrac{1}{14} & 1 \\ \tfrac{2}{7} & -\tfrac{1}{14} & 0 \end{bmatrix} \begin{bmatrix} 5 & 0 & 6 \\ 20 & 0 & 10 \\ 0 & 1 & 1 \end{bmatrix} = \begin{bmatrix} 1 & 0 & 0 \\ 0 & 1 & 0 \\ 0 & 0 & 1 \end{bmatrix},$$

which, in symbolic form, is $B^{-1}B = I$. The information contained in the coefficients of the slack variables is then the *inverse* of the current basis, since the multiplication produces the identity matrix. In general, to identify the basis corresponding to the inverse, it is only necessary to order the variables so that they correspond to the rows in which they are basic. In this case, the order is x_2, x_6, and x_1.

In matrix notation, the coefficients $\bar{A}_j$ of any column in the current tableau can then be determined from their coefficients A_j in the initial tableau and the basis inverse by $B^{-1}A_j$. For example,

$$\bar{A}_3 = B^{-1}A_3 = \begin{bmatrix} -\tfrac{1}{7} & \tfrac{3}{35} & 0 \\ -\tfrac{2}{7} & \tfrac{1}{14} & 0 \\ \tfrac{2}{7} & -\tfrac{1}{14} & 0 \end{bmatrix} \begin{bmatrix} 8 \\ 10 \\ 0 \end{bmatrix} = \begin{bmatrix} -\tfrac{2}{7} \\ -\tfrac{11}{7} \\ \tfrac{11}{7} \end{bmatrix}.$$

If we now consider the righthand-side values as just another column b in the initial tableau, we can, by analogy, determine the current righthand-side values by:

$$\bar{b} = B^{-1}b = \begin{bmatrix} -\tfrac{1}{7} & \tfrac{3}{35} & 0 \\ -\tfrac{2}{7} & \tfrac{1}{14} & 0 \\ \tfrac{2}{7} & -\tfrac{1}{14} & 1 \end{bmatrix} \begin{bmatrix} 60 \\ 150 \\ 8 \end{bmatrix} = \begin{bmatrix} 4\tfrac{2}{7} \\ 1\tfrac{4}{7} \\ 6\tfrac{3}{7} \end{bmatrix}.$$

* A discussion of vectors and matrices is included in Appendix A.

Hence, had x_3 been a promising variable to enter the basis, we could have easily computed the variable to drop from the basis by the minimum-ratio rule, once we had determined $\bar{A}_3$ and $\bar{b}$.

In performing the revised simplex method, then, we need not compute all of the columns of the tableau at each iteration. Rather, we need only keep track of the coefficients of the slack variables in all the equations including the objective function. These coefficients contain the simplex multipliers y and inverse of the current basis B^{-1}. Using the original data and the simplex multipliers, the reduced costs can be calculated easily and the entering variable selected. Using the original data and the inverse of the current basis, we can easily calculate the coefficients of the entering variable in the current tableau and the current righthand-side values. The variable to drop from the basis is then selected by the minimum-ratio rule. Finally, the basis inverse and the simplex multipliers are updated by performing the appropriate pivot operation on the current tableau, as will be illustrated, and the procedure then is repeated.

B.2 FORMALIZING THE APPROACH

We can formalize the ideas presented in the previous section by developing the revised simplex method in matrix notation. We will assume from here on that the reader is familiar with the first three sections of the matrix material presented in Appendix A.

At any point in the simplex method, the initial canonical form has been transformed into a new canonical form by a sequence of pivot operations. The two canonical forms can be represented as indicated in Fig. B.1.

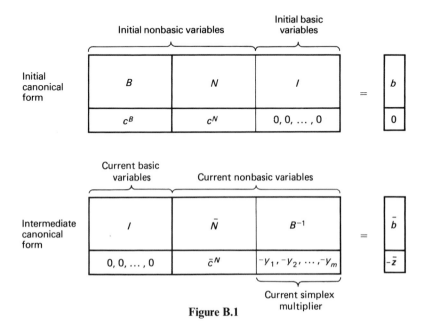

Figure B.1

The basic variables in the initial basis may include artificial variables, as well as variables from the original problem; the initial canonical form is determined by the procedures discussed in Chapter 2. The new basis may, of course, contain variables in common with the initial basis. To make Fig. B.1 strictly correct, we need to imagine including some of these variables *twice*.

We can derive the intermediate tableau from the initial tableau in a straight-forward manner. The initial system of equations in matrix notation is:

$$x^B \geqq 0, \qquad x^N \geqq 0, \qquad x^I \geqq 0,$$
$$Bx^B + Nx^N + Ix^I \quad = b, \qquad (1)$$
$$c^B x^B + c^N x^N \qquad - z = 0,$$

where the superscripts B, N, and I refer to basic variables, nonbasic variables, and variables in the initial identity basis, respectively. The constraints of the intermediate system of equations are determined by multiplying the constraints of the initial system (1) on the left by B^{-1}. Hence,

$$Ix^B + B^{-1}Nx^N + B^{-1}x^I = B^{-1}b, \qquad (2)$$

which implies that the updated nonbasic columns and righthand-side vector of the intermediate canonical forms are given by $\bar{N} = B^{-1}N$ and $\bar{b} = B^{-1}b$, respectively.

Since the objective function of the intermediate canonical form must have zeros for the coefficients of the basic variables, this objective function can be determined by multiplying each equation of (2) by the cost of the variable that is basic in that row and subtracting the resulting equations from the objective function of (1). In matrix notation, this means multiplying (2) by c^B on the left and subtracting from the objective function of (1), to give:

$$0x^B + (c^N - c^B B^{-1}N)x^N - c^B B^{-1}x^I - z = -c^B B^{-1}b. \qquad (3)$$

We can write the objective function of the intermediate tableau in terms of the simplex multipliers by recalling that the simplex multipliers are defined to be the *multiples of* the equations in the initial tableau that *produce zero* for the coefficients of the basic variables when *subtracted from* the initial objective function. Hence,

$$c^B - yB = 0 \qquad \text{which implies} \qquad y = c^B B^{-1}. \qquad (4)$$

If we now use (4) to rewrite (3) we have

$$0x^B + (c^N - yN)x^N - yx^I - z = -yb, \qquad (5)$$

which corresponds to the intermediate canonical form in Fig. B.1. The coefficients in the objective function of variables in the initial identity basis are the negative of the simplex multipliers as would be expected.

Note also that, since the matrix N is composed of the nonbasic columns A_j from the initial tableau, the relation $\bar{N} = B^{-1}N$ states that each updated column $\bar{A}_j$ of $\bar{N}$ is given by $\bar{A}_j = B^{-1}A_j$. Equivalently, $A_j = B\bar{A}_j$ or

$$A_j = B_1 \bar{a}_{1j} + B_2 \bar{a}_{2j} + \cdots + B_m \bar{a}_{mj}.$$

This expression states that the column vector A_j can be written as a linear combination of columns $B_1, B_2, \ldots, B_m$ of the basis, using the weights $\bar{a}_{1j}, \bar{a}_{2j}, \ldots, \bar{a}_{mj}$. In vector terminology, we express this by saying that the column vector

$$A_j = \langle \bar{a}_{1j}, \bar{a}_{2j}, \ldots, \bar{a}_{mj} \rangle$$

is the *representation* of A_j in terms of the basis B.

Let us review the relationships that we have established in terms of the simplex method. Given the current canonical form, the current basic feasible solution is obtained by setting the nonbasic variables to their lower bounds, in this case zero, so that:

$$x^B = B^{-1}b = \bar{b} \geq 0, \qquad x^N = 0.$$

The value of the objective function associated with this basis is then

$$z = yb = \bar{z}.$$

To determine whether or not the current solution is optimal, we look at the reduced costs of the nonbasic variables:

$$\bar{c}_j = c_j - yA_j.$$

If $\bar{c}_j \leq 0$ for all j nonbasic, then the current solution is optimal. Assuming that the current solution is not optimal and that the maximum $\bar{c}_j$ corresponds to x_s, then, to determine the pivot element, we need the representation of the entering column A_s in the current basis and the current righthand side,

$$\bar{A}_s = B^{-1}A_s \qquad \text{and} \qquad \bar{b} = B^{-1}b, \qquad \text{respectively.}$$

If $\bar{c}_s > 0$ and $\bar{A}_s \leq 0$, the problem is unbounded. Otherwise, the variable to drop from the basis x_r is determined by the usual minimum-ratio rule. The new canonical form is then found by pivoting on the element $\bar{a}_{rs}$.

Note that, at each iteration of the simplex method, only the column corresponding to the variable entering the basis needs to be computed. Further, since this column can be obtained by $B^{-1}A_s$, only the initial data and the inverse of the current basis need to be maintained. Since the inverse of the current basis can be obtained from the coefficients of the variables that were slack (or artificial) in the initial tableau, we need only perform the pivot operation on these columns to obtain the updated basis inverse. This computational efficiency is the foundation of the revised simplex method.

Revised Simplex Method

STEP (0): An initial basis inverse B^{-1} is given with $\bar{b} = B^{-1}b \geq 0$. The columns of B are $[A_{j_1}, A_{j_2}, \ldots, A_{j_m}]$ and $y = c^B B^{-1}$ is the vector of simplex multipliers.

STEP (1): The coefficients of $\bar{c}_j$ for the nonbasic variables x_j are computed by pricing out the original data A_j, that is,

$$\bar{c}_j = c_j - yA_j = c_j - \sum_{i=1}^{m} y_i a_{ij} \qquad \text{for } j \text{ nonbasic.}$$

If all $\bar{c}_j \leq 0$ then *stop*; we are optimal. If we continue, then there exists some $\bar{c}_j > 0$.

STEP (2): Choose the variable to introduce into the basis by

$$\bar{c}_s = \underset{j}{\text{Max}} \{\bar{c}_j | \bar{c}_j > 0\}.$$

Compute $\bar{A}_s = B^{-1}A_s$. If $\bar{A}_s \leq 0$, then *stop*; the problem is unbounded. If we continue, there exists $\bar{a}_{is} > 0$ for some $i = 1, 2, \ldots, m$.

STEP (3): Choose the variable to drop from the basis by the minimum-ratio rule:

$$\frac{\bar{b}_r}{\bar{a}_{rs}} = \underset{i}{\text{Min}} \left\{ \frac{\bar{b}_i}{\bar{a}_{is}} \,\middle|\, \bar{a}_{is} > 0 \right\}.$$

The variable basic in row r is replaced by variable s giving the new basis $B = [A_{j_1}, \ldots, A_{j_{r-1}}, A_s, A_{j_{r+1}}, \ldots, A_{j_m}]$.

STEP (4): Determine the new basis inverse B^{-1}, the new righthand-side vector $\bar{b}$, and new vector of simplex multipliers $y = c^B B^{-1}$, by pivoting on $\bar{a}_{rs}$.

STEP (5): Go to STEP (1).

We should remark that the initial basis in STEP (0) usually is composed of slack variables and artificial variables constituting an identity matrix, so that $B = I$ and $B^{-1} = I$, also. The more general statement of the algorithm is given, since, after a problem has been solved once, a good starting feasible basis is generally known, and it is therefore unnecessary to start with the identity basis.

	B^{-1}			$\bar{b}$
$\bar{a}_{1s}$	β_{11}	β_{12} $\cdots$	β_{1m}	$\bar{b}_1$
$\vdots$	$\vdots$	$\vdots$	$\vdots$	$\vdots$
$\textcircled{$\bar{a}_{rs}$}$	β_{r1}	β_{r2}	β_{rm}	$\bar{b}_r$
$\vdots$	$\vdots$	$\vdots$	$\vdots$	$\vdots$
$\bar{a}_{ms}$	β_{m1}	β_{m2} $\cdots$	β_{mm}	$\bar{b}_m$
$\bar{c}_s$	$-y_1$	$-y_2$ $\cdots$	$-y_m$	$-\bar{z}$

Fig. B.2 Updating the basis inverse and simplex multipliers.

The only detail that remains to be specified is how the new basis inverse, simplex multipliers, and righthand-side vector are generated in STEP (4). The computations are performed by the usual simplex pivoting procedure, as suggested by Fig. B.2. We know that the basis inverse for any canonical form is always given by the coefficients of the slack variables in the initial tableau. Consequently, the new basis inverse will be given by pivoting in the tableau on $\bar{a}_{rs}$ as usual. Observe that whether we compute the new columns $\bar{A}_j$ for $j \neq s$ or not, pivoting has the same effect upon B^{-1} and $\bar{b}$. Therefore we need only use the reduced tableau of Fig. B.2.

After pivoting, the coefficients in place of the β_{ij} and $\bar{b}_i$ will be, respectively, the new basis inverse and the updated righthand-side vector, while the coefficients in the place of $-y$ and $-\bar{z}$ will be the new simplex multipliers and the new value of the objective function, respectively. These ideas will be reinforced by looking at the example in the next section.

B.3 THE REVISED SIMPLEX METHOD—AN EXAMPLE

To illustrate the procedures of the revised simplex method, we will employ the same example used at the end of Chapter 2. It is important to keep in mind that the revised simplex method is merely a modification of the simplex method that performs fewer calculations by computing only those quantities that are essential to carrying out the steps of the algorithm. The initial tableau for our example is repeated as Tableau B.3. Note that the example has been put in canonical form by the addition of artificial variables, and the necessary Phase I objective function is included.

Tableau B.3 Initial data tableau.

Artificial variables

Basic variables	Current values	x_1	x_2	x_3	x_4	x_5	x_6	x_7	x_8	x_9	x_{10}	x_{11}
x_9	4	1	-1	1	-1	-4	2	-1		1		
x_8	6	-3	3	1	-1	-2	0	0	1			
x_{10}	1	0	0	-1	1	0	1	0			1	
x_{11}	0	1	-1	1	-1	-1	0	0				1
$(-z)$	0	-3	3	2	-2	-1	4	0				
$(-w)$	5	2	-2	1	-1	-5	3	-1				

At each iteration of the revised simplex method, the current inverse of the basis, and a list of the basic variables and the rows in which they are basic, must be maintained. This information, along with the initial tableau, is sufficient to allow us to carry out the steps of the algorithm. As in Chapter 2, we begin with the Phase I objective, maximizing the negative of the sum of the artificial variables, and carry along the Phase II objective in canonical form. Initially, we have the identity basis consisting of the slack variable x_8, the artificial variables x_9, x_{10}, and x_{11}, as well as $-z$ and $-w$, the Phase II and Phase I objective values, respectively. This identity basis is shown in Tableau B.4. Ignore for the moment the column labeled x_6, which has been appended.

Tableau B.4 Initial basis.

Basic variables	Current values	x_9	x_8	x_{10}	x_{11}	x_6	Ratio
x_9	4	1				2	$\frac{4}{2}$
x_8	6		1			0	
x_{10}	1			1		①	$\frac{1}{1}$
x_{11}	0				1	0	
$(-z)$	0					4	
$(-w)$	5					3	

Now, to determine the variable to enter the basis, find $\bar{d}_s = \text{Max } \bar{d}_j$ for j nonbasic. From the initial tableau, we can see that $\bar{d}_s = d_6 = 3$, so that variable x_6 will enter the basis and is appended to the current basis in Tableau B.4. We determine the variable to drop from the basis by the minimum-ratio rule:

$$\frac{\bar{b}_r}{\bar{a}_{r6}} = \underset{i}{\text{Min}} \left\{ \frac{\bar{b}_i}{\bar{a}_{i6}} \,\middle|\, \bar{a}_{i6} > 0 \right\} = \text{Min} \left\{ \frac{4}{2}, \frac{1}{1} \right\} = 1.$$

Since the minimum ratio occurs in row 3, variable x_{10} drops from the basis. We now perform the calculations implied by bringing x_3 into the basis and dropping x_{10}, but only on that portion of the tableau where the basis inverse will be stored. To obtain the updated inverse, we merely perform a pivot operation to transform the coefficients of the incoming variable x_6 so that a canonical form is maintained. That is, the column labeled x_6 should be transformed into all zeros except for a *one* corresponding to the circled pivot element, since variable x_6 enters the basis and variable x_{10} is dropped from the basis. The result is shown in Tableau B.5 including the list indicating which variable is basic in each row. Again ignore the column labeled x_3 which has been appended.

Tableau B.5 After iteration 1.

Basic variables	Current values	x_9	x_8	x_{10}	x_{11}	x_3	Ratio
x_9	2	1		-2		3	$\frac{2}{3}$
x_8	6		1	0		1	$\frac{6}{1}$
x_6	1			1		-1	
x_{11}	0			0	1	①	$\frac{0}{1}$
$(-z)$	-4			-4		6	
$(-w)$	2			-3		4	

We again find the maximum reduced cost, $\bar{d}_s = \text{Max } \bar{d}_j$, for j nonbasic, where $\bar{d}_j = d_j - yA_j$. Recalling that the simplex multipliers are the negative of the coefficients of the slack (artificial) variables in the objective function, we have $y = (0, 0, 3, 0)$. We can compute the reduced costs for the nonbasic variables from: $\bar{d}_j = d_j - y_1a_{1j} - y_2a_{2j} - y_3a_{3j} - y_4a_{4j}$, which yields the following values:

$\bar{d}_1$	$\bar{d}_2$	$\bar{d}_3$	$\bar{d}_4$	$\bar{d}_5$	$\bar{d}_7$	$\bar{d}_{10}$
2	-2	4	-4	-5	-1	-3

Since $\bar{d}_3 = 4$ is the largest reduced cost, x_3 enters the basis. To find the variable to drop from the basis, we have to apply the minimum-ratio rule:

$$\frac{\bar{b}_r}{\bar{a}_{r3}} = \underset{i}{\text{Min}} \left\{ \frac{\bar{b}_i}{\bar{a}_{i3}} \,\middle|\, \bar{a}_{i3} > 0 \right\}.$$

Now to do this, we need the representation of A_3 in the current basis. For this calculation, we consider the Phase II objective function to be a constraint, but never allow $-z$ to drop from the basis.

$$
\bar{A}_3 = B^{-1}A_3 = \begin{bmatrix} 1 & -2 & & & \\ & 1 & 0 & & \\ & & 1 & & \\ & & 0 & 1 & \\ & -4 & & & 1 \end{bmatrix} \begin{bmatrix} 1 \\ 1 \\ -1 \\ 1 \\ 2 \end{bmatrix} = \begin{bmatrix} 3 \\ 1 \\ -1 \\ 1 \\ 6 \end{bmatrix}
$$

Hence,

$$
\frac{\bar{b}_1}{\bar{a}_{13}} = \frac{2}{3}, \qquad \frac{\bar{b}_2}{\bar{a}_{23}} = \frac{6}{1}, \qquad \frac{\bar{b}_4}{\bar{a}_{43}} = \frac{0}{1},
$$

so that the minimum ratio is zero and variable x_{11}, which is basic in row 4, drops from the basis. The updated tableau is found by a pivot operation, such that the column $\bar{A}_3$ is transformed into a column containing all zeros except for a *one* corresponding to the circled pivot element in Tableau B.5, since x_3 enters the basis and x_{11} drops from the basis. The result (ignoring the column labeled x_2) is shown in Tableau B.6.

Tableau B.6 After iteration 2.

Basic variables	Current values	x_9	x_8	x_{10}	x_{11}	x_2	Ratio
x_9	2	1		-2	-3	②	$\frac{2}{2}$
x_8	6		1	0	-1	4	$\frac{6}{4}$
x_6	1			1	1	-1	
x_3	0			0	1	-1	
$(-z)$	-4			-4	-6	9	
$(-w)$	2			-3	-4	2	

Now again find $\bar{d}_s = \text{Max } \bar{d}_j$ for j nonbasic. Since $y = (0, 0, 3, 4)$ and $\bar{d}_j = d_j - yA_j$, we have:

$\bar{d}_1$	$\bar{d}_2$	$\bar{d}_4$	$\bar{d}_5$	$\bar{d}_7$	$\bar{d}_{10}$	$\bar{d}_{11}$
-2	2	0	-1	-1	-3	-4

Since $\bar{d}_2 = 2$ is the largest reduced cost, x_2 enters the basis. To find the variable to drop from the basis, we find the representation of A_2 in the current basis:

$$
\bar{A}_2 = B^{-1}A_2 = \begin{bmatrix} 1 & -2 & -3 & & \\ & 1 & 0 & -1 & \\ & & 1 & 1 & \\ & & 0 & 1 & \\ & -4 & -6 & & 1 \end{bmatrix} \begin{bmatrix} -1 \\ 3 \\ 0 \\ -1 \\ 3 \end{bmatrix} = \begin{bmatrix} 2 \\ 4 \\ -1 \\ -1 \\ 9 \end{bmatrix},
$$

and append it to Tableau B.6. Applying the minimum-ratio rule gives:

$$\frac{\bar{b}_1}{\bar{a}_{12}} = \frac{2}{2}, \qquad \frac{\bar{b}_2}{\bar{a}_{22}} = \frac{6}{4},$$

and x_9, which is basic in row 1, drops from the basis. Again a pivot is performed on the circled element in the column labeled x_2 in Tableau B.6, which results in Tableau B.7 (ignoring the column labeled x_5).

Tableau B.7 After iteration 3.

Basic variables	Current values	x_9	x_8	x_{10}	x_{11}	x_5
x_2	1	$\frac{1}{2}$		-1	$-\frac{3}{2}$	$-\frac{1}{2}$
x_8	2	-2	1	4	5	①(1)
x_6	2	$\frac{1}{2}$		0	$-\frac{1}{2}$	$-\frac{3}{2}$
x_3	1	$\frac{1}{2}$		-1	$-\frac{1}{2}$	$-\frac{3}{2}$
$(-z)$	12	$-\frac{9}{2}$		5	$\frac{15}{2}$	$\frac{19}{2}$
$(-w)$	0	-1		-1	-1	

Since the value of the Phase I objective is equal to zero, we have found a feasible solution. We end Phase I, dropping the Phase I objective function from any further consideration, and proceed to Phase II. We must now find the maximum reduced cost of the Phase II objective function. That is, find $\bar{c}_s = \text{Max}\ \bar{c}_j$, for j nonbasic, where $\bar{c}_j = c_j - yA_j$. The simplex multipliers to initiate Phase II are the negative of the coefficients of the slack (artificial) variables in the $-z$ equation. Therefore, $y = (\frac{9}{2}, 0, -5, -\frac{15}{2})$, and the reduced costs for the nonbasic variables are:

$\bar{c}_1$	$\bar{c}_4$	$\bar{c}_5$	$\bar{c}_7$	$\bar{c}_9$	$\bar{c}_{10}$	$\bar{c}_{11}$
0	0	$\frac{19}{2}$	$\frac{9}{2}$	$-\frac{9}{2}$	5	$\frac{15}{2}$

Since $\bar{c}_5$ is the maximum reduced cost, variable x_5 enters the basis. To find the variable to drop from the basis, compute:

$$\bar{A}_5 = B^{-1}A_5 = \begin{bmatrix} \frac{1}{2} & & -1 & -\frac{3}{2} \\ -2 & 1 & 4 & 5 \\ \frac{1}{2} & & 0 & -\frac{1}{2} \\ \frac{1}{2} & & -1 & -\frac{1}{2} \end{bmatrix} \begin{bmatrix} -4 \\ -2 \\ 0 \\ -1 \end{bmatrix} = \begin{bmatrix} -\frac{1}{2} \\ 1 \\ -\frac{3}{2} \\ -\frac{3}{2} \end{bmatrix}.$$

Since only $\bar{a}_{25}$ is greater than zero, variable x_8, which is basic in row 2, drops from the basis. Again a pivot operation is performed on the circled element in the column labeled x_5 in Tableau B.7 and the result is shown in Tableau B.8.

Tableau B.8 Final reduced tableau.

Basic variables	Current values	x_9	x_8	x_{10}	x_{11}
x_2	2	$-\frac{1}{2}$	$\frac{1}{2}$	1	1
x_5	2	-2	1	4	5
x_6	5	$-\frac{5}{2}$	$\frac{3}{2}$	6	7
x_3	4	$-\frac{5}{2}$	$\frac{3}{2}$	5	7
$(-z)$	32	$\frac{29}{2}$	$-\frac{19}{2}$	-33	-40

We again find $\bar{c}_s = \text{Max } \bar{c}_j$ for j nonbasic. Since $y = (-\frac{29}{2}, \frac{19}{2}, 33, 40)$ and $\bar{c}_j = c_j - yA_j$, we have:

$\bar{c}_1$	$\bar{c}_4$	$\bar{c}_7$	$\bar{c}_8$	$\bar{c}_9$	$\bar{c}_{10}$	$\bar{c}_{11}$
0	0	$-\frac{29}{2}$	$-\frac{19}{2}$	$\frac{29}{2}$	-33	-40

The only positive reduced cost is associated with the artificial variable x_9. Since artificial variables are never reintroduced into the basis once they have become nonbasic, we have determined an optimal solution $x_2 = 2$, $x_5 = 2$, $x_6 = 5$, $x_3 = 4$, and $z = -32$.

Finally, it should be pointed out that the sequence of pivots produced by the revised simplex method is exactly the same as that produced by the usual simplex method. (See the identical example in Chapter 2).

*B.4 COMPUTER CONSIDERATIONS AND THE PRODUCT FORM

The revised simplex method is used in essentially all commercial computer codes for linear programming, both for computational and storage reasons.

For any problem of realistic size, the revised simplex method makes fewer calculations than the ordinary simplex method. This is partly due to the fact that, besides the columns corresponding to the basis inverse and the righthand side, only the column corresponding to the variable entering the basis needs to be computed at each iteration. Further, in pricing out the nonbasic columns, the method takes advantage of the low density of nonzero elements in the initial data matrix of most real problems, since the simplex multipliers need to be multiplied only by the nonzero coefficients in a nonbasic column. Another reason for using the revised simplex method is that roundoff error tends to accumulate in performing these algorithms. Since the revised simplex method maintains the original data, the inverse of the basis may be recomputed from this data periodically, to significantly reduce this type of error. Many large problems could not be solved without such a periodic reinversion of the basis to reduce roundoff error.

Equally important is the fact that the revised simplex method usually requires less storage than does the ordinary simplex method. Besides the basis inverse B^{-1} and the current righthand-side vector $\bar{b}$, which generally contain few zeros, the revised simplex method must store the original data. The original data, on the other hand, generally contains many zeros and can be stored compactly using the following methods. First, we eliminate the need to store zero coefficients, by packing the nonzero coefficients in an array, with reference pointers indicating their location. Second, often the number of significant digits in the original data is small—say, three or fewer—so that these can be handled compactly by storing more than one coefficient in a computer word. In contrast, eight to ten significant digits must be stored for every nonzero coefficient in a canonical form of the usual simplex method, and most coefficients will be nonzero.

There is one further refinement of the revised simplex method that deserves mention, since it was a fundamental breakthrough in solving relatively large-scale problems on second-generation computers. The *product form* of the inverse was developed as an efficient method of storing and updating the inverse of the current basis, when this inverse has to be stored on a peripheral device.

$$
\begin{array}{c}
\overbrace{\hspace{5cm}}^{B^{-1}} \qquad \overbrace{\hspace{1cm}}^{\bar{b}} \\[4pt]
\begin{array}{c|ccc|c}
\bar{a}_{1s} & \beta_{11} & \beta_{12} & \cdots & \beta_{1m} & \bar{b}_1 \\
\vdots & \vdots & & & \vdots & \vdots \\
\boxed{\bar{a}_{rs}} & \beta_{r1} & \beta_{r2} & \cdots & \beta_{rm} & \bar{b}_r \\
\vdots & \vdots & & & \vdots & \vdots \\
\bar{a}_{ms} & \beta_{m1} & \beta_{m2} & \cdots & \beta_{mm} & \bar{b}_m
\end{array}
\end{array}
$$

Fig. B.3 Reduced tableau.

When recomputing B^{-1} and $\bar{b}$ in the revised simplex method, we pivot on $\bar{a}_{rs}$, in a reduced tableau, illustrated by Fig. B.3. The pivot operation first multiplies row r by $1/\bar{a}_{rs}$, and then subtracts $\bar{a}_{is}/\bar{a}_{rs}$ times row r from row i for $i = 1, 2, \ldots, m$ and $i \neq r$. Equivalently, pivoting premultiplies the above tableau by the elementary matrix.

$$
E = \begin{bmatrix}
1 & & & & \eta_1 & & \\
& 1 & & & \eta_2 & & \\
& & \ddots & & \vdots & & \\
& & & 1 & & & \\
& & & & \eta_r & & \\
& & & & 1 & & \\
& & & & \vdots & \ddots & \\
& & & & \eta_m & & 1
\end{bmatrix}
$$
$$
\underset{\text{Column } r}{\uparrow}
$$

where $\eta_r = 1/\bar{a}_{rs}$ and $\eta_i = -\bar{a}_{is}/\bar{a}_{rs}$ for $i \neq r$.

An elementary matrix is defined to be an identity matrix except for one column. If the new basis is B^*, then the new basis inverse $(B^*)^{-1}$ and new righthand-side vector $\bar{b}^*$ are computed by:

$$(B^*)^{-1} = EB^{-1} \qquad \text{and} \qquad \bar{b}^* = E\bar{b}. \tag{6}$$

After the next iteration, the new basis inverse and righthand-side vector will be given by premultiplying by another elementary matrix. Assuming that the initial basis is the identity matrix and letting E_j be the elementary matrix determined by the jth pivot step, after k iterations the basis inverse can be expressed as:

$$B^{-1} = E_k E_{k-1} \cdots E_2 E_1.$$

This *product form* of the inverse is used by almost all commercial linear-programming codes. In these codes, $\bar{b}$ is computed and maintained at each step by (6), but the basis inverse is not computed explicitly. Rather, the elementary matrices are stored and used in place of B^{-1}. These matrices can be stored compactly by recording only the special column $\langle \eta_1, \eta_2, \ldots, \eta_m \rangle$, together with a marker indicating the location of this column in E_j. Using a pivoting procedure for determining the inverse, B^{-1} can always be expressed as the product of no more than m elementary matrices. Consequently, when k is large, the product form for B^{-1} is recomputed. Special procedures are used in this calculation to express the inverse very efficiently and, consequently, to cut down on the number of computations required for the revised simplex method. The details are beyond the scope of our coverage here.

Since the basis inverse is used only for computing the simplex multipliers and finding the representation of the incoming column in terms of the basis, the elementary matrices are used only for the following two calculations:

$$y = c^B B^{-1} = c^B E_k E_{k-1} \cdots E_1 \tag{7}$$

and

$$\bar{A}_s = B^{-1} A_s = E_k E_{k-1} \cdots E_1 A_s. \tag{8}$$

Most commercial codes solve problems so large that the problem data cannot be kept in the computer itself, but must be stored on auxiliary storage devices. The product form of the inverse is well suited for sequential-access devices such as magnetic tapes or drums. The matrices $E_k, E_{k-1}, \ldots, E_1$ are stored sequentially on the device and, by accessing the device in one direction, the elementary matrices are read in the order $E_k, E_{k-1}, \ldots, E_1$ and applied sequentially to c^B for computing (7). When rewinding the device, they are read in opposite order $E_1, E_2, \ldots, E_k$ and applied to A_s for computing (8). The new elementary matrix E_{k+1} is then added to the device next to E_k. Given this procedure and the form of the above calculations, (7) is sometimes referred to as the *b-tran* (backward transformation) and (8) as the *f-tran* (forward transformation).

B.5 SENSITIVITY ANALYSIS REVISITED

In Chapter 3 we gave a detailed discussion of sensitivity analysis in terms of a specific example. There the analysis depended upon recognizing certain relationships between the initial and final tableaus of the simplex method. Now that we have introduced the revised simplex method, we can review that discussion and make some of it more rigorous. Since the revised simplex method is based on keeping track of only the original data tableau, the simplex multipliers, the inverse of the current basis, and which variable is basic in each row, the final tableau for the simplex method can be computed from this information. Therefore, all the remarks that we made concerning sensitivity analysis may be derived formally by using these data.

We will review, in this section, varying the coefficients of the objective function, the values of the righthand side, and the elements of the coefficient matrix. We will not need to review our discussion of the existence of alternative optimal solutions, since no simplifications result from the introduction of matrices. Throughout this section, we assume that we have a maximization problem, and leave to the reader the derivation of the analogous results for minimization problems.

To begin with, we compute the ranges on the coefficients of the objective function so that the basis remains unchanged. Since only the objective-function coefficients are varied, and the values of the decision variables are given by $x^B = B^{-1}b$, these values remain unchanged. However, since the simplex multipliers are given by $y = c^B B^{-1}$, varying any of the objective-function coefficients associated with basic variables will alter the values of the simplex multipliers.

Suppose that variable x_j is *nonbasic*, and we let its coefficient in the objective function c_j be changed by an amount Δc_j, with all other data held fixed. Since x_j is currently not in the optimal solution, it should be clear that Δc_j may be made an arbitrarily large negative number without x_j becoming a candidate to enter the basis. On the other hand, if Δc_j is increased, x_j will not enter the basis so long as its new reduced cost $\bar{c}_j^{\text{new}}$ satisfies:

$$\bar{c}_j^{\text{new}} = c_j + \Delta c_j - yA_j \leq 0,$$

which implies that

$$\Delta c_j \leq yA_j - c_j = -\bar{c}_j,$$

or that

$$-\infty < c_j + \Delta c_j \leq yA_j. \tag{9}$$

At the upper end of the range, x_j becomes a candidate to enter the basis.

Now suppose that x_j is a *basic* variable and, further, that it is basic in row i. If we let its coefficient in the objective function $c_j = c_i^B$ be changed by an amount Δc_i^B, the first thing we note is that the value of the simplex multipliers will be affected, since:

$$y = (c^B + \Delta c_i^B u^i)B^{-1}, \tag{10}$$

where u^i is a row vector of zeros except for a *one* in position i. The basis will not change so long as the reduced costs of all nonbasic variables satisfy:

$$\bar{c}_j^{\text{new}} = c_j - yA_j \leq 0. \tag{11}$$

Substituting in (11) for y given by (10),

$$\bar{c}_j^{\text{new}} = c_j - (c^B + \Delta c_i^B u^i)B^{-1}A_j \leq 0,$$

and noting that $B^{-1}A_j$ is just the representation of A_j in the current basis, we have

$$\bar{c}_j^{\text{new}} = c_j - (c^B + \Delta c_i^B u^i)\bar{A}_j \leqq 0. \tag{12}$$

Condition (12) may be rewritten as:

$$\bar{c}_j - \Delta c_i^B u^i \bar{A}_j \leqq 0,$$

which implies that:

$$\Delta c_i^B \geqq \frac{\bar{c}_j}{\bar{a}_{ij}} \qquad \text{for } \bar{a}_{ij} > 0,$$

and $$\tag{13}$$

$$\Delta c_i^B \leqq \frac{\bar{c}_j}{\bar{a}_{ij}} \qquad \text{for } \bar{a}_{ij} < 0,$$

and no limit for $\bar{a}_{ij} = 0$.

Finally, since (13) must be satisfied for all nonbasic variables, we can define upper and lower bounds on Δc_i^B as follows:

$$\underset{j}{\text{Max}} \left\{ \frac{\bar{c}_j}{\bar{a}_{ij}} \,\middle|\, a_{ij} > 0 \right\} \leqq \Delta c_i^B \leqq \underset{j}{\text{Min}} \left\{ \frac{\bar{c}_j}{\bar{a}_{ij}} \,\middle|\, \bar{a}_{ij} < 0 \right\}. \tag{14}$$

Note that, since $\bar{c}_j \leqq 0$, the lower bound on Δc_i^B is nonpositive and the upper bound is nonnegative, so that the range on the cost coefficient $c_i^B + \Delta c_i^B$ is determined by adding c_i^B to each bound in (14). Note that Δc_i^B may be unbounded in either direction if there are no $\bar{a}_{ij}$ of appropriate sign.

At the upper bound in (14), the variable producing the minimum ratio is a candidate to enter the basis, while at the lower bound in (14), the variable producing the maximum ratio is a candidate to enter the basis. These candidate variables are clearly not the same, since they have opposite signs for $\bar{a}_{ij}$. In order for any candidate to enter the basis, the variable to drop from the basis x_r is determined by the usual minimum-ratio rule:

$$\frac{\bar{b}_r}{\bar{a}_{rs}} = \underset{i}{\text{Min}} \left\{ \frac{\bar{b}_i}{\bar{a}_{is}} \,\middle|\, \bar{a}_{is} > 0 \right\}. \tag{15}$$

If $\bar{A}_s \leqq 0$, then variable x_s can be increased without limit and the objective function is unbounded. Otherwise, the variable corresponding to the minimum ratio in (15) will drop from the basis if x_s is introduced into the basis.

We turn now to the question of variations in the righthand-side values. Suppose that the righthand-side value b_k is changed by an amount Δb_k, with all other data held fixed. We will compute the range so that the basis remains unchanged. The values of the decision variables will change, since they are given by $x^B = B^{-1}b$, but the values of the simplex multipliers, given by $y = c^B B^{-1}$, will not. The new values of the basic variables, x^{new}, must be nonnegative in order for the basis to remain feasible. Hence,

$$x^{\text{new}} = B^{-1}(b + u^k \Delta b_k) \geqq 0, \tag{16}$$

where u^k is a column vector of all zeros except for a *one* in position k. Noting that $B^{-1}b$ is just the representation of the righthand side in the current basis, (16) becomes

$$\bar{b} + B^{-1}u^k \, \Delta b_k \geqq 0;$$

and, letting β_{ij} be the elements of the basis inverse matrix B^{-1}, we have:

$$\bar{b}_i + \beta_{ik} \, \Delta b_k \geqq 0 \qquad (i = 1, 2, \ldots, m), \tag{17}$$

which implies that:

$$\Delta b_i \geqq \frac{-\bar{b}_i}{\beta_{ik}} \qquad \text{for } \beta_{ik} > 0,$$

and

$$\tag{18}$$

$$\Delta b_i \leqq \frac{-\bar{b}_i}{\beta_{ik}} \qquad \text{for } \beta_{ik} < 0,$$

and no limit for $\beta_{ik} = 0$.

Finally, since (18) must be satisfied for all basic variables, we can define upper and lower bounds on Δb_i as follows:

$$\text{Max}_i \left\{ \frac{-\bar{b}_i}{\beta_{ik}} \,\middle|\, \beta_{ik} > 0 \right\} \leqq \Delta b_k \leqq \text{Min}_i \left\{ \frac{-\bar{b}_i}{\beta_{ik}} \,\middle|\, \beta_{ik} < 0 \right\}. \tag{19}$$

Note that since $\bar{b}_i \geqq 0$, the lower bound on Δb_k is nonpositive and the upper bound is nonnegative. The range on the righthand-side value $b_k + \Delta b_k$ is then determined by adding b_k to each bound in (19).

At the upper bound in (19) the variable basic in the row producing the minimum ratio is a candidate to be dropped from the basis, while at the lower bound in (19) the variable basic in the row producing the maximum ratio is a candidate to be dropped from the basis. The variable to enter the basis in each of these cases can be determined by the ratio test of the dual simplex method. Suppose the variable basic in row r is to be dropped; then the entering variable x_s is determined from:

$$\frac{\bar{c}_s}{\bar{a}_{rs}} = \text{Min}_j \left\{ \frac{\bar{c}_j}{\bar{a}_{rj}} \,\middle|\, \bar{a}_{rj} < 0 \right\}. \tag{20}$$

If there does not exist $\bar{a}_{rj} < 0$, then no entering variable can be determined. When this is the case, the problem is infeasible beyond this bound.

Now let us turn to variations in the coefficients in the equations of the model. In the case where the coefficient corresponds to a nonbasic activity, the situation is straightforward. Suppose that the coefficient a_{ij} is changed by an amount Δa_{ij}. We will compute the range so that the basis remains unchanged. In this case, both the values of the decision variables and the shadow prices also remain unchanged. Since x_j is assumed nonbasic, the current basis remains optimal so long as the new reduced cost $\bar{c}_j^{\text{new}}$ satisfies

$$\bar{c}_j^{\text{new}} = c_j - y(A_j + u^i \, \Delta a_{ij}) \leqq 0, \tag{21}$$

where u^i is a column vector of zeros except for a *one* in position i. Since $\bar{c}_j = c_j - yA_j$, (21) reduces to:

$$\bar{c}_j^{\text{new}} = \bar{c}_j - y_i \, \Delta a_{ij} \leqq 0. \tag{22}$$

Hence, (22) gives either an upper or a lower bound on Δa_{ij}. If $y_i > 0$, the appropriate range is:

$$\frac{\bar{c}_j}{y_i} \leqq \Delta a_{ij} < +\infty, \tag{23}$$

and if $y_i < 0$, the range is:

$$-\infty < \Delta a_{ij} \leqq \frac{\bar{c}_j}{y_i}. \tag{24}$$

The range on the variable coefficient $a_{ij} + \Delta a_{ij}$ is simply given by adding a_{ij} to the bounds in (23) and (24). In either situation, some x_s becomes a candidate to enter the basis, and the corresponding variable to drop from the basis is determined by the usual minimum-ratio rule given in (15).

The case where the coefficient to be varied corresponds to a basic variable is a great deal more difficult and will not be treated in detail here. Up until now, all variations in coefficients and righthand-side values have been such that the basis remains unchanged. The question we are asking here violates this principle. We could perform a similar analysis, assuming that the basic variables should remain unchanged, but the basis and its inverse will necessarily change. There are three possible outcomes from varying a coefficient of a basic variable in a constraint. Either (1) the basis may become singular; (2) the basic solution may become infeasible; or (3) the basic solution may become nonoptimal. Any one of these conditions would define an effective bound on the range of Δa_{ij}. A general derivation of these results is beyond the scope of this discussion.

B.6 PARAMETRIC PROGRAMMING

Having discussed changes in individual elements of the data such that the basis remains unchanged, the natural question to ask is what happens when we make simultaneous variations in the data or variations that go beyond the ranges derived in the previous section. We can give rigorous answers to these questions for cases where the problem is made a function of one parameter. Here we essentially compute ranges on this parameter in a manner analogous to computing righthand-side and objective-function ranges.

We begin by defining three different parametric-programming problems, where each examines the optimal value of a linear program as a function of the scalar (*not* vector) parameter θ. In Chapter 3, we gave examples and interpreted the first two problems.

Parametric righthand side

$P(\theta) = \text{Max } cx,$

   subject to:

$$Ax = b^1 + \theta b^2, \tag{25}$$

$$x \geq 0.$$

Parametric objective function

$Q(\theta) = \text{Max } (c^1 + \theta c^2)x,$

   subject to:

$$Ax = b, \tag{26}$$

$$x \geq 0.$$

Parametric rim problem

$R(\theta) = \text{Max } (c^1 + \theta c^2)x,$

   subject to:

$$Ax = b^1 + \theta b^2, \tag{27}$$

$$x \geq 0.$$

Note that, when the parameter θ is fixed at some value, each type of problem becomes a simple linear program.

We first consider the parametric righthand-side problem. In this case, the feasible region is being modified as the parameter θ is varied. Suppose that, for $\underline{\theta}$ and $\bar{\theta}$, (25) is a feasible linear program. Then, assuming that $\underline{\theta} < \bar{\theta}$, (25) must be feasible for all θ in the interval $\underline{\theta} \leq \theta \leq \bar{\theta}$. To see this, first note that any θ in the interval may be written as $\theta = \lambda\underline{\theta} + (1 - \lambda)\bar{\theta}$, where $0 \leq \lambda \leq 1$. (This is called a convex combination of $\underline{\theta}$ and $\bar{\theta}$.) Since (25) is feasible for $\underline{\theta}$ and $\bar{\theta}$, there must exist corresponding $\underline{x}$ and $\bar{x}$ satisfying:

$$A\underline{x} = b^1 + \underline{\theta}b^2, \qquad \underline{x} \geq 0. \qquad\qquad A\bar{x} = b^1 + \bar{\theta}b^2, \qquad \bar{x} \geq 0.$$

Multiplying the former by λ and the latter by $(1 - \lambda)$ and adding yields:

$$\lambda A\underline{x} + (1 - \lambda)A\bar{x} = \lambda(b^1 + \underline{\theta}b^2) + (1 - \lambda)(b^1 + \bar{\theta}b^2), \qquad \lambda\underline{x} + (1 - \lambda)\bar{x} \geq 0,$$

which may be rewritten as:

$$A(\lambda\underline{x} + (1 - \lambda)\bar{x}) = b^1 + (\lambda\underline{\theta} + (1 - \lambda)\bar{\theta})b^2, \qquad \lambda\underline{x} + (1 - \lambda)\bar{x} \geq 0. \tag{28}$$

Equation (28) implies that there exists a feasible solution for any θ in the interval $\underline{\theta} \leq \theta \leq \bar{\theta}$.

The implication for the parameteric righthand-side problem is that, when increasing (decreasing) θ, once the linear program becomes infeasible it will remain infeasible for any further increases (decreases) in θ.

Let us assume that (25) is feasible for $\theta = \theta_0$ and examine the implications of varying θ. For $\theta = \theta_0$, let B be the optimal basis with decision variables $x^B = B^{-1}(b^1 + \theta_0 b^2)$ and shadow prices $y = c^B B^{-1}$. The current basis B remains optimal as θ is varied, so long as the current solution remains feasible; that is,

$$\bar{b}(\theta) = B^{-1}(b^1 + \theta b^2) \geq 0,$$

or, equivalently,

$$\bar{b}^1 + \theta \bar{b}^2 \geq 0. \tag{29}$$

Equation (29) may imply both upper and lower bounds, as follows:

$$\theta \geq \frac{-\bar{b}_i^1}{\bar{b}_i^2} \qquad \text{for } \bar{b}_i^2 > 0,$$

$$\theta \leq \frac{-\bar{b}_i^1}{\bar{b}_i^2} \qquad \text{for } \bar{b}_i^2 < 0;$$

and these define the following range on θ:

$$\text{Max}_i \left\{ \frac{-\bar{b}_i^1}{\bar{b}_i^2} \,\middle|\, \bar{b}_i^2 > 0 \right\} \leq \theta \leq \text{Min}_i \left\{ \frac{-\bar{b}_i^1}{\bar{b}_i^2} \,\middle|\, \bar{b}_i^2 < 0 \right\}. \tag{30}$$

If we now move θ to either its upper or lower bound, a basis change can take place. At the upper bound, the variable basic in the row producing the minimum ratio becomes a candidate to drop from the basis, while at the lower bound the variable producing the maximum ratio becomes a candidate to drop from the basis. In either case, assuming the variable to drop is basic in row r, the variable to enter the basis is determined by the usual rule of the dual simplex method:

$$\frac{\bar{c}_s}{\bar{a}_{rs}} = \text{Min}_j \left\{ \frac{\bar{c}_j}{\bar{a}_{rj}} \,\middle|\, \bar{a}_{rj} < 0 \right\}. \tag{31}$$

If $\bar{a}_{rj} \geq 0$ for all j, then no variable can enter the basis and the problem is infeasible beyond this bound. If an entering variable is determined, then a new basis is determined and the process is repeated. The range on θ such that the *new* basis remains optimal, is then computed in the same manner.

On any of these successive intervals where the basis is unchanged, the optimal value of the linear program is given by $P(\theta) = y(b^1 + \theta b^2)$, where the vector $y = c^B B^{-1}$ of shadow prices is not a function of θ. Therefore, $P(\theta)$ is a straight line with slope yb^2 on a particular interval.

Further, we can easily argue that $P(\theta)$ is also a concave function of θ, that is,

$$P(\lambda \underline{\theta} + (1 - \lambda)\bar{\theta}) \geq \lambda P(\underline{\theta}) + (1 - \lambda)P(\bar{\theta}) \qquad \text{for } 0 \leq \lambda \leq 1.$$

Suppose that we let $\underline{\theta}$ and $\bar{\theta}$ be two values of θ such that the corresponding linear programs defined by $P(\underline{\theta})$ and $P(\bar{\theta})$ in (25) have finite optimal solutions. Let their respective optimal solutions be $\underline{x}$ and $\bar{x}$. We have already shown in (28) that $\lambda \underline{x} + (1 - \lambda)\bar{x}$ is a feasible solution to the linear program defined by $P(\lambda \underline{\theta} + (1 - \lambda)\bar{\theta})$ in (25). However, $\lambda \underline{x} + (1 - \lambda)\bar{x}$ may not be optimal to this linear program, and hence

$$P(\lambda \underline{\theta} + (1 - \lambda)\overline{\theta}) \geqq c(\lambda \underline{x} + (1 - \lambda)\overline{x}).$$

Rearranging terms and noting that $\underline{x}$ and $\overline{x}$ are optimal solutions to $P(\underline{\theta})$ and $P(\overline{\theta})$, respectively, we have:

$$c(\lambda \underline{x} + (1 - \lambda)\overline{x}) = \lambda c\underline{x} + (1 - \lambda)c\overline{x} = \lambda P(\underline{\theta}) + (1 - \lambda)P(\overline{\theta}).$$

The last two expressions imply the condition that $P(\theta)$ is a concave function of θ. Hence, we have shown that $P(\theta)$ is a concave piecewise-linear function of θ. This result can be generalized to show that the optimal value of a linear program is a concave polyhedral function of its righthand-side vector.

Let us now turn to the *parametric objective-function* problem. Assuming that the linear program defined in (26) is feasible, it will remain feasible regardless of the value of θ. However, this linear program may become unbounded. Rather than derive the analogous properties of $Q(\theta)$ directly, we can determine the properties of $Q(\theta)$ from those of $P(\theta)$ by utilizing the duality theory of Chapter 4. We may rewrite $Q(\theta)$ in terms of the dual of its linear program as follows:

$$Q(\theta) = \text{Max } (c^1 + \theta c^2)x \qquad = \qquad \text{Min } yb, \qquad\qquad (32)$$

$$\text{subject to:} \qquad\qquad\qquad \text{subject to:}$$

$$Ax = b, \qquad\qquad yA \geqq c^1 + \theta c^2,$$

$$x \geqq 0$$

and, recognizing that a minimization problem can be transformed into a maximization problem by multiplying by minus one, we have

$$Q(\theta) = -\text{Max} - yb \qquad\qquad = -P'(\theta), \qquad\qquad (33)$$

$$yA \geqq c^1 + \theta c^2.$$

Here $P'(\theta)$ must be a concave piecewise-linear function of θ, since it is the optimal value of a linear program considered as a function of its righthand side. Therefore,

$$Q(\lambda \underline{\theta} + (1 - \lambda)\overline{\theta}) \leqq \lambda Q(\underline{\theta}) + (1 - \lambda)Q(\overline{\theta}) \qquad \text{for all } 0 \leqq \lambda \leqq 1,$$

or

$$P'(\lambda \underline{\theta} + (1 - \lambda)\overline{\theta}) \geqq \lambda P'(\underline{\theta}) + (1 - \lambda)P'(\overline{\theta}) \qquad \text{for all } 0 \leqq \lambda \leqq 1,$$

which says that $Q(\theta)$ is a convex function.

Further, since the primal formulation of $Q(\theta)$ is assumed to be feasible, whenever the dual formulation is infeasible the primal must be unbounded. Hence, we have a result analogous to the feasibility result for $P(\theta)$. Suppose that for $\underline{\theta}$ and $\overline{\theta}$, (26) is a bounded linear program; then, assuming $\underline{\theta} < \overline{\theta}$, (26) must be bounded for all θ in the interval $\underline{\theta} \leqq \theta \leqq \overline{\theta}$. The implication for the parametric objective-function problem is that, when increasing (decreasing) θ, once the linear program becomes unbounded it will remain unbounded for any further increase (decrease) in θ.

Let us assume that (26) is bounded for $\theta = \theta_0$ and examine the implications of varying θ. For $\theta = \theta_0$, let B be the optimal basis with decision variables $x^B = B^{-1}b$ and shadow prices $y = (c^{1B} + \theta c^{2B})B^{-1}$. The current basis B remains optimal as θ is varied, so long as the reduced costs remain nonpositive; that is,

$$\bar{c}_j(\theta) = (c_j + \theta c_j^2) - yA_j \leq 0.$$

Substituting for the shadow prices y,

$$\bar{c}_j(\theta) = (c_j^1 + \theta c_j^2) - (c^{1B} + \theta c^{2B})B^{-1}A_j \leq 0;$$

and collecting terms yields,

$$\bar{c}_j(\theta) = c_j^1 - c^{1B}B^{-1}A_j + \theta(c_j^2 - c^{2B}B^{-1}A_j) \leq 0,$$

or, equivalently,

$$\bar{c}_j(\theta) = \bar{c}_j^1 + \theta\bar{c}_j^2 \leq 0. \tag{34}$$

Equation (34) may imply both upper and lower bounds, as follows:

$$\theta \geq \frac{-\bar{c}_j^1}{\bar{c}_j^2} \qquad \text{for } \bar{c}_j^2 < 0,$$

$$\theta \leq \frac{-\bar{c}_j^1}{\bar{c}_j^2} \qquad \text{for } \bar{c}_j^2 > 0;$$

and these define the following range on θ;

$$\text{Max}_j \left\{ \frac{-\bar{c}_j^1}{\bar{c}_j^2} \,\middle|\, \bar{c}_j^2 < 0 \right\} \leq \theta \leq \text{Min}_j \left\{ \frac{-\bar{c}_j^1}{\bar{c}_j^2} \,\middle|\, \bar{c}_j^2 > 0 \right\}. \tag{35}$$

If we move θ to either its upper or lower bound, a basis change can take place. At the upper bound, the variable producing the minimum ratio becomes a candidate to enter the basis, while at the lower bound, the variable producing the maximum ratio becomes a candidate to enter the basis. In either case, the variable to drop from the basis is determined by the usual rule of the primal simplex method.

$$\frac{\bar{b}_r}{\bar{a}_{rs}} = \text{Min}_i \left\{ \frac{\bar{b}_i}{\bar{a}_{is}} \,\middle|\, \bar{a}_{is} > 0 \right\}. \tag{36}$$

If $\bar{A}_s \leq 0$, then x_s may be increased without limit and the problem is unbounded beyond this bound. If a variable to drop is determined, then a new basis is determined and the process is repeated. The range on θ such that the *new* basis remains optimal is then again computed in the same manner.

On any of these successive intervals where the basis is unchanged, the optimal value of the linear program is given by:

$$Q(\theta) = (c^{1B} + \theta c^{2B})x^B,$$

where $x^B = B^{-1}b$ and is not a function of θ. Therefore, $Q(\theta)$ is a straight line with slope $c^{2B}x^B$ on a particular interval.

Finally, let us consider the *parametric rim problem*, which has the parameter in both the objective function and the righthand side. As would be expected, the optimal value of the linear program defined in (27) is neither a concave nor a convex function of the parameter θ. Further, $R(\theta)$ is not a piecewise-linear function, either. Let us assume that (27) is feasible for $\theta = \theta_0$, and examine the implications of varying θ. For $\theta = \theta_0$, let B be the optimal basis with decision variables $x^B = B^{-1}(b^1 + \theta b^2)$ and shadow prices $y = (c^{1B} + \theta c^{2B})B^{-1}$. The current basis remains optimal so long as the current solution remains feasible and the reduced costs remain nonpositive. Hence, the current basis remains optimal so long as both ranges on θ given by (30) and (35) are satisfied. Suppose we are increasing θ. If the upper bound of (30) is reached before the upper bound of (34), then a dual simplex step is performed according to (31). If the opposite is true, then a primal simplex step is performed according to (36). Once the new basis is determined, the process is repeated. The same procedure is used when decreasing θ.

For a given basis, the optimal value of the objective function is given by multiplying the basic costs by the value of the basic variables x^B; that is,

$$R(\theta) = (c^{1B} + \theta c^{2B})B^{-1}(b^1 + \theta b^2),$$

which can be rewritten as

$$R(\theta) = c^{1B}B^{-1}b^1 + \theta(c^{2B}B^{-1}b^1 + c^{1B}B^{-1}b^2) + \theta^2(c^{2B}B^{-1}b^2).$$

Hence, the optimal value of the parametric rim problem is a quadratic function of θ for a fixed basis B. In general, this quadratic may be either a concave or convex function over the range implied by (30) and (35).

It should be clear that the importance of parametric programming in all three cases is the efficiency of the procedures in solving a number of different cases. Once an optimal solution has been found for some value of the parameter, say $\theta = \theta_0$, increasing or decreasing θ amounts to successively computing the points at which the basis changes and then performing a routine pivot operation.

B.7 DUALITY THEORY IN MATRIX FORM

The duality theory introduced in Chapter 4 can be stated very concisely in matrix form. As we saw there, a number of variations of the basic duality results can be formulated by slightly altering the form of the primal problem. For ease of comparison with the Chapter 4 results, we will again employ the symmetric version of the dual linear programs:

Primal	*Dual*
Max $z = cx$,	Min $v = yb$,
subject to:	subject to:
$Ax \leq b$,	$yA \geq c$,
$x \geq 0$.	$y \geq 0$.

Note that y and c are row vectors while x and b are column vectors. Let $\bar{z}$ and $\bar{v}$ denote the optimal values of the objective functions for the primal and dual problems, respectively. We will review the three key results of duality theory.

Weak duality. If $\bar{x}$ is a feasible solution to the primal and $\bar{y}$ is a feasible solution to the dual, then $c\bar{x} \leq \bar{y}b$, and consequently, $\bar{z} \leq \bar{v}$.

Strong duality. If the primal (dual) has a finite optimal solution, then so does the dual (primal), and the two extremes are equal, $\bar{z} = \bar{v}$.

Complementary slackness. If $\bar{x}$ is a feasible solution to the primal and $\bar{y}$ is a feasible solution to the dual, then $\bar{x}$ and $\bar{y}$ are optimal solutions to the primal and dual, respectively, if and only if

$$\bar{y}(A\bar{x} - b) = 0 \quad \text{and} \quad (\bar{y}A - c)\bar{x} = 0.$$

The arguments leading to these results were given in Chapter 4 but will be briefly reviewed here. *Weak duality* is a consequence of primal and dual feasibility, since multiplying the primal constraints on the left by $\bar{y}$ and the dual constraints on the right by $\bar{x}$ and combining gives:

$$c\bar{x} \leq \bar{y}A\bar{x} \leq \bar{y}b.$$

Since the optimal values must be at least as good as any feasible value,

$$\bar{z} = \underset{x}{\text{Max}}\ c\bar{x} \leq \underset{y}{\text{Min}}\ \bar{y}b = \bar{v}.$$

Weak duality implies that if $c\bar{x} = \bar{y}b$ for some primal feasible $\bar{x}$ and dual feasible $\bar{y}$, then $\bar{x}$ and $\bar{y}$ are optimal solutions to the primal and dual, respectively.

The termination conditions of the simplex method and the concept of simplex multipliers provides the *strong duality property*. Suppose that the simplex method has found an optimal basic feasible solution $\bar{x}$ to the primal with optimal basis B. The optimal values of the decision variables and simplex multipliers are $\bar{x}^B = B^{-1}b$ and $\bar{y} = c^B B^{-1}$, respectively. The simplex optimality conditions imply:

$$\bar{z} = c^B x^B = c^B B^{-1}b = \bar{y}b, \tag{37}$$

$$\bar{c}^N = c^N - \bar{y}N \leq 0. \tag{38}$$

The latter condition (38), plus the definition of the simplex multipliers, which can be restated as:

$$\bar{c}^B = c^B - \bar{y}B = 0,$$

imply that $c - \bar{y}A \leq 0$, so that $\bar{y} \geq 0$ is a dual feasible solution. Now $\bar{x}^N = 0$ and (37) implies that:

$$c\bar{x} = c^B\bar{x}^B + c^N\bar{x}^N = \bar{y}b. \tag{39}$$

Since $\bar{x}$ and $\bar{y}$ are feasible solutions to the primal and dual respectively, (39) implies $\bar{z} = c\bar{x} = \bar{y}b = \bar{v}$ which is the desired result.

Finally, *complementary slackness* follows directly from the strong-duality property. First, we assume that complementary slackness holds, and show optimality.

If $\bar{x}$ and $\bar{y}$ are feasible to the primal and dual, respectively, such that

$$\bar{y}(A\bar{x} - b) = 0 \quad \text{and} \quad (\bar{y}A - c)\bar{x} = 0,$$

then

$$c\bar{x} = \bar{y}A\bar{x} = \bar{y}b. \tag{40}$$

Condition (40) implies the $\bar{x}$ and $\bar{y}$ are optimal to the primal and dual, respectively. Second, we assume optimality holds, and show complementary slackness. Let $\bar{x}$ and $\bar{y}$ be optimal to the primal and dual, respectively; that is,

$$c\bar{x} = \bar{y}b.$$

Then

$$0 = c\bar{x} - \bar{y}b \le c\bar{x} - \bar{y}A\bar{x} = (c - \bar{y}A)\bar{x} \tag{41}$$

by primal feasibility, and $\bar{y} \ge 0$. Now since $0 \ge c - \bar{y}A$ by dual feasibility, and $\bar{x} \ge 0$, we have

$$0 \ge (c - \bar{y}A)\bar{x}, \tag{42}$$

and (41) and (42) together imply:

$$(c - \bar{y}A)\bar{x} = 0.$$

An analogous argument using dual feasibility and $x \ge 0$ implies the other complementary slackness condition,

$$\bar{y}(A\bar{x} - b) = 0.$$

These results are important for applications since they suggest algorithms other than straightforward applications of the simplex method to solve specially structured problems. The results are also central to the theory of systems of linear equalities and inequalities, a theory which itself has a number of important applications. In essence, duality theory provides a common perspective for treating such systems and, in the process, unifies a number of results that appear scattered throughout the mathematics literature. A thorough discussion of this theory would be inappropriate here, but we indicate the flavor of this point of view.

A problem that has interested mathematicians since the late 1800's has been characterizing the situation when a system of linear equalities and inequalities does not have a solution. Consider for example the system:

$$Ax = b, \qquad x \ge 0.$$

Suppose that y is a row vector. We multiply the system of equations by y to produce a single new equation:

$$(yA)x = yb.$$

If x is feasible in the original system, then it will certainly satisfy this new equation. Suppose, though, that $yb > 0$ and that $yA \le 0$. Since $x \ge 0$, $yAx \le 0$, so that no $x \ge 0$ solves the new equation, and thus the original system is infeasible. We have determined a single inconsistent equation that summarizes the inconsistencies in the original system. The characterization that we are working towards states that such a summarizing inconsistent equation always can be found when the system has no solution. It is given by:

Farkas' Lemma. Exactly one of the following two systems has a solution:

$$\begin{array}{lll} \text{I} & Ax = b, & \quad \text{II} \quad yA \leq 0, \\ & x \geq 0. & \qquad\qquad yb > 0. \end{array}$$

or

The proof of the lemma is straightforward, by considering the Phase I linear-programming progam that results from adding artificial variables to System (I):

Minimize et,

subject to:

$$Ax + It = b, \tag{43}$$
$$x \geq 0, \quad t \geq 0,$$

where e is the sum vector consisting of all ones and t is the vector of artificial variables. Suppose that the simplex method is applied to (43), and the optimal solution is $(\bar{x}, \bar{t})$ with shadow prices $\bar{y}$. By the termination conditions of the simplex method, the shadow prices satisfy:

$$0 - \bar{y}A \geq 0 \quad\quad (\text{that is, } \bar{y}A \leq 0). \tag{44}$$

By the duality property, we know $e\bar{t} = \bar{y}b$, and since (43) is a Phase I linear program, we have:

$$\begin{array}{ll} \bar{y}b = e\bar{t} = 0 & \quad \text{if and only if } I \text{ is feasible,} \\ \bar{y}b = e\bar{t} > 0 & \quad \text{if and only if } I \text{ is infeasible.} \end{array} \tag{45}$$

Equations (44) and (45) together imply the lemma.

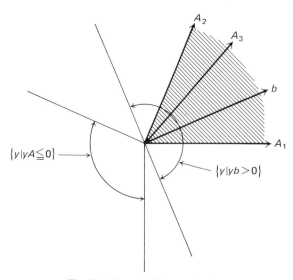

Fig. B.4 System I has a solution.

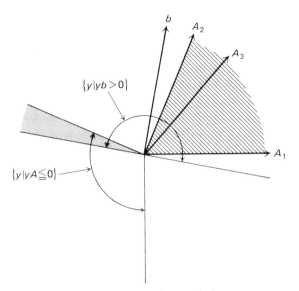

Fig. B.5 System II has a solution.

We can give a geometrical interpretation of Farkas' Lemma by recalling that, if the inner product of two vectors is positive, the vectors make an acute angle with each other, while if the inner product is negative, the vectors make an obtuse angle with each other. Therefore, any vector y that solves System II must make acute angles with all the column vectors of A and a strictly obtuse angle with the vector b. On the other hand, in order for System I to have a feasible solution, there must exist nonnegative weights x that generate b from the column vectors of A. These two situations are given in Figs. B.4 and B.5.

B.8 RESOLVING DEGENERACY IN THE SIMPLEX METHOD

In Chapter 2 we showed that the simplex method solves any linear program in a finite number of steps if we assume that the righthand-side vector is strictly positive for each canonical form generated. Such a canonical form is called *nondegenerate*.

The motivation for this assumption is to ensure that the new value of the entering variable x_s^*, which is given by:

$$x_s^* = \frac{\overline{b}_r}{\overline{a}_{rs}}$$

is strictly positive; and hence, that the new value of the objective function z^*, given by:

$$z^* = \overline{z} + \overline{c}_s x_s^*,$$

shows a strict improvement at each iteration. Since the minimum-ratio rule to determine the variable to drop from the basis ensures $\overline{a}_{rs} > 0$, the assumption that $\overline{b}_i > 0$ for all i implies that $x_s^* > 0$. Further, introducing the variable x_s requires that $\overline{c}_s > 0$ and therefore that $z^* > \overline{z}$. This implies that there is a strict improvement in the value of the objective function at each iteration, and hence that no basis is repeated. Since there is a finite number of possible bases, the simplex method must terminate in a finite number of iterations.

The purpose of this section is to extend the simplex method so that, without the nondegeneracy assumption, it can be shown that the method will solve any linear program in a finite number of iterations. In Chapter 2 we indicated that we would do this by perturbing the righthand-side vector. However, a simple perturbation of the righthand side by a scalar will not ensure that the righthand side will be positive for all canonical forms generated. As a result, we introduce a vector perturbation of the righthand side and the concept of lexicographic ordering.

An m-element vector $a = (a_1, a_2, \ldots, a_m)$ is said to be *lexico-positive*, written $a \succ 0$, if at least one element is nonzero and the first such element is positive. The term lexico-positive is short for lexicographically positive. Clearly, any positive multiple of a lexico-positive vector is lexico-positive, and the sum of two lexico-positive vectors is lexico-positive. An m-element vector a is *lexico-greater* than an m-element vector b, written $a \succ b$, if $(a - b)$ is lexico-positive; that is, if $a - b \succ 0$. Unless two vectors are identical, they are lexicographically ordered, and therefore the ideas of *lexico-max* and *lexico-min* are well-defined.

We will use the concept of lexicographic ordering to modify the simplex method so as to produce a unique variable to drop at each iteration and a strict lexicographic improvement in the value of the objective function at each iteration. The latter will ensure the termination of the method after a finite number of iterations.

Suppose the linear program is in canonical form initially with $b \geqq 0$. We introduce a unique perturbation of the righthand-side vector by replacing the vector b with the $m \times (m + 1)$ matrix $[b, I]$. By making the vector x an $n \times (m + 1)$ matrix X, we can write the initial tableau as:

$$X^B \geqq 0, \qquad X^N \geqq 0, \qquad X^I \geqq 0,$$
$$BX^B + NX^N + IX^I \quad = [b, I], \tag{46}$$
$$c^B X^B + c^N X^N \quad - Z = [0, 0],$$

where X^B, X^N, and X^I are the matrices associated with the basic variables, the non-basic variables, and the variables in the initial identity basis, respectively. The intermediate tableau corresponding to the basis B is determined as before by multiplying the constraints of (46) by B^{-1} and subtracting c^B times the resulting constraints from the objective function. The intermediate tableau is then:

$$X^B \geqq 0, \qquad X^N \geqq 0, \qquad X^I \geqq 0,$$
$$IX^B + \bar{N}X^N + B^{-1}X^I \quad = [\bar{b}, B^{-1}] = \bar{B}, \tag{47}$$
$$\bar{c}^N X^N - \quad yX^I - Z = [-\bar{z}, -y] = -\bar{Z},$$

where $\bar{N} = B^{-1}N$, $y = c^B B^{-1}$, $\bar{c}^N = c^N - yN$, and $\bar{z} = yb$ as before. Note that if the matrices X^N and X^I are set equal to zero, a basic solution of the linear program is given by $X^B = [\bar{b}, B^{-1}]$, where the first column of X^B, say X_1^B, gives the usual values of the basic variables $X_1^B = \bar{b}$.

We will formally prove that the simplex method, with a lexicographic resolution of degeneracy, solves any linear program in a finite number of iterations. An essential aspect of the proof is that at each iteration of the algorithm, each row vector of the righthand-side perturbation matrix is lexico-positive.

In the initial tableau, each row vector of the righthand-side matrix is lexico-positive, assuming $b \geq 0$. If $b_i > 0$, then the row vector is lexico-positive, since its first element is positive. If $b_i = 0$, then again the row vector is lexico-positive since the first nonzero element is a plus *one* in position $(i + 1)$.

Define $\bar{B}_i$ to be row i of the righthand-side matrix $[\bar{b}, B^{-1}]$ of the intermediate tableau. In the following proof, we will assume that each such row vector $\bar{B}_i$ is lexico-positive for a particular canonical form, and argue inductively that they remain lexico-positive for the next canonical form. Since this condition holds for the initial tableau, it will hold for all subsequent tableaus.

Now consider the simplex method one step at a time.

STEP (1): If $\bar{c}_j \leq 0$ for $j = 1, 2, \ldots, n$, then *stop*; the current solution is optimal. Since $X^B = \bar{B} = [\bar{b}, I]$, the first component of X^B is the usual basic feasible solution $X_1^B = \bar{b}$. Since the reduced costs associated with this solution are $\bar{c}^N \leq 0$, we have:

$$z = \bar{z} + \bar{c}^N X_1^N \leq \bar{z},$$

and hence $\bar{z}$ is an upper bound on the maximum value of z. The current solution $X_1^B = \bar{b}$ attains this upper bound and is therefore optimal. If we continue the algorithm, there exists some $\bar{c}_j > 0$.

STEP (2): Choose the column s to enter the basis by:

$$\bar{c}_s = \underset{j}{\text{Max}} \ \{\bar{c}_j \,|\, \bar{c}_j > 0\}.$$

If $\bar{A}_s \leq 0$, then *stop*; there exists an unbounded solution. Since $\bar{A}_s \leq 0$, then

$$X_1^B = \bar{b} - \bar{A}_s x_s \geq 0 \qquad \text{for all } x_s \geq 0,$$

which says that the solution X_1^B is feasible for all nonnegative x_s. Since $\bar{c}_s > 0$,

$$z = \bar{z} + \bar{c}_s x_s$$

implies that the objective function becomes unbounded as x_s is increased without limit. If we continue the algorithm, there exists some $\bar{a}_{is} > 0$.

STEP (3): Choose the row to pivot in by the following modified ratio rule:

$$\frac{\bar{B}_r}{\bar{a}_{rs}} = \underset{i}{\text{lexico-min}} \ \left\{ \frac{\bar{B}_i}{\bar{a}_{is}} \,\middle|\, \bar{a}_{is} > 0 \right\}.$$

We should point out that the lexico-min produces a *unique* variable to drop from the basis. There must be a unique lexico-min, since, if not, there would exist two vectors $\bar{B}_i/\bar{a}_{is}$ that are identical. This would imply that two rows of $\bar{B}$ are proportional and, hence, that two rows of B^{-1} are proportional, which is a clear contradiction.

STEP (4): Replace the variable basic in row r with variable s and re-establish the canonical form by pivoting on the coefficient $\bar{a}_{rs}$.

We have shown that, in the initial canonical form, the row vectors of the righthand-side matrix are each lexico-positive. It remains to show that, assuming the vectors $\bar{B}_i$ are lexico-positive at an iteration, they remain lexico-positive at the next iteration. Let the new row vectors of the righthand-side matrix be $\bar{B}_i^*$. First,

$$\bar{B}_r^* = \frac{\bar{B}_r}{\bar{a}_{rs}} \succ 0,$$

since $\bar{B}_r \succ 0$, by assumption, and $\bar{a}_{rs} > 0$. Second,

$$\bar{B}_i^* = \bar{B}_i - \bar{a}_{is}\frac{\bar{B}_r}{\bar{a}_{rs}}. \tag{48}$$

If $\bar{a}_{is} \leq 0$, then $\bar{B}_i^* \succ 0$, since it is then the sum of two lexico-positive vectors. If $\bar{a}_{is} > 0$, then (48) can be rewritten as:

$$\bar{B}_i^* = \bar{a}_{is}\left[\frac{\bar{B}_i}{\bar{a}_{is}} - \frac{\bar{B}_r}{\bar{a}_{rs}}\right]. \tag{49}$$

Since $\bar{B}_r/\bar{a}_{rs}$ is the unique lexico-min in the modified ratio rule, (49) implies $\bar{B}_i^* \succ 0$ for $\bar{a}_{is} > 0$, also.

STEP (5): Go to STEP (1)

We would like to show that there is a strict lexicographic improvement in the objective-value vector at each iteration. Letting $-Z = [-z, -y]$ be the objective-value vector, and $\bar{Z}^*$ be the new objective-value vector, we have:

$$\bar{Z}^* = \bar{Z} + \bar{c}_s X_s^*;$$

then, since $\bar{c}_s > 0$ and $X_s^* = \bar{B}_r^* \succ 0$, we have

$$\bar{Z}^* = \bar{Z} + \bar{c}_s X_s^* \succ \bar{Z},$$

which states that $\bar{Z}^*$ is lexico-greater than $\bar{Z}$. Hence we have a strict lexicographic improvement in the objective-value vector at each iteration.

Finally, since there is a strict lexicographic improvement in the objective-value vector for each new basis, no basis can then be repeated. Since there are a finite number of bases and no basis is repeated, the algorithm solves any linear program in a finite number of iterations.

C

Network-flow problems can be solved by several methods. In Chapter 8 we introduced this topic by exploring the special structure of network-flow problems and by applying the simplex method to this structure. This appendix continues the analysis of network problems by describing the application of the labeling algorithm to the maximum-flow network problem. Labeling methods provide an alternative approach for solving network problems. The basic idea behind the labeling procedure is to systematically attach labels to the nodes of a network until the optimum solution is reached.

Labeling techniques can be used to solve a wide variety of network problems, such as shortest-path problems, maximal-flow problems, general minimal-cost network-flow problems, and minimal spanning-tree problems. It is the purpose of this appendix to illustrate the general nature of the labeling algorithms by describing a labeling method for the maximum-flow problem.

C.1 THE MAXIMAL-FLOW PROBLEM

The maximal-flow problem was introduced in Section 8.2 of the text. If v denotes the amount of material sent from node s, called the source, to node t, called the sink, the problem can be formulated as follows:

Maximize v,

subject to:

$$\sum_j x_{ij} - \sum_k x_{ki} = \begin{cases} v & \text{if } i = s, \\ -v & \text{if } i = t, \\ 0 & \text{otherwise,} \end{cases}$$

$$0 \leq x_{ij} \leq u_{ij}, \qquad (i = 1, 2, \ldots, n; \quad j = 1, 2, \ldots, n).$$

We assume that there is no arc from t to s. Also, $u_{ij} = +\infty$ if arc i–j has unlimited capacity. The interpretation is that v units are supplied at s and consumed at t.

Letting x_{ts} denote the variable v and rearranging, we see that the problem assumes the following special form of the general network problem:

Maximize x_{ts},

subject to:

$$\sum_j x_{ij} - \sum_k x_{ki} = 0 \qquad\qquad (i = 1, 2, \ldots, n)$$

$$0 \leq x_{ij} \leq u_{ij} \qquad\qquad (i = 1, 2, \ldots, n; \quad j = 1, 2, \ldots, n).$$

Here arc t–s has been introduced into the network with u_{ts} defined to be $+\infty$, x_{ts} simply returns the v units from node t back to node s, so that there is no formal external supply of material. Let us recall the example (see Fig. C.1) that was posed in Chapter 8 in terms of a water-pipeline system. The numbers above the arcs indicate flow capacity and the bold-faced numbers below the arcs specify a tentative flow plan.

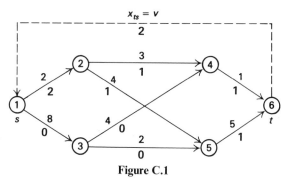

Figure C.1

The algorithm for finding maximal flow rests on observing two ways to improve the flow in this example. The following two "paths" appear in Fig. C.1.

In the first case, the directed path 1–3–5–6 has the capacity to carry 2 additional units from the source to the sink, as given by the capacity of its weakest link, arc 3–5. Note that adding this flow gives a feasible flow pattern, since 2 units are added as input as well as output to both of nodes 3 and 5.

The second case is not a directed path from the source to the sink since arc 2–4 appears with the wrong orientation. Note, however, that adding one unit of flow to the "forward arcs" from 1 to 6 and subtracting one unit from the "reverse arc" 2–4 provides a feasible solution, with increased source-to-sink flow. Mass balance is maintained at node 4, since the one more unit sent from node 3 cancels with the one less unit sent from node 2. Similarly, at node 2 the one additional unit sent to node 5 cancels with the one less unit sent to node 4.

The second case is conceptually equivalent to the first if we view decreasing the flow along arc 2–4 as sending flow from node 4 back to node 2 along the reverse arc 4–2. That is, the unit of flow from 2 to 4 increases the *effective capacity* of the "return" arc 4–2 from 0 in the original network to 1. At the same time, it decreases the usable or effective capacity along arc 2–4 from 3 to 2. With this view, the second case becomes:

Now both instances have determined a directed *flow-carrying path* from source to sink, that is, a directed path with the capacity to carry additional flow.

The maximal-flow algorithm inserts return arcs, such as 4–2 here, and searches for flow-carrying paths. It utilizes a procedure common to network algorithms by "fanning out" from the source node, constructing flow-carrying paths to other nodes, until the sink node is reached.

The procedure starts by labeling all nodes with the capacity to receive flow from the source node by a single arc. For the pipeline example, arc 1–2 is saturated and has a zero effective capacity, whereas arc 1–3 can carry 8 additional units. Thus, only node 3 is labeled from the source

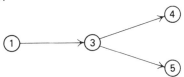

The procedure is repeated by labeling all nodes with the capacity to receive flow directly from node 3 by a single arc. In this case, nodes 4 and 5 are labeled and the labeled nodes become:

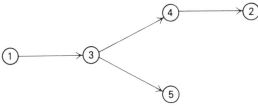

Additional nodes can now be labeled from either node 4 or 5. If node 4 is selected next, then node 2 is labeled, since the effective capacity (return capacity) of arc 4–2 is positive (node 6 cannot be labeled from node 4 since arc 4–6 is saturated). The labeled nodes are:

At any point in the algorithm, some of the labeled nodes, here nodes 1, 3 and 4, will already have been used to label additional nodes. We will say that these nodes have been *scanned*. Any unscanned labeled node can be selected and scanned next.

Scanning node 2 produces no additional labelings, since node 2 can only send flow directly to nodes 1, 4, and 5, and these have been labeled previously. Node 6 is labeled when node 5 is scanned, and the labeled nodes are:

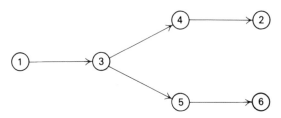

A flow-carrying path has been identified, since the sink has been labeled. In this case, the path discovered is 1–3–5–6, which was the first path given above. The flow along this path is updated by increasing the flow along each arc in the path by the flow capacity of the path, here 2 units. Effective capacities are now recomputed and the labeling procedure is repeated, starting with only the source node labeled.

Following is a formal description of the algorithm and a solution of the water-pipeline example. Notice that the flow value on the arcs need not be recorded from step to step in the algorithm since this information is computed easily from the effective (or usable) capacities of the arcs and their return arcs. For the example above, the effective capacity on arc 2–4 is less than its initial capacity.

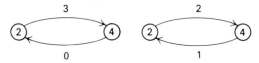

The difference $3 = 2 - 1$ must be the flow value on that arc that resulted in the reduced capacity. The effective capacity on arc 4–2 has been increased from 0 to 1 by adding return-flow capacity to that arc, not by physically adding flow to that arc. In general, this is the case whenever effective capacity exceeds the original capacity. These arcs, consequently, carry no flow.

When a flow-carrying path has been found from source to terminal, that is able to carry θ additional units, the effective capacity in every arc i–j of that path is reduced by θ. At the same time, the effective capacity of each reverse arc j–i increases by θ, since they can now be used to divert (or back up) these units of flow.

C.2 MAXIMAL-FLOW ALGORITHM—FORMAL STATEMENT

Initialization

Assume a given feasible flow plan x_{ij} (if none is given, use the feasible plan with all $x_{ij} = 0$). The initial effective capacity u^*_{ij} on arc i–j is given by calculating $u^*_{ij} = u_{ij} - x_{ij} + x_{ji}$ (i.e., unused capacity $u_{ij} - x_{ij}$ plus return capacity x_{ji}).

Path Search

Start with the source node s and label (mark) every node k with $u_{sk}^* > 0$. Then, in turn, select any labeled node i not already scanned (i.e., used to label other nodes) and label every node j with $u_{ij}^* > 0$ until either t has been labeled or no further labeling can take place.

Capacity Update

If t has been labeled, then a flow-carrying path P has been found from source to sink ($u_{ij}^* > 0$ for every arc i–j on the path), and

$$\theta = \text{Min } \{u_{ij}^* \mid i\text{–}j \text{ in } P\}$$

is the flow capacity of the path. For every arc i–j of P, change u_{ij}^* to $u_{ij}^* - \theta$, and change u_{ji}^* to $u_{ji}^* + \theta$; i.e., increase the effective capacity of the return path. (Adding or subtracting finite θ to any $u_{ij}^* = +\infty$ keeps the u_{ij}^* at $+\infty$. If every u_{ij}^* in P is $+\infty$, then $\theta = +\infty$ and the optimal flow is infinite.)

Termination

If the path search ends without labeling t, then terminate. The optimal flow pattern is given by:

$$x_{ij} = \begin{cases} u_{ij} - u_{ij}^* & \text{if } u_{ij} > u_{ij}^*, \\ 0 & \text{if } u_{ij} \leqq u_{ij}^*. \end{cases}$$

C.3 SAMPLE SOLUTION

Figure C.2 solves the water-pipeline example in Fig. C.1 by this algorithm. Checks next to the rows indicate that the node corresponding to that row has already been scanned. The first three tableaus specify the first application of the path-search step in detail. In Tableau 1, node 1 has been used to label node 3. In Tableau 2, node 3 was used to label nodes 4 and 5 (since $u_{34}^* > 0$, $u_{35}^* > 0$). At this point either node 4 or 5 can be used next for labeling. The choice is arbitrary, and in Tableau 3, node 5 has been used to label node 6. Since 6 is the sink, flow is updated. The last column in the tableau keeps track of how nodes have been labeled.[†] By backtracking, we get a flow-carrying path P from source to terminal. For instance, from Tableau 3, we know that 6 has been labeled from 5, 5 from 3, and 3 from 1, so that the path is 1–3–5–6.

[†] For computer implementation, another column might be maintained in the tableau to keep track of the capacity of the flow-carrying paths as they are extended. In this way, θ need not be calculated at the end.

Initial capacity u_{ij}

		1	2	3	4	5	6
Source	1		2	8			
	2				3	4	
	3				4	2	
	4						1
	5						5
Destination	6						

Tableau 1

		1	2	3	4	5	6	Labeled from	
Source	1			8				Start	✓
	2	2			2	3			
	3				4	2		1	
	4		1						
	5		1			4			
Destination	6				1	1			

Tableau 2

		1	2	3	4	5	6	Labeled from	
Source	1			8				Start	✓
	2	2			2	3			
	3				4	2		1	✓
	4		1					3	
	5		1			4		3	
Destination	6				1	1			

Fig. C.2 Tableaus for the maximal-flow algorithm.

Tableau 3

	1	2	3	4	5	6	Labeled from	
1			8				Start	✓
2	2		2	3				
3			4	2	1		1	✓
4		1			3		3	
5		1			4		3	✓
6				1	1		5	

Path P

①
↓
③ $\theta = \underset{i\text{-}j \text{ in } P}{\text{Min}} \{u_{ij}^*\} = \text{Min}\{8, 2, 4\} = 2$
↓
⑤ Subtract 2 from u_{13}^*, u_{35}^*, and u_{56}^*.
↓
⑥ Add 2 to u_{31}^*, u_{53}^*, and u_{65}^*.

Path 1–3–5–6

Tableau 4

	1	2	3	4	5	6	Labeled from	
1			6				Start	✓
2	2		2	3			4	✓
3	2		4				1	✓
4		1					3	✓
5		1	2		2		2	✓
6				1	3		5	

Path P

① $\theta = \underset{i\text{-}j \text{ in } P}{\text{Min}} \{u_{ij}^*\} = \text{Min}\{6,4,1,5,2\} = 1$
↓
③ Subtract 1 from u_{13}^*, u_{34}^*, u_{42}^*, u_{25}^*, u_{56}^*.
↓
Add 1 to u_{31}^*, u_{43}^*, u_{24}^*, u_{52}^*, u_{65}^*.
↓
Path 1–3–4–2–5–6

Tableau 5

	1	2	3	4	5	6	Labeled from	
1			5				Start	✓
2	2			3	2			
3	3			3			1	✓
4			1				3	✓
5		2	2			2		
6			1	4				

Key: Entry in ith row and jth column of each
tableau is u_{ij}^*. Blank entries denote zeros.

Fig. C.2 *(Cont.)*

Tableau 4 contains the updated capacities and a summary of the next path search, which used nodes 1, 3, 4, 2, and 5 for labeling. The fifth tableau contains the final updated capacities and path search. Only nodes 1, 3, and 4 can be labeled in this tableau, so the algorithm is completed.

The flow at any point in the algorithm is obtained by subtracting effective capacities from initial capacities, $u_{ij} - u_{ij}^*$, and discarding negative numbers. For Tableaus 1, 2, or 3, the flow is given by Fig. C.1. After making the two indicated flow changes to obtain Tableau 4 and then Tableau 5, the solutions are given in two networks shown in Figs. C.3 and C.4. The optimal solution sends two units along each of the paths 1–2–5–6 and 1–3–5–6 and one unit along 1–3–4–6.

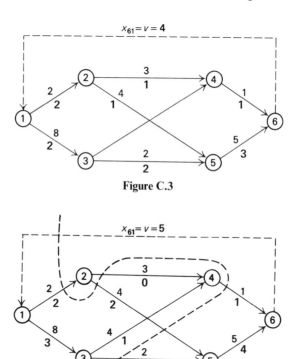

Figure C.3

Figure C.4

C.4 VERIFYING THE ALGORITHM—MAX-FLOW/MIN-CUT

The final solution to the sample problem shown in Fig. C.4 illustrates an important conceptual feature of the maximum-flow problem. The heavy dashed line in that figure "cuts" the network in two, in the sense that it divides the nodes into two groups, one group containing the source node and one group containing the sink node. Note that the most that can ever be sent across this cut to the sink is two units from node

1 to node 2, one unit from 4 to 6, and two units from 3 to 5, for a total of 5 units (arc 2–5 connects nodes that are both to the right of the cut and is not counted). Since no flow pattern can send more than these 5 units and the final solution achieves this value, it must be optimal.

Similarly, the cut separating labeled from unlabeled nodes when the algorithm terminates (see Fig. C.5) shows that the final solution will be optimal.

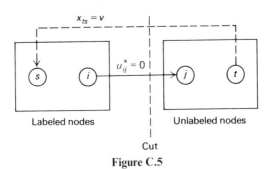

Figure C.5

By conservation of mass, v equals the net flow from left to right across this cut, so that v can be no greater than the total capacity of all arcs i–j pictured. In fact, this observation applies to any cut separating the source node s from the sink node t.

$$\begin{bmatrix} \text{Any} \\ \text{feasible flow} \\ v \text{ from} \\ \text{node } s \text{ to} \\ \text{node } t \end{bmatrix} \leq \begin{pmatrix} \text{Capacity of} \\ \text{any cut separating} \\ \text{node } s \text{ and node } t \end{pmatrix}.$$

As in the example discussed above, the capacity of any cut is defined as the total capacity of all arcs directed "from left to right" across the cut.

By virtue of the labeling scheme, however, every arc i–j directed from left to right across the cut separating the labeled and unlabeled nodes must have $u_{ij}^* = 0$; otherwise j would have been labeled from i. This observation implies that

$$x_{ij} = u_{ij} \quad \text{and} \quad x_{ji} = 0,$$

since if either $x_{ij} < u_{ij}$ or $x_{ji} > 0$, then $u_{ij}^* = (u_{ij} - x_{ij}) + x_{ji}$ would be positive. Consequently, the net flow across the cut, and thus v, equals the cut capacity. Since no flow can do any better, this flow pattern is optimal.

We can recast this observation in a slightly different way. The inequality stated above shows that any feasible flow v from source to sink is bounded from above by the capacity of any cut. Consequently, the maximum flow from source to sink is bounded by the capacity of *any* cut separating the source and sink. In particular, the maximum flow from source to sink is bounded from above by the minimum capacity of any cut. We have just shown, though, that the maximum flow v equals the capacity of the cut separating the labeled and unlabeled nodes when the algorithm terminates; therefore, we have verified the famous *max-flow/min-cut theorem*.

$$\begin{pmatrix} \text{The maximum flow} \\ v \text{ from node } s \text{ to} \\ \text{node } t \end{pmatrix} = \begin{pmatrix} \text{The minimum capacity} \\ \text{of any cut separating} \\ \text{node } s \text{ and node } t \end{pmatrix}.$$

There are many important ramifications of this theorem. It can be used, for example, to establish elegant results in combinatorial theory. Although such results lead to many useful applications of combinatorial analysis, their development is beyond the scope of our coverage here. We have merely shown how the max-flow/min-cut theorem arises, and used the theorem to show that when the maximum flow-labeling algorithm terminates, it has found the maximum possible flow from the source node to the sink node.

Finally, we should note that the maximum-flow procedure eventually terminates, as in the above example, and does not continue to find flow-carrying paths from s to t forever. For certain cases this is easy to show. If the capacity data u_{ij} is all integral and every x_{ij} in the original flow plan is an integer, then every u_{ij}^* and θ encountered will be integral. But then, if the maximal flow v is not $+\infty$, we approach it in integral steps and must eventually reach it. Similarly, if the original data and flow is fractional, v increases at each step by at least a given fractional amount and the procedure terminates in a finite number of steps. If the data is irrational, then it is known that the algorithm finds the maximum flow in a finite number of steps as long as the nodes are considered during path search on a first-labeled–first-scanned basis.

Note that starting with an integral flow plan x_{ij} when all u_{ij} are integral leads to integral u_{ij}^*. Consequently, the final (and optimal) flow plan will be integral. This is a special instance of the network integrality property introduced in Chapter 8.

References

Rather than attempting to provide a comprehensive bibliographic reference to all major publications related to the field of mathematical programming, we have selected a few references representing the major books in the field and the most significant survey articles. The interested reader can find in the following publications additional bibliographic material.

LINEAR PROGRAMMING

1. Beale, E. M. L., *Mathematical Programming in Practice*, Pitman, 1968.
2. Beale, E. M. L., (Ed.), *Applications of Mathematical Programming Techniques*, American Elsevier, New York, 1970.
3. Charnes, A., and W. W. Cooper, *Management Models and Industrial Applications of Linear Programming*, Vols. I and II, John Wiley & Sons, New York, 1961.
4. Dantzig, G. B., *Linear Programming and Extensions*, Princeton University Press, Princeton, N.J., 1963.
5. Dorfman, R., P. A. Samuelson, and R. M. Solow, *Linear Programming and Economic Analysis*, McGraw-Hill, New York, 1958.
6. Daellenbach, H. G., and E. J. Bell, *User's Guide to Linear Programming*, Prentice-Hall, Inc., Englewood Cliffs, N.J., 1970.
7. Driebeek, N. J., *Applied Linear Programming*, Addison-Wesley, Reading, Mass., 1969.
8. Gale, D., *The Theory of Linear Economic Models*, McGraw-Hill, New York, 1960.
9. Gass, S. I., *Linear Programming: Methods and Applications* (Third edition), McGraw-Hill, New York, 1969.
10. Gass, S. I., *An Illustrated Guide to Linear Programming*, McGraw-Hill, New York, 1970.
11. Hadley, G., *Linear Programming*, Addison-Wesley, Reading, Mass., 1962.

12. Orchard-Hays, W., *Advanced Linear-Programming Computing Techniques,* McGraw-Hill, New York, 1968.

13. Nicholson, T., *Optimization in Industry,* Vol. 2: *Industrial Applications,* Longman Group, London, 1971.

14. Pun, L., *Introduction to Optimization Practice,* John Wiley & Sons, New York, 1969.

15. Simmonnard, M., *Linear Programming,* Translated by W. Ś. Jewell, Prentice-Hall, Englewood Cliffs, N.J., 1966.

16. Simmons, D. M., *Linear Programming for Operations Research,* Holden-Day, San Francisco, 1972.

17. Spivey, W. A. and R. M. Thrall, *Linear Optimization,* Holt, Rinehart, and Winston, New York, 1970.

GAME THEORY

1. Luce, R. D., and H. Raiffa, *Games and Decisions,* John Wiley & Sons, New York, 1957.

2. May, F. B., *Introduction to Games of Strategy,* Allyn and Bacon, Boston, 1970.

3. McKinsey, J. C. C., *Introduction to the Theory of Games,* McGraw-Hill, New York, 1952.

4. Owen, G., *Game Theory,* W. B. Saunders, Philadelphia, 1968.

5. Williams, J. D., *The Compleat Strategyst,* McGraw-Hill, New York, revised edition, 1966.

NETWORK THEORY

1. Bradley, G., "Survey of deterministic networks," *AIIE Trans.* **7**(3), (1975) 222–234.

2. Busacker, R. G., and T. L. Saaty, *Finite Graphs and Networks: An Introduction with Applications,* McGraw-Hill, New York, 1965.

3. Christofides, N., *Graph Theory: An Algorithmic Approach,* Academic Press, New York, 1975.

4. Dreyfus, S., "An appraisal of some shortest-path algorithms," *Op. Res.* **17**(3), (1969) 395–412.

5. Elmaghraby, S., *Some Network Models in Operations Research,* Springer-Verlag, Berlin and New York, 1970.

6. Ford, L. R., Jr., and D. R. Fulkerson, *Flows in Networks,* Princeton University Press, Princeton, N.J., 1962.

7. Frank, H., and I. Frisch, *Communication, Transmission, and Transportation Networks,* Addison-Wesley, Reading, Mass., 1971.

8. Frank, H., and I. Frisch, "Network analysis," *Sci. Amer.,* July 1970.

9. Fulkerson, D., "Flow networks and combinational operations research," *Amer. Math. Monthly,* **73**(2), (1966) 115–138.

10. Fulkerson, D. R. (Ed.), *Studies in Graph Theory*, Parts I and II, Vols. 11 and 12 of the MAA series Studies in Mathematics, The Mathematical Association of America, 1975.

11. Harary, F., *Graph Theory*, Addison-Wesley, Reading, Mass., 1969.

12. Ore, O., *Graphs and their Uses*, Random House, The L. W. Singer Co., New York, 1963.

13. Pierce, A., "Bibliography on algorithms for shortest-path, shortest spanning tree, and related circuit-routing problems (1956–1974)," *Networks* **5**(2), (1975) 129–150.

14. Potts, R. B., and R. M. Oliver, *Flows in Transportation Networks*, Academic Press, New York, 1972.

INTEGER PROGRAMMING

1. Abadie, J. (Ed.), *Integer and Nonlinear Programming*, American Elsevier, New York, 1970.

2. Balinsky, M. L., "Integer Programming: Methods, uses, computation," *Man. Sci.* **12**(3), (November 1965) 253–313.

3. Beale, E. M. L., "Survey of integer programming," *Operational Research Quarterly* **16**(2), (June 1965) 219–228.

4. Garfinkel, R. S., and G. L. Nemhauser, *Integer Programming*, John Wiley & Sons, New York, 1972.

5. Geoffrion, A. M., "A guided tour of recent practical advances in integer linear programming," Working Paper No. 220, Western Management Science Institute, UCLA, September 1974.

6. Geoffrion, A. M., and R. E. Marsten, "Integer programming algorithms: A framework and state-of-the-art survey," *Man. Sci.* **18**(7), (March 1972).

7. Greenberg, H., *Integer Programming*, Academic Press, New York, 1971.

8. Harris, B. (Ed.), *Graph Theory and its Applications*, Academic Press, New York, 1970.

9. Hu, T. C., *Integer Programming and Network Flows*, Addison-Wesley, Reading, Mass., 1969.

10. Ivanescu, P. L., *Pseudo-Boolean Programming Methods for Bivalent Programming*, Springer-Verlag, Berlin and New York, 1966.

11. Lawler, E. L., *Combinatorial Optimization: Networks and Matroids*, Holt, Rinehart, and Winston (forthcoming).

12. Lawler, E. L., and D. E. Wood, "Branch-and-bound methods: A survey," *Op. Res.* **14**(4), (July-August 1966) 699–719.

13. Liu, C., *Introduction to Combinatorial Mathematics*, McGraw-Hill, New York, 1968.

14. Plane, D. R., and C. McMillan, *Discrete Optimization: Integer Programming and Network Analysis for Management Decisions*, Prentice-Hall, Englewood Cliffs, N.J., 1971.

15. Saaty, T. L., *Optimization in Integers and Related Extremal Problems*, McGraw-Hill, New York, 1970.

16. Salkin, H., *Integer Programming,* Addison-Wesley, Reading, Mass., 1975.

17. Shapiro, J. F., "Dynamic-programming algorithms for integer programming, **I:** The integer-programming problem viewed as a knapsack-type problem," *Op. Res.* **16**(1), (January-February 1968) 103–121.

18. Taha, H. A., *Integer Programming Theory, Applications, and Computations,* Academic Press, New York, 1975.

19. Zionts, S., *Linear and Integer Programming,* Prentice-Hall, Englewood Cliffs, N.J., 1973.

DYNAMIC PROGRAMMING

1. Beckmann, M. J., *Dynamic Programming of Economic Decisions,* Springer-Verlag, Berlin and New York, 1968.

2. Bellman, R. E., *Dynamic Programming,* Princeton University Press, Princeton, N.J., 1957.

3. Bellman, R. E., and S. E. Dreyfus, *Applied Dynamic Programming,* Princeton University Press, Princeton, N.J., 1962.

4. Denardo, E. V., *Dynamic Programming: Theory and Application,* Prentice-Hall, Englewood Cliffs, N.J., 1975.

5. Howard, R. A., *Dynamic Programming and Markov Processes,* The Massachusetts Institute of Technology Press, Cambridge, Mass., 1960.

6. Kaufmann, A., and R. Cruon, *Dynamic Programming: Sequential Scientific Management* (trans. H. C. Sneyd), Academic Press, New York, 1967.

7. Nemhauser, G. L., *Introduction to Dynamic Programming,* John Wiley & Sons, New York, 1966.

8. White, D. J., *Dynamic Programming,* Holden-Day, San Francisco, 1969.

LARGE-SCALE SYSTEMS THEORY

1. Baumol, W. J., and T. Fabian, "Decomposition pricing for decentralization and external economies," *Man. Sci.* **11**(1), (September 1964) 1–32.

2. Dantzig, G. B., "Large-Scale Linear Programming," *Mathematics of the Decision Sciences,* Part I (G. Dantzig and A. Veinott, Eds.) Amer. Math. Soc., Providence, R.I., 1968, 77–92.

3. Dantzig, G. B., and P. Wolfe, "Decomposition principle for linear programs," *Op. Res.* **8**(1), (January-February 1960), 101–111. See also *Econometrica* **29**(4), (October) 767–778.

4. Geoffrion, A. M., "Elements of large-scale mathematical programming, Parts I and II," *Man. Sci.* **16** (1970) 652–691.

5. Geoffrion, A. M., "Primal resource-directive approaches for optimizing nonlinear decomposable systems," *Op. Res.* **18**(3), (1970) 375–403.

6. Grinold, R. C., "Steepest ascent for large-scale linear program," *SIAM Review* **14** (1972) 447–464.

7. Himmelblau, D. M. (Ed.), *Decomposition of Large-Scale Problems*, North-Holland Publishing Co., Amsterdam, 1973.

8. Lasdon, L. S., *Optimization Theory for Large Systems*, MacMillan, New York, 1970.

9. Lasdon, L. S., "A survey of large-scale mathematical programming," *Handbook of Operations Research* (Elmaghraby and Moler, Eds.), Van Nostrand Reinhold, New York, 1977.

10. Reid, J. K., (Ed.), *Large Sparse Sets of Linear Equations*, Academic Press, New York, 1971.

11. Tomlin, J. A., "Survey of computation methods for solving large-scale systems," Technical Report No. 72–25, Department of Operations Research, Stanford University, Stanford, Calif., October 1972.

12. Wismer, D. A. (Ed.), *Optimization Methods for Large-Scale Systems*, McGraw-Hill, New York, 1971.

NONLINEAR PROGRAMMING

1. Abadie, J. (Ed.), *Nonlinear Programming*, John Wiley & Sons, New York, 1967.

2. Arrow, K. J., L. Hurwicz, and H. Uzawa, *Studies in Linear and Nonlinear Programming*, Stanford University Press, Stanford, Calif., 1958.

3. Boot, J., *Quadratic Programming Algorithms: Anomalies and Applications*, North-Holland Publishing Co., Amsterdam, Rand McNally and Co., Chicago, 1964.

4. Bracken, J., and G. P. McCormick, *Selected Applications of Nonlinear Programming*, John Wiley & Sons, New York, 1968.

5. Brent, R. P., *Algorithms for Minimization without Derivatives*, Prentice-Hall, Englewood Cliffs, N.J., 1973.

6. Canon, M., C. Culum, Jr., and E. Polak, *Theory of Optimal Control and Mathematical Programming*, McGraw-Hill, New York, 1970.

7. Converse, A., *Optimization*, Holt, Rinehart, and Winston, New York, 1970.

8. Dennis, J., *Mathematical Programming and Electrical Networks*, John Wiley & Sons, New York, 1959.

9. Duffin, R. J., E. L. Peterson, and C. Zener, *Geometric Programming: Theory and Application*, John Wiley & Sons, New York, 1967.

10. Fiacco, A. V., and G. P. McCormick, *Nonlinear Programming, Sequential Unconstrained Minimization Techniques*, John Wiley & Sons, New York, 1968.

11. Geoffrion, A. M., "Duality in nonlinear programming: A simplified application-oriented development," *SIAM Review* **13**(1), (January 1971) 1–37.

12. Graves, R. L., and P. Wolfe, *Recent Advances in Mathematical Programming*, McGraw-Hill, New York, 1963.

13. Hadley, G., *Nonlinear and Dynamic Programming*, Addison-Wesley, Reading, Mass., 1964.

14. Hestenes, M. R., *Optimization Theory: The Finite-Dimensional Case*, John Wiley & Sons, New York, 1975.

15. Himmelblau, D. M., *Applied Nonlinear Programming*, McGraw-Hill, New York, 1972.

16. Jacoby, S., J. Kowalik, and J. Pizzo, *Iterative Methods for Nonlinear Optimization Problems*, Prentice-Hall, Englewood Cliffs, N.J., 1972.

17. Karlin, S., *Mathematical Methods and Theory in Games, Programming, and Economics*, Vols. I and II, Addison-Wesley, Reading, Mass., 1959.

18. Kowalik, J., and M. Osborne, *Methods for Unconstrained Optimization Problems*, American Elsevier, New York, 1968.

19. Künzi, H. P., W. Krelle, and W. Oettli, *Nonlinear Programming*, Blaisdell, 1966.

20. Luenberger, D. G., *Optimization by Vector-Space Methods*, John Wiley & Sons, New York, 1969.

21. Luenberger, D. G., *Introduction to Linear and Nonlinear Programming*, Addison-Wesley, Reading, Mass., 1973.

22. Mangasarian, O. L., *Nonlinear Programming*, McGraw-Hill, New York, 1969.

23. Martos, B., *Nonlinear Programming: Theory and Methods*, North-Holland Publishing Co., Amsterdam, 1975.

24. Powell, M. J. D., "A survey of numerical methods for unconstrained optimization," *SIAM Review* 12(1), (January 1970).

25. Reldaitis, G. V., "A survey of nonlinear programming," *AIIE Transactions*, 7(3), September 1975.

26. Roberts, A., and D. Varberg, *Convex Functions*, Academic Press, New York, 1973.

27. Rockafellar, R. T., *Convex Analysis*, Princeton University Press, Princeton, N.J., 1969.

28. Scarf, H., *The Computation of Economic Equilibria*, Yale University Press, New Haven, Conn., 1973.

29. Simmons, D. M., *Nonlinear Programming for Operations Research*, Prentice-Hall, Englewood Cliffs, N.J., 1975.

30. Sposito, V. A., *Linear and Nonlinear Programming*, Iowa State University Press, Ames, Iowa, 1975.

31. Stoer, J., and C. Witzgall, *Convexity and Optimization in Finite Dimensions*, **I**, Springer-Verlag, Berlin and New York, 1970.

32. Varaiya, P., *Notes on Optimization*, Van Nostrand Reinhold, New York, 1972.

33. Whittle, P., *Optimization under Constraints*, John Wiley & Sons, New York, 1971.

34. Wilde, D. J., *Optimum Seeking Methods*, Prentice-Hall, Englewood Cliffs, N.J., 1964.

35. Wilde, D. J., and C. S. Beightler, *Foundations of Optimization*, Prentice-Hall, Englewood Cliffs, N.J., 1967.

36. Zoutendijk, G., *Methods of Feasible Directions*, Elsevier, Amsterdam, 1960.

Index